The University of Chicago School Mathematics Project

Geometry

Second Edition

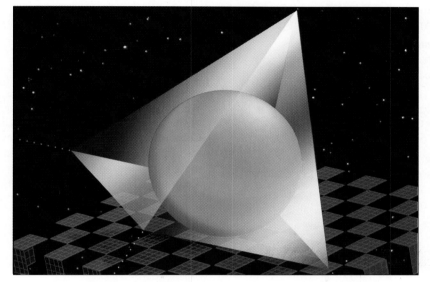

About the Cover The computer-generated art on the cover connects shapes studied by ancient geometers to the present and to the future. *UCSMP Geometry* integrates geometry with algebra, functions, and discrete mathematics.

Authors

Zalman Usiskin   Daniel Hirschhorn
Arthur Coxford   Virginia Highstone   Hester Lewellen
Nicholas Oppong   Richard DiBianca   Merilee Maeir

ScottForesman

Editorial Offices: Glenview, Illinois
Regional Offices: San Jose, California • Atlanta, Georgia
Glenview, Illinois • Oakland, New Jersey • Dallas, Texas

ACKNOWLEDGMENTS

Authors

Zalman Usiskin
Professor of Education, The University of Chicago, Chicago, IL

Daniel B. Hirschhorn
Assistant Professor of Mathematics, Illinois State University, Normal, IL

Arthur Coxford
Professor of Mathematics Education, University of Michigan, Ann Arbor, MI (First Edition only)

Virginia Highstone
Mathematics Teacher, York High School, Elmhurst, IL. (Second Edition only)

Hester Lewellen
Assistant Professor of Mathematics, Baldwin-Wallace College, Berea, OH (Second Edition only)

Nicholas Oppong
Assistant Professor of Mathematics Education University of Georgia, Athens, GA (Second Edition only)

Richard DiBianca
USCMP (Second Edition only)

Merilee Maeir
UCSMP (Second Edition only)

UCSMP Production and Evaluation

Series Editors: Zalman Usiskin, Sharon L. Senk
Director of Second Edition Studies: Gurcharn Kaeley
Director of First Edition Studies: Sandra Mathison (State University of New York, Albany), Penelope Flores
Technical Coordinator: Susan Chang
Second Edition Editors: Scott Anderson, Wei Tang
Second Edition Teacher's Edition Editors: Lianghuo Fan, Xinmei Cui

We wish to thank the many editors, production personnel, and design personnel at ScottForesman for their ongoing magnificent assistance.

We wish also to acknowledge the generous support of the **Amoco Foundation** and the **Carnegie Corporation of New York** for the development, testing, and distribution of the First Edition of these materials, and the continuing support of the **Amoco Foundation** for the Second Edition.

Design Development

Curtis Design

Multicultural Reviewers for ScottForesman

Cynthia Felton
Chicago Public Schools, Chicago, IL

Victor Gee
Oakland Public Schools, Oakland, CA

Ada Settle
Winter Park High School, Winter Park, FL

Linda Skinner
Edmond Public School, Edmond, OK

It is impossible for UCSMP to thank all the people who have helped create and test these books. We wish particularly to thank Carol Siegel, who coordinated the use of the test materials in schools; Suzanne Levin, David Witonsky, Maryann Kannappan, Alev Yalman, Kate Fahey, and Eric Wildt of our editorial staff; Eric Chen, Tim Day, Thao Do, Tony Ham, Jeong Moon, Adil Moiduddin, Antoun Nabhan, Young Nam, and Wei Tang of our technical staff; and Pushpam Jain and Kwai Ming Wa of our evaluation staff. We also wish to thank Peter Appelbaum, who helped at the University of Michigan with the first edition writing.

A first draft of *Geometry* was written and piloted during the 1986–1987 school year. The following schools participated in the first pilots of the first edition.

Collins High School
Chicago, Illinois

Corliss High School
Chicago, Illinois

Taft High School
Chicago, Illinois

Glenbrook South High School
Glenview, Illinois

Rich South High School
Richton Park, Illinois

Lake Park High School
Roselle, Illinois

The materials were revised and tested during the 1987–1988 school year, and again revised and tested during the 1988–1989 school year. The following schools participated in these studies.

Chaparral High School
Scottsdale, Arizona

Irvine High School
Irvine, California

Mendocino High School
Mendocino, California

Marietta High School
Marietta, Georgia

Hyde Park Career Academy
Chicago, Illinois

Taft High School
Chicago, Illinois

Lyons Township High School
La Grange, Illinois

Rich South High School
Richton Park, Illinois

Niles Township High School West
Skokie, Illinois

Fruitport High School
Fruitport, Michigan

North Hunterdon High School
Annandale, New Jersey

Aiken High School
Cincinnati, Ohio

Cincinnati Academy of Mathematics and Science
Cincinnati, Ohio

Walnut Hills High School
Cincinnati, Ohio

Carrick High School
Pittsburgh, Pennsylvania

Taylor Allderice High School
Pittsburgh, Pennsylvania

Since the ScottForesman publication of the First Edition of *Geometry* in 1991, thousands of teachers and schools have used the materials and have made additional suggestions for improvements. The materials were again revised, and a large field study was undertaken in 1993–1994. We appreciate the assistance of the following teachers who taught these preliminary second edition versions, participated in the research, and contributed ideas to help improve the text.

Rosalie McVay
Fayette County High School
Fayetteville, Georgia

Cheryl Dukes
Sandy Creek High School
Tyrone, Georgia

Paul Yu
Olympia High School
Stanford, Illinois

Rita Belluomini
Rich South High School
Richton Park, Illinois

Vicki Miller
Arsenal Technical High School
Indianapolis, Indiana

Leona Davidson
Franklin High School
Somerset, New Jersey

Brenda Paustain
Ashland High School
Ashland, Oregon

Bob Landwehr
Astoria High School
Astoria, Oregon

Rich Pearson
Lakeridge High School
Lake Oswego, Oregon

Betty Sharpe
Wando High School
Mt. Pleasant, South Carolina

Mel Noble
Olympic High School
Silverdale, Washington

Julie Milazzo
Lincoln High School
Wisconsin Rapids, Wisconsin

THE UNIVERSITY OF CHICAGO SCHOOL MATHEMATICS PROJECT

The University of Chicago School Mathematics Project (UCSMP) is a long-term project designed to improve school mathematics in grades K–12. UCSMP began in 1983 with a 6-year grant from the Amoco Foundation. Additional funding has come from the National Science Foundation, the Ford Motor Company, the Carnegie Corporation of New York, the General Electric Foundation, GTE, Citicorp/Citibank, and the Exxon Education Foundation, and from royalties from the sales of UCSMP materials by ScottForesman.

UCSMP is centered in the Departments of Education and Mathematics of the University of Chicago. The project has translated dozens of mathematics textbooks from other countries, held three international conferences, developed curricular materials for elementary and secondary schools, formulated models for teacher training and retraining, conducted a large number of large and small conferences, engaged in evaluations of many of its activities, and through its royalties has supported a wide variety of research projects in mathematics education at the University. UCSMP currently has the following components and directors:

Resources	Izaak Wirszup, Professor Emeritus of Mathematics
Elementary Materials	Max Bell, Professor of Education
Elementary Teacher Development	Sheila Sconiers, Research Associate in Education
Secondary	Sharon L. Senk, Associate Professor of Mathematics, Michigan State University Zalman Usiskin, Professor of Education
Evaluation Consultant	Larry Hedges, Professor of Education

From 1983 to 1987, the overall director of UCSMP was Paul Sally, Professor of Mathematics. Since 1987, the overall director has been Zalman Usiskin.

Geometry

The text *Geometry* has been developed by the Secondary Component of the project, and constitutes the core of the third year in a six-year mathematics curriculum devised by that component. The names of the six texts around which these years are built are:

Transition Mathematics
Algebra
Geometry
Advanced Algebra
Functions, Statistics, and Trigonometry
Precalculus and Discrete Mathematics

The content and questions of this book integrate algebra and some discrete mathematics together with geometry. Pure and applied mathematics are also integrated throughout. It is for these reasons that the book is deemed to be part of an integrated series. However, geometry is the trunk from which the various branches of mathematics studied in this book emanate, and which constitutes the main subject matter throughout. It is for this reason that we call this book simply *Geometry*.

Since the ScottForesman publication of the first edition of *Geometry* in 1991, the entire UCSMP secondary series has been completed and published. Thousands of teachers and schools have used the first edition and some made suggestions for improvements. There have been advances in technology and in thinking about how students learn. We attempted to utilize these ideas in the development of the second edition. Every bit of text and every question was examined. We moved lessons and reorganized others. We added new applications and updated others. With a reorganization of some material, the second edition has one less chapter than the first edition without any deletion of significant content.

The First Edition of *Geometry* introduced many features that are retained in this edition. The value of some of these features has influenced other geometry courses and integrated mathematics courses as well. There is **wider scope,** including significant amounts of algebra employed to motivate, justify, extend, and otherwise enhance the geometry. The coordinate and transformation approaches are particularly important because coordinates connect geometry with algebra, and transformations are functions which allow all figures to be considered as geometric. These two features enable this text to be particularly beneficial in any further study of algebra and functions. A **real-world orientation** has guided both the selection of content and the approaches used in working out exercises and problems, because being able to do mathematics is of little value to most individuals unless they can apply that content. We require **reading mathematics,** because students must read to understand mathematics in later courses and technical matter in the world at large.

We continue to use two unique overall features designed to maximize the acquisition of both skills and concepts. **Four dimensions of understanding** are emphasized: skill in drawing, visualizing, and following algorithms; understanding of properties, mathematical relationships and proofs; using geometric ideas in real situations; and representing geometric concepts with coordinates, networks, or other diagrams. We call this the SPUR approach: **S**kills, **P**roperties, **U**ses, **R**epresentations.

The **book organization** is designed to maximize performance. Ideas introduced in a lesson are reinforced through Review questions in succeeding lessons. This feature allows students several nights to learn and practice important concepts and skills. At the end of each chapter, a Progress Self-Test and a Chapter Review, each keyed to objectives in all the dimensions of understanding, are used to solidify performance of skills and concepts from the chapter so that they may be applied later with confidence. Finally, to increase retention, important ideas are reviewed in later chapters.

As in all UCSMP texts, **up-to-date technology** is integrated throughout. There is a rather substantial increase in the use of technology in this edition. In addition to the scientific calculators students are expected to have with them at all times—and, increasingly these calculators are automatic graphers—we encourage students to have continual access to *automatic drawers* that enable geometric figures to be drawn and constructions to be done automatically.

There are a number of other features new to this edition. **Activities** are incorporated between and inside lessons to help students develop concepts before or as they read. The In-class activities often require automatic drawers, so students cannot be expected to be able to do them as homework. There are **projects** at the end of each chapter because in the real world much of the mathematics done requires a longer period of time than is customarily available to students in daily assignments, and because teachers who have tried this idea are enthusiastic about it. There are many more questions requiring **writing** and a special writing font, because writing helps students clarify their own thinking, and writing is important in communicating mathematical ideas to others. The writing font is particularly useful in showing students what needs to be written in proof arguments.

Draft second edition materials were tested in a dozen schools. A summary of results is found in the Teacher's Edition. Refinements were made as a result of the testing.

Comments about these materials are welcomed. Please write: UCSMP, The University of Chicago, 5835 S. Kimbark, Chicago, IL 60637.

CONTENTS

CHAPTER 1   4

POINTS AND LINES

CHAPTER 2   62

THE LANGUAGE AND LOGIC OF GEOMETRY

To the Student:

GETTING STARTED

Welcome to Geometry.
We hope you enjoy this book; it was written for you.

Studying Mathematics

A goal of this book is to help you learn mathematics on your own, so that you will be able to deal with the mathematics in newspapers, magazines, on television, at work, and in school. The authors, who are all experienced teachers, offer the following advice on studying geometry.

1 You can watch basketball often on television. Still, to learn how to play basketball, you must have a ball in your hand and actually dribble, shoot, and pass it. Mathematics is no different. You cannot learn much geometry just by watching other people do it. You must participate. You must draw pictures, measure lengths, and think through problems. Some teachers have a slogan:

Mathematics is not a spectator sport.

2 You should read each lesson. You may do this as a class or in a small group; or you may do the reading on your own. No matter how you do the reading, it is vital for you to understand what you have read. In *Geometry*, there are also diagrams and symbols that are necessary for this understanding. Here are some ways to improve your reading comprehension.

Read slowly and thoughtfully, paying attention to each word, diagram, and symbol.

Look up the meaning of any word you do not understand. Work examples yourself as you follow the steps in the text.

Draw your own pictures when following a proof or other discussion.

Reread sections that are unclear to you.

Make a note of difficult ideas to discuss later with a fellow student or your teacher.

3 Writing is a tool for communicating solutions and thoughts and can help you understand mathematics. In *Geometry* you will often have to justify your solution to a problem and you will learn to write formal arguments. Writing good explanations takes practice. You should use solutions to the examples to guide your writing.

4 If you cannot answer a question immediately, don't give up! Read the lesson again. Read the question again. Look for examples. If you can, go away from the problem and come back to it a little later. Do not be afraid to ask questions in class and to talk to others when you do not understand something.

Equipment Needed for This Book

Geometry is the study of visual patterns. These patterns are found in physical objects. These patterns are also found in representations of numbers and other ideas whose origin is not visual. Good drawings can help you to learn geometry. Poor drawings hide patterns in geometry just as poor computation can hide patterns in arithmetic. Thus, for this course, you should have good drawing equipment.

In addition to the notebook paper, sharpened pencil, and erasers that you should always have, you need to have some drawing equipment.

Ruler (marked in both centimeters and inches)
Protractor
Compass
Graph paper
Scientific calculator

It is best if the ruler and protractor are made of transparent plastic. We recommend compasses that tighten with a screw and can hold regular pencils.

Geometry is the study of shape and size. Every physical object has a shape and its size can be measured in various ways. You will need a scientific calculator. Since a good calculator will last for many years, you may wish to obtain a graphics calculator. Such a calculator is needed for the mathematics courses which follow this one and may be able to create tables useful in this course. It also helps to have a dictionary.

Getting Acquainted with UCSMP Geometry

It is always helpful to spend some time getting acquainted with your textbook. The questions that follow are designed to help you become familiar with *Geometry*.

By the end of this year you will have had experience drawing, constructing, measuring, visualizing, comparing, transforming, and classifying geometric figures. You will have had experience reasoning about those figures and separating fact from opinion. We hope you will have had a chance to use what you already know and also to stretch your imagination.

We hope you join the hundreds of thousands of students who have enjoyed this book. We wish you much success.

Covering the Reading

1. What is geometry?

2. Name four things you will experience by the end of the year in geometry.

3. List four items of drawing equipment besides paper and pencil that you will need for your work in *Geometry*.

4. How can the statement "Mathematics is not a spectator sport" be applied to the study of geometry?

5. What two things other than words are important to understand when reading a geometry text.

Knowing Your Textbook

In 6–13, answer the questions by looking at the Table of Contents, the lessons and chapters of the textbook, or the end-of-book material.

6. Refer to the Table of Contents. What lesson would you read to learn about an automatic drawer?

7. Look at several lessons.
   a. What are the four categories of questions at the end of each lesson?
   b. What word is formed by the first letters of these categories?

8. Suppose you have just finished the questions in Lesson 4-3. On what page can you find answers to check your work? What answers are given?

9. In the vocabulary sections, why are some terms marked with an asterisk?

10. What is in the Glossary?

11. What should you do after taking a Progress Self-Test at the end of a chapter?

12. Use the Index.
    a. In what lesson is Euclid first mentioned?
    b. Who was Euclid?

13. This book has some Appendices. What do they cover?

CHAPTER

1

POINTS AND LINES

From 1884 to 1886, Georges Seurat, a French artist, worked on the painting reproduced here, entitled *Sunday Afternoon on the Island of La Grande Jatte.* This painting is entirely made up of small dots of about equal size. In it Seurat showed that a painting could be made without long brushstrokes. Even delicate figures and shadows could be formed from dots. The original painting, worth millions of dollars today, hangs in the Art Institute of Chicago.

The geometry that you will study in this book involves many different ideas, but almost all these ideas concern visual patterns of figures or shapes. As in Seurat's painting, any figure in geometry may be thought of as being made up of points. Geometry, being the study of visual patterns, can be thought of as the study of sets of points.

A surprising fact is that points themselves are not always the same. Nor are lines. In this chapter, you will first learn about different ways of describing points and lines. Then you will study two of these descriptions in more detail because they are the most important in this book.

Making Pictures from Dots

Seurat's actual painting, pictured on the previous page, is over 6 feet high and about 10 feet wide. A closer view of a part of the painting is shown here. If you look at the actual painting from a few feet away, you can see the individual dots that make it up. But if you are far away, or if the picture is reduced in size, you do not see the individual dots.

Seurat probably did not realize that his idea of making pictures from dots would become commonplace within 100 years. Today, television screens, computer monitors, and graphics calculator displays are made up of tiny dots, called **pixels.** Combinations of the pixels make up the pictures you see. The pixels are so numerous and so close together that what you see appears connected. The pixels are arranged in a rectangular array of rows and columns, called a **matrix.** Some Apple Macintosh screens have 640 columns and 480 rows. An IBM PC computer commonly uses a screen with 1024 columns and 768 rows. If the screens are the same size, the IBM PC screen will have more pixels per square inch. It would have sharper pictures. We say the IBM PC allows better *resolution.*

Some computer printers and most cash registers are *dot-matrix printers.* The more rows and columns in the matrix, the better the resolution of the printer, and the better the letters and figures look. For example, a printer might use a matrix with 9 rows and 8 columns. That matrix has 9 · 8, or 72, cells. To print a letter, the printer puts dots in the centers of particular cells. One way of making a capital (upper-case) "A" and a small (lower-case) "j" is shown on the following page.

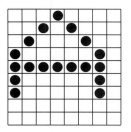 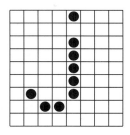

Not all dots are small. Signs can be formed by light bulbs arranged in a large matrix. By turning light bulbs on and off quickly, the letters can look as though they are moving. In this way, a long message can be put in a little space. Part of a message is shown below.

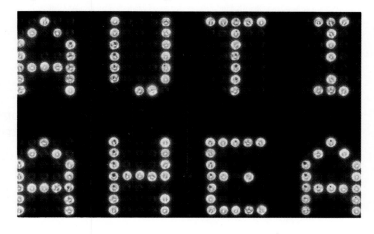

Marching bands form letters and simple pictures. When creating a new band formation, the drillmaster treats each band member as a dot. The more people in a band, the more complicated the pictures that can be created. Below is a band formation from a half time at a University of Michigan football game.

A Discrete Geometry

In this chapter, you will see four descriptions of *points.* The previous examples are all examples of points as dots. Since *lines* are made up of points, for each description of points, there is a corresponding description of lines.

Different descriptions of points and lines lead to different types of geometry. When a point is a dot, a line is made up of points with space between their centers. Lines made from dots are called **discrete lines.** The study of points as dots and discrete lines is one type of **discrete geometry.**

Discrete Geometry
Description of a point
A point is a dot.
Description of a line
A line is a set of dots in a row.

There are discrete geometries where points do not have size. In this book, when we refer to discrete geometry, points have size and shape.

Every line is either **horizontal, vertical,** or **oblique.** Discrete lines go on forever, so we can draw only parts of them.

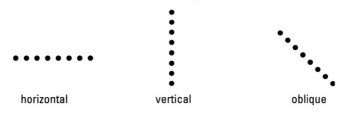

horizontal   vertical   oblique

Points and Lines in Discrete Geometry

When a point is thought of as a dot, lines have some thickness, and between two points on a line there may or may not be other points. There may be more than one line through two dots, depending on whether you require that the line go through their centers. It is possible for two of these lines to cross without having any points in common. The overhead view of an airport showing neat arrangements of lights provides examples of lines that cross with or without having any points in common.

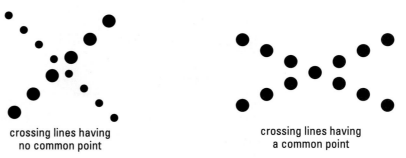

crossing lines having
no common point

crossing lines having
a common point

Lines drawn on a calculator or computer screen must be discrete, because the screens are composed of square pixels. When you use drawing programs to investigate geometric figures and their properties, oblique lines on the screen may look like steps up close. This is because they are composed of discrete points.

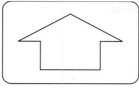

how the drawing of an
arrow appears on the screen

how the peak of the arrow
might appear close up

QUESTIONS

Covering the Reading

Use these questions to check your understanding of the reading. If you cannot answer a question, you should go back to the reading for help in obtaining an answer.

1. Who painted *Sunday Afternoon on the Island of La Grande Jatte?*

2. What is a pixel?

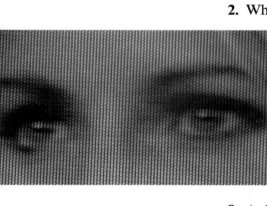

3. *True or false.* If a computer screen has more pixels than another screen of the same size, it has better resolution.

4. *True or false.* Two given points can be on several discrete lines.

5. A dot is one description of a(n) __?__.

6. In the "light bulb" road sign pictured in this lesson, part of a message is given. What might be the complete message?

7. What is *discrete geometry?*

8. A discrete line may be __?__, __?__, or oblique.

9. Oblique lines on a computer often look like __?__ up close.

10. What two different shapes are used for points in this lesson?

11. Draw two discrete lines, one horizontal and one vertical, with a point in common.

12. Draw two discrete lines which cross, but have no points in common.

Applying the Mathematics

These questions extend the content of the lesson. You should study the examples and explanations if you cannot get an answer. For some questions, you can check your answers with the ones in the back of the book.

13. **a.** How many total pixels are on a commonly used IBM PC screen?
b. How many total pixels are on the Macintosh screen mentioned in the lesson?

14. If two screens have the same dimensions, which has the better resolution, one with 200 rows and 300 columns of pixels, or one with 150 rows and 310 columns of pixels?

15. Pictured at the right is a horse's head as drawn with the help of a computer. The pixels are square.
a. Which are used to form this picture, black squares or white squares?
b. How are some parts made to look darker than others?

16. Use a 9-row by 8-column dot matrix like the one at the left to create a zero and a capital letter O that are different.

Getting to the point.
Shown is a detail of a 1972 painting by artist Albert Nagele. Nagele used the pointillism painting style created by Georges Seurat.

Review

Every lesson contains review questions to give you practice on ideas you have studied earlier.

17. What is the area in square feet of the Seurat painting shown on page 5? *(Previous course)*

18. Suppose $4x - 3y = 12$. What is the value of x if $y = 6$? *(Previous course)*

19. Graph on a coordinate plane: (6, 5), (6, -4), (2.5, 0.9), and $\left(-\frac{1}{2}, -\frac{1}{2}\right)$. *(Previous course)*

20. Evaluate each expression. *(Previous course)*
a. $|-23|$ **b.** $|4 - 19|$ **c.** $|19.3 - 11|$ **d.** $|7 - -5|$

Exploration

These questions ask you to explore mathematics topics related to the chapter. Sometimes you will need to use dictionaries or other sources of information.

21. Find an example of a picture made up of dots in a newspaper or other reading matter.

22. Use a dictionary to find the meaning of the term *pointillism*.

LESSON

1-2

Locations as Points

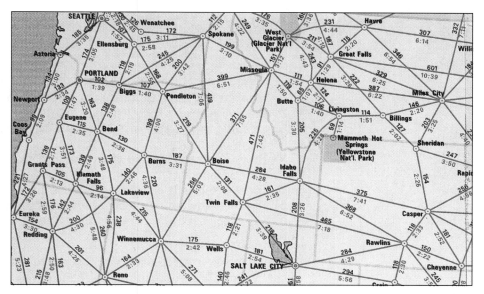

Road trip. *Many atlases contain maps that show the distances and travel times between major cities. Travelers can use these maps to estimate travel time.*

Measuring Distance Between Cities

You can find tables of distances from cities to other cities in atlases, maps, and almanacs. For instance, the *1995 World Almanac* gives the road mileage from New York to Los Angeles as 2786 miles. But New York and Los Angeles are both large cities, each many miles across. A location from which to calculate distances has to be chosen in each city. It may be the city hall, the control tower of an airport, or the intersection of two main streets.

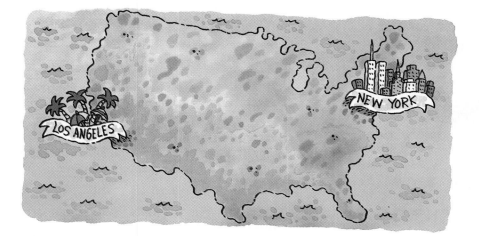

The road distance from New York to Los Angeles is not the same as the air distance, which the *1995 Information Please Almanac* lists as 2451 miles. Distance between two points depends on the route or path you take.

● ●
I II

1. Measure the distance in centimeters between these two dots.
2. Did you measure from left side of I to left side of II? Right side of I to left side of II? Center of I to center of II? Some other way?
3. How would you explain your choice to someone who had done it differently?
4. Draw a segment connecting the two dots.
5. Describe how you decided where to begin and end your segment.

As you can see from this activity, even straight-line measurements of distance can differ, depending on the accuracy of the measuring tools and exactly which locations you're measuring between. In fact, if the points have any size at all, your own measurements taken at different times may not agree.

Points as Exact Locations

A great many mathematical discoveries were made in the Greek empire from about 550 B.C. to 150 A.D. Greek mathematicians of that period made significant advances in number theory and geometry. They considered points not as actual physical dots, but as idealized dots with no size. For them, a point represented an *exact location* of this idealized dot.

When two points are exact locations, there is exactly one line containing them. This line contains the shortest path connecting them. The study of this second description of points and lines is called **synthetic geometry.**

Exactly! *Shown is Elizabeth Felton, a college student who repositioned the marker for the South Pole in January, 1995. Glacial ice shifts about 11 yards a year. Moving the marker compensates for the shift.*

Synthetic Geometry
Description of a point
A point is an exact location.

Description of a line
A line is a set of points extending in both directions containing the shortest path between any two points on it.

Number Lines

To find the unique distance between two locations on a line, a number can be assigned to each location. This creates a **number line.** You can choose any point you want for a zero point and either side of zero can be the positive side. Every number identifies a point and every point is identified with exactly one number, its **coordinate.** The line is said to be **coordinatized.** On the oblique line below, point A has coordinate 3; point B has coordinate -1.

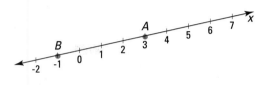

Number lines are **dense.** That is, between any two points you can always find another point, and there are points as close together as you wish.

Distance on a Number Line

The distance between two points can be calculated from their coordinates. Recall from algebra that the absolute value of a number n is written $|n|$. When n is positive, it equals its absolute value. For instance, $|6.2| = 6.2$. The absolute value of a negative number is the opposite of that number: $|-2500| = 2500$. The absolute value of 0 is 0: $|0| = 0$.

Using the ideas of a coordinatized line and absolute value, *distance* is defined as follows.

> **Definition**
> The **distance** between two points on a coordinatized line is the absolute value of the difference of their coordinates. In symbols, the distance between two points with coordinates x and y is $|x - y|$.

When we want to emphasize what you should write in answer to a question, **we use printing that looks like this.**

The distance between points A and B is written $\boldsymbol{AB}$.

> **Example**
> Find AB on the number line above.
>
> **Solution**
> The coordinate of A is 3. The coordinate of B is -1. Find the difference of the coordinates by subtracting. Then take the absolute value.
> $$AB = |3 - (-1)| = |4| = 4.$$
>
> **Check**
> Count the units to verify that there are 4 units between A and B.

Caution: When A and B are points, AB always means the distance from A to B, not their product. You cannot multiply points.

Jump start. *Shown is a long jump at the National Junior Olympics in San Antonio. The jump is measured from the near edge of the jumping board to the closest mark in the sand. The official checks that the jumper did not cross the edge.*

Because of properties of absolute value, if you switch the order of the points, the distance will be the same:

$$BA = |\text{-}1 - 3| = |\text{-}4| = 4.$$

For any two points A and B, $AB = BA$.

Tape measures (when stretched) and rulers resemble coordinatized lines. For example, if you wish to buy draperies or blinds for a window, you need the dimensions of the window. Suppose a tape measure crosses the top of a window at the 6-inch mark and the bottom at the 81-inch mark. Then the window is $|6 - 81|$, or $|\text{-}75|$, or 75, inches tall. When you connect two locations with a ruler, the edge of the ruler lies on a line. Whenever you use a ruler, you are applying the definition of distance.

QUESTIONS

Covering the Reading

1. **a.** According to the *1995 World Almanac,* what is the road mileage from New York to Los Angeles?
   b. According to the *1995 Information Please Almanac,* what is the air distance from New York to Los Angeles?

2. Show your answer for the activity in the lesson.

3. What did Greek mathematicians consider a point to be?

4. What is a description of *point* in synthetic geometry?

5. What does it mean for a line to be *dense?*

6. Define the distance between two points on a coordinatized line.

7. The symbol for the distance between two points A and B is __?__.

In 8–10, give the distance between two points with the given coordinates.

8. -321 and 32   9. 3 and -4   10. x and y

11. A person stretches a tape measure over a table. At one edge of the table the tape measure reads 1″. At the other edge the tape measure reads 45″. How long is the table from edge to edge?

Applying the Mathematics

12. The *1995 Information Please Almanac* gives 2825 miles as the road mileage from New York to Los Angeles. How is it possible that this distance differs from that in Question 1?

13. Why is the road distance from New York to Los Angeles greater than the air distance?

14. **a.** On a drive to Dinosaur, Colorado, the site of Dinosaur National Monument where over 350 tons of dinosaur bones were found, you pass a road marker that reads "Dinosaur National Monument, 88 miles." Along the same road you pass a later marker that reads "Dinosaur National Monument, 52 miles." How far have you traveled between markers?
   b. If the first road marker reads "x km to Dinosaur National Monument" and the second reads "y km to Dinosaur National Monument," how far have you traveled?

15. Use the number line below.

   a. Calculate AB, BC, and AC.
   b. *True or false.* $AB + BC = AC$.

16. Is a line in discrete geometry dense? Give an explanation for your answer.

The Parthenon, an ancient Greek temple built around 438 B.C. to honor Athena, the patron goddess of the city of Athens, is constructed entirely of white marble.

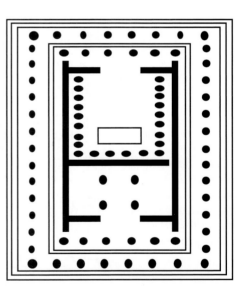

17. Above right is a floor plan for the Parthenon which contains both discrete and dense segments.
   a. What do the discrete segments represent?
   b. What do the dense segments represent?
   c. Do you believe that these representations give a clear picture of the actual building?
   d. Think of another situation where a plan would include both discrete and dense segments.

18. Use the following road mileage chart for four cities in Ohio.

	CINCINNATI	CLEVELAND	COLUMBUS	TOLEDO
CINCINNATI		244	108	200
CLEVELAND	244		139	111
COLUMBUS	108	139		133
TOLEDO	200	111	133	

How much farther is the drive from Cincinnati to Cleveland if you stop in Toledo?

Review

A

19. Draw an oblique discrete line. *(Lesson 1-1)*

B

20. How many different discrete lines contain points *A* and *B* at the left? Explain your answer. *(Lesson 1-1)*

Shown is the Senate Tower and the Lenin Mausoleum at the Kremlin in Moscow. The Kremlin is the center of the Russian government.

21. Graph the line containing (-2, 1) and (5, -6) on a coordinate plane. *(Previous course)*

22. For the equation $x - 3y = 5$, find the value of y for each given value of x. *(Previous course)*

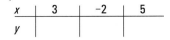

x	3	-2	5
y			

Exploration

23. To the nearest 100 miles, how far do you live from each of the following cities?
   a. New York   **b.** Los Angeles
   c. Honolulu   **d.** Moscow

24. Look up the word *synthetic* in a dictionary. Why do you think that the type of geometry in this lesson is called *synthetic geometry*?

1-3

Ordered Pairs as Points

Just "plane" wood. *A carpenter's plane, like the one shown, is used to smooth the edge of the wood so that it lies in one plane.*

A *plane* is a set of points thought of as something flat, like a tabletop. In small spaces, like a classroom, the floor can be considered a model of a plane. An unbent sheet of paper is part of a plane. The surface of Earth is *not* a plane because it is curved.

What is Coordinate Geometry?

Around the year 1630, the French mathematicians Pierre de Fermat and René Descartes realized that a location in the plane can be identified by an **ordered pair** of real numbers. At the right, the three points (0, 0), (3, 2), and (-5.3, 4.8) are graphed. The study of points as ordered pairs of numbers is called **plane coordinate geometry.**

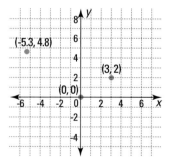

The plane containing these points is called the **Cartesian plane** (named after Descartes' Latin name *Cartesius*) or, more simply, the **coordinate plane.** For the point (*a*, *b*), *a* is the *x-coordinate* and *b* is the *y-coordinate*. The coordinates of a point are determined by the ***x*-axis** and the ***y*-axis.**

Lines as Sets of Ordered Pairs

In algebra you graphed lines and points in the coordinate plane. Here we review that skill.

Example 1

Graph the set of ordered pairs (x, y) satisfying $3x - y = 5$, and check your answer.

Solution

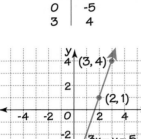

Find two points on the line. You can choose any value for x that you want. Substitute that value for x into the equation and solve for y.

If x = O,   If x = 3,
then $3 \cdot 0 - y = 5$.   then $3 \cdot 3 - y = 5$.
                  $-y = 5$ $9 - y = 5$
                   $y = -5$ $-y = -4$
                                                           $y = 4$
Graph (0, -5).   Graph (3, 4).

Draw the line through (0, -5) and (3, 4). Put arrows at both ends to indicate that the line continues in each direction.

Check

To check the graph, find a third point on the line.
When x = 2, $3 \cdot 2 - y = 5$
                   $6 - y = 5$
                    $-y = -1$
So $y = 1$.
The point (2, 1) is on the line. So it checks.

The Standard Form of an Equation for a Line

The form of the equation of the line in Example 1, $Ax + By = C$, where A and B cannot both be zero, is called the *standard form of an equation for a line.* It can be used for all lines in the plane. The values of A, B, and C determine the tilt and location of the line. When $A = 0$, the equation of the line is $By = C$ and the line is horizontal. When $B = 0$, the equation of the line is $Ax = C$ and the line is vertical. When neither A nor B is zero, the line is oblique (neither horizontal nor vertical).

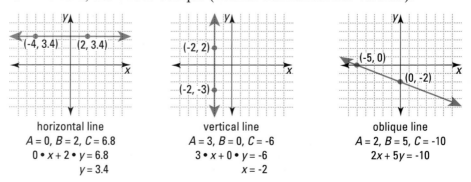

horizontal line
$A = 0, B = 2, C = 6.8$
$0 \bullet x + 2 \bullet y = 6.8$
$y = 3.4$

vertical line
$A = 3, B = 0, C = -6$
$3 \bullet x + 0 \bullet y = -6$
$x = -2$

oblique line
$A = 2, B = 5, C = -10$
$2x + 5y = -10$

The Slope-Intercept Form of an Equation for a Line

As Example 1 shows, it is useful to solve an equation of a line for y. This can always be done if the line is not a vertical line. The resulting form, $y = mx + b$, is called the *slope-intercept form of an equation for a line*, since m is the slope of the line and b is the y-intercept. But slope-intercept form cannot be used for the equation of every line in the plane. Vertical lines have an undefined slope and no single y-intercept. Lines with equations in slope-intercept form are easy to graph.

Example 2

Graph the line with equation $y = 3x - 1$ on the coordinate plane, and check your answer.

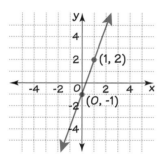

Solution 1

When $x = 0$, $y = 3 \cdot 0 - 1 = -1$. This gives the point $(0, -1)$.
When $x = 1$, $y = 3 \cdot 1 - 1 = 2$. This gives the point $(1, 2)$.
Draw the line through the two points.

Check

To check the graph, find a third point on the line.
When $x = 2$, $y = 3 \cdot 2 - 1 = 5$.
The new point $(2, 5)$ is on the line so it checks.

Solution 2

Use an automatic grapher. Here is a key sequence for one graphing calculator. Others may require a different sequence.

[y=] 3 [x] [x, T, θ] [–] 1 [GRAPH]

At the right is the output we get on the standard window for this calculator.

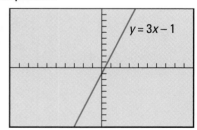

$-10 \leq x \leq 10$, x-scale $= 1$
$-10 \leq y \leq 10$, y-scale $= 1$

Check

Use the TRACE feature of the calculator to verify that when $x = 0$, $y = -1$, and when $x = 1$, $y = 2$. This checks two points, which is enough to check the line.

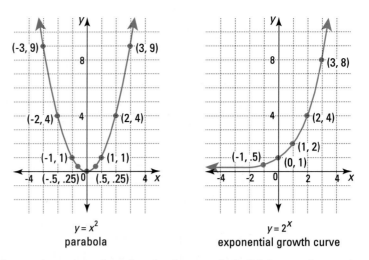

$y = x^2$
parabola

$y = 2^x$
exponential growth curve

Describing points as ordered pairs is very helpful in mathematics. Sets of ordered pairs may be a curve, and equations can describe that curve. The graphs above depict a parabola and an exponential growth curve, two curves that you may have studied or will learn about in other courses. Each curve can be described by an equation.

QUESTIONS

Covering the Reading

1. **a.** Which two mathematicians developed the idea of using ordered pairs of numbers to represent points?
   b. About how many years ago was this done?

2. In plane coordinate geometry, a point can be described as __?__.

3. **a.** $Ax + By = C$ is called the __?__ form for an equation of a line.
   b. Why is this form for a line useful?

In 4–7, classify the line with the given equation as vertical, horizontal, or oblique.

4. $x - 3y = 5$

5. $y = -1.234$

6. $x = 8$

7. $14 - 9y = 32x$

In 8–11, graph the set of ordered pairs satisfying each equation.

8. $x + y = 0$

9. $x + 2y = 4$

10. $y = -2x + 1$

11. $y = \frac{3}{2}x - 2$

12. *Multiple choice.* Which is an equation for an exponential growth curve?
   (a) $y = x^2$
   (b) $y = 2^x$
   (c) $y = 2x$

13. Give an equation for the horizontal line that passes through $(0, -5)$.

14. Give an equation for the vertical line that passes through $(7, -10)$.

15. **a.** Graph the line with equation $x = 3$.
   b. Graph the line with equation $4x + 3y = 6$ on the same coordinate plane.
   c. At what point(s) do the lines intersect?

16. Refer to the graphs of the parabola and the exponential growth curve on page 20. What labeled ordered pair is in both graphs? Do you think there are others? Explain why or why not.

17. Elaine has $12 in quarters and dimes.
   a. Name five ordered pairs (q, d) of quarters and dimes she might have.
   b. Graph these points.
   c. Do these points lie on the same line?

Review

18. Give the distance between two points on a number line with coordinates 1 and 10.   *(Lesson 1-2)*

19. If the ruler at the left is marked in centimeters, about how long is the segment AB?   *(Lesson 1-2)*

20. Suppose C has coordinate -7 and D has coordinate -211.
   a. Find CD **b.** Find DC.   *(Lesson 1-2)*

21. Use the road mileage chart at the right for the four largest cities in Georgia. Suppose you want to travel from Columbus to Savannah.
   a. How much further is the drive if you stop in Atlanta?
   b. At 55 mph, about how much longer would it take you if you go through Atlanta?
   (Lesson 1-2, Previous course)

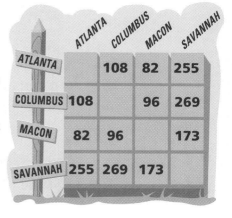

	ATLANTA	COLUMBUS	MACON	SAVANNAH
ATLANTA		108	82	255
COLUMBUS	108		96	269
MACON	82	96		173
SAVANNAH	255	269	173	

22. Draw two discrete vertical lines. *(Lesson 1-1)*

Exploration

23. The latitude and longitude of a location on Earth are like coordinates of a point. Look on a map to find the latitude and longitude of the place where you live. Compare these to the latitude and longitude of a place you would like to visit.

The XXVIth Olympiad.
Shown is project director, Scott Braley with a model for the 1996 Olympic Stadium in Atlanta.

Points in Networks

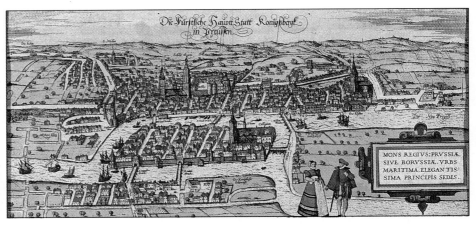

Bridging the gap. *This historic etching, picturing Königsberg, predates Euler. Note that only six bridges existed and were shown. The seventh bridge was built later.*

The Königsberg Bridge Problem

Through the city of Kaliningrad, in Russia, flows the Pregol'a River. There are two islands in this river, and seven bridges connect the islands to each other and to the shores. In the drawing below, which first appeared in an article by the great mathematician Leonhard Euler (pronounced "Oiler"), the islands are *A* and *D*. The bridges are *a, b, c, d, e, f,* and *g*. The shores of the river are *B* and *C*.

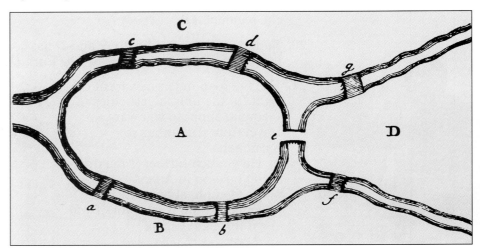

In the 1700s, this city was part of East Prussia and was known as Königsberg. It was common on Sunday for people to take walks over the bridges. These walks and bridges led to a problem.

Activity 1

Königsberg Bridge Problem: Find a way to walk across the seven bridges of Königsberg so that each bridge is crossed exactly once. Before you read the next page, take a few minutes to try to solve the Königsberg Bridge Problem.

The Königsberg Bridge Network

Euler solved the Königsberg Bridge Problem in 1736. First he named the islands, shores, and bridges as shown on the previous page. He then redrew the map with the islands *A* and *D* very small, and with longer bridges. This does not change the problem.

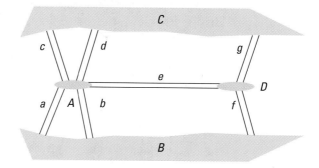

Then he realized that the shores *B* and *C* could be small. This again distorts the picture but it does not change the problem.

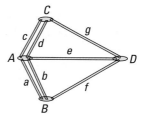

Finally—and this was the big step—he thought of the land areas *A, B, C,* and *D* as points and the bridges *a* through *g* as **arcs** connecting them. The result, shown below, is a **network** of points and arcs. In this network there is a path (though not necessarily direct) from any point to any other point.

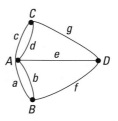

Networks are sometimes called **graphs,** and the geometry of networks is called **graph theory.** In a network, the *only* points are the endpoints of arcs. These endpoints have no size and are called **nodes** or **vertices.** The arcs are the lines of graph theory.

Graph Theory
Description of a point
A point is a node of a network.
Description of a line
A line is an arc connecting either two nodes or one node to itself.

Properties of Graph Theory

Euler was able to rephrase the Königsberg Bridge Problem in this way: *Without lifting a pencil off the paper, can one trace over **all** the arcs of this network exactly once?* If the answer is yes, this kind of network is called **traversable.** Euler noticed that the number of arcs at each node provided a clue as to whether or not a network was traversable. In the Königsberg Bridge Problem, there are 5 arcs at vertex A (c, d, e, b, and a) and 3 arcs at each of vertices B, C, and D. Below are some other networks showing different combinations of arcs and nodes.

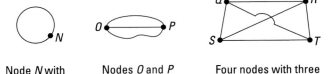

| Node N with arc to itself | Nodes O and P with 3 arcs each | Four nodes with three arcs each (arcs QT and SR do not intersect) |

There are no points in the middle of an arc, but between two nodes there can be many different arcs connecting them.

Before we show you Euler's answer to the Königsberg Bridge Problem, here are three more networks to consider. In these networks, the arcs are drawn as segments.

Example 1

How many arcs are at each node of networks I, II, and III?

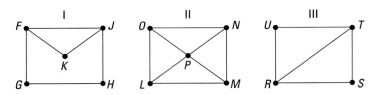

Solution

Network I: There are 2 arcs at nodes G, H, and K, and 3 at nodes F and J.

Network II: There are 3 arcs at nodes L, M, N, and O, and 4 arcs at node P.

Network III: There are 2 arcs at S and U, and 3 arcs at R and T.

Activity 2

Trace network III above. Is the network traversable if you start at
a. node R?   **b.** node S?   **c.** node T?   **d.** node U?

If the number of arcs at a node is even, the node is called an **even node.** Otherwise, it is an **odd node.** In Example 1, nodes G, H, K, P, U, and S are even nodes while nodes F, J, L, M, N, O, R, and T are odd nodes.

Euler's Solution to the Königsberg Bridge Problem

Euler realized that when a path goes through a node, it uses two arcs, one to the node, and one away from it. This led him to realize that when a network has an odd node, it *must* be the starting or finishing point for a traversable path. In Activity 2, you should have found that Network III is traversable only if you start at *R* or *T*. If you started at *R*, then you ended at *T*, and vice versa.

Euler saw that all four nodes in the Königsberg network are odd. Since a traversable network can have only one starting point and one finishing point, the Königsberg network is *not* traversable. Whenever a network has more than two odd nodes, it is not traversable.

Example 2

Which of the networks in Example 1 are traversable? Explain your answer.

Solution

Network I: Since it has only 2 odd nodes, it is traversable. A path must begin at *F* or *J* and end at the other. One such path is *F* to *K* to *J* to *H* to *G* to *F* to *J*.

Network II: Since it has more than 2 odd nodes, network II is not traversable.

Network III: It has only two odd nodes, so it is traversable. A possible path is *R* to *S* to *T* to *U* to *R* to *T*.

Problems like the Königsberg Bridge Problem may seem frivolous or silly, but there are important real situations that are similar. For example, companies, security agents, and bus drivers can improve their effectiveness by using networks to answer the following questions:

Can a repair crew for a telephone company inspect all the lines without going over any section twice?

What is the most efficient way for a security agent to patrol a museum?

Can a bus route be set up so that the bus drives on each street only once?

QUESTIONS

Covering the Reading

1. How many bridges connected the two islands to each other and to the shores of Königsberg in the 1700s?

2. What is the Königsberg Bridge Problem?

3. Who solved the Königsberg Bridge Problem and when?

In 4 and 5 refer to the Königsberg network.

4. What do the nodes in this network represent?

5. What do the arcs in this network represent?

6. What is the *geometry of networks* called?

7. What is the description of a point in this lesson?

8. What is the description of a line in this lesson?

9. What does it mean for a network to be *traversable?*

10. What did Euler notice about traversable networks that enabled him to solve the Königsberg Bridge Problem?

11. For Network I in Example 1, describe a traversable path different from the one given in the solution to Example 2.

12. Describe a real situation that uses networks.

13. Show your answers to Activity 2.

Applying the Mathematics

14. The network below is traversable.

a. Describe a traversable path.
b. At what vertices can a traversable path begin?

In 15 and 16, a network is given. **a.** Determine the number of even nodes and the number of odd nodes. **b.** Tell whether the network is traversable.

15.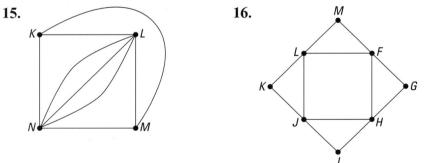

16.

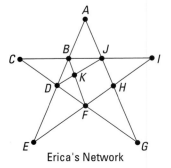
Erica's Network

17. Erica, a student who used the first edition of UCSMP *Geometry*, created the network at the left. Is her network traversable? Explain why or why not.

18. Refer to the etching at the top of page 22. Before the seventh bridge was built, could you walk across the six bridges so that each bridge is traversed exactly once?

19. Use the route map for Scandinavian Airlines pictured here.
   a. What is represented by the nodes?
   b. What is represented by the arcs?
   c. Is the network traversable?

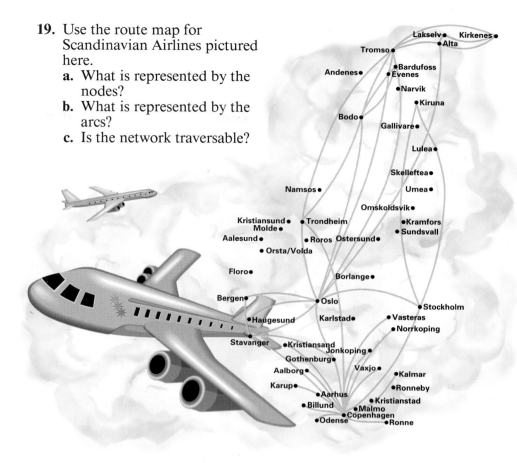

20. a. Draw a network with more arcs than nodes.
   b. Draw a network with more nodes than arcs.

21. An architect is planning a museum. A sketch is shown. Can a security guard walk through each aisle only once without going through any aisle twice? (Hint: Represent the intersections of the aisles as nodes, and the aisles themselves as arcs.)

Shown is an interior photo of the Orsay Museum in Paris, France.

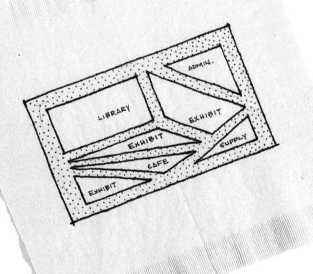

In 22 and 23, graph the set of ordered pairs satisfying each equation.

22. $4x - 3y = 6$ **23.** $x = -3$ *(Lesson 1-3)*

24. *True or false.* In plane coordinate geometry, points have size. *(Lessons 1-1, 1-3)*

25. Using the number line below, find *CD*.   *(Lesson 1-2)*

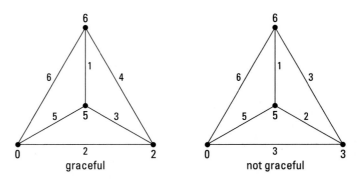

26. In measuring the width of a picture, Raina found that the top left corner was aligned to the 4 in. mark and the bottom left corner was aligned to the $21\frac{1}{2}$ in. mark. How wide is the picture?   *(Lesson 1-2)*

27. *True or false.* In a discrete line, between two points there is always a third point.   *(Lesson 1-1)*

28. There are many puzzles using networks. Here is one. Start with a network that has *n* arcs. (The networks below have 4 nodes and 6 arcs.) Name each *node* with a different number from 0 to *n*. Then number each arc by the absolute value of the difference of the nodes it connects. For example, in the network below at the right, the arc connecting nodes 5 and 3 is named 2 because $|5 - 3| = 2$. The goal is to name the nodes in such a way that the *n* arcs are numbered with all the integers from 1 to *n*. Such a network is called a *graceful* network.

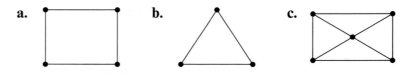

Number the nodes in these three networks so as to make them graceful.

a.   **b.**   **c.**

Drawing a Box in Perspective

IN-CLASS
ACTIVITY

At the right is a box that is not drawn in perspective. Drawing a figure in perspective gives a feeling of realistic depth.

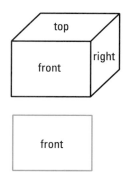

top
right
front

1 Copy the rectangle that represents the front face of the box. Make sure the sides are horizontal and vertical.

front

2 Mark a point (the vanishing point) outside your rectangle. Draw all segments that connect the vanishing point to a vertex of the rectangle. Each segment must lie entirely outside the rectangle except for the vertex. So you will draw only 3 segments.

Vanishing point

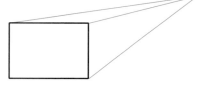

3 Draw a vertical segment between the lower two oblique lines to locate an edge of the back face of the box. The placement of this segment shows the depth of the box.

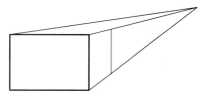

4 Draw a horizontal segment between the top two oblique lines, connecting with the segment you just drew.

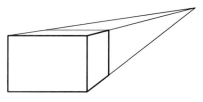

5 Shade in the box.

6 Using the steps above, redraw the "shoebox" at the right in perspective.

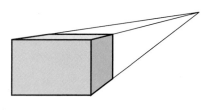

Courtesy of the Regis Collection Minneapolis, MN.

What is Perspective Drawing?

Drawings are done on a two-dimensional surface, but the world is three-dimensional. Refer to the picture of the railroad tracks on page 4. We know that railroad tracks are parallel and thus do not intersect, but in life and in many drawings they look as if they will meet if extended. The point where they seem to meet is called a **vanishing point.**

In the painting titled *Overseas Highway* shown above, artist Ralston Crawford has used a technique called **perspective drawing** to create a feeling of depth. You can see that the sides of the floor of the bridge meet at the vanishing point. Likewise, the railings on both sides of the bridge are drawn to the same vanishing point. Artists create a sense of three-dimensional space by choosing a horizon line and vanishing point(s) on the horizon line.

Below is an example of perspective drawing in block lettering.

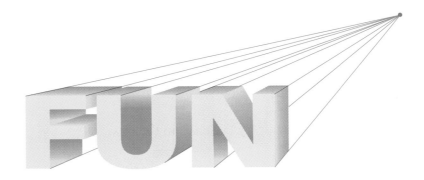

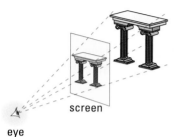

How Perspective Drawings Work

Renaissance artists imagined a clear screen between the viewer and the object they wanted to draw. The artist's eye was the vanishing point. Points on the object were connected to the eye with lines called **lines of sight**, as shown at the left. The points where the lines of sight intersected the screen showed where to draw the object. In the In-class Activity, you drew a box in perspective by first drawing the vanishing points and then lines of sight.

Perspective drawings show objects and relationships on a plane in a way that make the objects look three-dimensional. In perspective drawings, lines that are perpendicular to the viewer's line of sight do not change. So vertical lines remain vertical, and some horizontal lines remain horizontal. But oblique parallel lines will intersect if extended. Sometimes there is more than one vanishing point. The box below has two vanishing points, *P* and *Q*. Vertical lines on the box remain vertical and parallel, but the box has been tilted and has no horizontal edges.

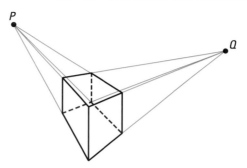

The box drawn below is tilted to have neither horizontal nor vertical edges. It has three vanishing points, *R, S,* and *T,* which are on the same line. They lie on the **vanishing line** for this drawing. In realistic drawings, the vanishing line is the horizon line and it corresponds to the height of the viewer's eye.

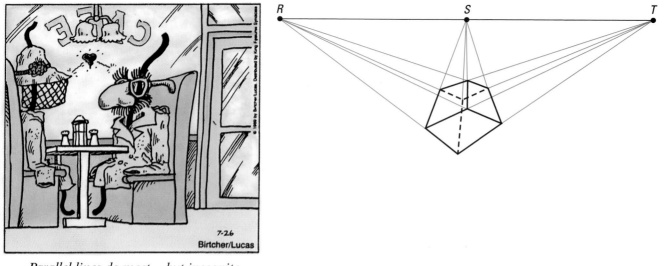

Parallel lines do meet—but incognito

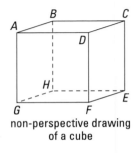

non-perspective drawing of a cube

Non-perspective Drawings

The artist uses perspective drawings in order to create the illusion of three-dimensional space. The mathematician, on the other hand, sometimes uses less realistic drawings in order to show clearly the geometric properties of figures. At the left, a cube is drawn as most mathematicians would draw it. Notice that the back of the cube, *BCEH*, is the same size as the front, *ADFG*. In the real world, the back of the cube is farther away than the front and it would look smaller. Artists trying to achieve realism would not like this drawing.

Geometers (mathematicians who study geometry) use another technique that preserves a feeling of depth. *Hidden lines,* such as the three back edges of the cube, are shown as dotted or dashed lines to indicate that they would not normally be seen.

In this book, you often will be asked to draw three-dimensional figures. You should follow the rules that geometers follow, and dot or dash hidden lines to show depth. You are not expected to use perspective unless asked.

The School of Athens *by Raphael (1483–1520), a fresco in the Vatican Palace.*

QUESTIONS

Covering the Reading

1. **a.** In a perspective drawing, do railroad tracks meet in the distance?
   b. In the real world, do railroad tracks actually meet in the distance?

2. What is a *vanishing point?*

3. What are *lines of sight?*

4. In a perspective drawing, tell whether these lines might intersect if extended.
   a. horizontal parallel lines
   b. vertical parallel lines
   c. oblique parallel lines

5. In general, do mathematicians use perspective in drawings?

6. How are hidden lines drawn in non-perspective drawings?

7. Draw the cube at the left in perspective.

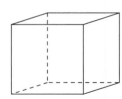

8. Consider these four drawings.

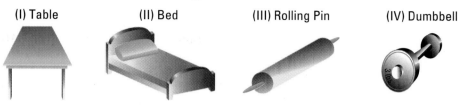

(I) Table   (II) Bed   (III) Rolling Pin   (IV) Dumbbell

    a. Which are drawn in perspective?
    b. Trace each perspective drawing and find its vanishing point.

9. Make a perspective drawing of a square floor tiled with square tiles shown at the left. Start with just the front edge of the floor and put the vanishing point on your paper far above the floor edge.

10. Consider the drawings of the two buildings below.
    a. Which appears greater, *BL* or *DG?*
    b. Measure to determine which is greater, *BL* or *DG?*
    c. Think of a reason that explains your answers to parts **a** and **b**.

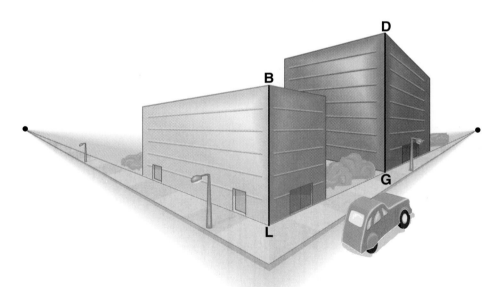

11. Use block lettering in perspective to print your initials.

12. Compare the figure given in Question 7 with your answer to the question.
    a. How many edges can you see in the original figure?
    b. How many edges can you see in your answer figure?
    c. How many other edges are drawn equal in length to the front left edge in the original figure?
    d. How many other edges are drawn equal in length to the front left edge in your answer figure?

In 13 and 14, is the network traversable? If so, give a path; if not, tell why not. *(Lesson 1-4)*

13.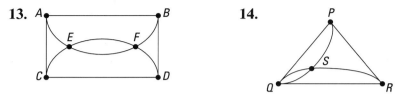

14.

15. **a.** Represent the floor plan at the left with a network.
    b. Is it traversable? *(Lesson 1-4)*

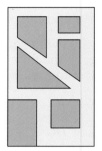

In 16 and 17, select one of the following as your answer.
(Lessons 1-1, 1-3, 1-4)
   (a) plane coordinate geometry
   (b) discrete geometry
   (c) graph theory

16. In which geometry do lines not contain an infinite number of points?

17. In which geometry do points have thickness?

18. Is the line with equation $4x + 0y = 31$ vertical, horizontal, or oblique? *(Lesson 1-3)*

19. **a.** Graph the lines $y = 3x - 2$ and $x + 2y = 10$ on the same coordinate plane.
    b. From the graph, estimate the coordinates of the point of intersection. *(Lesson 1-3)*

20. A tape measure was placed against a door. The left side was aligned to the 17 cm mark, and the right side to the 73 cm mark. How wide is the door? *(Lesson 1-2)*

21. Sally's baby brother has a block with ◖◗, ☼, ▲, ✳, ●, and ✜ on the six faces. Three views of the block are shown here. *(Previous course)*

Which pictures are on opposite sides of the block?

Exploration

22. In a newspaper, magazine, or book, find a picture that illustrates perspective.

The Need for Undefined Terms

The letter of the law. *Pictured here are lawyers and a judge in a courtroom in Portland, Oregon.*

Definitions are useful because they state precisely what words mean. Careful definitions are not always needed. However, in some fields, definitions are not just useful, but necessary. Law, economics, philosophy, science, and labor relations, in addition to mathematics, are some of the fields where precise definitions are necessary. For example, disputes may occur because individuals have no definitions or have different definitions for "overtime" or "freedom" or "force" or "obscene."

What Is the Meaning of Point?

Mathematicians try to define terms carefully. However, notice that the meanings of the words "space" and "figure" depend on what a point means. Trying to define "point" presents two problems. First, as you have seen in the previous lessons, there are many possible meanings for "point." A point may be a dot, or a location, or an ordered pair, or a node in a network.

Second, you can get into trouble if you try to select a meaning for "point." For instance, suppose you defined a point to be a *spot*. What about the word "spot"? Well, a spot is a *place*. But a place is an *exact location*. Trying to define this term, you might find that the best description of an exact location is a "point"!

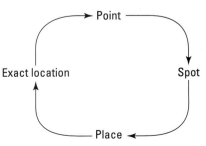

Thus you have returned to the original word which you were trying to define. You have circled back to where you started. When this "circling back" occurs, it usually means you are trying to define a basic term. It is called *circularity*. **Circularity** means that you have circled back to any word previously defined, not necessarily the original one. To avoid circularity, certain basic geometric terms must be *undefined*.

In this book, following a tradition begun by the German mathematician David Hilbert about 100 years ago, we choose *point, line,* and *plane* as **undefined terms.** One strength of this is that it allows geometry to be applied to different kinds of points and lines. For instance, we can talk about the distance between (4, 5) and (4, 2), because we can think of (4, 5) and (4, 2) as points, and not just pairs of numbers.

Also, very common English words, such as articles, prepositions, and conjunctions, are not defined. We assume that you are familiar with many words used in algebra or arithmetic, such as "equation," "number," "equal," "is less than," and so on. Other words, such as *if* and *then,* are the basic terms of logic, and are left undefined, too.

> **Undefined terms (in this book)**
> Geometric terms: point, line, plane
> Algebraic and arithmetic terms: number, equal, set, addition, . . .
> Logical terms: if, then, and, or, . . .
> Common English words: the, a, of, in, with, . . .

Definitions Using Undefined Terms

Definitions are made using the undefined terms. Here are definitions that apply to any of the kinds of geometries that you have studied.

> A **figure** is a set of points.
>
> The **space** of a geometry is the set of all points in that geometry.
>
> Three or more points are **collinear** if and only if they are on the same line.
>
> Four or more points are **coplanar** if and only if they are in the same plane.

When all points in a geometry are collinear, then the space and the geometry, and the figures in it, are called **one-dimensional.** A number line is one-dimensional. When the points in a geometry are not all collinear, but they are all coplanar, then the space and the geometry are called **two-dimensional.** A plane is two-dimensional. Figures that lie in a plane, such as squares, circles, and triangles, are two-dimensional.

Spheres, boxes, cubes, and most other real objects do not lie in a single plane. They are **three-dimensional** figures.

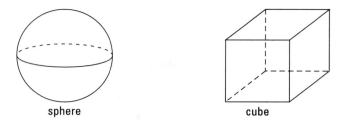

sphere   cube

For instance, the three-dimensional figure drawn at the left below is called a *cone*. Its circular base is a set of coplanar points. Points *A, D,* and *B* are collinear. In the box at the right below, points *F, G, L,* and *K* are coplanar.

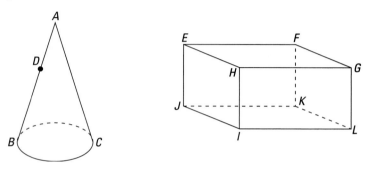

Comparing Different Descriptions of Points and Lines

If a word is not defined, it could mean anything. Point might mean "elephant"! So, to clarify that we are talking about a particular kind of point or line, we need to assume that points and lines have certain properties.

In this chapter, you have seen four different descriptions of points and lines. They are given below. Other descriptions are possible but are not discussed in this book.

Type of geometry	Description of point	Description of line	Example
discrete	dot	row of dots	computer screens
synthetic	location	straight path	maps
plane coordinate	ordered pair (x, y)	$Ax + By = C$	coordinate graphs
graph theory	node of network	arc of network	bridge network

As you have seen throughout this chapter, the type of geometry you work in affects the properties of figures and how figures are drawn. Even something as simple as representing a square region depends on how you describe terms such as point and line. Below are six different descriptions of a square region. Obviously, the figures are different. Figure (a) looks like a set of points in the shape of a square region, but the points are not connected. Figure (b) looks like the kind of square we would expect. Figure (c) seems to have only the vertices and is missing everything else. Figure (d) is a square region looked at from a different direction. Figure (e) is a square region graphed on the coordinate plane, and its algebraic description is Figure (f).

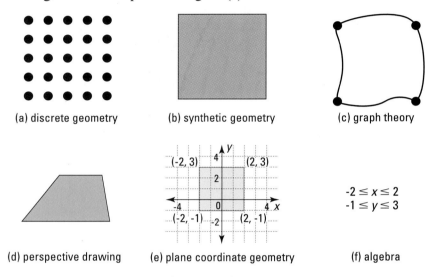

(a) discrete geometry　　(b) synthetic geometry　　(c) graph theory

(d) perspective drawing　　(e) plane coordinate geometry　　(f) algebra

These examples show that points and lines do not have the same properties when different descriptions are used. This is another reason why *point* and *line* are undefined.

QUESTIONS

Covering the Reading

1. In geometry, space is the __?__.

2. In this book, a figure is a(n) __?__.

3. In an attempt to define a simple word, such as "point," you may find the original word is used in the definition. This is called __?__.

4. Name three geometry terms undefined in this book.

5. Name two undefined words or phrases from algebra.

In 6–9, give the number of dimensions in each figure.

6. segment   **7.** square   **8.** point   **9.** suitcase

10. *True or false.*   Points, lines, and planes have the same properties in any geometry.

Applying the Mathematics

In 11 and 12, consider the clock pictured at the left.

11. a. Name three points that are collinear.
   b. Name three points that are not collinear, but are coplanar.

12. a. The actual clock is a(n) _____-dimensional figure.
   b. The drawing is _____-dimensional.

13. Look up the word *concord* in a dictionary. Try to find a one-word synonym. Look up this synonym. Find a one-word synonym for this second word. Continue this process. How many words did you look up before circularity happened?

In 14 and 15, follow the directions for Question 13 but begin with the following words.

14. inundate   **15.** satire

16. Define the word *number* on your own without a dictionary.

17. Refer to the six different views of a square region on page 38. Draw five views of a rectangular region.

18. Give the two meanings of "point" in the sign at the left.

Review

19. Do mathematicians generally draw in perspective? *(Lesson 1-5)*

20. Draw the railroad tracks at the left in perspective. *(Lesson 1-5)*

21. Graph the line with equation $5x + 3y = 15$. *(Lesson 1-3)*

22. Trace the cookie sheet below and find the vanishing point. *(Lesson 1-5)*

23. *Multiple choice.*   For which geometry is it true that through two points there is exactly one line? *(Lessons 1-1, 1-2, 1-4)*
(a) discrete geometry
(b) synthetic geometry
(c) graph theory

24. *The New York City Bridge Problem.* Below is a drawing of the five boroughs of New York City, Randall's Island, and New Jersey. (No water separates Brooklyn and Queens.) Bridges and tunnels connect the regions. Draw a network like the Königsberg bridge network to represent New York and determine if you could take a driving tour of New York going over each bridge and through each tunnel exactly once. *(Lesson 1-4)*

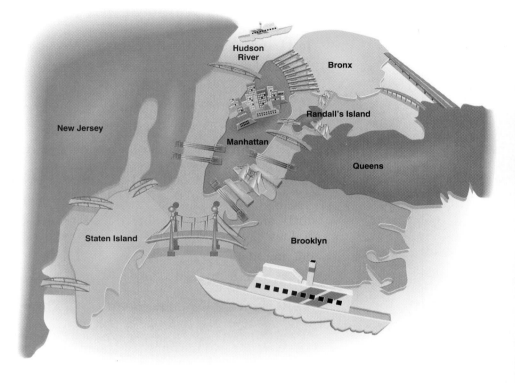

25. Calculate *AB.* *(Lesson 1-2)*

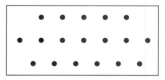

26. Pictured at the left is a plan for a garden. *(Lesson 1-1)*
   a. What is represented by discrete segments?
   b. What is represented by the dense segments?

27. On a number line, graph all solutions to the inequality $x \geq -10$.
(Previous course)

28. On a number line, graph all numbers t with $\frac{1}{2} \leq t \leq \frac{5}{2}$.
(Previous course)

Exploration

29. Some words have many definitions.
   a. Look in a dictionary. How many different definitions are given for the word "line"?
   b. Find another word with as many definitions.

Mathematician and teacher. *Shown is a detail of Raphael's* The School of Athens *showing Euclid with some of his students. Euclid founded and taught at a school in Alexandria, Egypt, which was then part of the Greek Empire.*

Different descriptions of points and lines lead to very different types of geometry. To make clear which description of *point* and *line* is being followed, assumptions or **postulates** are made about them. The first purpose served by postulates is to explain undefined terms. A second purpose, just as important, is to serve as a starting point for logically deducing, or proving, other statements. The postulates in this book were picked to fit the geometry known as **Euclidean geometry,** named after the Greek mathematician Euclid. The postulates for Euclidean geometry fit both the synthetic geometry and the plane coordinate geometry that have been described in earlier lessons.

The Point-Line-Plane Postulate of Euclidean Geometry

In this lesson we state three assumptions about points and lines. Together they are called the *Point-Line-Plane Postulate.* These assumptions fit Euclidean geometry but may not fit other geometries.

The first assumption of the Point-Line-Plane Postulate is sometimes read, "Two points determine a line."

Point-Line-Plane Postulate
a. Unique Line Assumption
Through any two points, there is exactly one line.

At the left, line ℓ is the unique line through points A and B. The symbol $\overleftrightarrow{AB}$ is read "line AB." $\overleftrightarrow{AB}$, $\overleftrightarrow{BA}$, and ℓ are three different names for the same line.

The Unique Line Assumption does not apply to graph theory, where there can be more than one line (arc) connecting two points (nodes). It also does not apply to lines in discrete geometry, as two points could be part of different lines that are near one another.

The second part of the Point-Line-Plane Postulate assures that lines in Euclidean geometry contain infinitely many points. So lines are *not* made up of dots placed next to one another.

> **Point-Line-Plane Postulate**
> **b. Number Line Assumption**
> Every line is a set of points that can be put into a one-to-one correspondence with the real numbers, with any point on it corresponding to 0 and any other point corresponding to 1.

Therefore, any line in Euclidean geometry can be made into a real number line. At the left, line ℓ is reproduced with point A corresponding to 0 and point B corresponding to 1.

The Number Line Assumption does not apply to graph theory as there are only 1 or 2 points on a given line in a graph.

The third part of the Point-Line-Plane Postulate assures that Euclidean geometry can study one-, two-, or three-dimensional figures.

> **Point-Line-Plane Postulate**
> **c. Dimension Assumption**
> (1) Given a line in a plane, there is at least one point in the plane that is not on the line.
> (2) Given a plane in space, there is at least one point in space that is not in the plane.

The surface of this page, which is two-dimensional, cannot contain a three-dimensional figure. So you have to use your imagination. A plane is usually drawn to look like a tabletop even though it goes on forever. Points not in the plane are drawn outside the tabletop.

In the figure below, line ℓ is in plane M, and point C is in M but not on line ℓ. Point D is in space but not in plane M.

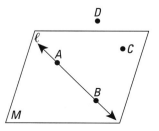

Intersecting and Parallel Lines

From the postulates, we can prove other properties of points and lines. A reasonable question is, "In how many different points can two different lines intersect?"

In algebra, if you solved the system $\begin{cases} 3x + 2y = 6 \\ 2x - y = 11 \end{cases}$ on the coordinate plane, you found (4, -3) as the point of intersection of the two lines. However, you also know that sometimes lines do not intersect, and sometimes two equations can describe the same line.

Suppose m and n are different lines. Can they intersect in two points, P and Q, as shown below? If they could, there would be two different lines through P and Q; yet from the Unique Line Assumption of the Point-Line-Plane Postulate, there can be only one line. We call this result the *Line Intersection Theorem*. A **theorem** is a statement which follows from postulates, definitions, or previously proved theorems.

P and Q are different points, and m and n are different lines. This is impossible in Euclidean geometry.

> **Line Intersection Theorem**
> Two different lines intersect in at most one point.

When coplanar lines do not intersect in exactly one point, they are called *parallel*. The idea behind parallel is "going in the same direction."

> **Definition**
> Two coplanar lines m and n are **parallel lines,** written $m \ /\!/ \ n$, if and only if they have no points in common or they are identical.

Why Is This Geometry Called Euclidean Geometry?

Euclid wrote the most famous organization of postulates and theorems (and perhaps the earliest) of all time, around 300 B.C., in a set of books called *Elements*. Euclid's *Elements* was still being used as a geometry text into the twentieth century. For these reasons, most of the world calls the geometry you study this year "Euclidean geometry."

Euclid's *Elements*. *The first Latin translation of* Elements *was made in about 1120 by Adelard of Bath who obtained a copy of an Arabic version. The first direct translation from the Greek without the Arabic intermediary was made by Bartolomeo Zamberti and published in Vienna in Latin in 1505.*

QUESTIONS

Covering the Reading

1. What is a *postulate?*

2. What two purposes do postulates serve?

3. What two views of points from this chapter are found in Euclidean geometry?

4. How is the Unique Line Assumption sometimes read?

5. If *m* and *n* are two lines, which assumption of the Point-Line-Plane Postulate is violated by this figure?

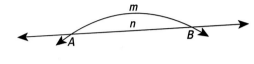

6. For the line at the left, which assumption of the Point-Line-Plane Postulate allows you to make zero the coordinate of *A* and 1 the coordinate of *B?*

7. *True or false.* A plane and space have the same dimensions.

8. If two coplanar lines do not intersect, then they are ___?___.

9. *True or false.* According to the definition of *parallel* in this lesson, every line is parallel to itself.

10. What is Euclid's *Elements?*

Applying the Mathematics

In 11 and 12, a statement is given. **a.** Determine if the statement is *true* or *false* in Euclidean geometry. **b.** Tell which assumption of the Point-Line-Plane postulate justifies your answer to part **a.**

11. A line contains infinitely many points.

12. For line ℓ in plane *M,* all points in *M* are also on line ℓ.

In 13 and 14, tell whether the statement is *true* or *false* in Euclidean geometry.

13. If two coplanar lines cross, then they have a point in common.

14. A line has thickness.

15. **a.** Graph $y = x$ and $y = x + 1$ on the same set of axes.
    b. Are the lines parallel?

16. Give an example of two different equations that describe the same line. Such lines are called **coincident lines.**

17. a. Draw 4 points *A, B, C,* and *D* that are coplanar with no three of them collinear.
  b. Draw a 5th point *E* that is not on the same plane as *A, B, C,* and *D.* *(Lesson 1-6)*

18. Draw the brick at the right in perspective.
  (Lesson 1-5)

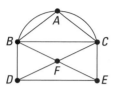

19. Refer to the network at the left.
  a. How many nodes does it have?
  b. How many arcs does it have?
  c. Is the network traversable? If so, give a route; if not, explain why not. *(Lesson 1-4)*

20. a. Represent the floor plan below with a network.
  b. Is it traversable? *(Lesson 1-4)*

21. Classify the line with the given equation as vertical, horizontal, or oblique. *(Lesson 1-3)*
  a. $y = 7$
  b. $2x - y = 9$

22. Lilly took a plane from Albuquerque to Boise and flew 764 miles. Nicole took a train from Albuquerque to Boise and rode 1831 miles. Kathryn drove 940 miles from Albuquerque to Boise. Why can all these distances be different? *(Lesson 1-2)*

23. If $CD = 12$ and *C* has coordinate of -4, what are the possible coordinates for *D?* *(Lesson 1-2)*

24. Use an encyclopedia or book on the history of mathematics and find the number of books in Euclid's *Elements.* How many have survived the centuries since 350 B.C.?

25. Physicists sometimes speak of space-time. How many dimensions does space-time have?

Lighting the way. *Shown is a Los Angeles studio experimenting with lasers as a possible game source of the future. Laser beams, powerful rays of light, are used in entertainment, business, medicine, and science.*

Activity

Find and then trace the two points *A* and *B* on this page. Draw the shortest path from *A* to *B*.

•*A*

The Idea of Betweenness

The path that you drew in the Activity above contains point *A,* point *B,* and all the points *between A and B.* To define precisely what is meant for a *point* to be between two other points, we start by considering numbers.

A *number* is **between** two others if it is greater than one of them and less than the other. For example, 5 is between -1 and 6.5, since $5 > -1$ and $5 < 6.5$. The number -1 is *not* considered to be between itself and 6.5.

A point is **between** two other points if it is on the same line and its coordinate is between their coordinates. For example, point *Q* above, with coordinate 5, is between the other two points *P* and *R*. Note that point *P* is *not* considered to be between *P* and *R* since the number -1 is not between -1 and 6.5.

Example 1

Suppose *D* and *F* have coordinates –53 and 670, respectively. If point *E* is between *D* and *F,* describe the possible values of the coordinate *x* of point *E*.

•
B

▶

▶ **Solution**

First draw a picture.

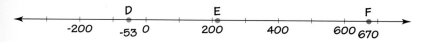

Since *x* has to be greater than -53 and less than 670, the range for *x* is expressed by the inequality -53 < x < 670.

Line Segments

The set of points that you drew in the Activity on page 46 is a *segment* with *endpoints A* and *B*.

> **Definition**
> The **segment** (or **line segment**) with **endpoints** A and B, denoted $\overline{AB}$, is the set consisting of the distinct points A and B and all points between A and B.

On a number line, the graph of two numbers, such as -1 and 6.5, and *all* points having coordinates between them, is a segment. The points for -1 and 6.5 are the endpoints of the segment.

The symbol $\overline{AB}$ is read "segment *AB*". On this number line, $\overline{AB}$ consists of all points on the line whose coordinates satisfy the inequality $-1 \le x \le 6.5$.

Rays

A *ray* is like a laser beam. A laser beam starts at a point and, if not blocked, continues forever in a particular direction. In geometry, a ray consists of an endpoint and all points of a line on one side of that endpoint. The idea of betweenness helps to give a precise definition of ray.

> **Definition**
> The **ray** with endpoint A and containing a second point B, denoted $\overrightarrow{AB}$, consists of the points on $\overline{AB}$ and all points for which B is between each of them and A.

A ray is named by its endpoint and any other point on the ray. The symbol $\overrightarrow{AB}$, read "ray AB", has an arrowhead above the point on the ray which is *not* the endpoint.

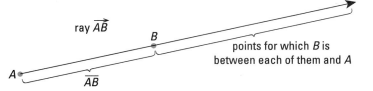

ray $\overrightarrow{AB}$

points for which B is
between each of them and A

$\overrightarrow{AB}$

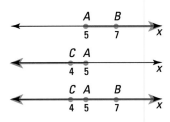

The graphs of the simplest inequalities in algebra are rays. For example, if A has coordinate 5 and B has coordinate 7, then $\overrightarrow{AB}$ consists of all points with coordinates $x \geq 5$.

If C has coordinate 4, then $\overrightarrow{AC}$ points in the opposite direction of $\overrightarrow{AB}$. $\overrightarrow{AC}$ consists of all points with coordinates $x \leq 5$. $\overrightarrow{AB}$ and $\overrightarrow{AC}$ are called *opposite rays*.

> **Definition**
> $\overrightarrow{AB}$ and $\overrightarrow{AC}$ are **opposite rays** if and only if A is between B and C.

Notice that, for any two points A and B, $\overrightarrow{BA}$ and $\overrightarrow{AB}$ are *not* opposite rays. They have many points in common, but $\overrightarrow{BA}$ has endpoint B, while $\overrightarrow{AB}$ has endpoint A.

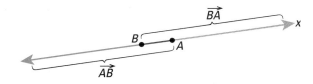

$\overrightarrow{BA}$

$\overrightarrow{AB}$

Assumptions About Distance

In the Activity, you were asked to draw the "shortest" path from A to B. In order to have the idea of "shortest" in our geometry, we need some assumptions about distance. The first two parts of the Distance Postulate repeat information from Lesson 1-2.

> **Distance Postulate**
> **a. Uniqueness Property** On a line, there is a unique distance between two points.
> **b. Distance Formula** If two points on a line have coordinates x and y, the distance between them is $|x - y|$.

You often have measured *lengths* of objects. We define the *length* of $\overline{AB}$ to be the distance from A to B. (Recall that AB with no bar above the A or B is the symbol for this distance.)

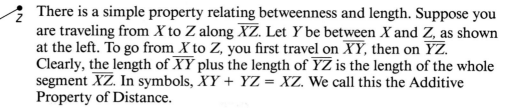

There is a simple property relating betweenness and length. Suppose you are traveling from X to Z along $\overline{XZ}$. Let Y be between X and Z, as shown at the left. To go from X to Z, you first travel on $\overline{XY}$, then on $\overline{YZ}$. Clearly, the length of $\overline{XY}$ plus the length of $\overline{YZ}$ is the length of the whole segment $\overline{XZ}$. In symbols, $XY + YZ = XZ$. We call this the Additive Property of Distance.

Distance Postulate
c. Additive Property If B is on $\overline{AC}$, then $AB + BC = AC$.

Example 2

In the figure below, Y is between X and Z. If $XY = 17$ and $XZ = 42$, what is YZ?

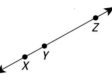

Solution

Sketch and mark the picture. Then use the Additive Property of Distance.
$$XY + YZ = XZ$$
$$17 + YZ = 42$$
$$YZ = 42 - 17 = 25$$

Caution: Be careful to distinguish the following symbols. They look similar, but their meanings are quite different.

$\overleftrightarrow{AB}$ is the *line* determined by points A and B.
$\overrightarrow{AB}$ is the *ray* with endpoint A and containing B.
$\overline{AB}$ is the *segment* with endpoints A and B. It is a set of points.
AB is the *distance* between A and B, or the length of $\overline{AB}$. It is a number.

QUESTIONS

Covering the Reading

1. **a.** Show your answer to the Activity in this lesson.
   b. What property justifies that there is a single answer to part **a**?

2. Suppose A and C have coordinates -3 and 82. If point B is between A and C, give the range of coordinates x for point B.

3. Define *segment*.

4. A laser beam is like the geometric figure called a(n) __?__.

5. $\overrightarrow{MN}$ has endpoint __?__ and contains a second point __?__.

6. a. On a number line, graph the set of numbers satisfying $y \leq 1$.
  b. What figure is the graph?

7. a. On a number line, graph the set of numbers satisfying $7 \leq x \leq 8.3$.
  b. What figure is the graph?

8. If Y is between X and Z, what is the relationship between lengths XY, XZ, and YZ?

9. In the figure at the left, Q is between P and R. If $PR = 110$ and $PQ = 48$, what is QR?

10. Match each symbol with the correct description.
  a. $\overline{AB}$ (i) line
  b. AB (ii) length
  c. $\overleftrightarrow{AB}$ (iii) ray
  d. $\overrightarrow{AB}$ (iv) segment

11. *Multiple choice.* Which sentence is *not* true?
  (a) $\overline{AB}$ and $\overline{BA}$ are the same figure.
  (b) $AB = BA$
  (c) $\overleftrightarrow{AB}$ and $\overleftrightarrow{BA}$ are the same figure.
  (d) $\overrightarrow{AB}$ and $\overrightarrow{BA}$ are the same figure.

Applying the Mathematics

12. *True or false.* If Q is between P and R, then P, Q, and R are collinear.

13. If M is between A and B in the figure below, give a possible coordinate for M.

14. Your height is a length. On your body, what are the endpoints of a segment whose length is your height?

In 15 and 16, use the figure at the left. Points R, L, D, and E are collinear.

15. If $ED = 15$, $ER = 97$, and $LR = 22$, find LD.

16. If $DE = 19x$, $DL = 62x$, and $LR = 12y$, find ER.

17. Write a sentence or two telling why the graph of $x > 1$ on a number line is not a ray.

18. The famous Wall Drug Store is between Rapid City and Sioux Falls, South Dakota, on Interstate 90. It is 56 miles from Rapid City to Wall Drug Store. If it is 294 miles to Sioux Falls from the Wall Drug Store, how far is it from Rapid City to Sioux Falls?

19. The drawing at the right shows 15 desks in a classroom. Think of the desks as discrete points. If the idea of betweenness were to exist in discrete geometry, do you think these statements would be true? Explain why or why not.

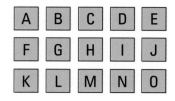

  a. *B* is between *A* and *E*.
  b. *I* is between *B* and *O*.

Pit stop. *In 1936, Ted and Dorothy Hustead's one room Wall Drug Store began advertising free ice water. Today, it is a three-generation family business occupying 50,000 square feet, with yearly sales of over $5.5 million.*

Review

20. *True or false.* *m* and *n* are coplanar lines.
  a. If *m* = *n*, then *m* // *n*.
  b. If *m* does not intersect *n*, then *m* // *n*.
  c. If *m* intersects *n* in exactly one point, then *m* // *n*.   *(Lesson 1-7)*

21. Lines on the surface of Earth do not behave like lines in a plane. Suppose only perfectly north-south streets are considered as lines. (These are lines of longitude.) What part or parts of the Point-Line-Plane Postulate do these lines violate?   *(Lesson 1-7)*

22. *True or false.*   A point has size in Euclidean geometry.   *(Lesson 1-7)*

23. Which geometric terms are undefined in this course?   *(Lesson 1-6)*

24. Graph $3x - y = 4$ and $8x = 24$ on the same axes.   *(Lesson 1-3)*

25. Three views of the same cube are given. Which symbols are on opposite faces of the cube?   *(Previous course)*

Exploration

26. Often graphs of number lines are used when the coordinates are dates or years. These number lines are called *time lines.* Find an example of a time line in a book, magazine, or newspaper. What do the terms *between, segment,* and *ray* mean on the time line you found?

PROJECTS

1 CHAPTER ONE

A project presents an opportunity for you to extend your knowledge of a topic related to the material of this chapter. You should allow more time for a project than you do for typical homework questions.

1 Latitude and Longitude

Latitude and longitude represent every point on Earth as an ordered pair of numbers.

a. Look on a globe or in an atlas to find (approximately) the latitude and longitude of your town or city.

b. Locate the country on your map or globe that is closest to latitude 0° and longitude 0°. Name the country.

c. Make a poster for your classroom that shows a map of the world indicating the location of your town, the location of (0°, 0°), and the latitude and longitude of one place on each of the seven continents.

2 Circularity in Songs

Find a copy of the song "There's a Hole in the Bucket." Write a paragraph about the circularity of the lyrics and make up your own poem with a circular theme.

3 Famous Artists

Write a report on an artist who uses dots of color to create paintings. Some possibilities are Eugene Delacroix, Georges Seurat, and Roy Lichtenstein. Include information on the artist's life, method of work, and famous paintings.

4 Computer Graphics

Look at the drawing of the horse in Lesson 1-1. Design a realistic picture of your own on graph paper, using the grid to provide shaded areas. Then reproduce your picture using a computer program such as MacPaint® or SuperPaint®.

MacPaint® is a registered trademark of Claris Corporation.
SuperPaint® is a registered trademark of Silicon Beach Software, Inc.

5 Resolution of Monitors

Pick four computer monitors of differing sizes. Calculate the resolution of each in terms of pixels per square inch. Use the cost of each monitor to decide which is the best buy. Display your information in a chart and write a short paper showing your work and explaining your decision.

6 Perspective Drawing

Find out how to make a one-point perspective drawing of a street. Pick a horizon line and a vanishing point and create your own drawing of an imaginary city street, including some buildings by it.

7 Reading Braille

The Braille Alphabet consists of discrete figures composed of raised dots. Write a report on the history of the Braille Alphabet and make a poster explaining how the letters are composed.

SUMMARY

Geometry is the study of visual patterns. In two-dimensional geometry, the basic building blocks of these patterns are points and lines. Four conceptions of points and lines are studied in this chapter. (1) In discrete geometry, points are dots and lines are collections of dots in a row. It is possible that between two dots, there might be no other dots. (2) In synthetic geometry, points are locations and lines are shortest paths between the locations. (3) In coordinate geometry, points are ordered pairs and lines are sets of ordered pairs (x, y) satisfying $Ax + By = C$. (4) In graph theory, points are nodes in networks and lines are arcs joining the nodes. Then lines have only two points on them and there may be many lines connecting two points.

In three-dimensional geometry, planes join points and lines as basic building blocks. It is impossible to show a three-dimensional figure on a two-dimensional page exactly as it is. A person can choose to draw the figure in perspective or not. Mathematicians usually do not use perspective.

It is impossible to define all terms in any system because of circularity. The terms left undefined in geometry are *point, line,* and *plane.* So any of the four conceptions of point and line might be possible. However, in this book we choose the assumptions in the Point-Line-Plane Postulate that fit Euclidean geometry. They apply only to the geometry of points either as locations or as ordered pairs, and they form a starting point for deducing properties of figures in Euclidean geometry.

Because the properties of points and lines as ordered pairs and locations are the same, everything that you learned in algebra about points and lines can be used in geometry.

Using the undefined terms in geometry, terms from algebra and logic, and common English words, the basic ideas in geometry can be precisely defined. Among the most basic is the idea of one point being between two others. Then, by using the definition of betweenness, precise definitions of segment and ray can be given. The length of a segment is the distance between its endpoints and that distance is unique. It is given by the formula $|x - y|$ if the endpoints have coordinates x and y on a number line. Distances satisfy the additive property $AB + BC = AC$, when B is between A and C.

VOCABULARY

Below are the most important terms and phrases for this chapter. For the starred (*) terms you should be able to give a definition of the term. For the other terms you should be able to give a general description and a specific example of each.

Lesson 1-1
pixel, matrix, resolution
discrete line
discrete geometry
oblique line
horizontal line
vertical line

Lesson 1-2
synthetic geometry
number line
coordinate, coordinatized
dense line
*distance, AB

Lesson 1-3
ordered pair
plane coordinate geometry
Cartesian plane
coordinate plane
x-coordinate, y-coordinate
x-axis, y-axis
Standard Form of an equation
    for a line
Slope-Intercept Form of an
    equation for a line

Lesson 1-4
Königsberg Bridge Problem
arc, network, graph theory
node, vertex (vertices)
traversable network
even node
odd node

Lesson 1-5
vanishing point
perspective drawings
line of sight, vanishing line

Lesson 1-6
circularity
undefined terms
*figure, *space
collinear
coplanar
one-dimensional
two-dimensional
three-dimensional

Lesson 1-7
*postulate
Euclidean geometry
$\overleftrightarrow{AB}$
Point-Line-Plane Postulate
    *Unique Line Assumption
    *Number Line Assumption
    *Dimension Assumption
*theorem
*Line Intersection Theorem
*parallel lines, //
coincident lines

Lesson 1-8
betweenness of numbers
betweenness of points
*segment, line segment, $\overline{AB}$
endpoint
*ray, $\overrightarrow{AB}$
*opposite rays
Distance Postulate
    *Uniqueness Property
    Distance Formula
    *Additive Property
length of a line segment, AB

PROGRESS SELF-TEST

Directions: Take this test as you would take a test in class. Then check your work with the solutions in the Selected Answers section in the back of the book. You will need graph paper and a ruler. Calculators are allowed.

1. Using the number line below, find AB.

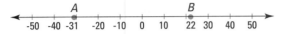

2. Draw this box in perspective.

3. Trace the table below. Then find the vanishing point.

4. On a number line Q is between S and T. If $QS = 5.3$ and $ST = 14.4$, find TQ.

5. Ignoring small thicknesses, how many dimensions does a flat sheet of paper have?

6. If, in defining a word, you return to the original word, what has occurred?

7. A tape measure is stretched along the edge of a shelf. If one end reads 4″ and the other 32″, how long is the shelf?

8. The airport distance from Chicago to Duluth is 450 miles, but the road distance is 473 miles. Why are the distances different?

9. Draw two discrete lines that cross but do not have a point in common.

10. In graph theory, the statement *Through any two points there is exactly one line* is false. Draw a diagram showing why the statement is false and explain your diagram.

11. *True or false.* In Euclidean geometry a point has size.

12. *Multiple choice.* Consider the statement: *A line has infinitely many points.* In which geometry is the statement not true?
(a) graph theory
(b) Euclidean geometry
(c) discrete geometry

13. Graph the lines $y = 4$ and $-4x + y = 2$ on the same set of axes.

In 14 and 15, an equation for a line is given. Classify the line as vertical, horizontal, or oblique.

14. $x = \frac{3}{2}$ **15.** $11x + y = 3$

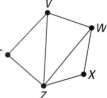

16. a. Is the network drawn above traversable?
b. If so, give a route. If not, explain why not.

17. The graph on a number line of the set of points satisfying $x \geq 40$ is the geometric figure called a(n) __?__.

In 18 and 19, use the line $\overleftrightarrow{BC}$ below.

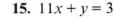

18. Give another name for $\overleftrightarrow{BC}$.

19. If $AB = 11$, $CD = 18$, and $AD = 41$, find BC.

20. Are $\overrightarrow{PQ}$ and $\overrightarrow{QP}$ opposite rays? Explain why or why not.

PROGRESS SELF-TEST

21. A face of a watch is pictured at the right.

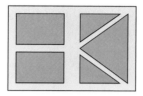

  a. What is represented by the discrete circle?

  b. What is represented by the dense circle?

22. Use this road mileage chart for three cities in North Carolina.

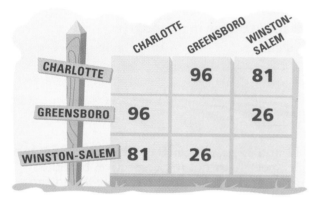

	CHARLOTTE	GREENSBORO	WINSTON-SALEM
CHARLOTTE		96	81
GREENSBORO	96		26
WINSTON-SALEM	81	26	

If you go from Charlotte to Greensboro by way of Winston-Salem, how much farther must you travel than if you had gone directly from Charlotte to Greensboro?

23. Below is a floor plan for one wing of a museum.

  a. Draw a diagram of this floor plan using nodes and arcs.

  b. Explain why a security guard cannot patrol each hallway without retracing his steps.

CHAPTER REVIEW

Questions on SPUR Objectives

SPUR stands for **S**kills, **P**roperties, **U**ses, and **R**epresentations. The Chapter Review questions are grouped according to the SPUR Objectives for this chapter.

SKILLS DEAL WITH THE PROCEDURES USED TO GET ANSWERS.

Objective A: *Draw discrete lines.* *(Lesson 1-1)*
1. Draw an oblique discrete line.
2. Draw a pair of horizontal discrete lines.
3. Draw a horizontal discrete line which intersects a vertical discrete line at exactly one point.
4. Draw a pair of oblique discrete lines which cross but do not have a point in common.

Objective B: *Analyze networks.* *(Lesson 1-4)*
In 5 and 6, refer to the network below.

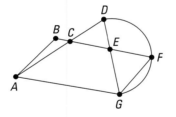

5. **a.** How many even nodes are there?
   b. How many odd nodes are there?
6. Is the network traversable?
7. **a.** Is the network below traversable?
   b. If so, give a path. If not, explain why not.

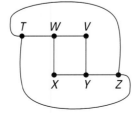

8. What must be true if a network is traversable?

Objective C: *Make and analyze perspective drawings.* *(Lesson 1-5)*
In 9 and 10, draw in perspective.
   9. a cube   **10.** a table

In 11 and 12, **a.** tell whether the picture is drawn in perspective or not in perspective. **b.** If the figure is drawn in perspective, trace it and show a vanishing point.

11.   **12.**

Objective D: *Recognize and use notation for lines, segments, and rays.* *(Lesson 1-8)*
In 13 and 14, use $\overleftrightarrow{SZ}$ below.

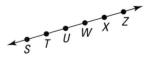

13. Give another name for ray $\overrightarrow{US}$.
14. Name $\overleftrightarrow{SZ}$ in two other ways.
15. Do $\overline{PQ}$ and $\overline{QP}$ consist of the same set of points? Explain why or why not.
16. Do $\overrightarrow{LM}$ and $\overrightarrow{ML}$ consist of the same set of points? Explain why or why not.

PROPERTIES DEAL WITH THE PRINCIPLES BEHIND THE MATHEMATICS.

Objective E: *Give the dimensions of figures and objects.* *(Lesson 1-6)*

In 17 and 18, tell how many dimensions each object has. Ignore small thicknesses.

17. a mirror   **18.** a tightrope

In 19 and 20, write the number of dimensions for each figure.

19. line segment   **20.** triangle

Objective F: *Given a property of points and lines, tell whether it is true in each of the four geometries: discrete geometry, synthetic geometry, plane coordinate geometry, graph theory.* *(Lessons 1-1, 1-2, 1-3, 1-4)*

In 21 and 22, *multiple choice.*

21. A point has size and shape in which geometry?
 (a) discrete geometry
 (b) plane coordinate geometry
 (c) synthetic geometry

22. A line does *not* contain infinitely many points in which geometry?
 (a) discrete geometry
 (b) graph theory
 (c) synthetic geometry

In 23 and 24, *true or false.*

23. In plane coordinate geometry, between two points on a line there is a third point.

24. In synthetic geometry, between two points on a line there is a third point.

25. The statement, *Two points determine a line,* is not true in discrete geometry. Explain why.

26. The statement, *Two different lines can have two points in common,* is true for graph theory. Explain why.

Objective G: *Recognize the use of undefined terms and postulates.* *(Lessons 1-6, 1-7)*

27. In an attempt to define a simple word, you may find the original word used in the definition. What is this called?

28. Name three undefined geometric terms.

29. What are the two major reasons for having postulates?

In 30 and 31, *multiple choice.*

30. Which assumption of the Point-Line-Plane Postulate allows a line to be coordinatized?
 (a) Unique Line Assumption
 (b) Number Line Assumption
 (c) Dimension Assumption

31. Which assumption of the Point-Line-Plane Postulate assures that not all points in space are coplanar?
 (a) Unique Line Assumption
 (b) Number Line Assumption
 (c) Dimension Assumption

In 32 and 33, tell whether the statement is true or false in Euclidean geometry.

32. If P and Q are points, there is only one line $\overleftrightarrow{PQ}$.

33. Lines have thickness.

Objective H: *Apply the Distance Postulate properties of betweenness.* *(Lesson 1-8)*

34. If Q is between P and R, then what is the relationship among PQ, QR, and PR?

35. Point C lies on $\overline{AB}$. If the coordinate of A is 14 and the coordinate of B is -17, what is the range of coordinates x for point C?

36. In the figure at the right, Z is between X and Y. If $XY = 10$ and $YZ = 6$, then $XZ = \underline{\ ?\ }$.

37. a. On a number line, graph the set of numbers satisfying $x \geq$ -3.
 b. What figure is the graph?

38. a. On a number line, graph the set of numbers satisfying $4 \leq x \leq 31$.
 b. What figure is the graph?

39. On the number line below, if $BC = 13$, $AB = 2$, and $AD = 31.8$, find CD.

USES DEAL WITH APPLICATIONS OF MATHEMATICS IN REAL SITUATIONS.

Objective I: *Apply the definition of distance to real situations.* *(Lesson 1-2)*

40. A student placed a meter stick on a desk. The front of the desk was aligned to the 13 cm mark, and its back to the 56 cm mark. How wide is the desk?

41. A thermometer reading was -6° at Billings and 2° at Fargo. How much colder was Billings than Fargo?

42. Jason took a plane from St. Louis to Kansas City and flew 237 miles. Zach took a train from St. Louis to Kansas City and rode 281 miles. Jessica drove 256 miles from St. Louis to Kansas City. Why can all of these distances be different?

43. Use the road mileage chart for three cities in Florida shown below. If you drive from Jacksonville to Miami through Tampa, how much longer is it than going directly from Jacksonville to Miami?

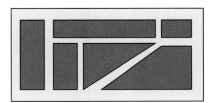

	JACKSONVILLE	MIAMI	TAMPA
JACKSONVILLE		356	198
MIAMI	356		254
TAMPA	198	254	

Objective J: *Use discrete geometry and graph theory to model real-world situations.* *(Lessons 1-1, 1-4)*

44. a. Represent the floor plan below with a network.
b. Could a security guard patrol all the hallways without retracing steps? Explain your answer.

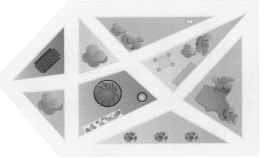

45. Below is a plan for paths through a garden in a park.
a. Draw a diagram of the paths as a network of nodes and arcs.
b. Find a path for visitors to take to cover all the paths without retracing their steps.

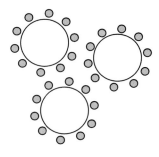

46. The diagram below uses discrete and dense lines.

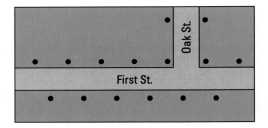

a. What might be represented by the discrete segments?
b. What might be represented by the dense segments?

47. Pictured at the right is the seating arrangement in a room at the Oakwood restaurant.
a. What might the dense circles represent?
b. What might the discrete circles represent?

REPRESENTATIONS DEAL WITH PICTURES, GRAPHS, OR OBJECTS THAT ILLUSTRATE CONCEPTS.

Objective K: *Determine distance on a number line.* *(Lesson 1-2)*

In 48 and 49, find *AB*.

48.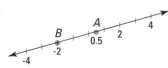

49.

In 50–53, give the distance between two points with the given coordinates.

50. 2 and 9

51. -31 and 47

52. -14 and -90

53. *x* and *y*

54. If *P* and *Q* are on a number line, *P* has coordinate 18, and *PQ* = 25, what could the coordinate of *Q* be?

55. If *R* and *S* are on the number line, *S* has coordinate $\frac{1}{2}$, and *RS* = $\frac{1}{4}$, what could the coordinate of *R* be?

Objective L: *Graph points and lines in the coordinate plane.* *(Lesson 1-3)*

In 56 and 57, graph the set of points satisfying the equation.

56. $4x - y = 8$ 57. $y = -2x + 1$

58. Graph $y = -2x$ and $4x - 3y = 10$ on the same set of axes.

59. Graph $x = -3$ and $x - y = 5$ on the same set of axes.

In 60–63, from the equation classify the line as vertical, horizontal, or oblique.

60. $5x + 3y = -19$ 61. $y = 3x$

62. $x = 11$ 63. $y = 3$

CHAPTER

2

THE LANGUAGE AND LOGIC OF GEOMETRY

What is a "cookie"? According to this newspaper article from *USA Weekend,* May 23–25, 1986, the answer is not so simple.

You could look it up, but that might not help

If it looks like a cookie and tastes like a cookie, it is a cookie, right? Not so fast . . .

Some of the simplest, everyday things turn out to be not so simple when you try to define them. Everybody knows what "time" is, for example. But if your life depended on coming up with a clear definition of time, you would be in a lot of trouble.

In the news, there are plenty of references to "terrorists." But anybody who tries to spell out what is meant by the word "terrorist" runs into difficulty. We all know what a "terrorist" is, but the United Nations has been unable to come up with a working definition. The State Department and the Pentagon have different definitions, and at least one Congressional Committee finally decided that there is no way of defining the word "terrorist" without making value judgments that not everybody is going to agree with.

One man's "terrorist" is another's "freedom fighter." It's impossible to pass laws against terrorism if you can't spell out with some precision what it is you are talking about.

Definitions are important in the law, of course. In Wilmington, Del., right now, there is a big legal battle being fought in the U.S. District Court. Several giant cookie companies are fighting over the recipe for so-called "dual textured" cookies. That means cookies that are crispy on the outside and soft and chewy on the inside. Procter & Gamble claims it discovered the process, patented it in 1983, and that Nabisco has infringed on the patent.

Nabisco, Keebler and Frito-Lay claim they were making cookies that were crispy on the outside and chewy on the inside before P&G got its patent. So the Nabisco, Keebler and Frito-Lay lawyers asked P&G to define their terms. Among the terms they wanted defined were "cookie" and "dough."

Now I know what the word cookie means and so do you. My 2-year-old Jamie knows what a cookie is and can ask for it by name.

But the definition turns out to be so important in this case that here are these high-priced lawyers, these learned counselors asking the judge, Joseph Longobardi, to please tell them what a "cookie" is, and what "dough" is.

Judge Longobardi is not a man who shies away from an intellectual exercise, but he declined to oblige the opposing lawyers in their request.

"It should not be the court's burden to supply definitions of the terms," he told them in a memo.

If a wise jurist like Judge Longobardi doesn't want to have to render definitions for such relatively simple concepts as "cookie" and "dough," no wonder the U.N. is bogged down with "terrorist."

If we truly don't know the meaning of "dough," if "cookie" truly is something mysterious, it is surely no wonder that we so often blunder when we're dealing with much more serious matters. ❏

CHARLES OSGOOD

Careful definitions are also one of the features of mathematical reasoning. In this chapter, the language and logic of geometry are applied to simple geometric figures, to some of the other mathematics you know, and to everyday situations.

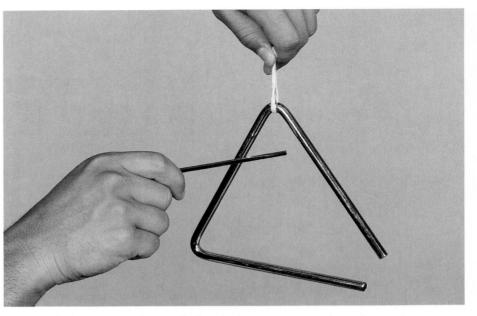

Sound definition. *The word* triangle *has many meanings. In music,
a* triangle *is a percussion instrument that consists of a steel bar bent in
a triangular shape. A player varies the tone quality by striking it with
different beaters.*

In the legal case described on page 63, millions of dollars were at stake.
This made a careful definition of "cookie" very important to the companies involved. The dispute over the meaning of "dual-textured cookie"
resulted in a $125,000,000 penalty that had to be paid to Procter and
Gamble by Nabisco, Keebler, and Frito-Lay, because they infringed on
Procter & Gamble's patent rights to that idea. At the time, this amount
was the largest penalty ever given in a United States patent trial.

You may wonder why the column was put in a book on geometry. The
reason is that it discusses definitions. Careful definitions are found
throughout mathematics. When ideas are not carefully defined, people
may not agree with what is written about them. Did Nabisco and
Procter & Gamble agree on what was meant by a "dual-textured
cookie"? Do you think everyone in your class would agree on what is
(and what is not) a "triangle"? Do you think everyone in the world
would agree on what a "circle" is?

Activity 1

Which of the following figures do you think should be called
"rectangles"? Write your choice or choices before reading on.

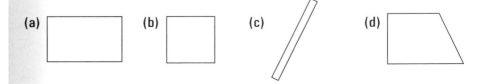

(a)   (b)   (c)   (d)

Almost everyone believes (a) is a rectangle. So do we. Many people think (b) is not a rectangle, because it is a square. Many people think (c) is not a rectangle because it is tilted, or because it is long and thin. Some people think (d) is a rectangle because it has right angles. Our view, which agrees with the view of most mathematicians, is that (a), (b), and (c) are rectangles, but (d) is not. That is, all squares are rectangles. A rectangle may be very thin. A rectangle tilted is still a rectangle. But (d) is not a rectangle, because it does not have four right angles.

Defining Geometric Terms

You have seen and used many geometric terms in previous courses. The terms may not have been carefully defined. In this book, however, geometry terms are carefully defined. Six terms from geometry are written below. The drawing beside each term pictures one example that will be included by the definition of that term.

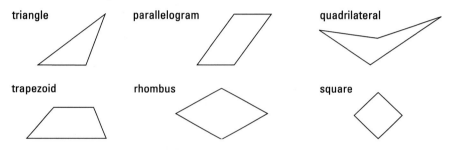

All of the above figures share certain characteristics. They all have straight sides, and they all fence in a certain area. Such figures are called *polygons.* Polygons are not limited to three or four sides. Below are three different nine-sided polygons.

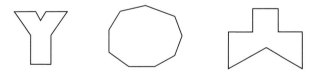

Yet these drawings do not tell us exactly what a polygon is.

Activity 2

Which figures below do you think are polygons? Why?

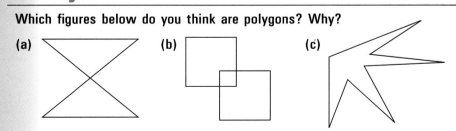

Your answers may not agree with those of others because *polygon* has not been carefully defined. Later, you will learn a careful definition of *polygon* that will be used throughout this book. With careful definitions, questions like the one posed in Activity 2 have a correct answer.

Convex Sets

In this lesson, there is only one definition. The term defined, *convex set,* helps us to distinguish between sets of points that have "dents" and those that do not. We want sets of points that do not have "dents" in them to be called convex sets.

> **Definition**
> A **convex set** is a set in which every segment that connects points of the set lies entirely in the set.

A set that is not convex is called, quite appropriately, a **nonconvex set.**

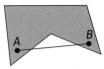 

A nonconvex set; part of $\overline{AB}$ lies outside the set.

A convex set; all segments connecting points in the set lie in the set.

A convex pentagonal region; all segments connecting points in the region lie in the region.

The word *convex* has the same meaning when it describes lenses. Lenses which are not convex are called *concave.*

convex lens

concave lens

Many other words used in mathematics, such as "angle" or "line" or "plane," have nonmathematical meanings that are quite different from their mathematical meaning. The definitions of terms in this book apply only to the mathematical meanings of the terms.

On the left is Danielle Mitterrand, wife of François Mitterrand, then president of France, attending a congressional committee meeting on human rights.

QUESTIONS

Covering the Reading

In 1–3, refer to the article on page 63.

1. What word did a Congressional Committee have trouble defining?

2. What does "dual-textured" mean?

3. What is the problem that led to a lawsuit involving the cookie companies?

4. How much was the definition of "cookie" worth to Procter & Gamble?

5. **a.** What were your answers to Activity 1 in this lesson?
   b. Draw a figure that you think is almost, but not quite, a rectangle.
   c. Write a sentence explaining why your drawing is not a rectangle.

6. What were your answers to the questions in Activity 2 in this lesson?

7. Draw two figures that you would classify as polygons, and one you would not.

8. Give a reason why careful definitions are needed in geometry.

In 9 and 10, consider the shaded set of points. Is the set convex?

9.

10.

Applying the Mathematics

11. Draw a nonconvex 4-sided region.

12. Draw a convex 8-sided region.

13. *True or false.* A segment is a convex set.

14. The word *midpoint* is carefully defined in Lesson 2-4. Before reading that lesson, tell whether you think point S is the midpoint of $\overline{RT}$ in the figure.

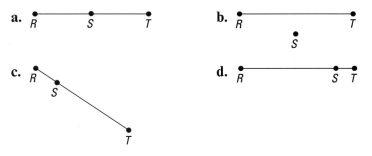

15. The word *quadrilateral* has two parts: "quadri" and "lateral." They come from the Latin *quattuor,* meaning "four," and *latus,* meaning "side." So a quadrilateral is meant to be a four-sided polygon.
    a. Which of the figures I, II, III, IV, and V do you think are quadrilaterals?
    b. If you think the figure is not a quadrilateral, give a reason for your decision.

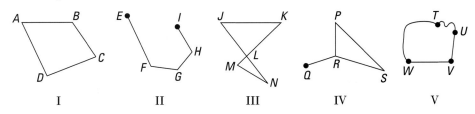

16. a. Which of the figures I, II, III, and IV do you think are circles?

I.

II.

III.

IV.

b. Describe what you think a circle is.

17. The points (1, 4), (1, 5), (3, 5), (3, 4), and (1, 4) are connected in the order named. Do you think the result should be called a rectangle? Why or why not?

18. An orange is cut with one slice of a sharp knife into two pieces. Do you think each piece should be called "half an orange"? Why or why not?

19. A cube is drawn at the right. The points *A, B, C,* and *D* are connected in order. They form a four-sided figure *ABCD* in which each angle is a right angle. Should *ABCD* be called a rectangle? Why or why not?

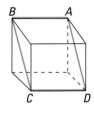

Review

20. Draw a single figure satisfying all of the following conditions:
(1) *P* is between *T* and *V*.
(2) $PM = 2 \cdot PV$.
(3) *Q* is not on $\overleftrightarrow{TV}$.
(4) $QP = PT$.   *(Lesson 1-8)*

21. The line $3x + 4y = 6$ contains the point (10, *a*). Find *a*.   *(Lesson 1-3)*

In 22–24, solve.   *(Previous course)*

22. $x - 23 = 180 - x$

23. $y = 6(90 - y)$

24. Solve for *m*: $225z = 15m$.

Exploration

25. Consider the article on page 63.
a. How would you define "cookie"?
b. How would you define "terrorist"?

Logical directions. *Implied if-then statements can be found in many places. This sign for the Blue Ridge Parkway in North Carolina means: If you want to go south, then turn right, and if you want to go north, then turn left.*

It is obvious that having precise meanings for words such as "rectangle" or "circle" is important for understanding geometry. What is not so obvious is that you also must know the meanings of small words such as *and* and *or,* and symbols such as $+$, $<$, and $\overrightarrow{AB}$. As you read this book, you often will have to slow down so that you do not ignore such words and symbols.

The Parts of a Conditional

The small word "if" is among the most important words in the language of logic and reasoning. It is used in everyday language, but not always carefully. In mathematics it is often paired with the word "then" and used *very carefully.* A sentence with an "if" clause and a "then" clause is called an **if-then statement,** or a **conditional.** Here is an example.

If a figure is a segment, then it is a convex set.

In an if-then statement, the clause following the "if" is called the **antecedent.** The clause following the "then" is the **consequent.** (**Hypothesis** and **conclusion** are alternate names for antecedent and consequent, respectively). In the conditional above, the antecedent is *a figure is a segment.* The consequent is *it is a convex set.* Notice that both the antecedent and the consequent could stand alone as complete sentences.

If <u>a figure is a segment</u>, then <u>it is a convex set</u>.
    antecedent   *consequent*

Example 1

Write the antecedent and consequent of the following conditional: *If the equation y = 3x − 4 is graphed on a coordinate plane, then its graph is a line.*

Solution

The antecedent follows the word *if.* Antecedent: The equation y = 3x − 4 is graphed on a coordinate plane.
The consequent follows the word *then.* Consequent: Its graph is a line.

Abbreviating Conditionals

It is tedious to write out long conditionals. So single letters are often used to stand for the antecedent and the consequent. For instance, for the conditional,

If a figure is a segment, then it is a convex set,

you can let *s = a figure is a segment* and let *c = it is a convex set.* Then the conditional may be rewritten:

If s, then c.

A still shorter way of writing this is to use the symbol ⇒, read "implies."

$$s \Rightarrow c$$

This is read, "*s* implies *c.*"

Instances of Conditionals

Consider again the conditional *If a figure is a segment, then it is a convex set.* In the drawing at the left, the figure is a segment $\overline{XY}$. So, for this drawing the antecedent of the conditional is true. $\overline{XY}$ is also a convex set, so the consequent of the conditional is also true. We say that the drawing is an *instance* of the conditional.

> **Definition**
> An **instance of a conditional** is a specific case in which both the antecedent (*if* part) and the consequent (*then* part) of the conditional are true.

This definition can be rewritten using variables. An instance of the conditional $p \Rightarrow q$ is a specific case in which both p and q are true.

Truth and Falsity of Conditionals

You might think that if you can find one instance of a conditional, then the conditional is true. Unfortunately, in mathematics, a conditional cannot be called true by finding one instance or 100 instances or even a million instances in which the antecedent and consequent are both true. A conditional is true only if for *every* specific case in which the antecedent is true, the consequent is also true. Thus, for a conditional to be true, it must be *proved* that the consequent is true for every specific case in which the antecedent is true. Later in this course, you will learn how to prove certain conditionals true.

Counterexamples to Conditionals

On the other hand, proving a conditional *false* is often easy. All you need to do is find *just one* specific case where the antecedent is true and the consequent is false. This one specific case proves that the conditional is false. Such a case serves as a *counterexample* to the conditional.

> **Definition**
> A **counterexample to a conditional** is a specific case for which the antecedent (*if* part) of the conditional is true and its consequent (*then* part) is false.

Thus, a counterexample to the conditional $p \Rightarrow q$ is a case in which p is true and q is false.

Example 2

Prove that the conditional *If February 28 falls on a Wednesday, then March 1 of that year falls on a Thursday* is false.

Solution

We need to find a counterexample. So we look for a situation in which the antecedent *February 28 falls on a Wednesday* is true but the consequent *March 1 of that year falls on a Thursday* is false. Recall that "leap years" contain a February 29. We consulted an almanac that has calendars for many years. *A counterexample to the conditional is in 1996, a leap year. In that year, February 28 is on a Wednesday, but March 1 falls on a Friday.*

In Example 2, despite the fact that you could find many instances of the conditional, such as the years 1990 and 2001, only one counterexample was needed to prove the conditional false.

Conditionals in Computer Programs

Animal years. *The Chinese calendar has repeating 12-year periods, each named after an animal. The year 2000 is the year of the dragon.*

Conditionals are commonly used in computer programs. Consider the following **BASIC** (Beginners All-purpose Symbolic Instructional Code) computer program.

```
10  INPUT N
20  IF N > 0 THEN PRINT "POSITIVE"
30  PRINT "DONE"
40  END
```

The 10, 20, 30, and 40 are line numbers. The computer performs the instructions in the order of the line numbers. In line 10, INPUT N tells the computer to take the value typed in by a user as the value of the variable N. Line 20 is a conditional. The computer considers a conditional in a program to be true. Thus, every time the antecedent is true, the computer performs the consequent. When the program is run, a question mark appears to let the user know the computer is waiting for input. If the user inputs 7, the following will be printed.

```
? 7
POSITIVE
DONE
```

A false antecedent causes the computer to ignore the *then* part of the conditional. If the user types in -2, then in line 20, the antecedent is false because -2 > 0 is false. So the word *positive* is not printed. The screen will look like this.

```
? -2
DONE
```

Later in this book, false antecedents will be discussed. But at this time, you do not need to worry about them.

Rewriting Sentences as Conditionals

There are sentences that have the same meaning as a conditional, but do not contain the words *if* or *then*. The following sentences are equivalent to the conditional *If a figure is a segment, then it is a convex set.*

All segments are convex sets.
Every segment is a convex set.

When statements follow the pattern *All A are B,* or *Every A is a B,* they can be rewritten as the conditional *If something is an A, then it is a B.* In geometry, often the "something" is "a figure."

Example 3

Rewrite the statement *All triangles have three sides* as a conditional.

Solution

Rewrite "All triangles" as "a figure is a triangle." Rewrite "have three sides" as "it has three sides." The final form is If a figure is a triangle, then it has three sides.

Covering the Reading

1. What is a *conditional*?

In 2 and 3, copy the statement. Underline the antecedent once and the consequent twice.

2. If a parallelogram has a right angle, then it is a rectangle.

3. If you can read this you're too close.

Drawing by Gahan Wilson; © 1993 *The New Yorker Magazine*, Inc.

In 4 and 5, let s = A figure is a square.
p = A figure is a polygon.
q = A figure is a quadrilateral.
Write the sentence symbolized by each statement.

4. $s \Rightarrow p$

5. q implies p

6. An instance of a conditional is a situation for which the antecedent is __?__ and the consequent is __?__.

7. Under what circumstances is a conditional true?

8. How many counterexamples are needed to show that a conditional is false?

9. Prove that the conditional *If February 1 of a year falls on a Thursday, then March 1 of that year also falls on Thursday* is false.

10. How does a BASIC computer program act when the antecedent of a conditional is true?

11. How does a BASIC computer program act when the antecedent of a conditional is false?

In 12 and 13, rewrite as a conditional.

12. Every rectangle is a quadrilateral.

13. All Irish setters are dogs.

Applying the Mathematics

14. *Multiple choice.* Consider the if-then statement *If a figure is a hexagonal region, then it is convex.* For each figure, tell if it
   (i)   is an instance of the statement,
   (ii) is a counterexample to the statement, or
   (iii) is neither an instance of nor a counterexample to the statement.

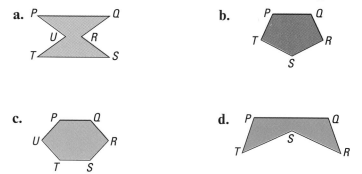

a. P Q U R T S

b. P Q T R S

c. P Q U R T S

d. P Q T S R

In 15 and 16, a conditional is given. **a.** Find an instance of the conditional. **b.** Find a counterexample to the conditional. **c.** Is the conditional true, or is it false?

15. If you are in Springfield, then you are in Massachusetts.

16. If a line contains the point (2, 3), then it is horizontal.

17. Write a true conditional with the antecedent $x \geq 9$.

18. This program calculates values from a formula you have studied in previous years.

```
10  INPUT R
20  LET A = 3.14159 * R * R
30  IF R > 0 THEN PRINT R, A
40  END
```

   a. Identify the formula and tell what A and R stand for.
   b. For what values of R will nothing be printed?

19. Heather read an ad: "If you want to improve your game, then you'll wear Nyebox Shoes." Heather wants to improve her game. If she believes the ad is true, what will she do?

20. a. Hervé heard an ad: "If you listen to WRNG, then you'll become more popular." He listened to WRNG, but he did not become more popular. Is the statement in the ad true? Explain your answer.

b. Howard heard the same ad as Hervé did. He listened to WRNG, and he became more popular. Is the statement in the ad true? Explain your answer.

21. Consider the following antecedent: *A person is 14 years old.*
a. Write a true conditional with this antecedent.
b. Write a false conditional with this antecedent.

Review

22. Which of figures I, II, III, and IV below do you think picture triangles? For each figure that you think is not a triangle, explain your reasons. *(Lesson 2-1)*

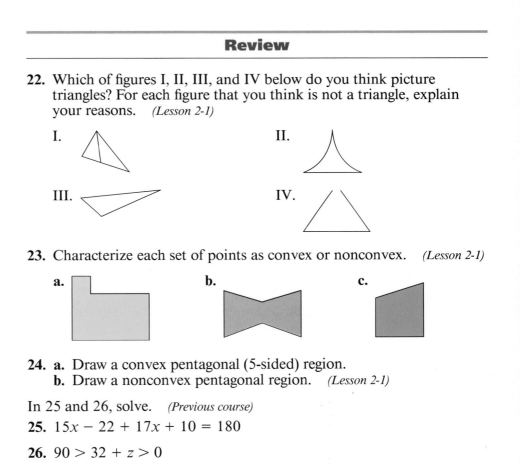

I.

II.

III.

IV.

23. Characterize each set of points as convex or nonconvex. *(Lesson 2-1)*

a. **b.** **c.**

24. a. Draw a convex pentagonal (5-sided) region.
b. Draw a nonconvex pentagonal region. *(Lesson 2-1)*

In 25 and 26, solve. *(Previous course)*

25. $15x - 22 + 17x + 10 = 180$

26. $90 > 32 + z > 0$

Exploration

27. The words *antecedent* and *consequent* are derived in part from the Latin words *ante* and *sequens*. What do these Latin words mean, and how are their meanings related to the mathematical meanings of *antecedent* and *consequent*?

If you are in Maracaibo . . . *Founded in 1529, Maracaibo is Venezuela's second largest city. Today it is one of the world's leading exporters of oil and coffee.*

Here is a true conditional.

> *If you are in Venezuela, then you are in South America.*

$$V \Rightarrow S$$

Every time you are in the country of Venezuela, you are also in South America.

In this example, switching the antecedent and consequent results in a false conditional.

> *If you are in South America, then you are in Venezuela.*

$$S \Rightarrow V$$

One counterexample is the case when you are in Peru. Then you would be in South America, but not in Venezuela.

The statements $V \Rightarrow S$ and $S \Rightarrow V$ are called *converses* of each other.

Definition
The **converse** of $p \Rightarrow q$ is $q \Rightarrow p$.

To write the converse of a statement, first write the original statement as a conditional. Then switch its antecedent and consequent.

Example 1

Consider the conditional *If $3x + 1 = 7$, then $x = 2$.*
a. Write its converse.
b. Is the original conditional true? Is its converse true?

▶

Solution

a. Since the statement is already a conditional, identify the antecedent, $3x + 1 = 7$, and the consequent, $x = 2$. Switch these to form the converse. Its converse is: If $x = 2$, then $3x + 1 = 7$.

b. Both the original conditional and its converse are true.

Truth and Falsity of Converses

Knowing that $p \Rightarrow q$ is true *does not* tell you whether $q \Rightarrow p$ is true or false. Here are two examples.

If you are in Venezuela, then you are in South America.

$p \Rightarrow q$: true
$q \Rightarrow p$: false

If $x = 2$, then $3x + 1 = 7$.

$p \Rightarrow q$: true
$q \Rightarrow p$: true

Sometimes $p \Rightarrow q$ is false. When this happens, the converse $q \Rightarrow p$ may be true or it may be false. Here are two examples.

If you are in California, then you are in Los Angeles.

$p \Rightarrow q$ false
$q \Rightarrow p$ true

If a figure is a segment, then it is a ray.

$p \Rightarrow q$ false
$q \Rightarrow p$ false

Unless you have other evidence, accepting the converse of a true statement as true is using incorrect reasoning. Accepting the converse of a false statement as false is also using incorrect reasoning. You must examine converses as independent statements.

Example 2

Let p be the statement $x \geq 9$.
Let q be the statement $x > 8$.
a. Write out $p \Rightarrow q$.
b. Is $p \Rightarrow q$ true? Explain your reasoning.
c. Is the converse true? Explain your reasoning.

Solution

a. $p \Rightarrow q$ is the statement: If $x \geq 9$, then $x > 8$.
b. $p \Rightarrow q$ is true. This can be shown with a number-line graph of $x \geq 9$. All the coordinates of the points on the graph of $x \geq 9$ are also on the graph of $x > 8$.

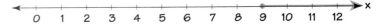

c. $q \Rightarrow p$ is the statement: if $x > 8$, then $x \geq 9$. Here is a graph of $x > 8$.

Any number between 8 and 9, such as 8.1, provides a counterexample to show that $q \Rightarrow p$ is false. So the converse is not true.

In Example 3, notice that even though there is a true instance of the conditional (a floor that is 5 feet by 4 feet), a single counterexample is enough to show that the conditional and its converse are false.

Example 3

Suppose p = A rectangular floor has perimeter 18 feet.
$\qquad q$ = A rectangular floor has area 20 square feet.
Show that both $p \Rightarrow q$ and its converse are false.

Solution

$p \Rightarrow q$ is the conditional: If a rectangular floor has perimeter 18 feet, then its area is 20 square feet. A counterexample will show it is false. The figure below has a perimeter of 18 feet, but the area is 18, not 20, square feet.

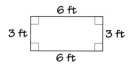

6 ft
3 ft   3 ft
6 ft

The converse is the conditional $q \Rightarrow p$: If a rectangular floor has area 20 square feet, then its perimeter is 18 feet. Here is a counterexample: The figure below has an area of 20 square feet, but the perimeter is 24, not 18, feet.

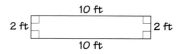

10 ft
2 ft   2 ft
10 ft

Distinguishing between a statement and its converse can be tricky. It helps to identify the antecedent and consequent carefully in order to reason logically.

QUESTIONS

Covering the Reading

1. Define *converse.*

In 2 and 3, the original conditional is true. **a.** Write the converse.
b. Tell whether the converse is true.

2. If points *A, B,* and *C* are collinear, then they all lie on the same line.

3. If you are a teenager, then you are at least 13 years old.

4. *Multiple choice.* Suppose a statement is true. Then its converse
(a) must be true.   (b) must be false.   (c) may be either true or false.

5. *Multiple choice.* Suppose a statement is false. Then its converse
(a) must be true.   (b) must be false.   (c) may be either true or false.

6. Suppose p = A rectangular floor has perimeter 26 feet.
q = A rectangular floor has area 42 square feet.
  a. Show by counterexample that $p \Rightarrow q$ is false.
  b. Show by counterexample that the converse of $p \Rightarrow q$ is false.

In 7 and 8, two statements are given. **a.** Write $p \Rightarrow q$. Is it true?
b. Write $q \Rightarrow p$. Is it true?

7. p: $x \leq -3$
q: $x < 0$

8. p: The perimeter of square $ABCD$ is 40 cm.
q: The area of square $ABCD$ is 100 cm^2.

Applying the Mathematics

In 9 and 10, use this fact. Sometimes a conditional is written with the word "if" in the middle of the statement. For example,

We'll go to the beach if the weather is hot.

The antecedent is *the weather is hot,* even though it is the latter phrase, because this phrase follows the "if." The consequent is *we'll go to the beach.* So the above statement can be rewritten as

If the weather is hot, then we'll go to the beach.

A statement is given. **a.** Rewrite the statement as a conditional. **b.** Write the converse of your answer to part **a.**

9. A polygon is a quadrilateral if it has four sides.

10. You may go to the movies if your homework is finished.

11. Refer to Example 3. Find an instance of $p \Rightarrow q$.

12. Lee saw an advertisement which read, "Everyone who attends the dance will receive Eau D'Or perfume." Lee received some Eau D'Or perfume.
  a. Rewrite the ad as a conditional.
  b. Did Lee attend the dance? Write a sentence or two explaining your answer.

Shown at the top is a perfume vessel from the tomb of Tutankhamun. About 3000 years ago the Egyptians soaked fragrant woods and resins in water and oil, then rubbed their bodies with the liquid. The second photo pictures perfume bottles from today. Since the late 1800s synthetic chemicals have been used extensively in perfumes.

Review

13. Consider the conditional *If an animal is a bird, then it can fly.*
(Lesson 2-2)
  a. Give an instance of this conditional.
  b. Give a counterexample to this conditional.

14. Find a counterexample to show that the conditional statement *If $x \geq -4$, then $x > -1$* is false. Draw a number-line graph to display your counterexample. *(Lesson 2-2)*

15. Refer to the following computer program. The symbol $>=$ means "$\geq$."

```
10  INPUT T
20  IF T >= 6 THEN PRINT "MATH IS FUN!"
30  END
```

Use the language of antecedents and consequents to explain why MATH IS FUN! will be printed if $T = 10$.   *(Lesson 2-2)*

16. Rewrite as a conditional: *All people born in New York City are U.S. citizens.*   *(Lesson 2-2)*

17. Give one reason for having undefined terms in geometry.   *(Lesson 1-6)*

In 18 and 19, Alpha, Beta, and Gamma are tollbooths on a highway. Pictured below are three mileage markers. They are like coordinates on a number line; for example, Alpha is 10.3 miles from one end of the highway.   *(Previous course, Lesson 1-2)*

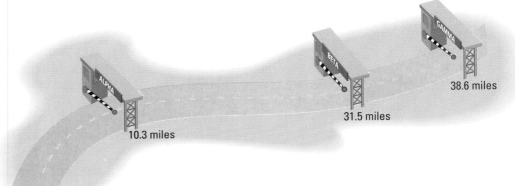

18. a. What is the mean of the mileages printed on the markers?
  b. What is the median of the mileages printed on the markers?

19. A tollbooth headquarters is planned with one of its functions being to collect money from the booths. If the headquarters is built at the median, how far would a courier have to travel to pick up the money from all the booths? (Assume the courier starts and finishes at headquarters.)

In 20 and 21, solve for W.   *(Previous course)*

20. $A = LW$ **21.** $P = 2L + 2W$

Exploration

---

22. Make up two nonmathematical conditionals that are false, but whose converses are true.

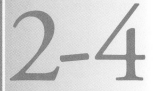

LESSON
2-4

Good Definitions

Tug of war. *Teams in a tug of war try to pull the midpoint of the rope, the red flag, over to their side. Often the red flag is placed over water holes or mud patches, making it very messy to be the losing team.*

Properties of a Good Definition

In this book, you have seen good definitions for several words, including "convex" and "segment." A good definition must

  I. include only words either commonly understood, defined earlier, or purposely undefined;

 II. accurately describe the idea being defined; and

III. include no more information than is necessary.

In this lesson, we give other definitions and examine definitions in more detail.

A Good Definition for the Midpoint of a Segment

Consider the term *midpoint.* We want "midpoint" to refer to a special point on a segment, the "halfway point." Such points are important. In a commercial flight over water, if there is trouble before the halfway point, the plane goes back to its point of origin. If there is trouble after that point, the plane goes on to its destination. Here is a good definition.

Definition
The **midpoint** of a segment $\overline{AB}$ is the point M on $\overline{AB}$ with $AM = MB$.

The definition of midpoint uses only words commonly understood (The, of, is, . . .), previously defined (segment $\overline{AB}$, distance AM, . . .), or purposely undefined (point). As seen below, the phrase "midpoint of a segment" can be traced back to these earlier or undefined terms.

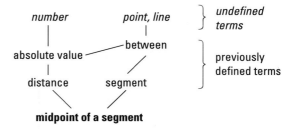

The definition of midpoint distinguishes the midpoint of a segment from other points that are not midpoints. From the definition, the midpoint must be *on* the segment. Below, $AM = MB$ but M is not on $\overline{AB}$. So M is *not* the midpoint of $\overline{AB}$. However, since $AM = MB$, we say that M is **equidistant** from A and B.

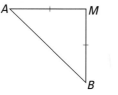

The tick marks on the segments show $AM = MB$.

Some true statements about midpoints are not good definitions of *midpoint.* You should be able to tell why they are not good.

Example 1

Why is each of these statements not a good definition of *midpoint*?
a. The midpoint of a segment is a point between the endpoints.
b. The midpoint M of $\overline{AB}$ is the point M on $\overline{AB}$ between A and B, the same distance from A and B, so that $MA = MB$.
c. The midpoint M of $\overline{AB}$ is the intersection of $\overline{AB}$ and a bisector of $\overline{AB}$.

Solution

a. The statement does not accurately describe midpoint (violates property II).
b. The statement contains too much information (violates property III).
c. The statement uses a term, "bisector," not previously defined or understood (violates property I).

Definitions as Two Conditionals

A good definition consists of a true conditional and its converse, which also must be true. So a good definition can be expressed symbolically as

$$p \Rightarrow q \text{ and } q \Rightarrow p.$$

This is abbreviated $p \Leftrightarrow q$, read "p **if and only if** q." Since $p \Leftrightarrow q$ is the combination of two conditionals, it is called a **biconditional.**

Example 2

Let $p = A$ is between C and D.
Let $q = CA + AD = CD$.
Write $p \Leftrightarrow q$ in words.

Solution

A is between C and D if and only if CA + AD = CD.

While biconditionals can appear in many situations, every good definition can be written as a biconditional. Below are the two true conditionals that make up the definition of midpoint. The first conditional goes from the defined *term* to tell you its *characteristics*.

If *M is the midpoint of $\overline{AB}$*	then	*M is on $\overline{AB}$ and AM = MB.*
term	$\Rightarrow$	characteristics

The second conditional in a good definition starts with the characteristics that allow you to use the defined term. It is the converse of the first conditional.

If *M is on $\overline{AB}$ and AM = MB*	then	*M is the midpoint of $\overline{AB}$.*
characteristics	$\Rightarrow$	term

Written as a biconditional, the definition of midpoint is as follows:

M is the midpoint of $\overline{AB}$	$\Leftrightarrow$	*M is on $\overline{AB}$ and AM = MB.*

M is the midpoint of $\overline{AB}$ if and only if M is on $\overline{AB}$ and AM = MB.

A Good Definition of Circle

We use the if-and-only-if form of a definition when we want to stress the two directions of a definition. However, most of the time we use the word "is" between the term and its defining characteristics because it is shorter.

Consider a familiar figure, the circle. Here is a good definition of *circle*.

> **Definition**
> A **circle** is the set of all points in a plane at a certain distance, its **radius**, from a certain point, its **center**.

The definition of a circle can be reworded to describe each point on it. The **circle with center C and radius r** is the set of all points P in a plane with $PC = r$. A circle with center C is often called **circle C** or $\odot$ **C**.

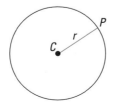

Circle with center C, radius r
$PC = r$
P is on the circle.
C, the center, is not on the circle.

Circle of friends. *These students are sitting in a circle to have a discussion. The mathematical definition of* circle *is more precise.*

Example 3

a. Write the conditional part of the definition of circle in the direction *term* ⇒ *characteristics.*

b. Write the conditional part of the definition of circle in the direction *characteristics* ⇒ *term.*

c. Write the definition of circle as a biconditional.

Solution

a. The *term* ⇒ *characteristics* direction is a conditional with the defined term "circle" in the antecedent. If a figure is a circle, then it is the set of all points in a plane at a certain distance from a certain point.

b. The *characteristics* ⇒ *term* direction is a conditional with the defined term "circle" in the consequent. If a figure is the set of all points in a plane at a certain distance from a certain point, then it is a circle.

c. Connect the term and characteristics with the words *if and only if.* A figure is a circle if and only if it is the set of all points in a plane at a certain distance from a certain point.

As in English, terms in mathematics may have more than one meaning. The term *radius of a circle* is one such term. It can mean *the distance* from the center to a point on the circle. It also can mean *a segment* connecting the center with a point on the circle. The situation almost always will indicate which meaning is appropriate. The same is true for the term **diameter of a circle,** which can mean either *a segment* connecting two points on the circle and containing the center of the circle, or *the length of that segment,* which is twice the radius.

QUESTIONS

Covering the Reading

1. List three properties of a good definition.

2. a. Define *midpoint of a segment.*
   b. What two terms that are defined earlier are contained in the definition?

3. To combine $a \Rightarrow b$ and $b \Rightarrow a$, we write a __?__ b.

4. The symbol "⇔" is read __?__.

5. Write the definition of midpoint as a biconditional.

6. If point Q is on circle C with radius r, then $QC =$ __?__.

7. What two meanings does the word *diameter* have?

Applying the Mathematics

8. Q is the midpoint of $\overline{RT}$ and $RT = 14$. Draw a diagram and then fill in each blank with a number.
  a. $RQ = \underline{\ ?\ }$.
  b. $TQ = \underline{\ ?\ }$.
  c. $TQ = \underline{\ ?\ } \cdot RT$.
  d. $\frac{RQ}{RT} = \underline{\ ?\ }$.

9. Why is each of these *not* a good definition of *circle*?
  a. A circle is the set of points in a plane at a certain distance from a certain point and it goes around the center.
  b. A circle is a plane section of a sphere.
  c. A circle is the set of points away from a certain point.

In 10 and 11, one of the two conditionals of a previous definition is given. Tell whether the statement goes in the direction *term* ⇒ *characteristics* or *characteristics* ⇒ *term*.

10. If two coplanar lines are parallel, then they have no points in common or they are identical.

11. If S is the set of all possible points, then S is space.

12. Let $p = $ *it is raining*, and $q = $ *drops of water are falling from the sky*. Write out $p \Leftrightarrow q$ in words.

13. Write the definition of *convex set* from Lesson 2-1 as a biconditional.

14. *Multiple choice.* The shaded portion of the figure at the right is the **interior of the circle** with center A and radius r. Which of these is a good definition of *interior of a circle*?
  (a) The interior of a circle with center A is the set of points inside the circle.
  (b) The interior of the circle with center A, radius r, is the set of points whose distance from A is less than r.

15. In the figure below, T is the midpoint of $\overline{DQ}$. The tick marks show $DT = TQ$. Let m be any point, line, ray, or segment other than $\overline{DQ}$ containing T. m is called a **bisector** of $\overline{DQ}$. Write a good definition of *bisector of a segment*.

16. M is equidistant from A and B. $AM = 4x + 10$ and $BM = 5x - 7$. Find x.

Rainy, Rainier, Rainiest. *Shown is part of a rain forest in Mount Rainier National Park near Seattle, Washington.*

The dodo was about the size of a large turkey, with short legs, stubby wings, and an enormous beak. It became extinct in 1681.

17. A dictionary includes this definition:

dodo *n.* An extinct flightless bird, *Raphus cucullatus,* once found on the island of Mauritius.

  a. Write the definition in biconditional form.

  b. What words in the definition are exact synonyms for dodo? Why do you think that the dictionary does not define dodo using only those words?

  c. Is this a good definition by the standard in this book? Do you think the standards of this dictionary are different from ours? Why or why not?

Review

In 18 and 19, a true statement is given. **a.** Write its converse. **b.** Is the converse true? *(Lesson 2-3)*

18. When you are reading this book, you are studying geometry.

19. If $\overrightarrow{AB}$ and $\overrightarrow{AC}$ are opposite rays, then A is between B and C.

20. Write as a conditional: *All elephants live in Africa.* Prove that the conditional is not true. *(Lesson 2-2)*

21. Draw a counterexample to the conditional *If P, Q and R are on the same line and PQ = 20 and QR = 6, then PR = 26.* *(Lesson 2-2)*

In 22 and 23, solve. *(Previous course)*

22. $50z + 3 = 67z + 1$

23. $180 > q + 19 > 90$

Exploration

24. Sometimes the meaning of a word can be found by looking at its parts. For instance, *geometer* originally meant "earth measurer," from *geo* meaning "earth" and *meter* meaning "measure." What do these words mean?

  a. geology

  b. geothermal

  c. geography

  d. geocentric

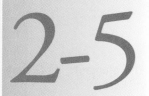

Unions and Intersections of Figures

Points, lines, segments, and rays are the building blocks of more complex figures. The two most common ways of combining figures or any other sets is to take their *union* or their *intersection*.

> **Definition**
> The **union of two sets** *A* and *B*, written *A* ∪ *B*, is the set of elements which are in *A*, in *B*, or in both *A* and *B*.

Written as a biconditional, the definition of *union of two sets* is as follows:

$C = A \cup B$ if and only if each element of *C* is in *A*, in *B*, or in both *A* and *B*.

> **Definition**
> The **intersection of two sets** *A* and *B*, written *A* ∩ *B*, is the set of elements which are in both *A* and *B*.

> **Activity**
> Write the definition of *intersection of two sets* as a biconditional.

The following three examples give unions and intersections of different types of sets.

Example 1

Let $P = \{4, 6, 8\}$ and $Q = \{10, 6, 4\}$.
a. Describe $P \cup Q$.
b. Describe $P \cap Q$.

Solution

a. First list the elements of P: 4, 6, 8. Then list the elements of Q that are not on the list for P: 10. Combining the two lists,
$P \cup Q = \{4, 6, 8, 10\}$.
b. Since 4 and 6 are in both sets, $P \cap Q = \{4, 6\}$.

Example 2

Let $C =$ the set of numbers x with $x \geq 3$ and let $D =$ the set of numbers x with $x \leq 7$.
a. Describe $C \cup D$.
b. Describe $C \cap D$.

Solution

It helps to draw a picture.

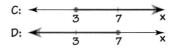

a. The graph of C is a ray including the point for 3 and all points to the right of 3. The graph of D includes the point for 7 and all points to the left of 7. So the graph of $C \cup D$ is the entire number line.

$C \cup D =$ set of all real numbers

b. To find the intersection, imagine the graphs on top of each other.

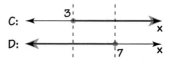

The points of the number line common to both graphs make up the segment with endpoints 3 and 7.

$C \cap D =$ set of numbers x with $3 \leq x \leq 7$, or
$C \cap D = \{x: 3 \leq x \leq 7\}$. Read $\{x: 3 \leq x \leq 7\}$ as "the set of all numbers x from 3 to 7."

Example 3

Let E = rectangle $PQRS$ and $F = \overline{PR} \cup \overline{QS}$. Draw
a. $E \cup F$ **b.** $E \cap F$.

Solution

First draw E.

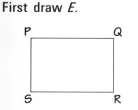

Then draw F.

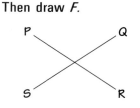

a. $E \cup F$ consists of all the points on each figure.
$E \cup F$:

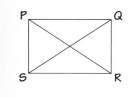

b. $E \cap F$ consists of points on both figures; that is,
$E \cap F = \{P, Q, R, S\}$.

Unions and intersections are important in describing figures accurately. For example, a triangle is the union of three segments. The symbol for triangle is $\triangle$. $\triangle ABC$ is read "triangle ABC."

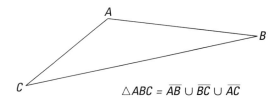

$\triangle ABC = \overline{AB} \cup \overline{BC} \cup \overline{AC}$

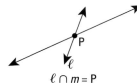

$\ell \cap m = P$

Often we speak of *intersections of lines*. At the left, the intersection of lines ℓ and m is point P. When a set of points contains just one point, we usually drop the braces { }. At the left, we may write $\ell \cap m = \{P\}$ or $\ell \cap m = P$.

If figures have no points in common, then their intersection is the **null set,** or **empty set.** That set is written ø, or { }. Below, rays $\overrightarrow{PQ}$ and $\overrightarrow{RT}$ do not have any points in common. You could write: $\overrightarrow{PQ} \cap \overrightarrow{RT} = $ ø.

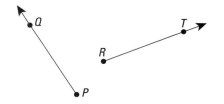

Both unions and intersections of figures occur in practical situations. Below is a map of part of the 1995 transit system for the city of Boston and some of its surrounding communities. This part consists of four *routes,* each named by a color: Red Line, Blue Line, Green Line, and Orange Line. (Note: North Station and Haymarket are on both the Green and Orange Lines.)

Shown is the Beacon Street Trolley in Boston.

Example 4

Refer to the map of part of the Boston transit system. Let R = Red Line, B = Blue Line, G = Green Line, and O = Orange Line.
a. What does $R \cup B \cup G \cup O$ mean?
b. What does $R \cap G$ mean?

Solution

a. $R \cup B \cup G \cup O$ consists of all the routes, so R ∪ B ∪ G ∪ O means this entire part of the transit system.
b. $R \cap G$ is where the two routes share the same station. According to the map, R ∩ G is the Park Street station. If you had to go from the Harvard station on the Red Line to the Boylston station on the Green Line, you could switch lines at the Park Street station.

QUESTIONS

Covering the Reading

1. What did you write for the Activity in this lesson?

In 2 and 3, describe $A \cup B$ and $A \cap B$.

2. $A = \{-3, 2, 5, 8\}$, $B = \{-3, 0, 2\}$

3. A = the set of numbers x with $x \geq 40$, B = the set of numbers x with $x \leq 50$.

4. Refer to Example 2.
   a. Why is $C \cup D$ "the set of all real numbers"?
   b. Write in words: $\{x: 3 \leq x \leq 7\}$.

5. Suppose m is the line and n is the circle pictured below. What are the elements of $m \cap n$?

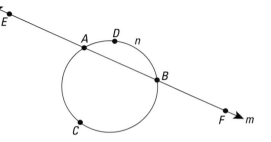

In 6 and 7, *SPED* and *PACE* are rectangles. Describe the named figure.

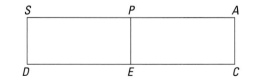

6. *SPED* $\cap$ *PACE*

7. *SPED* $\cup$ *PACE*

8. $\triangle PST$ is the __?__ of __?__, __?__, and __?__.

9. Symbols for the null set are __?__ and __?__.

In 10 and 11, refer to Example 4.

10. Choose the correct word from those in parentheses. The Boston transit system is the (union, intersection) of its four routes.

11. What is $G \cap O$?

Stairways to Heaven. *Rice terraces in southeast Asia are called* stairways to heaven *because they go so high up the mountains. Shown is a rice terrace in Bali, Indonesia. Rice is the chief food crop in Indonesia.*

12. Line ℓ has equation $4x - 3y = 12$. Line m has equation $2x + 5y = 6$. Graph ℓ and m on the same pair of axes. From the graph, estimate the coordinates of $\ell \cap m$.

In 13–15, describe $G \cup H$ and $G \cap H$.

13. $G =$ set of residents of Indonesia.
$H =$ set of residents of Bali, Indonesia.

14. $G =$ ages of people who are eligible to drive in your state.
$H =$ ages of people who are eligible to vote in your state.

15. $G =$ set of students in your geometry class.
$H =$ set of students in other geometry classes.

16. Refer to the figure at the right. Describe the named figure.
a. $\triangle GHI \cap \triangle GHJ$
b. $\triangle GHI \cup \triangle GHJ$

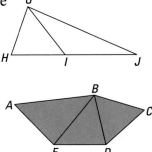

17. Refer to the two-dimensional figure $ABCDE$ at the right. Draw Region $ABDE \cap$ Region $BCDE$.

18. Let $A =$ the set of numbers y with $y \geq 10$, and let $B =$ set of numbers y with $y \geq 5$.
a. Draw a number-line graph of $A \cap B$.
b. Draw a number-line graph of $A \cup B$.

FoxTrot

by Bill Amend

19. *Multiple choice.* Refer to the cartoon above. Consider A, B, and C in the third frame as circular regions. Which of these describes the shaded parts?
(a) $(A \cup C) \cap (B \cup C)$ (b) $(A \cap C) \cup (B \cap C)$
(c) $(A \cup C) \cup (B \cup C)$ (d) $(A \cap C) \cap (B \cap C)$

20. Tell why each of the following is not a good definition of *speed limit*.
(Lesson 2-4)
   a. A speed limit tells how fast you must go.
   b. A speed limit gives the highest speed you can go. If you go faster the police may give you a speeding ticket. On some roads it may be dangerous to go faster than the speed limit.
   c. A speed limit is the maximum legal speed to avoid prosecution under the state's legislative sanctions.

21. Write the two conditionals that make up the biconditional *P is on circle O with radius r if and only if PO = r.* *(Lessons 2-2, 2-4)*

22. If the radius of a circle is $15x$, what is its diameter? *(Lesson 2-4)*

23. a. Write the converse of the statement *If B is on $\overrightarrow{AC}$, but not between A and C, then AC = AB − BC.*
   b. Is the statement true?
   c. Is its converse true? *(Lesson 2-3)*

24. Give the antecedent and the consequent of this statement: *When you work more than 40 hours a week, you will receive time-and-a-half for overtime.* *(Lesson 2-2)*

25. Draw a convex hexagonal (6-sided) region. *(Lesson 2-1)*

26. Draw a nonconvex decagonal (10-sided) region. *(Lesson 2-1)*

27. Solve for x and check: $x = 2(180 − x)$. *(Previous course)*

28. Graph all possibilities for q on a number line: $22 + q > 180$. *(Previous course)*

29. Here is a puzzle that combines some ideas of union and intersection (from *Mathematical Puzzles of Sam Loyd,* edited by Martin Gardner). It is told that three neighbors who shared a small park, as shown in the sketch, had a falling out. The owner of the large house, complaining that his neighbors' chickens annoyed him, built an enclosed pathway from his door to the middle gate at the bottom of the picture. Then the man on the right built a path to the gate on the left, and the man on the left built a path to the gate on the right. None of the paths crossed. Draw the three paths correctly. (Note: No path can go behind the large house.)

*Defining
Polygons*

IN·CLASS
ACTIVITY

Work in a group of two or three.

In 1–5, use figures that are unions of segments and consider the
following conditions:

    A. There are three or more segments.
    B. The figure lies entirely in a plane.
    C. Each segment intersects exactly two others in the figure.
    D. Each segment intersects other segments only at its endpoints.

1 Draw a figure satisfying **A, B, C,** and **D.**

2 Draw a figure satisfying **A, B,** and **D** but not **C.**

3 Draw a figure satisfying **B, C,** and **D** but not **A.**

4 Draw a figure satisfying **A, C,** and **D** but not **B.**

5 Draw a figure satisfying **A, B,** and **C** but not **D.**

6 If you think any of the figures in Questions 1–5 are impossible to
draw, explain why you think they are impossible.

Shayply manor. *The Shay hexagon house in Harbor Springs, Michigan, was built by Ephraim Shay in 1876–77. The central core and six pods are all hexagonal.*

Even a figure as simple as a triangle is hard to define. The last lesson contained the statement *a triangle is the union of three segments.* This statement by itself is not a good definition, because there are unions of three segments that do not look like figures we want to be triangles.

unions of three segments that are not triangles

Clearly, a triangle is not just any union of three segments. Each segment must intersect the others. Also, the intersections must be at endpoints. These criteria help to get a good definition not only of *triangle,* but also of the more general term *polygon.*

> **Definition**
> A **polygon** is the union of segments in the same plane such that each segment intersects exactly two others, one at each of its endpoints.

The In-class Activity on page 94 shows that this definition does not contain more information than is necessary. All parts of the definition are needed.

Indicate why each of the four unions of three segments pictured on the previous page is not a polygon. Tell which part of the definition each figure violates.

Parts of Polygons

To describe polygons, terminology is needed. The segments which make up a polygon are its **sides.** The endpoints of the sides are the **vertices** of the polygon. The singular of *vertices* is **vertex.** A polygon can be named by giving its vertices in order. Many names are possible; two names for the polygon shown below are *POLYGN* and *GYLOPN*.

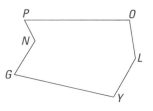

$$POLYGN = \overline{PO} \cup \overline{OL} \cup \overline{LY} \cup \overline{YG} \cup \overline{GN} \cup \overline{NP}$$

Consecutive (or **adjacent**) **vertices** are endpoints of a side. For instance, *G* and *Y* are consecutive vertices of *POLYGN*. **Consecutive** (or **adjacent**) **sides** are sides which share an endpoint. For instance, $\overline{PO}$ and $\overline{OL}$ are consecutive sides of *POLYGN*. A **diagonal** is a segment connecting nonadjacent vertices. For example, $\overline{NY}$ and $\overline{PG}$ are diagonals of *POLYGN*.

Names of Polygons

Polygons with *n* sides are called ***n*-gons.** When *n* is small, the polygons have the following special names: **triangle** (3 sides), **quadrilateral** (4 sides), **pentagon** (5 sides), **hexagon** (6 sides), **heptagon** (7 sides), **octagon** (8 sides), **nonagon** (9 sides), and **decagon** (10 sides). The boundaries of many objects are polygons; at the left is a picture of a house whose foundation's boundary is an 8-gon or octagon.

Polygonal Regions

From its definition, every polygon lies entirely in one plane. It separates the plane into two other sets, its interior and its exterior. The union of a polygon and its interior is a **polygonal** (puh-lig′-uh-null) **region.**

The Octagon House, owned by the Historical Society of Watertown, Wisconsin, has 5 floors and 57 rooms. Built around 1850–1854, its exterior is composed of 3 layers of brick with air spaces between each layer.

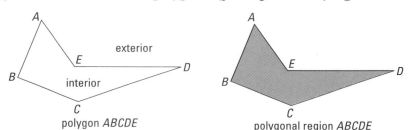

A polygon is **convex** if and only if its corresponding polygonal region is convex. Many commonly used polygons, such as squares and triangles, are convex. Some other polygons are also convex.

convex pentagon   nonconvex nonagon

Types of Triangles

Triangles with special characteristics are given specific names. When lengths of the sides of a triangle are considered, three possibilities occur: all are equal, two are equal, none is equal to the others. An **equilateral triangle** has all three sides equal. An **isosceles** (eye-soss′-suh-lees) **triangle** has at least two sides of equal length. This means that an equilateral triangle is also isosceles. A triangle with no sides of the same length is called **scalene** (skay-leen).

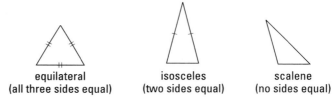

equilateral   isosceles   scalene
(all three sides equal)   (two sides equal)   (no sides equal)

The classification of triangles by sides is shown below in a **family tree** or **hierarchy.** Each type of figure is a special case of all types above it to which it is connected. Thus an equilateral triangle is a special type of isosceles triangle, a triangle, a polygon, and a figure, but an equilateral triangle is not a scalene triangle.

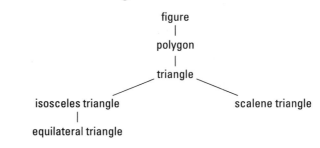

Example

According to the hierarchy, a scalene triangle is a special type of what figures?

Solution

Find "scalene triangle" at the lower right and go up the hierarchy:
A scalene triangle is a triangle, a polygon, and a figure.

QUESTIONS

Covering the Reading

In 1–3, why is each figure *not* a polygon?

1.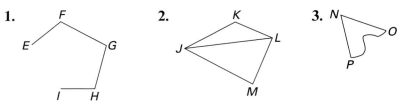

2.

3.

4. What was your answer to the Activity on page 96?

5. Use polygon *ABCDE* below.
   a. Name its vertices.
   b. Name a pair of adjacent sides.
   c. Name a pair of consecutive vertices.

6. a. How many vertices does an octagon have?
   b. How many vertices does an *n*-gon have?

7. How do polygons differ from polygonal regions?

In 8 and 9, characterize the polygonal regional as convex or nonconvex.

8.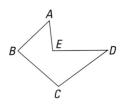

9.

In 10–12, classify the polygon by the number of its sides and by whether or not it is convex.

10. **11.** **12.**

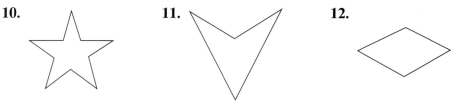

13. Draw a convex hexagon.

14. According to the hierarchy, what other names apply to an isosceles triangle?

15. Draw a union of five segments that is *not* a pentagon.

16. How many diagonals does a pentagon have?

17. Arrange in a hierarchy: isosceles triangle, figure, polygon, triangle, quadrilateral.

In 18 and 19, a real land area is shown. The boundary is very much like a polygon. Give the name of the polygon.

18.

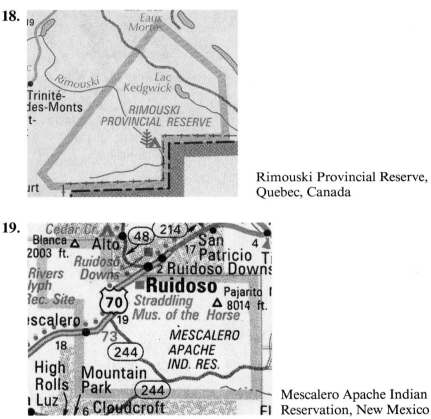

Rimouski Provincial Reserve, Quebec, Canada

19.

Mescalero Apache Indian Reservation, New Mexico

20. Draw a scalene triangle.

In 21–24, choose the figure that is *not* a polygon and indicate why it is not.

21.

22.

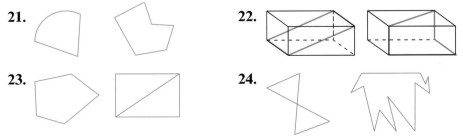

23.

24.

25. $S = \{-5, 0, 5, 10\}$ and $T = \{-3, 0, 3\}$.
   a. Find $S \cup T$.
   b. Find $S \cap T$.   *(Lesson 2-5)*

26. Refer to the figure at the right. Describe the named figure.
   a. $ABCD \cup CXZY$
   b. $ABCD \cap CXZY$.   *(Lesson 2-5)*

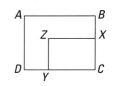

27. Let M be the midpoint of $\overline{AB}$.
   a. On a number line, suppose A has coordinate 12, and B has coordinate 8. What is the coordinate of M?
   b. If A has coordinate -43 and B has coordinate 39, then what is the coordinate of M?
   c. If A has coordinate x and B has coordinate y, then what is the coordinate of M?   *(Lessons 1-2, 2-4)*

28. On Interstate 80, Des Moines is 1793 miles from Sacramento. Salt Lake City is between them, 648 miles from Sacramento. How far is it to drive from Des Moines to Salt Lake City on this highway? *(Lesson 1-8)*

29. Give the symbol for
   a. the line through points X and Y.
   b. the line segment with endpoints X and Y.
   c. the distance between X and Y.   *(Lessons 1-7, 1-8)*

30. Some names for polygons are no longer used or are used only rarely. For each of the following names of polygons, guess how many sides the polygon has, and then check your guesses by looking in a large dictionary.
   a. dodecagon   **b.** duodecagon
   c. enneagon   **d.** pentadecagon
   e. quadrangle   **f.** tetragon
   g. trigon   **h.** undecagon

Using an Automatic Drawer: The Triangle Inequality

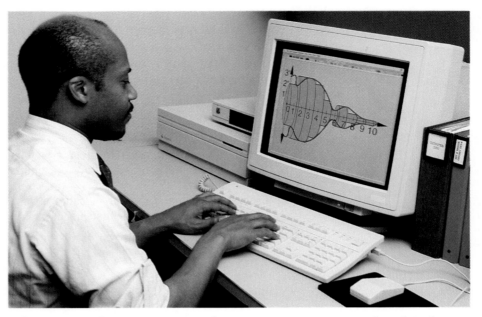

A leg to stand on. *Many manufacturers use computers to aid in their designs. This plan for a decorative chair leg will be used to create a mold. The actual chair leg will be made of plastic.*

Drawing and measuring figures are important parts of the study of geometry, but drawing figures can be time-consuming and difficult. If you do not have a good compass, your circles may start in one place and end in another. Pencil lines have thickness, and if a ruler slips just a little, the entire drawing may be off.

Computers have changed the way many people draw. Architects and draftsmen put their designs on computers. Virtually all animated cartoons are drawn with the aid of computers. Many commercials on TV use computer graphics. Manufacturers use computer drawings to help design new products. Even photographs can be enhanced with the help of computers.

Geometry Computer Software

We call any computer software or calculator that enables figures to be drawn or constructed an **automatic drawing tool** or **automatic drawer.** Most automatic drawers enable you to accurately construct, draw, or measure segments, angles, polygons, and circles of various sizes.

Geometry software may operate in a *static* or *dynamic mode.* Both modes allow the user to draw and measure figures. In **static mode,** a point that is moved changes its name, and other parts of the figure do not move with it. In **dynamic mode,** some parts of a figure can be moved while leaving other parts fixed, and lengths or angle measures displayed on the screen change automatically to reflect the changes in the figure. Each mode enables you to examine examples in a fraction of the time it would take to draw and measure them by hand.

Windows and Menus

The area in which an automatic drawer can draw is called the **window.** Outside the window is a **menu** of shapes and options. A screen from one such piece of software is pictured below with Activity 1. The computer screen will likely also show you a menu of *tools* which can be used to perform the drawings.

Activity 1

a. Use an automatic drawer to draw a segment and label its endpoints.
b. Have the automatic drawer draw the midpoint and one other point on the segment.
c. Have the automatic drawer measure all six possible distances on your screen. To how many decimal places are the measures given?

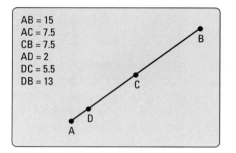

d. Your screen might look like the one above. Record your screen including the figure and the data. Save your screen for Activity 2.
e. Have the drawer compute $AD + DB$. (If your letters are different, use the lengths from each endpoint to the point that is not a midpoint.) Compare this measure to AB.
f. Do your data verify the Additive Property of Distance? Explain.

Dynamic Capabilities

Activity 2 utilizes the dynamic capabilities of an automatic drawer to generate many instances quickly.

Activity 2

a. Use your screen as it appears at the end of Activity 1. Select an endpoint of the segment. Move the endpoint to expand, to contract, and to tilt the segment. Stop after each move and record the figure and the measurements.
b. Explain what is happening to the measurements on the screen as you move the endpoint.
c. Does $AC = CB$ for all your cases? Why or why not?
d. One by one, select the two points that are not endpoints and try to slide them along the segment. Which point will not slide along the segment? Why?
e. Slide point *D.* Does $AD + DB = AB$ every time you move *D*? Explain.

Activity 3 illustrates a property of triangles that you may already know.

Activity 3

a. Place a triangle *ABC* on the screen.
b. Measure *AB*, *AC*, *AB* + *AC*, and *BC* on the screen. Record your screen; it may look like the one below. Which measure is greater, *AB* + *AC* or *BC*?

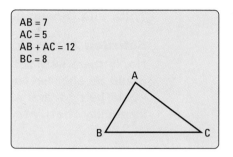

For parts **c–f**, answer the following three questions: Which lengths change? How do they change? Which is greater, *AB* + *AC* or *BC*?
c. Move point *A* far away from $\overline{BC}$ and record the screen.
d. Move point *A* very close to $\overline{BC}$ and record the screen.
e. Move point *A* to a point on $\overline{BC}$ and record the screen.
f. Move point *A* below $\overline{BC}$ and record the screen.

The Triangle Inequality

In Activity 3, you should have found that when *A* is not on $\overline{BC}$, $AB + AC > BC$. This result, which we take to be a postulate, is called the Triangle Inequality.

> **Triangle Inequality Postulate**
> The sum of the lengths of any two sides of a triangle is greater than the length of the third side.

For any triangle *ABC* such as the one drawn below, this relationship means three inequalities are true.

$$AB + BC > AC$$

$$BC + AC > AB$$

$$AB + AC > BC$$

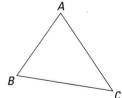

Shown is the bridge Severinsbrücke *in Cologne, Germany.*

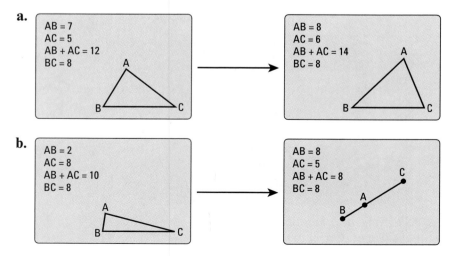

Q 5 R
7
15
P

Can this
triangle exist?

Example

Can 7, 15, and 5 be the sides of a triangle? Why or why not?

Solution 1

The Triangle Inequality Postulate says that the sum of any two sides of a triangle is greater than the third side.
7 + 15 > 5 and 15 + 5 > 7, but 5 + 7 is not greater than 15. So, 7, 15, and 5 cannot be the sides of a triangle.

Solution 2

No. If there were a triangle PQR with sides 7, 15, and 5 then it would be shorter to travel from P to Q and then to R than it would be to travel along $\overline{PR}$. This would violate the Triangle Inequality Postulate.

QUESTIONS

Covering the Reading

1. What is an automatic drawer?

2. In measuring segments, the automatic drawer you used gave lengths to how many decimal places?

3. In Activity 1, what was the procedure you had to use with your automatic drawer to draw a segment on the screen?

4. What was your answer to part **e** of Activity 2?

5. In Activity 3, you chose one vertex of the triangle and moved it. Which length remained constant during this activity?

6. The screens below were drawn by a student doing Activity 3. For each screen, on which part (**c, d, e,** or **f**) of Activity 3 was the student working?

a.

AB = 7
AC = 5
AB + AC = 12
BC = 8

→

AB = 8
AC = 6
AB + AC = 14
BC = 8

b.

AB = 2
AC = 8
AB + AC = 10
BC = 8

→

AB = 8
AC = 5
AB + AC = 8
BC = 8

7. State the Triangle Inequality Postulate.

8. Give three inequalities satisfied by the lengths of the sides of any triangle *ABC*.

In 9–11, tell whether the numbers can be lengths of the three sides of a triangle.

9. a. 2, 2, 2   **b.** 2, 2, 3   **c.** 2, 2, 4   **d.** 2, 2, 5

10. a. 15, 28, 12   **b.** 12, 15, 10   **c.** 15, 2, 12

11. a. 1, 2, 3   **b.** 4, 3, 2   **c.** 102, 103, 101

Applying the Mathematics

12. Cali performed the following procedure on an automatic drawer. She first put a circle with center *A* on the screen. She then placed 2 points *B* and *C* on the circle and connected them to make △*ABC*. What type of triangle is △*ABC*?

13. Clay wants to move point *C* to the midpoint of $\overline{AB}$. How will Clay know when *C* is at the midpoint?

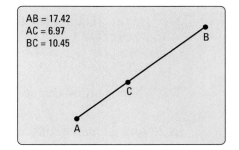

AB = 17.42
AC = 6.97
BC = 10.45

14. Suzette placed 4 points on the screen and connected them to form quadrilateral *ABCD*. If she has a dynamic drawer and she moves point *A*, which lengths of *ABCD* will change?

15. *A*, *B*, and *C* are points on a plane. For what positions of *A*, *B*, and *C* are the following true?
a. $AB + BC = AC$
b. $AB + BC > AC$
c. $AB + BC < AC$

16. Suppose two sides of a triangle have lengths 16 inches and 11 inches. What are the possible lengths *x* of the third side?

17. A hotel in the Houston area is 35 miles from the Houston Intercontinental Airport and 20 miles from Hobby Airport. With this information, you know that the airports are between __?__ and __?__ miles apart.

On formal occasions Thai dancers act out stories wearing traditional costumes. Jewelry and embroidery decorate the costumes.

18. In the *1995 World Almanac,* the air distance from Lima, Peru, to Bangkok, Thailand, is given as 12,244 miles. The air distance from Los Angeles to Lima is given as 4,171 miles. The air distance from Bangkok to Los Angeles is given as 7,637 miles. Choose one of the following and explain your choice.
  (a) This information violates the Triangle Inequality Postulate, so one of the numbers is in error.
  (b) The distances could be correct because the Triangle Inequality Postulate does not work for the surface of the earth.
  (c) If the distances were given in kilometers instead of miles, the Triangle Inequality Postulate would not be violated.

Review

19. Draw a nonconvex heptagon. *(Lesson 2-6)*

20. Consider the statement, *All equilateral triangles are isosceles triangles.*
  a. Rewrite the statement as a conditional.
  b. Write the converse of the conditional.
  c. Draw a counterexample to show that the converse of the conditional is false. *(Lessons 2-2, 2-3, 2-6)*

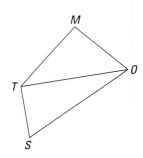

21. Use the diagram at the left. Describe the named figure. *(Lesson 2-5)*
  a. $\triangle MTO \cap \triangle STO$
  b. $\triangle MTO \cup \triangle STO$
  c. $\triangle MTO \cap MOST$

22. When is an if-then statement true? *(Lesson 2-2)*

In 23 and 24, solve for *y*. *(Previous course)*
23. $3x + 6y = 12$ **24.** $8x + 2y = 4y - 9$

Exploration

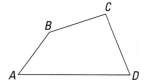

25. The lengths 2″, 2″, 2″, and 2″ can be sides of a quadrilateral, but 2″, 4″, 8″, and 16″ cannot. Is there a "quadrilateral inequality?" That is, is there some general way you can tell when four positive numbers *a, b, c,* and *d* can be the lengths of the sides of a quadrilateral? (You may wish to use an automatic drawing tool for this.)

26. Use an automatic drawer to draw one of the figures below.

  a. 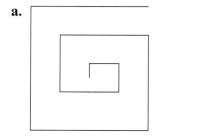   **b.**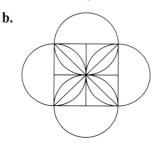

Famous conjecturer. *Sherlock Holmes, a fictional character invented by Sir Arthur Conan Doyle, was known for his ability to solve crimes using conjectures. This photo is from the movie* The Adventures of Sherlock Holmes.

Working with an automatic drawer, Maria and Luis created the screen shown below. They connected the midpoints of the sides of a triangle to form a new triangle. They noticed that the original triangle had 4 times the area of the new triangle. They tried this with the vertices of $\triangle ABC$ in different locations. The areas of the triangles changed, but the ratio of the areas was always 4.

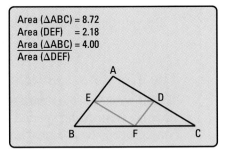

Area ($\triangle ABC$) = 8.72
Area (DEF)   = 2.18
Area ($\triangle ABC$) = 4.00
Area ($\triangle DEF$)

Activity 1

Repeat what Maria and Luis did.
1. On an automatic drawer, place a $\triangle ABC$.
2. Connect the midpoints of the sides to form $\triangle DEF$.
3. Have the drawer calculate Area($\triangle ABC$), Area($\triangle DEF$), and $\frac{\text{Area}(\triangle ABC)}{\text{Area}(\triangle DEF)}$.
4. Move the points *A, B,* and *C* around the screen. Do you find that $\frac{\text{Area}(\triangle ABC)}{\text{Area}(\triangle DEF)} = 4$ every time?

Conjectures

A **conjecture** is an educated guess or opinion. People use conjectures frequently in ordinary conversation, as well as in mathematics.

One type of conjecture is *specific,* made with reference to a particular situation. Here are some examples:

- The Sears Tower is taller than the Empire State Building.
- Caitlin has a better grade in history than Michael does.
- On Maria and Luis's screen, the area of the larger triangle is four times the area of the smaller one.

Each of these conjectures involves only one situation. Usually, specific conjectures can be proved or disproved easily with accurate data or an accurate drawing.

Another type of conjecture is a *generalization,* thought to be true for all situations of a particular type. Here are some examples:

- Juvenile delinquents grow up to become career criminals.
- Any man who has a bald maternal grandfather also will go bald.
- The triangle formed by the midpoints *D, E,* and *F* of the sides of any triangle *ABC* has one fourth the area of triangle *ABC.*

These conjectures apply to all cases of the type—juvenile delinquents, bald maternal grandfather, and triangles—not just to one situation. Generalizations always can be written in if-then form. Consequently, they are to be proved or disproved as you would any other conditional. Below, the pattern that Maria and Luis noticed is stated as a conjecture in if-then form.

If the midpoints *D, E,* and *F* of the sides of $\triangle ABC$ are connected, then

$$\frac{\text{Area}(\triangle ABC)}{\text{Area}(\triangle DEF)} = 4.$$

Activity 2

Do you think Maria and Luis's conjecture is true? Why?

The Truth of a Conjectured Generalization

To tell whether a generalization is true or false, mathematicians usually start by examining special cases in which the antecedent of the conjecture is true. For conjectures about geometric figures, this means that drawings are made and explored. If even one counterexample is found, the conjecture is not true. When instances of the generalization are found to be true, then there is evidence that the conjecture is true.

In Activity 1, you should have generated several instances of Maria and Luis's conjecture. But it is impossible to check every single triangle *ABC* that can be drawn; there are an infinite number of them! In mathematics, for a conjecture to be accepted in all cases, it must be **proved.**

You will learn in this course what it takes to prove statements true. In Chapter 3, you are asked to write one-step proofs. In Chapter 5, you will begin to write multi-step proofs. Throughout the book, you will see proofs of geometric theorems that we write and you will be asked to write your own proofs of certain statements.

Activity 3

Consider the following conjecture. If in a convex pentagon *ABCDE* the midpoints of the 5 sides, *F, G, H, I,* and *J,* are connected in order, then $\frac{\text{Area}(ABCDE)}{\text{Area}(FGHIJ)} = 1.5$.

1. On an automatic drawer, draw a convex pentagon *ABCDE*.

2. Connect the midpoints of the sides to form pentagon *FGHIJ*.

3. Calculate Area (*ABCDE*), Area (*FGHIJ*), and $\frac{\text{Area}(ABCDE)}{\text{Area}(FGHIJ)}$.

4. Move the vertices of *ABCDE* around the screen. Can you find a counterexample to the conjecture? Can you find instances of the conjecture? (Record the screens for Questions 7 and 8.)

Below is a *flow chart* which illustrates the process of testing a conjecture.

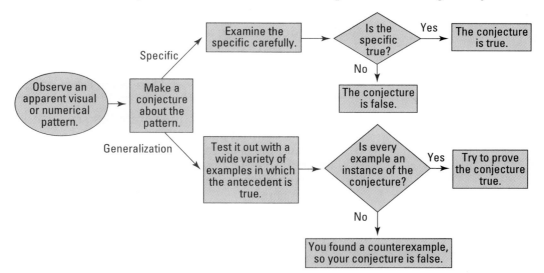

When a conjecture is not true, as in Activity 3, you may try to **refine** it. This means you change the statement slightly so that the conjecture is true. For instance, in Activity 3, it is true that $1 < \frac{\text{Area}(ABCDE)}{\text{Area}(FGHIJ)} < 2$. Is it true that $1.4 < \frac{\text{Area}(ABCDE)}{\text{Area}(FGHIJ)} < 1.6$? That idea is left for you to explore in Question 8.

QUESTIONS

Covering the Reading

1. What is a conjecture?

2. Refer to Activity 1. What was your answer in Step 4?

3. What was your answer to Activity 2?

4. What are the two types of conjectures?

5. To show that a conjecture is true in all cases, it must be __?__.

6. What is needed to show that a conjecture is false?

In 7 and 8, refer to Activity 3.

7. **a.** Give a figure which is an instance of the conjecture.
   b. Give a figure which is a counterexample to the conjecture.

8. Is it true that $1.4 < \dfrac{\text{Area}(ABCDE)}{\text{Area}(FGHIJ)} < 1.6$? Explain your answer.

9. Refer to the flow chart in the lesson. How can you test a generalization?

10. What does *refining a conjecture* mean?

Applying the Mathematics

In 11–13, a conjecture and three examples are given. For each example, tell if (i), (ii), or (iii) applies.
(i)   The example is an instance of the conjecture.
(ii) The example is a counterexample to the conjecture.
(iii) The example is neither an instance of nor a counterexample to the conjecture.

11. If $XM = MY$, then M is the midpoint of $\overline{XY}$.

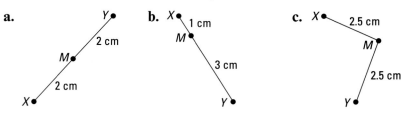

12. The square of any real number x is positive.
    a. $x = 7$ **b.** $x = \text{-}4$ **c.** $x = 0$

13. If △*ABC* is equilateral, then it is isosceles.

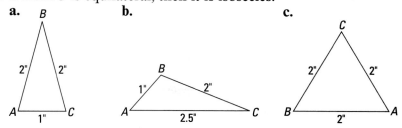

a.   **b.**   **c.**

14. a. Give your own example of a specific conjecture.
  b. Give your own example of a conjecture that is a generalization.

15. On an automatic drawer, draw a quadrilateral *ABCD*. Connect the midpoints *E, F, G, H* of the sides, to form a new quadrilateral. Compute Area(*ABCD*), Area(*EFGH*) and $\frac{\text{Area}(ABCD)}{\text{Area}(EFGH)}$. Make a conjecture and tell how you decided upon this conjecture.

In 16–18, a conjecture is stated. Choose the answer *a, b, c, d,* or *e* that best indicates your feeling about the statement. Draw pictures to help you choose.
(a) The conjecture is definitely true and in my mind needs no proof.
(b) The conjecture may be true, but I need a proof or a similar argument before I'd believe it.
(c) The conjecture doesn't seem true, but I am not sure. Some discussion would help.
(d) The conjecture is probably not true, but I'd be sure only if I had a counterexample.
(e) The conjecture is definitely false. No argument is needed to convince me that it is false.

16. If the midpoints of two sides of a triangle are joined, the segment is parallel to the third side.

17. If the midpoints of the four sides of a rectangle are connected, the resulting figure is a rectangle.

18. In the figure below, *PQR* is a semicircle with center *S*. Point *O* is the center of the larger circle and m∠*POR* = 90. Then the areas of the two shaded regions are equal.

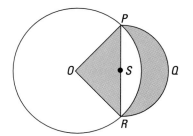

19. Can 1.3, 1.4, and 2 be the lengths of the sides of a triangle?
(Lesson 2-7)

20. It is a 12-minute bike ride from Peter's house to CJ's apartment. It is a 20-minute bike ride from CJ's apartment to Kate's home. At this rate, how long a bike ride is it from Peter's house to Kate's home? *(Lesson 2-7)*

21. Characterize each polygonal region as convex or nonconvex. *(Lesson 2-6)*

a.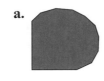

b.

22. Given $A = \{L, I, N, E, S\}$ and $B = \{P, O, I, N, T, S\}$. Find $A \cap B$. *(Lesson 2-5)*

23. Refer to the ad at the right. *(Lessons 2-2, 2-3)*
  a. Rewrite the idea of the ad as a conditional.
  b. Write the converse of the conditional in part **a.**

Looking for the 'Ultimate' New Baby Gift? Call 1-800-555-BABY For Details

24. Graph $4x - 3y = 15$. *(Lesson 1-3)*

25. A *simple closed curve* is a closed curve that does not intersect itself. Some examples are shown below.

I.   II.   III.

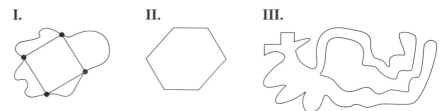

An unsolved conjecture (as of 1995) is that on every simple closed curve there are four points which are the vertices of a square. In curve **I,** four points are marked which might be the vertices of a square. Trace curves **II** and **III** and try to draw a square with vertices on each curve.

A project presents an opportunity for you to extend your knowledge of a topic related to the material of this chapter. You should allow more time for a project than you do for typical homework questions.

PROJECTS
2
CHAPTER TWO

1 Transportation Routes

Locate a map of a bus, subway, train, air, or other transportation system near where you live. Study the map and write a report which answers the following questions:

a. What is the purpose of the transportation system?
b. How does the configuration of the system reflect its purpose?
c. Is the system a union of several subsystems? If so, how?
d. Why do the intersections occur where they do?

2 Polygonal Buildings

Consider the house below. Its outline encloses a polygonal region. Find five different polygonal buildings near where you live or around your school. Draw and name each region accurately; show how each building is the union of smaller polygonal regions, as shown below. Name each polygonal building according to the number of sides it has.

3 Sayings and Adages

An old saying goes, "People who live in glass houses shouldn't throw stones." You can rewrite this as the conditional *If you live in a glass house, then you shouldn't throw stones.* Find five sayings or adages from different cultures. Write a report on your adages which addresses the following:

a. If an adage is not written as a conditional, express it as a conditional statement.
b. Tell the country of origin of each saying.
c. What does each saying mean? Which do you think are true sayings? Are any of the conditionals you wrote *literally* true?
d. Write the converse of each statement.
e. Does the converse make sense? Do you think it is true?

4 Word Roots

Many of the words in this chapter are formed using suffixes, prefixes, and word roots. Choose seven word roots from this chapter, and look up their meaning in a dictionary. For each word root, explain how it is used in the vocabulary of the lessons and find three new nonmathematical words that are derived from this root. Be sure to indicate how all these words are linked by the meaning of the root.

5 Conditionals in Advertisements

Find five examples in newspapers and magazines of advertisements that use conditionals. Tell whether you believe they are true and defend your opinion. Write the converse of each statement and tell whether you think it is true or false. Then write your own ad for a favorite product, using an appropriate conditional.

6 Tangram Polygons

Find a set of tangrams, or trace the set below, and use all seven tangram pieces to form the polygonal regions **a–g**. Draw and label your diagrams, clearly showing how you fit the tangram pieces together.

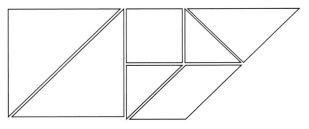

a. a square region
b. a quadrilateral region that is not a square
c. an isosceles triangle
d. a convex hexagonal region
e. a nonconvex hexagonal region
f. a heptagonal region
g. an octagonal region

SUMMARY

In Chapter 2, the language and logic of geometry are discussed and applied to segments and polygons. A common language is needed in geometry to name figures and to agree on their characteristics. This requires good definitions. A good definition is an accurate description of an idea which involves only words defined earlier, words commonly understood, or words purposely undefined. The logic of geometry is needed to express statements and determine whether or not they are true. Statements are written as conditionals. Every conditional has an antecedent, the "if" part, and a consequent, the "then" part. If the antecedent is p and the consequent is q, the conditional is $p \Rightarrow q$. The logic in this chapter dealt with three questions:

(1) How can you tell if $p \Rightarrow q$ is false? You need a counterexample. A counterexample is a situation for which p is true and q is false.

(2) How are $p \Rightarrow q$ and $q \Rightarrow p$ related? $q \Rightarrow p$ is the converse of $p \Rightarrow q$. The truth or falsity of one does not tell you anything about the truth or falsity of the other.

(3) What happens when both $p \Rightarrow q$ and $q \Rightarrow p$ are true? The statement $p \Rightarrow q$ and its converse $q \Rightarrow p$ together form the biconditional $p \Leftrightarrow q$. We then say "p if and only if q." Every definition can be reworded as an if-and-only-if statement and can be separated into two conditionals.

A conjecture is an educated guess or opinion. If it is specific, it applies only to one situation. If it is thought to be true for all situations, it is called a generalization and can be proved or disproved as you would any conditional.

A polygon is a particular union of segments in the same plane such that each segment intersects exactly two others, one at each of its endpoints.

VOCABULARY

Many terms were defined in this chapter. For the starred terms (*) below, you should be able to produce a *good* definition. For the other terms, you should be able to give a general description and a specific example, including a drawing where appropriate.

Lesson 2-1
*convex set, nonconvex set

Lesson 2-2
if-then statement, conditional
antecedent, consequent
hypothesis, conclusion
implies, $\Rightarrow$
*instance of a conditional
*counterexample to a
    conditional

Lesson 2-3
*converse

Lesson 2-4
*midpoint, equidistant
$\Leftrightarrow$, if and only if, biconditional
*circle, $\odot$, center, radius

diameter of a circle
interior of a circle

Lesson 2-5
*union of sets, $\cup$
*intersection of sets, $\cap$
$\triangle$
null set, empty set, ø, { }

Lesson 2-6
*polygon, side of polygon
vertex, vertices of a polygon
consecutive vertices, sides
adjacent vertices, sides
diagonal, *n*-gon
*triangle, quadrilateral
 pentagon, hexagon
 heptagon, octagon
 nonagon, decagon

polygonal region
convex polygon
*equilateral, isosceles,
 scalene triangles
hierarchy, family tree

Lesson 2-7
automatic drawing tool
automatic drawer
static mode, dynamic mode
window, menu
Triangle Inequality Postulate

Lesson 2-8
conjecture, specific
generalization
proof of a conjecture
refining a conjecture

PROGRESS SELF-TEST

Directions: Take this test as you would take a test in class. Use a ruler and a protractor. Then check your work with the solutions in the Selected Answers section in the back of the book.

1. Tell which property of a good definition is violated by this "bad" definition: The *midpoint of a segment AB is the point M on* $\overline{AB}$ *for which* $AM = MB$, $\frac{1}{2}AB = AM$, and $\frac{1}{2}AB = MB$.

2. *Multiple choice.* Match each term with the most appropriate drawing.
   a. hexagon   **b.** quadrilateral   **c.** octagon

   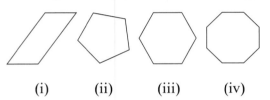

       (i)   (ii)   (iii)   (iv)

3. In the following conditional, underline the antecedent once and the consequent twice: *Two angles have equal measure if they are vertical angles.*

4. Rewrite as a conditional: *Every trapezoid is a quadrilateral.*

In 5 and 6, refer to the figure below.

5. Describe the named figure.
   a. rectangle $ABCD \cup \triangle ADC$
   b. rectangle $ABCD \cap \triangle ADC$

6. Name the points in the set $\overline{BC} \cap \overline{CD}$.

7. Let p = *There are over 50 books on that shelf* and let q = *The shelf falls.* Write $p \Leftrightarrow q$ in words.

8. Mr. Hawkins told his class, "If you do your homework every night, you will receive a passing grade." Assume Mr. Hawkins is telling the truth. Explain why the converse of this statement might not be true.

9. Tell whether the figure below is convex or nonconvex.

10. Draw a convex pentagonal region.

11. Write the following definition as two conditionals.
    A figure is an angle if and only if it is the union of two rays with a common endpoint.

12. On the number line below, point B has coordinate -10 and point O has coordinate 2. If O is the midpoint of $\overline{BT}$, what is the coordinate of T?

13. Consider the conditional *If* $x \geq 7$, *then* $x > 9$.
    a. Give an instance of the conditional.
    b. Give a counterexample to the conditional.

14. Can 11, 11, and 12 be the lengths of the sides of a triangle?

15. Along Highway 2A in Alberta, Canada, the distance from Calgary to Edmonton is 288 km and the distance from Edmonton to Red Deer is 140 km. From this information alone, what can you conclude about the distance from Calgary to Red Deer?

16. Use the definition of polygon to explain why the figure below is not a polygon.

17. Draw a hierarchy relating the following: polygon, equilateral triangle, scalene triangle, triangle.

PROGRESS SELF-TEST

18. P is the midpoint of $\overline{LT}$.
$PL = 3(7 - x)$
$PT = 2x - 5$
Find TL.

19. Pictured below is a floor plan of the Baha'i Temple in Wilmette, Illinois. What is the name given to the polygon that is the outside boundary of the temple (Assume that the outside boundary is connected all the way around the building.)?

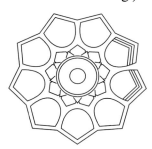

20. Use the polygon *OSTRICH* below.

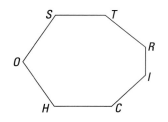

  a. Name a vertex adjacent to T.
  b. *True or false.* $\overline{OS}$ and $\overline{TR}$ are consecutive sides.
  c. Name the diagonals from vertex O.

For 21 and 22, use the following statement:
Every equilateral triangle is an isosceles triangle.

21. Write the statement as a conditional.

22. a. Write the converse of the conditional.
    b. Is the converse true? Explain your reasoning.

23. *Multiple choice.* Consider the conjecture *If M is between A and B, then M is the midpoint of $\overline{AB}$.*
For each figure, tell whether it is
(i) an instance of the conjecture.
(ii) a counterexample to the conjecture.
(iii) neither an instance of nor a counterexample to the conjecture.

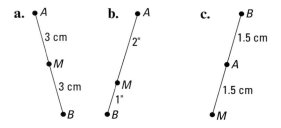

CHAPTER REVIEW

Questions on SPUR Objectives

SPUR stands for **S**kills, **P**roperties, **U**ses, and **R**epresentations. The Chapter Review questions are grouped according to the SPUR Objectives for this chapter.

SKILLS DEAL WITH THE PROCEDURES USED TO GET ANSWERS.

Objective A: *Distinguish between convex and nonconvex figures.*   *(Lessons 2-1, 2-6)*

In 1–4, characterize each figure as convex or nonconvex.

1. **2.**

3. **4.**

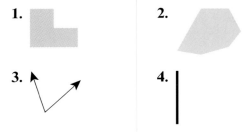

5. Draw a convex octagonal region.

6. Draw a nonconvex nonagonal region.

Objective B: *Draw and identify polygons.* *(Lesson 2-6)*

In 7 and 8, draw the polygon. **a.** Identify two consecutive vertices. **b.** Identify two nonadjacent sides. **c.** Name a diagonal.

7. quadrilateral   **8.** convex heptagon

9. *Multiple choice.*   Match each term with the appropriate drawing.

a. decagon   **b.** pentagon   **c.** quadrilateral

(i)   (ii)

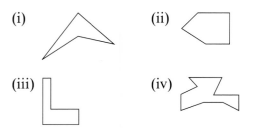

(iii)   (iv)

10. What is the difference between a pentagon and a pentagonal region?

In 11 and 12, use the definition of polygon to explain why the given figure is not a polygon.

11. **12.**

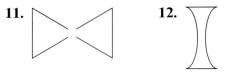

Objective C: *Use and interpret the symbols ⇒ and ⇔.*   *(Lessons 2-2, 2-3, 2-4)*

In 13 and 14, let $p = \triangle ABC$ *is equilateral* and $q = \triangle ABC$ *has three 60° angles*. Write each statement in words.

13. $q \Rightarrow p$ **14.** $p \Leftrightarrow q$

In 15 and 16, write each statement in symbols.

15. If *r*, then *p*.   **16.** *S* if and only if *t*.

Objective D: *Use the definition of midpoint to find lengths of segments.*   *(Lesson 2-4)*

17. On a number line, *T* has coordinate -6 and *Q* has coordinate 15. What is the coordinate of the midpoint of $\overline{TQ}$?

18. Point *A* on a number line has coordinate 235. It is the midpoint of a segment of length 76. What are the coordinates of the endpoints of the segment?

19. *S* is the midpoint of $\overline{LT}$.
$LS = 34 - 4a$
$ST = 3(a + 2)$
Find *LT*.

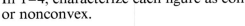

20. *R* is the midpoint of $\overline{PM}$.
$PR = 5y$
$PM = 2y + 40$
Find *RM*.

PROPERTIES DEAL WITH THE PRINCIPLES BEHIND THE MATHEMATICS.

Objective E: *Write the converse of a conditional.* *(Lesson 2-3)*

In 21 and 22, a conditional is given. **a.** Write the converse of the statement. **b.** Tell whether the converse is true.

21. If water freezes, then its temperature is less than 32° F.

22. If $AM = MB$, then M is the midpoint of $\overline{AB}$.

23. Write the converse of $r \Rightarrow t$.

24. *True or false.* If a statement is true, then its converse is also true.

Objective F: *Apply the properties of a good definition.* *(Lesson 2-4)*

In 25 and 26, tell which property of a good definition is violated by these "bad" definitions.

25. The midpoint M of $\overline{AB}$ is a point such that $AM = BM$.

26. A triangle is a closed path with three sides.

27. Here is a definition of a *secant to a circle: A secant to a circle is a line which intersects the circle in two points.*
This definition makes use of four undefined or previously defined terms. Name them.

28. Write the definition of *secant* in Question 27 as two conditionals.

Objective G: *Write conditionals and biconditionals.* *(Lessons 2-2, 2-4)*

In 29 and 30, rewrite as a conditional.

29. Every radius is a segment.

30. All Hawaiians live in the United States.

In 31 and 32, copy the statement and underline the antecedent once and consequent twice.

31. A figure is a rectangle if it is a square.

32. If p, then q.

33. Combine the following conditionals into a single biconditional: (1) If you are at 0° latitude, then you are at the equator, and (2) If you are at the equator, then you are at 0° latitude.

34. Separate the following biconditional into its two conditionals: *A polygon is a pentagon if and only if it has five sides.*

Objective H: *Evaluate conditionals and conjectures.* *(Lessons 2-2, 2-8)*

Multiple choice. In 35 and 36, a conjecture is given. For each figure determine whether
(i)   it is an instance of the conjecture.
(ii) it is a counterexample to the conjecture.
(iii) it is neither an instance of nor a counterexample to the conjecture.

35. If B is on $\overleftrightarrow{AC}$, then B is between A and C.

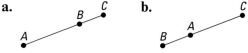

a.   **b.**

36. If $x > 12$, then $x > 13$.
    a. $x = 13.5$ **b.** $x = 12.5$

In 37 and 38, a conditional is given.
    a. Draw an instance of the conditional.
    b. Draw a counterexample to the conditional.

37. If a figure is a union of four segments, then it is a quadrilateral.

38. If a triangle is isosceles, then it is equilateral.

Objective I: *Determine the union and intersection of sets.* *(Lesson 2-5)*

39. If $A = \{2, 5, 9\}$ and $B = \{5, 13, 9\}$, find $A \cap B$.

40. Refer to the figure below. Describe the named figure.
    a. $\triangle MNO \cap \triangle MOP$ **b.** $\triangle MNO \cup \triangle MOP$

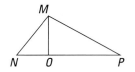

41. Draw line ℓ and line n so that $\ell \cap n = K$.

42. Use the figure below.
    a. What is $\overline{RM} \cap \overline{GM}$?
    b. What is $\overline{TR} \cap \overline{GM}$?

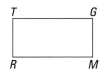

Objective J: *Determine whether a triangle can be formed with sides of three given lengths.*
(Lesson 2-7)

In 43 and 44, can the numbers be lengths of three sides of a triangle?

43. 14, 15, 30

44. $\frac{1}{3}, \frac{1}{4}, \frac{1}{5}$

45. Two sides of a triangle have lengths 4 cm and 7 cm. How long can a third side be?

46. You are told two sides of a triangle have lengths 1″ and 10″.
  a. Is this possible?
  b. If not, why not? If so, what are the possible lengths for the third side?

USES DEAL WITH APPLICATIONS OF MATHEMATICS IN REAL SITUATIONS.

Objective K: *Apply properties of conditionals in real situations.*   *(Lessons 2-2, 2-3)*

47. *If Bill hurries home, he will be able to watch his favorite TV program.* Call this statement $B \Rightarrow W$. Write the statement $W \Rightarrow B$.

48. Amanda's aunt said, "*If you help clean up, we'll make popcorn.*" Explain why the converse of her statement might not be true.

49. An ad states: "Everybody who wears Nyebox shoes can play like the pros." Rewrite the statement in the ad as a conditional.

50. *Webster's II New Revised Dictionary* defines *kimono* as "a long, loose, wide-sleeved Japanese robe worn with a broad sash." Rewrite this definition as a biconditional.

Objective L: *Identify polygons used in the real world.*   *(Lesson 2-6)*

51. An umpire has just swept off home plate in the picture at the right. What polygon is the shape of home plate?

52. A United States Susan B. Anthony dollar is pictured here. What polygon is on this coin?

53. Utah is the only state in the U.S. in the shape of a polygon other than a quadrilateral. What polygon is it?

UTAH

Objective M: *Apply the Triangle Inequality Postulate in real situations.*   *(Lesson 2-7)*

54. It is a 15-minute walk from DJ's place to Darlene's place. It is a 25-minute walk from Darlene's place to Becky's place. At this rate, by walking, how long would it take to get from DJ's place to Becky's place?

55. Vinh lives 3 blocks from the fire station and 12 blocks from school. How far apart can the fire station and the school be?

REPRESENTATIONS DEAL WITH PICTURES, GRAPHS, OR OBJECTS THAT ILLUSTRATE CONCEPTS.

Objective N: *Draw hierarchies of triangles and polygons.*   *(Lesson 2-6)*

56. Draw the hierarchy relating the following: figure, triangle, isosceles triangle, scalene triangle.

57. Draw the hierarchy relating the following: polygon, triangle, hexagon, isosceles triangle, equilateral triangle.

REFRESHER

Chapter 3, Angles and Lines, assumes that you have learned how to measure and draw angles with a protractor in earlier courses. Use the questions below to check your mastery of these skills.

A. Estimate the measure of an angle.

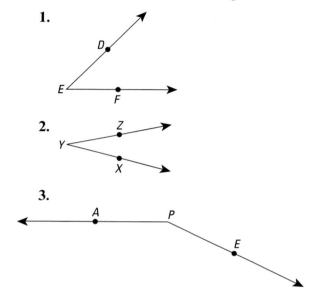

1.

2.

3.

B. Measure an angle with a protractor. (You are doing well if your estimate is within 1°.)

    4. angle in Question 1

    5. angle in Question 2

    6. angle in Question 3

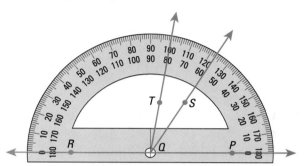

    7. a. $\angle PQS$
       b. $\angle SQR$

    8. a. $\angle SQT$
       b. $\angle TQR$

C. Draw an angle with a given measure.

    9. 55°

   10. 124°

   11. 90°

   12. 168°

D. Draw a circle with a given radius.

$A \bullet\!\!-\!\!-\!\!-\!\!-\!\!-\!\!-\!\!-\!\!-\!\!-\!\!-\!\!\bullet B$

   13. Draw a circle whose radius equals AB.

   14. Draw a circle with radius 6 cm.

CHAPTER 3

ANGLES AND LINES

Lines all look the same except for their direction, or *tilt*. Angles give a way of measuring and describing the tilt of a line, and also of measuring the differences in the tilts or directions of two lines. Angles occur everywhere, in both living and inanimate objects. An example is pictured here.

The leaf exhibits a network of veins. From the largest vein in the middle, smaller and smaller veins branch out. For a particular species of trees, these veins are set at characteristic angles, and often the smaller veins off a common vein are parallel.

Angles are found in all sorts of geometric figures. They occur whenever lines intersect. And they are found in all polygons. There are simple relationships among angles, parallel lines, and perpendicular lines, and surprising connections with the slopes of lines you studied in algebra. There are also angles between curves and between planes, but in this chapter we concentrate on the basic angles of the plane—those formed by rays or lines and those found in polygons.

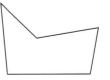

Shadows of time. *The first clocks were sundials. The sun hits the* gnomon— *the flat piece of metal in the center of the sundial—casting a shadow. The angle formed by the gnomon and the shadow determines the time of day.*

The points in a particular direction from a source all lie on the same ray. Two rays from the same source form an angle.

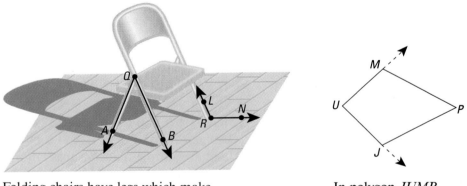

Folding chairs have legs which make angles with each other and with the plane of the floor.

In polygon *JUMP*, ∠*U* is the union of $\overrightarrow{UM}$ and $\overrightarrow{UJ}$.

> **Definition**
> An **angle** is the union of two rays that have the same endpoint.

The **sides** of an angle are the two rays that form it; the **vertex** of the angle is the common endpoint of the two rays. Angles may be determined by segments, as in polygon *JUMP* above, but you should still consider the sides of the angle to be rays. The symbol for angle is ∠. Notice the difference between the angle symbol (∠) and the symbol for "is less than" (<).

Names of Angles

Angles are named in various ways. When there is only one angle at a given vertex, the angle can be named by the vertex. On page 124, $\angle U$ (read "angle U") is shown.

When several angles have the same vertex, each one needs a unique name. In the figure below, there are three angles. Each can be named by giving a point on each side, with the vertex point in between: $\angle ABC$, $\angle CBD$, and $\angle ABD$. The smaller angles also can be named by numbers: $\angle ABC$ is $\angle 1$; $\angle CBD$ is $\angle 2$.

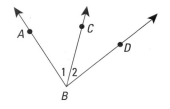

Pictured below are two special angles. $\angle STV$ is a *straight angle*, and $\angle XWY$ is a *zero angle*. They are special because some different properties apply to them. If you are asked to identify angles in a given figure or problem, you should ignore straight and zero angles unless they are specifically mentioned. Otherwise you would have to name infinitely many angles each time a ray or line appeared!

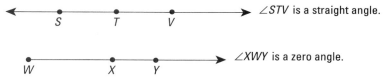

Every angle except a zero angle separates the plane into two nonempty sets other than the angle itself. If the angle is not a straight angle, exactly one of these sets is convex. The convex set is called the **interior** of the angle. The nonconvex set is the **exterior** of the angle. For a straight angle, both sets are convex and either set may be its interior.

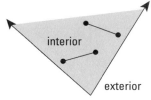

Measures of Angles

The **measure** of an angle indicates the amount of openness of the interior of the angle. For example, suppose you want to know how wide your angle of vision is. To compare it precisely with someone else's, you need a unit of measure for an angle. The unit of measure used in this book is the **degree,** denoted by °. Since no other units of measure for angles are used in this book, the degree symbol is sometimes omitted with angle measures. To indicate the measure of an angle, as opposed to the angle itself, the symbol **m$\angle ABC$,** read "the measure of angle ABC," is used.

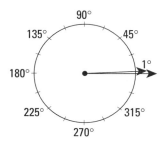

A 1° angle is determined by dividing a circle into 360 equal parts, and taking the angle determined by the center of the circle (vertex) and two consecutive subdivision points on the circle. The circle itself has 360°.

The division of the circle into 360° comes from the Babylonians, who lived in what is now Iraq. They noticed that a year was about 365 days long, and that stars appeared to move about $\frac{1}{365}$ of the way around the sky each day. The number 365 is not an easy number to work with; its only integer factors are 73 and 5. But 360 is divisible by 2, 3, 4, 5, 6, 8, 10, and 12. By splitting the circle into 360 equal parts, each part was approximately the same as the change in the position of a star from one night to the next.

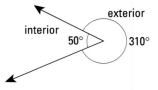

The measure of the exterior of an angle is 360° minus the measure of the interior. For instance, in the figure at the left, the measure of the exterior of the 50° angle is 310°. Sometimes in this book you will need to know this measure. But if we did this all the time, every angle would have two measures. In order to be able to speak of *the* measure of an angle, we assume that an angle has exactly one measure and that measure refers to the interior of the angle. This measure cannot be greater than 180°, which is half of the 360° encompassed by a circle.

Notice that we made choices regarding how to measure angles. We chose the number 360. We chose not to allow measures of the exterior of angles. We chose to allow straight angles and zero angles. These choices are summarized in the *Angle Measure Postulate.*

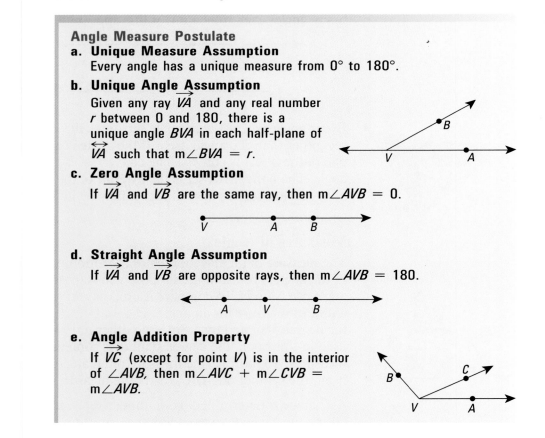

Angle Measure Postulate
a. **Unique Measure Assumption**
   Every angle has a unique measure from 0° to 180°.

b. **Unique Angle Assumption**
   Given any ray $\overrightarrow{VA}$ and any real number r between 0 and 180, there is a unique angle BVA in each half-plane of $\overleftrightarrow{VA}$ such that m$\angle BVA = r$.

c. **Zero Angle Assumption**
   If $\overrightarrow{VA}$ and $\overrightarrow{VB}$ are the same ray, then m$\angle AVB = 0$.

d. **Straight Angle Assumption**
   If $\overrightarrow{VA}$ and $\overrightarrow{VB}$ are opposite rays, then m$\angle AVB = 180$.

e. **Angle Addition Property**
   If $\overrightarrow{VC}$ (except for point V) is in the interior of $\angle AVB$, then m$\angle AVC +$ m$\angle CVB =$ m$\angle AVB$.

Parts **a** and **b** of the Angle Measure Postulate ensure that the interior of the angle is measured. Parts **c** and **d** allow for zero and straight angles. Part **e** states symbolically that when angles are next to each other, then their measures can be added to get the measure of the new, larger angle formed. This is like the Additive Property for distance. Later in this book you will see similar assumptions made for area and volume. An addition property is a fundamental property of all measures.

In drawings, an angle's measure is often written in its interior. For instance, in the figure of Example 1, m∠*APB* = 45.

Example 1

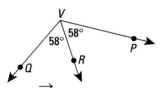

In the figure at the left, determine

a. m∠*APC*.   **b.** m∠*CPD*.

Solution

a. $\overrightarrow{PB}$ (except for point *P*) is in the interior of ∠*APC*. Use the Angle Addition Property.
$$\text{m}\angle APC = \text{m}\angle APB + \text{m}\angle BPC$$
$$= 45 + 80$$
$$= 125$$

b. Think of *P* as the center of a circle. The measures of the angles about *P* add to 360°. The given measures add to 215°.
So, m∠*CPD* = 360 − (90 + 45 + 80) = 360 − 215 = 145.

Every angle can be split by a ray into two angles of equal measure. This ray is its *bisector*.

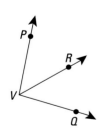

$\overrightarrow{VR}$ bisects ∠*PVQ*.

Definition

$\overrightarrow{VR}$ is the **bisector** of ∠*PVQ* if and only if $\overrightarrow{VR}$ (except for point *V*) is in the interior of ∠*PVQ* and m∠*PVR* = m∠*RVQ*.

Notice that the definition of angle bisector is similar in form to the definition of midpoint.

Example 2

In the figure at the left, $\overrightarrow{VR}$ bisects ∠*PVQ*. If m∠*PVR* = $5x - 11$ and m∠*RVQ* = $2x + 25$, find m∠*PVQ*.

Solution

First find *x*. Since $\overrightarrow{VR}$ is the angle bisector,
$$\text{m}\angle PVR = \text{m}\angle RVQ.$$
$$5x - 11 = 2x + 25$$
$$3x = 36$$
$$x = 12$$

Now substitute to find m∠*PVR* and m∠*RVQ*.
m∠*PVR* = $5x - 11 = 5 \cdot 12 - 11 = 49$
m∠*RVQ* = $2x + 25 = 2 \cdot 12 + 25 = 49$
m∠*PVQ* = m∠*PVR* + m∠*RVQ* = 49 + 49 = 98

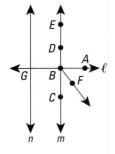

Assumptions from Drawings

There are limits to what information you can use from a drawing. Consider the figure at the left.

From a figure you *can* assume:

1. *Collinearity and betweenness of points drawn on lines.* For instance, E, D, B, and C all lie on line *m*, and D is between B and E.

2. *Intersections of lines at a given point.* For instance, *m* and *ℓ* intersect at point B.

3. *Points in the interior of an angle, on an angle, or in the exterior of an angle.* For instance, F is in the interior of ∠ABC, G is in the exterior of ∠ABC, and A is on ∠ABC.

From a figure, you *cannot* assume:

1. *Collinearity of three or more points that are not drawn on lines.* For instance, you cannot assume that F is between A and C, because there is no line drawn containing them.

2. *Parallel lines.* For instance, you cannot assume *m* // *n*, even though that may appear to be the case.

3. *Exact measures of angles and lengths of segments.* For instance, you cannot assume that m∠ABC = 90.

4. *Measures of angles or lengths of segments are equal.* For instance, you cannot assume that $\overrightarrow{BF}$ bisects ∠ABC or that DE = DB.

Often, information that you cannot assume from a figure is given elsewhere. Then, of course, you can use this information.

QUESTIONS

Covering the Reading

1. In this book, in what units are angles measured?

2. Define *angle*.

3. Estimate the measure of angle A shown at the left.

4. Draw angles with the following measures.
   a. 72   **b.** 135

In 5 and 6, use the angle shown at the left.

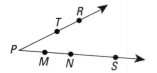

5. **a.** Name the vertex of the angle.
   b. Name its sides.
   c. Give five different names for this angle.

6. Copy the angle and shade its interior.

7. Which people developed the degree measure for an angle?

8. Use the drawing at the right.
   a. Name two straight angles.
   b. Name two angles with measure 0.
   c. m∠CED + m∠CEB = m∠ _?_.

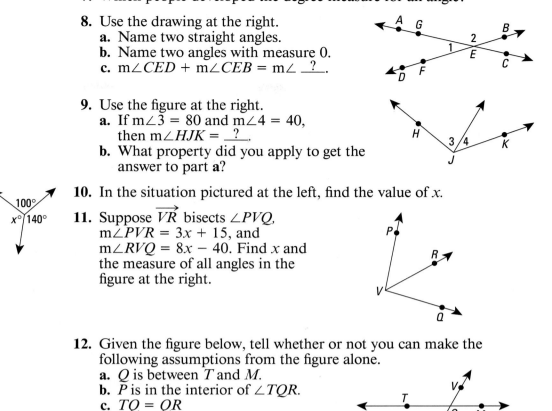

9. Use the figure at the right.
   a. If m∠3 = 80 and m∠4 = 40, then m∠HJK = _?_.
   b. What property did you apply to get the answer to part **a**?

10. In the situation pictured at the left, find the value of x.

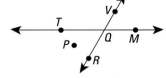

11. Suppose $\overrightarrow{VR}$ bisects ∠PVQ, m∠PVR = 3x + 15, and m∠RVQ = 8x − 40. Find x and the measure of all angles in the figure at the right.

12. Given the figure below, tell whether or not you can make the following assumptions from the figure alone.
    a. Q is between T and M.
    b. P is in the interior of ∠TQR.
    c. $TQ = QR$
    d. P is between T and R.

Applying the Mathematics

13. In this drawing of three hexagonal cells of a beehive, the three angles with vertex A have the same measure.
    a. Name these angles.
    b. What is their measure?

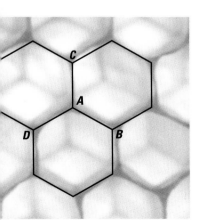

14. The figure at the right shows compass directions. For instance, N = north, SE = southeast, and WNW = west by northwest.
    a. Each small angle has the same measure. What is that measure?
    b. If you were to turn from going NE to go ESE, how many degrees would you turn?

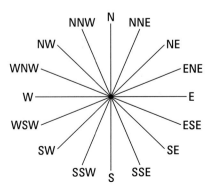

15. $\overrightarrow{ID}$ is a bisector of $\angle SIT$.
  a. Draw a picture of this situation.
  b. If $m\angle SIT = 9t - 14$ and $m\angle SID = 3t + 11$, find $m\angle SIT$.
  c. Is the picture you drew in part **a** accurate? If not, draw a more accurate picture.

Orchard Metro
Airport

N

16. The points of a compass are labeled clockwise in degrees. Airport runways are numbered with their compass direction, except that the final zero is dropped. North-south and east-west runways are labeled at left. The runway heading east is labeled 9 at the left, for 90°.
  a. The Orchard Airport Expansion Project planners want to add a runway labeled 32. Trace the drawing at the left and draw a possible runway.
  b. What number would be at the other end of the runway numbered 32?

Review

17. In at most how many points do the diagonals of a convex hexagon intersect? *(Lesson 2-6)*

18. In at least how many points do the diagonals of a convex hexagon intersect? *(Lesson 2-6)*

19. In this lesson, you learned that an angle is a union of two rays. What figures are the possible *intersections* of two rays? *(Lesson 2-5)*

20. a. Draw a circle with diameter 2″.
  b. What is its radius? *(Lesson 2-4)*

In 21 and 22, solve and graph solutions on a number line. *(Previous course)*
21. $3m + 81 \leq 180$ **22.** $0 < 83 - n < 90$

Exploration

23. Degrees can be divided into minutes and seconds.
  a. From a dictionary or other source, find out how many minutes are in one degree.
  b. How many seconds are in one degree?
  c. With your eye as the vertex, the width of the moon covers an angle of about 30 minutes in the sky. How many moons placed next to each other would extend from one point on the horizon to the point on the opposite side of the horizon?

24. Angles can be measured in units other than degrees. One of these units is the radian. Find out what a radian is and compare one radian to one degree.

Rotating a Triangle

IN·CLASS

ACTIVITY

Work on this activity with a partner. You will need a ruler, protractor, and compass.

Here is an algorithm for rotating △*ABC* 100° clockwise about point *O*.

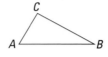

1   **a.** Trace △*ABC* and point *O* so that *O* is near the center of your paper.
    b. Draw $\overline{OA}$.
    c. Turning $\overline{OA}$ clockwise, use a protractor to draw an angle *DOA* so that m∠*DOA* = 100.
    d. Use a compass to draw a circle with center *O* and radius $\overline{OA}$.
    e. Mark the point *A′* (read "*A* prime") where $\overrightarrow{OD}$ intersects ⊙*O*.
    f. Repeat parts **b–e** for points *B* and *C* to find their images *B′* and *C′*.
    g. Connect *A′*, *B′*, and *C′* to form △*A′B′C′*.

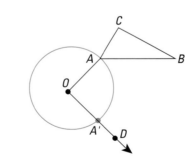

2   Draw a new triangle △*XYZ* and a point *P*.
Adapt the above procedure to rotate △*XYZ* 80° counterclockwise about *P*.

131

View from the top. *This Ferris wheel, located at Chicago's Navy Pier, is 15 stories tall. The 40 baskets, each of which seats six people, are placed at equal intervals around the circle. Each arc between baskets has a measure of nine degrees.*

Since the degree measure of an angle comes from a circle, it is not surprising that there is a close relationship between angles and circles. A circle with center O is drawn below. Suppose you walk along the circle counterclockwise from A to B. The part of the circle you have walked is **arc AB,** written $\overset{\frown}{AB}$.

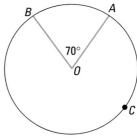

Like angles, arcs are usually measured in degrees. The measure of arc $\overset{\frown}{AB}$ is the same as the measure of the central angle, $\angle AOB$. Thus, the measure of $\overset{\frown}{AB}$ is 70°. That is, from A to B you have walked 70° around the circle. This tells how much you have turned and what part of the 360° circle you have traversed. If you walked the other way (clockwise) from A through C to B, you would have gone 290° around the circle.

Minor and Major Arcs

In general, a **central angle of a circle** is an angle whose vertex is the center of the circle. So, when A and B are points on a circle O, then $\angle AOB$ is a central angle. When $\angle AOB$ is not a straight angle, the points of $\odot O$ that are on or in the interior of $\angle AOB$ constitute the **minor arc** $\overset{\frown}{AB}$. The points A and B are the **endpoints** of the arc.

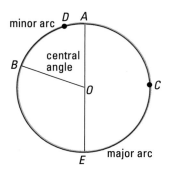

minor arc

central angle

major arc

The points of $\odot O$ which are on or in the exterior of $\angle AOB$ constitute a **major arc** of circle O. At the left, this arc is named $\overset{\frown}{ACB}$. The third point C is included to distinguish the major arc ($\overset{\frown}{ACB}$) from the minor arc ($\overset{\frown}{AB}$). In this book, major arcs are always denoted by three points. Hence, $\overset{\frown}{AB}$ will always refer to a minor arc. Sometimes, for clarity, you might want to use three letters to name a minor arc. In this instance $\overset{\frown}{AB}$ could be called $\overset{\frown}{ADB}$. The union of minor arc $\overset{\frown}{ADB}$ and major arc $\overset{\frown}{ACB}$ is the entire circle.

When a central angle is a straight angle, the arcs are called **semicircles.** In the figure above, $\angle AOE$ is a straight angle, and both $\overset{\frown}{ACE}$ and $\overset{\frown}{ABE}$ are semicircles.

> **Definitions**
>
> The **degree measure of a minor arc or semicircle** $\overset{\frown}{AB}$ of circle O, written m$\overset{\frown}{AB}$, is the measure of its central angle $\angle AOB$.
>
> The **degree measure of a major** $\overset{\frown}{ACB}$ of circle O, written m$\overset{\frown}{ACB}$, is $360° -$ m$\overset{\frown}{AB}$.

Example 1

Use $\odot O$ at the right.
a. Find m$\overset{\frown}{RS}$.
b. Find m$\overset{\frown}{RTS}$.

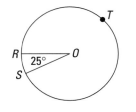

Solution
a. m$\overset{\frown}{RS}$ = m$\angle ROS$ = 25°
b. m$\overset{\frown}{RTS}$ = 360° – m$\overset{\frown}{RS}$ = 360° – 25° = 335°

Arc Length

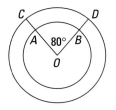

Circles are **concentric** if and only if they lie in the same plane and have the same center. At the left, in the two concentric circles with center O, arcs $\overset{\frown}{AB}$ and $\overset{\frown}{CD}$ have the same measure as m$\angle O$; m$\overset{\frown}{AB}$ = m$\overset{\frown}{CD}$ = 80°. Yet if you walked from C to D, you would walk a longer distance than if you walked from A to B. The arcs AB and CD have different lengths. Arc length is not the same as arc measure. Arc length indicates a distance; arc measure indicates an amount of a turn. The length of $\overset{\frown}{CD}$ is a distance that is measured in units such as centimeters or inches, but m$\overset{\frown}{CD}$ is measured in degrees. The units of measure are different. To avoid confusion, in this book we always put the degree symbol (°) with an arc measure. Here we discuss only arc measure; you will study arc length in Chapter 8.

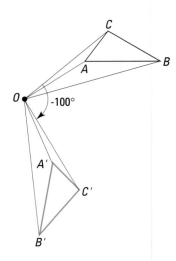

Rotations

Arc measure can be used to measure *rotations* or turns. In the figure at the left, $\triangle A'B'C'$ is the **image** of the **preimage** $\triangle ABC$ under a clockwise rotation of 100°.

Rotations always have a *center*. Here the center is O. The vertices of each figure have been connected to the center. Notice that $m\angle AOA' = m\angle BOB' = m\angle COC' = 100$. The turn has gone in a **clockwise direction**. In mathematics, **counterclockwise** is the positive direction and clockwise is the negative direction. Since the turn is clockwise, the rotation at the left has a **magnitude** of -100°. Every rotation can be expressed as having a magnitude from -180° to 180°. Notice that the magnitude of a rotation can be negative, but the magnitude of an angle (by the Angle Measurement Postulate) cannot be negative.

Rotations can also have magnitudes outside the range of -180° to 180°, such as 500°. This can always be converted to a rotation in the given range by adding or subtracting a multiple of 360°. A rotation of 500° yields the same images as a rotation of 140°.

Example 2

At the left, quadrilateral $A'B'C'D'$ is the image of $ABCD$ under a rotation about center Q. What is the magnitude of this rotation?

Solution

Draw $\overline{AQ}$ and $\overline{A'Q}$, and measure $\angle A'QA$. $m\angle A'QA = 75°$. Since the rotation from $\overline{QA}$ to $\overline{QA'}$ is counterclockwise as indicated by the arrow, the magnitude of the rotation is 75°.

Rotations can be useful in picturing regular patterns. For instance, on a piano keyboard, the notes repeat themselves as shown below. The three marked keys form a C-major chord.

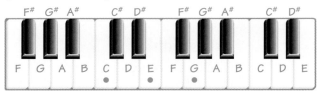

When the 12 different possible notes are placed on a circle as shown at the right, the notes of a C-major chord are represented by the vertices of triangle CEG. To find the notes of other major chords, you can rotate the triangle clockwise about the center O of the circle.

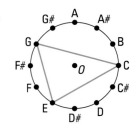

Example 3

Name the notes of a D-major chord.

Solution

Rotate △CEG so that the image of C is D. This is a rotation of 2 places in the clockwise direction. The notes are D, F#, and A.

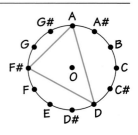

QUESTIONS

Covering the Reading

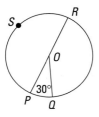

In 1–4, $\overline{PR}$ is a diameter of ⊙O at the left.

1. Name the minor arcs in the figure.

2. **a.** m$\widehat{PQ}$ = __?__   **b.** m$\widehat{PSQ}$ = __?__

3. **a.** m$\widehat{PSR}$ = __?__   **b.** m$\widehat{PQR}$ = __?__   **c.** m$\widehat{QR}$ = __?__

4. ∠ROQ is a __?__ angle of ⊙O.

5. What is the difference between *arc length* and *arc measure?*

6. Use the concentric circles pictured at the left.
   a. Find m$\widehat{WX}$.   **b.** Find m$\widehat{YUZ}$.

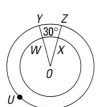

7. *True or false.* A clockwise rotation has a positive magnitude.

8. Trace the figure at the right. Rotate △DEF -70° about R.

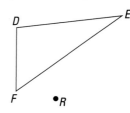

9. In the figure below, *LMNP* represents a flag and *L′M′N′P′* is its rotation image about center *C*. What is the magnitude of this rotation?

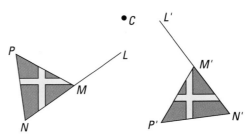

10. Name the notes of an A-major chord.

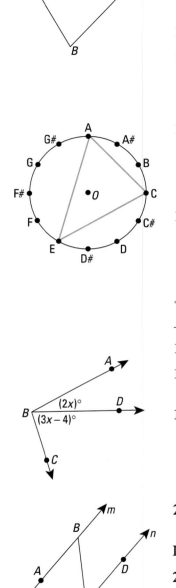

Applying the Mathematics

11. Trace △*ABC* at the left. Rotate △*ABC* 45° about *B*.

12. Refer to Example 2. What other angles have measure equal to m∠*AQA'*?

13. Draw two arcs with different lengths but the same measure, 120°.

14. Refer to the quotation below by basketball player Jason Kidd, the Dallas Mavericks' number one draft pick in 1994, about the Mavs' prospects. *"Now that I'm here, we'll turn the program around 360 degrees."* What do you think he meant to say? Explain.

15. At the left is a circle with an A-minor chord represented by triangle ACE.
   a. Find the notes of a C#-minor chord.
   b. What is the magnitude of the rotation you need to perform to answer part **a?**

16. In 1988, at the World Championships in Budapest, Hungary, Kurt Browning of Canada became the first figure skater to successfully complete a quadruple toe loop (four revolutions) in an international competition. What is the total number of degrees in four revolutions?

Review

17. If m∠*ABC* = 101 in the figure at the left, find m∠*DBC*.   *(Lesson 3-1)*

18. If two sides of a triangle are 7 and 15, write a sentence describing all possible lengths *t* for the third side.   *(Lesson 2-7)*

19. Use the figure at the right. If $\overrightarrow{BC}$ is the angle bisector of ∠*ABD*, find m∠*ABD*. *(Lesson 3-1)*

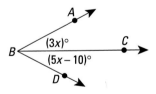

20. In the drawing at the left, can you assume line *m* is parallel to line *n*? *(Lesson 3-1)*

In 21 and 22, solve the system of equations.   *(Previous course)*

21. $\begin{cases} y = 4x \\ y = x + 20 \end{cases}$

22. $\begin{cases} 2x - y = 14 \\ x + y = 4 \end{cases}$

Exploration

23. Find a circle graph in a newspaper or magazine. Measure or estimate the measure of each central angle. In an accurate circle graph, the ratios of these measures should be the same as the ratios of the quantities graphed. Is the circle graph an accurate representation of the data?

Cool angles.   *Shown are fans from Indonesia, Spain, and China. The fans show angles of varying measures. Estimate the measure of each angle.*

Remember that every angle has a unique measure from 0 to 180 degrees. In this range, angles are classified into one of five types.

Definitions
If *m* is the measure of an angle, then the angle is
a. **zero** if and only if $m = 0$;
b. **acute** if and only if $0 < m < 90$;
c. **right** if and only if $m = 90$;
d. **obtuse** if and only if $90 < m < 180$;
e. **straight** if and only if $m = 180$.

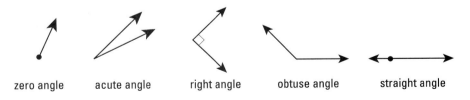

zero angle   acute angle   right angle   obtuse angle   straight angle

To identify a right angle, a " ⌐ " is drawn in the angle to form a square.

Complementary and Supplementary Angles

Some pairs of angles have special properties. For instance, two angles are given special names if the sum of their measures is 90 or 180.

If the measures of two angles are m_1 and m_2, then the angles are
a. **complementary** if and only if $m_1 + m_2 = 90$;
b. **supplementary** if and only if $m_1 + m_2 = 180$.

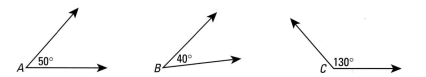

In the figures above, $\angle A$ and $\angle B$ are complementary angles, while $\angle A$ and $\angle C$ are supplementary angles. It is also said that $\angle A$ and $\angle B$ are **complements,** and $\angle A$ and $\angle C$ are **supplements.** If the measure of an angle is x, then any complement to it has measure $90 - x$ and any supplement to it has measure $180 - x$.

Example 1

The angle outlined in black on this Japanese eel basket is $\frac{1}{4}$ the measure of its supplement. Find the measure of the angle and check your answer.

Solution 1

Let the angle have measure x. Its supplement then has measure $180 - x$. From the given information,
$$x = \tfrac{1}{4}(180 - x).$$
$$4x = 180 - x$$
$$5x = 180$$
$$x = 36$$
The measure of the angle is 36.

Check
$180 - 36 = 144$. Since 36 is $\frac{1}{4} \cdot 144$, it checks.

Solution 2

Let x be the measure of the angle.
Let y be the measure of its supplement. Then
$$\begin{cases} x = \tfrac{1}{4}y & \text{(The angle is } \tfrac{1}{4} \text{ its supplement.)} \\ x + y = 180 & \text{(definition of supplementary angles)} \end{cases}$$
You can solve the system of equations by substitution. Substitute $\frac{1}{4}y$ for x in the second equation and solve for y.
$$\tfrac{1}{4}y + y = 180$$
$$\tfrac{5}{4}y = 180$$
$$y = 144$$
To find x, substitute 144 for y in the first equation.
$$x = \tfrac{1}{4}(144)$$
$$x = 36$$

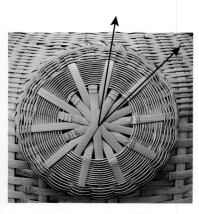

Eels for meals. *Eel baskets like this one are used in Japan for trapping and transporting eel.*

Adjacent Angles

Pairs of angles also have special names depending on their positions relative to each other.

> **Definition**
> Two non-straight and nonzero angles are **adjacent angles** if and only if a common side ($\overrightarrow{OB}$ in the figure at the right) is interior to the angle formed by the non-common sides ($\angle AOC$).

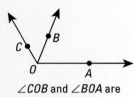

$\angle COB$ and $\angle BOA$ are adjacent angles.

Identification of adjacent angles can be tricky. Below, angles D and E are not adjacent. They do not have a common side since $\overrightarrow{DE} \neq \overrightarrow{ED}$. $\angle PQT$ and $\angle SQT$ are not adjacent because $\overrightarrow{QT}$, their common side, is not in the interior of $\angle PQS$. Angles 7 and 8 below at the right are not adjacent because $\overrightarrow{ZY}$, their common side, is not in the interior of $\angle XZW$.

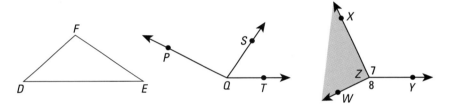

Example 2

Suppose $\angle 5$ and $\angle 6$ are complementary and adjacent angles, with $m\angle 5 = 47$.
a. Sketch a possible situation.   **b.** Find $m\angle 6$.

Solution

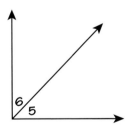

a. Since $\angle 5$ and $\angle 6$ are adjacent, they have a common side, so a figure could look like that drawn at the left.
b. Since $\angle 5$ and $\angle 6$ are complementary, $m\angle 5 + m\angle 6 = 90$.
Substitute 47 for $m\angle 5$. $47 + m\angle 6 = 90$.
So, $m\angle 6 = 43$.

Linear Pairs

Certain adjacent angles are *linear pairs*. In the figure at the right, $\angle 1$ and $\angle 2$ are a linear pair.

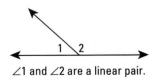

$\angle 1$ and $\angle 2$ are a linear pair.

> **Definition**
> Two adjacent angles form a **linear pair** if and only if their non-common sides are opposite rays.

Linear pairs of angles have a special property. Suppose $\overrightarrow{VA}$ and $\overrightarrow{VB}$ are opposite rays. Then, because of the Straight Angle Assumption in the Angle Measure Postulate, m∠AVB = 180. By the Angle Addition Property, m∠1 + m∠2 = 180. That is, ∠1 and ∠2 are supplementary.

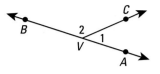

This argument works for any linear pair. It proves a theorem which will be used in later lessons, so we give it a name.

Linear Pair Theorem
If two angles form a linear pair, then they are supplementary.

Example 3

Refer to the figure above.
a. If m∠1 = 43, what is m∠2?
b. If m∠1 = x, what is m∠2?

Solution

a. By the Linear Pair Theorem, m∠1 + m∠2 = 180. So,
m∠2 = 180 − 43 = 137.
b. m∠2 = 180 − x.

Vertical Angles

When two lines intersect, four angles are formed. Each non-adjacent pair is a pair of *vertical angles.*

Definition
Two non-straight angles are **vertical angles** if and only if the union of their sides is two lines.

Focus on the vertical angles 3 and 5 at the right. Each forms a linear pair with ∠6. So, by the Linear Pair Theorem, m∠3 + m∠6 = 180 and m∠5 + m∠6 = 180. Thus, m∠3 + m∠6 = m∠5 + m∠6. Adding -m∠6 to each side, m∠3 = m∠5. So the vertical angles have the same measure. This argument works for any vertical angles and proves a second theorem.

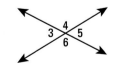

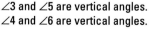
∠3 and ∠5 are vertical angles.
∠4 and ∠6 are vertical angles.

Hang your hat. *Vertical angles can be easily seen in this clothes rack.*

Vertical Angles Theorem

If two angles are vertical angles, then they have equal measures.

Example 4

Find the measures of as many angles as you can in the figure at the right, given m∠AEB = 62.

Solution

Using the Vertical Angles Theorem, m∠2 = 62. Since ∠AEB and ∠1 form a linear pair, by the Linear Pair Theorem they are supplementary. Thus m∠1 = 180 − 62 = 118. Lastly, since ∠1 and ∠3 are vertical angles, m∠3 = 118.

QUESTIONS

Covering the Reading

1. When classified by measure, there are five types of angles. Name them.

2. Suppose an angle has 4 times the measure of a supplement to it. Find the measure of the angle.

3. Suppose ∠7 and ∠8 are supplementary and adjacent angles, with m∠7 = 66.
   a. Sketch a possible situation.   **b.** Find m∠8.

4. What name is given to the pair of angles in Question 3?

5. In the figure at the right, ∠YVX and ∠WVX form a linear pair and m∠YVX = 103.
   a. Find m∠WVX.
   b. Is ∠YVX acute, obtuse, or right?
   c. Is ∠WVX acute, obtuse, or right?

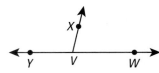

In 6–8, use the figure at the left.

6. ∠1 and ∠3 are __?__ angles.

7. If m∠1 = 121, find the measures of ∠2, ∠3, and ∠4.

8. If m∠1 = x, find the measures of ∠2, ∠3, and ∠4.

Applying the Mathematics

In 9 and 10, sketch a possible drawing of ∠1 and ∠2.

9. ∠1 and ∠2 are complementary and not adjacent.

10. ∠1 and ∠2 are adjacent, complementary, and have the same measure.

11. An angle's measure is 8 more than the measure of a complement to it. Find the measures of the angle and its complement.

12. In the figure below, ∠*AOC* and ∠*BOC* form a linear pair. Find m∠*COB*.

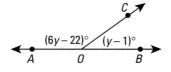

13. Some scissors are not formed by two straight pieces of metal, but by pieces of metal bent at the pivot *P*. When the scissors are open, as shown at the right below, are any two of the angles *APB*, *BPD*, *DPC*, and *CPA* a linear pair, adjacent, or vertical? Why or why not?

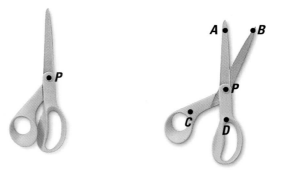

14. A triangle is called an **acute, right,** or **obtuse triangle** depending on the measure of its largest angle. Without measuring, tell whether each triangle appears to be acute, right, or obtuse.

a.   **b.**   **c.**

15. a. Write the converse of the Linear Pair Theorem.
b. Explain in words and with a drawing why the converse is not true.

16. The famous Leaning Tower of Pisa has been shifting on its foundation since it was built in the 12th century. At this time the smallest angle it makes with the ground measures about 85°. What is *x*, the measure of the largest angle it makes with the ground?

17. The measure *t* of an angle is less than the measure of its complement. Find all possible values of *t*.

18. The plane in the diagram at the right is heading in a direction 10° north of east, written 10° N of E. How many degrees would it have to turn in order to head due north?

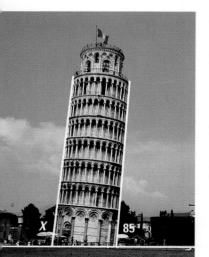

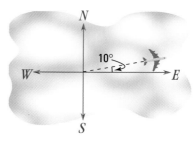

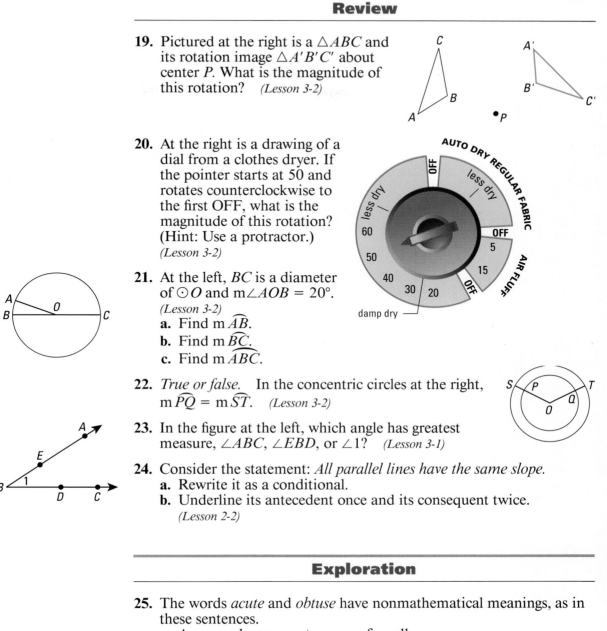

19. Pictured at the right is a △ABC and its rotation image △A'B'C' about center P. What is the magnitude of this rotation? *(Lesson 3-2)*

20. At the right is a drawing of a dial from a clothes dryer. If the pointer starts at 50 and rotates counterclockwise to the first OFF, what is the magnitude of this rotation? (Hint: Use a protractor.) *(Lesson 3-2)*

21. At the left, BC is a diameter of ⊙O and m∠AOB = 20°. *(Lesson 3-2)*
a. Find m $\widehat{AB}$.
b. Find m $\widehat{BC}$.
c. Find m $\widehat{ABC}$.

22. *True or false.* In the concentric circles at the right, m $\widehat{PQ}$ = m $\widehat{ST}$. *(Lesson 3-2)*

23. In the figure at the left, which angle has greatest measure, ∠ABC, ∠EBD, or ∠1? *(Lesson 3-1)*

24. Consider the statement: *All parallel lines have the same slope.*
a. Rewrite it as a conditional.
b. Underline its antecedent once and its consequent twice.
(Lesson 2-2)

Exploration

25. The words *acute* and *obtuse* have nonmathematical meanings, as in these sentences.
a. A person has an acute sense of smell.
b. That argument is obtuse.
What do these words mean in these sentences, and how do they relate to the corresponding names for angles?

26. The words *compliment* and *complement* sound alike but mean different things. They are called *homonyms*.
a. What is the meaning of *compliment*?
b. Give a homonym for *right*.
c. Find at least two other mathematical terms that have homonyms.

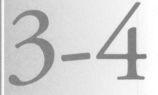

Algebra Properties Used in Geometry

The name of the game. *The properties of geometry are like the rules of a game. They tell you the moves you can make. This chess game in Zurich allows everyone to see each move.*

As you have seen in this chapter and the last, when you find measures in figures, you combine arithmetic, algebra, and geometry. Like geometry, arithmetic and algebra are also based on undefined terms (such as *number* or *set*), defined terms (such as *even number* or *integer*), and postulates. In this lesson, you will see some of the properties from arithmetic and algebra that are used in this course. All of these hold for *any* real numbers *a, b,* and *c.* These properties also hold for all algebraic expressions and for anything that stands for a real number.

Types of Properties

There are three types of properties in mathematics: *assumed properties, defining properties,* and *deduced properties.* Assumed properties are actual postulates. Defining properties are given in definitions. Deduced properties are concluded from theorems. The hierarchy below pictures this.

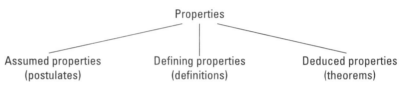

In this lesson, four sets of assumed properties from algebra are presented. After each set are instances of the properties from arithmetic, algebra, or geometry. Recall that postulates describe undefined terms and serve as starting points for logical deductions. In algebra, some of the undefined terms are *equality, addition,* and *multiplication.*

Describing Equality

The first set of properties describes what is commonly meant by the undefined equal sign. The names of the properties begin with the letters *R*, *S*, and *T*, which can help you remember them.

Postulates of Equality

For any real numbers *a, b,* and *c:*

Reflexive Property of Equality: $a = a$.

Symmetric Property of Equality: If $a = b$, then $b = a$.

Transitive Property of Equality: If $a = b$ and $b = c$, then $a = c$.

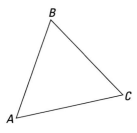

The Reflexive Property of Equality shows the obvious property that any number or expression is equal to itself.

You are applying the Symmetric Property when, to graph $4x - 1 = y$, you write it as $y = 4x - 1$.

The Transitive Property of Equality is used often. For instance, in $\triangle ABC$ at the left, if you know that $m\angle A = m\angle B$ and $m\angle B = m\angle C$, you can conclude that $m\angle A = m\angle C$.

Equality and the Operations of Addition and Multiplication

The second set of properties shows how equality is related to the operations of addition and multiplication. You have used these postulates often when solving equations.

Postulates of Equality and Operations

For any real numbers *a, b,* and *c:*

Addition Property of Equality: If $a = b$, then $a + c = b + c$.

Multiplication Property of Equality: If $a = b$, then $ac = bc$.

Example 1

Show the properties of equality and the operations used in solving $47x - 12 = 230$.

Solution

We indicate what was done by giving the steps at the left and writing the property at the right.

$$47x - 12 = 230 \quad \text{Given}$$
$$47x = 242 \quad \text{Addition Property of Equality}$$
$$\text{(12 added to each side)}$$
$$x = \frac{242}{47} \quad \text{Multiplication Property of Equality}$$
$$\text{(both sides multiplied by } \tfrac{1}{47})$$

These properties are also used in geometry. For instance, suppose that m∠*AVB* = m∠*CVD* in the figure at the left. By adding m∠*BVC* to both sides of this equation, you can conclude that m∠*AVC* = m∠*BVD*. The ﹀ marks mean that the measures of ∠*AVB* and ∠*CVD* are equal.

m∠*AVB* = m∠*CVD*		Given
m∠*AVB* + m∠*BVC* = m∠*CVD* + m∠*BVC*		Addition Property of Equality
m∠*AVC* = m∠*BVD*		Angle Addition Property

The Transitive Property does not always hold with relations other than *is equal to.* Consider the relationship *is a friend of.* If Anna is a friend of Brian and Brian is a friend of Consuelo, it does not mean that Anna is a friend of Consuelo!

Postulates of Inequality

Now we ask: If the equal sign = is replaced by the inequality sign < in the preceding postulates, are the resulting properties true?

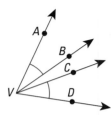

A friend, indeed! *If A is a friend of B and B is a friend of C, does it follow that A is a friend of C?*

Is there a Reflexive Property of Inequality? No, since $13 < 13$ is not true.

Is there a Symmetric Property of Inequality? If $3 < 20$, is $20 < 3$? No.

Is there a Transitive Property of Inequality? Given $-12 < 2$ and $2 < 9$, is it true that $-12 < 9$? Yes, at least in this case.

There is an Addition Property of Inequality. For example, if $2x + 9 < 13$, you can add -9 to both sides to get $2x < 4$.

In algebra, you also learned the Multiplication Properties of Inequality. Consider the true inequality $-6 < 10$. If you multiply both sides by a positive number such as $\frac{1}{2}$, the inequality does not change: $-3 < 5$. But if you multiply both sides of $-6 < 10$ by a negative number such as -2, the inequality sign does change: $12 > -20$.

These properties of inequality are formally stated below in the third set of postulates.

Postulates of Inequality and Operations
For any real numbers *a, b,* and *c:*
Transitive Property of Inequality: If $a < b$ and $b < c$, then $a < c$.
Addition Property of Inequality: If $a < b$, then $a + c < b + c$.
Multiplication Properties of Inequality: If $a < b$ and $c > 0$, then $ac < bc$.
If $a < b$ and $c < 0$, then $ac > bc$.

Below are three acute angles, P, Q, and R. If $m\angle P < m\angle Q$ and $m\angle Q < m\angle R$, then the Transitive Property allows you to conclude that $m\angle P < m\angle R$.

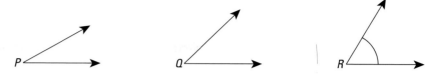

Suppose $m\angle R = m\angle BPE = m\angle DQF$. We can add angles with measures equal to $m\angle R$ to both $m\angle P$ and $m\angle Q$ as shown below. Then, by the Addition Property of Inequality, $m\angle APB < m\angle CQD$.

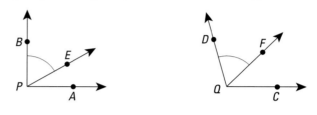

Example 2

Knowing that $m\angle X < m\angle Y$, what can you conclude about $2m\angle X$ and $2m\angle Y$ using the Multiplication Property of Inequality?

Solution

When both sides of $m\angle X < m\angle Y$ are multiplied by 2, a positive number, the order of the inequality does not change. So $2m\angle X < 2m\angle Y$.

Equations and Inequalities

A fourth set of properties applies to both equations and inequalities. The mathematics of these two properties should be familiar to you, but the formal statements of them may not be.

> **Postulates of Equality and Inequality**
> For any real numbers a, b, and c:
> **Equation to Inequality Property:** If a and b are positive numbers and $a + b = c$, then $c > a$ and $c > b$.
> **Substitution Property:** If $a = b$, then a may be substituted for b in any expression.

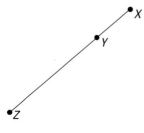

The Equation to Inequality Property is sometimes informally stated as "the whole is greater than any of its parts." For example, in the figure at the left, the length of the whole segment is greater than the length of any part of it. In symbols, since $XY + YZ = XZ$, then $XZ > XY$ and $XZ > YZ$.

You have already used the Substitution Property in this course. Suppose an angle has measure $90 - x$. If $x = 25$, then you substitute 25 for x in the expression to get $90 - 25$, or 65, as the measure of the angle.

QUESTIONS

Covering the Reading

1. Identify two undefined terms from arithmetic or algebra.

2. What are the three types of properties?

3. What undefined term do the Properties of Equality describe?

4. Name the three assumed Properties of Equality.

5. Since $\frac{1}{40} = 0.025$ and $0.025 = 2.5\%$, what can be concluded using the Transitive Property of Equality?

6. **a.** Solve $2x + 46 = 30$.
   b. What properties from this lesson did you use in finding your solution?

7. **a.** Solve $2x + 46 < 30$.
   b. What properties from this lesson did you use in finding your solution?

8. Given $m\angle WTX = m\angle ZTY$ in the figure at the left, what can you conclude using the Addition Property of Equality and the Angle Addition Property?

9. Suppose $m\angle A > m\angle B$. What can you conclude about $-3m\angle A$ and $-3m\angle B$ using the Multiplication Property of Inequality?

10. In the figure at the right, $PQ + QR = PR$. What can you conclude using the Equation to Inequality Property?

11. Suppose an angle has measure $180 - x$, and $x = 47$.
    a. What is the measure of the angle?
    b. What property did you use to answer part **a**?

Applying the Mathematics

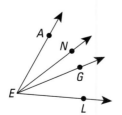

12. Consider the figure at the left. If you know that $m\angle AEN < m\angle GEL$, write an argument using the properties of this lesson to show that $m\angle AEG < m\angle NEL$.

13. *Multiple choice.* In an isosceles triangle ACE, $AC = CE$. Which postulate justifies $CE = AC$?
    (a) Symmetric Property of Equality
    (b) Addition Property of Equality
    (c) Equation to Inequality Property

14. Write an argument that you would use to convince a friend that there is no Symmetric Property of Inequality.

15. Match each angle type with an algebraic description. *(Lesson 3-3)*

  a. acute

  b. complementary

  c. obtuse

  d. right

  e. straight

  f. supplementary

  g. zero

  (i) $m_1 + m_2 = 90$

  (ii) $m_1 + m_2 = 180$

  (iii) $m = 0$

  (iv) $m = 90$

  (v) $m = 180$

  (vi) $90 < m < 180$

  (vii) $0 < m < 90$

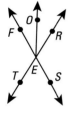

16. In the figure at the left, $\angle PSR$ is a straight angle. *(Lesson 3-3)*

  a. Find t.

  b. Find $m\angle VSP$.

17. The measure of a supplement to an angle is 20 more than the measure of the angle. Find the measure of the supplement. *(Lesson 3-3)*

18. Use the figure at the right.

  a. Find x.

  b. Find $m\angle EIT$.

  c. $\angle RIE$ and $\angle EIT$ are ___?___ of each other.

  (Lessons 3-1, 3-3)

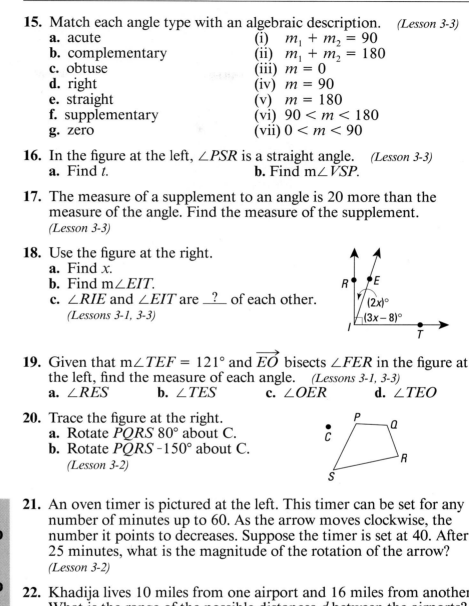

19. Given that $m\angle TEF = 121°$ and $\overrightarrow{EO}$ bisects $\angle FER$ in the figure at the left, find the measure of each angle. *(Lessons 3-1, 3-3)*

  a. $\angle RES$

  b. $\angle TES$

  c. $\angle OER$

  d. $\angle TEO$

20. Trace the figure at the right.

  a. Rotate $PQRS$ 80° about C.

  b. Rotate $PQRS$ -150° about C.

  (Lesson 3-2)

21. An oven timer is pictured at the left. This timer can be set for any number of minutes up to 60. As the arrow moves clockwise, the number it points to decreases. Suppose the timer is set at 40. After 25 minutes, what is the magnitude of the rotation of the arrow? *(Lesson 3-2)*

22. Khadija lives 10 miles from one airport and 16 miles from another. What is the range of the possible distances d between the airports? *(Lesson 3-2)*

Exploration

23. The transitive property is true for equality and for inequality. It is also true for *is an ancestor of:* If a is an ancestor of b and b is an ancestor of c, then a is an ancestor of c. But as you have seen, the transitive property is not always true for *is a friend of:* If a is a friend of b and b is a friend of c, then a is not necessarily a friend of c. Find two other relations, one that satisfies the transitive property and one that does not.

A two-step proof. *This security system reads fingerprints and matches them with access codes to* prove *that people are who they claim to be before allowing them access to files or entrance into certain locations.*

Simplifying expressions, solving equations, and applying algebraic techniques all represent a series of justified steps. If the given information p and the conclusion q following from it are important enough, then the statement $p \Rightarrow q$ is labeled as a theorem. Showing q follows from p *proves the conditional $p \Rightarrow q$.*

A proof-writer needs to indicate the given information p, the statement q which is the final conclusion, and the *proof argument* showing how q follows from p.

> **Definition**
> A **proof argument** for a conditional is a sequence of justified conclusions, starting with the antecedent and ending with the consequent.

For instance, here is a proof argument for the conditional *If $4r - 3 = 11$, then $r = 3.5$.* We number the antecedent 0 and the conclusions 1, 2, ... The justification or reason for each conclusion is shown to its right.

0. $4r - 3 = 11$ Given
1. $4r = 14$ Addition Property of Equality (3 was added to each side.)
2. $r = 3.5$ Multiplication Property of Equality (Both sides were multiplied by $\frac{1}{4}$.)

This is a *two-step proof argument.* You have written many arguments like this in algebra, though you did not usually write them in this way.

One-Step Proofs

When a proof has an argument in which there is a given and only one justified conclusion, it is often called a *one-step proof.*

A **justification** for a conclusion in a step is a general property for which the step is a special case. Other than the information given in a particular problem, all justifications are properties. Thus, there are three types of justifications: postulates, definitions, and previously proved theorems. In the algebra proof argument on page 150, both justifications are postulates. In Example 1, the justification is a definition.

Example 1

Given: *X* is the midpoint of $\overline{MT}$.
Conclusion: *XM* = *XT*.
Justify this conclusion.

Solution

The "Given" is the antecedent and the "Conclusion" is the consequent of the conditional *If X is the midpoint of MT, then XM = XT.* This is a restatement of one direction of the definition of midpoint. As a justification, you can write the name of the justification, **definition of midpoint,** or the statement, **If M is the midpoint of $\overline{AB}$, then AM = MB.**

When writing justifications, some teachers prefer that students write the names of definitions or theorems. Others prefer writing out the statements. Find out whether your teacher has a preference.

Definitions are biconditionals, so they can be used as justifications in two directions. To name or categorize figures, a common justification is to use the definition that goes in the direction of *characteristics ⇒ term.*

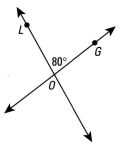

Example 2

Given: m∠*LOG* = 80. Justify the conclusion that ∠*LOG* is an acute angle.

Solution

Rewrite the given and conclusion as a conditional. *If m∠LOG = 80, then ∠LOG is an acute angle.* This is an instance of the definition of acute angle. Write the statement: **If an angle has measure less than 90, then it is acute,** or, more simply, **definition of acute angle.**

In Example 3, the justification is a previously proved theorem.

Example 3

Given: ∠*AEC* and ∠*DEB* are vertical angles. Justify the conclusion that m∠*AEC* = m∠*DEB*.

Solution

This example gives instances of the antecedent and consequent of the Vertical Angles Theorem. Write: **Vertical Angles Theorem** or **If two angles are vertical angles, then they are equal in measure.**

In Example 4, the justification is a postulate from algebra.

Example 4

Given: $x + y = 180$. Justify the conclusion $y = 180 - x$.

Solution

$-x$ has been added to both sides of $x + y = 180$ to get the conclusion. The Addition Property of Equality says you can add the same number to both sides of an equality without changing the solutions. So the justification is **Addition Property of Equality** or **If $a = b$, then $a + c = b + c$.**

To write justifications, you obviously need to be familiar with definitions, postulates, and theorems. In the back of the book you will find a list of postulates, a list of theorems, and a glossary. When you write a proof argument you can use only the justifications that have been presented prior to the statement you are proving. As each new postulate, definition, or theorem is presented in the text, it becomes a new possible justification for conclusions.

Why Are There Proofs?

In Lesson 3-3, two statements were proved—first the Linear Pair Theorem and then the Vertical Angles Theorem. These theorems may look rather obvious. So why are they proved? Mathematicians generally have three reasons for proving theorems.

The first reason is that people sometimes disagree; what is obvious to one person may not be so obvious to another person.

A second reason for proof is that unexpected results can be verified. For instance, in Chapter 1 you learned Euler's Theorem, which concerns traversable networks. The Pythagorean Theorem also is not an obvious theorem, but it will be proved later in this book. In algebra, the Quadratic Formula is proved, and that formula is not at all obvious.

A third reason for proving statements is also important: If a statement cannot be proved after many people have worked on it for a long time, it is quite possible that either (1) the statement cannot be proved or disproved from the postulates, or (2) the statement is not true even though it looks true.

An early proof? *This prehistoric hand print discovered by Heńri Cosquer in a cave in France may have been used to prove who painted the art or who visited the cave.*

The proofs that you write in this course are mostly of statements that appear to be true, but need justifications. As you learn more postulates, definitions, and theorems, you may be able to generate your own conjectures and then prove that they are true.

QUESTIONS

Covering the Reading

1. Showing that *q* follows from *p* is called __?__.

2. Define *proof argument.*

3. What is a proof with an argument that has only one justified conclusion often called?

4. Name the three kinds of statements which can be used as justifications in a proof argument.

5. Can a postulate from algebra be used as a justification in geometry?

6. Once a conditional is proved true, it is called a __?__.

7. Give three reasons why mathematicians feel it is important to have proofs.

8. *Multiple choice.* Most of the proofs you will do in this course will be of statements that
   (a) appear true, but need justification.
   (b) appear true, but are false.
   (c) appear false, but are true.

In 9 and 10, state the justification for the conclusion.

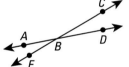

9. Given: In the figure at the left, $\angle ABC$ and $\angle EBD$ are vertical angles.
   Conclusion: $m\angle ABC = m\angle EBD$.

10. Given: $m\angle XYZ = 90$.
    Conclusion: $\angle XYZ$ is a right angle.

11. In going from $4x + 3 = 12$ to $4x = 9$, what is the justification?

Applying the Mathematics

12. *Multiple choice.* In the drawing at the left, $AB = BD$. You conclude that B is the midpoint of $\overline{AD}$. Which statement justifies this conclusion?
    (a) Addition Property of Distance   (b) definition of betweenness
    (c) Reflexive Property of Equality   (d) definition of midpoint

In 13 and 14, *multiple choice.* Use the figure at the left. Which statement justifies the conclusion?

13. Given: $m\angle 1 = 72$; $m\angle 2 = 74$. Conclusion: $m\angle FGH = 146$.
    (a) Angle Addition Property   (b) Linear Pair Theorem
    (c) definition of obtuse angle   (d) Vertical Angles Theorem

14. Given: $\angle 3$ and $\angle 4$ form a linear pair.
    Conclusion: $\angle 3$ and $\angle 4$ are supplementary angles.
    (a) Angle Addition Property   (b) Linear Pair Theorem
    (c) definition of supplementary angles (d) Vertical Angles Theorem

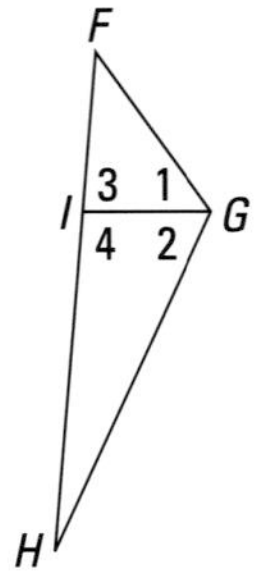

In 15–17, use △FGH at the left. Complete the one-step proofs.

15. Given: ∠3 and ∠4 are supplementary angles.
Conclusion: m∠3 + m∠4 = 180.

16. Given: $\overrightarrow{GI}$ bisects ∠FGH.
Conclusion: m∠1 = m∠2.

17. Given: m∠1 = m∠2.
Conclusion: $\overrightarrow{GI}$ bisects ∠FGH.

Review

18. State the Symmetric Property of Equality. *(Lesson 3-4)*

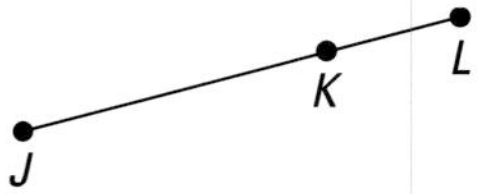

19. Use the figure at the left. Using the Equation to Inequality Property, what can you conclude about *JK* and *JL*? *(Lesson 3-4)*

20. ∠2 and ∠3 are vertical angles. If m∠2 = 12*q* and m∠3 = 4*z*, express *q* in terms of *z*. *(Previous course, Lesson 3-3)*

21. In the drawing at the right, $\overrightarrow{PN}$ and $\overrightarrow{PQ}$ are opposite rays. Find m∠NPO. *(Lesson 3-3)*

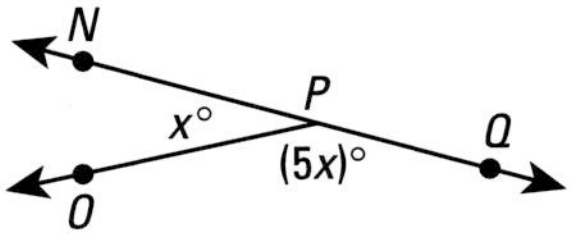

22. A complement of ∠*T* has fourteen times the measure of ∠*T*. What is m∠*T*? *(Lesson 3-3)*

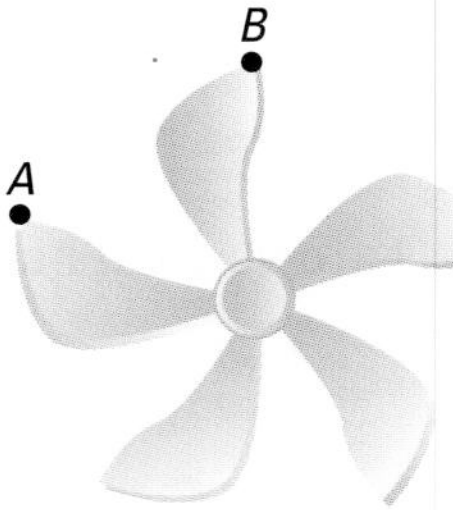

23. A fan with equally spaced propellers is pictured at the left. What is the magnitude of the rotation from *A* to *B*? *(Lesson 3-2)*

24. Raina and Lilly are sisters who share a square-shaped room. They have divided the room in half, as shown at the right. If $\overline{XW}$ bisects right angle *YXZ*, what is the measure of ∠*YXW*? *(Lesson 3-1)*

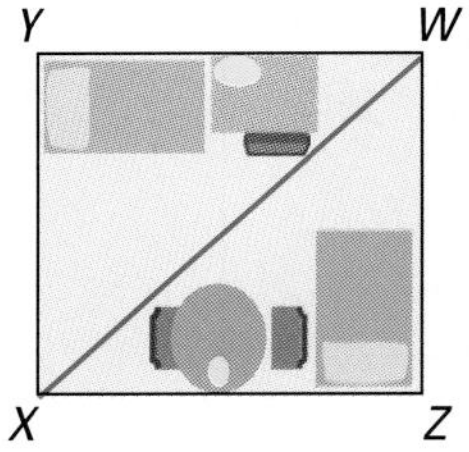

25. Let *v* be the measure of an obtuse angle. Graph all possibilities for *v* on a number line. *(Lesson 3-1)*

Exploration

In 26–28, the given information is from outside mathematics.
a. Make a conclusion from the given information.
b. How sure are you of your conclusion? Give your answer as a percent.

26. Today is Monday.

27. Last week the football team won its game.

28. You toss a coin nine times and it shows "heads" each time.

The bridge is up. *This bridge has a variable grade. As the bridge is opened, the grade becomes very steep. As it is lowered, the grade decreases to zero.*

Lines have *tilt.* The tilt of a line can be measured by the angle it makes with some line of reference. Often the reference line is horizontal. A road might be described as having a *grade* of 7°, which means that the angle it makes with the horizontal has a measure of 7°.

Corresponding Angles

Consider the angles formed when *two* lines *m* and *n* are intersected by a third line *t,* called a **transversal.** We say that the transversal "cuts" the lines. Eight angles are formed, as numbered below, four by *m* and the transversal and four by *n* and the transversal. Any pair of angles in similar locations with respect to the transversal and each line is called a pair of **corresponding angles.** In the figure, angles 1 and 5 are corresponding angles, because the interiors of the angles are to the left of the transversal and above line *m* and line *n,* respectively. In this figure, angles 2 and 6, 3 and 7, and 4 and 8 are also pairs of corresponding angles.

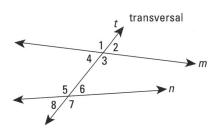

Recall the definition of parallel lines: two coplanar lines are parallel if and only if they are the same line or they do not intersect. (In symbols, $\ell \parallel m \Leftrightarrow \ell = m$ or $\ell \cap m = \emptyset$.) We say that segments or rays are parallel if the lines containing them are parallel.

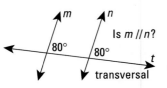

If two lines have the same tilt, that is, if they make the same angle with a transversal, you would definitely think they are parallel. Yet on Earth, two north-south streets make the same angle with any east-west street, but they would intersect at the North and South Poles if extended. Also, in perspective drawings, some parallel lines meet at a vanishing point. Assumptions about parallel lines are needed to ensure that the geometry you are studying is Euclidean geometry and not the geometry of the curved surface of Earth or the geometry of perspective drawings. We make the following assumption.

Corresponding Angles Postulate
Suppose two coplanar lines are cut by a transversal.
a. If two corresponding angles have the same measure, then the lines are parallel.
b. If the lines are parallel, then corresponding angles have the same measure.

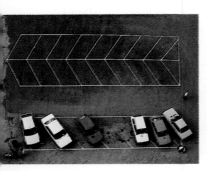

Looking for a place to park? *To enable more cars to fit in a parking lot, parallel segments are drawn to outline parking spaces.*

For instance, by part **a** of the Corresponding Angles Postulate, if segments are drawn creating equal angles with the line ℓ, then the segments are parallel.

In the figure at the right the small arrow marks "$\rightarrow$" on the lines m and n indicate that m and n are parallel. By part **b** of the Corresponding Angles Postulate, if $m \parallel n$, then $m\angle 1 = m\angle 2$.

When using this postulate as a justification, you can abbreviate each part with the following shorthand notation:

a. *corr.* $\angle s = \Rightarrow \parallel$ **lines.**
or **b.** $\parallel$ **lines** $\Rightarrow$ *corr.* $\angle s =$.

Properties of linear pairs and vertical angles can be used with the Corresponding Angles Postulate to determine the measures of angles formed by parallel lines.

Example 1

In the figure at the left, $s \parallel t$. Find $m\angle 8$.

Solution

$m\angle 7 = 105$ because $\parallel$ lines $\Rightarrow$ corr. $\angle s =$. Since $m\angle 7 + m\angle 8 = 180$, by the Linear Pair Theorem, $m\angle 8 = 75$.

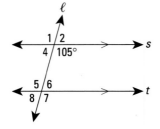

Slopes of Lines

In the coordinate plane, the tilt of a nonvertical line is indicated by a number called the *slope* of the line. You learned about slope in algebra; some ideas about slope are reviewed here.

> **Definition**
> The **slope** of the line through (x_1, y_1) and (x_2, y_2), with $x_1 \neq x_2$, is $\frac{y_2 - y_1}{x_2 - x_1}$.

The slope is the change in y-values divided by the corresponding change in x-values. It tells how many units the line goes up or down for every unit the line goes to the right. The slope of a horizontal line is equal to zero, while the slope of a vertical line is undefined.

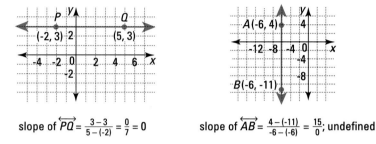

slope of $\overleftrightarrow{PQ} = \frac{3-3}{5-(-2)} = \frac{0}{7} = 0$

slope of $\overleftrightarrow{AB} = \frac{4-(-11)}{-6-(-6)} = \frac{15}{0}$; undefined

The slope of an oblique line is a single positive or negative number.

Example 2

Find the slope of the line through (7, 5) and (2, 4).

Solution

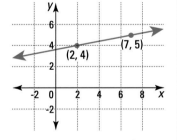

Use the definition of slope. Here $(x_1, y_1) = (7, 5)$ and $(x_2, y_2) = (2, 4)$. So $slope = \frac{4-5}{2-7} = \frac{-1}{-5} = \frac{1}{5}$.

Check

The line goes up $\frac{1}{5}$ unit for each unit you move to the right. Thus, as you move right 5 units (from 2 to 7), the lines goes up 1 unit (from 4 to 5). So it checks.

The slope of a line can be found from an equation for the line.

Example 3

Find the slope of the line with equation $4x + 3y = -12$.

Solution

Find two points on the line. We find (-3, 0) and (0, -4). Now use the definition of slope with these two points.

$Slope = \frac{-4-0}{0--3} = \frac{-4}{3}$

Slopes of Parallel Lines

On the automatic drawer screen at the right, two parallel lines were drawn using the parallel command. Their slopes were measured and found to be equal. Regardless of the tilt of the parallel lines, their slopes will always be equal to each other. This illustrates the following theorem. A general proof requires quite a bit of algebra and is omitted.

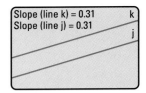

> **Parallel Lines and Slopes Theorem**
> Two nonvertical lines are parallel if and only if they have the same slope.

Thus, to determine whether nonvertical lines are parallel, you only have to know their slopes. All vertical lines, of course, are parallel.

The Parallel Lines and Slopes Theorem is a biconditional. When using it to justify that lines are parallel, you can abbreviate it: **= slopes ⇒ // lines.** When using it to justify equal slopes, you can write: **// lines ⇒ = slopes.**

Suppose two lines ℓ and n are each parallel to a third line m. Because // lines ⇒ = slopes, ℓ and m have the same slope, and so do m and n. By the Transitive Property of Equality, ℓ and n have the same slope. Then, since = slopes ⇒ // lines, ℓ and n are parallel. This sequence of justified conclusions proves a simple theorem.

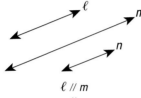

ℓ // m
n // m

> **Transitivity of Parallelism Theorem**
> In a plane, if line ℓ is parallel to line m and line m is parallel to line n, then line ℓ is parallel to line n.

In other words, if two lines are parallel to a third line, then they are parallel to each other.

QUESTIONS

Covering the Reading

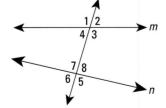

In 1–4, refer to the figure at the left.
1. ∠5 and __?__ are corresponding angles.

2. ∠1 and __?__ are corresponding angles.

3. If m∠4 = m∠6, redo the drawing correctly.

4. If m // n, name all pairs of angles with equal measures.

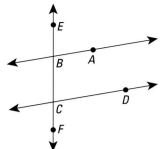

In 5 and 6, use the drawing at the left.

5. If $\overleftrightarrow{AB}$ // $\overleftrightarrow{CD}$ and m∠ABE = 80, find m∠DCB.

6. *True or false.* If m∠ABC = m∠FCD, then $\overline{AB}$ // $\overline{CD}$.

7. *True or false.* The Corresponding Angles Postulate is true on the surface of Earth.

8. Refer to Example 1.
   a. What was substituted for m∠7?
   b. What was added to both sides of m∠7 + m∠8 = 180?

In 9 and 10, find the slope of the line containing the two points.

9. (1, 4) and (6, 2)　　　　　**10.** (3, -7) and (-7, 3)

11. If two nonvertical lines are parallel, what can you say about their slopes?

In 12 and 13, find the slope of a line parallel to the line with the given equation.

12. $y = 4x - 5$　　　　　**13.** $12x - 3y = 10$

14. For which lines is slope undefined?

15. Suppose you write *When two lines are parallel to the same line they are also parallel to each other* as a justification for a step. What is the name of the justification?

Applying the Mathematics

In 16 and 17, use the part of a parking lot grid shown at the left.

16. a. Name all angles with the same measure as ∠6.
　　b. Name all angles supplementary to ∠6.

17. If m∠1 = 122, find the measures of as many other angles as you can.

In 18 and 19, use the figure below. *A*, *B*, *C*, and *D* are collinear and $\overline{BE}$ // $\overline{CF}$.

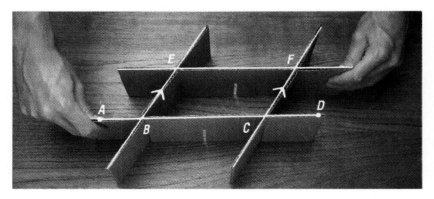

18. If m∠ABE = 106, then m∠__?__ = 106.

19. ∠EBC and __?__ are corresponding angles.

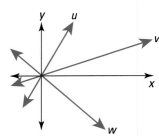

20. Refer to the diagram at the left.
   a. Which line has the greatest slope?
   b. Which line has a negative slope?

21. *True or false.* The slope of the line containing (x_1, y_1) and (x_2, y_2) equals the slope of the line containing (x_2, y_2) and (x_1, y_1).

22. Which would probably be easier to climb, a cliff with a grade of 65° or one with a grade of 80°?

23. State the justification for the conclusion.
   Use the figure at the right.
   Given: m∠1 = m∠6.
   Conclusion: ℓ // m.

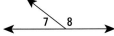

Review

In 24 and 25, state the justification for the conclusion.   *(Lessons 3-3, 3-5)*

24. Given: the figure at the left.
   Conclusion: m∠2 = m∠3.

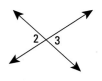

25. Given: the figure at the right.
   Conclusion: ∠7 and ∠8 are supplementary.

In 26–28, name the property that justifies each statement.   *(Lesson 3-4)*

26. $AB = AB$

27. If $x + y = 5$, then $x = 5 - y$.

28. If $\frac{1}{2}x = 7$, then $x = 14$.

29. Is the relationship *is the brother of* symmetric? Explain your answer.
   (Lesson 3-4)

30. Draw an obtuse scalene triangle.   *(Lessons 2-6, 3-3)*

31. M is the midpoint of $\overline{AB}$. N is the midpoint of $\overline{MB}$. Fill in the blanks with numbers. (Hint: Draw a figure.)
   a. If $AB = 5$, then $MN = \underline{\ ?\ }$.
   b. If $AN = x$, then $NB = \underline{\ ?\ }$.   *(Lesson 2-4)*

Exploration

32. With a ruler or an automatic drawing tool, draw three rays $\overrightarrow{OP}$, $\overrightarrow{OQ}$, and $\overrightarrow{OR}$. Let A be any point on $\overrightarrow{OP}$. Let B be any point on $\overrightarrow{OQ}$. Let C be any point on $\overrightarrow{OR}$. Draw the six lines $\overleftrightarrow{AB}$, $\overleftrightarrow{BC}$, $\overleftrightarrow{AC}$, $\overleftrightarrow{PQ}$, $\overleftrightarrow{PR}$, and $\overleftrightarrow{QR}$. Verify the incredible discovery of Girard Desargues in the early 1600s, that either (1) $\overleftrightarrow{AB}$ // $\overleftrightarrow{PQ}$, $\overleftrightarrow{AC}$ // $\overleftrightarrow{PR}$, or $\overleftrightarrow{BC}$ // $\overleftrightarrow{QR}$; or (2) the three points of intersection of these pairs of lines are collinear. (This result is known as Desargues' Theorem.)

Right angles are very common. The angles at the corners of this page are right angles. There are many right angles in the bookcase pictured above (though, due to perspective, some may not appear that way). As you know, the sides of right angles are *perpendicular*.

Definition
Two segments, rays, or lines are **perpendicular** if and only if the lines containing them form a 90° angle.

Perpendicular lines need not be horizontal and vertical. At the right, $\angle ABC$ is a right angle, so m and n are perpendicular. This is shown by the ⌐ sign in the drawing. You can write $m \perp n$, $\overline{AB} \perp \overline{BC}$, $\overleftrightarrow{AB} \perp \overleftrightarrow{BC}$, or $m \perp \overleftrightarrow{BC}$ to indicate perpendicularity. The symbol $\perp$ is read "is perpendicular to." You cannot assume drawn lines are perpendicular unless they are marked.

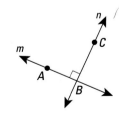

When intersecting lines form one right angle, the Vertical Angles and Linear Pair Theorems force the other three angles to be right angles. So you can put the ⌐ symbol by any one of the angles.

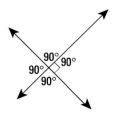

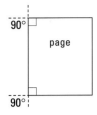

90°

page

90°

Pairs of Perpendicular Lines

The top and bottom edges of this page are each perpendicular to the left edge. If the edges are extended, corresponding 90° angles appear. By the Corresponding Angles Postulate, the top and bottom edges are parallel. This argument proves a useful theorem.

> **Two Perpendiculars Theorem**
> If two coplanar lines ℓ and m are each perpendicular to the same line, then they are parallel to each other.

In symbols, if $\ell \perp n$ and $m \perp n$, then $\ell \parallel m$. Notice that the relation *is perpendicular to* does *not* satisfy the transitive property.

Suppose you are given the lines as shown below, with one perpendicular relation ($\ell \perp m$) and one parallel relation ($m \parallel n$) among the lines. Then there are 90° angles where ℓ intersects m. Since $\parallel$ lines $\Rightarrow$ corr. $\angle$s =, there are also 90° angles where ℓ intersects n. So $\ell \perp n$.

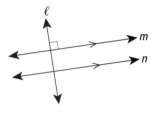

This simple argument proves that if $\ell \perp m$ and $m \parallel n$, then $\ell \perp n$. In words, this relation is as follows.

> **Perpendicular to Parallels Theorem**
> In a plane, if a line is perpendicular to one of two parallel lines, then it is also perpendicular to the other.

Slopes of Perpendicular Lines

Just as you think of parallel lines as having the same tilt, so you may think of two perpendicular lines as having the most different tilts imaginable. Since you know that parallel lines have the same slope, you might expect that the slopes of two perpendicular lines are as different as two related numbers can be. What numbers are most different? Opposites? Reciprocals? In fact, the slopes of perpendicular lines involve both ideas. Each slope is the opposite of the reciprocal of the other. In symbols, if the slope of a given line is m, the slope of any line perpendicular to it is $-\frac{1}{m}$. Since $m \cdot -\frac{1}{m} = -1$, the next theorem follows. A proof of it requires a bit of algebra and is omitted.

> **Perpendicular Lines and Slopes Theorem**
> Two nonvertical lines are perpendicular if and only if the product of their slopes is -1.

If one of two perpendicular lines is vertical (with undefined slope), it is perpendicular to a horizontal line (with 0 slope). Since one of the lines has undefined slope, there is no product of slopes in this case.

Example

A line has slope 5. What is the slope of any line perpendicular to it?

Solution

All coplanar lines perpendicular to a given line are parallel (Two Perpendiculars Theorem); thus they have the same slope. Let that slope be *m.* By the Perpendicular Lines and Slopes Theorem,

$$5 \cdot m = -1.$$
$$m = -\frac{1}{5}$$

Check

Notice that $-\frac{1}{5}$ is the opposite of the reciprocal of 5, which is the given slope.

An instance of the Example is shown at the right. On an automatic drawer two perpendicular lines, *j* and *m,* were created. The measures of their slopes were then put on the screen. A point on *j* was moved so that the slope of *j* would be 5. This made the slope of *m* equal to -0.20, which is $-\frac{1}{5}$.

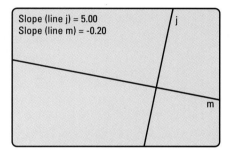

Slope (line j) = 5.00
Slope (line m) = -0.20

Summary of Parallel and Perpendicular Lines

In this and the preceding lesson, you have seen how parallel and perpendicular lines can be determined either by angle measures or by slopes. The table below relates parallelism and perpendicularity to the Euclidean descriptions of points and lines in synthetic and coordinate geometry that you saw in Chapters 1 and 2.

	synthetic geometry	coordinate geometry
point	location	ordered pair (x, y)
line	⟷	$Ax + By = C$
measure of tilt of line	angle measure	slope
parallel lines	corresponding angles have = measure	slopes equal
perpendicular lines	lines form 90° angles	product of slopes = -1

Covering the Reading

1. Define *perpendicular lines.*

2. Trace line ℓ below. Draw a line through point P perpendicular to ℓ.

3. What are the *two* symbols which indicate perpendicularity?

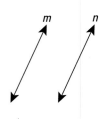

4. **a.** Trace the figure at the left. Draw a line t perpendicular to line m.
   b. If $m \parallel n$, must t also be perpendicular to n?

5. In the drawing at the right, $\overline{AD} \perp \overline{AB}$ and $\overline{BC} \perp \overline{AB}$. What theorem justifies the conclusion that $\overline{AD} \parallel \overline{BC}$?

6. In the drawing at the right, $\ell \parallel m$ and $m \perp n$. What theorem justifies the conclusion that $\ell \perp n$?

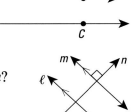

7. A line has slope $\frac{2}{3}$. What is the slope of a line perpendicular to it?

8. An oblique line has slope x. What is the slope of a line perpendicular to it?

Applying the Mathematics

9. Write the two conditionals that make up the Perpendicular Lines and Slopes Theorem.

10. Trace or copy the graph of line t at the right.
    a. Draw the line s through $(0, 1)$ that is perpendicular to t.
    b. What is the slope of s?

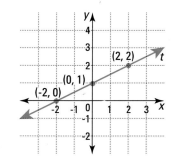

11. At the right, $\overrightarrow{AD} \perp \overrightarrow{AC}$. Answer with numbers.
    a. $m\angle DAC = \underline{\ ?\ }$
    b. $m\angle DAE = \underline{\ ?\ }$

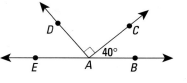

Meeting in the right way.
In this detail of a map of Sydney, Australia, you can see some streets that bend to make right-angle intersections.

12. Why are street intersections often planned so that the streets are perpendicular, even when one has to be bent to do so, as in the diagram at the right?

In 13 and 14, use the figure given below, with $\overline{QR} \parallel \overline{SU}$, $\overline{QT} \parallel \overline{PU}$, and $\overline{QT} \perp \overline{TU}$, as indicated.

13. Justify each conclusion using one of the following justifications.
  $\parallel$ lines $\Rightarrow$ corr. $\angle$s =
  corr. $\angle$s = $\Rightarrow \parallel$ lines
  $\ell \perp n$ and $m \perp n \Rightarrow \ell \parallel m$
  $\ell \perp m$ and $m \parallel n \Rightarrow \ell \perp n$
  definition of perpendicular lines

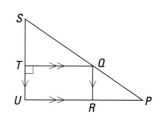

  a. $m\angle S = m\angle PQR$
  b. $\overline{QT} \perp \overline{QR}$
  c. $m\angle SQT = m\angle P$
  d. $m\angle QTU = 90$

14. If $m\angle PQR = 65$, give the measure of each angle.
  a. $\angle S$ **b.** $\angle TQR$ **c.** $\angle TQS$ **d.** $\angle P$

15. A line has equation $2x + 9y = 180$. What is the slope of any line perpendicular to this line?

16. Consider this statement: *If $\overline{AB} \perp \overline{BC}$ and $\overline{BC} \perp \overline{BD}$, then $\overline{AB} \perp \overline{BD}$.*
  a. Draw an instance of this statement in a plane, or tell why the drawing is impossible.
  b. Draw an instance of this statement in space, or tell why the drawing is impossible.

Review

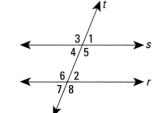

17. Lines *r, s,* and *t* intersect as shown at the left. $m\angle 1 = 70$ and $s \parallel r$. Find the measures of angles 2 through 8.   *(Lesson 3-6)*

18. A line contains $(4, -7)$ and $(-10, -9)$. Find its slope.   *(Lesson 3-6)*

In 19–21, use the figure at the right. In 19 and 20, state a justification for the conclusion.
(Previous course, Lessons 3-3, 3-5, 3-6)

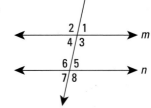

19. Given: $m\angle 3 = m\angle 8$.
  Conclusion: $m \parallel n$.

20. Given: the figure at the right.
  Conclusion: $\angle 7$ and $\angle 8$ are supplementary angles.

21. Use a protractor to find the measure of $\angle 6$.

22. In the figure below, describe ⊙ A ∩ ⊙ B.   *(Lesson 2-5)*

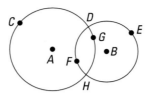

23. Draw a 24° angle.   *(Previous course)*

Amazeing garden. *This labyrinth located at Hever Castle in Great Britain is one of the most beautiful and best kept in England.*

Exploration

24. Mazes or labyrinths are puzzles in which you try to find a path from the start to the finish. Mazes are often designed so that the walls are horizontal or vertical. The maze below is from *The Dell Big Book of Crosswords and Pencil Puzzles #5.* Your teacher should have a copy of the maze.
  a. Try to find the way from the START to the FINISH.
  b. Design your own maze.

Drawing Parallel and Perpendicular Lines

Under construction. *Geometric constructions use the Unique Line Assumption of the Point-Line-Plane Postulate and the definition of circle. Often a combination of the two are used.*

It is useful to be able to draw figures accurately. There are many ways this can be done: with an automatic drawer; with careful measuring; or with special equipment, such as a straightedge and compass. Drawing is very easy with an automatic drawer.

Activity 1

Use an automatic drawer and follow the steps below to draw two parallel lines and a transversal.

Step 1: Choose two points *A* and *B*, and draw $\overleftrightarrow{AB}$. Choose a third point *C* not on $\overleftrightarrow{AB}$.

Step 2: Direct the drawer to draw a line through *C* parallel to $\overleftrightarrow{AB}$. Mark an additional point *D* on this line. Draw transversal $\overleftrightarrow{AC}$.

Step 3: Move point *B* around the screen. Which lines are moved? In what ways?

Step 4: Move point *C* around the screen. Which lines are moved? In what ways are they moved?

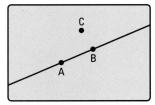

Step 1

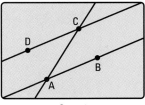

Step 2

The Perpendicular Bisector of a Segment

A **bisector of a segment** is its midpoint, or any line, ray, or segment which intersects the segment only at its midpoint. Every segment can have many bisectors, but in a given plane, only one line is a bisector and perpendicular to the segment. This is the **perpendicular bisector** or **⊥ bisector** of the segment.

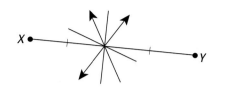

some of the many bisectors of $\overline{XY}$

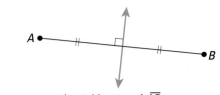

the ⊥ bisector of $\overline{AB}$

A geometric **construction** is a precise way of drawing which uses specific tools and follows specific rules. *Constructing* a geometric figure is like playing a game with rules telling you what is allowed and what is not. From the time of the ancient Greeks, only two tools have been permitted in making a construction. They are the **unmarked straightedge** and the **compass.**

In this book we describe a construction by showing you the steps in it. A sequence of steps leading to a desired end is called an **algorithm.** When we describe a construction, we show you the algorithm for it.

Activity 2

Follow this algorithm to construct (with compass and straightedge) the perpendicular bisector of a given segment $\overline{AB}$.

Step 1: $\odot A$ containing B

Step 2: $\odot B$ containing A

Step 3: $\odot A \cap \odot B = \{C, D\}$

Step 4: $\overleftrightarrow{CD}$

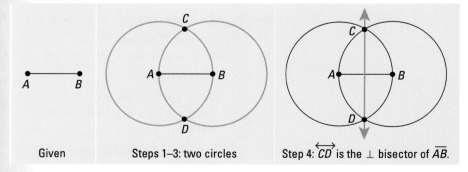

| Given | Steps 1–3: two circles | Step 4: $\overleftrightarrow{CD}$ is the ⊥ bisector of $\overline{AB}$. |

A bonus is that the midpoint of $\overline{AB}$ has also been constructed. It is the intersection of $\overline{AB}$ and $\overleftrightarrow{CD}$ in Step 4.

Another way to find a perpendicular bisector uses careful measuring with a ruler and a protractor. This algorithm is given in Activity 3.

Activity 3

Draw a segment *AB*. Measure $\overline{AB}$ to locate its midpoint, *M*. Use your protractor to measure and mark *X* above the 90° angle with *M* as its vertex. Draw $\overleftrightarrow{MX}$, the ⊥ bisector of $\overline{AB}$.

You should try to learn how to do accurate drawings with a variety of tools. Then, when they are all available, you can use the easiest way. When they are not all available, you will still be able to make accurate drawings.

QUESTIONS

Covering the Reading

1. Refer to Activity 1.
   a. Give your answers to the questions in Step 3.
   b. Give your answers to the questions in Step 4.

2. **a.** What tools are allowed in geometric constructions?
   b. What tools are allowed in geometric drawings?

3. What is an *algorithm*?

4. Define *perpendicular bisector*.

5. What is your answer to Activity 2?

6. What is your answer to Activity 3?

7. An automatic drawer screen showing $\overleftrightarrow{CD}$, the perpendicular bisector of $\overline{AB}$, is at the right.
   a. What is *BC*?
   b. What is m∠*DCB*?
   c. *C* is the ? of $\overline{AB}$.

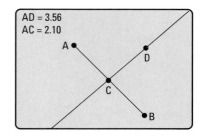

AD = 3.56
AC = 2.10

8. **a.** Trace $\overline{TR}$. Use a straightedge and compass to *construct* the perpendicular bisector of $\overline{TR}$.
   b. Trace $\overline{TR}$ again. This time use a ruler and protractor to *draw* the perpendicular bisector of $\overline{TR}$.

9. **a.** Trace $\overline{YZ}$ at the right. Draw a bisector that is not the perpendicular bisector of $\overline{YZ}$.
   b. Trace $\overline{YZ}$ again. Draw a perpendicular that is not the perpendicular bisector of $\overline{YZ}$.

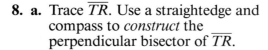

Many retirement communities have clubhouses where residents can get together to socialize or exercise.

10. a. Perform this construction, given two points A and B.

Step 1: $\odot A$ containing B

Step 2: $\overleftrightarrow{AB}$

Step 3: $\odot A \cap \overleftrightarrow{AB} = \{B, C\}$

Step 4: $\odot B$ containing C

Step 5: $\odot C$ containing B

Step 6: $\odot B \cap \odot C = \{D, E\}$

Step 7: $\overleftrightarrow{DE}$

b. How is $\overleftrightarrow{DE}$ related to the given points A and B?

11. The shortest path to a line ℓ from a point P not on ℓ is along the perpendicular to the line through the point.

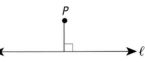

a. Trace the road and the clubhouse dot below. Draw the shortest path from the clubhouse to the road.

b. What tools did you use to answer part **a?**

c. Explain how you drew the path in part **b.**

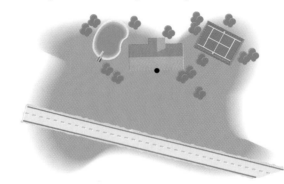

12. a. Use the figure below. Measure $\overline{GH}$, $\overline{GI}$, $\overline{GJ}$, $\overline{GK}$, $\overline{GL}$, and $\overline{GM}$ to the nearest millimeter.

b. Which segment is the shortest?

c. What do you think your results from part **b** show?

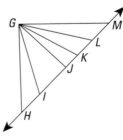

13. Use an automatic drawer if one is available.

a. Draw a triangle ABC.

b. Draw a line $\overleftrightarrow{DE}$ through C parallel to $\overline{AB}$. Label $\overleftrightarrow{DE}$ so that D and A are on the same side of CB, and C is between D and E.

c. Find m$\angle DCA$, m$\angle ACB$, m$\angle BCE$, m$\angle CAB$, and m$\angle CBA$.

d. Name two pairs of angles in your figure that have equal measures.

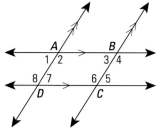

In 14 and 15, use an automatic drawer if one is available. Draw $\overleftrightarrow{AB} \parallel \overleftrightarrow{DC}$ and $\overleftrightarrow{AD} \parallel \overleftrightarrow{BC}$ and label the angles as shown at the left.

14. a. Find the measure of each of the angles 1 through 8.
   b. Name all pairs of corresponding angles among the angles 1 through 8.
   c. What postulate do your responses to parts **a** and **b** verify?

15. a. Find $m\angle 2 + m\angle 3 + m\angle 6 + m\angle 7$.
   b. Compare your answer to part **a** to those from another student. What seems to be true about $m\angle 2 + m\angle 3 + m\angle 6 + m\angle 7$ for any lines where $\overleftrightarrow{AB} \parallel \overleftrightarrow{DC}$ and $\overleftrightarrow{AD} \parallel \overleftrightarrow{BC}$?

Review

16. a. Consider the line with equation $x - 2y = 15$. What is its slope?
   b. What is the slope of a line perpendicular to it?   *(Lessons 3-6, 3-7)*

17. Given coplanar lines ℓ, m, and n, fill in each entry of the table with $\ell \parallel n$, $\ell \perp n$, or "can't tell."   *(Lessons 3-6, 3-7)*

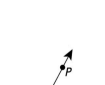

	$m \parallel n$	$m \perp n$
$\ell \parallel m$	**a.**	**b.**
$\ell \perp m$	**c.**	**d.**

18. Lincoln Avenue intersects the parallel streets Washington and Jefferson. Which of the eight angles at the corners have equal measures?   *(Lesson 3-6)*

19. In the figure at the left, $\ell \parallel m$. If $m\angle 6 = m\angle 1 + 12$, find $m\angle 1$.   *(Lessons 3-3, 3-6)*

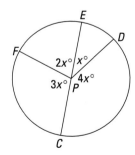

20. Given $\triangle TRI$, justify the conclusion $TR + RI > TI$.   *(Lessons 2-7, 3-5)*

21. Consider $\odot P$ with measures of central angles as indicated in the drawing at the right.
   a. What is the measure of the largest angle?
   b. Which pairs of angles, if any, are complementary?
   c. Which pairs of angles, if any, are supplementary?   *(Lessons 3-2, 3-3)*

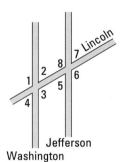

22. In the figure at the left, $\overrightarrow{OR}$ bisects $\angle POL$. $m\angle POR = 5x + 9$, $m\angle POL = 12x - 1$. Find $m\angle POL$.   *(Lesson 3-1)*

Exploration

23. Because the rules for constructions are quite specific, there are some figures that can be drawn but cannot be constructed. The Greeks were very puzzled by this, and three impossible constructions became famous. They are called "squaring the circle," "duplicating the cube," and "trisecting an angle." Do some research on one of these constructions, and draw an example of it.

A project presents an opportunity for you to extend your knowledge of a topic related to the material of this chapter. You should allow more time for a project than you do for typical homework questions.

1 Constructions

From another book, read about straightedge and compass constructions. Learn how to construct each of the following:

a. the perpendicular to a line at a given point on the line;

b. the perpendicular to a line from a point not on the line;

c. the parallel to a line through a given point not on the line.

Make a poster of these constructions and of the perpendicular bisector construction in Lesson 3-8.

2 Arcs in Rainbows

A rainbow is sometimes called an *arc* or *arch* in the sky. Look up *rainbow* in the encyclopedia and write a brief report on the science behind this phenomenon.

3 Mazes

Find two difficult mazes and decide why they are difficult. Then construct two difficult mazes of your own. Write a short report which includes your analysis of the mazes you found, your methods for creating your mazes, and the four mazes with solutions.

4 Turns in Ballet

In ballet one type of rotation is called a *pirouette*. Find out the names of at least three kinds of turns in ballet. Describe each turn with illustrations and prose, and determine their magnitudes.

5 Geometry in Sports

Pick a sport which uses angles and distance. (Possible sports include gymnastics, ice hockey, golf, football, basketball, sailing, bowling, and soccer.) Interview a player or a coach about how angles and distance are used and their importance to playing and scoring in the sport. Write a thorough summary of your interview.

6 Angles in Life

Opening and closing a pair of scissors is a dynamic illustration of the Linear Pair Theorem. Make a list of 10 real-world examples of vertical angles, supplementary or complementary angles, and adjacent angles. Include a drawing for each example.

7 Slopes and Staircases

The standard staircase, according to the National Association of Home Builders, has thirteen $8\frac{1}{4}''$ risers and twelve 9″ treads. A 1992 proposal to increase safety on stairs by the Building Officials and Code Administrators suggested a new standard of sixteen 7″ risers and fifteen 11″ treads.

a. On graph paper, make scale drawings of these two staircase models.

b. Find the slope of each staircase.

c. Write a paragraph explaining why the new staircase model might increase safety. What are its advantages? disadvantages? Do you think the suggestion is a good idea?

SUMMARY

This chapter discussed a variety of topics related to angles and lines. An angle is the union of two rays with the same endpoint. In degrees, the measure of an angle is a unique number from 0 to 180 and is subject to the Angle Measure Postulate. Angles are classified by their measures as zero, acute, right, obtuse, or straight.

The measure of an arc is a number between 0° and 360°. Arcs are classified by their measures as minor arcs or major arcs. In contrast, the magnitude of a rotation may be any real number. If its magnitude is positive, the rotation is counterclockwise; if its magnitude is negative, the rotation is clockwise. You can think of the measure of an angle as the magnitude of the counterclockwise rotation needed to turn one ray into the other, or as the measure of an arc of a circle where the angle is a central angle.

Two angles may be related by their measures. If their measures add to 90, they are complementary. If their measures add to 180, they are supplementary. Two angles may also be related by their positions. If they are vertical angles, their measures are equal. If they form a linear pair, the angles are supplementary.

Because the measures of angles and arcs are real numbers, the properties of real numbers are useful in geometry. In this chapter, postulates of equality and inequality, and the operations of addition and multiplication were discussed.

A proof argument of a conditional $p \Rightarrow q$ is a sequence of justified statements leading from p to q. Each step in the sequence is justified by finding a general postulate, definition, or theorem of which the instance is a special case. In this chapter, you were asked to consider one-step proofs, in which only one justification is needed.

When two lines are cut by a transversal, eight angles are formed. A pair of corresponding angles is equal in measure if and only if the lines are parallel. Parallelism is transitive: For any lines ℓ, m, and n, if ℓ // m and m // n, then ℓ // n. When nonvertical parallel lines are drawn on the coordinate plane, their slopes are equal.

Two segments, rays, or lines are perpendicular if and only if the lines containing them form a 90° angle. From the Corresponding Angles Postulate, one can prove that two lines perpendicular to the same line are parallel, and that if a line is perpendicular to one of two parallel lines, it is perpendicular to the other. When perpendicular lines are drawn on the coordinate plane and neither is a vertical line, then the product of their slopes is -1.

Parallel and perpendicular lines can be constructed with a straightedge and compass, or drawn freehand, or shown with the aid of automatic graphers. The construction of the perpendicular bisector of a segment is basic to many other constructions.

VOCABULARY

Below are the new terms and phrases for this chapter. You should be able to give a general description and specific example of each. For those terms that are starred, you should be able to give a *good* definition. You should also be able to rewrite each named theorem or postulate as a conditional or biconditional, as appropriate.

Lesson 3-1
*angle
sides of an angle
vertex of an angle
∠, ∠A, ∠ABC, ∠1
straight angle
zero angle
interior of an angle
exterior of an angle
measure of an angle, m∠ABC
degree, °
Angle Measure Postulate
  Unique Measure
    Assumption
  Unique Angle
    Assumption
  Zero Angle Assumption
  Straight Angle
    Assumption
  Angle Addition Property
*angle bisector

Lesson 3-2
arc of circle, $\overset{\frown}{AB}$
central angle of circle
minor arc, endpoints of arc
major arc
semicircle
*measure of minor arc
*measure of major arc
concentric circles
image, preimage
clockwise direction
counterclockwise direction
magnitude of rotation

Lesson 3-3
*zero angle, *acute angle,
*right angle, *obtuse angle,
*straight angle
*complementary angles
*supplementary angles
complements, supplements
*adjacent angles, *linear pair
Linear Pair Theorem
*vertical angles
Vertical Angles Theorem
acute, right, obtuse triangle

Lesson 3-4
Postulates of Equality
  Reflexive Property of
    Equality
  Symmetric Property of
    Equality
  Transitive Property of
    Equality
Postulates of Equality and
  Operations
  Addition Property of
    Equality
  Multiplication Property
    of Equality
Postulates of Inequality and
  Operations
  Transitive Property of
    Inequality
  Addition Property of
    Inequality
  Multiplication Properties
    of Inequality
Postulates of Equality and
  Inequality
  Equation to Inequality
    Property
  Substitution Property

Lesson 3-5
*proof argument
justification

Lesson 3-6
transversal, corresponding
  angles
Corresponding Angles
  Postulate
*slope
Parallel Lines and Slopes
  Theorem
Transitivity of Parallelism
  Theorem

Lesson 3-7
*perpendicular, ⌐, ⊥
Two Perpendiculars Theorem
Perpendicular to Parallels
  Theorem
Perpendicular Lines and
  Slopes Theorem

Lesson 3-8
bisector of a segment
perpendicular bisector, ⊥
  bisector
construction
unmarked straightedge,
  compass
algorithm

PROGRESS SELF-TEST

Directions: Take this test as you would take a test in class. Use a ruler, compass and a protractor. Then check your work with the solutions in the Selected Answers section in the back of the book.

1. Sketch two angles, ∠1 and ∠2, that are adjacent and complementary.

In 2 and 3, refer to the figure below.

2. If m∠3 = 77, find m∠4.
3. If m∠3 = 2x, then m∠5 = ___?___.

4. Draw an angle with measure 44°.
5. At a movie, a viewer has to look 15° above eye level to see the top of the movie screen. How many more degrees would the viewer have to bend his or her head to be staring straight up at the ceiling?

6. ∠1 and ∠2 are complementary.
m∠1 = 5x − 25 and m∠2 = 4x + 16.
a. Find x.   **b.** Find m∠1.

In 7 and 8, refer to the figure below.

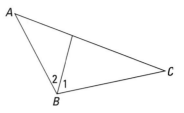

7. If m∠ABC = 105,
m∠2 = 6t, and m∠1 = 9t, find t.
8. *Multiple choice.* Tell which property justifies the conclusion m∠1 + m∠2 = m∠ABC.
(a) definition of complementary angles
(b) definition of supplementary angles
(c) Linear Pair Theorem
(d) Angle Addition Property

9. $\overrightarrow{BC}$ is the bisector of ∠ABD at the right.
If m∠ABC = 14y − 4 and m∠CBD = 37 − y, find y.

10. Trace the figure below and rotate ABCD 80° about point O.

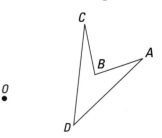

11. A spinner for a board game is divided into five equal sections. B is the image of A under a rotation with center O. What is the magnitude of this rotation?

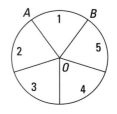

PROGRESS SELF-TEST

12. Refer to $\odot C$ below. $\overline{PT}$ is a diameter and $m\overset{\frown}{PR} = 76°$. Find each measure.
    a. $\overset{\frown}{TR}$ b. $\angle RCT$ c. $\overset{\frown}{RTP}$

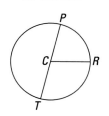

13. A bike path is to be built through G and made parallel to a road. Trace the figure at the right. Draw a line through G that is parallel to the road's edge.

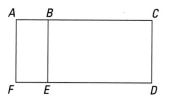

14. *Multiple choice.* Refer to the figure below.
    Given: $\overline{AF} \,/\!/\, \overline{BE}$ and $\overline{BE} \,/\!/\, \overline{CD}$.
    Conclusion: $\overline{AF} \,/\!/\, \overline{CD}$.
    Choose the correct justification from the list.

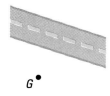

    (a) If $\ell \,/\!/\, m$ and $m \perp n$, then $\ell \perp n$.
    (b) If $\ell \perp m$ and $m \perp n$, then $\ell \,/\!/\, n$.
    (c) If $\ell \,/\!/\, m$ and $m \,/\!/\, n$, then $\ell \,/\!/\, n$.

15. Give the slope of the line through $(0, 0)$ and $(4, 1)$.

16. If a line has slope $\frac{4}{5}$, each line parallel to it has slope __?__ and each line perpendicular to it has slope __?__.

17. What is the slope of the line with equation $y = 3x - 7$?

18. What is the slope of any line perpendicular to the line $2x - y = 6$?

In 19 and 20, use the figure below where $\ell \,/\!/\, m$.

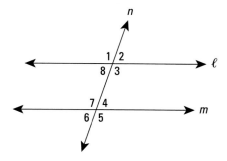

19. If $m\angle 6 = 5x - 12$ and $m\angle 8 = 3x + 14$, find $m\angle 6$.

20. If $n \perp \ell$ and $\ell \,/\!/\, m$, and $m\angle 5 = 11x + 5$, find x.

21. If Q is the midpoint of $\overline{OP}$, justify the conclusion that $OQ = QP$.

22. State the Reflexive Property of Equality.

23. Refer to the figure below. If $GP + PT = 14$ and $PT = RM$, make a conclusion using the Substitution Property of Equality.

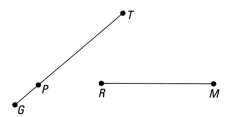

24. Trace the figure below. Construct the perpendicular bisector of $\overline{AB}$.

25. Find the measure of an angle that is 9 times the measure of its supplement.

CHAPTER REVIEW

Questions on SPUR Objectives

SPUR stands for **S**kills, **P**roperties, **U**ses, and **R**epresentations. The Chapter Review questions are grouped according to the SPUR Objectives for this chapter.

SKILLS DEAL WITH THE PROCEDURES USED TO GET ANSWERS.

Objective A: *Draw and analyze drawings of angles.*   *(Lessons 3-1, 3-3)*

In 1 and 2, use the drawing at the right.

1. **a.** Name a straight angle.
   b. Name an angle with measure 0.
   c. Name a linear pair.
   d. Give two other names for $\angle 1$.

2. Estimate $m\angle 2$.

In 3 and 4, give a possible sketch of two angles (label the angles 5 and 6) if

3. they form a linear pair.

4. they are complementary and have the same measure.

In 5 and 6, refer to the figure at the right. Find the measures of $\angle 1$, $\angle 3$, and $\angle 4$

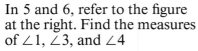

5. if $m\angle 2 = 78$.

6. if $m\angle 2 = 3x$.

7. Draw two different angles, each with measure 58°, that share a common side.

8. Draw an angle with measure 92°.

Objective B: *Use algebra to represent and find measures of angles.*   *(Lessons 3-1, 3-3)*

9. Point D is in the interior of $\angle ABC$ as shown at the right. If $m\angle ABD = 5t$, $m\angle DBC = 3t$, and $m\angle ABC = 72$, find the value of t.

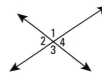

10. Let m be the measure of an acute angle. Graph all possible values of m on a number line.

11. In the figure below, $\angle 1$ and $\angle 2$ form a linear pair. If $m\angle 1 = 7x - 6$ and $m\angle 2 = 5x + 18$, find $m\angle 1$ and $m\angle 2$.

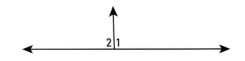

12. Lines ℓ and m intersect as in the diagram below. Find y.

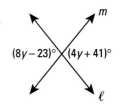

In 13 and 14, use the figure below.

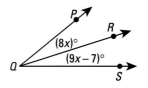

13. If $m\angle PQS = 40$, find x.

14. If $\overrightarrow{QR}$ is the bisector of $\angle PQS$, find $m\angle PQS$.

15. Find the measure of an angle that is $\frac{2}{3}$ the measure of its complement.

16. Let q be the measure of an angle that is less than its supplement. Find all possible values of q.

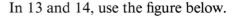

Objective C: *Determine measures of angles formed by parallel lines, perpendicular lines, and transversals.* *(Lessons 3-6, 3-7)*

In 17 and 18, use the figure at the right.

17. If $m\angle 1 = 83$, find the measures of angles 2 through 8.

18. If $m\angle 2 = 4x + 1$ and $m\angle 4 = 20x - 47$, find $m\angle 2$.

In 19 and 20, use the figure below, in which $\ell \perp m$.

19. If $m\angle 2 = 5x + 50$, find x.

20. If $m\angle 6 = m\angle 5$, is $m \parallel n$?

Objective D: *Draw parallel lines, $\perp$ bisectors, and perpendicular lines.* *(Lesson 3-8)*

21. Trace $\overline{AB}$. Then draw line ℓ, its $\perp$ bisector.

22. Trace $\overline{MP}$ at the right. Construct the perpendicular bisector of $\overline{MP}$.

23. Draw lines p, q, and r so that $p \parallel r$ and $p \perp q$.

24. Draw a segment $\overline{CD}$.
  a. Draw a bisector of $\overline{CD}$ which is not a $\perp$ bisector.
  b. Draw a line perpendicular to $\overline{CD}$ which is not a $\perp$ bisector.

Objective E: *Draw rotation images and find magnitudes of rotations.* *(Lesson 3-2)*

In 25 and 26, trace the figure at the right.

25. Draw the rotation image of $ABCD$ under a rotation about E of $-45°$.

26. Rotate $ABCD$ $180°$ about C.

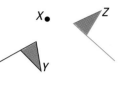

In 27 and 28, X is the center of rotation; Y and Z are pennants. Measure to determine the answer.

27. If Z is the image of Y, what is the magnitude of the rotation?

28. If Y is the image of Z, what is the magnitude of the rotation?

Objective F: *Find the measures of central angles and the degree measures of arcs.* *(Lesson 3-2)*

29. In the figure at the right, $\overline{AD}$ is a diameter of $\odot O$, $\overrightarrow{OC}$ bisects $\angle DOB$, and $m\angle AOB = 80$. Find
  a. $m\overset{\frown}{AB}$.   **b.** $m\overset{\frown}{BD}$.
  c. $m\overset{\frown}{BC}$.   **d.** $m\overset{\frown}{CDA}$.

30. Use $\odot P$ at the right.
  a. If $m\overset{\frown}{SZ} = 100°$, find $m\overset{\frown}{STZ}$.
  b. If $m\overset{\frown}{TZS} = 222°$, find $m\angle SPT$.

PROPERTIES DEAL WITH THE PRINCIPLES BEHIND THE MATHEMATICS.

Objective G: *Recognize and use the postulates of equality and inequality.* *(Lesson 3-4)*

31. State the Transitive Property of Equality.

32. If $m\angle 1 + m\angle 2 = 72$ and $m\angle 1 = m\angle 3$, use the Substitution Property of Equality to make a conclusion about $m\angle 3$ and $m\angle 2$.

In 33 and 34, use the figure below. M, R, S, and P are collinear.

33. If $MR = SP$, what can you conclude using the Addition Property of Equality?

34. Using the Equation to Inequality Property, what can you conclude about RS and RP?

Objective H: *Give justifications for conclusions involving angles and lines.* *(Lessons 3-1, 3-5, 3-6, 3-7)*

In 35 and 36, *multiple choice.* Which statement justifies the conclusion?

35. Given: m∠ABD = m∠DBC in the figure at the right.
Conclusion: $\overrightarrow{BD}$ bisects ∠ABC.
(a) Angle Addition Property
(b) Linear Pair Theorem
(c) definition of angle bisector
(d) definition of adjacent angles

36. Given: m∠B + m∠C = 180.
Conclusion: ∠B and ∠C are supplementary.
(a) Linear Pair Theorem
(b) Vertical Angles Theorem
(c) definition of straight angle
(d) definition of supplementary angles

In 37 and 38, *multiple choice.* Choose the correct justification from this list.
(a) If ℓ // m and m ⊥ n, then ℓ ⊥ n.
(b) If ℓ ⊥ m and m ⊥ n, then ℓ // n.
(c) If ℓ // m and m // n, then ℓ // n.

37. Given: $\overline{MJ} \perp \overline{JK}$ and $\overline{MJ}$ // $\overline{KL}$.
Conclusion: $\overline{KL} \perp \overline{JK}$

38. Given: $\overline{MJ} \perp \overline{JK}$ and $\overline{KL} \perp \overline{JK}$.
Conclusion: $\overline{JM}$ // $\overline{KL}$

In 39 and 40, justify the conclusion.

39. Given: ∠4 and ∠5 are a linear pair.
Conclusion: m∠4 + m∠5 = 180.

40. Given: the figure at the right.
Conclusion: m∠EFG + m∠GFH = m∠EFH.

USES DEAL WITH APPLICATIONS OF MATHEMATICS IN REAL SITUATIONS.

Objective I: *Apply angle and arc measures in real situations.* *(Lessons 3-2, 3-3)*

In 41 and 42, the latitude of a point P on Earth is the measure of the angle PCE, where C is the center of Earth and E is the point on the equator directly north or south of P. Suppose P is at 78° north latitude.

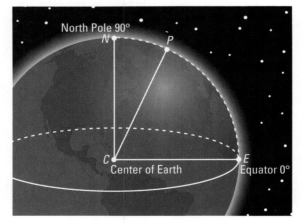

41. What is m∠PCE?

42. How many degrees of latitude are there between P and the North Pole?

In 43 and 44, use a diagram like the one at the right.

43. Draw $\overrightarrow{OP}$ going in the direction 20° S of W.

44. A plane takes off in the direction 30° E of S. How many degrees does it have to turn counterclockwise to fly due north?

45. If a circular pizza is cut into 8 equal pieces, what is the degree measure of the arc formed by the crust of one piece?

46. A child is swinging on a swing as pictured at the right. If A is the image of B under a rotation about point C, use a protractor to find the magnitude of the rotation.

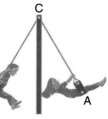

47. At the right is a dial with the numbers 1 through 8 equally spaced around it. What is the magnitude of the rotation needed to turn the dial from 4 to 2?

48. A nail has been driven into a wall to hang a picture. If the measure of one angle made by the nail and the wall is 45, what is the measure L of the other angle?

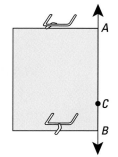

45°

Objective J: *Apply parallel and perpendicular lines in real situations.* (Lesson 3-8)

In 49 and 50, use the drawing below.

A

•C

B

49. A scale drawing of a football field is being made. The 50-yard line is the perpendicular bisector of $\overline{AB}$. Trace the figure and construct the 50-yard line.

50. A hash mark runs perpendicular to $\overleftrightarrow{AB}$. Draw a hash mark through point C.

In 51 and 52, use the drawing below. Some girls are having a race to the river.

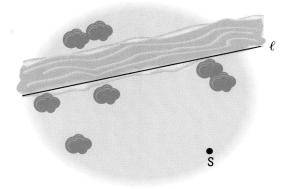

ℓ

•
S

51. Trace the river's edge ℓ and the starting point S. Draw a starting line through S parallel to ℓ.

52. Trace the river's edge ℓ and the starting point S. Draw the shortest path to run.

REPRESENTATIONS DEAL WITH PICTURES, GRAPHS, OR OBJECTS THAT ILLUSTRATE CONCEPTS.

Objective K: *Determine the slope of a line from its equation or given two points on it.* (Lesson 3-6)

53. Give the slope of the line through $(-5, -2)$ and $(6, 1)$.

54. Give the slope of the line through $(4, -3)$ and $(-3, 4)$.

In 55 and 56, find the slope of the line with the given equation.

55. $y = -\frac{3}{5}x + 11$ **56.** $5x - 4y = 45$

57. *Multiple choice.* Which line below has a negative slope?
(a) x (b) y
(c) m (d) n

y   m

x

n

58. Which line has a greater slope, $y = 3x$ or $x = 3y$?

Objective L: *Determine the slope of a line parallel or perpendicular to a given line.* (Lessons 3-6, 3-7)

59. If a line has slope 10, each line parallel to it has slope __?__ and each line perpendicular to it has slope __?__.

60. If a line has slope $-\frac{2}{3}$, each line parallel to it has slope __?__ and each line perpendicular to it has slope __?__.

61. Line m has equation $4x + 7y = 20$. If line n is perpendicular to m, what is the slope of n?

62. *Multiple choice.* The lines with equations $y = 2x + 1$ and $y = -2x + 1$ are
(a) parallel.
(b) perpendicular.
(c) neither parallel nor perpendicular.

CHAPTER 4

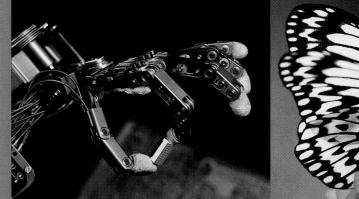

FROM REFLECTIONS TO CONGRUENCE

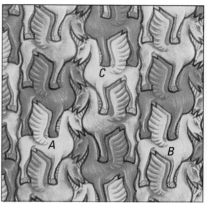

The drawing at the right is one of the many tessellations created by the Dutch artist Maurits Escher (1898–1972). This tessellation was designed by piecing together unicorns of the same size and shape. We say that the unicorns are all *congruent* to each other.

Unicorns A and B are related to each other by a *slide,* or *translation.* Each is a translation image of the other. On the other hand, unicorns A and C have different orientations. But they are not related by a single *flip,* or *reflection.* They are related by another kind of movement called a *walk,* or *glide reflection.*

It is natural to wonder what are all the possible ways in which Escher's congruent unicorns are related to each other. In this chapter, you will study the four types of *transformations* in the plane which yield congruent images of figures: reflections (flips), rotations (turns), translations (slides), and glide reflections (walks). These transformations, called *isometries,* were first identified in 1828 by the French mathematician Michel Chasles (1793–1880).

The first use of transformations dates back to the ancient Greeks around the time of Euclid. However, it was not until 1776 that Euler identified all the kinds of isometries in space. It is interesting that the three-dimensional analysis was accomplished before the two-dimensional analysis. This is probably because objects seen daily are three-dimensional.

Studying these various transformations helps a person to become more aware of the movements of objects such as gears (which rotate) and conveyer belts (which slide). More complicated movements, such as those done by robots, can be taken apart into their component movements and analyzed.

Which is the real house? *This house is located in northern Minnesota. Because the water is almost still, the reflection image is nearly perfect. Can you tell which is the house and which is the image?*

Reflection images of buildings, trees, and clouds can be seen in ponds, lakes, puddles, and streams. The surface of water, when still, reflects almost perfectly. Any time you use a mirror, you see your reflection image. You probably use a mirror when you wash your face and try on clothes, and babies are fascinated with the "other baby" in a mirror. In this lesson, we analyze the idea of reflection.

Reflection Images

Examine the figure below. Think of the picture of the wart hog at the left as the original. It is called the **preimage.** The reflection image at the right can be drawn by folding along the line m and then tracing the preimage. Line m is called the **reflecting line** or **line of reflection.**

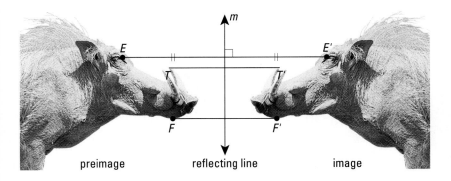

preimage   reflecting line   image

The prime symbol (′) indicates corresponding points; E' (read "E prime") corresponds to E. Notice that line m is the perpendicular bisector of the segments connecting the corresponding eyes E and E' and the tips of the tusks T and T'. This is the defining property of a reflection.

Definition
For a point P not on a line m, the **reflection image of P over line m** is the point Q if and only if m is the perpendicular bisector of $\overline{PQ}$. For a point P on m, the **reflection image of P over line m** is P itself.

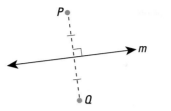

P is not on m.
Q is the reflection image
of P over m.
m is the $\perp$ bisector of $\overline{PQ}$.

P is on m.
P is its own reflection
image over m.

Drawing Reflection Images

Reflection images of individual points can be drawn using a variety of tools.

Example 1

Draw the reflection image P' of point P over line m in the figure at right.

Solution 1

Here is an algorithm that uses a ruler and protractor.
1. Place your protractor so that its 90° mark and the center of the protractor are on m.
2. Slide the protractor along m so that the base line of the protractor (the line through the 0° and 180° marks) goes through P.
3. Measure the distance from P to m along the base line. You may wish to draw this line lightly.
4. Locate P' on the other side of m along the base line, the same distance from m.

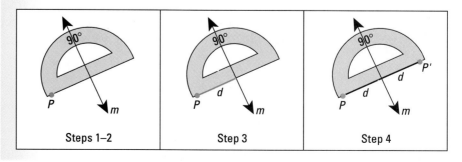

| Steps 1–2 | Step 3 | Step 4 |

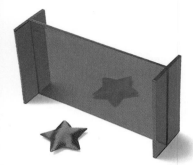

Solution 2

Here is an algorithm that uses a MIRA™ or some other plastic tool that shows reflection images.
1. Place the MIRA with the long thin part on line *m*.
2. Look through the MIRA from the side of *m* that contains point *P*. Look for the image of *P*.
3. Locate *P'* on the other side of the MIRA to coincide with the image of *P*.

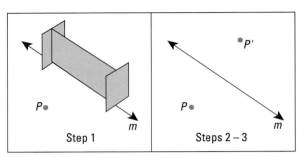

| Step 1 | Steps 2 – 3 |

Check

By the definition of reflection, $\overline{PP'}$ should be perpendicular to and bisected by *m*. Check by folding your paper along line *m*. *P* should land on *P'*.

Transformations

Every reflection is a transformation.

> **Definition**
> A **transformation** is a correspondence between two sets of points such that:
> (1) each point in the preimage set has a unique image, and
> (2) each point in the image set has exactly one preimage.

A transformation is often called a **mapping,** and it is said that a transformation *maps* a preimage onto an image. If the transformation is called T, then the image of a point *P* under that transformation is written **T(P)**, which is read "T of *P*."

Notation for Reflections

When we know that a transformation is a reflection, we use a lower-case letter "r" to refer to it. When discussing reflections in general, or when the reflecting line is obvious, we write

$$\mathbf{r}(A) = A'$$

and read "The reflection image of *A* is *A'*," or "r of *A* equals *A'*." When we want to emphasize the reflecting line *m*, we write

$$\mathbf{r}_m(P) = Q$$

which is read "The reflection image of *P* over line *m* is *Q*," or "r of *P* over line *m* equals *Q*."

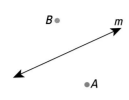

Example 2

At the left, the reflection image of *A* over line *m* is *B*. Name *B* in two ways, using reflection notation.

Solution

You can write B = r(A) since the reflecting line is obvious. Or you can specify the line: B = r$_m$(A).

Reflections are often performed over lines in the coordinate plane. If the reflecting line is one of the axes, coordinates of an image point can be found by using the definition of reflection image.

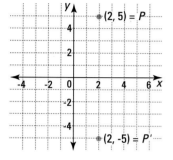

Example 3

Give the coordinates of the reflection image of (2, 5) over the *x*-axis.

Solution

Draw a coordinate plane and locate *P* = (2, 5). Since *P* is 5 units above the *x*-axis, its image will be 5 units below the *x*-axis. So *P'* = (2, –5). The image of (2, 5) is (2, -5).

In Example 3, r$_{x\text{-}axis}$ (*P*) = *P'*. So you could write r$_{x\text{-}axis}$ (2, 5) = (2, -5).

QUESTIONS

Covering the Reading

1. Name two common surfaces in which reflection images can be seen.

2. A figure that is to be reflected is called the __?__.

3. Suppose *B* is the reflection image of *A* over line *m* and *B* is not on *m*. How are *m*, *A*, and *B* related?

4. When a point *P* is on the reflecting line ℓ, then the reflection image of *P* is __?__.

In 5 and 6, trace the drawing at the left.

5. Use a ruler and protractor to draw *A'*, the reflection image of *A* over line ℓ.

6. Draw *B'*, the reflection image of point *B* over ℓ. Describe the method you used.

7. **a.** Define *transformation*.
   b. *True or false.* Every reflection is a transformation.
   c. *True or false.* Every transformation is a reflection.

8. If *P* is a point, write in words how each symbol can be read.
   a. r(*P*)   **b.** r$_m$(*P*)

9. Trace the drawing below. Then draw the reflection images of the points *U, V, W,* and *X* over line *m.*

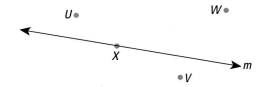

10. In the figure at the left, give the coordinates of each point.
    a. $r_{x\text{-}axis}(P)$ **b.** $r_{y\text{-}axis}(P)$

11. Find the reflection image of (-4, 1) over the given line.
    a. the *x*-axis   **b.** the *y*-axis

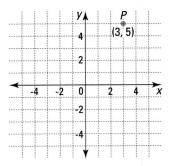

Applying the Mathematics

12. Repeat Question 11 for the point (c, d).

13. **a.** *B, C,* and *D* at the right are three reflection images of point *A.* Match each image with the correct reflecting line.
    b. Name each image of *A* using reflection notation.

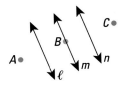

14. At the left, $r_{\ell}(P) = T.$ Trace *P* and *T,* and draw $\ell.$

15. Trace the figure below. Then find the reflection image of the giraffe over line *m* by folding and tracing.

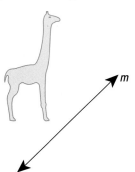

T•

•*P*

Not down for the count.
Because of the vulnerable position taken while watering, one giraffe usually stands guard while another is drinking.

16. The figures at the right are reflection images of each other. Trace the figures and then draw the line of reflection.

17. **a.** Decipher the message below.
    b. Which letter is written incorrectly?

    HELP! I'M TRAPPED
    INSIDE THIS PAGE!

18. *Multiple choice.* The lines with equations $x + 2y = 6$ and $2x - y = 8$ are
(a) parallel.
(b) perpendicular.
(c) neither parallel nor perpendicular. *(Lessons 3-6, 3-7)*

19. Use the coordinate plane at the left.
  a. Is ℓ // m? Justify your answer.
  b. Is $\ell \perp n$? Justify your answer. *(Lessons 3-6, 3-7)*

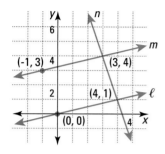

20. In the figure below, a // b. Find x. *(Lessons 3-3, 3-6)*

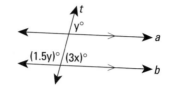

21. In the figure at the right, $m\angle G = 30$ and $m\angle GEO = 150$. Justify each conclusion.
  a. $m\angle VEO = 30$
  b. $\overline{EO}$ // $\overline{GL}$ *(Lessons 3-3, 3-6)*

22. *Multiple choice.* Consider the conjecture: If $|x| = 10$, then $x = 10$. For the value of x given, tell whether
  (i) it is an instance of the conjecture.
  (ii) it is a counterexample to the conjecture.
  (iii) it is neither an instance of nor counterexample to the conjecture.
  a. 10   **b.** 0   **c.** -10   *(Lessons 2-2, 2-8)*

23. Show that the conditional *If a network is traversable, then it has exactly two odd nodes* is false. *(Lessons 1-4, 2-2)*

24. Consider this statement: If D is on $\overline{XY}$, $XD = 11.2$, and $XY = 26.7$, then $DY = \underline{\ ?\ }$.
  a. Fill in the blank.
  b. Make a drawing of this situation. *(Lesson 1-8)*

Exploration

25. Consider this sentence: *If* $r_\ell(A) = B$ *and* $r_\ell(C) = D$, *then* $\underline{\ ?\ }$. Draw three possible pictures of its antecedent. (Make your pictures look as different from each other as you can.) Then fill in the blank with as many consequents as you think make the sentence true.

Reflecting Polygons

IN-CLASS
ACTIVITY

Work on this activity with a partner. You will need a ruler, protractor, and compass, or an automatic drawer for steps 1 through 5. Steps 6 through 8 require an automatic drawer.

1 Trace the figure at the right or draw one like it using an automatic drawer.

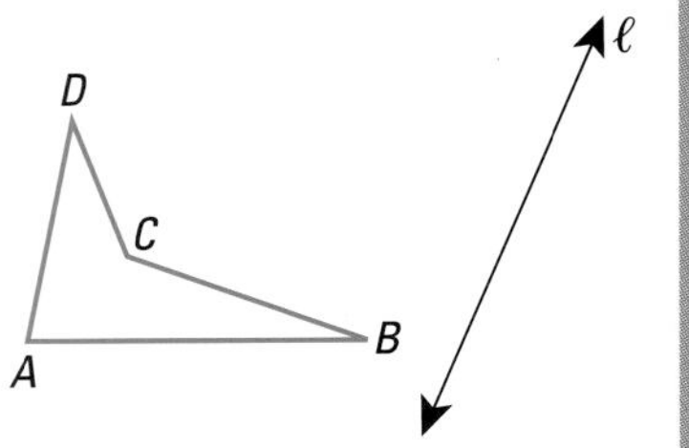

2 Reflect A, B, C, and D over line ℓ. Call them A', B', C', and D'. Connect the images to form polygon $A'B'C'D'$.

3 Find AB, $A'B'$, CD, $C'D'$, m$\angle DAB$, m$\angle D'A'B'$, m$\angle ADC$, and m$\angle A'D'C'$. List your answers in two columns on your paper.

$$AB = \underline{\quad} \qquad\qquad A'B' = \underline{\quad}$$
$$CD = \underline{\quad} \qquad\qquad C'D' = \underline{\quad}$$
$$m\angle DAB = \underline{\quad} \qquad m\angle D'A'B' = \underline{\quad}$$
$$m\angle ADC = \underline{\quad} \qquad m\angle A'D'C' = \underline{\quad}$$

4 Compare the lengths of the preimages $\overline{AB}$ and $\overline{CD}$ with the lengths of their images $\overline{A'B'}$ and $\overline{C'D'}$. Make a conjecture based on these lengths.

5 Compare the measures of the preimages $\angle DAB$ and $\angle ADC$ with the measures of their images $\angle D'A'B'$ and $\angle A'D'C'$. Make a conjecture using these angle measures.

Steps 6–8 require an automatic drawer.

6 Repeat steps 1–3 using a different quadrilateral $ABCD$. Place the measures of AB, CD, $A'B'$, $C'D'$, $\angle DAB$, $\angle ADC$, $\angle D'A'B'$, and $\angle A'D'C'$ on the screen.

7 Select point D and move it to two other positions. Stop at each position and observe the measures on the screen. Can you find any counterexamples to your conjectures in Steps 4 and 5?

8 Select line ℓ and move it to other positions.
a. When do you get polygons $ABCD$ and $A'B'C'D'$ to intersect?
b. Where do they intersect?
c. Continue to move line ℓ and stop at two different positions. Observe the measures on the screen. Can you find any counterexamples to your conjectures in Steps 4 and 5?

LESSON

4-2

Reflecting Figures

Majestic Reflection. *The Taj Mahal, shown with its reflection image, is located in Agra, India. It is made of white marble and was built between 1632 and 1653 as a tomb for the wife of the Indian ruler Shah Jahan.*

Preservation Properties of Reflections

The In-class Activity on page 190 illustrates that the reflection image of a polygon is a polygon whose sides are the same length and whose angles have the same measure as those of its preimage. Why do the reflected images of complex figures look so much like their preimages? The reason is that there are properties of all figures that are the same in both preimages and images. We say the property is preserved by reflections.

> **Reflection Postulate**
> Under a reflection:
> **a.** There is a 1-1 (one-to-one) correspondence between points and their images.
> **b.** Collinearity is preserved. If three points *A*, *B*, and *C* lie on a line, then their images *A'*, *B'*, and *C'* are collinear.
> **c.** Betweenness is preserved. If *B* is between *A* and *C*, then the image *B'* is between the images *A'* and *C'*.
> **d.** Distance is preserved. If $\overline{A'B'}$ is the image of $\overline{AB}$, then *A'B'* = *AB*.
> **e.** Angle measure is preserved. If ∠*A'C'E'* is the image of ∠*ACE*, then m∠*A'C'E'* = m∠*ACE*.

Each part of the Reflection Postulate is significant.

1. Property **a** means that each preimage has a unique image, and each image has a unique preimage.
2. Properties **b** and **c** imply that the image of a segment is a segment.
3. Properties **d** and **e** ensure that reflection images have the same size and shape as their preimages.
4. By assuming the Reflection Postulate, we are ensuring that in Euclidean geometry the two sides of a line act the same way.

The parts of the Reflection Postulate are in a logical order. Points must have images before there can be collinearity. Images of collinear points must be collinear before betweenness can be preserved. Betweenness precedes distance and the rays necessary to have angles.

Many people, however, remember this postulate by thinking of the letters A-B-C-D, as follows: Every reflection is a 1-1 correspondence that preserves Angle measure, Betweenness, Collinearity, and Distance.

Drawing Reflection Images of Figures

The **reflection image of a figure** is the set of reflection images of all the points in the figure. While a computer can reflect a figure point by point, it is easier to use a shortcut: simply reflect the points that determine the figure. Because reflections preserve betweenness and collinearity, the remaining points will fall into place. For instance, angles can be determined by a vertex and two points, one on each side. So, if $r(A) = X$, $r(B) = Y$, and $r(C) = Z$, the reflection image of $\angle ABC$ is $\angle XYZ$. We write $r(\angle ABC) = \angle XYZ$. This idea is generalized below.

> **Figure Reflection Theorem**
> If a figure is determined by certain points, then its reflection image is the corresponding figure determined by the reflection images of those points.

Example 1

Draw the reflection image $A'B'C'D'E'$ of pentagon $ABCDE$ over line m.

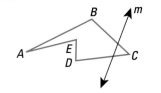

Solution

The five points A, B, C, D, and E determine pentagon $ABCDE$. Find the images of these points, and then connect them in order.

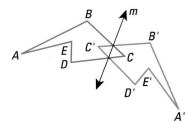

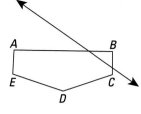

In Example 1, we write r(*ABCDE*) = *A′B′C′D′E′*, keeping preimage and image points in order. Notice that the image intersects the reflecting line *m* at two points. This is because *ABCDE* intersects *m* in two points and because the image of a point on the reflecting line is the point itself.

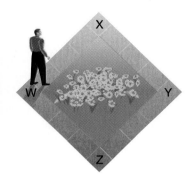

Activity

Repeat Example 1 using the figure at the left. Trace it first.

Orientation of Figures

Although reflections preserve angle measure, betweenness, collinearity, and distance, they do not preserve everything. The image *A′B′C′D′E′* in Example 1 looks reversed from the preimage *ABCDE*. Now we explore that reversal.

When you think of tracing the sides of a polygon, going from vertex to vertex, you are assigning an *orientation* to the polygon. Every polygon has two orientations, clockwise and counterclockwise.

At the left, imagine standing at point *W* and walking along the sides of polygon *WXYZ* from *W* to *X* to *Y* to *Z*. You are traveling clockwise and your right side is always towards the interior. If you walk counterclockwise, your left side is towards the interior.

Example 2

Refer to Example 1 on page 192.
a. Give the orientation of *ABCDE*.
b. Give the orientation of *A′B′C′D′E′*.

Solution
a. As you walk from *A* to *B* and continue around the polygon, the interior is on your right. So **The orientation is clockwise.**
b. As you walk from *A′* to *B′* to *C′* to *D′* to *E′* and back to *A′*, the interior is on your left. So **The orientation is counterclockwise.**

Orientation depends on the order in which the vertices are given. In Example 1, because the orientation of *ABCDE* is clockwise, the orientation of the image polygon, *taking the image vertices in the same order,* is counterclockwise. A figure and its reflection image always have opposite orientations. If a preimage is clockwise, its image is counterclockwise, and vice versa. We take the word *orientation* as undefined, and add the following part to the Reflection Postulate.

Reflection Postulate
Under a reflection:
f. Orientation is reversed. A polygon and its image, with vertices taken in corresponding order, have opposite orientations.

Covering the Reading

1. Can a point have two different reflection images over the same line?

2. Show your work from the Activity in this lesson.

3. In the figure at the left, V is between A and Z. If the reflection images of these points over some line are A', V', and Z', what is the relationship among A', V', and Z'?

4. Since reflections preserve distance, if $C' = r_\ell(C)$ and $D' = r_\ell(D)$, then $DC = \underline{\ \ ?\ \ }$.

5. Suppose $m\angle ABC = 50$ and $m\angle DEF = 100$.
   a. Can $\angle DEF$ be a reflection image of $\angle ABC$?
   b. Why or why not?

6. Name four properties that reflections preserve.

7. Trace the drawing at the left.
   a. Draw $r_\ell(\triangle DEF)$.
   b. Check by measuring the sides of $\triangle DEF$ and its image to see that distance has been preserved.

8. If $r_m(X) = Z$ and $r_m(T) = S$, then $r_m\overline{XT} = \underline{\ \ ?\ \ }$.

9. Refer to the figure at the right.
   a. Is the path $TXWA$ oriented clockwise or counterclockwise?
   b. What is the orientation of $XWAT$?
   c. Which way is $WXTA$ oriented?

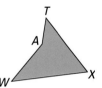

10. How are the orientations of a figure and its reflection image related?

11. *True or false.* Until you know the order of vertices, you cannot tell whether a polygon is oriented clockwise or counterclockwise.

12. How many image points are needed to determine the image of an angle?

13. Refer to the figure below.
    a. What is the fewest number of image points needed to determine the image of pentagon $ABCDE$ over line m?
    b. If $m\angle ABC = 130$, what is the measure of $\angle A'B'C'$ (its reflection image)?
    c. Trace the figure and draw $r_m(ABCDE)$.

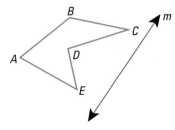

14. Copy the drawing at the right.
    a. Draw $r_{x\text{-}axis}(\triangle ABC)$.
    b. Give the coordinates of the images of the vertices.

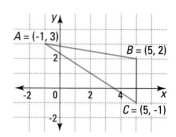

In 15–18, assume $r_\ell(M) = Q$, $r_\ell(N) = R$, $r_\ell(O) = S$, and that M, N, and O are not collinear. Choose the correct justification for each conclusion from the choices at the right.

15. $NO = RS$

16. $m\angle NMO = m\angle RQS$

17. $\ell \perp \overline{MQ}$

18. $r_\ell(\triangle MNO) = \triangle QRS$

(a) Reflections preserve angle measure.
(b) Reflections preserve betweenness.
(c) Reflections preserve collinearity.
(d) Reflections preserve distance.
(e) definition of reflection image
(f) Figure Reflection Theorem

In 19 and 20, trace the drawing. Then draw the reflecting line m so that XYZ is the reflection image of ABC over m.

19.   20.

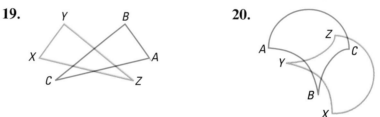

21. Draw and label a figure picturing $r_n(GLAD) = SLAT$.

22. What is the orientation of the path of a base runner around a baseball diamond?

23. In the figure at the right, Police Officer Lionel Hunter follows a clockwise path $ABCDXTZA$ when he walks his beat. If Officer Bernice Trapp walks the same beat, beginning at D, but with a counterclockwise orientation, name the path she takes.

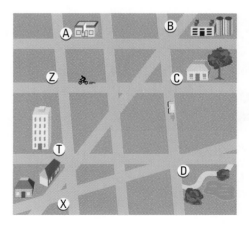

Take me out to the ball game. *Baseball is a very popular sport in Japan. The Tokyo Giants play their home games in the 56,000-seat Tokyo Dome sports arena.*

A

B

24. Trace *A* and *B* at the left. Draw the reflecting line *m* for which $r_m(A) = B$.   *(Lesson 4-1)*

25. In the figure at the right $r_m(E) = F$ and $r_m(C) = D$.
  a. *True or false.* $\overline{EF} \parallel \overline{CD}$.
  b. Justify your answer to part **a.**
    (Lessons 3-7, 4-1)

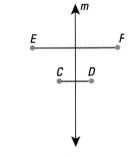

26. The **distance between two parallel lines** is defined as the length of a segment that is perpendicular to the lines with an endpoint on each of the lines.

Multiple choice.   In the figure below, which length appears to be the distance between ℓ and *m*?   *(Lesson 3-7)*
  (a) *PQ*   (b) *PR*   (c) *PS*

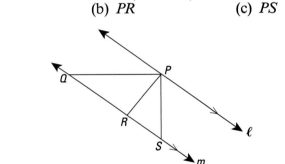

27. If a line has slope -3, each line parallel to it has slope __?__ and each line perpendicular to it has slope __?__.   *(Lessons 3-6, 3-7)*

28. Write the Transitive Property of Equality.   *(Lesson 3-4)*

29. If $p \Rightarrow q$ is true, what can be concluded about $q \Rightarrow p$?   *(Lesson 2-3)*

30. Many objects come in different orientations. For instance, there are right-handed golf clubs and left-handed golf clubs. Name at least three other objects that come in different orientations.

Eldrick (Tiger) Woods, the youngest player to win the U.S. Junior Amateur Golf title, won the U.S. Amateur Golf Championship in 1994, which earned him a place in the 1995 Masters tournament.

Bouncing off Surfaces

When a ball is rolled without spin against a wall, it bounces off the wall and travels in a ray that is the reflection image of the path of the ball going through the wall.

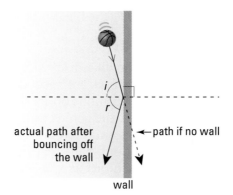

actual path after bouncing off the wall ←path if no wall

wall

This property is not limited to bouncing balls. It is true of any object traveling without spin which hits a surface, including sound, light, and radio waves. The angles marked i and r in the above figure are always of equal measure and are referred to in the study of light as the *angle of incidence* (i) and *angle of reflection* (r).

When you look in a mirror to see a person, your eye receives as images only those light waves which bounce off the mirror in your direction. This is shown in the diagram on the next page. You see the person's image in the mirror as if the person were reflected to the other side of the mirror. The image appears to be as far "behind" the mirror as the person is in front of it, and the orientation of the image is reversed. All of these results agree with the definition of reflection and the Reflection Postulate.

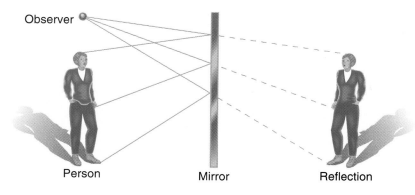

Observer

Person   Mirror   Reflection

Miniature Golf

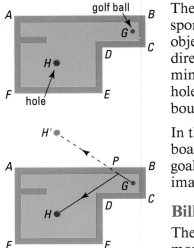

golf ball

The simple idea of bouncing off a wall also has applications in many sports, including miniature golf and billiards. In miniature golf, the object is to hit a golf ball into a hole. The hole is often placed so that a direct shot from the tee into it is impossible. At the left is a diagram of a miniature golf hole, as seen from above, with a golf ball at G and the hole at H. The segments represent wooden boards off which the ball can bounce. There is an obstructing board perpendicular to the wall $\overline{AF}$.

In this situation, a good strategy is to bounce (carom) the ball off a board, as shown at the left. To find where to aim the ball, start with the goal and work backward. Reflect the hole H over $\overline{AB}$. If you shoot at the image H', the ball will bounce off $\overline{AB}$ at P and go towards the hole.

Billiards

The following discussion of billiards shows how to aim when two or more caroms are needed. Billiards is a game played on a table with rubber cushions on its sides and no holes. In one version of 3-cushion billiards, the goal is to hit the cue ball so that it bounces off three cushions and then hits another ball. To simplify our discussion, we ignore other balls that may be on the table. Pictured below is a table with cushions w, x, y, and z.

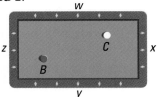

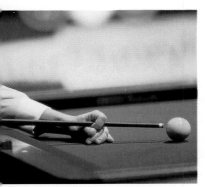

Most billiard tables are 5 feet by 10 feet. Billiard balls were once made of ivory, but today are made from a durable plastic material.

Suppose you want to shoot C off x, then y, then z, and finally hit B. To set up the shot, consider the walls in *reverse* order. First, reflect B over wall z to get B'. Then take the *image B'* and reflect it over wall y to get B'' (read "B double prime"). You may have to extend y as shown below.

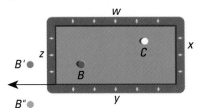

Now reflect the *image B″* over wall x to get $B‴$ (read "B triple prime"). Shoot C in the direction of $B‴$.

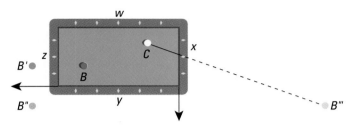

Study the diagram below to see what happens with the shot. On the way toward $B‴$, the cue ball C bounces off x in the direction of $B″$. On the way to $B″$, it bounces off y in the direction of $B′$. Finally it bounces off z, and hits B.

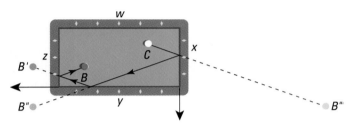

QUESTIONS

Covering the Reading

1. In the figure at the right, ball B is rolling toward the wall without spin. Trace the figure and draw the rest of the path showing how the ball will bounce off the wall.

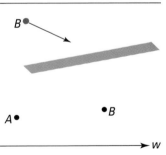

2. Trace the figure at the right. To shoot a ball from A to B off the wall w, where should you aim?

3. Trace the diagram of the miniature golf hole below. Where can you shoot in order to get the ball G into the hole H?

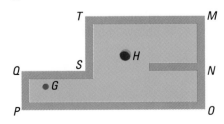

4. Trace the figure below. Where should you aim on wall x to shoot the golf ball G off sides x and y and into the hole at H?

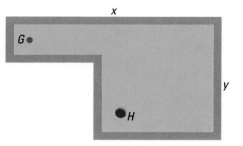

5. Trace the picture of the billiards table below. If you wish to bounce P off of the top, right, and bottom cushions, and then hit Q, show the path that P should travel.

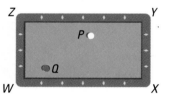

6. Trace the figure below.

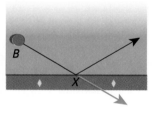

a. A billiard ball B travels along the path shown. Mark all angles that are equal in measure.
b. Draw a line through X perpendicular to the wall, and identify the angle of incidence and angle of reflection.

Applying the Mathematics

---

7. A laser beam sent from point S is to be reflected off two lines, ℓ and m, in such a way that it finally passes through point D. Trace the figure below. Then draw the path of the laser beam.

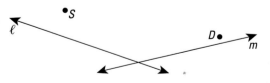

8. Trace the first miniature golf picture in this lesson. Find two other paths for getting the golf ball G in the hole H with one shot.

In 9 and 10, use this information. Pocket billiards is played on a table with rubber cushions on its sides and six holes called *pockets*. The goal is to hit the cue ball so that it hits an object ball into a pocket. Trace the figure below.

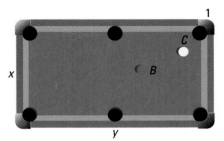

Behind the 8-ball. Pool *is the general term for pocket billiards in the U.S. The most popular pool game is 8-ball.*

9. To bounce the object ball *B* off wall *y* so that it goes into pocket 1, where on *y* should the ball bounce?

10. If you want to hit the cue ball *C* off side *x* before hitting object ball *B*, where should you aim?

11. Trace the diagram of the miniature golf hole below. Where can you shoot in order to get the ball *G* into the hole *H*?

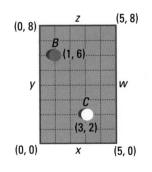

12. In the diagram below, a billiards table with sides *w*, *x*, *y*, and *z* is pictured on a coordinate plane. A player wants to hit a cue ball at $C = (3, 2)$ into a second ball at $B = (1, 6)$. At what point should the player aim in order to bounce the cue ball off sides *w*, *z*, and *y*, in that order?

Review

13. **a.** According to the Reflection Postulate, what four properties are preserved under reflections? (Hint: They start with the letters *A*, *B*, *C*, and *D*.)
   b. What property is not preserved under a reflection? *(Lesson 4-2)*

14. Trace the figure at the right. Draw the reflection image of quadrilateral *BCDE* over $\overleftrightarrow{AB}$.   *(Lesson 4-2)*

15. In the figure for Question 14, what is the orientation of *BCDE*?   *(Lesson 4-2)*

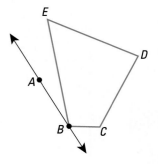

16. Trace the figure at the left. Then reflect the letter *R* over ℓ.   *(Lesson 4-2)*

In 17 and 18, $P = (-2, 0)$, $Q = (-4, 3)$, and $R = (2, 7)$.

17. Find the coordinates of the vertices of the reflection image of △*PQR* over the *y*-axis.   *(Lessons 4-1, 4-2)*

18. Show that $\overleftrightarrow{PQ} \perp \overleftrightarrow{QR}$.   *(Lessons 3-5, 3-7)*

19. In the figure below, *Z* is the reflection image of *A* over line ℓ. Match each conclusion with its justification.   *(Lessons 2-4, 3-7, 4-1)*

   a. $\overline{ZA} \perp \ell$ (i) definition of bisector
   b. ℓ bisects $\overline{ZA}$.   (ii) definition of midpoint
   c. *M* is the midpoint of $\overline{ZA}$.   (iii) definition of reflection
   d. $AM = MZ$

20. *True or false.*   A supplement of an acute angle is always an obtuse angle.   *(Lesson 3-3)*

21. Give a counterexample to the conjecture *If m∠A = x, then the measure of a complement to ∠A cannot be x.*   *(Lessons 2-2, 2-8, 3-3)*

22. In the picture of the toaster dial at the left, the numbers 2 and 6 are at the endpoints of a diameter. What is the magnitude of the rotation to turn the dial from 2 to 5?   *(Lesson 3-2)*

Exploration

23. Design a miniature golf hole in which smart golfers, but not those who don't apply reflections, could make a hole-in-one.

4-4

Composing Reflections over Parallel Lines

Mirror, mirror on the wall. *Shown are the preimage and many images of two children under composites of reflections over parallel planes. Composites over parallel lines work similarly.*

In some of the miniature golf and billiard table applications of Lesson 4-3, it was necessary to follow a reflection over one line by a reflection of the image over a second line. The general idea was to apply one transformation and then apply a second transformation to the image of the first. Following one transformation by another in this manner is called *composing* the transformations. Composing provides a fundamental way to analyze what can be done to a figure. In this lesson, we examine what happens when two reflections over parallel lines are composed.

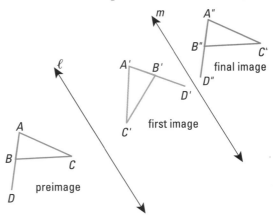

An Example of Composition of Transformations

In the figure above, flag *DBAC* has been reflected over line ℓ. Then its image $D'B'A'C'$ has been reflected over line m. The points A, A', and A'' are related as follows:

$$\mathrm{r}_\ell(A) = A' \text{ and } \mathrm{r}_m(A') = A''.$$

Substituting $r_\ell(A)$ for A' in the second equation (they are equal in the first equation),

$$r_m(r_\ell(A)) = A''.$$

This last equation is read "the reflection over m of the reflection over ℓ of A equals A''," or "r over m of r over ℓ of A is A''." The relationships among the three triangles that form the top of the flag can be stated in symbols.

$$r_\ell(\triangle ABC) = \triangle A'B'C'$$
$$r_m(\triangle A'B'C') = \triangle A''B''C''$$

When one transformation (here r_ℓ) is followed by a second transformation (here r_m), in this way, we say that the transformations have been composed. The result is called the *composite* of the transformations.

Definition

The **composite** of a first transformation S and a second transformation T, denoted **T ∘ S**, is the transformation that maps each point P onto $T(S(P))$.

T ∘ S is read "T following S." The operation denoted by the small circle ∘ is called *composition*. Any transformations can be composed. In the situation with the flags on the previous page, r_ℓ is the first transformation S and r_m is T. You can write either

$$r_m(r_\ell(\triangle ABC)) = \triangle A''B''C'' \text{ or } r_m \circ r_\ell(\triangle ABC) = \triangle A''B''C''.$$

You can say "reflection over m following reflection over ℓ of triangle ABC equals triangle $A''B''C''$."

Notice that the transformation applied first is written *on the right*. The reason for this is that with transformations, as in algebra, you must work inside parentheses first. Thus in $r_m(r_\ell(A))$, the reflection r_ℓ is applied before the reflection r_m.

Activity

In the figure at the left, $\ell \parallel m$. Trace the figure.
a. Draw $r_\ell \circ r_m(P)$. (Reflect P over m first, then reflect the image P' over ℓ.)
b. Measure the distance from P to your final image.
c. Measure the distance between ℓ and m.

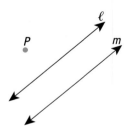

Translations

In the drawing on the previous page, notice the positions of the preimage $\triangle ABC$ and the final image $\triangle A''B''C''$. It seems that the composite of the two reflections over parallel lines has the effect of sliding a figure. This transformation is called a *translation*.

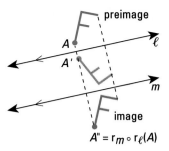

preimage

A

A'

ℓ

m

image

$A'' = r_m \circ r_\ell(A)$

At the left is another example of composing reflections over parallel lines. Again the composite seems to be a slide, or translation.

This is exactly the same sort of translation Escher used with unicorns A and B in the drawing at the beginning of the chapter and that you may have seen in earlier courses.

Now we do something that is often done by mathematicians to take advantage of what is already known. We use a not-so-obvious property as a defining property. We *define* a translation to be the transformation obtained by composing reflections over parallel lines. This enables us to take advantage of the properties of reflections when we study translations.

> **Definition**
> A **translation**, or **slide**, is the composite of two reflections over parallel lines.

That is, when $\ell \,/\!/\, m$, the transformation $r_m \circ r_\ell$ is a translation.

Properties of Translations

Here are some of the basic properties of every translation. Since each reflection in a translation preserves angle measure, betweenness, collinearity, and distance, so does the translation. Also, every translation preserves orientation, because the first reflection switches orientation, and the second switches it back.

You can describe a translation by telling how far a preimage is slid, and in what direction. The **direction** of a translation is given by any ray from a preimage point through its image point. The **magnitude** of a translation is the distance between any point and its image. For the translation above, $\overrightarrow{AA''}$ is the direction, and AA'' is its magnitude.

As you found in the Activity, the magnitude and direction of a translation are related to the reflecting lines in a surprisingly simple way.

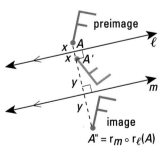

preimage

x A

x A'

ℓ

y

y

m

image

$A'' = r_m \circ r_\ell(A)$

In the picture at the left, the direction of the translation $r_m \circ r_\ell$ is $\overrightarrow{AA''}$, which is perpendicular to both ℓ and m. The translation is from ℓ toward m, and its magnitude AA'' is twice as long as the distance between ℓ and m. The picture and the following two paragraphs explain *why* this is true.

Look at A, A', and A''. We want to show that $\overline{AA''} \perp \ell$ and $\overline{AA''} \perp m$. By the definition of reflection, $\overline{AA''} \perp \ell$. Since $\ell \,/\!/\, m$, by the Perpendicular to Parallels Theorem, $\overleftrightarrow{AA''} \perp m$. Since $\overleftrightarrow{A'A''}$ is also perpendicular to m, $\overleftrightarrow{AA'}$ and $\overleftrightarrow{A'A''}$ must be the same line. So $\overleftrightarrow{AA''}$ is the same line, too. Thus, $\overline{AA''} \perp \ell$ and $\overline{AA''} \perp m$. So the direction of the translation is perpendicular to the reflecting lines.

Referring to the picture again, the distance between the parallel lines is $x + y$. From the definition of reflection, $AA' = 2x$ and $A'A'' = 2y$. By the Additive Property of Distance, $AA'' = 2x + 2y$ which is twice $x + y$. A similar argument can be given if A is in a different position relative to ℓ

and *m*. The magnitude of the translation is twice the distance between the reflecting lines. The Two-Reflection Theorem for Translations summarizes these conclusions.

> **Two-Reflection Theorem for Translations**
> If $m \parallel \ell$, the translation $r_m \circ r_\ell$ has magnitude two times the distance between ℓ and *m*, in the direction from ℓ perpendicular to *m*.

Translating Figures

Since a translation is defined by reflections, the Figure Reflection Theorem holds for translations. Thus to translate a figure, you only have to translate the points which determine the figure and then connect the image points to determine the image of the figure. You can locate the transformation image of a point in either of two ways: (1) slide the point the proper distance in the proper direction or (2) reflect the point over two parallel lines which are perpendicular to the line of direction and exactly half the distance between the preimage and image.

Example

In the figure below, $m \parallel n$. Use the Two-Reflection Theorem for Translations to draw $r_n \circ r_m(\triangle ABC)$.

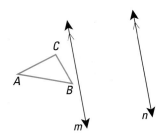

Solution

A composite of two reflections over parallel lines is a translation. So first draw any perpendicular from *m* toward *n* (colored green below). Double this length to get the magnitude of the translation. Translate points *A, B,* and *C,* and then connect them.

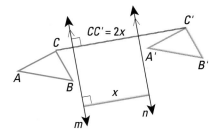

Have you ever been in a room with two mirrors on opposite walls? (Barber shops and beauty salons often have such rooms.) The mirrors apply the Two-Reflection Theorem for Translations.

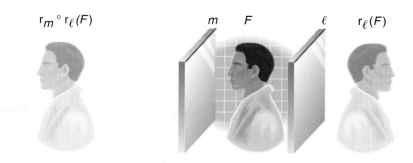

$$r_m \circ r_\ell(F) \qquad\qquad m \quad F \qquad\qquad \ell \qquad r_\ell(F)$$

If you look at an object in one of the mirrors, you see not only the usual reflection image but also many images of images. Some of these look like slides. This is why, by using two nearly parallel mirrors, you can see the back of your head.

QUESTIONS

Covering the Reading

1. *Multiple choice.* Which of the following symbols stands for the motion which results from first applying f and then applying s?
 (a) $f(s)$ (b) $s(f)$ (c) $f \circ s$ (d) $s \circ f$

2. Show your answers to the activity in this lesson.

3. Use the diagram at the left. Fill in each blank with a named point.
 a. $r_m(F) = \underline{\ ?\ }$ **b.** $r_\ell(r_m(F)) = \underline{\ ?\ }$ **c.** $r_\ell \circ r_m(F) = \underline{\ ?\ }$
 d. $r_\ell(A) = \underline{\ ?\ }$ **e.** $r_m \circ r_\ell(A) = \underline{\ ?\ }$

4. A translation is the $\underline{\ ?\ }$ of two reflections over parallel lines.

5. Name five properties preserved under translations.

6. a. Trace this drawing on a sheet of paper.

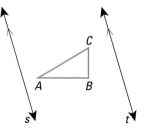

 b. Draw $r_s(\triangle ABC)$. Call it $A^*B^*C^*$.
 c. Draw $r_t \circ r_s(\triangle ABC)$. Call it $\triangle A'B'C'$.
 d. Verify in your drawing that $\overline{AA'}$, $\overline{BB'}$, and $\overline{CC'}$ are parallel segments of the same length.
 e. *Multiple choice.* How does the length of $\overline{AA'}$ compare with the distance between s and t?
 (i) It is the same.
 (ii) It is twice as long.
 (iii) It is half as long.

7. Horizontal lines ℓ and m are 1 cm apart, as shown. Consider the translation $r_m \circ r_\ell$.
   a. What is its magnitude?
   b. What is its direction?

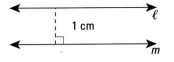

8. In the figure below, $\ell \; // \; m$. Use the Two-Reflection Theorem for Translations to draw $r_m \circ r_\ell(\triangle ABC)$.

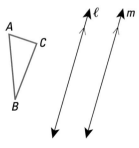

9. Where might you see parallel mirrors? Of what use are they?

Applying the Mathematics

---

10. When the letter R is reflected over line ℓ and then over line m, its final image is as shown. But someone erased line m. Trace the figure and put line m back.

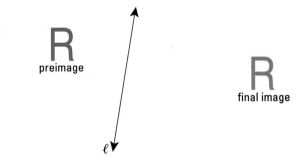

11. At the left, $r_m(B) = C$. What is $r_m \circ r_m(\overline{AB})$?

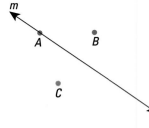

12. Generalize the result of Question 11.

13. Trace the figure below.
   a. Draw $r_m(\triangle ABC)$.
   b. Draw $r_n(r_m(\triangle ABC))$.
   c. Explain how the sides and angle measures of $r_n(r_m(\triangle ABC))$ are related to the sides and angle measures of $\triangle ABC$.

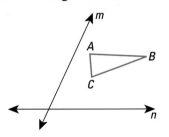

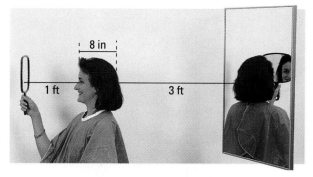

14. Rosita has just had her hair cut. With her back to a mirror on the wall, she stands and holds a hand mirror in front of her face at the distance shown in the photo above. About how far from her eyes will the image of the back of Rosita's head appear in the mirror?

Review

15. Trace the figure below. If you want to hit the cue ball *C* off of side *x* before *C* hits the object ball *B*, where should you aim? *(Lesson 4-3)*

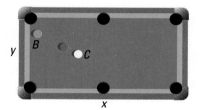

16. Trace the diagram of the miniature golf hole at the left. Where can you shoot in order to get ball *G* into hole *H*? *(Lesson 4-3)*

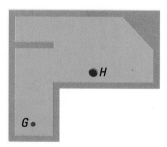

In 17 and 18, give the coordinates of the image of the point. *(Lesson 4-1)*

17. a. $r_{x\text{-}axis}(8, -5)$ **b.** $r_{y\text{-}axis}(8, -5)$

18. a. $r_{x\text{-}axis}(-4, 0)$ **b.** $r_{y\text{-}axis}(-4, 0)$

19. Justify the conclusion.
  Given: In $\triangle ABC$ at the right,
    $\overline{BE} \perp \overline{AC}$ and $\overline{AD} \perp \overline{DC}$.
  Conclusion: $m\angle AFE = m\angle BFD$. *(Lesson 3-5)*

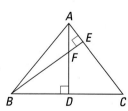

20. a. Write the converse of the statement, *If B is on $\overrightarrow{AC}$ but not between A and C, then AC = AB − BC.*
  b. Is the statement true?
  c. Is its converse true? *(Lessons 1-8, 2-3)*

Exploration

21. Suppose $\ell \,//\, m \,//\, n$. Describe $r_n \circ r_m \circ r_\ell$, the composite of three reflections over these parallel lines. (Hint: Draw some figures and lines ℓ, m, and n, and do the transformations.)

LESSON
4-5

Composing Reflections over Intersecting Lines

In clothing stores you often find intersecting mirrors. A shopper can see the reflection image of a reflection image in the mirrors. These mirrors allow the shopper to see how clothes look from the front, side, and back. This is an application of composing reflections over intersecting lines.

Pictured below is the composition of two reflections over intersecting lines. The result is amazing and not obvious: you can *turn* the preimage onto the final image. Moreover, the center of the turn is the intersection of the lines. The results are pictured on the next page with measurements.

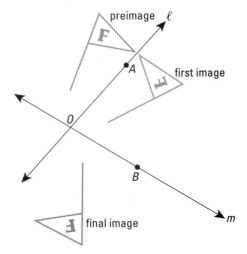

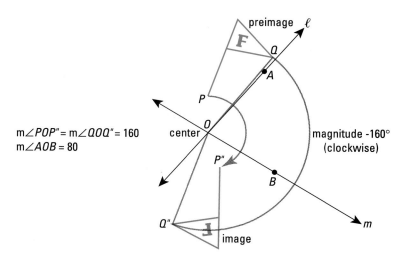

$m\angle POP'' = m\angle QOQ'' = 160$
$m\angle AOB = 80$

Rotations

As with translations, we define rotations in terms of reflection so that the properties of rotations can be more easily deduced.

> **Definition**
> A **rotation** is the composite of two reflections over intersecting lines.

That is, when ℓ intersects m, $\mathrm{r}_m \circ \mathrm{r}_\ell$ is a rotation.

Since each reflection in a rotation preserves angle measure, betweenness, collinearity, and distance, so does the rotation. Since a rotation is a composite of two reflections, orientation is also preserved.

Rotations and Reflections

There is a relationship between the magnitude of the rotation and the non-obtuse angle formed by the intersecting lines. In the figure above, the center of the turn is the intersection of the lines since that point coincides with its image under each reflection. Also $m\angle QOQ'' = 160$, which is twice the $m\angle AOB$. The rotation is clockwise, so the magnitude of the rotation is -160°.

> **Two-Reflection Theorem for Rotations**
> If m intersects ℓ, the rotation $\mathrm{r}_m \circ \mathrm{r}_\ell$ has center at the point of intersection of m and ℓ and has magnitude twice the measure of the non-obtuse angle formed by these lines, in the direction from ℓ to m.

Remember that one example does not prove that a conditional is true. The following argument explains why this theorem is true in general.

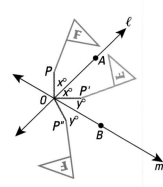

First, the point of intersection of ℓ and m is O. Point O is on both reflecting lines, so $r_m(r_\ell(O)) = r_m(O) = O$. This makes O the center of rotation of $r_m \circ r_\ell$.

Now, look at the points P, P', and P''. Remember that reflections preserve angle measure. From the first reflection over ℓ, $m\angle POA = m\angle P'OA = x$. From the second reflection over m, $m\angle P'OB = m\angle P''OB = y$. The measure of the angle between the reflecting lines ℓ and m is $m\angle AOB$, or $x + y$. The magnitude of the rotation is $m\angle POP''$, or $2x + 2y$, which is twice $m\angle AOB$.

If Q were located on one of these lines, or on the other side of ℓ or m, the argument would be a little different, but the result would still hold.

Rotating Figures

Since a rotation is defined by reflections, the Figure Reflection Theorem extends to rotations. To rotate a figure, you only have to rotate the points which determine the figure and then connect the image points to form the image.

Example

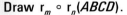

Draw $r_m \circ r_n(ABCD)$.

Solution

You could reflect *ABCD* over *n* and then reflect its image over *m,* but it is easier to use the Two-Reflection Theorem for Rotations. The magnitude of the rotation is twice the measure of the non-obtuse angle formed by the reflecting lines. Going from *n* to *m* is counterclockwise, so the magnitude is positive. Since the angle between the reflecting lines is 24°, the magnitude is 48°. The center of the rotation is *P.* Find the rotation images of *A, B, C,* and *D,* and connect them.

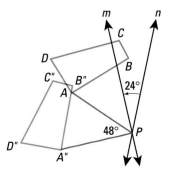

Now we can explain how intersecting mirrors in clothing stores work. In the drawing at the beginning of the lesson, a shopper is using the left and middle mirrors. This, in effect, turns the shopper twice the angle between the mirrors. The acute angle between the mirrors is 55°, and as a result, the final image of the shopper is rotated 110°. This is a large enough turn to give the shopper a view of the back of the clothing.

QUESTIONS

Covering the Reading

1. The composite of two reflections over two intersecting lines is a(n) __?__ .

2. The center of the rotation in Question 1 is the __?__ of the reflecting lines.

3. Identify the properties that rotations preserve from this list: angle measure, betweenness, collinearity, distance, orientation.

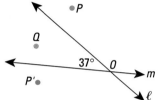

4. In the figure at the left, $r_\ell(P) = Q$, $r_m(Q) = P'$, and the acute angle between the lines is 37°.
   a. What is m∠POP'?
   b. What is the magnitude of the rotation $r_m \circ r_\ell$?
   c. Parts **a** and **b** are an instance of what theorem?

5. To rotate a figure -140°, you can reflect the figure over two lines, where the acute angle between the lines has measure __?__ .

6. Trace the figure below, leaving extra space around the figure for your work. Draw $r_m \circ r_\ell(WXYZ)$.

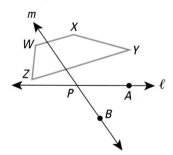

7. Name a place where you can find intersecting mirrors.

8. Trace the figure at the right.
   a. Draw $r_m \circ r_\ell(\triangle ABC)$.
   b. Use a protractor to determine the magnitude of this rotation.

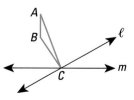

9. Draw the figure at the right.
   a. Reflect $\triangle ABC$ over the x-axis.
   b. Reflect its image over the y-axis.
   c. What is the magnitude of the rotation that has occurred?

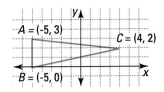

10. Explain why rotations preserve orientation.

In 11–13, use the figure at the right in which $r_\ell(\triangle LMN) = \triangle OPQ$ and $r_m(\triangle OPQ) = \triangle RST$.

11. Explain why $MN = ST$.

12. Explain why $m\angle N = m\angle T$.

13. Suppose $m\angle NCT = 128$.
   a. What is the magnitude of $r_m \circ r_\ell$?
   b. What is the magnitude of $r_\ell \circ r_m$?
   c. What is $m\angle ACB$?

14. The written alphabet of the Inupik dialect of the language family commonly referred to as Eskimo-Aleut uses symbols to denote sound combinations. One interesting feature of this alphabet is that it consists of 12 basic symbols, each of which is rotated or reflected to denote an additional sound.
   a. Trace the symbol for *va* at the right. Draw the Eskimo-Aleut letter *vai* by rotating this figure 90°.
   b. Reflect *va* on a vertical line to draw *vu*.
   c. Rotate *vu* 90° to draw *vi*.

Review

15. In the drawing below, $s \mathbin{/\!/} t$. Trace the drawing and draw $r_t \circ r_s(ABCD)$.   *(Lesson 4-4)*

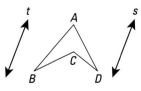

16. *Multiple choice.* Which is not preserved by translations?
   (a) angle measure
   (b) collinearity
   (c) orientation
   (d) All are preserved.   *(Lesson 4-4)*

17. Write an equation for the image of the line with equation $y = x + 1$ under $r_{x\text{-axis}}$. *(Lesson 4-2)*

18. Use *n*-gon $ABCD \ldots L$, which is pictured below.
  a. What is *n*?
  b. Is $ABCD \ldots L$ oriented clockwise or counterclockwise?
  (Lessons 2-1, 4-2)

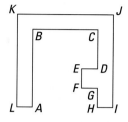

In 19 and 20, use this information about Oklahoma City. Britton Rd. does not curve. Western Ave. is perpendicular to Britton. Kelley Ave. is also perpendicular to Britton. Britton Road is parallel to Hefner Road. Hefner Road is parallel to Wilshire Blvd.

19. Make a conclusion using the Two Perpendiculars Theorem.
  (Lesson 3-7)

20. Make a conclusion using the Transitivity of Parallelism Theorem.
  (Lesson 3-6)

21. In 4 minutes, through how many degrees does
  a. the minute hand of a clock turn?
  b. the hour hand of a clock turn? *(Lesson 3-2)*

Exploration

22. Windshield wipers are objects that rotate. Check the windshield wipers of a car or other vehicle and determine the center and magnitude of the greatest possible rotation.

23. Point *H* is the image of point *G* under a rotation.
  a. Identify a point that could be the center of the rotation.
  b. Identify another point that could be the center of the rotation.
  c. Identify a third point that could be the center of the rotation.
  d. Generalize parts **a, b,** and **c.**

$G \bullet$

$\bullet H$

LESSON

4-6

Translations and Vectors

Translational architecture. *Shown is an audience hall in the Grand Palace in Bangkok, Thailand. Translations are illustrated in the roof tiling.*

In Lessons 4-4 and 4-5, translations and rotations were defined as composites of reflections. In Lesson 3-2, you had already applied rotations to figures. In this lesson, we further examine translations as individual transformations.

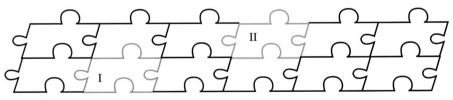

You can think of the tiling design above as one figure and many translation images. Consider region I to be the preimage and region II to be its image under a particular translation. We have taken the figures and connected corresponding points as shown below. You see five arrows, all with the same magnitude and pointing in the same direction. Since all the arrows are the same, a single arrow can describe the motion of the translation.

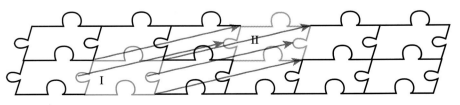

What Is a Vector?

The arrows in the drawing on page 216 picture a *vector*.

> **Definition**
> A **vector** is a quantity that can be characterized by its direction and magnitude.

Vectors are extremely important in many fields today. Airplanes travel on vectors that characterize their speed and direction. Forces such as gravity, pressure, and weight are all vectors because they have a magnitude and a direction. In this lesson, however, we focus only on the close relationship between vectors and translations.

The vector that starts at A and ends at B is denoted $\overrightarrow{AB}$. A is called the **initial point** of the vector and B its **terminal point.** An arrow above a single lower-case letter, as in $\vec{v}$, or a boldface single letter $\mathbf{v}$, is also used to note the many vectors that have the same magnitude and direction.

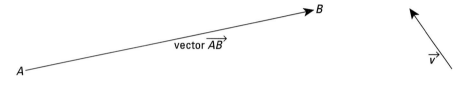

CAUTION: Drawings and notation for vectors and rays look identical, but vectors are not rays. Vectors have a fixed length and rays go on forever. The situation will usually tell you whether rays or vectors are being discussed.

Drawing Translation Images Given a Vector for the Translation

Suppose you are given a figure to translate, and the translation is described by a vector. With an automatic drawer, it is usually easy to draw the image of the figure. You just slide it the direction and amount of the vector. With paper and pencil, locating the image accurately requires drawing parallel lines and measuring lengths.

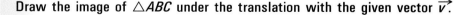

Example 1

Draw the image of $\triangle ABC$ under the translation with the given vector $\vec{v}$.

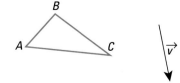

Solution

1. Trace △*ABC* and $\vec{v}$.
2. Very lightly draw line *ℓ* containing *C* and the initial point of $\vec{v}$. Measure an angle made by *ℓ* and the vector and also measure the length of the vector. We find the angle to have measure 112° and the length of the vector to be 1.6 cm.
3. Now draw the ray at *C* that makes the same corresponding angle with *ℓ* as the vector. Mark off a segment on that ray equal to the length of the vector. The endpoint of this segment is *C′*.

4. Repeat steps 2 and 3 for points *A* and *B*. The resulting image will be in the position shown here.

Vectors in the Coordinate Plane

It is often easier to work with vectors using coordinates. In coordinate geometry, vectors are described by ordered pairs. Each pair indicates how far the vector slides a figure horizontally and vertically. The **horizontal component** of the vector (a, b) is a; the **vertical component** is b. This is **the ordered-pair description of a vector.**

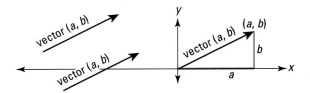

All of the vectors (a, b) in the figure above have horizontal component a and vertical component b. Note that each vector is written vector (a, b) even when its initial point is not at the origin.

When a figure is translated on the coordinate plane by the vector (a, b), the image of the point (x, y) is $(x + a, y + b)$. That is, a is added to the first coordinate and b is added to the second coordinate. You may have used this idea to slide figures in algebra.

Example 2

Quadrilateral *ABCD* has vertices *A* = (-7, 4), *B* = (0, 3), *C* = (5, -1), *D* = (-2, -2). It is translated by the vector (1, -3). Graph *ABCD* and its translation image *A'B'C'D'* on the coordinate plane.

Solution

First find the coordinates of the image vertices. Add 1 to each first coordinate and -3 to each second coordinate.

A' = (-7 + 1, 4 + -3) = (-6, 1)
B' = (0 + 1, 3 + -3)   = (1, 0)
C' = (5 + 1, -1 + -3) = (6, -4)
D' = (-2 + 1, -2 + -3) = (-1, -5)

Now graph *ABCD* and *A'B'C'D'*.

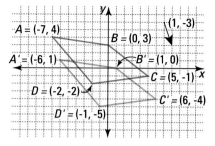

You have now studied three types of transformations in a plane: reflections, rotations, and translations. Here is a summary of what determines them.

Transformation in a Plane	Determined by	Description in terms of reflections
reflection	line	r_ℓ
rotation	center and magnitude	$r_m \circ r_n$ if m intersects n
translation	vector	$r_m \circ r_n$ if $m \parallel n$

QUESTIONS

Covering the Reading

In 1 and 2, trace the figure below. Find the image of $\triangle XYZ$ under the translation with the given vector.

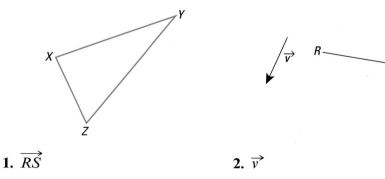

1. $\overrightarrow{RS}$

2. $\overrightarrow{v}$

3. What is a vector?

4. How are vectors and translations related?

5. Identify the horizontal and vertical components of the vector (-4, 8).

6. Quadrilateral $ABCD$ has vertices $A = (5, -2)$, $B = (-3, -3)$, $C = (0, 8)$, and $D = (3, 7)$. Let $A'B'C'D'$ be the image of $ABCD$ under a translation by vector (-1, 4). Graph $ABCD$ and $A'B'C'D'$ on the coordinate plane.

In 7–9, decide whether the transformation is a reflection, rotation, or translation.

7. The transformation is the composite of two reflections over intersecting lines.

8. The transformation is determined by a line.

9. The transformation is associated with a vector.

Applying the Mathematics

10. At the right is a preimage and translation image with corresponding points Q and Q' marked. What vector describes this translation?

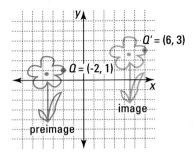

In 11–13, for each of the following real situations using the word *slide*, draw a picture with a preimage, image, and translation vector clearly marked.

11. a slide in baseball or softball

12. a playground slide

13. a rock slide

14. Let ℓ be the vertical line with equation $x = 3$. Let m be the vertical line with equation $x = 5$. Give an ordered pair for the vector associated with the translation
  a. $r_\ell \circ r_m$.
  b. $r_m \circ r_\ell$.

Review

15. Trace the figure at the right. Draw $r_\ell \circ r_m(ABCD)$.
  (Lesson 4-5)

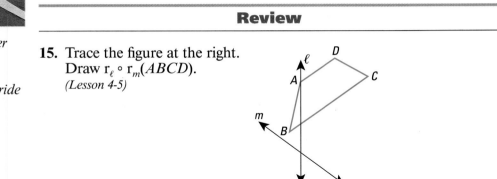

Slide rides. *Most water slides like this one in Anaheim, CA, offer a twisting, turning, cool ride into a pool.*

16. Let ℓ be the vertical line with equation $x = 3$. Let n be the horizontal line with equation $y = 2$. Describe the transformation $r_\ell \circ r_n$. *(Lesson 4-5)*

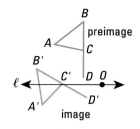

17. Trace the figure at the right. $A'B'C'D' = r_m \circ r_\ell(ABCD)$. Line m (which contains O) has been erased. Draw line m. *(Lesson 4-5)*

18. Suppose $s \parallel t$ and $r_s \circ r_t(DEFGH) = D'E'F'G'H'$. If $D'E'F'G'H'$ is clockwise oriented, what is the orientation of pentagon $DEFGH$? *(Lessons 4-2, 4-4)*

19. Trace the diagram of this billiards table. If you wish to bounce P off the top, right, and bottom sides, and then hit Q, in what direction should you shoot P? *(Lesson 4-3)*

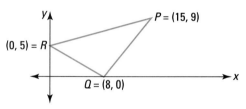

20. Reflect $\triangle PQR$ over the x-axis. Give the coordinates of the vertices of the image. *(Lesson 4-2)*

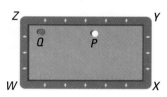

21. If a point P is reflected over the y-axis and its image is also P, what is its x-coordinate? Justify your answer. *(Lesson 4-1)*

22. $\overleftrightarrow{ZB}$ at the right is the perpendicular bisector of $\overline{MN}$. Justify each conclusion. *(Lessons 2-4, 3-8, 4-1)*
  a. $M = r_{\overleftrightarrow{BZ}}(N)$
  b. $MZ = NZ$
  c. B is the midpoint of $\overline{MN}$.

23. Give a good definition for *circle*. *(Lesson 2-4)*

Exploration

---

24. It is possible for the composite of two rotations with different centers to be a translation. Let R1 be a rotation with magnitude 90° with center point $(5, 0)$. Let R2 be a rotation with magnitude -90° with center $(0, 0)$. Describe R2 ∘ R1.

Isometries

A reflection or a composite of reflections is called an **isometry.** A reflection over just one line is, of course, a reflection. A composite of reflections over two parallel lines is a translation; a composite of reflections over two intersecting lines is a rotation. Since two lines in a plane are either parallel or they intersect, these are the only two-reflection isometries. A natural question is: What is the result of a composite of reflections over three coplanar lines?

Reflections over Three Lines

There are four possibilities for three coplanar lines. They intersect in either zero, one, two, or three points. When three coplanar lines intersect in zero or one point, the composite of three reflections equals a single reflection. For instance, in the figure at the left below, the composite $r_n \circ r_m \circ r_\ell$ over the parallel lines ℓ, m, and n is equal to the single reflection r_p. At the right below, the composite $r_s \circ r_t \circ r_v$ over the *concurrent* lines s, t, and v is equal to the single reflection r_q. Three or more lines are **concurrent** when they have a point in common.

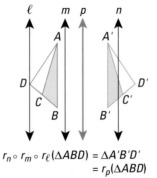

$$r_n \circ r_m \circ r_\ell (\triangle ABD) = \triangle A'B'D'$$
$$= r_p(\triangle ABD)$$

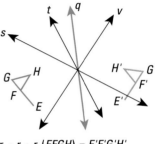

$$r_s \circ r_t \circ r_v(EFGH) = E'F'G'H'$$
$$= r_q(E'F'G'H')$$

In Question 14, you are asked to draw a composite of reflections over three lines that intersect in two points. Below is an example of a composite of reflections over three lines that intersect in three points. Let $G = r_n \circ r_m \circ r_\ell$. When the composite of three reflections is not equal to one reflection, an isometry G called a *glide reflection* occurs.

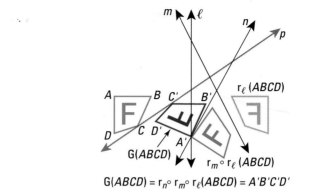

$$G(ABCD) = r_n \circ r_m \circ r_\ell(ABCD) = A'B'C'D'$$

Below is a drawing of G($ABCD$), G(G($ABCD$)), and G(G(G($ABCD$))). Notice how the successive quadrilaterals could be footsteps of a robot walking along the line p. The line p is the unique glide-reflecting line for G. In a later chapter, you will learn how to locate p.

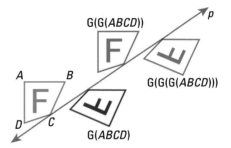

A glide reflection is the composite of a reflection over a line and a translation (the "glide" part) whose direction is parallel to the reflecting line. It makes no difference whether the reflection or the translation comes first; the composite is the same glide reflection.

To prove this, we locate a coordinate system so that the reflecting line is the x-axis. Then $r(x, y) = (x, -y)$. The translation T must be parallel to the x-axis, so $T(x, y) = (x + a, y)$.

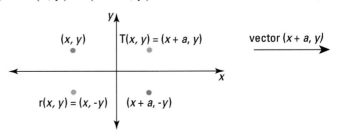

If the translation is done first: $r \circ T(x, y) = r(x + a, y) = (x + a, -y)$.
If the reflection is done first: $T \circ r(x, y) = T(x, -y) = (x + a, -y)$.
In either case, the final image is the same point, $(x + a, -y)$.

Successive footprints are very close to being glide-reflection images of one another, because each can be mapped onto the other by reflecting over a line and then translating in a direction parallel to that line. For this reason, a glide reflection is informally called a *walk*.

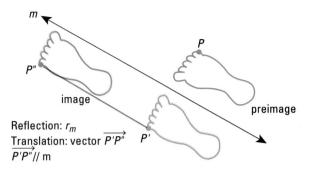

Reflection: r_m
Translation: vector $\overrightarrow{P'P''}$
$\overrightarrow{P'P''} \parallel m$

Since every glide reflection is a composite of three individual reflections, glide reflections preserve angle measure, betweenness, collinearity, and distance. A glide reflection reverses orientation since the first reflection reverses it, the second brings it back, and the third reflection leaves the orientation reversed.

Drawing Glide-Reflection Images

A glide reflection is determined when you know the reflecting line and the magnitude and direction of the translation. By the Figure Reflection Theorem, to glide-reflect a figure, you only have to glide-reflect the points that determine the figure.

Example 1

Given the figure at the right, let $G = T \circ r_\ell$ where ℓ is a line and T is the translation determined by the vector $\overrightarrow{AB}$. Draw $G(WXYZ)$.

Solution

1. First reflect polygon *WXYZ* over line ℓ by reflecting its vertices *W*, *X*, *Y*, and *Z*. The image is *W'X'Y'Z'* as shown in the drawing below.

2. Translate the images *W'*, *X'*, *Y'*, and *Z'* by the vector $\overrightarrow{AB}$ to find the glide-reflection images *W''*, *X''*, *Y''*, and *Z''*. Connect these points.

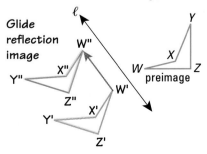

Types of Isometries

The word *isometry* comes from the Greek *isos* meaning "equal" and *metron* meaning "measure." The surprising fact is that reflections, translations, rotations, and glide reflections are all the isometries there are! Any composite of reflections over four or more lines can be reduced to a single reflection, translation, rotation, or glide reflection. Below is a hierarchy relating all the isometries.

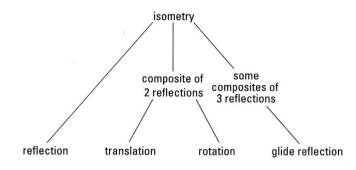

Preimages and images from reflections and glide reflections have opposite orientation; preimages and images from translations and rotations have the same orientation. By comparing the orientation of a figure and its image, you can identify the type of isometry relating the figures.

Example 2

Examine the tessellation at the left.
a. Which type of isometry maps figure *A* onto figure *B*?
b. Which type of isometry maps figure *A* onto figure *C*?

Solution

a. Since *A* and *B* have the same orientation, they are related by either a rotation or by a translation. *A* and *B* can't be related by a translation because they are tilted differently. So B is a *rotation image* of A.

b. Since *A* and *C* have opposite orientations, they are related by a reflection or glide reflection. They are not reflection images of each other, so C is a *glide reflection image* of A.

QUESTIONS

Covering the Reading

1. Name the transformation that results from a composite of reflections
   a. over three parallel lines.   **b.** over three concurrent lines.
   c. over three lines that intersect in three points.

2. Define *glide reflection.*

3. Name four properties preserved by glide reflections.

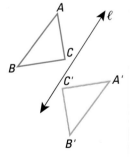

4. Name a property that glide reflections do not preserve.

5. Given the figure at the left, let G = T ∘ r$_\ell$ where ℓ is a line and T is the translation determined by the vector $\overrightarrow{XY}$. Trace the figure and draw G(*ABCD*).

6. Name the four types of isometries.

In 7–9, use the tessellation below.

7. Which type of isometry maps figure *A* onto figure *B*?

8. Which type of isometry maps figure *A* onto figure *C*?

9. Which type of isometry maps figure *C* onto figure *D*?

Applying the Mathematics

10. *True or false.* For any two figures, there is an isometry mapping one onto the other.

In 11 and 12, refer to the figure at the left. △*A'B'C'* is the image of △*ABC* under a glide reflection. Line ℓ is the glide-reflecting line.

11. Trace the figure and draw the translation vector.

12. a. What is the orientation of △*ABC*?
   b. What is the orientation of △*A'B'C'*?

13. The Sitka willow was used by Native Americans in the Pacific Northwest to dry fish, heal wounds, stretch skins, and make baskets. Notice that the placement of Sitka willow leaves looks like successive glide reflection images. What part of the plant is the glide-reflecting line?

14. **a.** Trace the drawing at the right.
    Then find $r_\ell \circ r_m \circ r_n$ (*FLAG*).
    b. Is $r_\ell \circ r_m \circ r_n$ a glide reflection?

15. Figure Q is the image of Figure P under
    the composite of six reflections.
    a. Name the two possible types of
    isometries which could be equivalent to
    this composite.
    b. Explain how you arrived at your
    answer.

16. Let $H = T \circ r_{x\text{-axis}}$ where T is the translation determined by the
    vector $(3, -1)$.
    a. Find $H(\triangle ABC)$ where $A = (0, 6)$, $B = (-5, 6)$, and $C = (-5, -2)$.
    b. Graph $\triangle ABC$ and $H(\triangle ABC)$ on the same axes.

Review

17. Trace $\triangle XYZ$ below and translate it with vector $\overrightarrow{XZ}$. *(Lesson 4-6)*

18. Trace the figure at the right. Draw its image under
    the composite $r_\ell \circ r_m$. *(Lesson 4-5)*

19. **a.** What properties are preserved by rotations?
    b. What properties are preserved by translations?
    (Lessons 4-4, 4-5)

20. The distance between two parallel lines is
    20 meters. What is the magnitude of the
    translation determined by a composite of
    reflections over those lines? *(Lesson 4-4)*

21. **a.** Trace the miniature golf hole at the
    right. If you want to hit the golf
    ball G off wall x and then off wall y
    to get it into the hole at H, where
    should you aim? *(Lesson 4-3)*
    b. Give the center and magnitude of
    the rotation $r_x \circ r_y (H)$. *(Lesson 4-5)*

Exploration

22. Draw a right triangle and its image under a composite of reflections
    $r_n \circ r_m \circ r_\ell$ in which ℓ, m, and n contain sides of a triangle. The
    composite is a glide reflection. Try to locate the glide-reflecting line
    p, and give the magnitude of the translation.

When Are Figures Congruent?

Congruent designs. *Stenciling uses the properties of congruent figures. A stencil contains cut-out shapes that can be copied repeatedly.*

Isometries and Congruent Figures

You first dealt with the idea of *congruent figures* when you were very young and copied figures. You may have used tracing paper to make an image of a drawing. With tracing paper you can slide (translate) or turn (rotate) the image. You can even flip (reflect) the paper over, and you can do these movements one after the other (compose them). With these transformations, whatever you do, you will get a figure that looks very much like the original, because under a reflection, rotation, or translation, corresponding angles have the same measure and corresponding distances are equal.

When you were a child, you also knew you could reverse the process. If you were given a figure and a tracing image of it, you could slide, turn, or flip the original figure to coincide with its image. You were showing that *if two figures are congruent, then there is a composite of translations, rotations, and reflections that map one onto the other.*

In this chapter, you have learned that translations, rotations, and glide reflections are composites of reflections. Since the word *isometry* refers to any reflection or composite of reflections, we can give a short definition of congruent figures.

> **Definition**
> Two figures F and G are **congruent figures,** written $F \cong G$, if and only if G is the image of F under an isometry.

Because of this definition, another term for isometry is **congruence transformation.** When figures are congruent and have the same orientation, we say they are **directly congruent.** Thus, translation or rotation images are directly congruent to their preimages. For instance, two stamps of the same kind are congruent regardless of where they are placed on an envelope. When figures are congruent and have opposite orientation, we say they are **oppositely congruent.** Reflection and glide-reflection images are oppositely congruent to their preimages. You and a mirror image of yourself are oppositely congruent.

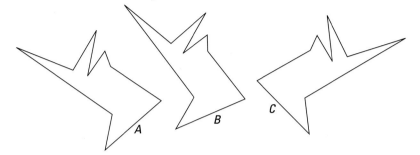

Figures *A* and *B* are directly congruent. Figures *B* and *C* are oppositely congruent.

Transformations that Do Not Yield Congruent Figures

Not all transformations are isometries. In each drawing below, there is a preimage and an image. The figure has been *transformed.* But the image and preimage are not congruent. In each case, however, a transformation has taken place because each point in the preimage corresponds to a unique image point, and vice versa.

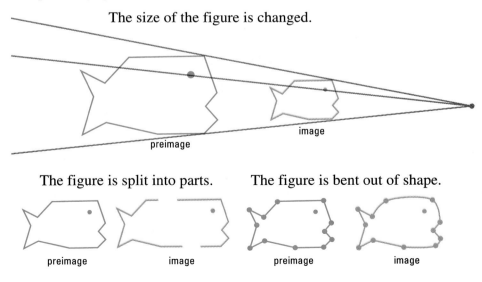

Even transformations with simple rules can give rise to distorted images. Drawn below is the front of a house and its image under the transformation S, where $S(x, y) = (2x, x + y)$. (To save space, only the outside vertices are named.) For instance, $B' = S(B) = S(12, 6) = (2 \cdot 12, 12 + 6) = (24, 18)$. Collinearity and betweenness are preserved, so you can still make out the house. But distance and angle measures are not preserved, so the image house is distorted.

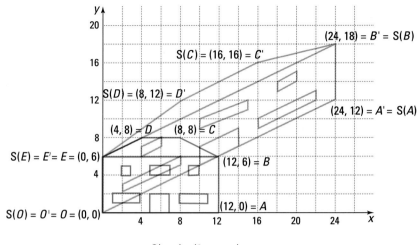

$S(x, y) = (2x, x + y)$

There are some important transformations that do not preserve distance, and you will study them in detail in a later chapter. But first, in the next few chapters, you will learn more about congruent figures.

QUESTIONS

Covering the Reading

1. Name three places outside of mathematics in which congruent figures may be found.

2. If $r(\triangle ABC) = \triangle DEF$, then according to the definition of congruence, $\triangle ABC \underline{\ ?\ } \triangle DEF$.

3. If $A \cong B$, then B is the image of A under what kind(s) of transformation(s)?

4. The door of the preimage house pictured above has vertices $(5, 0)$, $(5, 2)$, $(7, 2)$, and $(7, 0)$. Find the vertices of the door of the image house under the transformation S.

5. Copy the outline $ABCDEO$ of the preimage house above. Show its image under the transformation T, where $T(x, y) = (2x, x - y)$.

In 6–8, use the diagram below. Curve X is the image of curve W under a translation. Y is a rotation image of X. Z is a glide-reflection image of Y.

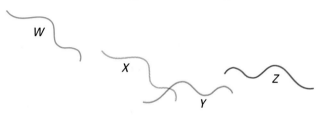

6. *True or false.* $W \cong Z$.

7. Name all curves that are directly congruent to Y.

8. Name all curves that are oppositely congruent to X.

9. Draw a figure and an image of that figure under a transformation that is not an isometry.

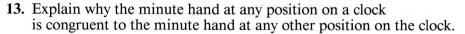

Applying the Mathematics

10. Draw a figure oppositely congruent to the figure at the right.

11. *True or false.*   Any two circles are congruent. Justify your answer with a drawing and an explanation.

12. Your great-grandfather looks at Question 6 and says, "The answer is false. Curves can't be congruent." What response might you give?

13. Explain why the minute hand at any position on a clock is congruent to the minute hand at any other position on the clock.

14. Tell whether or not the task uses the idea of congruence.
   a. making sure that the rear window glass for a car fits snugly into the window space
   b. constructing a circle with diameter equal to the length of $\overline{FY}$
   c. demonstrating that a tall thin glass and a short fat glass can hold the same amount of water
   d. explaining to a Martian that M and M are two ways to represent the same letter
   e. figuring out that two jigsaw puzzle pieces fit together

15. A transformation has the following rule: For all points (x, y), the image of (x, y) is $(x + 3, y - 4)$. Is this a congruence transformation? Why or why not?

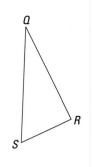

Review

16. Trace the figure at the left. Let T be the transformation described by vector $\overrightarrow{AB}$. Draw T($\triangle XYZ$).   *(Lesson 4-6)*

In 17 and 18, justify the conclusion.   *(Lessons 3-4, 3-5, 4-2)*

17. Given: m$\angle ABC$ = m$\angle DEF$ and
m$\angle DEF$ = m$\angle GHI$.
Conclusion: m$\angle ABC$ = m$\angle GHI$.

18. Given: The figure at the right with $r_\ell(A) = D$
and $r_\ell(B) = C$.
Conclusion: $r_\ell(AB) = CD$.

19. Name the three equivalence properties of equality.   *(Lesson 3-4)*

20. Trace $\triangle QRS$ at the left. Rotate $\triangle QRS$ -90° about R.   *(Lesson 3-2)*

Exploration

21. a. Find three characteristics that make Figure I *not* congruent to Figure II.

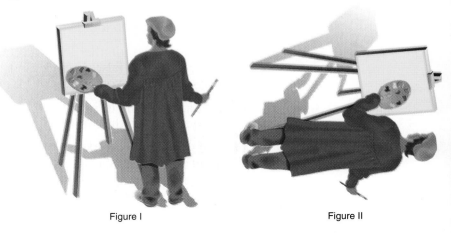

Figure I   Figure II

b. Make up a puzzle like the one in part **a,** or find such a puzzle in a newspaper or magazine.

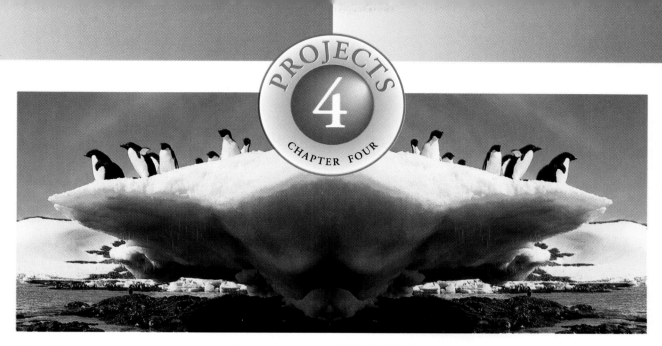

A project presents an opportunity for you to extend your knowledge of a topic related to the material of this chapter. You should allow more time for a project than you do for typical homework questions.

1 Composites of Three Reflections

Give a number of examples of the composite of reflections $r_u \circ r_t \circ r_s$ over three concurrent lines s, t, and u. Each composite equals the single reflection r_v. Explain how v is related to s, t, and u.

2 Miniature Golf

Design a five-hole miniature golf course. Each hole should include obstacles. Indicate an optimal shot for a hole-in-one for each hole, using reflections.

3 Real-World Isometries

Find eight real-world instances of isometries different from those in the chapter. Include at least one rotation, reflection, glide reflection, and translation. Display your examples on a poster. Label each one with its mathematical name.

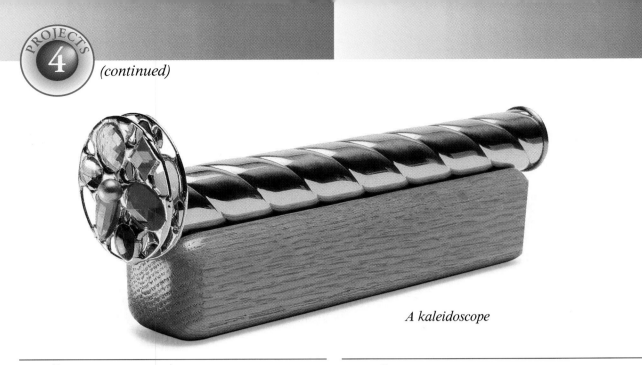

A kaleidoscope

4 Kaleidoscopes

Sir David Brewster (1781–1868) invented the first kaleidoscope. Find out how kaleidoscopic images use reflections and rotations, and then make your own. Prepare a report for your class on how a kaleidoscope works.

5 Letter Reflections

Pick four letters from the written alphabet of English or another language and perform reflections to create a pleasing design with each letter. (This project was suggested by John DiNicola, a Geometry teacher at the International School of Kuala Lumpur.)

6 House of Mirrors

Use small mirrors and cardboard to make a model of a room in a house of mirrors. Place two objects in the room. Draw a diagram of the room, showing several reflection images of each object.

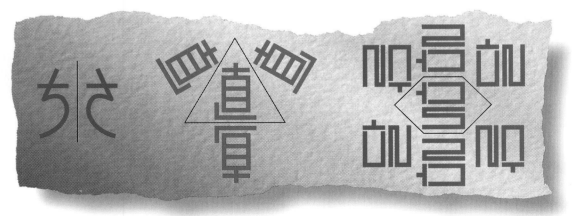

SUMMARY

A transformation is a one-to-one correspondence between sets of points. The first set is called the preimage, the second set the image. If the preimage point or figure is F and the transformation is T, we denote the image by T(F). The composite of two transformations, S followed by T, is the transformation T ∘ S that maps each point P onto T ∘ S(P). Four types of transformations are discussed in this chapter: reflections, translations, rotations, and glide reflections.

When point A is not on line m, the reflection image of A over line m, written $r_m(A)$, is the point B such that m is the perpendicular bisector of $\overline{AB}$. If A is on m, then $r_m(A) = A$. Reflections preserve collinearity, betweenness, distance, and angle measure, and switch orientation.

Composites of reflections are called isometries. A translation is a composite of two reflections over parallel lines; its magnitude is twice the distance between the lines and its direction is along the line perpendicular to those lines from the first line to the second. A rotation is a composite of two reflections over intersecting lines; its center is the intersection of the lines and its magnitude is twice the measure of the non-obtuse angle between the lines from the first line to the second. These composites of two reflections have the preservation properties of reflections and also preserve orientation.

A translation can be described by a vector. When this vector is parallel to a reflecting line, the composite of the translation and the reflection is a glide reflection. A composite of three reflections is either a single reflection or a glide reflection. Glide reflections have the preservation properties of reflections but reverse orientation.

Reflections have numerous applications. In billiards and miniature golf, using reflections and composites of reflections can help a player make more accurate shots. Combinations of mirrors are used in kaleidoscopes, telescopes, and other instruments, as well as in stores and homes. Two figures are congruent if and only if one is the image of the other under an isometry. Thus, isometries are called congruence transformations.

VOCABULARY

Below are the new terms and phrases for this chapter. For the starred terms (*) below, you should be able to produce a good definition. For the other terms, you should be able to give a general description and specific example of each.

Lesson 4-1
preimage
reflecting line, line of reflection
*reflection image of a point
*transformation, mapping
T(P), r, r(P), $r_m(P)$

Lesson 4-2
preserved property
Reflection Postulate
reflection image of a figure
Figure Reflection Theorem
orientation of a figure
distance between two parallel lines

Lesson 4-3
angle of incidence
angle of reflection

Lesson 4-4
*composite of two transformations
T ∘ S, T(S(P)), T ∘ S(P)
*translation, *slide
direction of translation
magnitude of translation
Two-Reflection Theorem for Translations

Lesson 4-5
*rotation
Two-Reflection Theorem for Rotations
center of rotation

Lesson 4-6
*vector, vector $\overrightarrow{AB}$, $\vec{v}$
initial point, terminal point
horizontal component, vertical component
ordered-pair description of a vector

Lesson 4-7
isometry
concurrent lines
*glide reflection

Lesson 4-8
*congruent figures, $F \cong G$
congruence transformation
directly congruent
oppositely congruent

PROGRESS SELF-TEST

Directions: Take this test as you would take a test in class. Use a ruler, compass, and graph paper. Then check your work with the solutions in the Selected Answers section in the back of the book.

1. Trace the figure below.

   a. Draw $r_\ell(A)$.   **b.** Draw $r_\ell(Q)$.

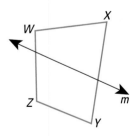

2. Trace W and V below. Draw the reflecting line m such that $r_m(W) = V$.

    •
   W

          •
         V

3. Trace the figure below. Draw the reflection image of $WXYZ$ over line m.

In 4 and 5, given $r_m(A) = B$ and $r_m(C) = D$. Justify each conclusion.

  4. $AC = BD$ **5.** $m \perp \overline{AB}$

6. The composite of three reflections can be which of the following?
    (a) reflection   (b) translation
    (c) rotation   (d) glide reflection

7. Give the image of $(4, -11)$ when translated by the vector $(3, -1)$.

8. Name five properties preserved by rotations.

9. $\triangle MNP$ below has vertices $M = (-2, 0)$, $N = (5, -1)$, and $P = (-3, 4)$. Give the vertices of the reflection image of $\triangle MNP$ over the x-axis.

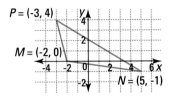

10. $\triangle ABC$ below was reflected over line m, then its image was reflected over line n. What angle of $\triangle JKL$ has the same measure as $\angle C$?

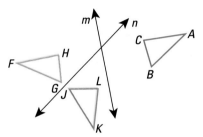

11. In the figure below, $\triangle X'Y'Z'$ is the image of $\triangle XYZ$ under a glide reflection. Line m is the glide-reflecting line. Trace the figure and draw the translation vector.

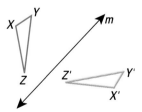

12. Trace the figure below. Draw the image and describe the transformation $r_t \circ r_s(\triangle PQR)$.

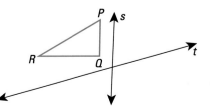

PROGRESS SELF-TEST

13. In the drawing below, ℓ // m.
  a. Draw $r_\ell \circ r_m(\triangle ABC)$.
  b. Name the composite transformation and describe its magnitude and direction.

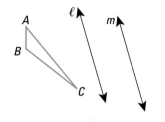

14. To rotate $\triangle WIN$ -50° about a point Q, you can successively reflect it over two lines that form an angle of __?__ whose vertex is __?__.

15. In the figure below, ℓ // m and $\ell \perp n$. Describe the transformation $r_\ell \circ r_m$.

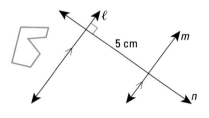

16. In the figure below, $r_{\overleftrightarrow{PR}}(Q) = S$. If $QR = 11$, find RS.

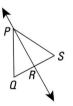

In 17 and 18, refer to the figure at the right.

17. What is the orientation of $ABCDEF$?

18. Trace the figure. Draw the translation image of $ABCDEF$ determined by vector $\overrightarrow{EC}$.

19. Trace the diagram of the miniature-golf hole below. Draw the path a ball at G must take if it is to bounce off of x and then y and go into the hole at H.

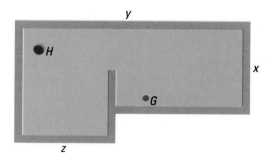

In 20 and 21, name the type of isometry that maps Figure I onto Figure II.

20.   **21.**

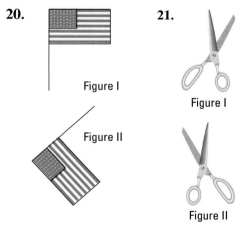

22. **a.** Draw and label a figure so that $r_p(MILD) = COLD$.
  b. From your figure give another name for p.

23. *Multiple choice.* Which letter below is not congruent to the others?

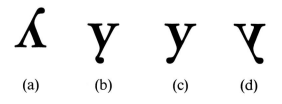

(a)   (b)   (c)   (d)

CHAPTER REVIEW

Questions on SPUR Objectives

SPUR stands for **S**kills, **P**roperties, **U**ses, and **R**epresentations. The Chapter Review questions are grouped according to the SPUR Objectives for this chapter.

SKILLS DEAL WITH THE PROCEDURES USED TO GET ANSWERS.

Objective A: *Draw figures by applying the definition of reflection image.*   *(Lessons 4-1, 4-2)*

In 1–4, trace the figure first.

1. **a.** Draw $r_\ell(E)$.
   b. Draw $r_\ell(B)$.

2. **a.** Draw $r_m(X)$.
   b. Draw $r_m(Y)$.

3. Draw the reflecting line ℓ for which $r_\ell(P) = Q$.

4. Draw the reflecting line m for which $r_m(\triangle ABC) = \triangle DEF$.

In 5 and 6, draw and label a figure matching the given conditions.

5. $r_p(\triangle CAT) = \triangle DOG$
6. $r_m(\triangle WHO) = \triangle WAT$

Objective B: *Draw reflection images of segments, angles, and polygons over a given line.* *(Lesson 4-2)*

In 7–10, trace the figure first. Then draw the reflection image of the given figure over the given line.

7.

8.

9.

10.

Objective C: *Draw translation and glide-reflection images of figures.*   *(Lessons 4-6, 4-7)*

In 11 and 12, trace the figure below.

11. Draw the translation image of *ABCDE* determined by vector $\overrightarrow{QT}$.

12. Draw the translation image of *ABCDE* determined by vector $\overrightarrow{AE}$.

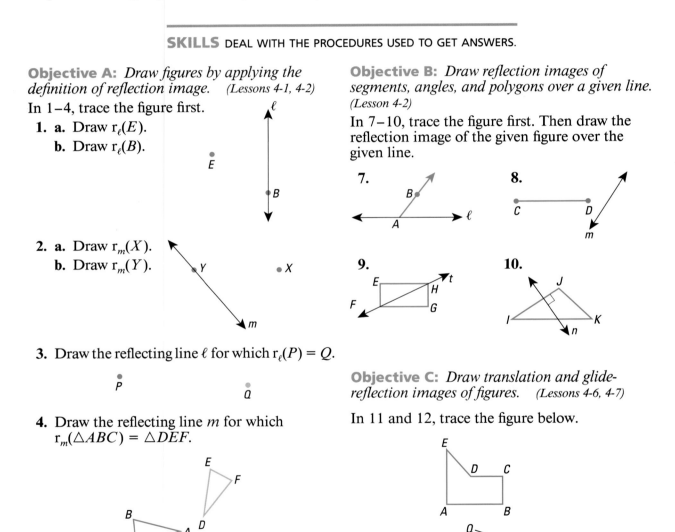

13. Trace the figure below. If $G = T \circ r_\ell$ where T is the translation determined by vector $\overrightarrow{RN}$, draw $G(ABCDE)$.

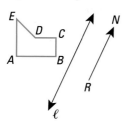

14. Trace the figure at the right. $\triangle A'B'C'$ is the image of $\triangle ABC$ under a glide reflection. Line m is the glide-reflecting line. Draw the translation vector $\vec{v}$.

Objective D: *Draw or identify images of figures under composites of two reflections.*
(Lessons 4-4, 4-5)

In 15 and 16, trace the diagram at the right. Then draw the image and describe the transformation.

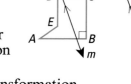

15. $r_a(r_b(\triangle DEF))$ **16.** $r_b \circ r_a(\triangle DEF)$

In 17 and 18, trace the drawing at the right.

17. a. Draw the image of $ABCDE$ under the transformation $r_n \circ r_m$.

    b. Describe this transformation.

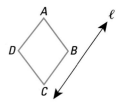

18. a. Draw $r_m \circ r_n (ABCDE)$.

    b. Describe this transformation.

PROPERTIES DEAL WITH THE PRINCIPLES BEHIND THE MATHEMATICS.

Objective E: *Apply properties of reflections to make conclusions, using one or more of the following justifications.*
definition of reflection
Reflections preserve distance.
Reflections preserve angle measure.
Reflections switch orientation.
Figure Reflection Theorem (Lessons 4-1, 4-2)

19. If $r_\ell(A) = B$ and A and B are different points, what conclusion follows due to the definition of reflection?

In 20–22, suppose $r(CDEF) = GHIJ$ as shown at the right. Justify each conclusion.

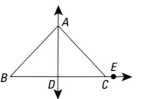

20. $r(E) = I$

21. $m\angle EDF = m\angle IHJ$

22. $FC = JG$

In 23 and 24, use the figure below in which $r_{\overleftrightarrow{AD}}(B) = C$.

23. If $m\angle BAD = 42$, find $m\angle BAC$.

24. If $m\angle B = x$, find $m\angle ACE$.

In 25 and 26, refer to the figure at the right.

25. What is the orientation of $ADCB$?

26. If $ADCB$ is reflected over line ℓ, what is the orientation of the image $A'D'C'B'$?

Objective F: *Apply properties of reflections to obtain properties of other isometries.*
(Lessons 4-4, 4-5, 4-7)

27. *Multiple choice.* Rotations do not preserve
(a) betweenness.
(b) distance.
(c) orientation.
(d) All of these properties are preserved.

28. *Multiple choice.* A composite of two reflections
(a) can never be a reflection.
(b) can never be a rotation.
(c) can never be a translation.
(d) can be any of the transformations mentioned in (a), (b), and (c).

29. *True or false.* Glide reflections preserve orientation.

30. What properties do all isometries preserve?

In 31 and 32, use the figure below. In the figure, $r_m(\triangle ABC) = \triangle FED$ and $r_\ell(\triangle DEF) = \triangle IGH$.

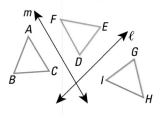

31. Which side of $\triangle GHI$ has the same length as $\overline{BC}$?

32. Which angle of $\triangle ABC$ has the same measure as $\angle G$?

Objective G: *Apply the Two-Reflection Theorems for Translations and for Rotations.* (Lessons 4-4, 4-5)

In 33 and 34, ℓ // m and $\ell \perp n$. Name the transformation, and describe it.

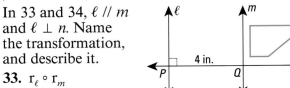

33. $r_\ell \circ r_m$

34. $r_m \circ r_n$

35. To translate a figure 6″ horizontally, you can reflect successively over two parallel __?__ lines where the distance between them is __?__.

36. To rotate $\triangle MNO$ 30° about a point C, you can reflect successively over two lines that form an angle of __?__ whose vertex is __?__.

USES DEAL WITH APPLICATIONS OF MATHEMATICS IN REAL SITUATIONS.

Objective H: *Determine the isometry which maps one figure onto another.* (Lesson 4-7)

In 37–40, name the type of isometry which maps Figure I onto Figure II.

37.

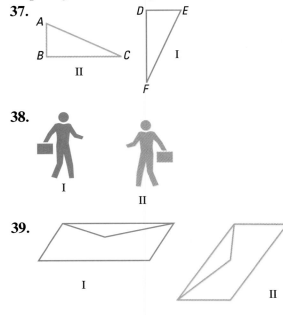

38.

39.

40.

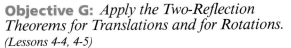

In 41 and 42, use the tessellation below.

41. Which type of isometry maps figure A onto figure B?

42. Which type of isometry maps figure B onto figure C?

Objective I: *Use reflections to find a path from an object to a particular point.* *(Lesson 4-3)*

In 43 and 44, find a path to shoot the ball *G* into the hole at *H* which would

43. bounce off $\overline{AB}$ only.

44. bounce off $\overline{AB}$ and then $\overline{AC}$.

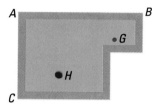

In 45 and 46, trace the drawing below.

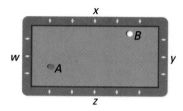

45. Draw the path of a ball that starts at *B* and bounces off sides *x*, *y*, and *z* in that order, and then hits *A*.

46. Draw a path from *B* that bounces off *x* and *z* in that order, and then hits *A*.

Objective J: *Use congruence in real situations.* *(Lesson 4-8)*

47. *Multiple choice.* Choose the process(es) that will create a copy that is congruent to the original.
(a) faxing
(b) photocopying at 200%
(c) photocopying at 100%
(d) taking a Polaroid photo

Multiple choice. In 48 and 49, the Folkwang and Erasmees light typefaces are used. Pick the letter that is not congruent to the other two.

48. (a)   (b)   (c)   **49.** (a)   (b)   (c)

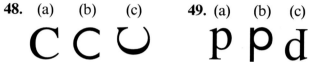

50. Use the mosaic pictured below. Find the smallest fundamental region that is copied to make this design.

REPRESENTATIONS DEAL WITH PICTURES, GRAPHS, OR OBJECTS THAT ILLUSTRATE CONCEPTS.

Objective K: *Find coordinates of reflection and translation images of points over the coordinate axes.* *(Lessons 4-1, 4-2, 4-6)*

51. Find the reflection image of (3, 7) over the *x*-axis.

52. Give the image of (*a*, *b*) when reflected over the *y*-axis.

53. Give the image of (8, 2) when translated by the vector (-6, 13).

54. Give the image of (*c*, *d*) when translated by the vector (-6, 3).

In 55 and 56, a quadrilateral has vertices *A* = (3, 7), *B* = (3, 1), *C* = (0, -5), and *D* = (-2, 8).

55. Graph *ABCD* and its reflection image over the *y*-axis.

56. Graph T(*ABCD*), where T is the translation determined by the vector (4, -1).

PROOFS USING CONGRUENCE

Congruent objects are everywhere. Teachers duplicate worksheets for students, and businesses photocopy pages for records and communication. Tool-and-die makers create molds (the "dies") for cutting and forging metal so that manufacturers can make identical parts. When computers copy files, a pattern from one microchip is copied onto another.

It is important to know when objects are congruent and when they are not. A copy of a document that has been changed slightly may change the meaning of words, sentences, paragraphs, and perhaps the entire document. If a die cut is slightly off, the objects it makes may not work. When there is a computer glitch and a file is not copied correctly, major problems can arise. Incorrect amounts of money might be transferred from one account to another, or entire documents may be lost.

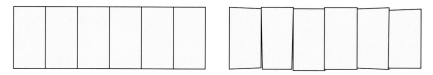

Congruence of geometric figures has the same importance. The rectangles above at the left are congruent and fit nicely together. The figures above at the right are very close to being congruent rectangles but that is not good enough. We see that there are problems with them. If these figures represented actual bricks or pieces of notebook paper or boxes to be placed next to each other, the result would not be as visually pleasing and perhaps would not be as useful.

In this chapter, you will take a close look at some properties of congruent figures. You will see how congruent figures are used and use them yourself to prove properties of angles and segments within particular figures. These properties and results have been key ideas for as long as people have studied geometry, so in this chapter you will be introduced to some of the history of the development of geometry.

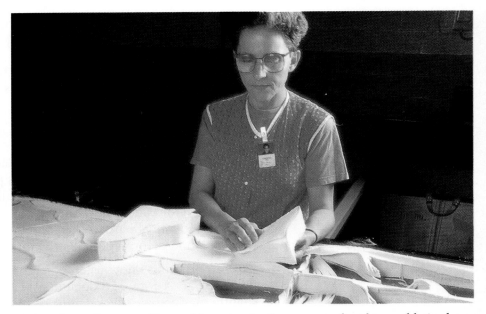

If the shoe fits . . . *Pictured is a step in the process of making athletic shoes. Corresponding parts must be congruent to ensure correct pairings and size.*

In the next three chapters, you will be looking at geometric figures and asking:

What parts of these figures do I know are congruent?

Can I use known congruent parts to determine that other parts are congruent?

These questions are easier to answer when you know that the figures themselves are congruent.

Angles and Sides of Congruent Figures

In the figure below, $\triangle ABC$ was translated onto $\triangle A'B'C'$. Then $\triangle A'B'C'$ was reflected over m onto $\triangle A''B''C''$. Lastly, $\triangle A''B''C''$ was rotated about O onto $\triangle XYZ$. Under this composite of reflections, call it T, $\triangle XYZ$ is the image of $\triangle ABC$. So, by the definition of congruence, $\triangle ABC \cong \triangle XYZ$. Specifically, $T(A) = X$, $T(B) = Y$, and $T(C) = Z$.

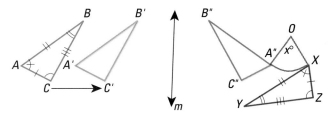

Angles and sides that are images of each other under a transformation are called **corresponding parts.** Since T maps the entire figure $\triangle ABC$ onto $\triangle XYZ$, it also maps each part of $\triangle ABC$ to its corresponding part of $\triangle XYZ$. Thus, all the corresponding parts of the triangles are congruent.

$$\triangle ABC \cong \triangle XYZ$$

$$\angle A \cong \angle X \qquad \angle B \cong \angle Y \qquad \angle C \cong \angle Z$$
$$\overline{AB} \cong \overline{XY} \qquad \overline{BC} \cong \overline{YZ} \qquad \overline{AC} \cong \overline{XZ}$$

In general, when two figures are congruent, pairs of corresponding sides, angles, and other parts of the figures are congruent. This result is widely used, and we state it as a theorem with an abbreviated name.

> **Corresponding Parts of Congruent Figures (CPCF) Theorem**
> If two figures are congruent, then any pair of corresponding parts is congruent.

Unless stated otherwise, corresponding parts of a figure refer to sides or angles. With a pair of congruent triangles, there are six pairs of corresponding parts: three pairs of sides and three pairs of angles. However, there may be other corresponding parts in a pair of congruent figures, such as diagonals or angle bisectors; these also are congruent. The order of vertices tells you which points are images of each other and, therefore, which parts correspond.

Example 1

$\triangle TOP \cong \triangle JKL$. List the six pairs of congruent parts. Sketch a possible situation and mark the congruent parts.

Solution

You can use the congruence statement to match the corresponding parts:

$$\triangle TOP \cong \triangle JKL$$

So $\qquad \angle T \cong \angle J, \angle O \cong \angle K,$ and $\angle P \cong \angle L.$

$$\triangle TOP \cong \triangle JKL$$

So $\qquad \overline{TO} \cong \overline{JK}, \overline{OP} \cong \overline{KL},$ and $\overline{TP} \cong \overline{JL}.$

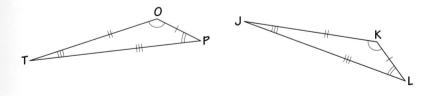

Preservation Properties and Congruence

By the Reflection Postulate, you know that every reflection preserves angle measure, betweenness, collinearity, and distance. So these properties are preserved by any isometry. As we have noted before, the names of these four properties happen to begin with the first four letters of the alphabet.

The preservation properties can tell us that certain segments have equal length and certain angles have equal measure.

Example 2

Suppose *MNOP* ≅ *RSMQ* in the figure at the right. Using the A-B-C-D Theorem,
a. which angle has measure equal to m∠*PMN*?
b. which segment has length equal to *NP* (not drawn)?

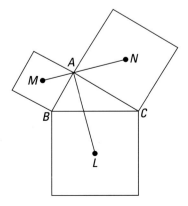

Solution

From the order of the vertices in *MNOP* ≅ *RSMQ*, there is a congruence transformation that maps *M* onto *R*, *N* onto *S*, *O* onto *M*, and *P* onto *Q*. Thus ∠*PMN* is mapped onto ∠*QRS*, and $\overline{NP}$ is mapped onto $\overline{SQ}$. So,
a. ∠QRS has measure equal to m∠PMN because angle measure is preserved.
b. $\overline{SQ}$ has length equal to NP because distance is preserved.

When figures are not known to be congruent, then it is not so easy to tell whether parts are congruent. For instance, consider the figure below. It was formed by drawing squares on the three sides of △*ABC*. Then the centers *M* and *N* of two of the squares were connected. The center *L* of the third square was connected to the opposite vertex *A*. Does $\overline{MN}$ always contain *A*? Is $\overline{MN} ≅ \overline{AL}$? Do the segments form congruent angles so that they are perpendicular? It isn't easy to answer any of these questions because the segments do not seem to be corresponding parts of any figure. To answer questions like these, we need to examine the properties of congruence more closely.

School days. *School photos are often sold in sheets which can be cut apart. The photos are congruent to each other.*

QUESTIONS

Covering the Reading

1. Name three places outside mathematics in which congruent figures may be found.

2. Unless otherwise stated, corresponding parts refer to __?__.

3. Suppose $\triangle ATV \cong \triangle MCI$. What part has measure equal to the given measure?
   a. $m\angle T$ **b.** $m\angle VAT$ **c.** IC

4. In a pair of congruent triangles, there are __?__ pairs of corresponding parts.

5. Suppose $\triangle ADE \cong \triangle BNO$.
   a. Sketch this situation.
   b. Which pairs of sides are congruent?
   c. Which pairs of angles are congruent?

6. Name four properties preserved by isometries.

7. Suppose $\triangle ATV \cong \triangle MCI$ and $AT = 5$ cm. What other distance can be found?

8. Suppose $\triangle MCI \cong \triangle GTE$ and $m\angle ICM = 94$. What other angle measures 94?

9. Begin with your own $\triangle ABC$ and draw a figure like that on page 246. From the figure in the book and your figure, guess at the answers to the three questions posed above the figure.

Applying the Mathematics

In 10–12, for each figure assume the two triangles that appear to be congruent are congruent. **a.** Write a congruence statement for each with the vertices in correct order. **b.** Tell whether the triangles are directly or oppositely congruent. **c.** List, in pairs, all corresponding congruent parts.

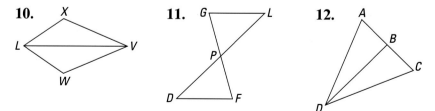

10.

11.

12.

13. Here is a puzzle by Robert Leighton from the May 1995 issue of *Games World of Puzzles,* by the editors of *Games Magazine.* There are 11 pairs of congruent figures in this scene. For instance, the crescent-shaped head of the pick held by one of the dwarfs is congruent to the collar of the balding man at the lower right. Identify the other ten pairs, but *do not put any marks in this book;* that would spoil the puzzle for the next reader. Your teacher will give you a copy of the puzzle.

Reprinted with permission from GAMES WORLD OF PUZZLES Magazine (19 West 21st Street, New York, NY 10010). Copyright © 1995. B. & P. Publishing Co., Inc.

14. Draw a figure oppositely congruent to the figure at the left. *(Lesson 4-8)*

15. Taking a photograph of the front of a building can be thought of as mapping points on the front of the building onto points of the photo.
a. *Multiple choice.* Which of the following are *not* preserved by this transformation?
(a) angle measure   (b) betweenness
(c) collinearity   (d) distance
b. Explain your answer to part **a.** *(Lessons 4-1, 4-8)*

In 16 and 17, refer to this statement: *If a figure is an image under an isometry, then it is directly congruent to its preimage.* *(Lessons 2-1, 2-2, 4-8)*

16. Draw an instance of the conditional.

17. Draw a counterexample to the conditional.

18. Let $F = S \circ r_n$ where n is the line below and S is the translation determined by vector $\vec{v}$. Trace the figure and draw $F(HIJKL)$. *(Lessons 4-6, 4-7)*

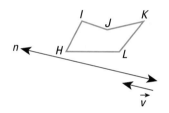

19. Trace the bow tie below.
a. Sketch line ℓ so that $ABCD = r_\ell(BADC)$.
b. Sketch line m so that $ABCD = r_m(DCBA)$.
c. Describe the rotation R so that $ABCD = R(CDAB)$.
(Lessons 4-1, 4-5)

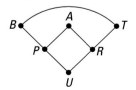

20. Is the network at the left traversable? If not, tell why not. If so, give a possible path. *(Lesson 1-4)*

21. Make up your own drawing like that in Question 13. Include at least 5 pairs of congruent figures.

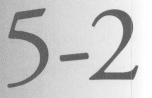

*Congruence
and
Equality*

The Equivalence Properties of Congruence

Some properties of congruence ($\cong$) are like properties of equality ($=$). Three of these properties are called the *equivalence properties of congruence.* Here we give these properties and arguments that explain why they are true.

1. Any figure is congruent to itself.

    Argument: Suppose the figure is F. Reflect it over any line m and then back over that line. Because $r_m \circ r_m(F) = F$, by the definition of congruence, $F \cong F$.

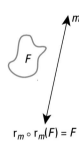

$r_m \circ r_m(F) = F$

2. For any figures F and G: if $F \cong G$, then $G \cong F$.

    Argument: If $F \cong G$, then there is a composite of reflections that maps F onto G. Using the reflecting lines in reverse order, you can map G onto F. So $G \cong F$. For example, in the figure below, $r_m \circ r_\ell(F) = G$. So, in reverse order, $r_\ell \circ r_m(G) = F$.

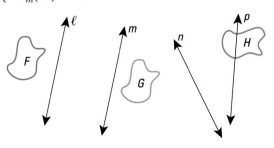

3. For any figures, F, G, and H: if $F \cong G$ and $G \cong H$, then $F \cong H$.

Argument: Since $F \cong G$ and $G \cong H$, there is a composite of reflections mapping F onto G and a composite of reflections mapping G onto H. Take all of the reflecting lines in these composites in order, and they will map F onto H. So $F \cong H$.

For example, in the figure at the bottom of page 250, $r_m \circ r_\ell(F) = G$ and $r_p \circ r_n(G) = H$ so $r_p \circ r_n \circ r_m \circ r_\ell(F) = H$.

The names of these properties are stated here.

Theorem (Equivalence Properties of $\cong$)
For any figures F, G, and H:
$F \cong F$.   (Reflexive Property of Congruence)
If $F \cong G$, then $G \cong F$.   (Symmetric Property of Congruence)
If $F \cong G$ and $G \cong H$, then $F \cong H$. (Transitive Property of Congruence)

Example 1

$G \cong H$ in the figure at the bottom of page 250. By the Symmetric Property of Congruence, $H \cong G$. What composite of reflections maps H onto G?

Solution

It is given that $r_p \circ r_n(G) = H$. So take the reflecting lines in reverse order:

$$r_n \circ r_p(H) = G.$$

Congruent Segments and Length

If you are given two congruent segments (or other figures), then you know that there is an isometry T which maps one onto the other. In the figure below, $T(\overline{CD}) = \overline{EF}$.

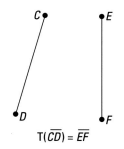

$T(\overline{CD}) = \overline{EF}$

By the A-B-C-D theorem, $CD = EF$. This argument proves the conditional:

If two segments are congruent, then they have the same length.

Is the converse of this statement true as well? If segments have equal lengths, then are they congruent? Yes, and a proof is given on the following page.

In Figure 1 below, $WX = YZ$. The proof argument shows that $\overline{YZ}$ is the image of $\overline{WX}$ under an isometry.

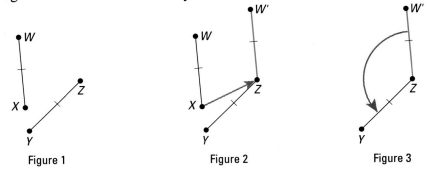

Figure 1   Figure 2   Figure 3

Translate $\overline{WX}$ by vector $\overrightarrow{XZ}$ to get image $\overline{W'Z}$. (See Figure 2.) Thus $\overline{WX} \cong \overline{W'Z}$, and by the Transitive Property of Equality, $W'Z = YZ$. Rotate $\overline{W'Z}$ by $m\angle W'ZY$ around center Z. (See Figure 3.) The image of Z is Z. The image of W' is Y because $W'Z = YZ$. So $\overline{W'Z} \cong \overline{YZ}$. By the Transitive Property of Congruence, $\overline{WX} \cong \overline{YZ}$.

Since both the conditional and its converse are true, we can state the theorem as a biconditional.

> **Segment Congruence Theorem**
> Two segments are congruent if and only if they have the same length.

Example 2

In the figure at the left, $\overline{AB} \cong \overline{CD}$. M and N are the midpoints of $\overline{AB}$ and $\overline{CD}$, respectively. If $AB = 12x$, what is CN?

Solution
Since $\overline{AB} \cong \overline{CD}$, $AB = CD$. Since $AB = 12x$, $CD = 12x$.
Since N is a midpoint, $CN = \frac{1}{2}CD$. By substitution, $CN = \frac{1}{2}(12x) = 6x$.

Congruent Angles and Angle Measure

The Segment Congruence Theorem, which connects segments and lengths, has a counterpart involving angles and angle measure.

> **Angle Congruence Theorem**
> Two angles are congruent if and only if they have the same measure.

The Angle Congruence Theorem is also a biconditional. The first conditional is *If two angles are congruent, then they have the same measure.* You are asked to give an argument for this conditional in Question 12. The second conditional states: *If two angles have the same measure, then they are congruent.* In Question 15 you are asked to describe an isometry that maps one angle onto another.

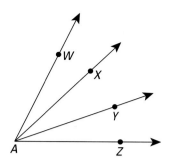

Example 3

In the figure at the left, $\angle WAX \cong \angle YAZ$. If $m\angle WAY = 44$, what is $m\angle ZAX$?

Solution

By the Angle Congruence Theorem $m\angle WAX = m\angle YAZ$. By the Addition Property of Equality (add $m\angle XAY$ to both sides), $m\angle WAY = m\angle ZAX$. So, $m\angle ZAX = 44$.

Notation of Congruence and Equality

For instance, the definition of an equilateral triangle given in Lesson 2-6 could be restated as: An equilateral triangle is a triangle with three congruent sides. In this book, we often use these substitutions.

Congruence	Equality
congruent segments	segments of equal length
$\overline{AX} \cong \overline{BY}$	$AX = BY$
congruent angles	angles with the same measure
$\angle ABC \cong \angle XYZ$	$m\angle ABC = m\angle XYZ$

QUESTIONS

Covering the Reading

1. What are the equivalence properties of congruence?

2. What does the transformation $r_\ell \circ r_\ell$ do to a figure?

3. If $r_\ell \circ r_m(F) = G$, what isometry maps G onto F?

4. According to the Segment Congruence Theorem, if $AX = BY$, then $\underline{\ \ ?\ \ }$.

5. Trace the two congruent segments $\overline{EF}$ and $\overline{GH}$ at the left.
   a. Translate $\overline{EF}$ by the vector $\overrightarrow{FG}$ and draw the image $\overline{E'G}$.
   b. What is the magnitude of rotation needed to map $\overline{E'G}$ onto $\overline{HG}$?

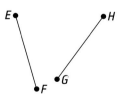

6. In the figure at the right, $\overline{AB} \cong \overline{CD}$. M and N are the midpoints of $\overline{AB}$ and $\overline{CD}$, respectively. If $AB = 38x$, what is ND?

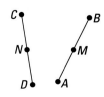

7. According to the Angle Congruence Theorem, if $\angle C \cong \angle T$, then $\underline{\ \ ?\ \ }$.

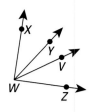

8. In the figure at the left, $\angle XWY \cong \angle ZWV$. If m$\angle XWV = 56$, what other angle has a measure of 56?

In 9–11, which equivalence property justifies each statement?

9. If $\overline{AB} \cong \overline{CD}$ and $\overline{CD} \cong \overline{GF}$, then $\overline{AB} \cong \overline{GF}$.

10. If $\triangle MPQ \cong \triangle ABC$, then $\triangle ABC \cong \triangle MPQ$.

11. $\angle QAM \cong \angle QAM$.

Applying the Mathematics

12. Give justifications to complete this argument proving one direction of the Angle Congruence Theorem.
Given: $\angle APD \cong \angle CQB$.
  a. Conclusion: There is an isometry T for which T($\angle APD$) = $\angle CQB$.
  b. Conclusion: m$\angle APD$ = m$\angle CQB$.

In 13 and 14, reword the statement using the word *congruent*.

13. An isosceles triangle has two sides of the same length.

14. If two lines are cut by a transversal and corresponding angles have the same measure, then the lines are parallel.

15. At the right, $\angle RXS$ can be mapped onto $\angle WYZ$ by a translation followed by a rotation.
  a. What is the translation vector?
  b. Trace the figure and draw the image of $\angle RXS$ under the translation.
  c. What is the magnitude of the rotation?
  d. Must m$\angle RXS$ be equal to m$\angle WYZ$? Why or why not?

16. In the figure at the right, $\triangle ABC$ is isosceles with $\overline{AB} \cong \overline{AC}$. $\triangle CAD$ is isosceles with $\overline{AC} \cong \overline{AD}$. Name all lengths in the figure that are equal to AB.

17. Suppose that $\triangle FRL \cong \triangle QXZ$.
  a. If m$\angle R = 32$, then __?__ = 32.
  b. If $FL = 9$, then __?__ = 9.

Review

18. Name four properties preserved by isometries. *(Lesson 5-1)*

19. According to the definition of congruence, when are triangles congruent? *(Lesson 4-8)*

20. In the figures below, $POLYGN \cong P'O'L'Y'G'N'$.
   a. Name the type of isometry that maps Figure 1 to Figure 2.
   b. Name the theorem that justifies this statement: $\angle N \cong \angle N'$.
   c. What is the orientation of $POLYGN$? *(Lessons 4-2, 4-7, 5-1)*

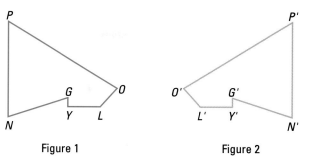

Figure 1   Figure 2

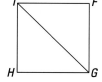

21. In the figure at the left, if $\triangle FGI \cong \triangle HGI$, then $\angle HIG \cong$ _?_ .
   (Lesson 5-1)

22. A taxi driver makes a copy of the key to a taxi. How can the driver know if the copy is congruent to the original key? *(Lesson 5-1)*

23. Trace the billiards table below.
   a. Find the path that bounces ball A off wall x, wall w, and wall z in that order, and then hits ball B.
   b. Find another path for ball A that uses three walls and then hits ball B. *(Lesson 4-3)*

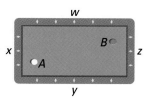

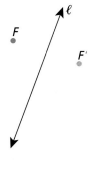

24. Refer to the figure at the left. If $r_\ell(F) = F'$, what justifies the conclusion that ℓ is the perpendicular bisector of $\overline{FF'}$?
   (Lessons 3-5, 4-1)

25. Justify the following conclusion.
   Given: $\odot O$ at the right.
   Conclusion: $OA = OB$.
   (Lessons 2-4, 3-5)

26. The measure x of an angle is greater than four times the measure of a supplement to it. Is this possible? If so, graph the possibilities for x on a number line. If not, explain why not. *(Lesson 3-3)*

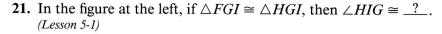

Exploration

27. On an automatic drawer, create two congruent segments $\overline{AB}$ and $\overline{CD}$ so that when point C is moved, AB will remain equal to CD.

In this lesson, you will work with statements that can be used as justifications in proofs that involve congruence. The proof arguments in this lesson are only one step long, but in the next lesson you will see that by putting a few of these steps together, you can prove some very important theorems that may not be so obvious.

Common Justifications for Congruence

Remember that every conclusion in a proof needs to be justified. Every justification is either a postulate, a definition, or a theorem that we have already proved. Sometimes a justification is one direction of a biconditional. In that case we name the biconditional but write only the correct direction as an if-then statement. Here is a table of some justifications that will be helpful.

Justifications that segments are congruent	Justifications that angles are congruent
definition of ⊥ bisector: If the perpendicular bisector of $\overline{AB}$ intersects $\overline{AB}$ at M, then $\overline{AM} \cong \overline{MB}$. *(Lesson 3-8)*	Corresponding Angles Postulate: If two parallel lines are cut by a transversal, then corresponding angles are congruent. *(Lesson 3-6)*
definition of midpoint: If M is the midpoint of $\overline{AB}$, then $\overline{AM} \cong \overline{MB}$. *(Lesson 2-4)*	definition of ∠ bisector: If $\overrightarrow{BD}$ is the bisector of $\angle ABC$, then $\angle ABD \cong \angle DBC$. *(Lesson 3-1)*
CPCF Theorem: Corresponding parts of congruent figures are congruent. *(Lesson 5-1)*	CPCF Theorem: Corresponding parts of congruent figures are congruent. *(Lesson 5-1)*
Segment Congruence Theorem: If $AB = CD$, then $\overline{AB} \cong \overline{CD}$. *(Lesson 5-2)*	Angle Congruence Theorem: If $m\angle 1 = m\angle 2$, then $\angle 1 \cong \angle 2$. *(Lesson 5-2)*
definition of circle: If A and B are on circle O, then $\overline{OA} \cong \overline{OB}$. *(Lesson 2-4)*	Vertical Angles Theorem: If $\angle 1$ and $\angle 2$ are vertical angles, then $\angle 1 \cong \angle 2$. *(Lesson 3-3)*
definition of congruence: If one segment is the image of another under an isometry, then the segments are congruent. *(Lesson 4-8)*	definition of congruence: If one angle is the image of another under an isometry, then the angles are congruent. *(Lesson 4-8)*

Making Justified Conclusions from Given Information

Suppose a friend told you he would not be at home, he was going to buy a snack, and he was walking. Then you may be able to make a conclusion about where your friend is. This process is called *making a justified conclusion from given information.*

The same process occurs in geometry. You may be given a figure or some other information and asked to prove whatever you can. Then you need to provide a conclusion and a justification for it. If there is no figure, you may want to draw one. Here is a typical example.

Example 1

Given: V is the midpoint of $\overline{AP}$. What can you deduce?

Solution

It is often helpful to draw a figure, such as the one shown here.

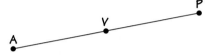

From the definition of midpoint you can conclude that $\overline{AV} \cong \overline{VP}$. You can write: $\overline{AV} \cong \overline{VP}$ **because of the definition of midpoint.** The more information that is given, the more that can usually be deduced.

Example 2

Given: $\overline{PR}$ and $\overline{SQ}$ are diameters of circle O. Make and justify a conclusion.

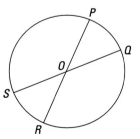

Solution

First consider the possibility of congruent segments. Since a circle is given, there will be congruent radii. Here is one conclusion.

$$\overline{OP} \cong \overline{OQ} \cong \overline{OR} \cong \overline{OS} \text{ by the definition of circle.}$$

Now consider angles. The two lines $\overleftrightarrow{PR}$ and $\overleftrightarrow{QS}$ are intersecting, so they form vertical angles. Here is another possible conclusion.

$$\angle POQ \cong \angle ROS \text{ because of the Vertical Angles Theorem.}$$

Many things can be deduced when you know that one figure is the image of another under an isometry.

Example 3

Given: $r(\triangle ABC) = \triangle ADE$. What can you deduce?

Solution

First, draw a figure.

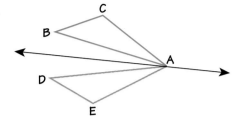

You can deduce that the triangles are congruent. You might write:
From the definition of congruence, $\triangle ABC \cong \triangle ADE$.
Or, if you wished to be clearer, you could write: Since $\triangle ADE$ is the image of $\triangle ABC$ under an isometry, $\triangle ABC \cong \triangle ADE$ by the definition of congruence. From this conclusion, you can make other conclusions about the corresponding parts. You might write: $\angle CBA \cong \angle EDA$ because of the CPCF Theorem.

Determining Justifications

Often you know what you need to prove, and you need to search for a justification. For instance, someone may tell you that your friend Jennie is at a basketball game. You ask: "How do you know?" The justification may be simple: "She told me she was going." Or the justification may be a little more complicated: "I saw Jennie walking with Tom. Tom told me he was going to the basketball game with someone." This is called *justifying a conclusion that has been asserted*.

When a conclusion is that one figure is congruent to another, there are only a few possible justifications. The most common justifications are given in the table on page 256.

Example 4

Given: $m \parallel n$. Justify the conclusion that $\angle 1 \cong \angle 5$.

Solution

$\angle 1$ and $\angle 5$ are corresponding angles. This suggests that the Corresponding Angles Postulate may be appropriate. That postulate is a biconditional, so be careful to use the correct order. If parallel lines are cut by a transversal, then corresponding angles are congruent. ($\parallel$ lines $\Rightarrow$ corr. $\angle$s $\cong$.)

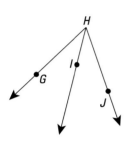

Example 5

In the figure at the left, $\angle GHI$ is the reflection image of $\angle JHI$ over $\overleftrightarrow{HI}$. Justify the conclusion that $\angle GHI \cong \angle JHI$.

Solution

Examine the list of possible justifications for angles to be congruent. Only one involves reflections, namely, *the definition of congruence.* To be clearer, you may write the appropriate direction of the definition: If *one angle is the image of another under an isometry, then the angles are congruent.*

The next example uses the language of proof. Notice that we continue to use "Given" and we replace "justify the conclusion that" with "To prove."

Example 6

Given: $r(\angle GHI) = \angle JHI$.
To prove: $\angle GHI \cong \angle JHI$.

Solution

This is the same problem as in Example 5, only with different wording. The answer is the same! You can write $\angle GHI \cong \angle JHI$ *because of the definition of congruence.* Or you could write $\angle GHI \cong \angle JHI$ *because if one angle is the image of another under an isometry (every reflection is an isometry), then the angles are congruent.*

QUESTIONS

Covering the Reading

1. What kinds of statements are allowed as justifications in a proof?

2. Name a definition that might be used to justify that two segments are congruent.

3. Name a theorem that might be used to justify that two angles are congruent.

In 4 and 5, make and justify a conclusion from the given information.

4. Given: m is the $\perp$ bisector of $\overline{WY}$.

5. Given: $\overrightarrow{AV}$ is the bisector of $\angle BAC$.

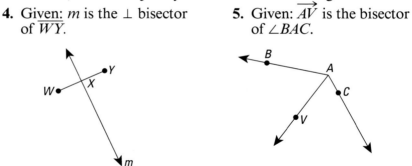

6. Given: $\triangle ABC$ is the image of $\triangle DEF$ under a rotation of 40°. Draw a figure, and make and justify a conclusion.

7. Given the figure below with lengths as indicated, justify the conclusion $\overline{FL} \cong \overline{MT}$.

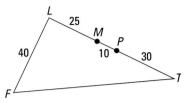

8. If quadrilateral $ABCD \cong$ quadrilateral $EFGH$, justify the conclusion that $\angle CDB \cong \angle GHF$.

Applying the Mathematics

9. Refer to Example 3. Make and justify two conclusions that are different from the ones shown there.

10. In the figure below, lines m and n are parallel and intersected by transversal t. Make and justify at least five different conclusions.

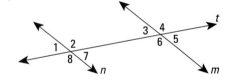

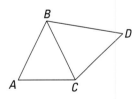

11. In the figure at the left, $AB = BC$ and $BC = CD$. Make and justify at least three congruence conclusions. (Hint: You will need to use a justification from Lesson 5-2).

In 12 and 13, draw a figure and state a possible conclusion for the justification used.

12. If A, B, and C are points and $3 \cdot AB = BC$, what conclusion can be made using the Multiplication Property of Equality?

13. Suppose $\overline{AB}$, $\overline{CD}$, and $\overline{EF}$ intersect at three points G, H, and I, with G between A and B, H between C and D, and I between E and F. Using the Vertical Angles Theorem as the justification, what conclusion can be made?

14. *Multiple choice.* Refer to the table of justifications in this lesson. Only one direction for each definition is written. For the definition of midpoint, in which direction is this?
(a) term $\Rightarrow$ characteristics
(b) characteristics $\Rightarrow$ term

15. a. Give a non-mathematical example, different from the one in this lesson, of making a justified conclusion from given information.
b. Give a non-mathematical example, different from the one in this lesson, of justifying a conclusion that someone has made.

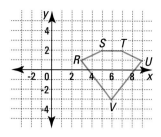

16. a. Consider pentagon *RSTUV* at the left. Give the coordinates of its image *R′S′T′U′V′* under the translation associated with the vector (-6, 6).
  b. Why is *RSTUV* ≅ *R′S′T′U′V′*?
  c. The original pentagon has a line of reflection. What is an equation for that line of reflection?
  d. What is an equation for the line of reflection of the image?
  (Lessons 4-1, 4-6, 5-1)

17. Trace $\overline{NM}$ at the right.

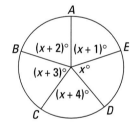

  a. Construct the perpendicular bisector of $\overline{NM}$. Label it $\overleftrightarrow{RS}$. *(Lesson 3-8)*
  b. Construct a line parallel to $\overleftrightarrow{RS}$ through point *M*. Label it $\overleftrightarrow{MQ}$. *(Lesson 3-8)*
  c. If the slope of $\overleftrightarrow{MQ}$ = *n*, what is the slope of $\overleftrightarrow{NM}$? *(Lesson 3-7)*

18. Write the direction of the Corresponding Angles Postulate that is not given in the table of justifications in this lesson. *(Lesson 3-6)*

19. On the game show "Wheel of Fortune," the wheel has 24 congruent regions. What is the measure of the arc along the outside of each region? *(Lesson 3-2)*

20. The measures of five central angles are shown in the circle at the right. Find
  a. m $\widehat{AE}$.
  b. m $\widehat{BDE}$.
  (Lessons 3-1, 3-2)

21. There are at least twenty different conclusions that can be made from this drawing. Find as many as you can.

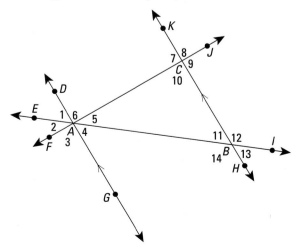

Circle Constructions

IN-CLASS
ACTIVITY

Work on this activity with a partner. Use a ruler, protractor, and compass, or use an automatic drawer.

1 **a.** Trace $\overline{AB}$ and complete the following algorithm.
    Step 1: Draw $\odot A$ with radius AB.
    Step 2: Draw $\odot B$ with radius BA.
    Step 3: Label the intersections of $\odot A$ and $\odot B$ as C and D.
    Step 4: Draw $\overline{AC}$ and $\overline{BC}$ to form $\triangle ABC$.

$A \bullet\text{———}\bullet B$

  b. *Multiple choice.* What type of triangle is $\triangle ABC$?
    (i)   equilateral
    (ii)   isosceles, but not equilateral
    (iii) scalene
  c. Justify your answer to part **b.**

2 **a.** Draw any three noncollinear points X, Y, and Z. Follow this algorithm.
    Step 1: Construct m, the perpendicular bisector of $\overline{XY}$.
    Step 2: Construct n, the perpendicular bisector of $\overline{YZ}$.
    Step 3: Label $m \cap n$ as O.
    Step 4: Draw the circle with center O and radius OX.

  b. *True or false.* The circle in step 4 of the construction contains points Y and Z.
  c. Justify your answer to part **b.**

LESSON
5-4

Proofs Using Transitivity

The Domino Effect. *These dominoes are spaced equally, so when one topples it knocks the next one over and so on. Each domino's fall relies on the one before it. Proofs using transitivity rely on this principle, too.*

The First Written Proofs

To our knowledge, the first person to write proofs like those used today was the Greek mathematician Thales, in the 6th century B.C. Three hundred years later, Euclid wrote *Elements,* which we mentioned in Lesson 1-7. *Elements* consists of 13 parts, called "books." Within these books 465 theorems are proved.

In writing a proof of a statement, it is helpful to rewrite the statement as a conditional. A proof of a conditional $p \Rightarrow q$ has four components:

> the *Given p,*
> the *To prove q,*
> a *Drawing* (which may be omitted), and
> the *Argument* by which q is shown to follow from p.

The very first proof in *Elements* involves the first construction in the In-class Activity on page 262. It is the proof that, by going through the algorithm, an equilateral triangle is constructed. On the next page, we state it as a theorem and give the complete proof, similar to what is found in *Elements,* but in modern style and language.

Theorem
If the algorithm at the top of page 262 is followed, an equilateral triangle is constructed.

Proof

Given: ⊙A with radius AB, ⊙B with radius BA,
⊙A ∩ ⊙B = {C, D}.
To prove: △ABC is equilateral.
Drawing:

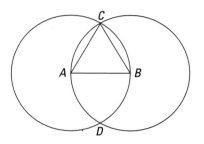

Argument:

Conclusions	Justifications
1. AC = AB	definition of circle (⊙A)
2. AB = BC	definition of circle (⊙B)
3. AC = BC	Transitive Property of Equality (Steps 1 and 2)
4. △ABC is equilateral.	definition of equilateral triangle

The argument of the proof contains conclusions and justifications. Notice that the argument for this proof has four steps. The first two steps are conclusions of congruent segments justified by the definition of circle. The third step follows from the first two, using the Transitive Property of Equality. So the first three steps show that all sides of the triangle are congruent. Then, in the fourth step, the definition of equilateral triangle justifies the conclusion that the triangle is equilateral.

Notice the parentheses in the justifications. These provide information helpful to the reader. In Steps 1 and 2, the parentheses identify the circle that is being used to make the conclusion. In Step 3, the parentheses identify the two conclusions that are used to apply the Transitive Property of Equality.

The Greek mathematicians wrote proofs in paragraphs. The argument above is written in **two-column form.** The two-column form was first used in geometry books about 100 years ago.

A proof of the Vertical Angles Theorem appears in this 1888 Geometry text.

Alternate Interior Angles

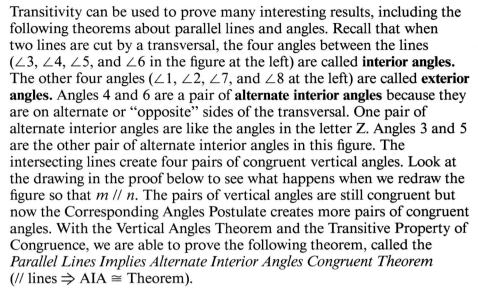

Transitivity can be used to prove many interesting results, including the following theorems about parallel lines and angles. Recall that when two lines are cut by a transversal, the four angles between the lines (∠3, ∠4, ∠5, and ∠6 in the figure at the left) are called **interior angles.** The other four angles (∠1, ∠2, ∠7, and ∠8 at the left) are called **exterior angles.** Angles 4 and 6 are a pair of **alternate interior angles** because they are on alternate or "opposite" sides of the transversal. One pair of alternate interior angles are like the angles in the letter Z. Angles 3 and 5 are the other pair of alternate interior angles in this figure. The intersecting lines create four pairs of congruent vertical angles. Look at the drawing in the proof below to see what happens when we redraw the figure so that *m // n.* The pairs of vertical angles are still congruent but now the Corresponding Angles Postulate creates more pairs of congruent angles. With the Vertical Angles Theorem and the Transitive Property of Congruence, we are able to prove the following theorem, called the *Parallel Lines Implies Alternate Interior Angles Congruent Theorem* (// lines ⇒ AIA ≅ Theorem).

> **// Lines ⇒ AIA ≅ Theorem**
> If two parallel lines are cut by a transversal, then alternate interior angles are congruent.

The proof of this theorem is given below. Examine it carefully. Some teachers like students to rewrite the given as part of the argument. When we put the *given* at the beginning of the argument, we call it step 0 (since it is not a new conclusion). Also, we have omitted some justifications. These are left for you as Question 7.

Proof

Given: *m // n* with angles as numbered in the diagram at the right.

To prove: ∠4 ≅ ∠6.

Argument:

Conclusions	Justifications
0. *m // n*	Given
1. ∠6 ≅ ∠2	?
2. ∠2 ≅ ∠4	?
3. ∠6 ≅ ∠4	?

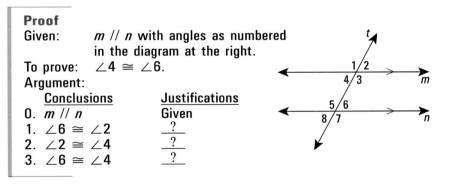

Example 1

In the figure at the left, *n // p*. If m∠3 = 43, find the measures of angles 1 and 2.

Solution

Angles 2 and 3 are alternate interior angles with *m* as a transversal. Thus, m∠2 = m∠3 = 43. Angles 1 and 2 form a linear pair, so m∠1 = 180 − 43 = 137.

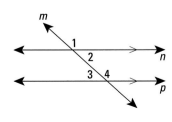

The converse of the // Lines $\Rightarrow$ AIA $\cong$ Theorem is also true. Not surprisingly, it is called the *Alternate Interior Angles Congruent Implies Parallel Lines Theorem* (AIA $\cong$ $\Rightarrow$ // Lines Theorem).

AIA $\cong$ $\Rightarrow$ // Lines Theorem
If two lines are cut by a transversal and form congruent alternate interior angles, then the lines are parallel.

Example 2

Prove the AIA $\cong$ $\Rightarrow$ // Lines Theorem, using the figure below as a drawing.

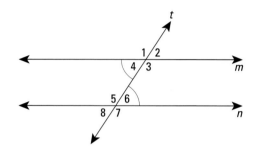

Solution

First write what is given and what is to be proved, using the figure.

Given: $\angle 4 \cong \angle 6$

To prove: m // n

You can prove lines parallel if corresponding angles are congruent. $\angle 6$ and $\angle 2$ are corresponding angles, while $\angle 4$ and $\angle 2$ are vertical angles. This is enough to show that the corresponding angles are congruent. We write this argument in a combination of paragraph form and two-column form.

Argument:

Since	$\angle 4 \cong \angle 6$	(Given)
and	$\angle 2 \cong \angle 4$	(Vertical Angles Theorem),
we have	$\angle 2 \cong \angle 6$.	(Transitive Property of Congruence)
Therefore, m // n		(Corr. $\angle$s $\cong$ $\Rightarrow$ // lines Postulate)

QUESTIONS

Covering the Reading

1. About how long ago did Thales live, and for what is he famous?

2. Trace $\overline{PQ}$. Construct an equilateral triangle with $\overline{PQ}$ as one side.

3. In $\odot O$ at the right, what is the justification for the conclusion $OA = OB$?

4. Name the four components of a proof.

5. What is the *argument* in a proof that $p \Rightarrow q$?

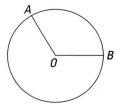

6. Name two ways in which proof arguments are written.

7. Supply the justifications for steps 1–3 in the proof of the // Lines ⇒ AIA ≅ Theorem.

In 8–10, use the figure at the right.

8. **a.** Name the interior angles.
   b. Name the pairs of alternate interior angles.

9. If ℓ // m and m∠7 = 21, find the measures of angles 1 through 6, and 8.

10. *True or false.* If m∠2 = m∠6, then ℓ // m.

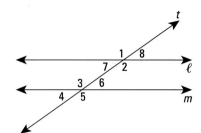

Applying the Mathematics

11. In the Navajo rug at the right, assume that segments that look parallel are parallel. Write an argument to convince a friend that m∠1 = m∠2.

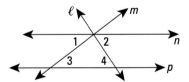

12. *True or false.* If the figure at the left (not drawn to scale) has angles as marked, then
    a. $\overline{AB}$ // $\overline{CD}$.   **b.** $\overline{AD}$ // $\overline{BC}$.

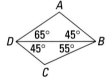

13. In the figure at the right, n // p. If m∠3 = 38 and m∠4 = 57, find the measures of angles 1 and 2.

14. In the figure at the left, ∠8 and ∠4 are a pair of **alternate exterior angles**. ∠1 and ∠5 are another pair of alternate exterior angles. Follow Example 2 to write a proof of the following theorem:

> **Theorem**
> If two lines are cut by a transversal and form congruent alternate exterior angles, then the lines are parallel.

15. Use the diagram below and complete the proof by writing the argument.
    Given: B is the midpoint of $\overline{AC}$.
    C is the midpoint of $\overline{BD}$.
    To prove: $\overline{AB} \cong \overline{CD}$.

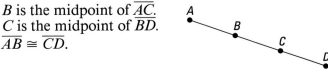

16. Given: $\triangle BIG \cong \triangle LOW$. Make and justify two conclusions. *(Lesson 5-3)*

17. The letter "R" at the left is to be reflected over line m. The same letter (not its image) is reflected over line ℓ. Are the two reflection images congruent? Why or why not? *(Lesson 4-8)*

18. State the Figure Reflection Theorem. *(Lesson 4-2)*

19. Consider the lines with equations $y = 2x + 1$ and $y = 2x - 1$.
a. Are these lines parallel?
b. Are these lines perpendicular? *(Lessons 3-6, 3-7)*

20. In the drawing at the left $\triangle LMN$, $\triangle LMO$, and $\triangle MOP$ are all equilateral. Which point could be the center of a circle containing all of the other named points? How do you know? *(Lessons 2-4, 5-1, 5-3)*

21. *True or false.* In Euclidean geometry, two points determine a line. *(Lesson 1-7)*

22. Is the network below traversable? Explain your answer. *(Lesson 1-4)*

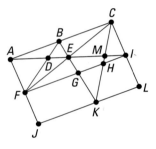

23. a. Other than Z, which printed capital letters usually contain parts of alternate interior angles?
b. Which printed capital letters usually contain parts of corresponding angles?

24. The saw below has teeth whose angles are equal. For example, $m\angle B = m\angle D$. In this saw, $AB = BC$, $\overline{AB} \parallel \overline{CD}$, and $m\angle C = m\angle B$. Sketch a picture of a saw for which none of these three relationships is true.

IN · CLASS
ACTIVITY

*Investigating
Perpendicular
Bisectors*

Work on this activity with a partner. You will need a ruler and a compass, or an automatic drawer.

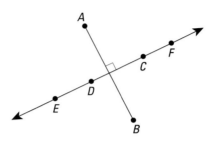

1 Draw a segment $\overline{AB}$ and construct its perpendicular bisector. (If available, you may use an automatic drawer for this Activity.) Label four points C, D, E, and F on the perpendicular bisector as shown above.

2 Measure these lengths on your drawing.

a. AC b. BC
c. AD d. BD
e. AE f. BE
g. AF h. BF

3 Remove C, D, and E from your drawing. Move F along the ⊥ bisector. Stop several times and draw $\overline{AF}$ and $\overline{BF}$.
a. What is the smallest length $\overline{AF}$ can have?
b. What is the largest length $\overline{AF}$ can have?
c. Are $\overline{AF}$ and $\overline{BF}$ always congruent?

Building up. *In order to construct sound buildings, like these in London, a proper foundation must be laid. Then, work on successive floors can begin.*

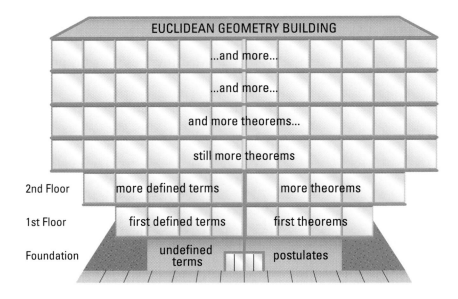

Euclidean geometry can be compared to a multi-story building. Its foundation is its undefined terms and postulates. In Chapter 2, you saw how a few definitions use undefined terms, and how other terms can be defined using previously defined terms. In this chapter, you are seeing how to deduce the first theorems from definitions and postulates, and how other theorems can be proved using definitions, postulates, and previously proved theorems. As you study this book, you are traveling through the lower floors to the higher floors of the building. Sometimes, you may see the scaffolding for a new floor. This is because the building is still growing as people discover and prove new theorems.

In this lesson, properties of reflections are used to prove some important theorems.

Using Preservation Properties of Reflections

When a point P is the same distance from A and B, P is said to be **equidistant** from A and B. In the In-class Activity on page 269, you found instances of the generalized conditional *If P is a point on the perpendicular bisector of a segment, then P is equidistant from the endpoints of the segment.* We call this generalization the *Perpendicular Bisector Theorem.*

As you know, instances indicate that a conditional might be true, but the only way to guarantee the truth of a statement with infinitely many instances is to prove it. The proof of the Perpendicular Bisector Theorem uses the definition of reflection and the fact that reflections preserve distance. You might have expected that reflections would be involved, since the definition of reflection mentions a perpendicular bisector.

Perpendicular Bisector Theorem
If a point is on the perpendicular bisector of a segment, then it is equidistant from the endpoints of the segment.

Proof

Given: m is the perpendicular bisector of $\overline{AB}$.
            P is on m.
To prove: $PA = PB$.
Drawing:

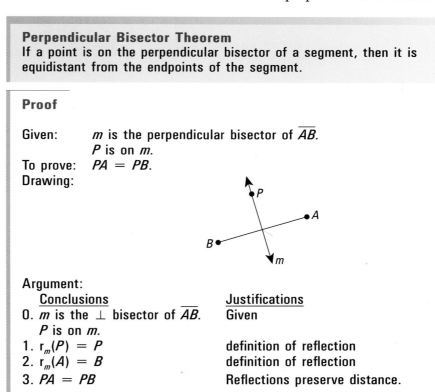

Argument:
Conclusions	Justifications
0. m is the $\perp$ bisector of $\overline{AB}$.	Given
P is on m.	
1. $r_m(P) = P$	definition of reflection
2. $r_m(A) = B$	definition of reflection
3. $PA = PB$	Reflections preserve distance.

Example 1

In the drawing at the left, $\overleftrightarrow{MB}$ is the $\perp$ bisector of $\overline{AC}$. If $AB = 10$, what is BC?

Solution

From the Perpendicular Bisector Theorem,
AB = BC. So BC = 10.

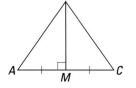

Locating the Center of a Circle

The Perpendicular Bisector Theorem has a surprising application. It helps prove that the second construction of the In-class Activity on page 262 results in the circle containing the three points. Example 2 exhibits a proof of this result with its argument written in paragraph form.

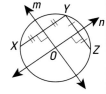

Example 2

Write an argument to complete the proof. Use the figure at the left.
Given: m is the perpendicular bisector of $\overline{XY}$.
             n is the perpendicular bisector of $\overline{YZ}$.
To prove: X, Y, and Z lie on $\odot O$ with radius OX.

Solution

We need to show that $OX = OY = OZ$ so that they can be radii of the same circle. The Perpendicular Bisector Theorem sets up the congruences.
Argument: Because of the Perpendicular Bisector Theorem with line m, OX = OY. With line n, this theorem also justifies the conclusion OY = OZ. By the Transitive Property of Equality, the three distances OX, OY, and OZ are all equal. Thus by the definition of circle, $\odot O$ with radius OX contains points X, Y, and Z.

Using the Figure Reflection Theorem

You have seen a variety of ways to prove segments congruent. You also have seen many ways to prove angles congruent. A two-step combination of the Figure Reflection Theorem and the definition of congruence can prove that figures of *any* type are congruent if there are images of enough points to determine the figures.

Example 3

Prove: If $r_\ell(A) = Y$, $r_\ell(B) = X$, and $r_\ell(C) = Z$, then $\triangle ABC \cong \triangle YXZ$.

Solution

Remember that in a proof of a conditional $p \Rightarrow q$, p is the "given" and q is the "to prove."
     Given: $r_\ell(A) = Y$, $r_\ell(B) = X$, $r_\ell(C) = Z$.
     To prove: $\triangle ABC \cong \triangle YXZ$.
It may be helpful to draw a figure such as the one at the left.
     Argument:

Drawing:

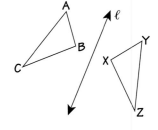

Conclusions	Justifications
0. $r_\ell(A) = Y$, $r_\ell(B) = X$, $r_\ell(C) = Z$	Given
1. $r_\ell(\triangle ABC) = \triangle XYZ$	Figure Reflection Theorem (from the Given)
2. $\triangle ABC \cong \triangle XYZ$	definition of congruence

The Figure Reflection Theorem is used in Example 4 to prove that two angles are congruent.

Example 4

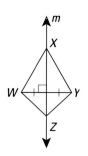

Write the argument to complete the proof.

Given: $\overleftrightarrow{XZ}$ is the perpendicular bisector of $\overline{WY}$.
To prove: $\angle WXZ \cong \angle YXZ$.

Solution

The angles will be congruent if it can be proved that they are reflection images of each other. Thus each point must be reflected in order. (Notice that we label $\overleftrightarrow{XZ}$ as m in step 0 to make it easier to write the reflections.)

Argument:

Conclusions	Justifications
0. $\overleftrightarrow{XZ}$ is the $\perp$ bisector of $\overline{WY}$. Let $m = \overleftrightarrow{XZ}$.	Given
1. $r_m(W) = Y$	definition of reflection
2. $r_m(X) = X$	definition of reflection
3. $r_m(Z) = Z$	definition of reflection
4. $r_m(\angle WXZ) = \angle YXZ$	Figure Reflection Theorem
5. $\angle WXZ \cong \angle YXZ$	definition of congruence

QUESTIONS

Covering the Reading

1. In this lesson, Euclidean geometry is compared to a building. What is found in its foundation?

2. Draw a figure in which point A is equidistant from points B and C.

3. Suppose t is the $\perp$ bisector of $\overline{MN}$ in the figure at the left. Justify each of these conclusions from the proof of the Perpendicular Bisector Theorem.
   a. P is the reflection image of P over t.
   b. $r_t(N) = M$
   c. $PN = PM$

4. Choose the correct word from each pair in parentheses. Any point on the (bisector, perpendicular bisector) of a segment is equidistant from the (endpoints, midpoint) of the segment.

5. Refer to the In-class Activity on page 269. If in Step 3 you had moved E along the $\perp$ bisector, would $\overline{AE}$ and $\overline{BE}$ always be congruent?

6. Write the argument to complete the proof.
   Given: In the figure at the right,
   $r_\ell(P) = S$, $r_\ell(Q) = T$,
   $r_\ell(R) = U$.
   To prove: $\triangle PQR \cong \triangle STU$.

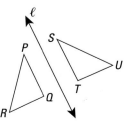

7. Given: $\overleftrightarrow{WY}$ is the perpendicular bisector of $\overline{XZ}$ at the right.

To prove: $\angle WXY \cong \angle WZY$.

a. Trace the figure and mark the given information.

b. Write the proof argument.

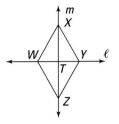

Applying the Mathematics

8. In the figure at the right, $\overleftrightarrow{BD}$ is the $\perp$ bisector of $\overline{AC}$.

a. If $CD = 12x$, what is AD?

b. If $AC = 6$, what is AB?

9. Rewrite the proof argument of Example 2 in two-column form.

10. A tree stands midway between two stakes. Guy wires are attached from the stakes to a narrow collar around the trunk. Explain why the guy wires must have the same length.

In 11 and 12, write the proof argument using the given figure.

11. Given: ℓ is the perpendicular bisector of $\overline{CD}$.

$r_\ell(E) = F$.

To prove: Quadrilateral $ABCE \cong$ Quadrilateral $ABDF$.

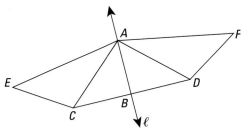

12. Given: $r_m(X) = Y$;

$r_m(Z) = W$.

To prove: $XZ = YW$.

13. Fill in the justifications in this proof argument. *(Lesson 5-4)*

Given: C is the midpoint of $\overline{BD}$.
A and C are on $\odot D$.

To prove: $AD = BC$.

Argument:

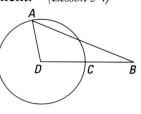

Conclusions	Justifications
1. $AD = DC$	**a.**
2. $DC = BC$	**b.**
3. $AD = BC$	**c.**

14. Suppose you were asked to prove the Linear Pair Theorem: *If two angles form a linear pair, then they are supplementary.*
a. Draw a figure.
b. What would be the "given"?
c. What would be the "to prove"?   *(Lessons 3-3, 5-4)*

In 15 and 16, $\triangle QZP \cong \triangle KRA$.

15. List six pairs of corresponding parts.

16. If m$\angle ZQP = 107$, what other angle measures 107?   *(Lessons 5-1, 5-2)*

17. The composite of two reflections over parallel lines is a(n) __?__ .
(Lesson 4-4)

18. The measure of an angle is greater than the measure of a complement to it. What can be deduced about the original angle?
(Lesson 3-3)

19. Two sides of a triangle are 91 cm and 38 cm. Give the possible lengths of the third side.   *(Lesson 2-7)*

20. Trace $\triangle VWX$ below.
a. Construct a circle through the three vertices of $\triangle VWX$.
b. Is the center of the circle inside or outside the triangle?
c. What determines whether the center will be inside or outside? (If you have an automatic drawer, experiment with triangles of different shapes.)
d. Is this circle unique? That is, can there ever be more than one circle through the three vertices of a triangle? Explain your answer.

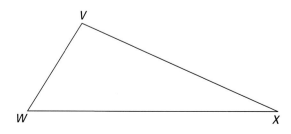

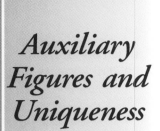

LESSON

5-6

Auxiliary Figures and Uniqueness

Some Examples of Uniqueness

The adjective *unique* means "exactly one." When exactly one thing satisfies some given conditions, we say the thing is **uniquely determined.** For instance, the clues and diagram of a crossword puzzle uniquely determine the answers to the puzzle. In algebra, the given condition $3x + 5 = 14$ uniquely determines x.

Example 1

Given a segment $\overline{AB}$, which of these things are uniquely determined?
a. midpoint of $\overline{AB}$
b. bisector of $\overline{AB}$
c. perpendicular bisector of $\overline{AB}$

Solution

a. Does a segment have exactly one midpoint? Yes
b. Does a segment have exactly one bisector? No. There can be many lines through the midpoint, and each is a bisector.
c. Does a segment have exactly one perpendicular bisector? Yes

We usually take uniqueness for granted. Yet the construction of *the* circle through three noncollinear points can be done only because (1) the perpendicular bisector of a segment is unique, (2) the two perpendicular bisectors intersect at a unique point, and (3) there is a unique circle with a particular center and radius.

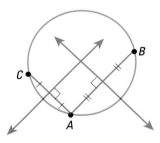

What Are Auxiliary Figures?

To obtain the circle through three noncollinear points we draw two perpendicular bisectors to locate the center of the circle. We call a segment, line, or other figure that is added to a diagram an **auxiliary figure.** The word *auxiliary* means "assisting" or "giving help."

When an auxiliary figure is not uniquely determined, then there are two possibilities. (1) There may be more than one figure satisfying the conditions. This is the case with bisectors of segments. (2) The other possibility is that there may be no figure satisfying the given conditions.

Example 2

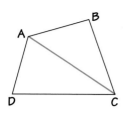

In quadrilateral *ABCD*, a student wished to draw as an auxiliary segment the diagonal $\overline{AC}$ that bisects ∠*A*. Is this possible? Why or why not?

Solution

It is not possible to do this in every quadrilateral. Draw a quadrilateral ABCD. Diagonal $\overline{AC}$ is uniquely determined because points A and C determine a line. However, $\overrightarrow{AC}$ does not have to be the angle bisector, as the figure shows.

Activity

Draw another quadrilateral *ABCD* in which the diagonal $\overline{AC}$ *does not* lie on the bisector of ∠*A*.

How Many Parallels Are There?

As Example 2 demonstrates, uniqueness is not always obvious. How many lines are there which are parallel to a given line ℓ, containing a point not on the given line? The answer to this question is given in the next theorem. In its proof, two auxiliary lines are drawn. We do this to create alternate interior angles that are congruent.

Uniqueness of Parallels Theorem (Playfair's Parallel Postulate)
Through a point not on a line, there is exactly one line parallel to the given line.

Proof
Given: Point *P* not on line ℓ.
              Points *R* and *Q* on line ℓ.
To prove: There is exactly one line
              parallel to ℓ through *P*.

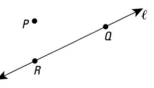

Argument:

Draw $\overleftrightarrow{PQ}$. By the Point-Line-Plane Postulate, this line is uniquely determined. We label ∠*PQR* as ∠1. (See the figure below.)

Now draw line $\overleftrightarrow{PA}$ so that *A* is on the other side of $\overleftrightarrow{PQ}$ from *R* and m∠*QPA* = m∠1. $\overleftrightarrow{PA}$ is unique because of the Unique Angle Assumption in the Angle Measure Postulate. Then, because AIA ≅ ⟹ // lines, $\overleftrightarrow{PA}$ // ℓ. So there is at least one line parallel to ℓ through *P*.

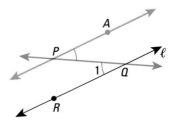

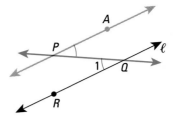

Can there be another parallel? Because // lines ⇒ AIA ≅, every line // to ℓ through *P* must contain $\overrightarrow{PA}$. So there cannot be more than one parallel. Thus, $\overleftrightarrow{PA}$ is unique, and there is exactly one line parallel to ℓ through *P*.

Playfair's Parallel Postulate

The Scottish mathematician and geologist John Playfair published his Elements of Geometry *in 1795.*

The Uniqueness of Parallels Theorem is important in the history of mathematics. In fact, it ultimately changed the entire nature of mathematics. In Euclid's *Elements,* the fifth and final geometric postulate states: *If two lines are cut by a transversal, and the interior angles on the same side of the transversal have a total measure of less than 180°, then the lines will intersect on that side of the transversal.* This postulate bothered mathematicians, who felt that such a complicated statement should not be assumed true. For 2000 years they tried to prove the fifth postulate from Euclid's other postulates.

After years of being unable to prove Euclid's fifth postulate, some mathematicians substituted simpler statements for it. The uniqueness of parallels on the preceding page was first suggested by the Greek mathematician Proclus about 450 A.D., but it is known as *Playfair's Parallel Postulate* because it was used by the Scottish mathematician John Playfair in 1795. We were able to prove it as a theorem in this lesson because we assumed the Corresponding Angles Postulate (// lines ⇔ corr. ∠s ≅) in Lesson 3-6. Using that postulate, we proved the // Lines ⇒ AIA ≅ Theorem and its converse.

Other mathematicians substituted different statements for Playfair's Parallel Postulate. When they assumed *there are no parallels to a line through a point not on it,* they were able to develop a spherical geometry that could apply to the surface of the Earth. When they assumed *there is more than one parallel to a line through a point not on it,* they developed types of geometries for other surfaces. All of these geometries are called **non-Euclidean geometries.**

These mathematicians greatly influenced *all* later mathematics with their work. For the first time, postulates were viewed as statements *assumed* true instead of statements definitely true. With this point of view, mathematicians experimented with a variety of algebras and geometries formed by modifying or changing postulates. Their experiments were at first thought to be merely a game, but now are considered quite important. Non-Euclidean geometries are important in physics in Einstein's theory of relativity. A useful algebra, with some postulates different from those you have studied, is applied in logic and in the operation of computers.

QUESTIONS

Covering the Reading

In 1–4, tell whether the figure is or is not uniquely determined.

1. midpoint of a given segment

2. bisector of a given segment

3. line parallel to a given line through a point not on it

4. diagonal bisecting a given angle of a given quadrilateral

5. Draw the figure you created in the Activity of this lesson.

6. State Playfair's Parallel Postulate.

7. Refer to the proof of the Uniqueness of Parallels Theorem. Give justifications for the following conclusions.
   a. There is a unique line through points P and Q.
   b. There is a unique line containing ray $\overrightarrow{PA}$ so that $\angle QPA$ and $\angle 1$ are alternate interior angles and m$\angle QPA$ = m$\angle 1$.

8. Which of Euclid's postulates most troubled mathematicians, and why?

9. Can Playfair's Parallel Postulate be proved from Euclid's other postulates?

10. Geometries in which Playfair's Parallel Postulate is *not* true are called __?__.

11. Since the discovery of non-Euclidean geometries, postulates have been viewed as __?__ rather than as statements which are definitely true.

Applying the Mathematics

In 12–14, use a ruler, compass, and protractor, or use an automatic drawer. Tell whether the auxiliary figure is uniquely determined. If so, trace $\triangle ABC$ and draw the auxiliary figure; if not, explain why not.

12. line perpendicular to $\overleftrightarrow{BC}$ through A

13. point E between A and B

14. point D on $\overrightarrow{AC}$ so that $AD = BC$

15. In quadrilateral $ABCD$, a student wished to draw as a single auxiliary line, the $\perp$ bisector of $\overline{AB}$ and $\overline{CD}$. Is this possible? Why or why not?

16. Draw a figure describing Euclid's fifth postulate. State the antecedent and consequent of that postulate in terms of your figure.

17. *Multiple choice.* Which of these lines contains the point (10, 13) and is parallel to the line with equation $3x + 2y = 5$?
(a) $3x + 2y = -19$ (b) $2x - 3y = -19$
(c) $3x + 2y = 56$ (d) $2x - 3y = 56$

18. a. In the figure at the right, $\overleftrightarrow{YZ}$ is parallel to $\overleftrightarrow{WX}$ and contains B. Find $m\angle ABC$.
b. Is $\overleftrightarrow{YZ}$ unique? Why or why not?

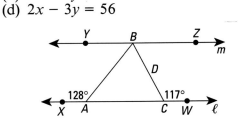

Review

In 19 and 20, write the proof argument.

19. Given: In the figure at the left, m is the $\perp$ bisector of $\overline{AB}$ and $\overline{CD}$.
To prove: $\triangle ACD \cong \triangle BDC$. *(Lesson 5-5)*

20. Given: $\odot O$ and $\odot P$ at the right.
          $OA = PC$.
To prove: $OB = PD$.   *(Lesson 5-4)*

21. In the figure at the right, $\ell \,/\!/\, m$. If $m\angle 5 = \frac{4}{5} m\angle 4$, find the measures of angles 1 to 8.   *(Lesson 5-4)*

22. Find $r_{y\text{-axis}}(-q, p)$.   *(Lesson 4-1)*

23. Use the figure at the right. All acute angles in the figure have measure 60°. Suppose a rotation R has center O.
a. What is its magnitude if R(B) = C?
b. What is its magnitude if R(A) = E?
(Lesson 3-2)

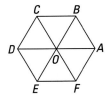

24. a. Graph the line with equation $12x - y = 6$.
b. What is its slope?   *(Lesson 1-2)*

Exploration

25. Draw a scalene triangle ABC. Let n be the line parallel to $\overleftrightarrow{AB}$ through C. Let ℓ be the line parallel to $\overleftrightarrow{BC}$ through A. Let m be the line parallel to $\overleftrightarrow{AC}$ through B. Make and try to prove conjectures about the figure formed when all these lines are drawn.

26. What does the word *auxiliary* mean outside mathematics?

27. a. Name five things outside mathematics which are unique.
b. Name five things outside mathematics which are not unique.
c. How is the word *unique* misused in casual speech? Give an example.

What's in a name? *Who is Paul Simon? Shown are Senator Paul Simon and the singer Paul Simon.*

LESSON

5-7

Sums of Angle Measures in Polygons

Triangular beauty. *The Bavarian Alps, located in Germany, are part of the largest mountain system in Europe.*

Activity

Refer to the figure below. With a protractor, determine each measure as accurately as you can.
a. m∠A
b. m∠B
c. m∠C
d. m∠A + m∠B + m∠C

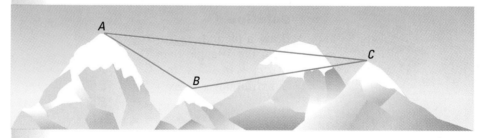

In the early 1800s, the great mathematician Karl Friedrich Gauss wondered whether the theorems of Euclidean geometry were true over long distances. And so he measured the angles between three mountaintops in Germany to see if they added to 180°. Gauss found that the sum of the angles in his measurements was very close to 180°, within the limits of the accuracy of his instruments. He was checking the truth of the next theorem, one that was known to mathematicians in ancient Greece. It probably also has been known to you for some time.

Triangle-Sum Theorem
The sum of the measures of the angles of a triangle is 180°.

To prove this theorem, we think of it as a conditional: *If a figure is a triangle, then the sum of the measures of its angles is 180°.* We then draw a triangle and restate the antecedent and consequent in terms of that triangle. The proof argument uses the auxiliary line through B parallel to $\overleftrightarrow{AC}$.

Proof
Given: $\triangle ABC$.
To prove: $m\angle A + m\angle B + m\angle C = 180$.

Argument:
Draw $\overleftrightarrow{BD}$ with $\overleftrightarrow{BD} \parallel \overleftrightarrow{AC}$. This line is uniquely determined, due to the Uniqueness of Parallels Theorem. By angle addition,

$$m\angle 1 + m\angle 2 + m\angle 3 = 180.$$

Since // lines $\Rightarrow$ AIA $\cong$, $m\angle 1 = m\angle A$ and $m\angle 3 = m\angle C$. So by substitution,

$$m\angle A + m\angle B + m\angle C = 180.$$

Drawing:

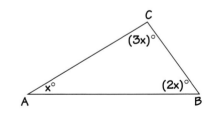

Example
In $\triangle ABC$, the angles are in the *extended ratio* 1:2:3. This means for some real number x, the angles have measures $1x$, $2x$, and $3x$. Find their measures.

Solution
Draw a picture.
From the Triangle-Sum Theorem,
$m\angle A + m\angle B + m\angle C = 180.$
Substituting, $x + 2x + 3x = 180.$
$$6x = 180$$
$$x = 30$$

So $m\angle A = \ \ x = 30,$
$\ \ \ m\angle B = 2x = 60,$ and
$\ \ \ m\angle C = 3x = 90.$

Check
The numbers 30, 60, and 90 are in the extended ratio 1:2:3.

Non-Euclidean Geometry
You might wonder why Gauss climbed the mountaintops in the first place. He was trying to imagine a geometry where the Triangle-Sum

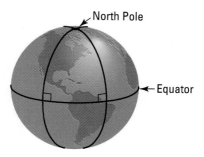

North Pole

Equator

Theorem is not true. It turned out that his suspicions were justified. In many non-Euclidean geometries, the sum of the measures of the angles of a triangle is not 180°.

The geometry of the surface of Earth is not Euclidean. The surface of Earth can be approximated as a sphere. In a plane, two perpendiculars to the same line cannot intersect to form a triangle, but this can happen on a sphere. A triangle formed by two longitudes (north-south lines connecting the North Pole and the South Pole) and the equator has two right angles! Since there is a third angle at the North Pole, the measures add to more than 180°. Thus neither the Two Perpendiculars Theorem nor the Triangle-Sum Theorem holds for triangles drawn on the surface of Earth.

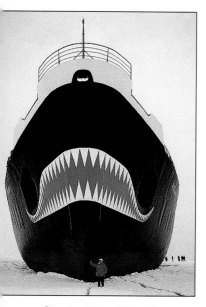

Crushed ice. *Yamal, a Russian-built nuclear powered icebreaker, is the twelfth surface ship in history to reach the North Pole. It has blazed trails—usually in total darkness—through ice as thick as 16 feet.*

Sums of Angles in Quadrilaterals

Now let us return to Euclidean geometry, where the sum of the measures of the angles of a triangle is 180°.

The Triangle-Sum Theorem allows you to calculate the sum of the measures of the angles of any convex polygon. Quadrilaterals are the obvious place to start.

Let S = the sum of the measures of the angles of *QUAD* at the right.

$$S = m\angle U + m\angle A + m\angle D + m\angle Q$$

Draw the auxiliary segment $\overline{AQ}$. (Do you know why it is uniquely determined?) This splits each of $\angle A$ and $\angle Q$ into two smaller angles. Now by the Angle Addition Postulate and substitution,

$$S = m\angle U + (m\angle 1 + m\angle 2) + m\angle D + (m\angle 3 + m\angle 4).$$

Rearrange the terms in this sum to put those in the same triangles together and apply the Triangle-Sum Theorem.

$$S = (m\angle U + m\angle 1 + m\angle 3) + (m\angle 2 + m\angle 4 + m\angle D)$$
$$= \qquad\quad 180 \qquad\quad + \qquad\quad 180$$
$$= \qquad\qquad\qquad 360$$

This argument proves the following theorem.

> **Quadrilateral-Sum Theorem**
> The sum of the measures of the angles of a convex quadrilateral is 360°.

Sums of Angles in Convex Polygons

The sum of the measures of the angles of a convex *n*-gon can be determined in a similar manner. Consider the convex polygons displayed below.

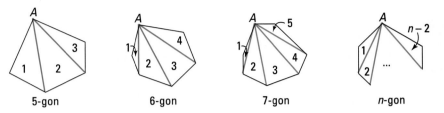

| 5-gon | 6-gon | 7-gon | *n*-gon |

Choose a vertex on each polygon. Call it *A*. Draw all the diagonals from *A*. This *triangulates* the polygon. For the 5-gon, there are two diagonals containing *A* and they form 3 triangles. The sum of the measures is $3 \cdot 180°$.

For the 6-gon, there are 4 triangles; for the 7-gon, there are 5 triangles; and for the *n*-gon, there are $n - 2$ triangles. The sums of the measures are thus

6-gon	$4 \cdot 180°$
7-gon	$5 \cdot 180°$
⋮	⋮
n-gon	$(n - 2) \cdot 180°$.

This argument proves the Polygon-Sum Theorem.

Polygon-Sum Theorem
The sum of the measures of the angles of a convex *n*-gon is $(n - 2) \cdot 180°$.

QUESTIONS

Covering the Reading

1. Give your answers to the Activity on page 281.

2. Here is a slightly different drawing for a proof of the Triangle-Sum Theorem. In the figure below,
   $\ell \mathbin{/\!/} AB$. Justify each statement.
   a. $m\angle A = m\angle 1$
   b. $m\angle B = m\angle 3$
   c. If $m\angle 1 + m\angle 2 + m\angle 3 = 180$,
   then $m\angle A + m\angle 2 + m\angle B = 180$.

3. Refer to $\triangle EFG$ below. If $m\angle F = 115$ and $m\angle G = 40$, find $m\angle E$.

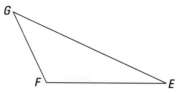

4. If the angles of a triangle are in the extended ratio 1:3:5, find their measures.

5. Suppose two north-south streets on the surface of Earth were extended.
   a. Would they ever intersect?
   b. If so, where? If not, why not?

6. a. Give an example of a triangle for which the measures of its angles do not add to 180°.
   b. What are geometries called in which this happens?

7. In the proof of the Quadrilateral-Sum Theorem, why is $\overline{AQ}$ uniquely determined?

8. In convex quadrilateral $HIJK$, $m\angle H + m\angle I + m\angle J + m\angle K = \underline{\ ?\ }$.

9. State the Polygon-Sum Theorem.

In 10–13, give the sum of the measures of the angles of the given figure.

10. convex quadrilateral

11. equilateral triangle

12. 10-gon

13. 20-gon

14. Refer to the triangulated polygons pictured on page 284.
    a. *True or false.* Each side of the polygon which does not contain vertex A is a side of exactly one triangle.
    b. How many sides of the n-gon do not contain vertex A?
    c. Into how many triangles does an n-gon triangulate?

Applying the Mathematics

Mathematical career.
Surveyors, like this one, use measuring tools like the theodolite. A theodolite measures angles and determines direction. Surveyors must know algebra, geometry, trigonometry, and calculus.

15. Surveyors often measure the angles of lots in degrees and minutes. There are 60 minutes (') in a degree. What is the value of x in degrees and minutes?

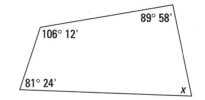

16. Given: Rectangle *ABCD* at the right.
Fill in each blank with a number.
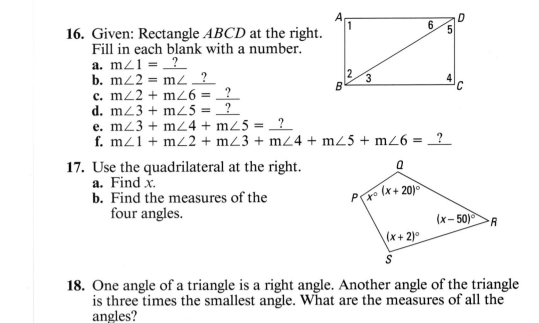
a. m∠1 = __?__
b. m∠2 = m∠ __?__
c. m∠2 + m∠6 = __?__
d. m∠3 + m∠5 = __?__
e. m∠3 + m∠4 + m∠5 = __?__
f. m∠1 + m∠2 + m∠3 + m∠4 + m∠5 + m∠6 = __?__

17. Use the quadrilateral at the right.
a. Find *x*.
b. Find the measures of the
four angles.

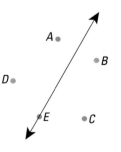

18. One angle of a triangle is a right angle. Another angle of the triangle
is three times the smallest angle. What are the measures of all the
angles?

Review

In 19–21, *A* and *B* are given points.
a. Draw the indicated figure.
b. Tell whether the figure is uniquely determined. *(Lesson 5-6)*

19. point *C* such that *AC* = *CB*

20. in a given plane, a perpendicular bisector of $\overline{AB}$

21. ray $\overrightarrow{BC}$ such that m∠*ABC* = 50

22. Use the drawing below.
Prove: If r(*A*) = *B* and r(*C*) = *D*, then ∠*AED* ≅ ∠*BEC*. Write a
complete proof (given, to prove, and argument). *(Lesson 5-5)*

23. Write a proof argument using the given figure.
Given: Lines ℓ, *m*, and *n* intersect at *O*.
$\overrightarrow{OA}$ bisects ∠*BOF*.
$\overrightarrow{OB}$ bisects ∠*AOC*.
To prove: m∠*AOF* = m∠*EOF*.
(Lesson 5-4)

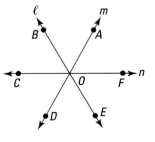

24. a. Name four properties that are preserved under a reflection.
b. Name one property that is not preserved under a reflection.
(Lesson 4-2)

25. *P*, *Q*, and *R* below are ships. From ship *Q*, ship *P* is 30°N of E and ship *R* is 10°S of E. From ship *R*, ship *P* is 15°W of N. From *P*, what are the positions of the other two ships? *(Lessons 3-1, 3-3)*

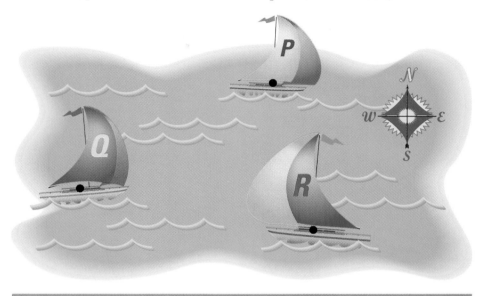

Exploration

26. Find a globe. Estimate the sum of the measures of the angles in the smallest triangle determined by Phoenix, London, and Rio de Janeiro.

27. Gauss is considered one of the greatest mathematicians of all times. Write about some of his accomplishments.

Cool globe. *This ice sculpture of a globe was created and displayed in Duluth, MN, as part of the annual winter carnival.*

A project presents an opportunity for you to extend your knowledge of a topic related to the material of this chapter. You should allow more time for a project than you do for typical homework questions.

1 Templates

Interview a carpenter to find out how templates are used to ensure congruence and uniformity in work. Write an illustrated report which includes examples of three actual templates and how they are used.

2 Who Was Euclid?

Write a report on Euclid including information on his life, the scope of his work, and his importance to mathematics.

3 Lawyers' Arguments

Interview a lawyer and find out how arguments are constructed to win a case on a point of law. What is the given information? What is to be proved? What kinds of justifications are used for conclusions? Give at least one example of an argument in detail.

4 Quilt Squares

Traditional American quilt patterns are made up of congruent figures. Find five quilt patterns and identify their congruent figures. Explain how the large quilt is formed from these smaller pieces, and how changing colors creates new designs from the same pieces. Finally, design your own quilt pattern.

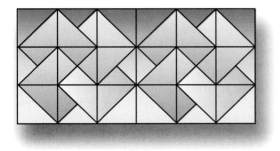

5 Non-Euclidean Geometries

Find out about the non-Euclidean geometries developed by Nicolai Lobachevsky and Georg Riemann. How does each geometry differ from Euclidean geometry?

6 Triangles on Curved Surfaces

You will need pieces of thin, bendable wire. Use (1) a large ball or other sphere, (2) a cylinder, (3) a large mixing bowl, and (4) another curved surface of your own choosing. Bend a thin wire "triangle" onto the surface so that the sides of the triangle hug the surface. Next, measure each angle of the triangle as close as you can to the vertex, as shown below. Record the sum of the measures of the angles of each triangle.

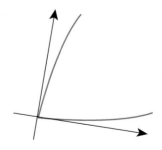

Do this with three triangles on each surface. Make some conjectures about the smallest and largest sum for each surface.

SUMMARY

Congruence has some properties which are like those of equality: the reflexive, symmetric, and transitive properties. There are three other basic properties of congruence: the CPCF Theorem, the Segment Congruence Theorem and the Angle Congruence Theorem.

A proof of a conditional statement $p \Rightarrow q$ usually has four parts: the "given" p, the "to prove" q, a drawing (sometimes), and an argument to show how q follows from p. Though mathematicians almost always write arguments in paragraphs, in elementary geometry, arguments are commonly written either in two columns or in paragraphs.

Most of the proofs in this chapter involve congruence. Definitions that involve segments of equal length or angles of equal measure, the congruence theorems, the Transitive Property of Congruence, and properties of reflections are common justifications.

The properties of reflections help to prove that any point on the perpendicular bisector of a segment is equidistant from the endpoints of the segment. This explains why the construction of a circle through three noncollinear points works.

From the Corresponding Angles Postulate we proved: Two lines cut by a transversal are parallel if and only if a pair of alternate interior angles are congruent. From this we also can deduce that there is exactly one line parallel to a given line through a point not on the line. This Uniqueness of Parallels Theorem helps deduce the Triangle-Sum Theorem, which is used to prove the Quadrilateral Sum Theorem and to develop the formula $S = 180(n - 2)$ for the sum, S, of the measures of the interior angles of any convex n-gon.

This chapter shows the power of proof to deduce many results from a few postulates and definitions. The geometry of the surface of Earth (or any other sphere-like surface) is non-Euclidean, because the Uniqueness of Parallels Theorem and the Triangle-Sum Theorem are not true on the surface of Earth.

VOCABULARY

Below are the new terms and phrases for this chapter. You should be able to give a general description and specific example of each.

Lesson 5-1
corresponding parts
Corresponding Parts of Congruent Figures (CPCF) Theorem
A-B-C-D Theorem

Lesson 5-2
Equivalence Properties of ≅ Theorem
Reflexive Property of ≅
Symmetric Property of ≅
Transitive Property of ≅
Segment Congruence Theorem
Angle Congruence Theorem

Lesson 5-4
Given, To prove
paragraph form
two-column form
interior angles
exterior angles
alternate interior angles
// Lines ⇒ AIA ≅ Theorem
AIA ≅ ⇒ // Lines Theorem
alternate exterior angles

Lesson 5-5
equidistant
Perpendicular Bisector Theorem

Lesson 5-6
uniquely determined
auxiliary figure
Uniqueness of Parallels Theorem
Playfair's Parallel Postulate
non-Euclidean geometries

Lesson 5-7
extended ratio
Triangle-Sum Theorem
Quadrilateral-Sum Theorem
Polygon-Sum Theorem

PROGRESS SELF-TEST

You will need a ruler and a compass for this test, and you may want to use a calculator.

In 1 and 2, $BLAST \cong CROST$.

1. If m$\angle LBS = 47$, what other angle measure(s) can be found?

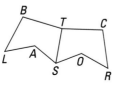

2. What segment has ___ the same length as $\overline{AL}$?

3. Trace $\overline{PQ}$. Construct an equilateral triangle with side length equal to PQ.

4. Use the figure below, in which $\ell \parallel m$. If m$\angle 5 = 9z - 52$ and m$\angle 3 = 2z + 45$,
   a. find z.   b. find m$\angle 3$.

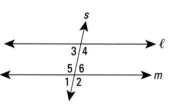

5. Justify this conclusion.
   Given: $AL = BK$.
   Conclusion: $AL \cong BK$.

6. Give the justification for each conclusion.
   Given: t is the $\perp$ bisector of $\overline{PQ}$;
                $r_t(S) = M$.
   To prove: $\angle PSL \cong \angle QML$.
   Drawing:

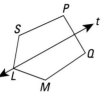

   Argument:
Conclusions	Justifications
0. $r_t(S) = M$	a.
1. $r_t(P) = Q$	b.
2. $r_t(L) = L$	c.
3. $r_t(\angle PSL) = \angle QML$	d.
4. $\angle PSL \cong \angle QML$	e.

7. Write the argument for this proof.
   Given: M is the midpoint of $\overline{AB}$.
                $\overline{AM} \cong \overline{BC}$.
   To prove: $\overline{MB} \cong \overline{BC}$

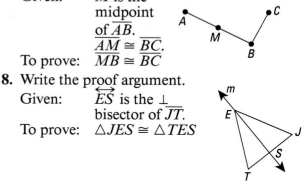

8. Write the proof argument.
   Given: $\overleftrightarrow{ES}$ is the $\perp$ bisector of $\overline{JT}$.
   To prove: $\triangle JES \cong \triangle TES$

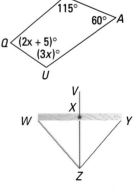

9. The angles of a triangle are in the extended ratio 2:3:4. Find the measure of each angle.

10. In $QUAD$ at the right, find x and m$\angle Q$.

11. What is the sum of the measures of the angles of a convex 30-gon?

12. To create a marionette, a wooden stick $\overline{WY}$ is pierced by $\overline{VZ}$ such that $\overline{VZ}$ is the $\perp$ bisector of $\overline{WY}$. Why are the strings $\overline{WZ}$ and $\overline{YZ}$ congruent?

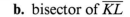

13. In a $\triangle JKL$ tell whether the figure is uniquely determined. Illustrate your argument with a diagram.
    a. bisector of $\angle J$ b. bisector of $\overline{KL}$

14. Trace $\triangle MNO$ at the right. Construct the circle which passes through the vertices of the triangle.

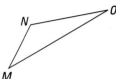

15. Until the discovery of non-Euclidean geometries, postulates were thought to be __?__ true. Now they are considered as __?__ true.

CHAPTER REVIEW

Questions on SPUR Objectives

SPUR stands for **S**kills, **P**roperties, **U**ses, and **R**epresentations. The Chapter Review questions are grouped according to the SPUR Objectives for this chapter.

SKILLS DEAL WITH THE PROCEDURES USED TO GET ANSWERS.

Objective A: *Identify and determine measures of parts of congruent figures.* *(Lessons 5-1, 5-2)*

1. Suppose $\triangle HAT \cong \triangle TOP$.
   a. Sketch a possible situation.
   b. If m$\angle THA = 102$, then __?__ = 102.
   c. If $PT = 6.9''$, then __?__ = 6.9''.

In 2–4, use the figure below, in which $ABCD \cong DEFA$.

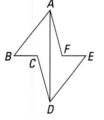

2. Name three pairs of congruent sides.
3. If m$\angle B = 42$, then __?__ = 42.
4. If m$\angle BAD = 50$ and m$\angle CDA = 19$, find as many other angle measures as you can.

In 5 and 6, use the figure below with $\overline{PL} \cong \overline{RT}$. M and N are midpoints of $\overline{PL}$ and $\overline{RT}$, respectively.

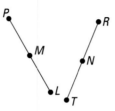

5. If $RT = 16$ cm, find the lengths of as many other segments as you can.
6. If $PM = 12x$, find the lengths of as many other segments as possible (in terms of x).

In 7 and 8, refer to the figure below with $\angle 5 \cong \angle 6$ and $\angle 7 \cong \angle 6$.

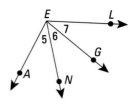

7. If m$\angle 5 = 38$, then
   a. m$\angle 6 = $ __?__
   b. m$\angle LEN = $ __?__
   c. m$\angle AEL = $ __?__
8. If m$\angle AEG = 30p$, then
   a. what other angle has measure $30p$?
   b. m$\angle AEN = $ __?__
   c. m$\angle AEL = $ __?__

Objective B: *Construct equilateral triangles and construct the circle through three noncollinear points.* *(Lesson 5-4)*

9. Trace $\overline{TA}$ and construct an equilateral triangle with side length *TA*.

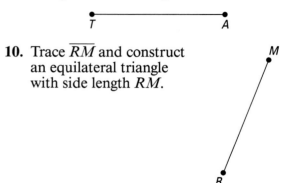

10. Trace $\overline{RM}$ and construct an equilateral triangle with side length *RM*.

11. Trace points X, Y, and Z, and construct the circle which passes through them.

Y
•

X
•

Z
•

12. Trace $\triangle ABC$ and construct the circle which contains its vertices.

A

B ——————— C

Objective C: *Find lengths and angle measures using properties of perpendicular bisector and alternate interior angles.* *(Lessons 5-4, 5-5)*

In 13 and 14, $\overline{PQ}$ is the perpendicular bisector of $\overline{RS}$.

13. If $QR = 12$, find QS.
14. If $PR = 4x$, $RS = 3x$, and $PS = 30$, find x.

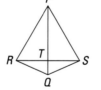

15. In the figure below, $\ell \,/\!/\, m$.
    a. If $m\angle 1 + m\angle 5 = 140$, find $m\angle 2$.
    b. If $m\angle 7 = 117$ and $m\angle 3 = 60$, find $m\angle 2$.
    c. If $m\angle 5 = m\angle 6$, what other pairs of angles have equal measures?

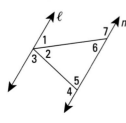

16. $s \,/\!/\, t$ as pictured below.
    a. If $m\angle 3 = 2y - 11$ and $m\angle 6 = -y + 46$, find $m\angle 3$.
    b. If $m\angle 1 = m\angle 7 + 54$, find $m\angle 1$.

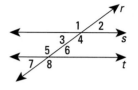

Objective D: *Use the Triangle-, Quadrilateral-, and Polygon-Sum Theorems to determine angle measures.* *(Lesson 5-7)*

17. The measures of the angles of a triangle are described by the extended ratio $5{:}8{:}12$. Find the measure of the largest angle.
18. Find the sum of the measures of the angles of a convex 18-gon.
19. Use $\triangle DEF$ at the right.
    a. Find x.
    b. Find the measures of the three angles of the triangle.
20. Two angles of a triangle have measures 43 and 91. Find the measure of the third angle.
21. Find the measures of all four angles of quadrilateral $RSTU$ below.

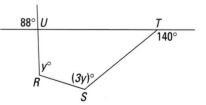

22. Use quadrilateral $EACH$ below. Find x and the measure of each angle.

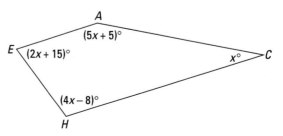

PROPERTIES DEAL WITH THE PRINCIPLES BEHIND THE MATHEMATICS.

Objective E: *Make and justify conclusions about congruent figures.* *(Lessons 5-1, 5-2, 5-3)*

23. If $\triangle PAT \cong \triangle LOG$, list all congruent pairs of segments and angles.

24. Suppose quadrilateral $BOWS \cong$ quadrilateral $HAIR$. What part has measure equal to the given measure?

  a. m$\angle SWO$ **b.** RH **c.** m$\angle H$

In 25–28, use the figure below.

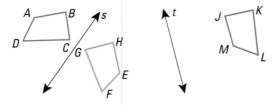

Given: $r_s(ABCD) = EHGF$;
    $r_t(EFGH) = MLKJ$.

Fill in the blanks (where necessary) and justify the conclusions.

25. $ABCD \cong \underline{\ ?\ }$.

26. $ABCD \cong EHGF$ and
    $EHGF \cong MJKL \Rightarrow ABCD \cong \underline{\ ?\ }$.

27. $AB = EH$

28. m$\angle G =$ m$\angle K$

29. In Questions 7 and 8, why is $\angle 5 \cong \angle 7$?

30. T$(\overline{XZ}) = (\overline{XW})$ and T is an isometry.

  a. Draw a possible figure.

  b. Is $\overline{XZ} \cong \overline{XW}$? Why or why not?

  c. Is X the midpoint of $\overline{ZW}$? Why or why not?

31. Justify this conclusion:
    Given: m$\angle XYZ =$ m$\angle LMN$.
    Conclusion: $\angle XYZ \cong \angle LMN$.

Objective F: *Write proofs using the Transitive Properties of Equality or Congruence.* *(Lesson 5-4)*

In 32 and 33, supply justifications in the proof argument.

32. Given: E is the midpoint of $\overline{SG}$ and $\overline{ET} \cong \overline{EG}$.
    To prove: $\overline{ET} \cong \overline{SE}$

    Drawing:

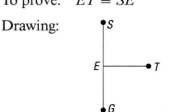

Argument:

Conclusions	Justifications
0. E is the midpoint of $\overline{SG}$ and $\overline{ET} \cong \overline{EG}$.	**a.**
1. $\overline{SE} \cong \overline{EG}$	**b.**
2. $\overline{ET} \cong \overline{SE}$	**c.**

33. Given: $\angle 10 \cong \angle 11$ and $\angle 11 \cong \angle 12$.
    To prove: $\ell \,/\!/\, p$.

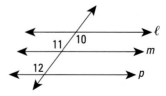

Argument:

Conclusions	Justifications
0. $\angle 10 \cong \angle 11$ $\angle 11 \cong \angle 12$	**a.**
1. $\angle 10 \cong \angle 12$	**b.**
2. $\ell \,/\!/\, p$	**c.**

In 34–37, write a proof argument.

34. Given: $\triangle PQT \cong \triangle RQS$,
$\triangle TSQ \cong \triangle RQS$.

To prove: $\triangle PQT \cong \triangle TSQ$.

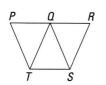

35. Given: $\angle 2 \cong \angle 3$.
To prove: $\angle 1 \cong \angle 3$.

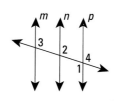

36. Given: $m \parallel n$;
$n \parallel p$.

To prove: $\angle 1 \cong \angle 3$.

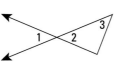

37. Given: $\overrightarrow{OB}$ bisects $\angle AOC$.
$\overrightarrow{OC}$ bisects $\angle BOD$.

To prove: $m\angle AOB = m\angle COD$.

Objective G: *Write proof arguments using properties of reflections.*   *(Lesson 5-5)*

In 38 and 39, give the justification for each conclusion.

38. Given: $r_{\overleftrightarrow{RC}}(M) = P$
To prove: $\triangle MRC \cong \triangle PRC$.

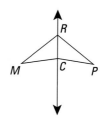

Argument:

Conclusions	Justifications
0. $r_{\overleftrightarrow{RC}}(M) = P$	**a.**
1. $r_{\overleftrightarrow{RC}}(R) = R$	**b.**
$r_{\overleftrightarrow{RC}}(C) = C$	
2. $r(\triangle RCM) = \triangle RCP$	**c.**
3. $\triangle RCM \cong \triangle RCP$	**d.**

39. Given: t is the $\perp$ bisector of $\overline{PC}$.
To prove: $\angle EPA \cong \angle ECA$.

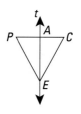

Argument:

Conclusions	Justifications
0. t is the $\perp$ bisector of $\overline{PC}$.	**a.**
1. $r_t(P) = C$	**b.**
2. $r_t(A) = A$	**c.**
3. $r_t(E) = E$	**d.**
4. $r_t(\angle EPA) = \angle ECA$	**e.**
5. $\angle EPA \cong \angle ECA$	**f.**

In 40–43, write a proof argument.

40. Given: $r_m(B) = S$, $r_m(A) = L$,
$r_m(R) = I$, $r_m(T) = P$.
To prove: $BART \cong SLIP$.

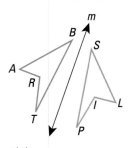

41. Given: $\overleftrightarrow{AJ}$ is the $\perp$ bisector of $\overline{RM}$.
To prove: $\angle JAM \cong \angle JAR$.

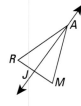

42. Given: $\overline{PT}$ is the $\perp$ bisector of $\overline{LS}$.
To prove: **a.** $\overline{PL} \cong \overline{PS}$.
b. $\triangle PLT \cong \triangle PST$.

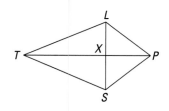

43. Given: $r_p(A) = C$ and
p is the $\perp$ bisector of $\overline{EB}$.
To prove: $\angle ABD \cong \angle CED$.

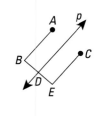

Objective H: *Tell whether auxiliary figures are uniquely determined.* *(Lesson 5-6)*

In 44–46, use the rectangle below. Tell whether each auxiliary figure is uniquely determined. Illustrate your argument with a diagram.

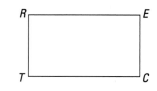

44. bisector of $\angle R$

45. diagonal $\overline{ET}$

46. bisector of segment $\overline{EC}$

47. In the figure for Questions 44–46, must $\overline{ET}$ lie on the bisector of $\angle REC$? Why or why not?

In 48 and 49, use the figure below.

48. *Multiple choice.* How many lines contain point A and are parallel to $\overleftrightarrow{BC}$?
(a) none   (b) one
(c) two   (d) more than two

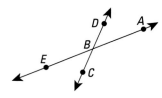

49. *Multiple choice.* Ray $\overrightarrow{AF}$ is located so that F is on the other side of $\overleftrightarrow{AE}$ from D and $m\angle FAB = m\angle DBA$. How many possible positions are there for this ray?
(a) none   (b) one
(c) two   (d) more than two

USES DEAL WITH APPLICATIONS OF MATHEMATICS IN REAL SITUATIONS.

Objective I: *Use the Perpendicular Bisector Theorem and theorems on alternate interior angles in real situations.* *(Lessons 5-4, 5-5)*

50. In the kite *ABCD* at the right, $\overline{AC}$ is the perpendicular bisector of $\overline{BD}$. Which edges of the kite are congruent?

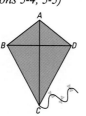

51. For the antenna pictured below, $\overline{WX} \parallel \overline{YZ}$. Justify why $\angle WLM \cong \angle ZML$.

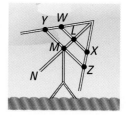

Pictured is a stunt event at the International Kite Festival in Long Beach, Washington.

CULTURE DEALS WITH THE PEOPLES AND THE HISTORY RELATED TO THE DEVELOPMENT OF MATHEMATICAL IDEAS.

Objective J: *Know the history and impact of postulates relating to parallel lines on the development of geometry.* *(Lessons 5-6, 5-7)*

52. *True or false.* In geometry that is not Euclidean, there can be triangles in which the sum of the measures of the angles is greater than 180°.

53. *True or false.* In Euclidean geometry, Playfair's Parallel Postulate is true.

54. How many geometric postulates were in Euclid's *Elements*?

55. How did the development of non-Euclidean geometries change the view of the truth of Euclid's postulates?

CHAPTER

6

POLYGONS AND SYMMETRY

Many companies have logos or trademarks that are reflection-symmetric. Can you name the company associated with each trademark below?

Symmetry also provides a powerful way to study polygons and other geometric figures. For example, the symmetry of a triangular roof lets you know that both sides have the same length and tilt. The symmetry of a rectangular table allows for the same number of people to sit on opposite sides. In this chapter, you will see how symmetry helps prove certain properties of triangles, quadrilaterals, other polygons, and circles.

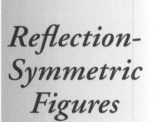
Reflection-Symmetric Figures

Kudos to the kudu. *The greater kudu is an antelope that lives in southern Africa. The male greater kudu has spiral horns that may be up to 1.5 meters long. Its head has almost perfect reflection symmetry.*

What Is Reflection Symmetry?

When a figure is reflected over a line, one of two situations arises. In the first case, as pictured below at the left, the preimage and image are distinct.

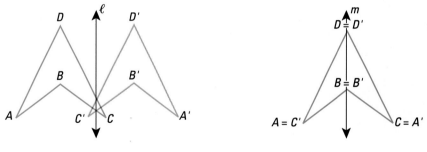

$$r_{\ell}(ABCD) = A'B'C'D'$$

$$r_m(ABCD) = A'B'C'D' = CBAD$$

In the second case, pictured at the right, the image of quadrilateral *ABCD* is itself. The preimage and image *coincide*. Note that $r_m(A) = C$, $r_m(B) = B$, $r_m(C) = A$, and $r_m(D) = D$. Thus, $A'B'C'D' = CBAD$ and $r_m(ABCD) = CBAD$. The figure is called *reflection-symmetric*, and the reflecting line is called a *symmetry line* for the figure.

> **Definition**
> A plane figure *F* is a **reflection-symmetric figure** if and only if there is a line *m* such that $r_m(F) = F$. The line *m* is a **symmetry line** for the figure.

We say reflection-symmetric figures possess **line symmetry**.

Trace each capital letter below, and draw all symmetry lines.

a. **A**   b. **C**   c. **X**   d. **N**

You should have drawn a total of two horizontal and two vertical lines in Activity 1. Also, you should have recognized that one letter is not reflection-symmetric.

Which Figure Is the Image?

The inkblot pictured at the left was made by placing dabs of ink on paper and then folding the paper onto the wet ink. It shows the union of a figure and its reflection image. From this information it is impossible to tell which is the preimage and which is the image, even if the reflecting line is identified. This is because of the following theorem. We call it the *Flip-Flop Theorem*, but you may wish to call it by some other name. Notice that this theorem applies to both points and figures, and that we use the first part of the theorem, after proving it, to prove the second part.

Flip-flop inkblot. *Pictured is a Rorschach-type inkblot, a reflection-symmetric figure. Herman Rorschach was a Swiss psychiatrist who used inkblots to investigate the workings of the human mind.*

Flip-Flop Theorem
(1) If F and G are points and $r_\ell(F) = G$, then $r_\ell(G) = F$.
(2) If F and G are figures and $r_\ell(F) = G$, then $r_\ell(G) = F$.

Proof
Given: $r_\ell(F) = G$.
To prove: $r_\ell(G) = F$.

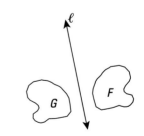

Argument:
(1) Since F and G are points and $r_\ell(F) = G$ (given), then ℓ is the perpendicular bisector of $\overline{FG}$, by the definition of reflection. But $\overline{FG}$ and $\overline{GF}$ are the same segment. So ℓ is the perpendicular bisector of $\overline{GF}$. Consequently, again using the definition of reflection, $r_\ell(G) = F$.

(2) For this part, it is given that F and G are figures and $r_\ell(F) = G$. This means each point on G is the reflection image of a point on F, and all points of F have been used. Using part (1) of this theorem, all points on F must be reflection images of the points of G. So $r_\ell(G) = F$.

Activity 2

Trace each figure below and draw all of its symmetry lines.

a.

b.

c.

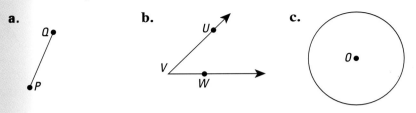

The Reflection Symmetry of Segments

The reflection symmetry of polygons is based on the reflection symmetry of segments and angles. Did you draw *both* symmetry lines for the segment in Activity 2?

> **Segment Symmetry Theorem**
> Every segment has exactly two symmetry lines:
> (1) its perpendicular bisector, and
> (2) the line containing the segment.

Example 1 gives the proof of part (1) of this theorem. The proof of part (2) is left for you as Question 6.

Example 1

Write the proof argument using the figure at the left.
Given: ℓ is the $\perp$ bisector of $\overline{AB}$.
To prove: $\overline{AB}$ is a reflection-symmetric figure with ℓ as its symmetry line.

Solution
Argument:

Conclusions	Justifications
0. ℓ is the $\perp$ bisector of $\overline{AB}$.	Given
1. $r_\ell(A) = B$, $r_\ell(B) = A$	definition of reflection
2. $r_\ell(\overline{AB}) = \overline{BA}$	Figure Reflection Theorem
3. $\overline{AB}$ is a reflection-symmetric figure with symmetry line ℓ.	definition of reflection-symmetric figure

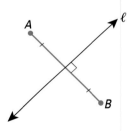

Reflection Symmetry of Angles

Angles also possess reflection symmetry. In the figure at the left, consider reflecting $\angle ABC$ over the line containing its bisector, $\overrightarrow{BD}$. Because $\overrightarrow{BD}$ is the angle bisector, $m\angle ABD = m\angle CBD$. We name these angles $\angle 1$ and $\angle 2$.

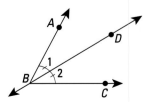

Now reflect $\angle 1$ over $\overleftrightarrow{BD}$. One side of $\angle 1$, $\overrightarrow{BD}$, coincides with its image. The other side, $\overrightarrow{BA}$, has $\overrightarrow{BC}$ as its reflection image, because reflections preserve angle measure. Thus the reflection image of $\angle 1$ is $\angle 2$.

> **Side-Switching Theorem**
> If one side of an angle is reflected over the line containing the angle bisector, its image is the other side of the angle.

This theorem gives the angle its symmetry. Since $r_{\overleftrightarrow{BD}}(\overrightarrow{BA}) = \overrightarrow{BC}$, by the Flip-Flop Theorem, $r_{\overleftrightarrow{BD}}(\overrightarrow{BC}) = \overrightarrow{BA}$. Thus the sides of the angles reflect onto each other, so the image of $\angle ABC$ is the angle itself.

> **Angle Symmetry Theorem**
> The line containing the bisector of an angle is a symmetry line of the angle.

Thus you should have drawn one symmetry line for the angle of Activity 2.

Reflection Symmetry of Circles

In Activity 2, you should have had great difficulty drawing *all* the symmetry lines of a circle. There are infinitely many symmetry lines!

> **Circle Symmetry Theorem**
> A circle is reflection-symmetric to any line through its center.

Proof
Given: $\odot O$ with line m through O.
To prove: $r_m(\odot O) = \odot O$.
Drawing:

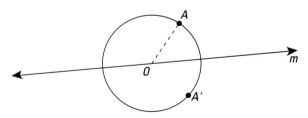

Argument:
We need to show that the reflection image of any point on the circle also lies on the circle, and that every point on the circle is an image. Let A be any point on $\odot O$, and let $A' = r_m(A)$. This means that OA is the radius of the circle.

Since O is on m, $r_m(O) = O$ by the definition of reflection.
So $OA = OA'$, because reflections preserve distance.
Thus, A' is on $\odot O$, by the definition of circle.

So the reflection image of each point on $\odot O$ is also on $\odot O$. By the Flip-Flop Theorem, $r_m(A') = A$. So every point on the circle is the image of some point on the circle. Consequently, $\odot O$ coincides with its image.

Big wheel. *The design on the wheel of this Costa Rican oxcart has many symmetry lines, but not as many as the wheel does.*

Why Are Symmetric Figures Important?

By the Reflexive Property of Congruence, every figure is congruent to itself. So, by the CPCF Theorem, each of its parts is congruent to itself. These are called *trivial* congruences because they are so obvious. A reflection-symmetric figure is also congruent to itself. But since a reflection-symmetric figure is its own image under a reflection, it is congruent to itself in a *non-trivial* way. Its parts correspond to different parts.

Symmetric Figures Theorem
If a figure is symmetric, then any pair of corresponding parts under the symmetry are congruent.

Example 2

In the figure at the left, line m is a symmetry line for figure *ABCD*.
a. Find $r_m(ABCD)$.
b. What can be deduced using the Symmetric Figures Theorem?

Solution
a. A and B are images of each other, as are C and D.
   So $r_m(ABCD) = BADC$.
b. There is one pair of congruent corresponding sides: $\overline{AD} \cong \overline{BC}$.
   There are two pairs of congruent corresponding angles:
   $\angle A \cong \angle B$ and $\angle C \cong \angle D$.
   There is one pair of congruent corresponding diagonals (not drawn):
   $\overline{AC} \cong \overline{BD}$.

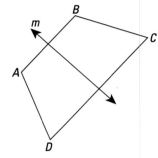

Because congruent parts can be found so easily using the Symmetric Figures Theorem, it is very useful to know that figures are symmetric. For this reason, polygons are examined for their symmetry throughout much of this chapter.

QUESTIONS

Covering the Reading

1. Why did Hermann Rorschach present reflection-symmetric inkblots to his patients?

2. Define *reflection-symmetric figure*.

3. Show your answers for Activity 1 of this lesson.

4. Justify the following conclusion.
   Given: $r_\ell(F) = G$, and F is a triangle.
   Conclusion: $r_\ell(G) = F$.

5. Show your answers to Activity 2 of this lesson.

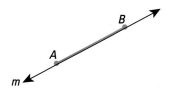

6. Complete the proof of part 2 of the Segment Symmetry Theorem.
   Given: m contains $\overline{AB}$.
   To prove: $\overline{AB}$ is a reflection-symmetric figure with m as a symmetry line.

In 7 and 8, trace the figure and draw all lines of symmetry.

7. $\overline{AB}$ **8.** $\angle CDE$

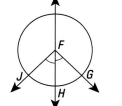

9. Use the drawing at the left. Identify the reflection image of each figure over $\overleftrightarrow{FH}$.
   a. $\overrightarrow{FJ}$ **b.** $\angle JFG$ **c.** $\odot F$

10. How many symmetry lines does a circle have?

11. Triangle ABC at the right has symmetry line m.
    a. $r_m(\triangle ABC) = \underline{\ ?\ }$
    b. What can be deduced by the Symmetric Figures Theorem?

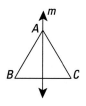

Applying the Mathematics

12. Alchemists used these symbols during the Middle Ages. How many symmetry lines does each have?
    a. Arsenic   **b.** Lead   **c.** Platinum   **d.** Sulphur

13. Trace each figure and draw all symmetry lines.
    a. urn   **b.** triskelion   **c.** union of two circles

          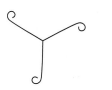

14. **a.** Trace the figure below and reflect it over line m.
    b. What natural condition might produce this kind of figure?

15. Copy the figure below and complete its shape so that the result is symmetric to the line.

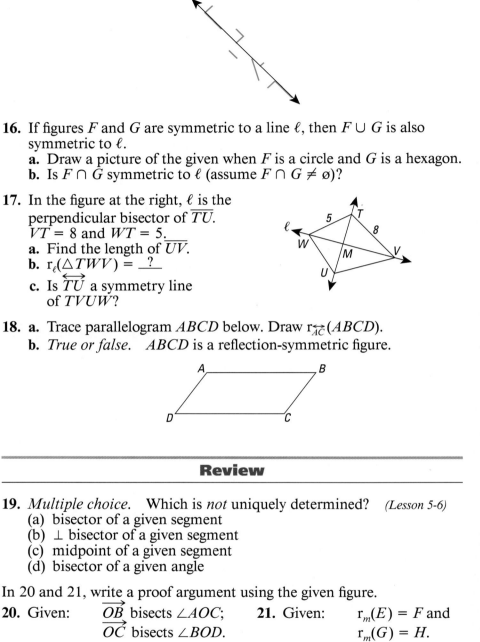

16. If figures F and G are symmetric to a line ℓ, then $F \cup G$ is also symmetric to ℓ.
a. Draw a picture of the given when F is a circle and G is a hexagon.
b. Is $F \cap G$ symmetric to ℓ (assume $F \cap G \neq \emptyset$)?

17. In the figure at the right, ℓ is the perpendicular bisector of $\overline{TU}$. $VT = 8$ and $WT = 5$.
a. Find the length of $\overline{UV}$.
b. $r_\ell(\triangle TWV) = \underline{\ ?\ }$
c. Is $\overleftrightarrow{TU}$ a symmetry line of $TVUW$?

18. a. Trace parallelogram $ABCD$ below. Draw $r_{\overleftrightarrow{AC}}(ABCD)$.
b. *True or false.* $ABCD$ is a reflection-symmetric figure.

Review

19. *Multiple choice.* Which is *not* uniquely determined? *(Lesson 5-6)*
(a) bisector of a given segment
(b) ⊥ bisector of a given segment
(c) midpoint of a given segment
(d) bisector of a given angle

In 20 and 21, write a proof argument using the given figure.

20. Given: $\overrightarrow{OB}$ bisects $\angle AOC$; $\overrightarrow{OC}$ bisects $\angle BOD$.
To prove: $m\angle AOB = m\angle COD$.
(Lesson 5-4)

21. Given: $r_m(E) = F$ and $r_m(G) = H$.
To prove: $EH = FG$.
(Lesson 5-5)

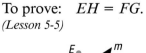

In 22 and 23, trace △*PAT* below.

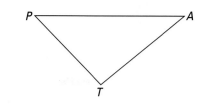

22. Construct the perpendicular bisector of $\overline{PA}$. *(Lesson 3-8)*

23. Draw the angle bisector of ∠*T*. *(Lesson 3-1)*

24. At the right, *ABCDEF* is a hexagon
with diagonals $\overline{AD}$, $\overline{BF}$, and $\overline{CE}$.
m∠1 = 70 and m∠2 = 70.
 a. Which numbered angles
 3 through 8 also measure 70?
 b. *True or false.* $\overline{BF}$ // $\overline{CE}$.
 c. *True or false.* ∠4 and ∠6 are
 supplementary angles. *(Lessons 3-3, 3-6)*

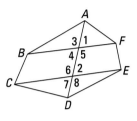

25. Line *m* goes through points (3, -1) and (-7, 4). Find the slope of
line *n* if *m* ⊥ *n*. *(Lesson 3-7)*

26. An ad states: "Try Roboflex for just two weeks and lose 5 pounds!"
Kalie lost 5 pounds in two weeks. Did she use Roboflex? *(Lesson 2-3)*

Exploration

27. a. Draw two non-perpendicular intersecting lines. The resulting
 figure has two symmetry lines. Draw these lines.
 b. Repeat part **a** with two other lines.
 c. In parts **a** and **b,** how are these symmetry lines related to each
 other?

28. The word *trivial* is used in this lesson.
 a. What does *trivial* mean?
 b. In medieval universities, students studied seven liberal arts.
 Three of these subjects were called the *trivium*. Find out what
 subjects formed the trivium.
 c. The other four subjects were called the *quadrivium*. Name them.

*Oxford University, located
in Oxford, England, is the
oldest university in Great
Britain. It was founded in
the 1100s and received its
first official recognition in
1214. Today it has over
13,000 students.*

*Angle
Bisectors,
Medians,
⊥ Bisectors
of a
Triangle*

IN-CLASS
ACTIVITY

Work on this activity with a partner. You will need an automatic drawer.

1 On an automatic drawer, place a scalene triangle *ABC* and the measures *AB* and *AC*.

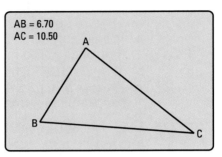

AB = 6.70
AC = 10.50

2 Draw *m*, the bisector of ∠*BAC*.

3 Draw the midpoint *D* and the perpendicular bisector of $\overline{BC}$.

4 A segment connecting a vertex of a triangle to the midpoint of the opposite side is called a **median** of the triangle. Draw the median from *A* to $\overline{BC}$.

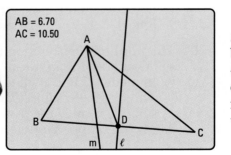

AB = 6.70
AC = 10.50

m contains the bisector of ∠*BAC*. *ℓ* is the ⊥ bisector of $\overline{BC}$. $\overline{AD}$ is the median from vertex *A*.

5 Move point *A* until the three figures drawn in Steps 2–4 coincide. Record *AB* and *AC*.

6 Move point *A* to two other places in which the three figures drawn in Steps 2–4 coincide. Record *AB* and *AC*.

7 What can you conclude about △*ABC* when the angle bisector of ∠*BAC*, the median from *A* to $\overline{BC}$, and the perpendicular bisector of $\overline{BC}$ coincide?

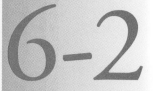

Recall that isosceles triangles have (at least) two sides of equal length. They are the outlines of many rooftops, right cones, and other objects that taper to a point. They occur when two radii of a circle are drawn and their endpoints are joined.

The angle determined by the equal sides in an isosceles triangle is called the **vertex angle.** The side opposite the vertex angle is called the **base.** The two angles whose vertices are the endpoints of the base are the **base angles.** The base angles are opposite the equal sides. In the triangle below, $\angle A$ is the vertex angle; $\overline{BC}$ is the base; $\angle B$ and $\angle C$ are base angles.

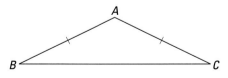

Symmetry of an Isosceles Triangle

Every isosceles triangle is reflection-symmetric. The proof of this statement is long, but this symmetry is useful for easily deducing some other properties of isosceles triangles.

> **Isosceles Triangle Symmetry Theorem**
> The line containing the bisector of the vertex angle of an isosceles triangle is a symmetry line for the triangle.

Proof

Given: Isosceles triangle ABC with vertex angle A bisected by line m.

To prove: m is a symmetry line for $\triangle ABC$.

Drawing:

Argument:
We first show that $r_m(B) = C$. Since m is an angle bisector, the Side-Switching Theorem tells us that, when $\overrightarrow{AB}$ is reflected over m, its image is $\overrightarrow{AC}$. Thus $r_m(B)$ is on $\overrightarrow{AC}$. Let $B' = r_m(B)$. Since reflections preserve distance, $AB' = AB$. It is given that $\triangle ABC$ is isosceles with vertex angle A, so $AB = AC$. Thus, by the Transitive Property of Equality, $AB' = AC$. So B' and C are points on $\overrightarrow{AC}$ at the same distance from A, and therefore $B' = C$. That is, $r_m(B) = C$. By the Flip-Flop Theorem, $r_m(C) = B$. So, by the Figure Reflection Theorem, $r_m(\triangle ABC) = \triangle ACB$, which is the characteristic required for line m to be a symmetry line for the triangle.

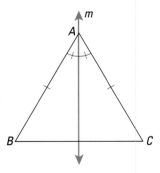

In the In-class Activity, you should have seen that in a scalene triangle, the median, angle bisector, and $\perp$ bisector are on different lines. But at the left is isosceles triangle ABC with bisector m of $\angle A$. Since $r_m(B) = C$, m is the $\perp$ bisector of $\overline{BC}$. And so m contains the midpoint of $\overline{BC}$. Thus m also contains the median from vertex A. This proves the following theorem.

> **Isosceles Triangle Coincidence Theorem**
> In an isosceles triangle, the bisector of the vertex angle, the perpendicular bisector of the base, and the median to the base determine the same line.

Base Angles in an Isosceles Triangle

In $\triangle ABC$ above, since m is a symmetry line, $\angle ABC \cong \angle ACB$ by the Symmetric Figures Theorem. This conclusion is an important theorem.

> **Isosceles Triangle Base Angles Theorem**
> If a triangle has two congruent sides, then the angles opposite them are congruent.

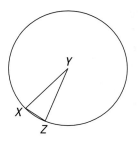

Example 1

In $\odot Y$ at the left, $m\angle Y = 23$. What is $m\angle X$?

Solution

By the Triangle-Sum Theorem,
$$m\angle Y + m\angle X + m\angle Z = 180.$$
Substituting, $\qquad 23 + m\angle X + m\angle Z = 180.$
So, $\qquad\qquad\qquad m\angle X + m\angle Z = 157.$
But $\triangle XYZ$ is isosceles ($XY = ZY$) with vertex angle Y. So, from the Isosceles Triangle Base Angles Theorem, $m\angle X = m\angle Z$.
Thus, $\qquad\qquad\qquad m\angle X + m\angle X = 157.$
$$2m\angle X = 157$$
$$m\angle X = 78.5$$

The Isosceles Triangle Base Angles Theorem is useful in proofs in which you must go from equal sides to equal angles.

Example 2

Write a proof argument.
Given: The figure at the left, with $PQ = QR$.
To Prove: $m\angle 1 = m\angle 3$.

Solution

Argument:

Conclusions	Justifications
0. $PQ = QR$	Given
1. $m\angle 1 = m\angle 2$	Isosceles Triangle Base Angles Theorem
2. $m\angle 2 = m\angle 3$	Vertical Angles Theorem
3. $m\angle 1 = m\angle 3$	Transitive Property of Equality (steps 1 and 2)

Equilateral Triangles

In an equilateral triangle, all sides have the same length, so any side of an equilateral triangle may be considered as a base, with the angle opposite as its vertex angle. Notice how this idea can be used to deduce many properties special to equilateral triangles.

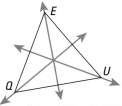

equilateral triangle *EQU*
with its three symmetry lines

Example 3

Prove the following theorem.

Equilateral Triangle Symmetry Theorem
Every equilateral triangle has three symmetry lines, which are the bisectors of its angles (or equivalently, the perpendicular bisectors of its sides).

Proof
First, consider this theorem as a conditional: If a triangle is equilateral, then it has three symmetry lines. This clarifies what is given and what is to be proved.

Given: Equilateral △EQU.
To Prove: △EQU has three symmetry lines, the bisectors of its angles.

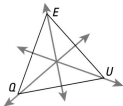

Now think of the triangle as being isosceles in three different ways.

Argument:
Thinking of △EQU as isosceles with base $\overline{QU}$, the bisector of ∠E is a symmetry line by the Isosceles Triangle Symmetry Theorem. Similarly, thinking of △EQU as isosceles with base $\overline{EU}$, the bisector of ∠Q is a symmetry line. Thinking of △EQU as isosceles with base $\overline{QE}$, the bisector of ∠U is a symmetry line. Thus there are three symmetry lines.

If all angles of a polygon have the same measure, then the polygon is called **equiangular.** It is rather easy to deduce that an equilateral triangle is equiangular. This is left to you in Question 10.

Equilateral Triangle Angle Theorem
If a triangle is equilateral, then it is equiangular.

A **corollary** (pronounced core′-oh-larry or cor-ahl′ur-ee) is a theorem that follows immediately from another theorem. The following corollary follows from the Equilateral Triangle Angle Theorem.

Corollary
Each angle of an equilateral triangle has measure 60°.

Have you lost your marbles? *There are many equilateral triangles on this Chinese checkers board. Each player places ten marbles in a triangle and moves them across the board to the opposite triangle.*

Covering the Reading

1. Write your answer to Step 7 of the In-class Activity on page 308.

In 2 and 3, refer to △*WIN*, in which *IW* = *IN*.

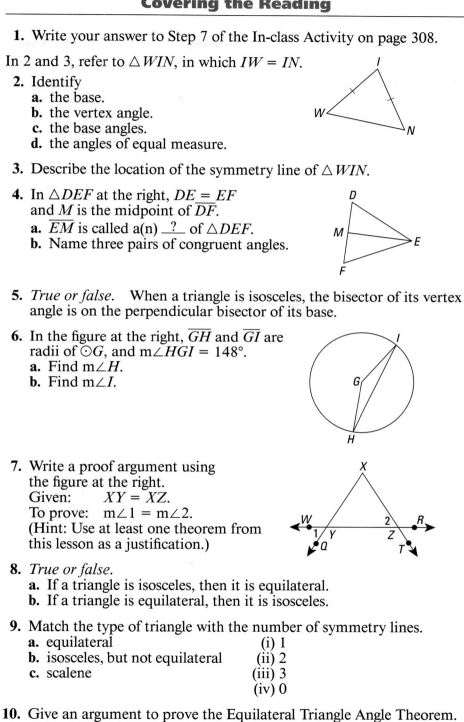

2. Identify
   a. the base.
   b. the vertex angle.
   c. the base angles.
   d. the angles of equal measure.

3. Describe the location of the symmetry line of △*WIN*.

4. In △*DEF* at the right, *DE* = *EF* and *M* is the midpoint of $\overline{DF}$.
   a. $\overline{EM}$ is called a(n) __?__ of △*DEF*.
   b. Name three pairs of congruent angles.

5. *True or false.* When a triangle is isosceles, the bisector of its vertex angle is on the perpendicular bisector of its base.

6. In the figure at the right, $\overline{GH}$ and $\overline{GI}$ are radii of ⊙*G*, and m∠*HGI* = 148°.
   a. Find m∠*H*.
   b. Find m∠*I*.

7. Write a proof argument using the figure at the right.
   Given: *XY* = *XZ*.
   To prove: m∠1 = m∠2.
   (Hint: Use at least one theorem from this lesson as a justification.)

8. *True or false.*
   a. If a triangle is isosceles, then it is equilateral.
   b. If a triangle is equilateral, then it is isosceles.

9. Match the type of triangle with the number of symmetry lines.
   a. equilateral   (i) 1
   b. isosceles, but not equilateral   (ii) 2
   c. scalene   (iii) 3
                        (iv) 0

10. Give an argument to prove the Equilateral Triangle Angle Theorem. (Hint: Use an idea similar to that found in the proof of Equilateral Triangle Symmetry Theorem, and use transitivity.)

Applying the Mathematics

11. *Multiple choice.* In the figure of Question 6, suppose m∠G = x. What is m∠H?
   (a) 180 − x
   (b) 180 − $\frac{x}{2}$
   (c) 90 − x
   (d) 90 − $\frac{x}{2}$

12. *Multiple choice.* The horizontal beam $\overline{RS}$ helps support this roof.

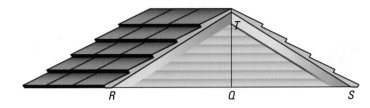

To keep the roof from collapsing, the support $\overline{QT}$ is used, where Q is the midpoint of $\overline{RS}$. If the parts $\overline{RT}$ and $\overline{ST}$ of the roof are of equal length, then
   (a) $\overrightarrow{TQ}$ bisects ∠RTS.
   (b) $\overline{QT}$ is vertical.
   (c) $\overrightarrow{TQ} \perp \overline{RS}$.
   (d) All of the above.

13. In the figure at the left, △AOB and △BOC are each isosceles. $\overrightarrow{OM}$ and $\overrightarrow{ON}$ are bisectors of the vertex angles AOB and BOC, respectively. If m∠AOC = 83, what is m∠MON?

14. Write a proof argument.
   Given: $\overline{AD}$ and $\overline{BE}$ intersect at C, and
            AB = AC, as shown at the right.
   To prove: m∠B = m∠ECD.

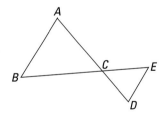

Review

15. $\overleftrightarrow{AD}$ is a symmetry line for polygon ABCDEF at the right.
   a. r$_{\overleftrightarrow{AD}}$(ABCDEF) = __?__
   b. Which sides of ABCDEF have the same length?
   c. Which angles in the figure have the same measure? *(Lesson 6-1)*

16. Trace each figure below and draw all symmetry lines. *(Lesson 6-1)*

   a.   **b.**

17. Is it possible for the four angles of a quadrilateral to have the extended ratio 1 : 2 : 3 : 4? If so, determine the angle measures and use a protractor to draw such a quadrilateral. If this is not possible, tell why not.   *(Lesson 5-7)*

18. In the figure below, ⊙*A* and ⊙*H* each have radius *AH* and ⊙*A* ∩ ⊙*H* = {*M*, *T*}. Is △*AHM* equilateral, isosceles but not equilateral, or scalene? Justify your answer.   *(Lesson 5-4)*

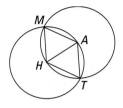

19. Give the reflection image of (*x*, *y*) over the *y*-axis.   *(Lesson 4-1)*

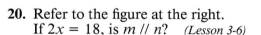

20. Refer to the figure at the right. If 2*x* = 18, is *m* // *n*?   *(Lesson 3-6)*

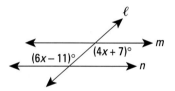

21. The age difference between Ahmed and Bjorn is 12 years. The age difference between Bjorn and Crystal is 15 years. What can you say about the age difference between Ahmed and Crystal?   *(Lesson 2-7)*

22. Arrange from least to greatest number of sides: rectangle, octagon, pentagon, equilateral triangle, nonagon, 7-gon.   *(Lesson 2-6)*

Exploration

23. Find two meanings of the word *median* other than the one in this lesson.

24. Use a ruler, compass, and protractor, or use an automatic drawer.
   a. Draw an equilateral triangle. Draw all three medians. They should all intersect in a point called the **centroid** of the triangle.
   b. Measure the lengths of the medians and the distances from the centroid to each of the vertices. What can you conclude?
   c. Draw an isosceles triangle that is not equilateral. Repeat the steps in parts **a** and **b** for your triangle.
   d. Draw a scalene triangle. Repeat the steps in parts **a** and **b** again.
   e. Summarize your work in parts **a–d** in a sentence or two.

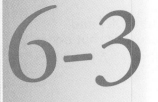
Types of Quadrilaterals

Pastureized. *This horse farm is sectioned into quadrilateral fields where horses exercise or graze.*

The three-sided polygons, the triangles, are usually classified by the number of equal sides as scalene, isosceles, or equilateral. Or they are classified by their largest angle as acute, right, or obtuse. The four-sided polygons, the quadrilaterals, are more diverse and the classification is more complicated. There are seven major types: parallelogram, rhombus, rectangle, square, kite, trapezoid, and isosceles trapezoid. Here are definitions and examples of four of them.

Definition
A quadrilateral is a **parallelogram** if and only if both pairs of its opposite sides are parallel.

$\overline{AB} \parallel \overline{CD}, \overline{BC} \parallel \overline{AD}$

Definition
A quadrilateral is a **rhombus** if and only if its four sides are equal in length.

$EF = FG = GH = HE$

Definition
A quadrilateral is a **rectangle** if and only if it has four right angles.

$\angle I, \angle J, \angle K,$ and $\angle L$ are right angles.

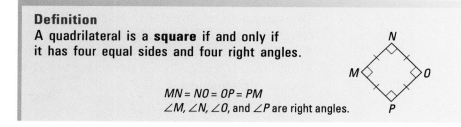

$MN = NO = OP = PM$
$\angle M$, $\angle N$, $\angle O$, and $\angle P$ are right angles.

From the definitions, you can see that every square is a rhombus, since every square has four equal sides, but not every rhombus is a square. You also can conclude that every square is a rectangle, since every square has four right angles. This information is summarized in the network below. This network shows part of a **hierarchy** of quadrilaterals.

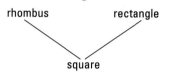

If a figure is of any type in the hierarchy, it is also a figure of all types connected above it in the hierarchy.

Refer to the rectangle at the left. Because two perpendiculars to the same line are parallel, the opposite sides of a rectangle are parallel. Thus, every rectangle is a parallelogram. So we can add "parallelogram" to the hierarchy.

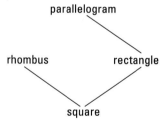

Since all squares are rectangles, and all rectangles are parallelograms, you can conclude that all squares are parallelograms.

A fifth type of quadrilateral is formed by the union of two isosceles triangles having the same base, with the base removed. The result is a quadrilateral that resembles a *kite* or arrowhead. Pictured here are the convex kite *ABCD* and the nonconvex kite *FORM*.

$AB = BC$
$AD = DC$

$FO = OR$
$FM = MR$

Two quadrilaterals are drawn below: *EFGH* with exactly three sides congruent, and *IJKL* with four sides congruent (a rhombus). *EFGH is not* a kite, since the pairs of consecutive equal sides are not distinct; while *IJKL is* a kite since it does have two distinct pairs of consecutive sides of the same length.

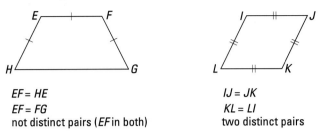

EF = HE
EF = FG
not distinct pairs (*EF* in both)

IJ = JK
KL = LI
two distinct pairs

Thus, every rhombus is a special kite. This information is added to the hierarchy. You also can conclude now that every square is a kite by reading up the hierarchy from square to rhombus to kite.

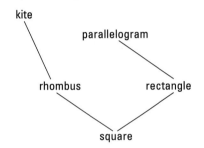

A figure that is more general than the parallelogram is the *trapezoid*.

> **Definition**
> A quadrilateral is a **trapezoid** if and only if it has at least one pair of parallel sides.

trapezoids:

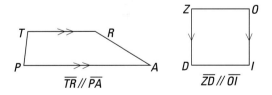

$\overline{TR} \mathbin{/\mkern-5mu/} \overline{PA}$

$\overline{ZD} \mathbin{/\mkern-5mu/} \overline{OI}$

Parallel sides of a trapezoid are called **bases**. In the figures above, $\overline{TR}$ and $\overline{PA}$ are bases and $\overline{ZD}$ and $\overline{OI}$ are bases. Two consecutive angles whose vertices are endpoints of a single base constitute a **pair of base angles.** For example, the angles $\angle TPA$ and $\angle RAP$ are a pair of base angles of trapezoid *TRAP*. Another pair of base angles are $\angle PTR$ and $\angle ART$. This terminology enables us to define a special type of trapezoid.

> **Definition**
> A trapezoid is an **isosceles trapezoid** if and only if it has a pair of base angles equal in measure.

Isosceles trapezoids:

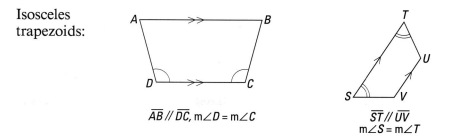

$\overline{AB} \text{ // } \overline{DC}$, m$\angle D$ = m$\angle C$

$\overline{ST} \text{ // } \overline{UV}$
m$\angle S$ = m$\angle T$

Because a rectangle has opposite sides parallel and all angles congruent, every rectangle is an isosceles trapezoid. Now you can relate all of these seven types of quadrilaterals in the same hierarchy. We show this with the solid lines below. One other hierarchy relationship will be deduced in the next lesson: every rhombus is a parallelogram. We show it now with a dashed line. At the right below is the hierarchy with a drawing of a representative figure surrounding the word.

We call the complete network of connections among the seven types of quadrilaterals the *Quadrilateral Hierarchy Theorem.*

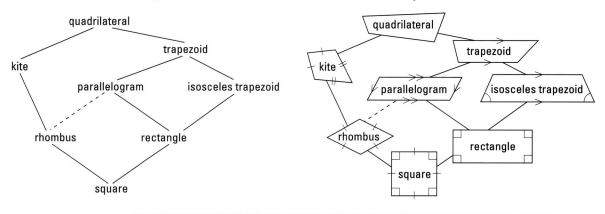

> **Quadrilateral Hierarchy Theorem**
> The seven types of quadrilaterals are related as shown in the hierarchy pictured above.

The Quadrilateral Hierarchy Theorem is very useful because it allows you to relate the properties of specific quadrilaterals. The general rule is:

Every property true of all figures of one type in the hierarchy is also true of all figures of all the types below it to which the first type is connected.

For example, *square* is below *rhombus* in the hierarchy. Thus any square has all the properties of a rhombus. Squares and rhombuses are below *kite*. Thus they have all the properties of kites. In the next few lessons this rule will be used to identify many properties of specific quadrilaterals.

Caution: The converse of the above rule is not true. For example, all parallelograms share some properties which are not true of all trapezoids.

Covering the Reading

In 1–7, a quadrilateral is named. **a.** Give a definition. **b.** Draw an example that is *not* an example of a figure below it (if there is one) in the quadrilateral hierarchy.

1. parallelogram
2. rhombus
3. rectangle
4. square
5. kite
6. trapezoid
7. isosceles trapezoid

8. Draw a hierarchy of the following quadrilaterals: kite, square, rhombus, rectangle, parallelogram.

In 9–15, *true or false.*

9. Every square is a rhombus.

10. Every rhombus is a square.

11. Every square is a kite.

12. Every kite is a rhombus.

13. If a quadrilateral has three sides of equal length, then it is a kite.

14. A property of every square is a property of every trapezoid.

15. A property of every trapezoid is a property of every parallelogram.

Applying the Mathematics

In 16–19, a figure is given. **a.** Write (in symbols) the information marked on the figure. **b.** Use this information to name the figure. Be as specific as you can, but do not be fooled by the looks of the diagram.

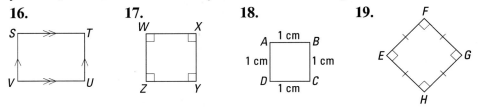

16. 17. 18. 19.

20. Let A = set of all rectangles and B = set of all rhombuses. Describe $A \cap B$.

21. In the figure at the left, circles O and Q intersect at N and P. Justify each conclusion in the proof that $NOPQ$ is a kite.

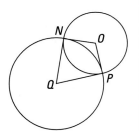

Conclusions	Justifications
1. $QN = QP$	**a.**
2. $ON = OP$	**b.**
3. $NOPQ$ is a kite.	**c.**

22. Which of the types of quadrilaterals in the hierarchy can be nonconvex? Support your answer with drawings.

In 23–25, a real object that has a quadrilateral boundary is given. Name all of the types of quadrilaterals that apply to each object.

23.   poster   **24.**   warning sign   **25.**   cross-section of
                                                                               a water trough

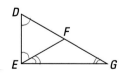

FLOUR SACK TOWELS
Popular Square Size 29" x 28"

26. Examine this ad for towels made by Excello, Ltd.
    a. What is mathematically wrong with the ad?
    b. Why do you think these towels are advertised in this way?

Review

27. One angle of an isosceles triangle has measure 84. Give *two* possible pairs of measures for the other two angles.   *(Lesson 6-2)*

28. Trace the diagram of the hockey rink below and draw the lines of symmetry.   *(Lesson 6-1)*

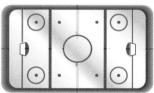

29. In the figure at the right, ℓ is a symmetry line for *ABCDEF*. What can be deduced from the Symmetric Figures Theorem?   *(Lesson 6-1)*

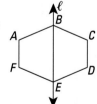

30. Given: In the figure at the left, $\triangle DEF$ is isosceles with vertex angle *F*. $\triangle EFG$ is isosceles with vertex angle *F*.
     To prove: $DF = FG$.   *(Lesson 5-4)*

Exploration

31. Biologists classify living things in a hierarchy. Show a hierarchy containing the following terms: human, cat, animal, mammal, organism, primate, chimpanzee, lion, feline, plant.

32. Describe another hierarchy outside of mathematics different from that in Question 31.

*Conjectures
about
Quadrilaterals*

IN·CLASS
ACTIVITY

Work on this activity in a group of 3 or 4. You will need an automatic drawer.

Pictured at the right is a kite with two distinct pairs of congruent sides. By measuring with a protractor, you can verify that m∠A = 118, m∠C = 60, and m∠B = m∠D = 91.

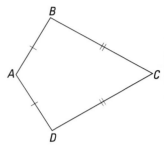

Does *every* kite have two angles of equal measure, or is *ABCD* in some way an unusual kite? To answer this question, examine other kites. Below are some kites drawn with an automatic drawing tool.

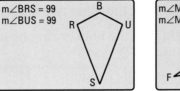

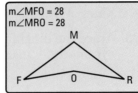

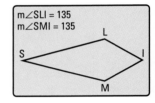

Each of the three kites has two angles with equal measures. These instances support the conjecture *Every kite has at least two angles of equal measure.*

1 **a.** Test this conjecture: *If a figure is a kite, then its diagonals are perpendicular.*
   b. Do you think the conjecture is true?

2 **a.** Test this conjecture: *If a figure is a parallelogram, then its diagonals have equal length.*
   b. Do you think the conjecture is true?

3 Make a conjecture about the diagonals of a rhombus.

4 Make a conjecture about the opposite angles of an isosceles trapezoid.

Properties of Kites

City on the lake. *Chicago's famous skyline is host to the Associates Building (center) as well as many other famous structures. The outline of the roof of the Associates Building is very nearly a kite.*

Paper kites and arrowheads are two of the places in which the quadrilateral called a kite appears. Kites are formed when four radii are drawn in intersecting circles, as in Figure I. And they are formed when a triangle is reflected over one of its sides, as in Figure II.

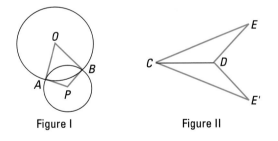

Figure I   Figure II

Kites also occupy an important position in the hierarchy of quadrilaterals. Remember that any property held by all figures of one type in the hierarchy is held by all types below it which are connected to it. Since all seven types of special quadrilaterals are either kites, trapezoids, or both, it is particularly useful to know the properties of these two types of figures.

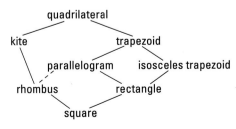

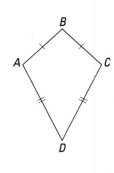

The common endpoints of the equal sides of a kite are the **ends** of the kite. In the kite at the left, B and D are the ends. For a rhombus, all four of the vertices are ends.

Reflection Symmetry of a Kite

Of all properties of kites, the most powerful is the kite's reflection symmetry. As usual, we begin the proof by drawing a figure and stating the "Given" and "To prove" in terms of that figure. The key to the proof is to use the symmetries of two isosceles triangles.

Kite Symmetry Theorem
The line containing the ends of a kite is a symmetry line for the kite.

Proof
Given: $ABCD$ is a kite with ends B and D.
To prove: $\overleftrightarrow{BD}$ is a symmetry line for $ABCD$.

Drawing:

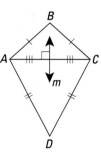

Argument:

Conclusions	Justifications
0. $ABCD$ is a kite with ends B and D.	Given
1. Let m be the $\perp$ bisector of $\overline{AC}$.	A segment has exactly one perpendicular bisector.
2. $r_m(A) = C$, $r_m(C) = A$	definition of reflection
3. $AB = BC$, $AD = DC$	definition of ends of kite
4. $\triangle ABC$ and $\triangle ADC$ are isosceles.	definition of isosceles triangle
5. m contains B and D.	The $\perp$ bisector of the base of an isosceles triangle is the angle bisector of the vertex angle (so it contains the vertex).
6. $r_{\overleftrightarrow{BD}}(B) = B$, $r_{\overleftrightarrow{BD}}(D) = D$	definition of reflection image
7. $r_{\overleftrightarrow{BD}}(ABCD) = CBAD$	Figure Reflection Theorem
8. $\overleftrightarrow{BD}$ is a symmetry line for $ABCD$.	definition of symmetry line

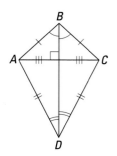

The diagonal determined by the ends ($\overline{BD}$ at the left) is called the **symmetry diagonal** of the kite. The line $\overleftrightarrow{BD}$ is the ⊥ bisector of the other diagonal of the kite. By the Figure Reflection Theorem, $r_{\overleftrightarrow{BD}}(\angle ABD) = \angle CBD$ and $r_{\overleftrightarrow{BD}}(\angle ADB) = \angle CDB$. Thus $\overline{BD}$ bisects $\angle ABC$ and $\angle ADC$. This proves the following theorem.

> **Kite Diagonal Theorem**
> The symmetry diagonal of a kite is the perpendicular bisector of the other diagonal and bisects the two angles at the ends of the kite.

Example 1

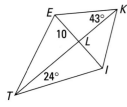

Given kite *KITE* at the left with ends *K* and *T*, *EL* = 10, m∠*EKT* = 43, and m∠*ITK* = 24. Find
a. m∠*IKT*.   **b.** m∠*ETK*.   **c.** *LI*.

Solution
a. By the Kite Diagonal Theorem, $\overleftrightarrow{KT}$ is a symmetry line for the kite. So r($\angle EKT$) = $\angle IKT$. Thus, m∠IKT = 43.
b. Also, r($\angle ITK$) = $\angle ETK$, so m∠ETK = 24.
c. Since $\overleftrightarrow{KT}$ is the ⊥ bisector of $\overline{EI}$, LI = 10.

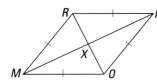

Batplane? *The shape of this stealth bomber resembles the kite in Figure II on page 323.*

Symmetry of Rhombuses

The Kite Diagonal Theorem applies to rhombuses and squares as well as to kites because rhombuses and squares are below and connected to kites in the hierarchy. A rhombus is a kite, either of whose opposite vertices can be a pair of ends. Thus a rhombus has two symmetry diagonals. Consequently, the diagonals of a rhombus are perpendicular and bisect each other.

> **Rhombus Diagonal Theorem**
> Each diagonal of a rhombus is the perpendicular bisector of the other diagonal.

In rhombus *RHOM* at the left, $\overline{RO}$ is the perpendicular bisector of $\overline{HM}$ and $\overline{HM}$ is the perpendicular bisector of $\overline{RO}$.

Because of the symmetry of *RHOM*, many angles have the same measure, and many segments have the same length. For instance, *X* is the midpoint of $\overline{HM}$ and $\overline{RO}$. Also, because of the two symmetry lines, opposite angles of a rhombus are congruent. That is, $\angle MRH \cong \angle MOH$ and $\angle RHO \cong \angle RMO$.

Example 2

In rhombus *RHOM* below, if m∠*RHM* = 23, find
a. m∠*MHO*.　　**b.** m∠*RMH*.　　**c.** m∠*OMH*.　　　**d.** m∠*XRH*.

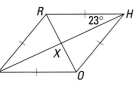

Solution

a. Each diagonal is an angle bisector of the vertex angles.
Thus m∠MHO = 23.

b. △*RMH* is isosceles with base $\overline{MH}$. So, m∠RMH = m∠RHM = 23.

c. Since $\overleftrightarrow{MH}$ bisects ∠*RMO*, m∠OMH = 23.

d. Since $\overline{MH}$ ⊥ $\overline{RO}$, ∠*RXH* is a right angle.
By the Triangle-Sum Theorem, m∠XRH = m∠XOH = 67.

Completing the Hierarchy of Quadrilaterals

Notice in Example 2 that there are two pairs of congruent alternate interior angles: ∠*RHM* and ∠*OMH*, and ∠*RMH* and ∠*OHM*. From the first pair $\overline{RH}$ // $\overline{OM}$ and from the second pair $\overline{OH}$ // $\overline{MR}$. This allows us to prove the final connection of the hierarchy of quadrilaterals.

Theorem

If a quadrilateral is a rhombus, then it is a parallelogram.

Proof
Given:　　*RHOM* is a rhombus.
To prove:　*RHOM* is a parallelogram.

Drawing:

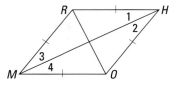

Argument:
By the Kite Symmetry Theorem, both $\overline{RO}$ and $\overline{HM}$ are symmetry diagonals of *RHOM*. By the Kite Diagonal Theorem using symmetry diagonal $\overline{HM}$, ∠1 ≅ ∠2 and ∠4 ≅ ∠3. Using symmetry diagonal $\overline{RO}$, ∠2 ≅ ∠4. Applying the Transitive Property of Congruence, ∠1 ≅ ∠4 and ∠2 ≅ ∠3. By the AIA ≅ ⇒ // Lines Theorem, $\overline{RH}$ // $\overline{MO}$ and $\overline{MR}$ // $\overline{OH}$. This makes *RHOM* a parallelogram, by the definition of parallelogram.

Covering the Reading

1. Which types of quadrilaterals in the hierarchy of quadrilaterals are always kites?

2. State the Kite Symmetry Theorem.

3. Refer to kite *PLAY* at the right.
   $PL = LA$ and $AY = YP$.
   a. Name the ends of *PLAY*.
   b. Name its symmetry line.
   c. $r_{\overleftrightarrow{LY}}(\angle LPY) = $ ___?___

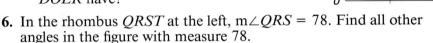

4. Use kite *ABCD* at the left. If $BE = 15$, $m\angle BCA = 30$, and $m\angle DAC = 50$, find as many other angle measures and lengths as you can.

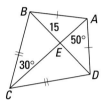

5. At the right, $DO = OL = LR = RD$.
   a. Is *DOLR* a kite?
   b. If so, name its ends. If not, why not?
   c. How many lines of symmetry does *DOLR* have?

6. In the rhombus *QRST* at the left, $m\angle QRS = 78$. Find all other angles in the figure with measure 78.

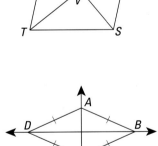

7. Rewrite the argument from the proof that all rhombuses are parallelograms, using two-column form.

Applying the Mathematics

8. Complete the following paragraph proof that the opposite angles of a rhombus are equal. Use the diagram at the left.

   The reflection image of $\angle ADC$ over $\overleftrightarrow{AC}$ is __a.__ , so $m\angle ADC = $ __b.__ because reflections preserve __c.__ . Similarly, the reflection image of $\angle DAB$ over __d.__ is __e.__ , so $m\angle DAB = $ __f.__

9. Given that $E' = r_{\overleftrightarrow{CD}}(E)$ in the figure on page 323, explain why *ECE'D* is a kite.

10. a. Draw a nonconvex kite *ABCD* with ends *B* and *D*.
    b. Are the Kite Symmetry and Kite Diagonal theorems true for nonconvex kites? Explain why or why not.

11. Two sticks $\overline{XZ}$ and $\overline{WY}$ are lashed together to form the kite at the left. If $XZ = 20''$, $WY = 30''$, and $VY = 7''$, find each measure.
    a. XV b. VZ c. WV d. $m\angle YVZ$

12. In rhombus *BUSY* at the right, m∠*YBU* = 76. Find each angle measure.

    a. m∠*YBS*   **b.** m∠*UBS*

    c. m∠*USB*   **d.** m∠*YSB*

    e. Is $\overleftrightarrow{BU}$ // $\overleftrightarrow{YS}$? Explain your answer.

Review

In 13–15, true or false. *(Lesson 6-3)*

13. Every parallelogram is a kite.   **14.** All rectangles are trapezoids.

15. If a figure is an isosceles trapezoid, then it is a square.

In 16 and 17, define. *(Lesson 6-3)*

16. trapezoid   **17.** isosceles trapezoid

18. Name all the special types of quadrilaterals which are isosceles trapezoids. *(Lesson 6-3)*

In 19–21, refer to the hierarchy of quadrilaterals. Answer always, sometimes, *or* never. *(Lesson 6-3)*

19. A figure is a parallelogram if it is a rectangle.

20. Every square is a trapezoid.

21. If a figure is a rhombus, then it is a square.

22. One side of an equilateral triangle has length 3″. What is the measure of each angle of the triangle? *(Lesson 6-2)*

23. In the figure at the right, *m* // *n*.

    a. Find *x*.

    b. Is the figure accurately drawn? Explain your answer. *(Lesson 3-6)*

$(7x + 10)°$ *m*

$(6x + 31)°$ *n*

24. At the left is pictured a clock. What is the measure of the arc that the minute hand travels from 3:00 to 3:01? *(Lesson 3-2)*

Exploration

25. Let *B* and *D* be ends of a kite, as shown at the right. Let *M*, *N*, *P*, and *Q* be midpoints of the sides of the kite. If m∠*A* = 115 and m∠*B* = 92, find the measures of as many other angles in this figure as you can.

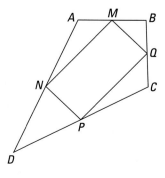

Properties of Trapezoids

Trapezoidal property. *Flood irrigation in the Arizona desert helps farmers grow crops. The darker green fields are alfalfa.*

Think of two parallel pieces of wood or metal used in building a structure. If the ends of the pieces are connected, a trapezoid is formed. So trapezoids are exceedingly common. Trapezoids also occupy a high place in the hierarchy of quadrilaterals.

Any property of all trapezoids holds for all parallelograms, rhombuses, rectangles, squares, and isosceles trapezoids. This makes it very valuable and efficient to look for properties of trapezoids.

Angles in Trapezoids

In the trapezoid at the right, $\overline{AB} \parallel \overline{DC}$. $\overline{AD}$ has been extended beyond A to point E. This forms the linear pair $\angle 1$ and $\angle 2$. From this information, we can deduce the following:

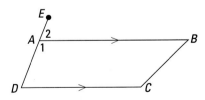

Conclusions	Justifications
1. $m\angle 1 + m\angle 2 = 180$	Linear Pair Theorem
2. $m\angle 2 = m\angle D$	$\parallel$ lines $\Rightarrow$ corr. $\angle$s $=$
3. $m\angle 1 + m\angle D = 180$	Substitution (step 2 into step 1)
4. $\angle 1$ and $\angle D$ are supplementary.	definition of supplementary angles

This argument could be repeated with $\angle B$ and $\angle C$ and with any trapezoid. The result is the following theorem.

> **Trapezoid Angle Theorem**
> In a trapezoid, consecutive angles between a pair of parallel sides are supplementary.

Example 1

In trapezoid *TRAP* at the right, $\overline{TR} \,/\!/\, \overline{PA}$.
If m∠*A* = 82, find the measures of as many other angles as you can.

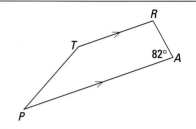

Solution

From the Trapezoid Angle Theorem, angles *A* and *R* are supplementary.
82 + m∠R = 180. So m∠R = 98.
No other specific angle measures can be found. However, the Trapezoid Angle Theorem tells you that m∠*P* + m∠*T* = 180.

Recall that an isosceles trapezoid is defined to be a trapezoid with a pair of base angles of equal measure. If *TRAP* in Example 1 were an isosceles trapezoid, then m∠*P* = m∠*A* = 82. (The diagram would need to be changed.) Then m∠*T* = m∠*R* = 98. Thus, in an isosceles trapezoid, each pair of base angles is equal in measure.

Symmetry of Isosceles Trapezoids

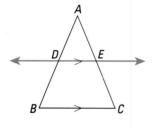

Isosceles trapezoids are also related to isosceles triangles. Let △*ABC* be isosceles with vertex angle *A* and base $\overline{BC}$. Draw $\overleftrightarrow{DE} \,/\!/\, \overline{BC}$. Then you can prove that *DECB* is an isosceles trapezoid. (See Question 7.) Since an isosceles triangle is reflection-symmetric, you would expect that an isosceles trapezoid is reflection-symmetric as well. This is the case.

> **Isosceles Trapezoid Symmetry Theorem**
> The perpendicular bisector of one base of an isosceles trapezoid is the perpendicular bisector of the other base and a symmetry line for the trapezoid.

Drawing:

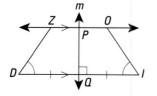

Proof

Given: Isosceles trapezoid *ZOID* with m∠*I* = m∠*D*.
    m is the perpendicular bisector of $\overline{DI}$.

To prove: *m* is the ⊥ bisector of $\overline{ZO}$.
    m is a symmetry line for *ZOID*.

Argument:

To show that *m* is a symmetry line for *ZOID*, we need to show that $r_m(ZOID) = OZDI$. Obviously, $r_m(D) = I$ and $r_m(I) = D$ since *m* is the ⊥ bisector of $\overline{DI}$. Now we want to show that $r_m(Z) = O$. Since *ZOID* is a trapezoid, $\overline{ZO} \,/\!/\, \overline{ID}$; thus $\overline{ZO} \perp m$ by the Perpendicular to Parallels

▶

330

Theorem. Thus, by the definition of reflection, $r_m(Z)$ lies on $\overleftrightarrow{ZO}$. Since reflections preserve angle measure, you know that $r_m(\overrightarrow{DZ}) = \overrightarrow{IO}$, so $r_m(Z)$ lies on $\overrightarrow{IO}$. Since $\overrightarrow{IO}$ and $\overleftrightarrow{ZO}$ intersect at O, $r_m(Z)$ must be the point O. This proves, by the definition of reflection, that m is the $\perp$ bisector of $\overline{OZ}$.

By the Flip-Flop Theorem, $r_m(O) = Z$. Thus, by the Figure Reflection Theorem, $r_m(ZOID) = OZDI$, and m is a symmetry line of $ZOID$.

The Isosceles Trapezoid Symmetry Theorem has a corollary. By the Symmetric Figures Theorem, $\overline{ZD} \cong \overline{OI}$. This proves:

Isosceles Trapezoid Theorem
In an isosceles trapezoid, the non-base sides are congruent.

Example 2

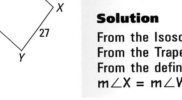

$WXYZ$ at the left is an isosceles trapezoid with bases $\overline{WX}$ and $\overline{YZ}$, $XY = 27$, and $m\angle Z = 83$. Find as many other lengths and angle measures as possible.

Solution

From the Isosceles Trapezoid Theorem, WZ = XY, so WZ = 27.
From the Trapezoid Angle Theorem, $m\angle W = 180 - 83 = 97$.
From the definition of isosceles trapezoid,
$m\angle X = m\angle W = 97$ and $m\angle Y = m\angle Z = 83$.

Rectangles

A rectangle can be considered an isosceles trapezoid in two ways. Either pair of parallel sides can be the bases. This yields another corollary of the Isosceles Trapezoid Symmetry Theorem.

Rectangle Symmetry Theorem
The perpendicular bisectors of the sides of a rectangle are symmetry lines for the rectangle.

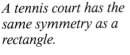

A tennis court has the same symmetry as a rectangle.

Rectangle $RECT$ and its two symmetry lines ℓ and m are pictured below.

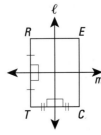

Covering the Reading

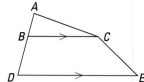

1. In the figure at the left, B is between A and D. $BCED$ is a trapezoid with $\overline{BC}$ // $\overline{DE}$. Justify each statement in this proof of the Trapezoid Angle Theorem.
   a. $m\angle ABC = m\angle D$
   b. $m\angle ABC + m\angle CBD = 180$
   c. $m\angle D + m\angle CBD = 180$
   d. $m\angle D$ and $m\angle CBD$ are supplementary.

2. To which special types of quadrilaterals does the Trapezoid Angle Theorem apply?

3. In trapezoid $PART$ at the right, $\overline{PA}$ // $\overline{TR}$, $m\angle P = 68$, and $m\angle R = 95$.
   a. Name the bases.
   b. Name a pair of base angles.
   c. Find the measure of as many angles in $PART$ as possible.

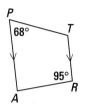

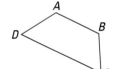

4. Isosceles trapezoid $ABCD$ at the left has bases $\overline{AB}$ and $\overline{DC}$. Trace $ABCD$ and construct m, a symmetry line for the trapezoid.

5. Refer to isosceles trapezoid $ZOID$ with bases $\overline{ZO}$ and $\overline{ID}$ at the right. If q is the $\perp$ bisector of $\overline{DI}$, $DZ = 12$ cm, and $m\angle I = 125$, find as many other lengths and angle measures as possible.

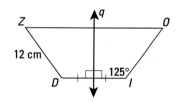

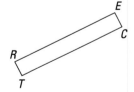

6. Trace rectangle $RECT$ at the left and draw all of its symmetry lines.

Applying the Mathematics

7. Given: $\triangle ISO$ at the right is isosceles with $IS = SO$, and $\overline{DZ}$ // $\overline{IO}$. Fill in each justification in this argument that $ZOID$ is an isosceles trapezoid.

Conclusions	Justifications
0. $\triangle ISO$ is isosceles with $IS = SO$, $\overline{DZ}$ // $\overline{IO}$.	Given
1. $m\angle I = m\angle O$	**a.**
2. $ZOID$ is a trapezoid.	**b.**
3. $ZOID$ is an isosceles trapezoid.	**c.**

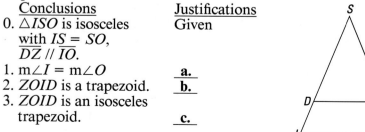

8. Use the figure at the right. Quadrilateral *PARL* is a parallelogram. If m∠*R* = 27.3, find the measures of as many other angles as you can.

9. Given: *ABCD* is an isosceles trapezoid with bases $\overline{AB}$ and $\overline{DC}$.
  a. Prove: *AC* = *BD*. (Hint: use symmetry and reflections.)
  b. State the result of part **a** in words as a theorem.

10. Refer to Question 9. *True or false.*
  a. The diagonals of an isosceles trapezoid are equal in measure.
  b. The diagonals of a rectangle are equal in measure.
  c. The diagonals of a square are equal in measure.

11. In trapezoid *ABCD* with bases $\overline{AB}$ and $\overline{DC}$ at the right, m∠*D* = *x* and m∠*B* = 2*y*. Find each measure.
  a. m∠*A*
  b. m∠*C*

12. a. Which theorem of this lesson tells you that the top of most mattresses has two symmetry lines?
  b. Why do you think most mattresses are shaped this way?

13. Describe all the symmetry lines of a square *ABCD*.

Review

14. From memory, try to draw the complete quadrilateral hierarchy. *(Lesson 6-3)*

In 15 and 16, use the kite at the right with ends *I* and *E*.
15. If *KI* = 4, *KR* = 3, *ET* = 8, m∠*RIT* = 49, and m∠*KET* = 41, find as many other lengths and angle measures as possible. *(Lesson 6-4)*

16. Justify each conclusion. *(Lessons 4-1, 6-4)*
  a. $\overleftrightarrow{IE}$ is a symmetry line for the kite.
  b. $\overleftrightarrow{IE}$ is the ⊥ bisector of $\overline{KT}$.
  c. $r_{\overleftrightarrow{IE}}(R) = R$

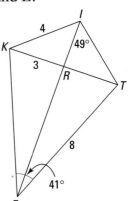

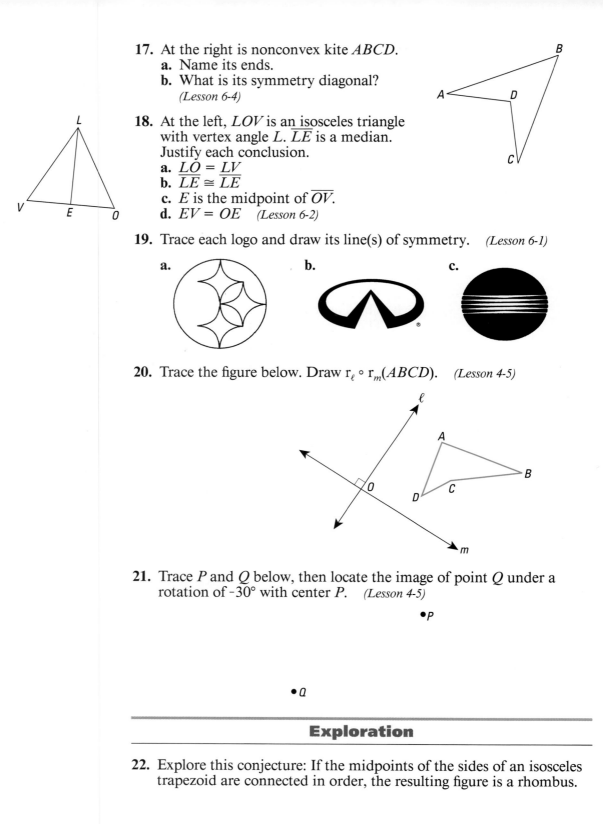

17. At the right is nonconvex kite *ABCD*.
  a. Name its ends.
  b. What is its symmetry diagonal?
  (Lesson 6-4)

18. At the left, *LOV* is an isosceles triangle with vertex angle *L*. $\overline{LE}$ is a median. Justify each conclusion.
  a. $LO = LV$
  b. $\overline{LE} \cong \overline{LE}$
  c. *E* is the midpoint of $\overline{OV}$.
  d. $EV = OE$ *(Lesson 6-2)*

19. Trace each logo and draw its line(s) of symmetry. *(Lesson 6-1)*

  a.   **b.**   **c.**

20. Trace the figure below. Draw $r_\ell \circ r_m(ABCD)$.   *(Lesson 4-5)*

21. Trace *P* and *Q* below, then locate the image of point *Q* under a rotation of $-30°$ with center *P*.   *(Lesson 4-5)*

• *P*

• *Q*

Exploration

22. Explore this conjecture: If the midpoints of the sides of an isosceles trapezoid are connected in order, the resulting figure is a rhombus.

LESSON

6-6

Rotation Symmetry

The cutting edge. *Saw blades have many shapes. Circular saw blades vary in the number and the size of teeth, but all have rotation symmetry.*

When Are Figures Rotation-Symmetric?

Reflection symmetry is not the only type of symmetry. Just as reflection symmetry means that there is a reflection mapping a figure onto itself, *rotation symmetry* means that there is a rotation that maps a figure onto itself.

For instance, if this saw blade is rotated 15° (clockwise or counterclockwise), then it will look as it does now. Its *center of symmetry* is the point *O*. Here is the general definition.

> **Definition**
> A plane figure *F* is a **rotation-symmetric figure** if and only if there is a rotation R with magnitude between 0° and 360° such that R(*F*) = *F*. The center of R is a **center of symmetry** for *F*.

The blade on page 335 has *24-fold rotation symmetry* because 24 rotations of magnitude 15° bring the saw blade back to its original position. In general, if *m* is the smallest positive magnitude for a rotation that maps a figure onto itself and if $n = \frac{360}{m}$, then *n* rotations will bring it back to its original position. We say that the figure has ***n*-fold rotation symmetry.**

Activity

Trace the following figures. Mark the center of rotation and give the *n*-fold rotation symmetry.

a.   b.   c.

Reflection and rotation symmetry do not always occur together, as the saw blade and these designs show.

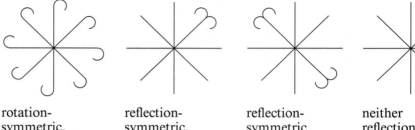

| rotation-symmetric, not reflection-symmetric | reflection-symmetric, not rotation-symmetric | reflection-symmetric and rotation-symmetric | neither reflection-symmetric nor rotation-symmetric |

When Does Reflection Symmetry Imply Rotation Symmetry?

A rotation is a composite of reflections over intersecting lines. So whenever a figure has intersecting symmetry lines, it has rotation symmetry.

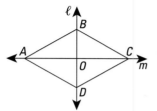

Consider rhombus *ABCD* at the left, and its symmetry lines ℓ and *m*. Both reflections, r_ℓ and r_m, map *ABCD* onto itself. Thus, so does their composite $r_m \circ r_\ell$. Specifically,

$$r_m \circ r_\ell \,(ABCD) = r_m(CBAD) = CDAB.$$

Since the angle between ℓ and *m* has measure 90°, the Two-Reflection Theorem for Rotation tells us the rotation $r_m \circ r_\ell$ has measure 180°. Its center is *O*, the intersection of the diagonals. Since $\frac{360}{180} = 2$, *ABCD* has 2-fold rotation symmetry.

This argument can be repeated whenever a figure has two intersecting symmetry lines.

Rotor rooter. *This helicopter rotor has 2-fold rotation symmetry. Other helicopter rotors may have 3-fold or 4-fold symmetry.*

Theorem

If a figure possesses two lines of symmetry intersecting at a point *P*, then it is rotation-symmetric with a center of symmetry at *P*.

Line segments, equilateral triangles, and circles have two or more intersecting symmetry lines, so they are rotation-symmetric.

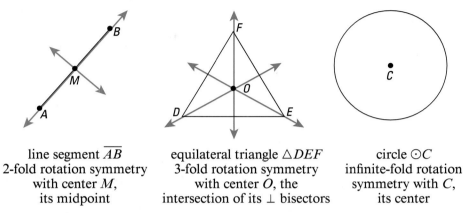

| line segment $\overline{AB}$ | equilateral triangle $\triangle DEF$ | circle $\odot C$ |
| 2-fold rotation symmetry with center *M*, its midpoint | 3-fold rotation symmetry with center *O*, the intersection of its $\perp$ bisectors | infinite-fold rotation symmetry with *C*, its center |

From the Rectangle Symmetry Theorem, rectangles have perpendicular symmetry lines, so rectangles, and thus squares, have rotational symmetry. The only other quadrilaterals with rotation symmetry are parallelograms.

You will examine the properties of parallelograms in the next chapter.

QUESTIONS

Covering the Reading

1. *ABCD* is a rhombus. Describe its reflection and rotation symmetry.

2. Define *rotation-symmetric figure*.

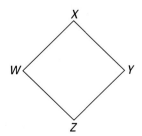

3. Trace square *WXYZ* at the left.
   a. Locate its center of symmetry.
   b. This square has *n*-fold rotation symmetry. What is the value of *n*?

4. At the right is a ceiling fan.
   a. What is the magnitude of the rotation about *O* which maps *A* to *B*?
   b. The fan has __?__-fold rotation symmetry.

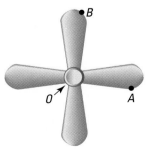

5. Give your answers for the Activity in this lesson.

6. Rectangle *RECT* below has its two symmetry lines ℓ and m drawn.
 a. What point is its center of symmetry?
 b. It has *n*-fold rotation symmetry. What is *n*?

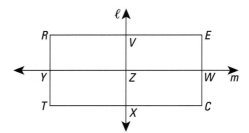

7. Draw a figure that has 3-fold rotation symmetry but no reflection symmetry.

Applying the Mathematics

8. Below is rectangle *ABCD* with its diagonals. From the symmetry of the rectangle, list as many properties of the diagonals as you can.

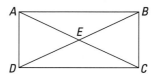

9. At the left is equilateral triangle *PQR* with symmetry lines ℓ, *m*, and *n*. Find *x*.

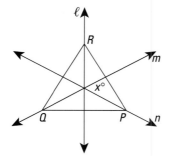

10. Draw a hexagon that has rotation symmetry but no reflection symmetry.

11. Draw a hexagon that has reflection symmetry but no rotation symmetry.

12. Explain why normal playing cards (for games like bridge) have rotation symmetry.

13. What kind of symmetry does the word below have?

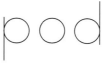

Review

14. Draw a quadrilateral hierarchy with a representative figure by each word. Mark the angles of the figures with tick marks to show all congruent angles. *(Lessons 6-4, 6-5)*

In 15 and 16, *WXYZ* is an isosceles trapezoid with base angles *X* and *Y*. *(Lesson 6-5)*

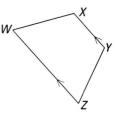

15. Trace the figure and construct its symmetry line.

16. If $m\angle X = -2q + 71$ and
$m\angle Y = -5q + 32$, find $m\angle Z$ and $m\angle W$.

17. Graph points $A = (0, 0)$, $B = (10, 0)$, $C = (7, 3)$, and $D = (2, 3)$. What kind of figure is *ABCD*? Explain how you know. *(Lesson 6-5)*

18. Write an argument to complete the proof.
Given: *ABCD* is a kite with ends *B* and *D*.
To prove: $\angle BAC \cong \angle BCA$.   *(Lesson 6-4)*

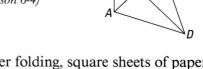

19. In origami, the Japanese art of paper folding, square sheets of paper are repeatedly folded. If you take a square sheet of paper and fold it onto itself along a diagonal, what kind of figure is formed? *(Lesson 6-2)*

In 20–22, use the figure below, in which $\ell \, / \! / \, m$.   *(Lessons 3-3, 3-4, 3-6, 5-7)*

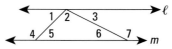

20. a. $m\angle 1 + m\angle 2 + m\angle 3 = \underline{\ ?\ }$
   b. If $m\angle 1 = 35$, what is $m\angle 4$?
   c. If $m\angle 1 + m\angle 2 = 145$, what is $m\angle 6$?

21. If $m\angle 5 = 45$ and $m\angle 6 = 40$, find the measures of as many other angles as possible.

22. *Multiple choice.*   If $m\angle 4 = x$, then $m\angle 2 + m\angle 3 = \underline{\ ?\ }$.
   (a) x (b) $90 - x$ (c) $180 - x$ (d) $\dfrac{180 - x}{2}$

23. Give a counterexample to this conditional: *If a triangle is isosceles, then its largest angle is acute.*   *(Lessons 2-2, 2-6)*

Exploration

24. a. Draw a hexagon with no symmetry lines.
   b. Draw a hexagon with exactly 1 symmetry line.
   c. Draw a hexagon with exactly 2 symmetry lines.
   d. Draw a hexagon with exactly 3 symmetry lines.
   e. Draw a hexagon with 6 symmetry lines.
   f. Kichiro did parts **a** through **e** and made the following conjecture: An *n*-gon can have *m* symmetry lines only if *m* is a factor of *n*. Do you think Kichiro's conjecture is true?

To life! *The origami crane is a symbol of happiness and long life. Chains of 1000 cranes are made by and for people who are ill.*

*Symmetry
Lines in
Polygons*

IN-CLASS
ACTIVITY

Work with a partner. You will need a ruler, compass, and protractor.

Below are two *regular polygons,* one with seven sides and one with eight sides.

regular heptagon   regular octagon

1 Trace each figure. Draw all symmetry lines.

2 Consider the regular octagon.
a. How many symmetry lines does this figure have?
b. How many of these lines bisect two angles of the octagon?
c. How many of these lines are ⊥ bisectors of two sides of the octagon?
d. Are there any other symmetry lines? If so, describe them.

3 Repeat Step 2 for the regular heptagon.

4 For each figure, let O be the point of concurrence of the symmetry lines. Using a compass, draw ⊙O with radius OA for each figure. Which other vertices of the figure lie on ⊙O?

5 Retrace the original figures and put point O on each figure. Draw $\overline{OA}$, $\overline{OB}$, $\overline{OC}$, and $\overline{OD}$. Draw in the one symmetry line ℓ where $r_\ell(B) = C$.
a. Which point is $r_\ell(A)$?
b. Which point is $r_\ell(O)$?
c. What kind of triangle is $\triangle ABC$?
d. What kind of quadrilateral is $OABC$?
e. What kind of quadrilateral is $ABCD$?

Lighten up. *The tips of the lights of this chandelier are vertices of a regular hexagon.*

Chandeliers usually need to light a room in all directions, so they are designed to have as much symmetry as possible. You have seen that the triangle with the most symmetry is the equilateral triangle. The quadrilateral that has the most symmetry is the square. In general, the polygons that exhibit the most reflection and rotation symmetry for their number of sides are the *regular polygons*.

> **Definition**
> A **regular polygon** is a convex polygon whose angles are all congruent and whose sides are all congruent.

You can see regular polygons in the structure of many chandeliers. Only the regular polygons with 3 and 4 sides have special names. Otherwise they are simply called **regular pentagons, regular hexagons,** and so on.

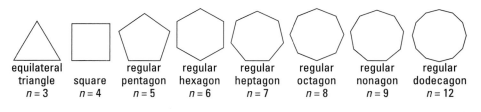

| equilateral triangle $n = 3$ | square $n = 4$ | regular pentagon $n = 5$ | regular hexagon $n = 6$ | regular heptagon $n = 7$ | regular octagon $n = 8$ | regular nonagon $n = 9$ | regular dodecagon $n = 12$ |

Equilateral and Equiangular Polygons

If all sides of a polygon have the same length, the polygon is called **equilateral.** If all angles of a polygon have the same measure, the polygon is called **equiangular.** Rhombuses and squares are equilateral quadrilaterals. Rectangles and squares are equiangular polygons. Regular polygons are both equilateral and equiangular.

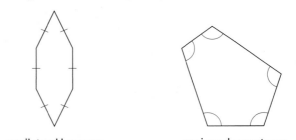

equilateral hexagon   equiangular pentagon

The Polygon-Sum Theorem gives a quick way to determine the measure of an individual angle in an equiangular polygon.

Example 1

What is m∠ZUV in regular hexagon *UVWXYZ* below?

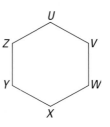

Solution

By the Polygon-Sum Theorem, the sum of all the angle measures in a hexagon is $(6 - 2) \cdot 180 = 720$. Since a regular hexagon is equiangular, there are 6 angles of equal measure. So m∠ZUV = $\frac{720}{6}$ = 120°.

Regular Polygons and Circles

Notice that the chandelier pictured on page 341 has a center, and each of its lights is equidistant from that center. In Question 4 of the In-class Activity, you should have found that you can draw a single circle which contains the vertices of a regular polygon. This is an important property of regular polygons, and we call it the *Center of a Regular Polygon Theorem*. In the proof of this theorem, which is on the next page, notice that we do not draw the entire regular polygon, since regular polygons with different numbers of sides look different. The proof applies many properties you have learned, so you should read it carefully. We suggest that you draw a figure like the one in the proof. As you read through the argument, mark each justified conclusion on your figure.

One can only hope! *The famous blue Hope diamond, which weighs 45.52 carats, was cut from a diamond found in India. Diamonds are cut to have many faces, called* facets. *The more facets the diamond has, the greater its brilliance. Often the facets are equilateral polygons.*

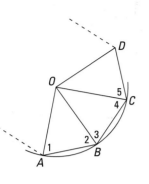

Center of a Regular Polygon Theorem
In any regular polygon there is a point (its **center**) which is equidistant from all of its vertices.

Proof

Given: Regular polygon $ABCD$ . . . .

To prove: There is a point O equidistant from A, B, C, D, ...

Drawing: $ABCD$ . . . at the left.

Argument:

There is a circle containing the three vertices A, B, and C. Call its center O. By the definition of circle, $OA = OB = OC$. Our goal is to prove that $OC = OD$ because this will mean that D is also on the circle.

Since $AB = BC$ by the definition of regular polygon, quadrilateral $OABC$ is a kite and $\overline{OB}$ is its symmetry diagonal. From this, and the fact that $\triangle OBC$ is isosceles,

$$m\angle 2 = m\angle 3 = m\angle 4.$$

The angles of a regular polygon are all congruent, so

$$m\angle ABC = m\angle BCD.$$

By Angle Addition and Substitution, $m\angle 2 + m\angle 3 = m\angle 4 + m\angle 5$. Subtracting two of the equal measures, $m\angle 2 = m\angle 5$, and so by transitivity, $m\angle 5 = m\angle 4$. This makes $\overrightarrow{CO}$ an angle bisector. By the definition of regular polygon, $\triangle BCD$ (not drawn) is isosceles, and the angle bisector $\overleftrightarrow{CO}$ is its symmetry line. Over that line, the reflection image of B is D, and O is on $\overleftrightarrow{CO}$, so the reflection image of O is O itself. Since reflections preserve distance, $OB = OD$. This means that D is on the circle with center O and radius OB. Repeating this process, the circle that contains B, C, and D also contains the next vertex of the regular polygon, and so on. So one circle contains all the vertices of any regular polygon, and the center of that circle is equidistant from all the vertices of the polygon.

The Center of a Regular Polygon Theorem enables regular polygons to be drawn. Draw the circle first. Then equally space the vertices of the polygon around the circle.

Activity

Follow this algorithm to draw a regular pentagon.

Step 1: Draw a circle with center O and mark a point A on it.

Step 2: Divide 360 by 5 (the number of sides in the pentagon).

Step 3: Find a point B on the circle so that $m\angle AOB$ equals your answer to Step 2.

Step 4: Find points C, D, and E so that $m\angle BOC$, $m\angle COD$, and $m\angle DOE$ each equal $m\angle AOB$.

Step 5: Connect A, B, C, D, and E to form the regular pentagon.

Regular Polygons and Symmetry

All regular polygons possess reflection symmetry. To the diagram for the Center of a Regular Polygon Theorem we add ℓ, the perpendicular bisector of $\overline{AB}$.

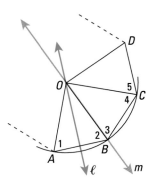

The line ℓ is a symmetry line for isosceles $\triangle ABO$, and it can be proved it is a symmetry line for the entire regular polygon. The line m that contains the bisector of $\angle ABC$ is a symmetry line for the kite $OABC$ and also for the entire regular polygon.

If a regular polygon has an odd number of sides, a bisector of one angle is a perpendicular bisector of the opposite side. This happens in equilateral triangles. If the number of sides is even, such as the square, a perpendicular bisector of one side is also a perpendicular bisector of the opposite side. So it may not be clear at first how many symmetry lines a regular n-gon has. But there is a nice way to determine the number. All the symmetry lines are either angle bisectors, perpendicular bisectors of sides, or both. Since there are n angles and n sides in an n-gon, there would seem to be $2n$ lines. But each line is counted twice, and so there are n lines of symmetry.

Regular polygons also possess rotation symmetry. By the Two-Reflection Theorem, $r_m \circ r_\ell$ is a rotation with center O and magnitude $m\angle AOB$. By taking individual points, you can verify that $r_m \circ r_\ell\,(\overline{AB}) = \overline{BC}$ and $r_m \circ r_\ell\,(\overline{AB}) = \overline{CD}$. So, under this rotation, each side of the regular polygon is mapped onto the next side. This indicates that the regular polygon is rotation-symmetric. Since there are n sides, there is n-fold rotation symmetry. The following theorem summarizes these ideas.

Regular Polygon Symmetry Theorem
Every regular n-gon possesses
(1) n symmetry lines, which are the perpendicular bisectors of each of its sides and the bisectors of each of its angles;
(2) n-fold rotation symmetry.

Example 2

At the left is regular heptagon *ABCDEFG*. What is the smallest positive magnitude of rotation which maps *ABCDEFG* onto itself?

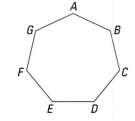

Solution

Since *ABCDEFG* is a regular 7-gon, it has 7-fold rotation symmetry. So the least positive magnitude of rotation is $\frac{360°}{7}$, or $51\frac{3}{7}°$.

QUESTIONS

Covering the Reading

1. Why are the structures of many chandeliers based on regular polygons?

2. A regular polygon with four sides is usually called a(n) __?__.

In 3 and 4, draw the figure.

3. an equilateral pentagon that is not a regular pentagon

4. an equiangular hexagon that is not a regular hexagon

In 5–7, refer to the In-class Activity on page 340.

5. Give your answers to Step 2.

6. Give your answers to Step 3.

7. Give your answers to Step 5.

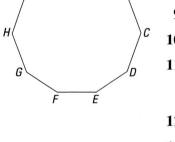

In 8 and 9, refer to regular decagon *ABCDEFGHIJ* at the left.

8. What is m∠*DEF*?

9. Trace the figure and draw the circle containing its vertices.

10. Give your answer to the Activity in the lesson.

11. **a.** Draw a regular hexagon *PQRSTU*.
    b. What is the smallest positive magnitude of rotation which maps *PQRSTU* onto itself?

12. A regular nonagon has __?__-fold rotation symmetry.

13. A regular nonagon has __?__ symmetry lines.

14. When is the perpendicular bisector of a side of a regular *n*-gon also an angle bisector?

Applying the Mathematics

In 15 and 16, an object is given. For each object, explain why its shape is a regular polygon.

15. a nut on a car wheel   16. an open umbrella

17. Construct equilateral triangles *AOB*, *BOC*, *COD*, *DOE*, *EOF*, and *FOA*, where *A*, *B*, *C*, *D*, *E*, and *F* are all different points.
    a. What kind of figure is *ABCDEF*?
    b. Justify your answer to part **a.**

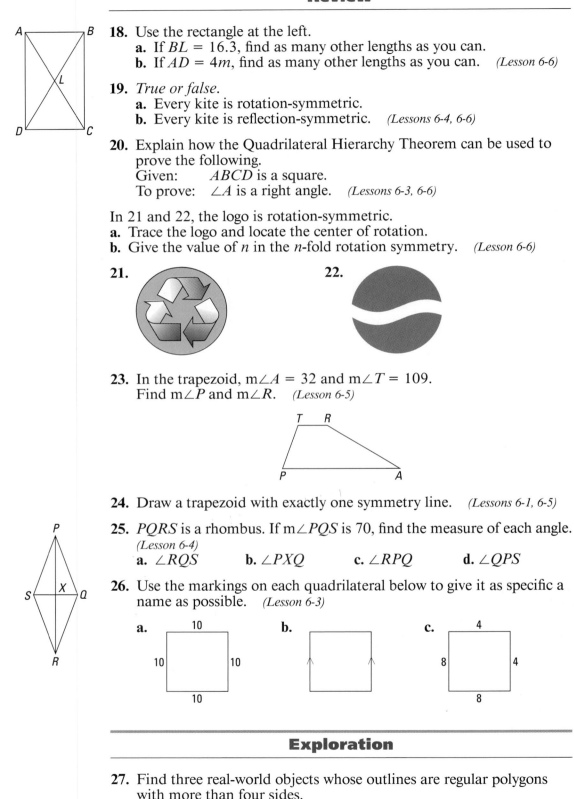

18. Use the rectangle at the left.
  a. If $BL = 16.3$, find as many other lengths as you can.
  b. If $AD = 4m$, find as many other lengths as you can. *(Lesson 6-6)*

19. *True or false.*
  a. Every kite is rotation-symmetric.
  b. Every kite is reflection-symmetric. *(Lessons 6-4, 6-6)*

20. Explain how the Quadrilateral Hierarchy Theorem can be used to prove the following.
  Given: $ABCD$ is a square.
  To prove: $\angle A$ is a right angle. *(Lessons 6-3, 6-6)*

In 21 and 22, the logo is rotation-symmetric.
a. Trace the logo and locate the center of rotation.
b. Give the value of n in the n-fold rotation symmetry. *(Lesson 6-6)*

21.

22.

23. In the trapezoid, $m\angle A = 32$ and $m\angle T = 109$.
  Find $m\angle P$ and $m\angle R$. *(Lesson 6-5)*

24. Draw a trapezoid with exactly one symmetry line. *(Lessons 6-1, 6-5)*

25. $PQRS$ is a rhombus. If $m\angle PQS$ is 70, find the measure of each angle.
  (Lesson 6-4)
  a. $\angle RQS$ **b.** $\angle PXQ$ **c.** $\angle RPQ$ **d.** $\angle QPS$

26. Use the markings on each quadrilateral below to give it as specific a name as possible. *(Lesson 6-3)*
  a.   **b.**   **c.**

27. Find three real-world objects whose outlines are regular polygons with more than four sides.

Regular Polygons and Schedules

Spare time. *Bowling-league schedules are usually round-robin with each team playing every other team the same number of times.*

In this lesson, we show a surprising application of regular polygons and circles to a discrete geometry situation in which points stand for teams.

We can shorten the discussion by introducing a few new terms relating to circles. A segment that connects two points on a circle is called a **chord.** If A and B are points on circle O, then $\overset{\frown}{AB}$ is the **minor arc of the chord.**

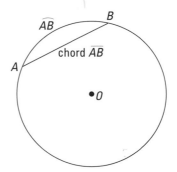

For this application, we use a theorem that we prove in Chapter 14: *In a circle, the greater the measure of the minor arc of a chord, the longer the chord.*

A Round-Robin Tournament

In many tournaments, from baseball to chess to soccer to bowling, each competitor (or team) plays all the others. When each competitor plays each of the other competitors exactly once, it is called a **round-robin tournament.**

Suppose there are seven teams to be scheduled so that each plays the other six. The first thing you might do is number the teams 1 through 7. Now schedule a first week. One team doesn't play. That team gets a **bye.**

| 1 plays 2 | 3 plays 4 | 5 plays 6 | 7 bye |

What about a second week? Try 1 playing 3. Here's a possibility.

| 1 plays 3 | 2 plays 4 | 5 plays 7 | 6 bye |

Now a third week. We just try to have teams play teams they haven't already played.

| 1 plays 4 | 2 plays 5 | 6 plays 7 | 3 bye |

It seems easy. Here is a fourth-week schedule.

| 1 plays 5 | 2 plays 6 | 3 plays 7 | 4 bye |

Another week. It's starting to get complicated. We keep looking back so as not to repeat.

| 1 plays 6 | 2 plays 3 | 4 plays 7 | 5 bye |

How many weeks to go? Team 1 still has to play 7. So does 2. We make 2 the bye.

| 1 plays 7 | 3 plays 6 | 4 plays 5 | 2 bye |

There's one week left. Can you see what it should be? (See Question 2.)

| ? | ? | ? | ? |

Scheduling an Odd Number of Teams

It doesn't seem hard. But perhaps we were lucky. What if there were more teams? It would be nice to have an algorithm that automatically creates the schedule. The algorithm described here is surprising in that it uses properties of regular polygons and circles.

Step 1: Let the 7 teams be vertices of a regular 7-gon (heptagon).

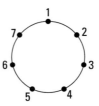

Step 2: (The first week) Draw a chord and all chords parallel to it. Because the polygon has an odd number of sides, the minor arcs of each chord have different measures. So no two chords have the same length. This is the first week's schedule.

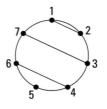

1st week: 1-2   7-3   6-4   5 bye

Since the top chord connects 1 and 2, team 1 plays team 2 the first week. This is called a **pairing** and is written 1-2. Also in the first week, team 7 plays team 3, team 6 plays team 4, and team 5 gets a bye. The full schedule will be completed when all sides and diagonals of the heptagon have been drawn.

Step 3: (The second week) Rotate the chords $\frac{1}{7}$ of a revolution. For example, the chord pairing 7-3 the first week rotates into the pairing 1-4 for the second week.

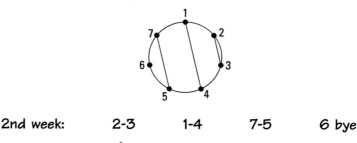

2nd week: 2-3   1-4   7-5   6 bye

Step 4: Continue rotating $\frac{1}{7}$ of a revolution for each week. Because in a week no two chords have the same length, no pairing repeats. (For instance, look at the chord that forms the pair 7-3 in Week 1. No other chord has the same length, so no other chord will pair team 7 with team 3.)

In a total of seven weeks, the schedule is complete.

3rd week:	3-4	2-5	1-6	7 bye
4th week:	4-5	3-6	2-7	1 bye
5th week:	5-6	4-7	3-1	2 bye
6th week:	6-7	5-1	4-2	3 bye
7th week:	?	?	?	?

Again we leave the 7th week for you to figure out. It is easier this time. (See Question 6.)

With 7 teams there are 3 games for each of 7 weeks, or 21 pairings. Of these, 7 are sides of the regular heptagon, 7 are congruent shorter diagonals, and 7 are congruent longer diagonals. Scheduling a league of 7 thus has revealed some properties of the regular heptagon!

Scheduling an Even Number of Teams

The above procedure will not work with an even number of teams. If parallel chords are drawn using the vertices of a regular octagon, some will have the same length. You can see that as you rotate, you will repeat pairings.

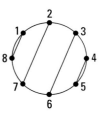

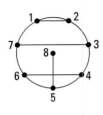

However, instead of putting the 8th team on the circle, it can be placed at the center of the circle. The radius joins team 8 and the team receiving the bye in the schedule for 7 teams. As you rotate the chords to make the schedule, rotate the radius too! This shows the surprising result: it takes as many weeks for a round-robin schedule of 7 teams as it does for a schedule of 8 teams. In general, when n is even, it takes as many weeks for a schedule of $n - 1$ teams as for a schedule of n teams.

In the Questions, you should assume that all schedules are round-robin.

QUESTIONS

Covering the Reading

1. Draw a circle and a chord in it.

2. Complete the last week of pairings in the first schedule discussed in this lesson.

In 3–8, refer to the use of the regular heptagon to schedule teams.

3. How is a game between teams 4 and 6 pictured?

4. To find the pairings for the first week, all chords __?__ to a given chord are drawn.

5. How are the chords for one week related to the chords for the next week?

6. Complete the last week of pairings.

7. Where is an 8th team placed so that this idea can be used to schedule 8 teams?

8. Write a complete schedule for 8 teams.

9. A regular heptagon has diagonals of how many different lengths?

10. *True or false.* It takes as many weeks for a schedule of 9 teams as it does for a schedule of 10 teams.

In the Olympics, small groups of teams play round-robin tournaments to decide who advances. This picture is from a volleyball match between the USSR and the USA at the 1988 Summer Olympics in Seoul, Korea.

Applying the Mathematics

11. a. To schedule 9 teams, what should be done first?
    b. Give the first two weeks of a schedule for 9 teams.
    c. Complete a schedule for 9 teams.
    d. How many diagonals does a nonagon have?
    e. Indicate what you can do to convert the schedule in part c to one with 10 teams.

12. Explain why you cannot begin with the diagram at the left and use the ideas of this lesson if you wish to schedule 6 teams.

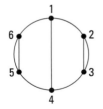

13. Make a complete schedule for a tournament with 6 teams.

14. Which two of the following descriptions of a point are used in the idea of scheduling?
(a) dot   (b) ordered pair   (c) location   (d) node of network

Review

In 15 and 16, consider the regular heptagon from this lesson.

15. How many symmetry lines does it have? *(Lesson 6-7)*

16. What is the measure of each of its angles? *(Lesson 5-7)*

17. Many nuts have hexagonal boundaries, like the one pictured here.
a. A wrench is placed on the nut in order to turn it. In how many positions can the wrench be placed?
b. How is the answer to part **a** related to the symmetry of the nut?
(Lessons 6-6, 6-7)

18. Consider the following statement: *If A, B, C, and D are consecutive vertices of a regular polygon,* then *ABCD is an isosceles trapezoid.*
a. Draw a picture of this situation.
b. Use the Regular Polygon Symmetry Theorem to prove it.
c. Show that the converse of the statement is false.   *(Lessons 2-3, 6-5, 6-7)*

19. Draw the hierarchy of quadrilaterals. Next to the name of each quadrilateral, write what is special about the sides of the quadrilateral.   *(Lessons 6-3, 6-4, 6-5)*

20. A triangle has two sides of length 50 cm and 40 cm.
a. What are the possible lengths for the third side t of the triangle?
b. What are the possible lengths for the third side t of the triangle if the triangle has a symmetry line?   *(Lessons 2-7, 6-1)*

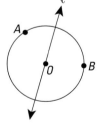

21. Trace the drawing at the left. Draw line m so that $r_m \circ r_\ell(A) = B$.
(Lesson 4-5)

22. The circle with center C and radius r is defined to be the set of all points in a plane whose distance from C is equal to r. Give a good definition of the *interior* of this circle.   *(Lesson 2-4)*

23. Solve for n: $s = 180(n - 2)$.   *(Previous course)*

Exploration

24. a. Find a schedule for teams in a league involving your school or community.
b. What factors affect schedules that this lesson does not mention?

A project presents an opportunity for you to extend your knowledge of a topic related to the material of this chapter. You should allow more time for a project than you do for typical homework questions.

PROJECTS
6
CHAPTER SIX

1 Symmetric Faces: Artist's Version

Find three large photographs or magazine pictures of faces (human or otherwise) looked at straight on. Cut each one in half along its "symmetry line." Attach one half of each to a large piece of poster board, then use your artistic skills to draw the missing half. Then do the same with the other half. Write a paragraph comparing the two faces you have now created from each photograph. How are they alike? How are they different?

2 Constructing Quadrilaterals

a. Use an automatic drawer to make a series of drawings of each type of quadrilateral in the hierarchy. By each quadrilateral, put the properties of its sides, angles, diagonals, and symmetries. Have the drawer measure angles and lengths for your drawings. Print them all and present them in a notebook.

b. Using a straightedge and compass, construct an isosceles trapezoid, a kite, a rhombus, a parallelogram, a rectangle, and a square. Present your constructions in a notebook.

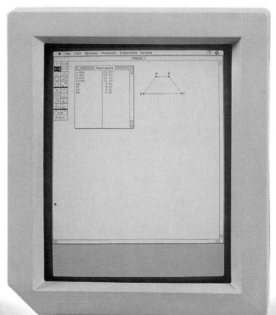

PROJECTS 6 *(continued)*

3 Logo Collection

a. Find ten logos of teams or companies. Photocopy each logo and draw any lines of symmetry. Locate the center of rotational symmetry. Then write a sentence or two for each logo that describes its symmetries. Tell why you think each company chose its logo.

b. Design an original logo. Use reflection and rotation symmetries. Make a colorful rendering of your logo on unlined $8\frac{1}{2}''$ by $11''$ paper.

4 Upside-Down Names

In his book *Inversions*, Scott Kim writes words and names in such a way that when they are rotated 180°, they spell the same word or name, or a related word or name. Here are two examples.

Experiment with writing your own name in such a way that when viewed upside down it spells either your name again or some related idea. Show your best work. Then invent a few other examples of upside-down words or names.

5 Symmetries in Rugs

Find rug patterns designed by people living in different parts of the world. For example, you might include rugs made in Asia and the Middle East. You also

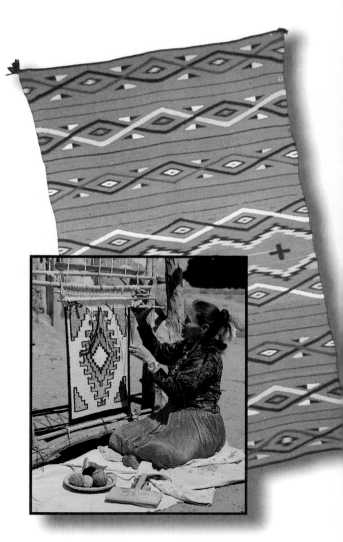

can research designs by Native Americans. Choose eight different rugs and examine them for symmetric designs. Look for symmetric motifs as well as at the symmetry of the whole rug. Photocopy the designs and mark lines and centers of symmetry. Write a brief report on rug-making. Include your ideas about the symmetries you found.

SUMMARY

In this chapter you studied many polygons and their properties. The simplest polygon is the triangle. Triangles can be classified by the number of congruent sides (shown below at the left) or by the size of the largest angle (shown below at the right).

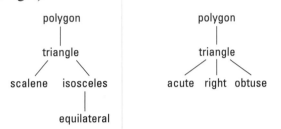

Every isosceles triangle has at least one line of symmetry. That line contains the bisector of the vertex angle, the perpendicular bisector of the base, and the median to the opposite side. An equilateral triangle has three lines of symmetry and 3-fold rotation symmetry. A scalene triangle has no symmetry.

Quadrilaterals were also examined in this chapter. A hierarchy for quadrilaterals is given below.

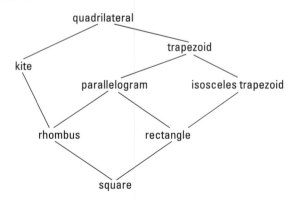

The importance of these hierarchies is that any property true for figures of one type is true for figures of every type below it to which it is connected. Thus the Kite Symmetry Theorem applies to rhombuses and squares. The Trapezoid Angle Theorem can be applied to all special types of quadrilaterals except kites. Similarly, the Isosceles Triangle Symmetry Theorem applies to equilateral triangles.

Many properties of polygons can be deduced from their symmetries. Every isosceles trapezoid has a line of symmetry, the ⊥ bisector of its bases. Also, every kite has a line of symmetry, the line containing its ends. From these two symmetries all other line symmetries of quadrilaterals are derived. When a figure has two intersecting lines of symmetry, as rectangles and rhombuses do, then it possesses rotation symmetry.

The polygons with the most reflection and rotation symmetry for their number of sides are the regular polygons. Every regular n-gon has n symmetry lines which contain all the bisectors of its angles and the perpendicular bisectors of its sides. Every regular n-gon also has n-fold rotation symmetry. This rotation symmetry can be applied to produce a schedule for the teams in a round-robin tournament.

VOCABULARY

Below are the new terms and phrases for this chapter. (Some you have studied in previous years, but we repeat them here for reference.) You should be able to give a general description and specific example of each. For those terms that are starred (*), you should be able to give a precise definition.

Lesson 6-1
*reflection-symmetric figure
*symmetry line
line symmetry
Flip-Flop Theorem
Segment Symmetry Theorem
Side-Switching Theorem
Angle Symmetry Theorem
Circle Symmetry Theorem
Symmetric Figures Theorem

Lesson 6-2
median of a triangle
vertex angle
base angles
base of an isosceles triangle
Isosceles Triangle Symmetry Theorem
Isosceles Triangle Coincidence Theorem
Isosceles Triangle Base Angles Theorem
Equilateral Triangle Symmetry Theorem
Equilateral Triangle Angle Theorem
corollary
centroid

Lesson 6-3
*parallelogram
*rhombus
*rectangle
*square
hierarchy of quadrilaterals
*kite
*trapezoid
bases of a trapezoid, pair of base angles
*isosceles trapezoid
Quadrilateral Hierarchy Theorem

Lesson 6-4
ends of a kite
Kite Symmetry Theorem
symmetry diagonal
Kite Diagonal Theorem
Rhombus Diagonal Theorem

Lesson 6-5
Trapezoid Angle Theorem
Isosceles Trapezoid Symmetry Theorem
Isosceles Trapezoid Theorem
Rectangle Symmetry Theorem

Lesson 6-6
rotation-symmetric figure
*center of symmetry
n-fold rotation symmetry

Lesson 6-7
*regular polygon
regular pentagon
regular hexagon
equilateral polygon
equiangular polygon
Center of a Regular Polygon Theorem
center of a regular polygon
Regular Polygon Symmetry Theorem

Lesson 6-8
chord
minor arc of a chord
round-robin tournament
bye
pairing

PROGRESS SELF-TEST

Directions: Take this test as you would take a test in class. You will need a ruler and a calculator for this test. Then check your work with the solutions in the Selected Answers section in the back of the book.

In 1 and 2, trace the figure. Then draw all lines that seem to be symmetry lines.

1.

2. The figure is a regular heptagon.

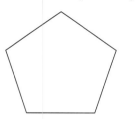

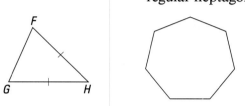

3. Trace the regular pentagon below. Locate and label its center of symmetry.

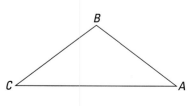

4. Draw an isosceles trapezoid with a horizontal line of symmetry.

5. Draw a quadrilateral that is equilateral but not regular.

6. $\triangle ABC$ below is isosceles with vertex $\angle CBA$. If $m\angle A = 6x$ and $m\angle C = 50 - 2x$, find $m\angle B$.

7. *DEFGH* below is a regular pentagon. Find $m\angle HDG$.

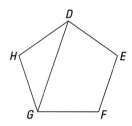

8. Angle *W* of an isosceles trapezoid *WXYZ* with base $\overline{WX}$ has measure 100°. Find the measure of as many other angles as you can.

9. *PQRS* is a kite with ends *Q* and *S*. Find each length.

   a. *SR*   **b.** *QR*

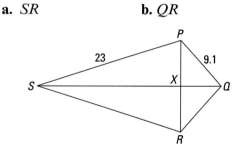

10. *RHOM* is a rhombus. If $m\angle 1 = 78$, find each measure.

   a. $m\angle 2$ **b.** $m\angle 3$

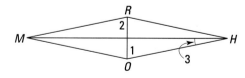

PROGRESS SELF-TEST

In 11 and 12, polygon *ABCD* below is symmetric to line *n*.

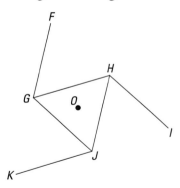

11. $r_n(ABCD) = \underline{\quad?\quad}$.

12. Name all angles congruent to $\angle DBC$.

13. Figure *FGHIJK* below has 3-fold rotation symmetry with center of symmetry *O*. Name all segments congruent to $\overline{FG}$.

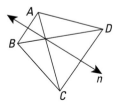

In 14 and 15, *multiple choice*.

14. A triangle cannot have exactly which number of symmetry lines?
  (a) 0   (b) 1   (c) 2   (d) 3

15. A regular polygon has *n* lines of symmetry. Each one is a perpendicular bisector of a side of the polygon. Then
  (a) *n* is even.   (b) *n* is odd.
  (c) You cannot tell whether *n* is even or odd.
  (d) There is no *n* for which this is true.

16. From the information given in the drawing, what quadrilateral is pictured? Be as specific as possible.

In 17 and 18, *true or false*.   If false, give a counterexample.

17. Every square is a rectangle.

18. Every kite has two lines of symmetry.

In 19 and 20, write a proof argument using the given figure.

19. Given: $\overline{AD}$ intersects $\overline{BE}$ at *C*. $AB = BC$.
  To prove: $m\angle A = m\angle ECD$.

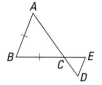

20. Given: $\odot C$ and $\odot T$.
  To prove: *CITY* is a rhombus.

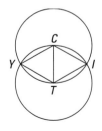

21. Trace the hubcap below.
  a. If this hubcap is reflection-symmetric, draw the line(s) of symmetry.
  b. If this hubcap has *n*-fold rotation symmetry, mark the center of symmetry.

22. Draw the hierarchy relating the following: polygon, square, parallelogram, rectangle, rhombus.

23. Schedule five teams for a round-robin tournament so that teams 1 and 2 play each other during the first round.

CHAPTER REVIEW

Questions on SPUR Objectives

SPUR stands for **S**kills, **P**roperties, **U**ses, and **R**epresentations. The Chapter Review questions are grouped according to the SPUR Objectives for this chapter.

SKILLS DEAL WITH THE PROCEDURES USED TO GET ANSWERS.

Objective A: *Locate symmetry lines and centers of symmetry of geometric figures.*
(Lessons 6-1, 6-2, 6-4, 6-5, 6-6, 6-7)

In 1–6, trace each figure and draw its symmetry line(s).

1.

2.

3.

4.

5. *PENTA* is regular.   6.

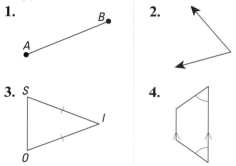

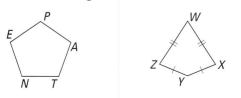

In 7 and 8, trace each figure and locate its center of symmetry.

7. rectangle *RECT*   8.

Objective B: *Draw polygons satisfying various conditions.* *(Lessons 6-2, 6-3, 6-7)*

In 9–14, draw an example of the figure using a ruler, compass, or protractor.

9. a scalene right triangle

10. an isosceles acute triangle

11. a kite that is not a rhombus

12. a trapezoid that is not isosceles

13. a regular hexagon

14. a hexagon that is equilateral but not regular

Objective C: *Apply theorems about isosceles triangles to find angle measures and segment lengths.* *(Lessons 6-2, 6-7)*

In 15 and 16, use isosceles △*GHK* with base $\overline{GH}$.

15. If m∠*K* = 34, find m∠*H* and m∠*G*.

16. If *KH* = 2*x*, *KG* = 20, and *HG* = 12, find *x*.

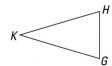

17. Refer to △*ISO* at the right with m∠*I* = m∠*S*. If *IS* = 35 and *IO* = 34, find *OS*.

18. In isosceles △*ABC*, m∠*A* = 72. Draw two possible situations for △*ABC* and find m∠*B* and m∠*C* in each situation.

19. In regular hexagon *ABCDEF* at the right, find m∠*AFB*.

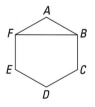

20. The figure at the right is a regular 12-gon. Find m$\angle X$ and m$\angle XZY$.

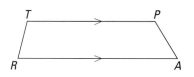

Objective D: *Apply theorems about quadrilaterals and regular polygons to find angle measures and segment lengths.*
(Lessons 6-3, 6-4, 6-5, 6-7)

In 21 and 22, use parallelogram *PQRS* at the right.

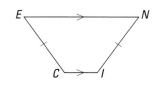

21. If m$\angle S = 43$, find all the other angle measures of the figure.

22. If $PQ = 9$ and $PS = 15$, find SR.

In 23 and 24, use rhombus *RHOM* at the right.

23. If m$\angle 1 = 63$, find the measures of all the numbered angles in the figure.

24. If $MR = 18$ and $MO = 12s$, find s.

25. Use trapezoid *TRAP* below. If m$\angle A = 8x - 12$ and m$\angle P = 15x - 15$, find m$\angle A$.

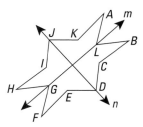

26. Use isosceles trapezoid *NICE* below. Suppose m$\angle N = 5t - 7$ and m$\angle E = 11t - 79$.

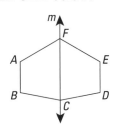

   a. Find m$\angle E$.
   b. Find m$\angle C$.

27. Find the measure of each interior angle of a regular 24-gon.

28. Find the measure of each interior angle of a regular octagon.

PROPERTIES DEAL WITH PRINCIPLES BEHIND THE MATHEMATICS.

Objective E: *Apply properties of symmetry to assert and justify conclusions about symmetric figures.*   *(Lessons 6-1, 6-6)*

In 29 and 30, $\overleftrightarrow{XZ}$ is a symmetry line of polygon *WXYZ* at the right.

29. Which sides of the polygon have the same length?

30. Which angles of the figure are congruent?

In 31–33, r$_m$(*FABCDE*) = *FEDCBA* below.

31. r$_m$($\angle ADE$) = __?__

32. Name all angles with the same measure as $\angle D$.

33. *True or false.* *m* bisects $\angle AFE$.

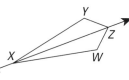

34. In the figure below, *m* and *n* are symmetry lines. Name three other angles congruent to $\angle A$.

In 35 and 36, consider the seven types of quadrilaterals in this chapter.

35. Which types always possess rotation symmetry?

36. For which types is the perpendicular bisector of a side always a line of symmetry?

In 37 and 38, hexagon *GHIJKL* at the right has 2-fold rotation symmetry with center of symmetry at *O*.

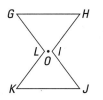

37. Name all angles congruent to ∠*G*.

38. Name all segments congruent to $\overline{IJ}$.

Objective F: *Know the properties of the various types of triangles and regular polygons.*
(Lessons 6-2, 6-7)

39. *Multiple choice.* Suppose that in △*QRS* the bisector of ∠*Q* coincides with the median to side *RS*. What type of triangle is *QRS*?
  (a) acute   (b) equilateral
  (c) isosceles   (d) obtuse

40. Draw a triangle that has three lines of symmetry.

In 41 and 42, *true or false*.

41. A regular *n*-gon has *n* symmetry lines.

42. The bisector of an angle in a regular polygon is also the perpendicular bisector of a side.

Objective G: *Know the properties of the seven special types of quadrilaterals.*
(Lessons 6-3, 6-4, 6-5)

In 43 and 44, use the information given in the drawing to name the type of quadrilateral. Be as specific as possible.

43.   **44.**

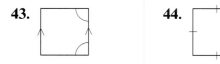

In 45 and 46, *true or false*. If false, give a counterexample.

45. Every square is a parallelogram.

46. Every parallelogram is a rectangle.

In 47 and 48, refer to the quadrilateral below in which *AB* = *BC* and *DA* = *DC*.

47. Does the figure have any symmetry lines? If so, describe them.

48. Name two angles which must have equal measures.

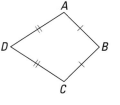

49. Name a property of all rhombuses that is not a property of all parallelograms.

50. Explain why every square is a kite.

In 51–54, *true or false*.

51. Every trapezoid has at least one symmetry line.

52. The diagonals of a rhombus bisect each other.

53. The perpendicular bisector of one base of an isosceles trapezoid is the perpendicular bisector of the other base.

54. The perpendicular bisector of the base of an isosceles triangle contains the angle bisector of the vertex angle.

Objective H: *Write proofs using properties of triangles and quadrilaterals.*
(Lessons 6-2, 6-3, 6-4, 6-5, 6-7)

55. Give justifications for each conclusion.
  Given: △*ABC* is isosceles with vertex ∠*BAC*. △*ACD* is isosceles with vertex ∠*ACD*.
  To prove: *AB* = *CD*.

Conclusions	Justifications
0. △*ABC* is isosceles with vertex ∠*BAC*. △*ACD* is isosceles with vertex ∠*ACD*.	**a.**
1. *AB* = *AC* *AC* = *CD*	**b.**
2. *AB* = *CD*	**c.**

In 56–60, write a proof argument using the given figure.

56. Given: △*ACH* is isosceles with base $\overline{AH}$.
  To prove: ∠1 ≅ ∠*H*.

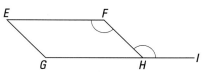

57. Given: m∠*F* = m∠*FHI*.
  To prove: *EFHG* is a trapezoid.

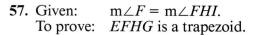

58. Given: Isosceles trapezoid *ABCD* with bases $\overline{AB}$ and $\overline{CD}$.
To prove: *AC = BD*.
(Hint: Use symmetry.)

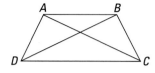

59. Given: $\odot O$ and $\odot P$ intersect at *Q* and *R*.
To prove: *OQPR* is a kite.

60. Given: *MLKJ* is a square.
To prove: $\overline{LK} /\!/ \overline{MJ}$.

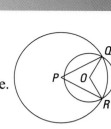

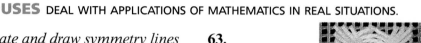

USES DEAL WITH APPLICATIONS OF MATHEMATICS IN REAL SITUATIONS.

Objective I: *Locate and draw symmetry lines in real-world designs.* *(Lessons 6-1, 6-6)*

In 61–64, trace the figure. **a.** If the figure is reflection-symmetric, draw the line(s) of symmetry. **b.** If the figure has *n*-fold rotation symmetry, mark the center of symmetry and give the value of *n*.

61.

carved mask from Indonesia

62.

airplane propeller

63.

large petal stitch

64.

Flag of Trinidad and Tobago

Objective J: *Make a schedule for a round-robin tournament.* *(Lesson 6-8)*

65. Four teams need to be scheduled for a round-robin tournament. Picture the schedule using chords of circles, and write a complete schedule.

66. Schedule a round-robin tournament for 11 teams A–K.

67. Schedule a round-robin tournament for 12 teams A–L.

REPRESENTATIONS DEAL WITH PICTURES, GRAPHS, OR OBJECTS THAT ILLUSTRATE CONCEPTS.

Objective K: *Draw and apply hierarchies of polygons.* *(Lesson 6-3)*

68. Draw the hierarchy relating the following: isosceles △, polygon, triangle, scalene △, equilateral △.

69. Draw the hierarchy relating the following: figure, quadrilateral, isosceles trapezoid, trapezoid, parallelogram, rectangle.

70. Arrange from most general to most specific: rhombus, quadrilateral, polygon, square, parallelogram.

71. Arrange from least to greatest number of sides: hexagon, heptagon, kite, scalene triangle, octagon.

CHAPTER

7

TRIANGLE CONGRUENCE

Triangles, the simplest of all polygons, have many uses. Triangles are rigid, so the supporting frameworks for many structures, such as the ones pictured at the left, use triangles.

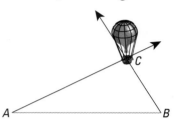

Triangles can be used to determine the location of an object. If a person at a point A sights a hot air balloon or other object in the sky, that person can only determine the direction of the balloon, not how far away it is. But if a person at a second point B sights the balloon at the same time, then the balloon is at the intersection, C, of the two rays. This method is called *triangulation*, because A, B, and C are vertices of a triangle. For hundreds of years, until cameras on satellites provided accurate maps of the earth, triangulation was the best way to determine the precise locations of mountain peaks, harbors, and other places not easily reached.

In this chapter, you will learn what the minimum conditions are to determine a triangle uniquely. These conditions are found in the triangle congruence theorems, which have been utilized from the time of ancient Greece to deduce properties of polygons. Thus, in this chapter, you will study an old subject with many modern applications.

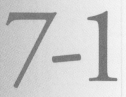

If you join two boards with a nail, you can turn the boards and change the measure of the angle between the boards. If you join four boards with four nails, the four angles can still change. In both cases, the structure is not rigid because its shape and size are not constant.

What if three boards are joined? Consider this situation. Ramon was building a roof for a small storage shed. In order for the roof to lie in one plane, and to fit snugly with the walls, the triangular supports must be congruent. He needed four triangular frames for his shed, each with sides 2, 4, and 5 feet.

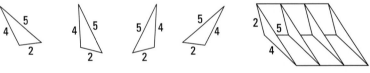

Ramon's Frames

Since the supports were to be made of wood, Ramon had to make the triangles one at a time. Before buying and cutting the wood, Ramon wondered whether his frames would be congruent, as required by the plan for the shed. Was it possible that the angles might have different measures in these triangles?

Drawing a Triangle Given Three Sides

We can represent Ramon's triangles using lengths 2 cm, 4 cm, and 5 cm, but drawing a triangle on a sheet of paper is different from constructing one out of wood. You can't pick up, cut, or nail the sides when they are on paper! Some automatic drawers enable you to enter side lengths and angle measures for a triangle and they then draw the triangle. With others you construct triangles as you would with a compass, protractor, and ruler. Activity 1 gives an algorithm for constructing a triangle with three sides of given lengths that satisfy the Triangle Inequality.

Carry out this construction of a triangle given sides of lengths 2, 4, and 5. You can pick the unit.

Step 1: Construct any line and call it *t*. Choose point *X* on *t*.

Step 2: Construct $\odot X_1$ with radius 5. ($\odot X_1$ identifies the first circle with center *X*.) $\odot X_1 \cap t = \{Y, V\}$.

Step 3: Construct $\odot X_2$ with radius 4. ($\odot X_2$ is a second circle with center *X*.)

Step 4: Construct $\odot Y$ with radius 2.

Step 5: $\odot X_2 \cap \odot Y = \{Z, W\}$. (The circles will not intersect if the Triangle Inequality is violated.)

Step 6: Both $\triangle XYZ$ and $\triangle XYW$ are triangles with sides 2, 4, and 5.

Step 7: Measure all the angles in $\triangle XYZ$ and $\triangle XYW$ to see if they are the same.

The Academy Cadet Chapel at the United States Air Force Academy in Colorado Springs makes use of congruent triangles.

Here is what an automatic drawer screen could look like after completing Activity 1 and asking for lengths of sides and measures of angles in $\triangle XZY$ and $\triangle XWY$.

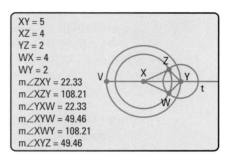

XY = 5
XZ = 4
YZ = 2
WX = 4
WY = 2
m∠ZXY = 22.33
m∠XZY = 108.21
m∠YXW = 22.33
m∠XYW = 49.46
m∠XWY = 108.21
m∠XYZ = 49.46

The angle measures in your triangles should be close to those shown above. This is because, given the lengths of three segments that satisfy the Triangle Inequality, a unique triangle is determined. *Triangles are rigid.* Their rigidity is why triangles are used as supports for structures. In Activity 1 you verified this property for one set of lengths. In Lesson 7-2, you will see a proof of this property for all triangles.

Drawing a Triangle Given Three Angles

Now suppose you are given the measures of the three angles of a triangle. Do you think your triangle will be congruent to the triangle drawn by the other students?

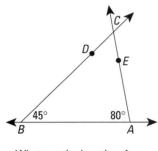

What are the lengths of $\overline{AB}$, $\overline{BC}$, and $\overline{AC}$?

Draw a triangle *ABC* in which m∠*A* = 80, m∠*B* = 45, and m∠*C* = 55. To start, identify two points as *A* and *B*, and draw $\overleftrightarrow{AB}$. Now draw $\overrightarrow{BD}$ so that ∠*ABD* has measure 45°. On the same side of $\overleftrightarrow{AB}$ as *D*, draw $\overrightarrow{AE}$ so that ∠*BAE* has measure 80°. $\overrightarrow{BD} \cap \overrightarrow{AE} = C$. $\triangle ABC$ should be the desired triangle. Check to see if m∠*C* = 55.

Notice that the measures of only two angles were needed to do Activity 2. The measure of the third angle is the check. This is because all triangles with two pairs of congruent angles have congruent third angles.

Theorem
If two angles in one triangle are congruent to two angles in another triangle, then the third angles are congruent.

Proof
Given: Triangles *ABC* and *DEF* with $\angle A \cong \angle D$ and $\angle B \cong \angle E$.
To prove: $\angle C \cong \angle F$.
Drawing:

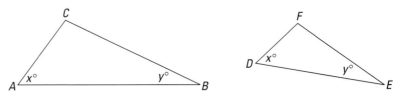

Argument:
Congruent angles have equal measures. Let $x = m\angle A = m\angle D$ and let $y = m\angle B = m\angle E$. Because of the Triangle-Sum Theorem, $x + y + m\angle C = 180$ and $x + y + m\angle F = 180$. Adding $-x + -y$ to each side of each equation,

$$m\angle C = 180 - x - y$$
$$\text{and } m\angle F = 180 - x - y.$$

Therefore, by the Transitive Property of Equality, $m\angle C = m\angle F$. So the third angles are congruent.

Drawing a Triangle Given Two Sides and an Angle

In Activity 3, you are asked to draw a triangle given the lengths of two sides and the measure of the angle between the sides. Do you think everyone's triangle will be congruent to yours?

Activity 3

In a triangular sail *ABC*, $m\angle A = 70$, $AB = 5$ meters, and $AC = 3.5$ meters. Make a scale drawing using centimeters instead of meters as the unit.
1. First, draw a 70° angle. Call it $\angle A$.
2. Let 1 cm in the drawing equal 1 meter of the actual sail. Mark off 5 cm on one side of $\angle A$ and 3.5 cm on the other. Label the points *B* and *C*, and draw $\overline{BC}$.
3. Find $m\angle B$, $m\angle C$, and BC in your drawing.

Below is a screen from an automatic drawer that shows m∠A = 70, AC = 3.5 and AB = 5. Do m∠B, m∠C, and BC have the same measures as those you found in step 3 of Activity 3?

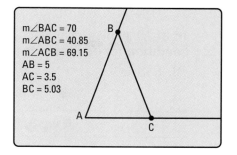

Sufficient Conditions for a Triangle

Whenever a set of given conditions is enough to determine a figure, we call the set a **sufficient condition** for the figure.

In the Activities in this lesson you have been asked to draw a triangle given three different conditions.

Activity	Given	Name for given condition
1	three sides	SSS
2	three angles	AAA
3	two sides and the angle they form	SAS

In part **a** of Questions 11–15, you are asked to draw triangles given other conditions. Part **b** of these Questions asks whether you think everyone else's triangles will be congruent to yours. In other words, are these given conditions sufficient conditions?

QUESTIONS

Covering the Reading

1. What is *triangulation*?

2. **a.** Show your triangle and angle measures in Step 7 of Activity 1.
   b. Do you think everyone else's triangle should be congruent to yours?

3. Construct a triangle with sides of 3 cm, 4 cm, and 5 cm.

4. **a.** Show your answer to Activity 2.
   b. Do you think everyone else's triangle should be congruent to yours?

5. If two angles of a triangle have measures 32 and 102, what is the measure of the third angle?

6. If two angles of a triangle have measures x and y, what is the measure of the third angle?

7. In Activity 3, did your measures of $\angle B$, $\angle C$, and BC agree with the measures shown?

In 8–10, explain why you think the condition is or is not a sufficient condition for determining a unique triangle.

8. SSS   **9.** AAA   **10.** SAS

Applying the Mathematics

In 11–15, you may use an automatic drawer or any other drawing tools.
a. Accurately draw a triangle ABC with the given information.
b. Conjecture whether the condition is a sufficient condition for determining a unique triangle.

11. AA condition: $m\angle A = 70$, $m\angle B = 38$.

12. SA condition: $AB = 4$ cm, $m\angle A = 60$.

13. AAS condition: $m\angle A = 40$, $m\angle B = 60$, $BC = 3$ cm.

14. ASA condition: $m\angle A = 40$, $m\angle B = 60$, $AB = 2$ cm.

15. SSA condition: $AB = 2.0''$, $BC = 1.75''$, $m\angle C = 70$.

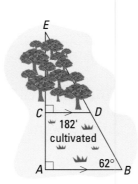

16. The Foresters own a triangular plot of land. Part they have cultivated, and part is densely wooded as shown at the left. They want to know DE.
a. Use the measurements in the figure, and make a scale drawing.
b. Use your drawing to estimate DE.
c. What name is given to the set of conditions known about $\triangle CDE$?

Review

17. Stop signs have the shape of regular octagons. What is the measure of each angle in a stop sign?   *(Lesson 6-7)*

In 18 and 19, suppose $\triangle BRI = r_m(\triangle GHT)$.   *(Lessons 4-1, 6-5)*
18. a. $\overline{IR} \cong \underline{\;?\;}$ **b.** $\angle GHT \cong \underline{\;?\;}$

19. If B, G, H, and R are not collinear, then these points are vertices of a quadrilateral called a(n) $\underline{\;?\;}$.

20. Given $\triangle FGH \cong \triangle LMN$, what can be concluded by the definition of congruence?   *(Lesson 4-8)*

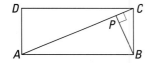

21. In the figure at the left, $\overline{AC}$ is a diagonal of rectangle $ABCD$ and $\overline{AC} \perp \overline{PB}$. If $m\angle PAB = 40$, find as many angle measures as you can.   *(Lesson 6-5)*

22. In the figure at the right, $r_m \circ r_n(\overline{AB}) = \overline{CD}$.
  a. Trace the figure and locate possible positions of m and n.
  b. What kind of isometry is $r_m \circ r_n$?
  (Lesson 5-2)

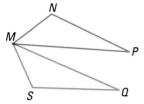

23. At the left, $r_\ell(\triangle MNP) = \triangle MSQ$. Trace the drawing and draw in line ℓ.   *(Lesson 4-2)*

Exploration

24. Each of the objects at the left is hidden in the picture. Give an ordered pair to describe the location of each object.

book
telephone receiver
parrot
car
roller skate
bell
airplane
bugle
mouse
rooster's head
wool cap
bathrobe
yo-yo
Puritan's hat

"Hidden Pictures" puzzle by Christopher Wray from *Highlights for Children,* July-August, 1986.

Triangle Congruence Theorems

Make no bones about it. *This Byzantine motif, made of wood with bone inlay, decorates the side of a blanket chest made around 1775.*

In Lesson 7-1, you drew or constructed a triangle given some lengths of sides and measures of angles. After drawing each triangle, you were asked if your triangle would be congruent to the triangles drawn by everyone else who followed the same instructions.

The SSS Condition

The first condition considered in Lesson 7-1 was the SSS condition. We asked if all triangles with sides of lengths 2 cm, 4 cm, and 5 cm are congruent. You should have found that they are. The generalization is called the *SSS Congruence Theorem*. It is proved by showing that whenever two triangles have congruent sides, then there is an isometry that maps one triangle onto the other.

> **SSS Congruence Theorem**
> If, in two triangles, three sides of one are congruent to three sides of the other, then the triangles are congruent.

Proof
Given: $\overline{AB} \cong \overline{DE}$, $\overline{BC} \cong \overline{EF}$, and $\overline{AC} \cong \overline{DF}$.
To prove: $\triangle ABC \cong \triangle DEF$.
Drawing:

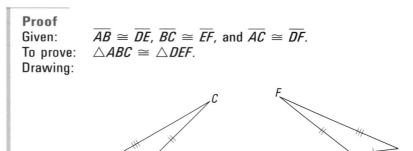

Argument:

It is given that $\overline{AB} \cong \overline{DE}$, $\overline{BC} \cong \overline{EF}$, and $\overline{AC} \cong \overline{DF}$. Two figures are congruent if and only if one is the image of the other under an isometry. Since $\overline{AB} \cong \overline{DE}$, there is an isometry T with $T(\overline{AB}) = \overline{DE}$ so that $T(A) = D$ and $T(B) = E$. Furthermore, T can be chosen so that $T(C)$ is on the other side of $\overleftrightarrow{DE}$ from F. Label the image of $\triangle ABC$ as $T(\triangle ABC)$ or $\triangle A'B'C'$.

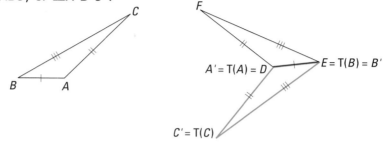

We write the rest of the argument in two-column form. You should examine the figure after each step.

Conclusions	Justifications
1. $\triangle A'B'C' \cong \triangle ABC$	definition of congruence
2. $\overline{BC} \cong \overline{B'C'}$, $\overline{AC} \cong \overline{A'C'}$	CPCF Theorem
3. $\overline{B'C'} \cong \overline{EF}$, $\overline{A'C'} \cong \overline{DF}$	Transitive Property of Congruence (using Given and step 2)
4. $C'DFE$ is a kite with ends D and E.	definition of kite
5. $C'DFE$ is reflection-symmetric to $\overleftrightarrow{DE}$.	Kite Symmetry Theorem
6. $r_{\overleftrightarrow{DE}}(\triangle A'B'C') = \triangle DEF$	definition of reflection-symmetric figure
7. $\triangle A'B'C' \cong \triangle DEF$	definition of congruence
8. $\triangle ABC \cong \triangle DEF$	Transitive Property of Congruence

Because of the SSS Congruence Theorem, if the sides of a triangle are fixed in length, the shape and size of the triangle cannot change. The rigidity of triangles in construction means that walls with carefully measured triangular supports will remain perpendicular to the floor and that roofs with triangular supports will be stable.

Triangular supports are needed for these basketball hoops, constructed for a park district competition in Chicago in 1990.

When three boards are joined, the structure is rigid.

The SSS Congruence Theorem is the first of four triangle congruence theorems proved in this lesson. Three of the theorems are proved in much the same way, by mapping $\triangle ABC$ onto a conveniently located congruent image $\triangle A'B'C'$. Then $\triangle A'B'C'$ and $\triangle DEF$ are congruent, so $\triangle ABC$ and $\triangle DEF$ are congruent. Just how this is done varies from proof to proof.

The SAS Condition

Now consider the *SAS condition.* This condition refers to two sides and the angle they form (called the **included angle**). The proof uses the symmetry of an isosceles triangle.

SAS Congruence Theorem
If, in two triangles, two sides and the included angle of one are congruent to two sides and the included angle of the other, then the triangles are congruent.

Proof
Given: $\overline{AB} \cong \overline{DE}$, $\overline{AC} \cong \overline{DF}$, and $\angle A \cong \angle FDE$.
To prove: $\triangle ABC \cong \triangle DEF$.

Drawing:

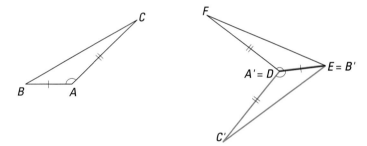

Argument:
Since $\overline{AB} \cong \overline{DE}$, there is, as in the SSS proof, an isometry T mapping $\overline{AB}$ onto $\overline{DE}$. T($\triangle ABC$) = $\triangle A'B'C'$, and so $\triangle A'B'C' \cong \triangle ABC$. (Look carefully at the tick and angle marks. They are the only things making this drawing different from the drawing for the SSS Congruence Theorem.)

In this case, the defining conditions for a kite do not appear. However, $\triangle C'DF$ (not drawn) is isosceles and $\overleftrightarrow{DE}$ contains the bisector of its vertex angle. (You are asked to prove this in Question 19.) Because of the Isosceles Triangle Symmetry Theorem, the reflection image of C' over $\overleftrightarrow{DE}$ is F. So the reflection image of $\triangle A'B'C'$ over $\overleftrightarrow{DE}$ is $\triangle DEF$. This makes $\triangle A'B'C' \cong \triangle DEF$. So by the Transitive Property of Congruence, $\triangle ABC \cong \triangle DEF$.

For instance, if two triangles both have sides with lengths 4″ and 6″, and their included angles each measure 50°, then the triangles are congruent.

The ASA Condition

The *ASA condition* refers to two angles and the side between their vertices (called the **included side**). This condition also determines a unique triangle.

> **ASA Congruence Theorem**
> If, in two triangles, two angles and the included side of one are congruent to two angles and the included side of the other, then the two triangles are congruent.

Proof
Given: $\overline{AB} \cong \overline{DE}$, $\angle A \cong \angle FDE$, and $\angle B \cong \angle FED$.
To prove: $\triangle ABC \cong \triangle DEF$.

Drawing:

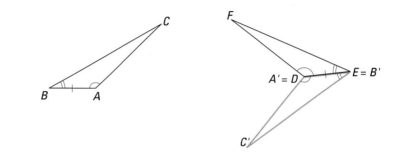

Argument:
Again consider the image $\triangle A'B'C'$ of $\triangle ABC$ under an isometry mapping $\overline{AB}$ onto $\overline{DE}$. $\triangle A'B'C'$ and $\triangle DEF$ form a figure much like that in the previous two proofs, but now with two pairs of congruent angles, as shown in the diagram.

Again, think of reflecting $\triangle A'B'C'$ over line $\overleftrightarrow{DE}$. Since $\overleftrightarrow{DE}$ contains the bisector of $\angle C'DF$ (see Question 19) and of $\angle C'EF$, you can apply the Side-Switching Theorem. The image of $\overrightarrow{A'C'}$ is $\overrightarrow{DF}$. Applying the Side-Switching Theorem to $\angle C'EF$, the image of $\overrightarrow{B'C'}$ is $\overrightarrow{EF}$. This forces the image of C' to be on both $\overrightarrow{DF}$ and $\overrightarrow{EF}$, and so the image of C' is F. Therefore the image of $\triangle A'B'C'$ is $\triangle DEF$. This makes $\triangle A'B'C' \cong \triangle DEF$ and, by the Transitive Property of Congruence, $\triangle ABC \cong \triangle DEF$.

Shown is part of a window of the Techny Towers Conference Center in Techny, Illinois. Which triangles do you think are congruent?

The AAS Condition

There is one other general condition that leads to congruent triangles: the AAS condition. This means two pairs of angles and one pair of corresponding non-included sides of two triangles are congruent. This is easily proved. If two pairs of angles are congruent, such as $\angle A \cong \angle X$ and $\angle B \cong \angle Y$ below, then the third pair is congruent. Thus, $\angle C \cong \angle Z$. This makes $\triangle ABC \cong \triangle XYZ$ by the ASA Congruence Theorem.

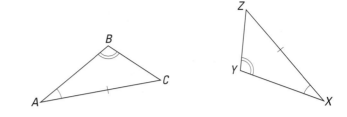

AAS Congruence Theorem
If, in two triangles, two angles and a non-included side of one are congruent respectively to two angles and the *corresponding* non-included side of the other, then the triangles are congruent.

Example

Using only the information marked, which pairs of triangles are congruent? Justify each choice with a triangle congruence theorem.

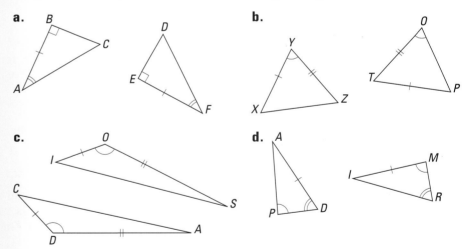

Solution
a. $\triangle ABC \cong \triangle FED$ by ASA Congruence Theorem.
b. You cannot conclude that the triangles are congruent. The congruent angles are not included by the corresponding sides in both triangles.
c. $\triangle CAD \cong \triangle ISO$ by the SAS Congruence Theorem.
d. You cannot conclude that the triangles are congruent. The congruent sides are not opposite corresponding congruent angles.

374

Covering the Reading

1. List four conditions that lead to triangle congruence.

2. In Lesson 7-1, Ramon built four triangular frames, each with sides 2 meters, 4 meters, and 5 meters. Which triangle congruence theorem justifies the conclusion that the frames are congruent?

3. In the proof of the SSS Congruence Theorem, the symmetry of what figure is used?

4. In the proof of the SAS Congruence Theorem, the symmetry of what figure is used?

5. State the ASA Congruence Theorem.

6. How does the AAS Congruence Theorem differ from the ASA Congruence Theorem?

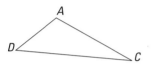

7. Use the figure at the left. In △*ACD*, what side is included by ∠*CAD* and ∠*ADC*?

In 8–13, if the given triangles are congruent, justify with a triangle congruence theorem and indicate corresponding vertices. Otherwise, write *not enough information to know.*

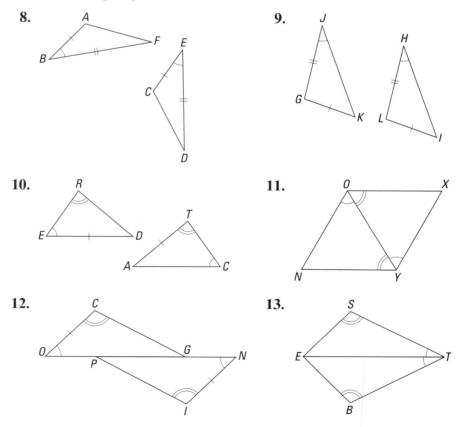

8.

9.

10.

11.

12.

13.

14. Which triangle congruence theorem explains why all triangles are rigid?

Applying the Mathematics

15. Given: $\overline{DP} \cong \overline{JK}$, $\overline{DO} \cong \overline{JL}$, and $\overline{OP} \cong \overline{LK}$
  a. $\triangle DPO \cong \underline{\ ?\ }$.
  b. If $OD = 5$ yards, then $\underline{\ ?\ }$ is 5 yards.
  c. $\angle O \cong \angle \underline{\ ?\ }$.
  d. If m$\angle P = 73$, then $\underline{\ ?\ } = 73$.

16. In the figures below, $\overline{AB} \cong \overline{FD}$ and $\angle B \cong \angle D$. State the additional information that must be known in order to use the congruence theorem to show that the figures are congruent.
  a. AAS $\cong$ Theorem
  b. ASA $\cong$ Theorem
  c. SAS $\cong$ Theorem
  d. SSS $\cong$ Theorem

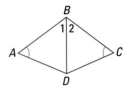

17. In the sail for a hang glider at the right, the seam $\overline{BD}$ bisects $\angle ABC$, and $\angle A \cong \angle C$. Why can you be sure that the cloth used for the left side can be cut from the same pattern design as the right side?

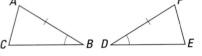

18. Sally attached three sticks together as shown below. She rotated stick A until it made an angle of 23° with the meter stick. Then she rotated stick B so that it made an angle of 49° with the meter stick. Lastly, she secured A and B where they crossed to make a triangle. Why will anyone get a triangle congruent to hers by following the same procedure?

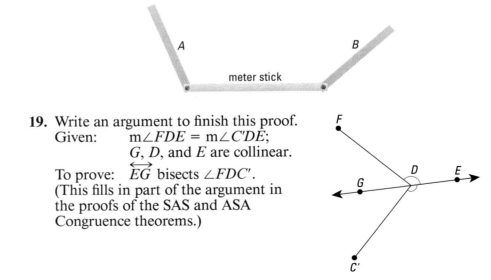

meter stick

19. Write an argument to finish this proof.
  Given: m$\angle FDE =$ m$\angle C'DE$;
         G, D, and E are collinear.

  To prove: $\overleftrightarrow{EG}$ bisects $\angle FDC'$.
  (This fills in part of the argument in the proofs of the SAS and ASA Congruence theorems.)

Hanging out. *The pilot of a hang glider hangs from a harness and steers and regulates the speed of the glider by shifting his or her body weight.*

In 20–22, the figure has *n*-fold rotation symmetry.
a. Determine *n*.
b. Trace the figure and draw all the symmetry lines.   *(Lesson 6-6)*

20.   **21.**   **22.**

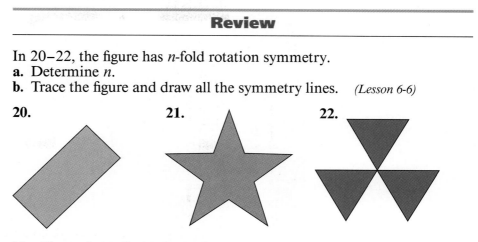

23. Given: $\triangle ABC \cong \triangle DEF$. If m$\angle A$ = 32 and m$\angle B$ = 64, what is m$\angle F$?
(Lessons 4-9, 6-3)

24. Given: $\triangle OLD \cong \triangle NEW$. Make and justify four conclusions from this information.   *(Lesson 5-1)*

25. In the miniature golf hole below, where should you aim on wall *x* to shoot a golf ball *G* off sides *x*, *y*, and *z*, and into the hole at *H*?
(Lesson 4-3)

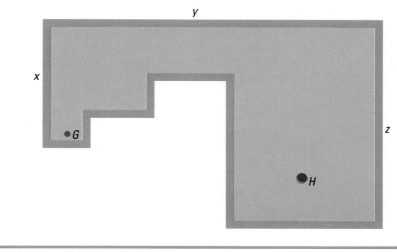

26. Draw two noncongruent triangles *ABC* and *DEF* such that $AB = DE$ and each angle in $\triangle ABC$ is congruent to a different angle in $\triangle DEF$.

LESSON

7-3

*Proofs
Using
Triangle
Congruence
Theorems*

To use any of the triangle congruence theorems of the preceding lesson, you need to know that three parts (SSS, SAS, ASA, or AAS) of one triangle are congruent to the corresponding three parts of another. The theorems then enable you to conclude that the triangles are congruent.

Examine Example 1 carefully. Notice that from the given information, two pairs of angles and their included sides are congruent. This is the ASA condition. The task is to write a clear argument.

Example 1

Write a proof argument.
Given: In the figure at the right,
          ∠*EBA* ≅ ∠*DBC*, *B* is the midpoint
          of $\overline{AC}$, and ∠*A* ≅ ∠*C*.
To prove: △*ABE* ≅ △*CBD*.

Solution

It helps to copy the figure and mark it with the given information.

Argument:

Conclusions	Justifications
0. ∠EBA ≅ ∠DBC. B is the midpoint of $\overline{AC}$. ∠A ≅ ∠C	Given
1. $\overline{AB}$ ≅ $\overline{BC}$	definition of midpoint
2. △ABE ≅ △CBD	ASA Congruence Theorem (Given and step 1)

In the justification for step 2, the part in parentheses refers to the given or the steps in which the corresponding parts were shown to be congruent. The congruence of these three parts must be stated explicitly, either in the given or in the proof, before applying a triangle congruence theorem as a justification.

Often, information is not given as directly as in Example 1. Even so, from given information you sometimes can deduce the conditions necessary to prove triangles congruent.

Example 2

Write a proof argument.
Given: *M* is the midpoint of $\overline{CD}$ and $\overline{EF}$.
To prove: $\triangle CME \cong \triangle DMF$.

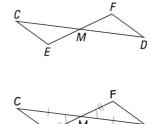

Solution

Again, it helps to copy the figure and mark it with the given information.

Argument:

	Conclusions	Justifications
	0. M is the midpoint of $\overline{CD}$. M is the midpoint of $\overline{EF}$.	Given
Because *M* is given to be a midpoint of two segments, there are two pairs of congruent sides.	1. $\overline{MC} \cong \overline{MD}$, $\overline{ME} \cong \overline{MF}$	definition of midpoint
The intersecting lines form vertical angles, which are also congruent.	2. $\angle CME \cong \angle DMF$	Vertical Angles Theorem
Now two sides and the included angle of each triangle are congruent. This is the SAS condition.	3. $\triangle CME \cong \triangle DMF$	SAS $\cong$ Theorem (steps 1 and 2)

Parts of Congruent Triangles

When triangles are congruent, all their corresponding parts are congruent due to the CPCF Theorem. Thus the SSS, SAS, ASA, and AAS theorems enable you to conclude that the three other pairs of parts are congruent, when you only knew about three pairs before. That makes these theorems quite powerful. For instance, in Example 2 you could conclude $\angle E \cong \angle F$, $\angle D \cong \angle C$, and $\overline{CE} \cong \overline{DF}$ by the CPCF Theorem.

Often a side or angle is shared by two triangles. The Reflexive Property of Congruence is the justification for their congruence. These ideas are Illustrated in Example 3 on the next page.

Example 3

Write a proof argument.
Given: $AB = CD$;
           $BC = AD$.
To prove: $\angle B \cong \angle D$.

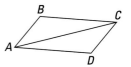

Solution

We copy the figure and mark it with the given information.

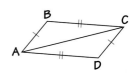

$\overline{AC}$ is a side of both triangles.

Argument:

Conclusions	Justifications
0. $AB = CD$, $BC = AD$	Given
1. $\overline{AC} \cong \overline{AC}$	Reflexive Property of Congruence
2. $\triangle ABC \cong \triangle CDA$	SSS $\cong$ Theorem (Given and step 1)
3. $\angle B \cong \angle D$	CPCF Theorem

Important theorems can be proved using the triangle congruence theorems. The AAS Congruence Theorem helps prove the converse of the Isosceles Triangle Base Angles Theorem. The argument below uses an auxiliary line. You are asked to complete the proof in Question 5.

> **Isosceles Triangle Base Angles Converse Theorem**
> If two angles of a triangle are congruent, then the sides opposite them are congruent.

Proof
Given: $\angle B \cong \angle C$ in the figure at the right.
To prove: $\overline{AB} \cong \overline{AC}$.

Argument:

Conclusions	Justifications
0. $\angle B \cong \angle C$	Given
1. Draw the bisector of $\angle A$, intersecting $\overline{BC}$ at D.	An angle has exactly one bisector.
2. $\angle BAD \cong \angle CAD$	definition of angle bisector
3. $\overline{AD} \cong \overline{AD}$	**a.**
4. $\triangle ABD \cong \triangle ACD$	**b.** Congruence Theorem (Given, steps 2 and 3)
5. $\overline{AB} \cong \overline{AC}$	**c.**

Notice the isosceles triangles in the barn gate of this farm in the Blue Ridge Mountains of North Carolina.

The triangle congruence theorems are helpful in deducing properties of many figures, as you will see. Writing proofs using these theorems takes some practice; it often takes a while to become proficient.

Covering the Reading

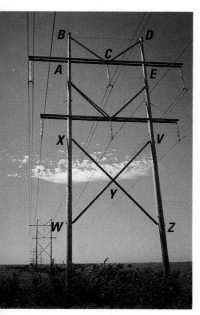

In 1 and 2, use the figures named on the picture of the electrical transmission tower.

1. Given: C is the midpoint of $\overline{AE}$, $\angle BAC \cong \angle DEC$, and $\angle BCA \cong \angle DCE$ in triangles ABC and EDC.
   a. Copy the triangles and mark the given on them.
   b. What theorem justifies the congruence of the triangles?
   c. Write the argument of a proof that $\triangle ABC \cong \triangle EDC$.

2. Write a proof argument.
   Given: Y is the midpoint of $\overline{WV}$.
   $\quad\quad$ m$\angle Z$ = m$\angle X$.
   To prove: $\triangle WXY \cong \triangle VZY$.

3. Using the given of Example 3, prove $\angle BAC \cong \angle DCA$.

4. a. With angle measures as given in the figure at the right, which sides of $\triangle ABC$ are congruent?
   b. What is the justification for your answer to part a?

5. In the proof of the Isosceles Triangle Base Angles Converse Theorem, give a justification for the conclusion of each step.
   a. Step 3 $\quad\quad$ b. Step 4 $\quad\quad$ c. Step 5

Applying the Mathematics

In 6–8, write a proof argument using the given figure.

6. Given: $XA = YA$ and
   $\quad\quad$ $XC = YC$.
   To prove: $\angle XAC \cong \angle YAC$.

7. Given: $\overrightarrow{NK}$ bisects $\angle JNM$.
   $\quad\quad$ $\overline{NM} \cong \overline{NK}$ and $\overline{NL} \cong \overline{NJ}$.
   To prove: $\triangle JNK \cong \triangle LNM$.

8. Given: $\overline{BC} \parallel \overline{AD}$ and
   $\quad\quad$ $BC = AD$.
   To prove: $\triangle ABC \cong \triangle CDA$.

9. Write a proof argument using the figure at the right.
   Given: m∠*ADB* = m∠*CDB*,
            m∠*ABD* = m∠*CBD*.
   To prove: *AB* = *CB*.

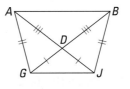

10. a. List all eight triangles in the figure below.
   b. From the marked information, there are three pairs of triangles
      which can be proved congruent. Name them, with vertices in
      corresponding order.

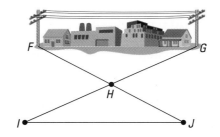

Review

11. The figure below shows electric wires that run from post *F* to post *G*
   over the buildings. People can't measure the length of wire needed
   because the buildings are in the way. Here is how it can be done
   using triangle congruence. Place a stake at *H*. Then mark off *J* on
   $\overrightarrow{FH}$ so that *FH* = *HJ*. Now mark off *I* on $\overrightarrow{GH}$ so that *GH* = *HI*.

   a. What triangle congruence theorem indicates that △*FGH* ≅ △*JIH*?
   b. Why does *IJ* = *FG*?   *(Lesson 7-2)*

In 12 and 13, if the given triangles are congruent, justify with a triangle
congruence theorem and indicate all the corresponding vertices.
Otherwise, write *not enough information to know*.   *(Lesson 7-2)*

12.

A

D *C* *B*

13. *E* *F*

G

I *H*

14. *True or false.* If Trip and Trisha each make a triangle out of straws with lengths 3 cm, 4 cm, and 6 cm, then the two triangles
a. must be congruent.
b. must have the same orientation. *(Lessons 4-8, 7-1)*

15. At the right is a triangle *JKL* with measures as indicated.
a. Are all triangles with these dimensions congruent?
b. Why or why not? *(Lessons 7-1, 7-2)*

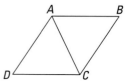

In 16 and 17, use the figure at the right.

16. If *ABCD* is a rhombus and m∠*DAC* = 50, find as many other angle measures as you can. *(Lesson 6-4)*

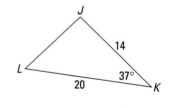

17. Trace the figure and translate it by the vector $\overrightarrow{AB}$. *(Lesson 4-6)*

18. The sum of the measures of the acute angles in a right triangle is __?__. *(Lesson 5-7)*

19. Draw a right triangle *ABC* with right angle at *C*. Find the image of △*ABC* under a rotation of 180° about the midpoint of $\overline{AB}$. *(Lesson 3-2)*

20. Why is there no triangle with sides 3 cm, 11 cm, and 6 cm? *(Lesson 2-7)*

Exploration

21. Quadrilateral *ABCD* in Example 3 looks like a parallelogram. With the information as given in the example, must *ABCD* be a parallelogram? Explain why or why not.

LESSON

7-4

Overlapping Triangles

Folding forms figures. *Origami figures are created without cutting, pasting, or decorating. Japanese works fall into two categories: natural shapes and ceremonial designs.*

How many triangles do you see in the figure at the right? At first, many people see only the small triangles *BFD* and *CFE*. But there are also *overlapping triangles ACD* and *ABE*. **Overlapping figures** have some part of their interiors in common. (Figures that do not overlap are called **nonoverlapping figures**.) Given appropriate information, overlapping triangles can be proved congruent just as other triangles can.

Example 1

In the figure above, if *AC* = *AB* and *AD* = *AE*, prove ∠*D* ≅ ∠*E*.

Solution

Given: *AC* = *AB* and *AD* = *AE*.
To prove: ∠*D* ≅ ∠*E*.
It may seem that the triangles to use are the small triangles *BFD* and *CFE*. But no sides or angles of these triangles are given as congruent, and even though the vertical angles ∠*BFD* and ∠*CFE* are congruent, one pair of angles is not enough to prove any triangles congruent. So try the overlapping triangles, △*ACD* and △*ABE*.

Drawing: We draw the figure twice to see the overlapping triangles. (You may not have to do this.) Mark the figure with the given information. Notice that although only two pairs of sides are given as congruent, there is a common angle, ∠*A*.

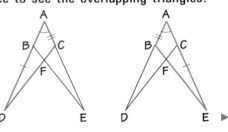

Argument:

Conclusions	Justifications
1. ∠CAD ≅ ∠BAE	Reflexive Property of Congruence
2. △ADC ≅ △AEB	SAS ≅ Theorem (Given and step 1)
3. ∠D ≅ ∠E	CPCF Theorem

With overlapping triangles, keeping track of corresponding vertices can be tricky. As you deduce congruent sides or angles, mark the figure. Some students find it helpful to use colored pencils and mark each triangle with a different color.

Overlapping triangles occur when intersecting diagonals of a polygon are drawn.

Example 2

Given: Regular hexagon *ABCDEF* below.
To prove: $\overline{BF} \cong \overline{AC}$.

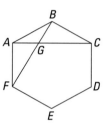

Solution

You may want to redraw △*ABF* and △*ABC* for clarity.
Drawing:

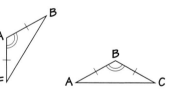

Argument:

Conclusions	Justifications
0. ABCDEF is a regular hexagon.	Given
1. $\overline{AF} \cong \overline{BC}$	definition of regular polygon
2. ∠FAB ≅ ∠ABC	definition of regular polygon
3. $\overline{AB} \cong \overline{AB}$	Reflexive Property of ≅
4. △FAB ≅ △ABC	SAS Congruence Theorem (steps 1, 2, and 3)
5. $\overline{BF} \cong \overline{AC}$	CPCF Theorem

Covering the Reading

1. Use the figure at the right.
   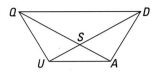
   a. How many triangles are in the figure?
   b. It looks as if △QUA is congruent to what other triangle?
   c. Find a second pair of overlapping triangles that seem congruent.

2. Using the diagram and the given of Example 1, fill in the justifications in this proof argument that ∠DCA ≅ ∠EBA.

Conclusions	Justifications
0. $AC = AB$ and $AD = AE$	**a.**
1. ∠DAC ≅ ∠EAB	**b.**
2. △ADC ≅ △AEB	**c.**
3. ∠DCA ≅ ∠EBA	**d.**

3. Given: Regular pentagon $ALIVE$ at the right.
   To prove: ∠EAV ≅ ∠VIE.

Applying the Mathematics

In 4–6, write a proof argument using the given figure.

4. Given: $\overline{QU}$ ≅ $\overline{AD}$ and $\overline{QA}$ ≅ $\overline{UD}$.
   To prove: ∠1 ≅ ∠2.
   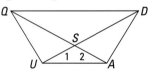

5. Given: $AD = AE$ and m∠D = m∠E.
   To prove: $EB = CD$.
   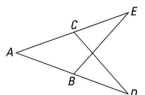

6. Given: $PR = QS$ and $PS = QR$.
   To prove: m∠P = m∠Q.
   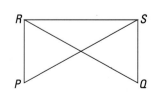

7. Refer to Example 2. Write an argument showing that $\overline{AC}$ ≅ $\overline{BF}$ using symmetry instead of triangle congruence.

Review

In 8 and 9, write a proof argument using the given figure.

8. Given: A, D, E, and B are collinear.
$\overline{AD} \cong \overline{BE}$ and $\overline{AC} \cong \overline{BC}$.
To prove: $\overline{DC} \cong \overline{CE}$. *(Lesson 7-3)*

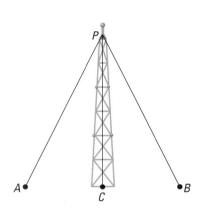

9. Given: $\overline{QR} \parallel \overline{TU}$, and S is the midpoint of $\overline{QU}$.
To prove: S is the midpoint of $\overline{RT}$.
(Lesson 7-3)

10. $\overline{PC}$ is a vertical radio tower on level land supported in part by the taut guy wires $\overline{PA}$ and $\overline{PB}$. Explain why the guy wires will have the same length if they are attached to the ground at the same distance from C. *(Lessons 7-2, 7-3)*

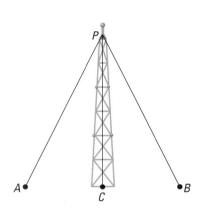

11. Given that $\overline{DF}$ and $\overline{AG}$ intersect at E with $\overline{AE} \cong \overline{FE}$ and $\overline{BE} \cong \overline{GE}$ as marked at the right, justify each conclusion in this proof that $\angle FGE \cong \angle DBC$. *(Lessons 7-2, 7-3)*

Conclusions	Justifications
1. $\angle AEB \cong \angle FEG$	**a.**
2. $\triangle ABE \cong \triangle FGE$	**b.**
3. $\angle FGE \cong \angle ABE$	**c.**
4. $\angle ABE \cong \angle DBC$	**d.**
5. $\angle FGE \cong \angle DBC$	**e.**

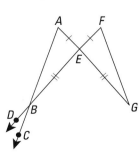

12. a. Draw a $\triangle ABC$ with m$\angle A = 32$ and m$\angle B = 110$.
   b. Should everyone's answers to part **a** be congruent? Why or why not?   *(Lesson 7-1)*

13. What is the measure of each angle of a regular pentagon? *(Lesson 6-7)*

In 14 and 15, *true or false.*   If *QRST* is a translation image of *MNOP*,

14. $MNOP \cong QRST$.

15. $MQ = OS$.   *(Lesson 5-1)*

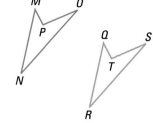

16. If y is the measure of an angle and $0 < 180 - 2y$, is the angle acute, obtuse, or right?   *(Lesson 3-3)*

17. In the figure of Example 1, describe $\triangle ADC \cap \triangle AEB$.   *(Lesson 2-5)*

18. In the figure below, $WX = 13$, $WZ = 45$ and $YZ = 17$. Find XY. *(Lesson 1-8)*

Exploration

19. In Example 1 of the lesson, the outer edges of the overlapping congruent triangles determine the quadrilateral $ADFE$. In Example 2, the outer edges determine the pentagon $AFGCB$. For the questions below, note that overlapping figures do not have to share a side or angle.
   a. Draw two overlapping congruent triangles whose outer edges determine a hexagon.
   b. Draw two overlapping congruent triangles whose outer edges determine a 12-gon.
   c. What other polygons can be determined by the outer edges of two overlapping congruent triangles? Show an example of each.

LESSON

7-5

The SSA Condition and HL Congruence

Catching the angle. *The Tennessee Aquarium in Chattanooga is the world's largest freshwater aquarium, containing more than 3,500 specimens.*

The SSA Condition

Since there are AAS and ASA Congruence Theorems, and there is a congruence theorem with two sides and an included angle (SAS), it is natural to ask what happens if the angle is not the included angle. We call this the **SSA condition.**

Examine $\triangle ABC$ and $\triangle XYZ$ below. There are two pairs of congruent sides, $\overline{AB} \cong \overline{XY}$ and $\overline{BC} \cong \overline{YZ}$. Also, there is a pair of congruent corresponding nonincluded angles, $\angle A$ and $\angle X$.

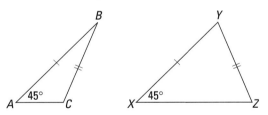

But clearly the triangles are not congruent. In fact, a translation image of $\triangle ABC$ fits snugly into one corner of $\triangle XYZ$. Thus, in general, the SSA condition is not sufficient for determining a unique triangle.

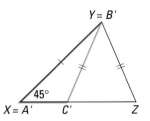

The HL Condition

However, when the corresponding nonincluded angles are right angles, then the situation is different. Recall that in a right triangle, the **legs** are the sides that include the right angle, while the **hypotenuse** is the side opposite the right angle. Suppose $BC = YZ$, $AB = XY$, and $\angle C$ and $\angle Z$ are right angles as shown in the triangles below. This is the **hypotenuse-leg** or **HL condition.** This condition is enough to guarantee congruence.

> **HL Congruence Theorem**
> If, in two right triangles, the hypotenuse and a leg of one are congruent to the hypotenuse and a leg of the other, then the two triangles are congruent.

Proof

Given: $\overline{BC} \cong \overline{YZ}$.
             $\overline{AB} \cong \overline{XY}$.
             $\angle C$ and $\angle Z$ are right angles.
To prove: $\triangle ABC \cong \triangle XYZ$.
Drawing:

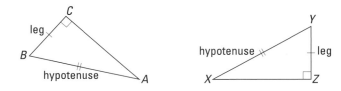

Argument:
We use the symmetry of an isosceles triangle. Since $\overline{BC} \cong \overline{YZ}$, there is an isometry T with $T(\overline{BC}) = \overline{YZ}$, and with $T(A)$ on the other side of $\overline{YZ}$ from X. This gives the figure below, in which $\triangle ABC \cong \triangle A'B'C'$.

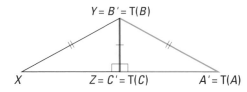

$\angle C$ and $\angle Z$ are right angles. So $m\angle A'ZX = 180$ by Angle Addition, which means A', Z, and X are collinear. This makes $\triangle A'YX$ an isosceles triangle. Applying the Isosceles Triangle Base Angles Theorem, $\angle A' \cong \angle X$. With two pairs of congruent angles and $C'B' = ZY$, $\triangle A'B'C' \cong \triangle XYZ$ by the AAS $\cong$ Theorem. So $\triangle ABC \cong \triangle XYZ$ by the Transitive Property of Congruence.

Constructing the right way. *Right triangles can often be seen while buildings are being constructed.*

Example 1

Given: $\overline{AC} \perp \overline{BD}$ and $AD = AB$.

To prove: $\triangle ABC \cong \triangle ADC$.

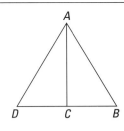

Solution

If you mark the figure, you will see that you may apply the HL condition.

Conclusions	Justifications
0. $\overline{AC} \perp \overline{BD}$, $AD = AB$	Given
1. $\overline{AC} \cong \overline{AC}$	Reflexive Property of $\cong$
2. $\triangle ABC \cong \triangle ADC$	HL Congruence Theorem (Given and Step 1)

Once triangles are proved congruent, many deductions can follow if you use the congruent parts.

Example 2

Write a proof argument.

Given: $\overline{AB} \perp \overline{BD}$, $\overline{CD} \perp \overline{BD}$, and $AD = BC$ in the figure below.

To prove: $\overline{AD} \parallel \overline{BC}$.

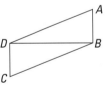

Solution

Drawing: Mark the figure. The triangles are congruent by HL congruence. Corresponding angles allow you to deduce parallel lines.

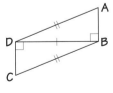

Argument:

Conclusions	Justifications
1. $\overline{BD} \cong \overline{BD}$	Reflexive Property of $\cong$
2. $\triangle ABD \cong \triangle CDB$	HL Congruence (step 1 and Given)
3. $\angle ADB \cong \angle CBD$	CPCF Theorem
4. $\overline{AD} \parallel \overline{BC}$	AIA $\cong \Rightarrow \parallel$ Lines

SsA Condition

The HL condition is the special case of the SSA condition when the congruent angles are right angles. Because there is an HL Congruence Theorem, the SSA condition works sometimes. A natural question is: Does the SSA condition give congruent triangles at any other time? The answer is "Yes." We write it as the *SsA Congruence Theorem*. In order for it to work, the sides opposite the congruent angles in each triangle must be longer than the other pair of congruent sides. The proof is long, so we number the steps even though they are in a paragraph.

SsA Congruence Theorem
If two sides and the angle opposite the longer of the two sides in one triangle are congruent, respectively, to two sides and the corresponding angle in another triangle, then the triangles are congruent.

Proof
Given: $\overline{AB} \cong \overline{XY}$, $\overline{AC} \cong \overline{XZ}$, $\angle C \cong \angle Z$, and $XY > XZ$.
To prove: $\triangle ABC \cong \triangle XYZ$.
Drawing:

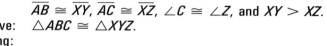

Argument:
The argument is like the arguments for the other triangle congruence theorems. To begin with, in steps 1–3 we map $\triangle ABC$ onto a conveniently located congruent image $\triangle A'B'C'$.

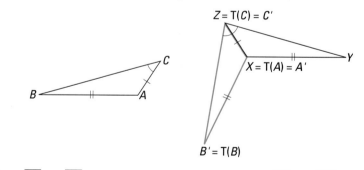

1. Since $\overline{AC} \cong \overline{XZ}$, there is an isometry T with $T(\overline{AC}) = \overline{XZ}$ and with $T(A) = A' = X$ and $T(C) = C' = Z$. Furthermore, T can be chosen so that $T(B) = B'$ is on the other side of XZ from Y.
2. By the definition of congruence, $\triangle ABC \cong \triangle A'B'C'$.
3. It is given that $\angle BCA \cong \angle YZX$. By the CPCF Theorem, $\angle BCA \cong \angle B'C'A'$. And so, by the Transitive Property of Congruence, $\angle B'C'A' \cong \angle YZX$. This makes $\overrightarrow{ZX}$ the bisector of $\angle B'ZY$.

In steps 4–8, we show that $\triangle A'B'C'$ is congruent to $\triangle XYZ$. This is done by showing that $\triangle XYZ$ is the reflection image of $\triangle A'B'C'$ over the line $\overleftrightarrow{ZX}$.

▶

4. C' and A' are the points Z and X on the reflection line, so by the definition of reflection, $r(C') = Z$ and $r(A') = X$.

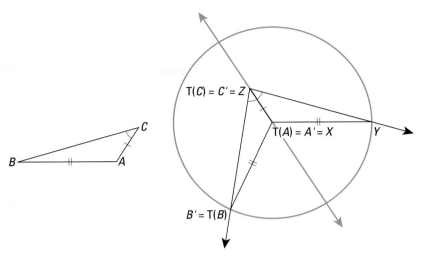

5. To find $r(B')$, consider the auxiliary circle with center X and radius XY. B' is on this circle because $XB' = XY$. $r(B')$ is on the circle because $\overleftrightarrow{XZ}$ contains a diameter, and any diameter is a symmetry line for a circle.

6. Z is in the interior of this circle because $XY > XZ$. Thus the ray $\overrightarrow{ZY}$ intersects the circle in exactly one point. $r(B')$ must also be on $\overrightarrow{ZY}$ because of the Side-Switching Theorem. Consequently, $r(B') = Y$, the only point of intersection of the circle and $\overrightarrow{ZY}$.

7. By the Figure Reflection Theorem (steps 4 and 6), $r(\triangle A'B'C') = \triangle XYZ$.

8. By the definition of congruence, $\triangle A'B'C' \cong \triangle XYZ$.

9. By the Transitive Property of Congruence (steps 2 and 8), $\triangle ABC \cong \triangle XYZ$.

Notice that if $XZ > XY$, then the given congruent angles are no longer opposite the longer congruent sides. In this case, the point Z is outside the circle and the ray intersects the circle in two points. Then there are two choices for B' and we cannot conclude that $r(B') = Y$. This is why XY must be greater than XZ.

The argument in the proof of the SsA Congruence Theorem is the longest in this book. But there are much longer arguments in mathematics. Some proofs take many pages of writing, and a few proofs are the lengths of books. The SsA Congruence Theorem is not in Euclid's *Elements*, nor is it found in many geometry books.

You now have studied the five triangle congruence theorems: SSS, SAS, ASA, AAS, and SsA, and HL, a special case of SsA.

QUESTIONS

Covering the Reading

1. Draw two noncongruent triangles satisfying the SSA condition.

2. **a.** What kind of triangle has sides represented by H and L?
   b. What does H stand for? What does L stand for?

3. The proof of the HL Congruence Theorem makes use of the symmetry of which figure?

4. Write a proof argument using the figure at the right.
   Given: $\overline{BD} \perp \overline{AC}$ and $AB = BC$.
   To prove: $\triangle ABD \cong \triangle CBD$.

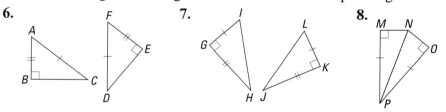

5. When does the SSA condition lead to congruence?

In 6–8, use the information given in the figure. **a.** What triangle congruence theorem tells you that the two triangles are congruent? **b.** Name the congruent triangles with vertices in corresponding order.

6.   7.   8.

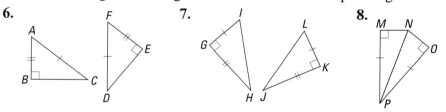

Applying the Mathematics

9. Follow the steps to draw an instance of a triangle given the SSA condition.
   a. Draw $\overrightarrow{XY}$.
   b. Draw $\angle ZXY$ with measure 50 and with $XZ = 11$ cm.
   c. Draw circle Z with radius 9 cm. Let W be a point of intersection of $\odot Z$ and $\overrightarrow{XY}$.
   d. Consider $\triangle XZW$. Will everyone else who follows these steps correctly have a triangle XZW congruent to yours?

In 10 and 11, write a proof argument using the given figure.

10. Given: $\overline{VX} \perp \overline{WY}$, $WZ = VY$, and $XZ = XY$.
    To prove: $\angle W \cong \angle V$.

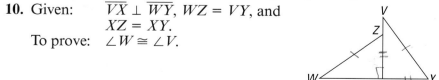

394

11. Given: $AP = AR$, $\angle P$ and $\angle R$ are
right angles.
To prove: $PBRA$ is a kite.

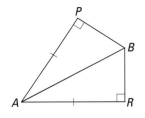

12. Recall that two circles are concentric if they have the same
center. The two concentric circles at the left have center O, with
$OQ > OP$. If $\angle OAB \cong \angle OPQ$, complete the proof argument that
$\triangle ABO \cong \triangle PQO$.

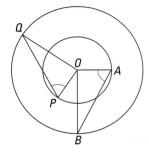

Argument:

Conclusions	Justifications
0. $OQ > OP$,	**a.**
$\angle OAB \cong \angle OPQ$	
1. $OP = OA$	**b.**
2. $OQ = OB$	**c.**
3. $\triangle ABO \cong \triangle PQO$	**d.**

13. June and April are dancing around a
maypole. The streamers they are holding are
the same length. If June's and April's hands
are the same height and the ground is level,
explain why their hands are the same
distance from the maypole.

Review

In 14 and 15, arguments are shown to prove that the diagonals of an
isosceles trapezoid are congruent.
Given: $WXYZ$ is an isosceles trapezoid with bases $\overline{WX}$ and $\overline{ZY}$.
To prove: $XZ \cong WY$.

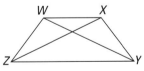

14. Fill in the justifications in this argument using congruent triangles.
(Lesson 7-4)

Conclusions	Justifications
1. $\angle WZY \cong \angle XYZ$	**a.**
2. $\overline{WZ} \cong \overline{XY}$	**b.**
3. $\overline{ZY} \cong \overline{ZY}$	**c.**
4. $\triangle WZY \cong \triangle XYZ$	**d.**
5. $\overline{XZ} \cong \overline{WY}$	**e.**

15. Fill in the justifications using symmetry.   *(Lesson 6-5)*

Conclusions	Justifications
1. $WXYZ$ is symmetric to ℓ,	**f.**
the bisector of $\overline{WX}$ and $\overline{ZY}$.	
2. $\overline{XZ} \cong \overline{WY}$	**g.**

May days may daze. *In
medieval times, villagers
would dance around the
maypole holding ends of
ribbons. Today that
tradition continues.*

16. For Questions 14 and 15, which argument do you prefer? Why?

17. a. Draw $\triangle QRS$ with m$\angle Q = 45$, m$\angle R = 80$, and $RS = 2''$.
  b. Will everyone else's correct drawing be congruent to yours? Explain why or why not. *(Lesson 7-1)*

18. Write a proof argument using the given figure.
  Given: $AB = AC$ and $BE = DC$.
  To prove: **a.** $\angle ADE \cong \angle AED$.
    b. $\triangle ADE$ is isosceles. *(Lesson 6-2)*

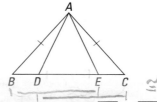

19. Given the figure below as marked with $\overline{AB} \perp \overline{XY}$, name all triangles congruent to $\triangle AGX$. *(Lesson 6-2)*

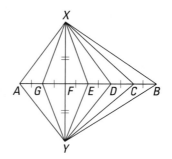

20. Suppose $r_m(\triangle ABC) = \triangle DEF$. State a conclusion which follows due to each justification. *(Lessons 4-2, 4-8, 6-1)*
  a. definition of congruence
  b. Flip-Flop Theorem
  c. Figure Reflection Theorem

Exploration

21. Explore this conjecture: If, in quadrilaterals $ABCD$ and $EFGH$, angles A, C, E, and G are right angles, $AB = EF$, and $BC = FG$, then the quadrilaterals are congruent.

Mt. Shuksan is located in the state of Washington.

Many floors and walls are covered with congruent polygons. Pictured above is a bamboo pattern formed from congruent parallelograms. Many other regions could also be used to cover the plane.

Covering the Plane with Congruent Triangles

Here are two tiling patterns using congruent copies of $\triangle ABC$. Notice that the patterns are different. Pattern I contains parallelograms, while pattern II contains kites.

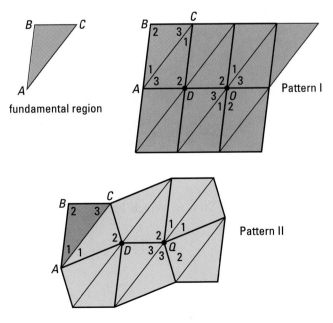

fundamental region

Pattern I

Pattern II

A covering of a plane with nonoverlapping congruent regions and no holes, is called a **tessellation.** The region is called a **fundamental region** for the tiling. On page 397, $\triangle ABC$ is a fundamental region for each tessellation shown. We say that $\triangle ABC$ **tessellates** the plane. A key question is whether a given region can tessellate the plane. The two activities that follow show how to tessellate the plane with a triangle forming the patterns shown. You will need a full sheet of paper.

Activity 1

Tessellate part of a plane using Pattern I.
Use a full sheet of paper.

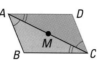

Step 1: Draw a scalene triangle ABC, with sides two or three times larger than the one above.
Step 2: Find M, the midpoint of $\overline{AC}$, and rotate $\triangle ABC$ 180° about M.

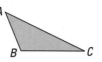

Step 3: Repeatedly translate the region $ABCD$ by the vectors $\overrightarrow{BC}$ and $\overrightarrow{CB}$.

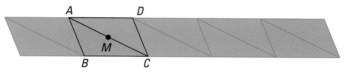

Step 4: Repeatedly translate the entire figure by the vectors $\overrightarrow{AB}$ and $\overrightarrow{BA}$. The result covers as much of the plane as you wish.

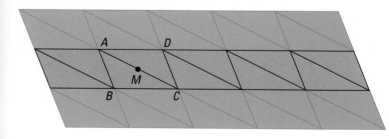

Step 5: Which kind of special quadrilateral is $ABCD$? How do you know?

Activity 2

Tessellate part of a plane with any triangle *XYZ* using Pattern II.

Step 1: Draw a triangle *XYZ* not necessarily congruent to the one in the box below.

Step 2: Reflect △*XYZ* over $\overleftrightarrow{YZ}$. The image of △*XYZ* is △*WYZ*. *XYWZ* is a kite.

Step 3: Rotate the entire figure 180° about *M*, the midpoint of $\overline{XY}$.

Step 4: Repeatedly translate the four-triangle region by the vectors $\overrightarrow{YZ}$ and $\overrightarrow{ZY}$.

Step 5: Translate the entire figure by the vectors $\overrightarrow{XW}$ and $\overrightarrow{WX}$. The result can cover as much of the plane as you wish to cover.

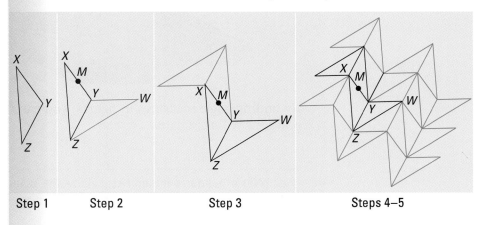

Step 1   Step 2   Step 3   Steps 4–5

The idea of a tessellation is an old one. The word *tessellate* comes from a Latin word meaning "small stone." Small stones, put together into mosaics, covered the floors of many Roman buildings. The Moors, whose religion (Islam) does not allow any pictures in their places of worship, used many different tessellations in decorating their mosques. Below is a photograph of tessellations in the Alhambra, a museum in Granada, Spain, that was built in the 1300s as a mosque.

Tessellations with Other Polygons

It is not as easy to see, but every quadrilateral region will tessellate!
Since the sum of the angles in a quadrilateral is 360°, it is possible to
have a different angle from each of four congruent quadrilaterals
meeting at a single point. A start of one such tessellation is shown below.
This tessellation was generated using the procedure of Question 14.

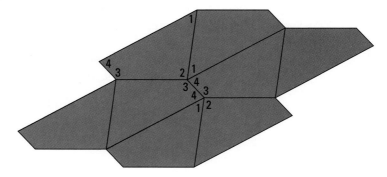

By finding angle measures in figures, you can tell whether certain regular
polygons will tessellate.

Example

a. Will a regular pentagon tessellate the plane?
b. Will a regular hexagon tessellate the plane?

Solution

To answer these questions, the measure of an individual angle in each
figure must be determined.

a. The sum of the angle measures in a pentagon is $(5 - 2) \cdot 180 = 540$, so each angle in a regular pentagon has measure 108. Since 360 is not evenly divisible by 108, a regular pentagon will not tessellate. At the left three regular pentagons are pictured around a point. Notice that there is a gap which cannot be filled with a regular pentagon.

b. The sum of the angle measures in a hexagon is $(6 - 2) \cdot 180 = 720$, so each angle in a regular hexagon has measure 120. Since 360 is divisible by 120, a regular hexagon will tessellate. At the right is a sample of the start of such a tessellation.

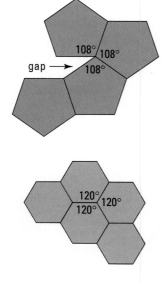

While a regular pentagon will not tessellate, other pentagons do. But no
one yet has proved exactly how many types of pentagons tessellate the
plane! It is an unsolved problem in which there have been some
advances in recent years. In 1975, Marjorie Rice, a homemaker from
California, used her knowledge of high school geometry to discover a
new type of pentagon that tessellates the plane. As recently as 1985, Rolf
Stein of the University of Dortmund in Germany found still another
new type of tessellating pentagon. So an old application of geometry is
still yielding new mathematics. Perhaps *you* will be able to discover a
new tessellation.

Technicolored tessellation. *This quilt was made by Acadienne women in Abram Village, Prince Edward Island, Canada. Quilt patterns often are tessellations.*

QUESTIONS

Covering the Reading

1. What is a tessellation?

2. Give two examples of tessellations seen where you live.

3. Show your results from Activity 1.

4. Show your results from Activity 2.

In 5 and 6, trace the figure repeatedly to show part of a tessellation using the figure as a fundamental region.

5.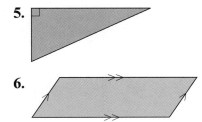

6.

7. What key question concerning tessellations is discussed in this lesson?

8. One museum where many tessellations can be found is the ___?___.

9. Can quadrilateral *ABCD* below tessellate the plane? Why or why not?

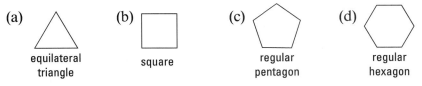

10. *Multiple choice.* Pictured here are regular polygons with 3, 4, 5, and 6 sides. Which one cannot be a fundamental region for a tessellation?

   (a) equilateral triangle

   (b) square

   (c) regular pentagon

   (d) regular hexagon

11. Will a regular heptagon tessellate the plane? Explain your answer.

12. A new type of tessellating pentagon was discovered as recently as ___?___.

13. The tessellation below is by Sheila Haak, a teacher and mathematics textbook editor. Trace a possible fundamental region.

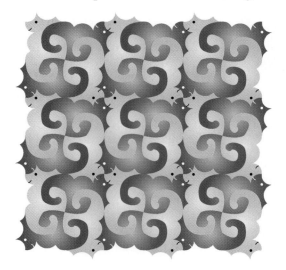

14. **a.** Follow this algorithm to tessellate the plane with quadrilateral $ABOC$ below, where X, Y, and Z are the midpoints of $\overline{BO}$, $\overline{AB}$, and $\overline{AC}$, respectively.

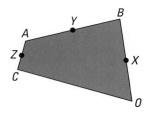

    Step 1: Trace quadrilateral $ABOC$ or place a figure like $ABOC$ on the screen of an automatic drawer.

    Step 2: Rotate $ABOC$ 180° about X (image $A'OBC'$).

    Step 3: Rotate $A'OBC'$ 180° about Y', the image of Y (image $OA'B''C''$).

    Step 4: Rotate $OA'B''C''$ 180° about Z'', the image of Z' (image $C''A^*CO$).

  b. There are now 4 angles with vertex O. How do you know that these angles fit around O exactly?

  c. What needs to be done to complete the tessellation?

  d. If you have an automatic drawer, repeat the tessellation with a nonconvex quadrilateral $ABOC$.

15. Refer to Pattern I on page 397. Prove that $ABCD$ is a parallelogram.

16. Refer to Pattern II on page 397. Prove that $ABCD$ is a kite.

In 17–19, write a proof argument using the given figure.

17. Given: $AD = BC$, $\overline{DA} \perp \overline{AB}$, and $\overline{BC} \perp \overline{CD}$.
To prove: $\overline{AB} \parallel \overline{DC}$. *(Lesson 7-5)*

18. Given: $\overline{AC} \cong \overline{BD}$ and $\overline{AD} \cong \overline{BC}$.
To prove: $\angle ADB \cong \angle BCA$. *(Lesson 7-4)*

19. Given: $m\angle ABD = m\angle BDC$ and $m\angle ADB = m\angle DBC$.
To prove: $AB = CD$. *(Lesson 7-3)*

20. In the figure at the right, $\overline{AB} \cong \overline{XY}$ and $\angle A \cong \angle Y$. Name three additional pieces of information, each of which would be enough to guarantee congruence of the triangles, and name the appropriate congruence theorem to justify that conclusion. *(Lesson 7-2)*

21. In $\triangle PQR$, a student wanted to draw a segment from P perpendicular to $\overline{QR}$ at its midpoint.
a. Is this segment uniquely determined?
b. If so, give the justification. If not, tell why not. *(Lesson 5-6)*

22. State Playfair's Parallel Postulate. *(Lesson 5-6)*

23. Write a paragraph discussing the Moors and what happened to them.

24. Below is part of a tessellation using the pentagon discovered by Rolf Stein. Trace the part below and continue it to fill a sheet of paper.

*Congruent
Parts of
Parallelograms*

IN-CLASS
ACTIVITY

Work with a partner. You will need a ruler, compass, and protractor.

Steps 1–3   Steps 4–6

Step 1. Draw a parallelogram *ABCD* and its diagonal $\overline{AC}$. Identify all angles that are congruent.

Step 2. Prove that the two triangles formed are congruent.

Step 3. What pairs of segments and angles are congruent due to proving the triangles congruent?

Step 4. Draw $\overline{BD}$. Let $\overline{BD} \cap \overline{AC} = E$. Prove as many pairs of triangles congruent as you can.

Step 5. What additional pairs of segments and angles are congruent due to proving the triangles congruent in Step 4?

Step 6. What symmetry does the parallelogram possess?

LESSON

7-7

Properties of Parallelograms

Fenceful figures. *Many parallelograms can be seen in this bamboo fence surrounding a tea house at the Japanese Embassy in Washington D.C.*

The In-class Activity on the previous page shows that when the diagonals of a parallelogram are drawn, there are many pairs of congruent triangles, segments, and angles. You should be able to find four pairs of congruent triangles in Step 4.

From these triangle congruences, the following theorem can be proved. We show the proof of part (a) and leave the proofs of parts (b) and (c) to you in Questions 6 and 7.

Properties of a Parallelogram Theorem
In any parallelogram:
(a) opposite sides are congruent;
(b) opposite angles are congruent;
(c) the diagonals intersect at their midpoints.

Proof
Given: parallelogram *ABCD*.
To prove: $\overline{AB} \cong \overline{CD}$ and $\overline{AD} \cong \overline{CB}$.

Drawing:

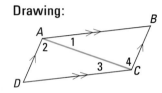

Argument:
Draw auxiliary diagonal $\overline{AC}$ and number the angles formed.

Now consider $\triangle ACD$ and $\triangle CAB$.
Since $\overleftrightarrow{AD}$ // $\overleftrightarrow{BC}$ and // Lines $\Rightarrow$ AIA $\cong$, $\angle 2 \cong \angle 4$.
Since $\overleftrightarrow{AB}$ // $\overleftrightarrow{CD}$ and // Lines $\Rightarrow$ AIA $\cong$, $\angle 1 \cong \angle 3$.
Also $\overline{AC} \cong \overline{AC}$, so the triangles are congruent by ASA.
Using the CPCF Theorem, $\overline{AB} \cong \overline{CD}$ and $\overline{AD} \cong \overline{CB}$.

Because of the Quadrilateral Hierarchy Theorem, you can further conclude that the properties of parallelograms apply to all rhombuses, rectangles, and squares.

Example

In parallelogram *WXYZ* at the right, $WQ = 4$, $XQ = 6$, and $YZ = 7$.
a. Find *QY*.
b. Find *WX*.

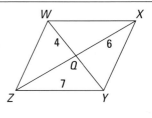

Solution

a. *Q* is the midpoint of $\overline{WY}$, so
   $QY = 4$.
b. $\overline{WX}$ and $\overline{YZ}$ are opposite sides of the parallelogram, so $WX = YZ$.
   $WX = 7$.

Distance Between Parallel Lines

A corollary of the Properties of a Parallelogram Theorem involves the distance between parallel lines. Recall that the distance between two parallel lines is the length of a segment that is perpendicular to them and connects two points on them. This result is important for deducing area formulas, as you will see in the next chapter.

Theorem
The distance between two given parallel lines is constant.

Proof
Given: $\ell \parallel m$, $\overline{AB} \perp \ell$, and $\overline{XY} \perp \ell$.
To prove: $AB = XY$.

Drawing:

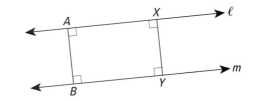

Argument:
By the Perpendicular to Parallels Theorem, $\overline{AB} \perp m$ and $\overline{XY} \perp m$. So *AB* and *XY* are distances between ℓ and *m*. With four right angles, *ABYX* is a rectangle, by the definition of rectangle, so *ABYX* is also a parallelogram. Since opposite sides of any parallelogram are congruent, opposite sides of this rectangle are congruent. So $\overline{AB} \cong \overline{XY}$, from which $AB = XY$.

Rotation Symmetry of a Parallelogram

You may already have conjectured that a parallelogram is rotation-symmetric. We now prove that this conjecture is true.

Parallelogram Symmetry Theorem
Every parallelogram has 2-fold rotation symmetry about the intersection of its diagonals.

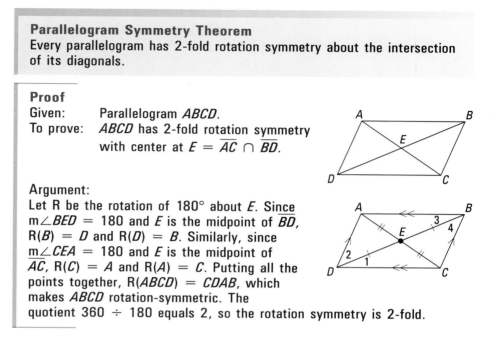

Proof
Given: Parallelogram *ABCD*.
To prove: *ABCD* has 2-fold rotation symmetry
                 with center at $E = \overline{AC} \cap \overline{BD}$.

Argument:
Let R be the rotation of 180° about *E*. Since
m∠*BED* = 180 and *E* is the midpoint of $\overline{BD}$,
R(*B*) = *D* and R(*D*) = *B*. Similarly, since
m∠*CEA* = 180 and *E* is the midpoint of
$\overline{AC}$, R(*C*) = *A* and R(*A*) = *C*. Putting all the
points together, R(*ABCD*) = *CDAB*, which
makes *ABCD* rotation-symmetric. The
quotient 360 ÷ 180 equals 2, so the rotation symmetry is 2-fold.

Do All Parallelograms Possess Reflection Symmetry?

Are parallelograms reflection-symmetric? Do this activity.

Activity

Make three tracings of *ABCD* at the right.
1. Reflect the first tracing over $\overleftrightarrow{AC}$.
2. Reflect the second tracing over $\overleftrightarrow{BD}$.
3. Reflect the third tracing over the line
   containing the midpoints of $\overline{AB}$ and $\overline{CD}$.

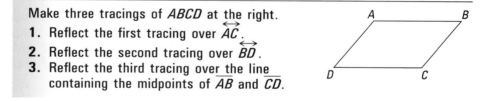

You should find that none of these reflection images coincide with
ABCD. Parallelograms are not necessarily reflection-symmetric.

QUESTIONS

Covering the Reading

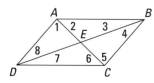

1. In the figure at the left, the diagonals of parallelogram *ABCD*
   intersect at *E*.
   a. Which segments are congruent to $\overline{AD}$?
   b. Which angles are congruent to ∠1?
   c. The midpoints of __?__ and __?__ are the same point.

2. Repeat Question 1, but suppose $ABCD$ is a rhombus.

In 3–5, use the diagram at the right.
   Given: $ABCD$ is a parallelogram and
            E is the midpoint of $\overline{AC}$.

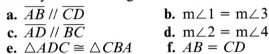

3. Justify the following conclusions.
   a. $\overline{AB} \parallel \overline{CD}$ **b.** $m\angle 1 = m\angle 3$
   c. $\overline{AD} \parallel \overline{BC}$ **d.** $m\angle 2 = m\angle 4$
   e. $\triangle ADC \cong \triangle CBA$ **f.** $AB = CD$

4. What theorem is proved by the sequence of conclusions in Question 3?

5. Under a rotation of $180°$ with center E, what is the image of $ABCD$?

6. Prove: In any parallelogram, opposite angles are congruent.

7. Prove: In any parallelogram, the diagonals intersect at their midpoints.

In 8 and 9, use parallelogram $ABCD$ at the right.

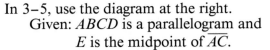

8. If $AE = 7$ and $AD = 12$, find
   a. CE.   **b.** BC.

9. If $BE = x$ and $BC = 2y$, find
   a. ED.   **b.** AD.   **c.** BD.

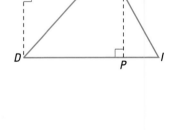

10. $ZOID$ at the left is a trapezoid with bases $\overline{ZO}$ and $\overline{ID}$. $\overleftrightarrow{ZO} \perp \overline{DR}$ and $\overleftrightarrow{ID} \perp \overline{PO}$.
   a. Which two segments are congruent?
   b. State the theorem that justifies your answer to part **a.**

11. Give your answers to the Activity in this lesson.

12. *Multiple choice.* Let A = Parallelograms have reflection symmetry. Let B = Parallelograms have rotation symmetry.
   (a) A is true, B is false.   (c) A and B are both true.
   (b) A is false, B is true.   (d) A and B are both false.

Applying the Mathematics

13. In a 100-meter race, each runner is in a different lane, as shown at the right. What theorem in this lesson justifies that each runner has the same distance to run?

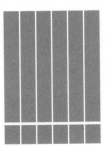

14. a. Place the figure names *kite, parallelogram, rectangle, rhombus,* and *square* in this hierarchy.

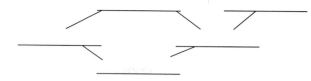

b. Next to each name, write (rd) if the figure is reflection-symmetric to a diagonal, (rpb) if the figure is reflection-symmetric to a perpendicular bisector of a side, and (R) if the figure is rotation-symmetric.

15. The figure below shows a sidewalk constructed so that *ABDC*, *CDFE*, and *EFHG* are parallelograms. Explain why *AB* = *GH*.

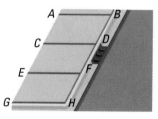

16. Write a proof argument using the given figure.

Given: *ABCD* is a parallelogram and *M* is the midpoint of $\overline{AB}$.

To prove: If *MD* = *MC*, then *ABCD* is a rectangle.

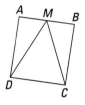

Review

17. At the right is a pattern from a Bushongo sewn mat. Thinking of this as part of a tessellation, trace a possible fundamental region. *(Lesson 7-6)*

18. In trapezoid *ABCF* below, $\overline{AE} \perp \overline{CF}$ and $\overline{BD} \perp \overline{CF}$. Also, *AF* = *BC* and $\overleftrightarrow{AB}$ // $\overleftrightarrow{FC}$. Prove that $\angle F \cong \angle C$. *(Lesson 7-5)*

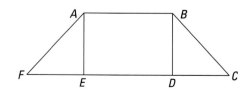

19. In right triangles ABC and XYZ below, two sides and an angle of one are congruent to two sides and an angle of the other. Are the triangles congruent? Why or why not? *(Lesson 7-5)*

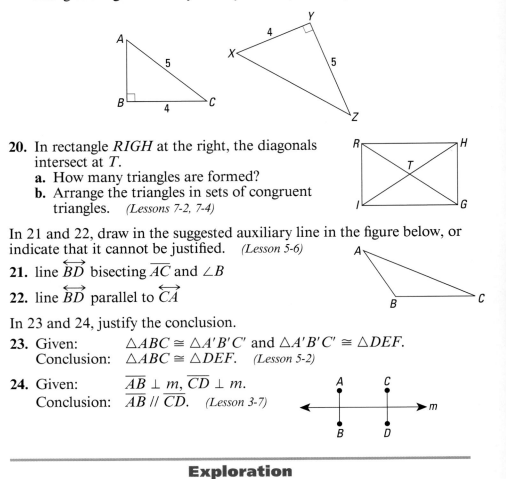

20. In rectangle $RIGH$ at the right, the diagonals intersect at T.
   a. How many triangles are formed?
   b. Arrange the triangles in sets of congruent triangles. *(Lessons 7-2, 7-4)*

In 21 and 22, draw in the suggested auxiliary line in the figure below, or indicate that it cannot be justified. *(Lesson 5-6)*

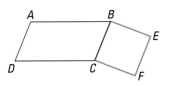

21. line $\overleftrightarrow{BD}$ bisecting $\overline{AC}$ and $\angle B$

22. line $\overleftrightarrow{BD}$ parallel to $\overleftrightarrow{CA}$

In 23 and 24, justify the conclusion.

23. Given: $\triangle ABC \cong \triangle A'B'C'$ and $\triangle A'B'C' \cong \triangle DEF$.
   Conclusion: $\triangle ABC \cong \triangle DEF$. *(Lesson 5-2)*

24. Given: $\overline{AB} \perp m$, $\overline{CD} \perp m$.
   Conclusion: $\overline{AB} \,/\!/\, \overline{CD}$. *(Lesson 3-7)*

Exploration

25. Draw a parallelogram $ABCD$. On each side of $ABCD$, draw a square. One square has been drawn in the figure at the right. Connect the centers of symmetry of the four squares. What type of figure is formed?

26. Below is a pentagon that tessellates, of a type discovered by Marjorie Rice in 1975. Draw enough of a tessellation to show the pattern.

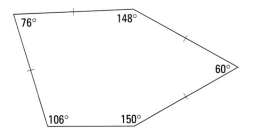

Sufficient Conditions for Parallelograms

Let's Swing! *The Rock-and-Roll Hall of Fame and Museum opened in Cleveland on September 1, 1995. The 150,000-square-foot building, designed by I. M. Pei, utilizes triangles and parallelograms.*

If a quadrilateral has two pairs of parallel sides, then by the definition of parallelogram, it must be a parallelogram. Consequently, "two pairs of parallel sides" is a sufficient condition for a quadrilateral to be a parallelogram. Because parallelograms have so many properties, it is useful to know whether there are other sufficient conditions.

Activity 1

a. Draw two parallel lines. On each line, draw a segment with length 6.5 cm, and connect the endpoints of the two segments to form a quadrilateral.
b. What special kind of quadrilateral is formed?
c. If you move one of the segments along its line, can you form any other kinds of quadrilaterals? If so, what kind(s)?

In your drawing for Activity 1, you should have an instance of the conditional proved in Example 1. Its proof argument uses congruent triangles.

Example 1

Prove: If a quadrilateral has one pair of sides both parallel and congruent, then the quadrilateral is a parallelogram.

Solution

First draw a figure and restate the "given" and "to prove" in terms of the figure.

Drawing:

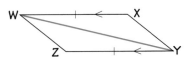

Given: WXYZ is a quadrilateral with $\overline{WX}$ // $\overline{YZ}$ and $\overline{WX} \cong \overline{YZ}$.
To prove: WXYZ is a parallelogram.
Argument:

Conclusions	Justifications
1. Draw $\overline{WY}$.	Two points determine a line.
2. $\overline{WY} \cong \overline{WY}$	Reflexive property of $\cong$
3. $m\angle XWY = m\angle ZYW$	// Lines $\Rightarrow$ AIA $\cong$
4. $\triangle WZY \cong \triangle YXW$	SAS Congruence Theorem (steps 2 and 3 and given)
5. $m\angle ZWY = m\angle XYW$	CPCF Theorem
6. $\overline{WZ}$ // $\overline{XY}$	AIA $\cong$ $\Rightarrow$ // Lines
7. WXYZ is a parallelogram.	definition of a parallelogram

Where Is a Good Place To Find Sufficient Conditions?

Defining conditions are always sufficient conditions. So if both pairs of opposite sides of a quadrilateral are parallel, then the quadrilateral is a parallelogram. Another place to look for possible sufficient conditions is with the properties of a figure. In parallelograms, opposite sides are congruent. Is one pair of opposite sides congruent enough to guarantee that a quadrilateral is a parallelogram? No, an isosceles trapezoid has a pair of opposite sides congruent but it does not have to be a parallelogram.

Is one pair of opposite angles congruent enough to guarantee that a quadrilateral is a parallelogram? Again the answer is "no," because a kite has one pair of opposite angles congruent but it does not have to be a parallelogram.

Flower frame. *Climbing and rambler roses are often trained on trellises and fences, like these on Cape Cod in Massachusetts. Notice the parallelograms in the trellis.*

quadrilateral with pair
of opposite sides congruent

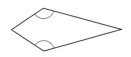

quadrilateral with one pair
of opposite angles congruent

However, in a quadrilateral, if *both* pairs of opposite sides are congruent, or if *both* pairs of opposite angles are congruent, then the quadrilateral is a parallelogram. And if the diagonals of a quadrilateral intersect at their midpoints, then the quadrilateral is a parallelogram.

Activity 2

a. Draw two nonperpendicular segments that have the same midpoint. Now connect the endpoints of the two segments.
b. What kind of figure is formed?

These statements are summarized in the following theorem. Example 1 has proved part (a). You are asked to prove parts (b) and (c) in the questions. Both proofs use triangle congruence and alternate interior angles just as part (a) does. The proof of part (d) uses algebra, and we show it here.

Sufficient Conditions for a Parallelogram Theorem
If, in a quadrilateral,
(a) one pair of sides is both parallel and congruent, or

(b) both pairs of opposite sides are congruent, or

(c) the diagonals bisect each other, or

(d) both pairs of opposite angles are congruent,

then the quadrilateral is a parallelogram.

Proof (d)

Given: In quadrilateral *WXYZ* at the right,
                $a = m\angle WXY = m\angle Z$
                and $b = m\angle W = m\angle Y.$
To prove: *WXYZ* is a parallelogram.

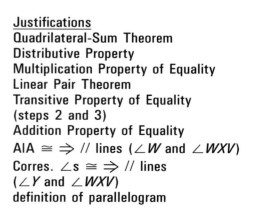

Argument:

Conclusions	Justifications
1. $a + b + a + b = 360$	Quadrilateral-Sum Theorem
2. $2a + 2b = 360$	Distributive Property
3. $a + b = 180$	Multiplication Property of Equality
4. $a + m\angle WXV = 180$	Linear Pair Theorem
5. $a + m\angle WXV = a + b$	Transitive Property of Equality (steps 2 and 3)
6. $m\angle WXV = b$	Addition Property of Equality
7. $\overline{WZ}\ //\ \overline{XY}$	AIA $\cong \Rightarrow$ // lines ($\angle W$ and $\angle WXV$)
8. $\overline{WX}\ //\ \overline{ZY}$	Corres. $\angle$s $\cong \Rightarrow$ // lines ($\angle Y$ and $\angle WXV$)
9. *WXYZ* is a parallelogram.	definition of parallelogram

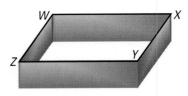

Example 2

Pictured at the left is a rectangular cardboard box with its top and bottom removed, so it is no longer rigid. Since the original box was rectangular, its opposite sides are known to be of the same length. But now the angles can change; they no longer must be right angles. If you are careful not to bend the sides, will the darkened edge WXYZ always be a parallelogram?

Solution

Given $WX = YZ$ and $WZ = XY$. By part (b) of the Sufficient Conditions for a Parallelogram Theorem, WXYZ will always be a parallelogram.

QUESTIONS

Covering the Reading

1. According to its definition, what makes a quadrilateral a parallelogram?

2. Give your answers to Activity 1 in this lesson.

3. Give your drawing and answer to Activity 2 in this lesson.

4. Give four sufficient conditions for parallelograms other than its defining characteristics.

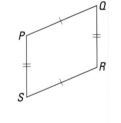

5. Write a proof argument for part (b) of the Sufficient Conditions for a Parallelogram Theorem: *If both pairs of opposite sides of a quadrilateral are congruent, then the quadrilateral is a parallelogram.*

    Given: Quadrilateral $PQRS$ at the left with $\overline{PQ} \cong \overline{RS}$ and $\overline{PS} \cong \overline{QR}$.
    To prove: $PQRS$ is a parallelogram.

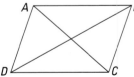

In 6–8, use the figure at left. Given the following information, is $ABCD$ a parallelogram?

6. $m\angle BAD = 60$, $m\angle ADC = 60$, $m\angle DCB = 120$, $m\angle CBA = 120$

7. $AB = 11$, $BC = 15$, $CD = 11$, $AD = 15$

8. $\overline{AB} \parallel \overline{CD}$, $AD = 8$, $BC = 8$

9. What figure is formed by the top edge of a box (when the top and bottom of the box are removed)?

Applying the Mathematics

10. Two yardsticks and two meter sticks are joined end to end to form a quadrilateral. Name all the quadrilaterals that can be formed.

11. Prove part (c) of the Sufficient Conditions for a Parallelogram Theorem.

12. In quadrilateral *NEST*, angles have measures as marked. Is *NEST* a parallelogram? Why or why not?

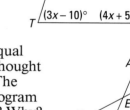

13. Two scouts lashed together two sticks of equal length at their midpoints. The first scout thought *ABCD* (not drawn) is always a rectangle. The second thought *ABCD* is always a parallelogram but not always a rectangle. Who was right? Why?

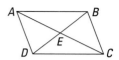

14. Tick marks of equal length are spaced around a watch, as shown at the left. Is *WXYZ* (not drawn) a parallelogram? Explain why or why not.

Review

In 15 and 16, refer to parallelogram *ABCD* at the right. *(Lesson 7-7)*

15. If m∠*BAD* = 70, find as many other angle measures as you can.

16. If *BE* = 11 and *BC* = 20, find as many other lengths as you can.

17. Model railroad tracks intersect each other as shown at the right. Explain why *PQ* = *RS*. *(Lesson 7-7)*

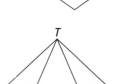

18. At the right is a badge in the shape of a regular pentagon. Can this shape tessellate the plane? Explain why or why not. If so, show part of the tessellation. *(Lesson 7-6)*

19. Write the proof argument using the given figure.
Given: △*PTS* is isosceles with vertex angle *T*, and m∠*PTQ* = m∠*STR*.
To prove: **a.** △*TPQ* ≅ △*TSR*.
    b. △*TQR* is isosceles. *(Lesson 7-3)*

20. Consider the figure in Question 19 as a network. Is it a traversable network? If so, give a path. If not, explain why not. *(Lesson 1-4)*

Exploration

21. Prove or disprove this conjecture: If, in quadrilateral *ABCD*, *AB* = *CD* and ∠*A* ≅ ∠*C*, then *ABCD* is a parallelogram.

Working on the railroad. *Model railroading requires knowledge about carpentry, electrical wiring, and making scenery. In the early 1990s, Model Railroader magazine estimated that about 250,000 adults operated model railroads as a hobby.*

Exterior Angles

An evening exterior. *The exterior angle that the Pyramid of Cheops forms with the ground is shown above.*

The proof of part (d) of the Sufficient Conditions for a Parallelogram Theorem on page 413 makes use of *exterior angle VXW.*

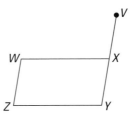

> **Definition**
> An angle is an **exterior angle** of a polygon if and only if it forms a linear pair with one of the angles of the polygon.

Exterior Angles of Triangles

A triangle is pictured below. At each vertex, one of its sides has been extended. The exterior angles are ∠*ABE*, ∠*ACD*, and ∠*FAC*.

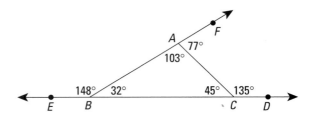

To distinguish exterior angles of a polygon from the polygon's own angles, the angles of the polygon sometimes are called **interior angles.** Measures of exterior angles of triangles are related in several ways to the interior angles of the triangles. Notice in the triangle on page 416 that each exterior angle measure is the sum of the measures of two of the triangle's interior angles: $148 = 103 + 45$, $77 = 45 + 32$, and $135 = 103 + 32$. The general property is quite easy to prove using algebra.

Exterior Angle Theorem
In a triangle, the measure of an exterior angle is equal to the sum of the measures of the interior angles at the other two vertices of the triangle.

Proof
Given: $\triangle XYZ$ with exterior angle $\angle 4$.
To prove: $m\angle 4 = m\angle 2 + m\angle 3$.

Drawing:

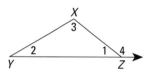

Argument:
Conclusions	Justifications
1. $m\angle 1 + m\angle 2 + m\angle 3 = 180$.	Triangle-Sum Theorem
2. $m\angle 1 + m\angle 4 = 180$.	Linear Pair Theorem
3. $m\angle 1 + m\angle 4 = m\angle 1 + m\angle 2 + m\angle 3$	Transitive Property of Equality (steps 1 and 2)
4. $m\angle 4 = m\angle 2 + m\angle 3$.	Addition Property of Equality

There is a conclusion which directly follows from the Exterior Angle Theorem. Since $m\angle 4 = m\angle 2 + m\angle 3$, we can use the Equation to Inequality Property to conclude

$$m\angle 4 > m\angle 2$$
and
$$m\angle 4 > m\angle 3.$$

The result is called the *Exterior Angle Inequality*.

Exterior Angle Inequality
In a triangle, the measure of an exterior angle is greater than the measure of the interior angle at each of the other two vertices.

Example 1

Refer to the figure below. If m∠*APE* = 35, what can be concluded about the other angle measures?

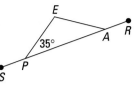

Solution

You can conclude something about each angle in the figure.
(1) m∠*SPE* = 145 by the Linear Pair Theorem.

Now, applying the Exterior Angle Inequality,
(2) m∠*E* < 145 and
(3) m∠*EAP* < 145.

The Exterior Angle Inequality also shows
(4) m∠*EAR* > 35.

Exterior Angles of Convex Polygons

The Exterior Angle Inequality does not extend to polygons other than triangles. In the figure below, ∠*BCE* is an exterior angle to polygon *ABCD*. But m∠*BCE* is not greater than the measure of any of the interior angles of *ABCD*.

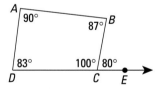

However, there is a surprising property of exterior angles that holds in all convex polygons.

Activity

a. Add the measures of the exterior angles shown for △*ABC* on page 416.
b. Find the measure of one exterior angle at each vertex of quadrilateral *ABCD* above. Then add the four measures.

Exterior angles also enable us to obtain properties of interior angles. You know that if two sides of a triangle are congruent, then the angles opposite those sides are congruent. That is the Isosceles Triangle Base Angles Theorem. But what if two sides are not congruent? The next theorem shows the answer.

Proof
First we draw a figure and state the given and to prove in terms of
that figure.

Given: $\triangle ABC$ with $BA > BC$.
Prove: $m\angle C > m\angle A$.

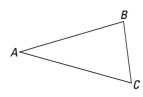

Drawing:

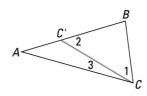

Argument:

Conclusions	Justifications
1. Identify point C' on $\overrightarrow{BA}$ with $BC' = BC$. Then C' is between A and B because $BA > BC'$.	On a ray, there is exactly one point at a given distance from an endpoint.
2. $m\angle 1 = m\angle 2$	Isosceles Triangle Base Angles Theorem
3. $m\angle 2 > m\angle A$	Exterior Angle Inequality (with $\triangle CC'A$)
4. $m\angle 1 > m\angle A$	Substitution (step 2 into step 3)

Now all that is left to show is that $m\angle BCA > m\angle 1$.

5. $m\angle 1 + m\angle 3 = m\angle BCA$	Angle Addition Postulate
6. $m\angle BCA > m\angle 1$	Equation to Inequality Property
7. $m\angle BCA > m\angle A$	Transitive Property of Inequality (steps 4 and 6)

Finally, we ask a question about unequal angles in a triangle. If no two
angles in a triangle are congruent, then the triangle is not isosceles, so
the sides opposite the angles are not congruent. But which side is
opposite a larger angle? Because of the Unequal Sides Theorem, a larger
side cannot be opposite a smaller angle. So a larger side must be
opposite a larger angle.

Example 2

In △QRS below, arrange the sides in order from shortest to longest.

Solution

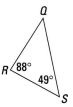

From the Triangle-Sum Theorem,
m∠Q + 88 + 49 = 180.
So,   m∠Q = 43.
Since ∠Q is the smallest angle, $\overline{RS}$ is the shortest
side. Since ∠R is the largest angle, $\overline{QS}$ is the
longest side. Thus the sides of △QRS from shortest
to longest are: $\overline{RS}$, $\overline{RQ}$, $\overline{QS}$.

QUESTIONS

Covering the Reading

In 1 and 2, use the figure at the right.

1. If m∠C = 50 and m∠D = 90, find
   a. x.   **b.** y.   **c.** z.

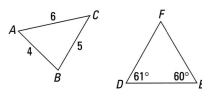

2. How is m∠ABD related to
   m∠C and m∠D?

In 3–5, the sides of △FGI have been extended, as shown in the drawing at the left. Give two angles that have measures less than

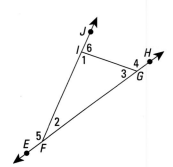

3. m∠4.   **4.** m∠5.   **5.** m∠6.

6. a. State the Exterior Angle Theorem.
   b. State the Exterior Angle Inequality.

In 7 and 8, use the triangles at the right.

7. a. Name the largest angle of
         △ABC.
   b. Name the smallest angle of
         △ABC.

8. a. Name the longest side of △DEF.
   b. Name the shortest side of △DEF.

9. Give your answers to the Activity in this lesson.

10. In △UVW, X is between U and V. Draw a picture. If m∠WXU = 70, what can be concluded about
   a. m∠V?   **b.** m∠U?   **c.** m∠UWX?

11. *True or false.* In an obtuse triangle, the longest side is opposite the obtuse angle.

12. In $\triangle LUV$, $LU > UV > LV$ and one angle has measure 60. Which angle is it?

13. Name the shortest segment in the figure below. (The figure is not necessarily drawn accurately.)

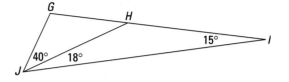

14. Segments $\overline{PS}$ and $\overline{PR}$ trisect $\overline{QT}$ in right triangle PQT.
Trisect means to divide into three congruent parts.
a. Prove: m$\angle 3 >$ m$\angle 1$.
b. Prove: $PQ < PS$.

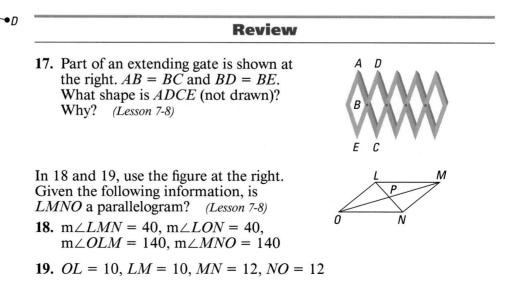

15. At the right, $\overrightarrow{CB}$ bisects $\angle ACD$. In terms of x and y, give the measure of each angle.
a. $\angle ACB$
b. $\angle CBD$
c. $\angle D$

16. At vertex C of $\triangle ABC$ at the left, there are two exterior angles. Explain why they have the same measure.

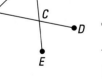

17. Part of an extending gate is shown at the right. $AB = BC$ and $BD = BE$. What shape is $ADCE$ (not drawn)? Why? *(Lesson 7-8)*

In 18 and 19, use the figure at the right. Given the following information, is $LMNO$ a parallelogram? *(Lesson 7-8)*

18. m$\angle LMN = 40$, m$\angle LON = 40$, m$\angle OLM = 140$, m$\angle MNO = 140$

19. $OL = 10$, $LM = 10$, $MN = 12$, $NO = 12$

20. Refer to parallelogram $QRST$ at the right. If $QV = 3x$ and $QR = y$, find as many other lengths as you can. *(Lesson 7-7)*

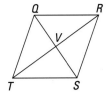

21. Given $\triangle ABC$ and $\triangle DEF$ with congruent parts marked.
   a. Name an additional piece of information that would be enough to guarantee congruence.
   b. Name the triangle congruence theorem used to justify your answer to part **a.** *(Lessons 7-2, 7-5)*

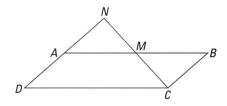

22. In the drawing below, M is the midpoint of $\overline{AB}$, and $\angle N \cong \angle MCB$. Prove: M is the midpoint of $\overline{NC}$. *(Lesson 7-3)*

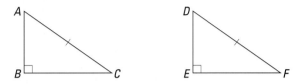

Exploration

23. Think of walking around a convex polygon. At each vertex, you must change direction. The amount of the change in direction is measured by the exterior angle at that vertex. The total amount, as you walk around, is the sum of the measures of the exterior angles of the polygon; one exterior angle is picked at each vertex.
   a. Draw a pentagon. Find the sum of the measures of the exterior angles of your pentagon, one at each vertex.
   b. Repeat part **a** with a hexagon.
   c. Generalize parts **a** and **b.**

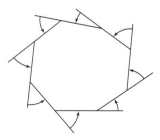

A project presents an opportunity for you to extend your knowledge of a topic related to the material of this chapter. You should allow more time for a project than you do for typical homework questions.

1 Triangles and Quadrilaterals in Architecture

Interview a carpenter, architect, or construction worker to find out about building the wooden frame for a house. Write an illustrated report which answers the following questions.

a. Where can rectangles, squares, and other quadrilaterals be found in a house frame?

b. Where can congruent triangles be found? Are there overlapping or isosceles triangles in a frame?

c. One technique for building a house is called *balloon-framing*. Where and when was this first used? What are its advantages and disadvantages?

2 The "Bridge of Fools"

The fifth theorem in Euclid's *Elements* was what we call the Isosceles Triangle Base Angles Theorem. Euclid's proof was quite different from the one in this book and used the diagram shown here. The proof was often a stumbling block when first encountered by students, and it was called the "Pons Asinorum," or "Bridge of Fools." Here are the given and what is to be proved in terms of the figure.

Given: $\overline{AB} \cong \overline{AC}$, $\overline{AF} \cong \overline{AG}$.
To prove: $\angle ABC \cong \angle ACB$.

Write a proof argument. (Hints: The fourth theorem in Euclid's Elements is the SAS Congruence Theorem. You will need to find two pairs of congruent overlapping triangles.)

3 Tessellations Everywhere

Make a collage or poster display using photographs and/or pictures in magazines of tessellations found in everyday life. In each picture, highlight the shape or shapes which tessellate.

4 Morley's Theorem

One of the most surprising theorems in all of geometry was discovered in 1904 by geometer Frank Morley. Morley's theorem concerns the trisectors of the angles of any triangle. Consider △*ABC* below. In Figure 1, the trisecting rays of ∠*B* have been drawn. When all six of these rays are drawn, as in Figure 2, points of intersection are found for adjacent rays. △*DEF* is the triangle formed by connecting these points. Morley's theorem is: *In any triangle, the points of intersection of adjacent trisectors of the angles are the vertices of an equilateral triangle.*

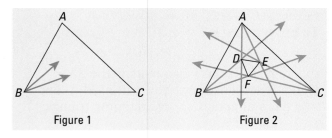

Figure 1   Figure 2

Illustrate this theorem with carefully drawn figures of three noncongruent triangles. Use a protractor and ruler or an automatic drawer. Include measures of all important angles and sides.

5 Sufficient Conditions for Special Quadrilaterals

The six statements **a–f** given below are true properties of quadrilaterals. (1) Write the converse of each statement. (2) If the converse is false, find a counterexample. (At least two converses are false.) If the converse is true, prove it.

a. *If a quadrilateral is a rhombus, then it is a parallelogram with perpendicular diagonals.*
b. *If a quadrilateral is a rectangle, then it is a parallelogram with a right angle.*
c. *If a quadrilateral is a square, then it is a kite with a right angle.*
d. *If a quadrilateral is a trapezoid, then two consecutive angles are supplementary.*
e. *If a quadrilateral is an isosceles trapezoid, then it is a trapezoid with congruent diagonals.*
f. *If a quadrilateral is a parallelogram, then it is a trapezoid that has a pair of congruent opposite sides.*

6 Drawing Polygons Using Logo

In the computer language Logo, there is an arrowed cursor on the screen. The command FORWARD moves the cursor forward, tracing a segment of an indicated length on the computer monitor. The command RIGHT turns the arrow clockwise the number of degrees you specify. The commands BACKWARD and LEFT are defined similarly. So, for instance, FORWARD 10 RIGHT 120 traces a segment of length 10 and then turns the arrow 120° clockwise in preparation for another tracing.

a. Write and run a program to trace a rectangle that is not a square, and print your results.
b. Write and run a program to trace a convex pentagon, and print your result.
c. Write and run a program to trace a convex hexagon, and print your result.
d. Write and run a program to trace a convex heptagon, and print your result.
e. Your programs should have something in common that relate to the RIGHT commands. What is this commonality?
f. Draw a many-sided polygon with appropriate angles so that it looks like a circle. How many sides does your polygon have?

SUMMARY

This chapter discusses triangle congruence theorems and their applications. Two triangles are congruent if the following parts of one are congruent to corresponding parts of the other: SSS (three sides); SAS (two sides and the included angle); ASA (two angles and the included side); AAS (two angles and a nonincluded side); SsA (two sides and the angle opposite the larger side); and HL (in a right triangle, hypotenuse and leg).

These theorems also indicate the conditions sufficient to determine a unique triangle. For instance, the SSS Congruence Theorem indicates that all triangles with three given side lengths are congruent.

From the triangle congruence theorems, we deduced a variety of theorems about geometric figures.

(1) If a triangle has two congruent angles, then it has two congruent sides.
(2) Parallelograms have 2-fold rotation symmetry; both pairs of opposite sides and opposite angles are congruent; and their diagonals have the same midpoint.

(3) If a quadrilateral has two pairs of congruent opposite sides, two pairs of congruent opposite angles, one pair of congruent and parallel opposite sides, or diagonals with the same midpoint, then the quadrilateral is a parallelogram.

When congruent copies of the same figure cover the plane without gaps or overlapping, a tessellation is formed. The figure is a fundamental region for the tessellation. Tessellations are important in manufacturing and building, because if a figure tessellates, then copies may be cut out from a larger piece without waste. Any triangle or quadrilateral can be a fundamental region, but not all pentagons can. No one knows if we have found all the pentagons which tessellate; it is an unsolved problem in mathematics.

It is also useful to know when figures are not congruent. The measure of an exterior angle of a triangle equals the sum of the measures of the nonadjacent interior angles. So an exterior angle has a measure greater than that of either nonadjacent interior angle. From this theorem, we proved that in a triangle, larger angles are opposite longer sides, and longer sides are opposite larger angles.

VOCABULARY

Below are the most important terms and phrases for this chapter. You should be able to give a definition of the starred (*) term. For the other terms you should be able to give a general description and a specific example of each. You should be able to state any theorem as a conditional and draw a picture.

Lesson 7-1
sufficient condition

Lesson 7-2
SSS, SAS, ASA, and AAS
    Congruence Theorems
included angle
included side

Lesson 7-3
Isosceles Triangle Base Angles
    Converse Theorem

Lesson 7-4
overlapping figures
nonoverlapping figures

Lesson 7-5
SSA condition
legs, hypotenuse
HL condition
HL Congruence Theorem
SsA Congruence Theorem

Lesson 7-6
tessellation, tessellates
fundamental region

Lesson 7-7
Properties of a Parallelogram
    Theorem
Parallelogram Symmetry
    Theorem

Lesson 7-8
Sufficient Conditions for a
    Parallelogram Theorem

Lesson 7-9
*exterior angle
interior angle
Exterior Angle Theorem
Exterior Angle Inequality
Unequal Sides Theorem
Unequal Angles Theorem
trisect

PROGRESS SELF-TEST

Directions: Take this test as you would take a test in class. You will need a ruler, compass, and protractor. Then check your work with the solutions in the Selected Answers section in the back of the book.

1. In the figure below, suppose m∠ABD = 60. What can you conclude about each angle?

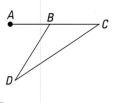

  a. m∠CBD

  b. m∠C

  c. m∠D

2. a. Draw a triangle LMN with LM = 10 cm, m∠M = 70, and m∠L = 30.

  b. If everyone else draws a △LMN so that it has the three measures given in part **a,** will their triangles be congruent to yours?

  c. Explain why or why not.

3. a. Draw a triangle ABC with AB = 1″, m∠B = 100, and m∠C = 45.

  b. Will everyone else's correct drawing be congruent to yours?

  c. Explain why or why not.

In 4 and 5, a triangle is drawn with certain parts marked. **a.** Is this a unique shape, given the measurements? **b.** Why or why not?

4.   **5.**

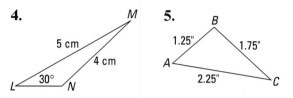

6. Give a complete statement of the AAS Triangle Congruence Theorem.

7. If the two triangles are congruent, justify with a triangle congruence theorem and indicate corresponding vertices. Otherwise, tell why the triangles are not congruent.

  a.   **b.**

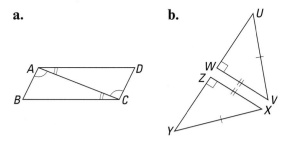

8. In the figure below, ∠1 ≅ ∠3 and ∠2 ≅ ∠4.

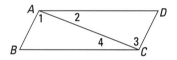

  a. Name the congruent triangles with vertices in correct order.

  b. What theorem guarantees that the triangles are congruent?

In 9 and 10, write a proof argument using the given figure.

9. Given: M is the midpoint of $\overline{AC}$ and $\overleftrightarrow{AB} \parallel \overleftrightarrow{CD}$.

   To prove: △MBA ≅ △MDC

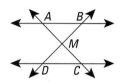

PROGRESS SELF-TEST

10. Given: $WX = WY$ and
                   $\angle WUY \cong \angle WVX$.
       To prove: $\triangle WUV$ is isosceles.

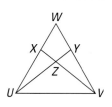

11. $\overline{AB}$ and $\overline{AC}$ are ropes that hold a tarpaulin in place over a picnic table. $\overline{AD}$ is perpendicular to the ground, which is level. If $AB = AC$, explain why B and C are each the same distance away from D.

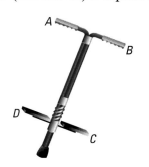

12. Walt bought Kelly a pogo stick for her birthday. The handlebar $\overline{AB}$ is parallel to and congruent to the footbar $\overline{CD}$. Explain why $ABCD$ (not drawn) is a parallelogram.

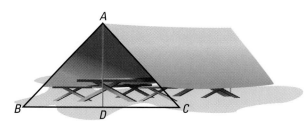

13. In parallelogram $PARL$ below, $PO = x$ and $AP = y$. Find as many other lengths as you can.

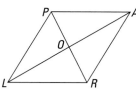

14. In the figure below $AE = 9$, $BE = 10$, $CE = 9$, and $DE = 10$. Is $ABCD$ a parallelogram?

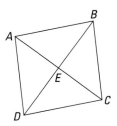

15. A picture-frame maker makes plastic triangles to put in the corners of frames. In order not to waste material, the frame maker uses a tessellation. Trace the figure and draw part of a tessellation.

16. *Multiple choice.*   Using the given angle measures in the figure below (not necessarily drawn accurately), which is the shortest segment?
   (a) $\overline{WX}$ (b) $\overline{WY}$ (c) $\overline{WZ}$
   (d) $\overline{XY}$ (e) $\overline{XZ}$ (f) $\overline{YZ}$

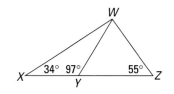

CHAPTER REVIEW

Questions on SPUR Objectives

SPUR stands for **S**kills, **P**roperties, **U**ses, and **R**epresentations. The Chapter Review questions are grouped according to the SPUR Objectives for this chapter.

SKILLS DEAL WITH THE PROCEDURES USED TO GET ANSWERS.

Objective A: *Draw triangles satisfying certain conditions and determine whether all such triangles are congruent.*   *(Lessons 7-1, 7-2, 7-5)*

In 1–4, a triangle is drawn with certain measures indicated. **a.** Are all triangles with these measures congruent? **b.** Why or why not?

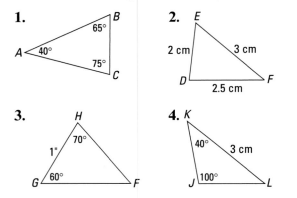

1.
65° B
A 40°
75° C

2.
E
2 cm   3 cm
D
2.5 cm F

3.
H
70°
1"
60°
G   F

4. K
40° 3 cm
J 100°   L

In 5–8, **a.** Draw a triangle satisfying the given condition. **b.** Will every correct drawing be congruent to yours? Why or why not?

5. Draw a triangle *MNO* with *MN* = 5 cm and m∠*N* = 90.

6. Draw a triangle *PQR* with *PQ* = 1 in., *QR* = 1.5 in., and m∠*R* = 30.

7. Draw a triangle *STU* with *ST* = 2 cm, *TU* = 4 cm, and m∠*S* = 60.

8. Draw a triangle *VWX* with m∠*W* = 55, *VW* = 2 in., and *WX* = 1 in.

Objective B: *Determine measures of angles in polygons using exterior angles.*   *(Lesson 7-9)*

9. In △*QRS* below, if m∠*QST* = 132, what can you conclude about
   a. m∠*QSR*?
   b. m∠*Q*?
   c. m∠*Q* + m∠*R*?

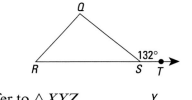

Q
R   132°
S T

10. Refer to △*XYZ* at the right.
   a. Find m∠*X*.
   b. Find m∠*Y*.

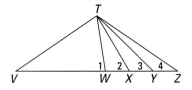

Y
(2q)°
q°   138°
X   Z

11. Refer to the figure below. Which angle is largest: 1, 2, 3, or 4? Explain your reasoning.

T
1 2 3 4
V   W X Y Z

12. ∠1 is an exterior angle of quadrilateral *ABCD*. What is its measure?

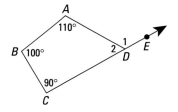

A
110°
B 100°
1
2 E
D
90°
C

PROPERTIES DEAL WITH THE PRINCIPLES BEHIND THE MATHEMATICS.

Objective C: *Determine whether triangles are congruent from given information.* *(Lessons 7-2, 7-5)*

In 13–18, if the two triangles are congruent, justify with a triangle congruence theorem and indicate corresponding vertices. Otherwise, write *not enough information to tell*.

Objective D: *Write proofs that triangles are congruent.* *(Lessons 7-3, 7-4, 7-5)*

In 21–26, write a proof argument using the given figure.

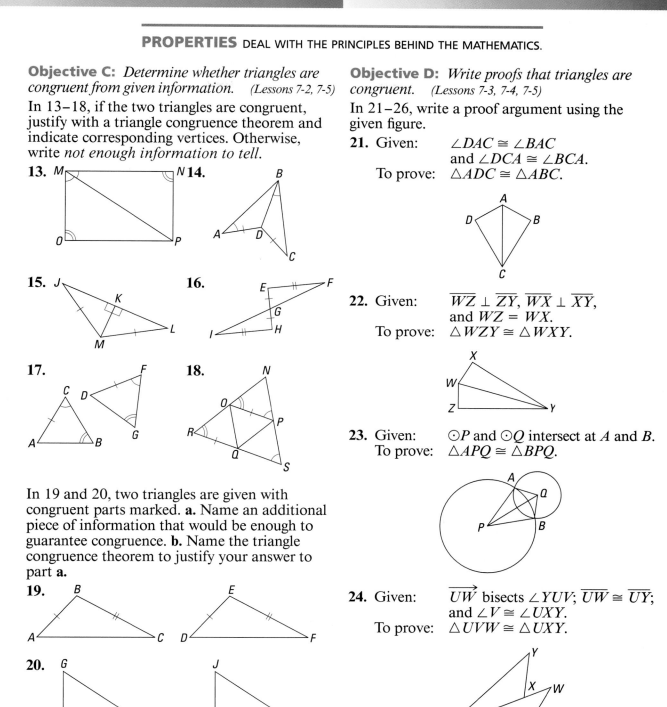

13.

14.

15.

16.

17.

18.

In 19 and 20, two triangles are given with congruent parts marked. **a.** Name an additional piece of information that would be enough to guarantee congruence. **b.** Name the triangle congruence theorem to justify your answer to part **a.**

19.

20.

21. Given: ∠DAC ≅ ∠BAC
            and ∠DCA ≅ ∠BCA.
    To prove: △ADC ≅ △ABC.

22. Given: $\overline{WZ} \perp \overline{ZY}$, $\overline{WX} \perp \overline{XY}$,
            and WZ = WX.
    To prove: △WZY ≅ △WXY.

23. Given: ⊙P and ⊙Q intersect at A and B.
    To prove: △APQ ≅ △BPQ.

24. Given: $\overrightarrow{UW}$ bisects ∠YUV; $\overline{UW} \cong \overline{UY}$;
            and ∠V ≅ ∠UXY.
    To prove: △UVW ≅ △UXY.

25. Given: Concentric circles with center O, $OB > OA$, and $\angle CDO \cong \angle ABO$.
  To prove: $\triangle DOC \cong \triangle BOA$.

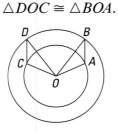

26. Given: $\overline{AB} \cong \overline{DC}$ and $\angle ABC \cong \angle DCB$.
  To prove: $\triangle ACB \cong \triangle DBC$.

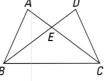

Objective E: *Apply the triangle congruence and CPCF theorems to prove that segments or angles are congruent.*   *(Lessons 7-3, 7-4, 7-5)*

In 27–32, write a proof argument using the given figure.

27. Given: $\overline{AB} \cong \overline{AC}$ and $\overline{BD} \cong \overline{DC}$.
  To prove: $\angle BAD \cong \angle CAD$.

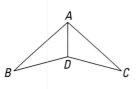

28. Given: *ABCDEFGH* is a regular octagon.
  To prove: $\overline{AC} \cong \overline{BD}$.

29. Given: $\overrightarrow{JK}$ bisects $\angle MJL$, and $\overline{MJ} \cong \overline{LJ}$.
  To prove: $\angle M \cong \angle L$.

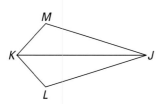

30. Given: $\overline{AD} \perp \overline{DC}$,
     $\overline{AB} \perp \overline{BC}$,
     $\overline{AD} \cong \overline{BC}$.
  To prove: $\overline{AB} \cong \overline{CD}$.

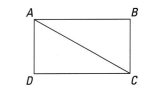

31. Given: N is the midpoint of $\overline{OE}$. $\ell \mathbin{/\!/} m$.
  To prove: $\overline{AE} \cong \overline{UO}$.

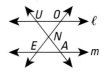

32. Given: $\angle SPQ \cong \angle RQP$ and $\angle S \cong \angle R$.
  To prove: $\overline{QS} \cong \overline{PR}$.

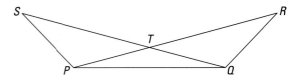

Objective F: *Apply properties of parallelograms.*   *(Lesson 7-7)*

In 33–35, refer to parallelogram *ABCD* below.

33. If $BE = 12$ and $BC = 19$, find as many other lengths as you can.

34. If $CE = 3x$ and $BC = y - 2$, find as many other lengths as you can.

35. If m$\angle DAB = 130$, find as many other angle measures as you can.

36. Trace parallelogram *WXYZ* below. Locate its center of rotation symmetry and label it as point C.

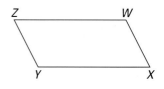

Objective G: *Determine whether conditions are sufficient for parallelograms.* *(Lesson 7-8)*

In 37–40, use the figure at the right. Given the following information, is *ABCD* a parallelogram?

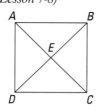

37. *AB* = 7, *BC* = 8, *CD* = 7, *AD* = 8
38. *AE* = *ED*, *CE* = *BE*
39. $\overline{AB}$ // $\overline{CD}$, *AD* = 9, *BC* = 9
40. m∠*ABC* = 80, m∠*ADC* = 80, m∠*BAD* = 100, m∠*BCD* = 100

Objective H: *From given information, deduce which sides or angles of triangles are smallest or largest.* *(Lesson 7-9)*

In 41 and 42, refer to △*ABC* below.

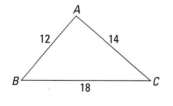

41. Name the largest angle.
42. Name the smallest angle.

In 43 and 44, the figures are not drawn accurately.

43. Name the sides of △*DEF* in order from shortest to longest.

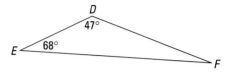

44. Name the shortest segment in the figure.

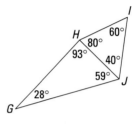

USES DEAL WITH APPLICATIONS OF MATHEMATICS IN REAL SITUATIONS.

Objective I: *Use theorems about triangles to explain real situations.* *(Lessons 7-2, 7-5)*

45. $\overline{RA}$ and $\overline{AB}$ are beams of the same length at the end of a roof. $\overline{AP}$ is perpendicular to $\overline{RB}$. Explain why $\overline{RP}$ and $\overline{PB}$ have the same length.

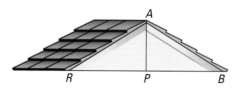

46. Tony and Maria each made pieces of triangular jewelry with sides 21 mm, 18 mm, and 26 mm. Explain why the pieces must have the same shape.

47. A vertical radio tower is supported by guy wires that hit the level ground at points *A*, *C*, and *D*. Explain why, if the angles at *A*, *C*, and *D* have the same measure, then these points are the same distance from *B*, the bottom of the tower.

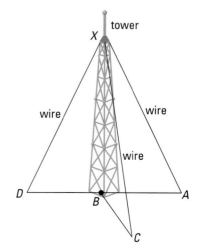

48. A scout used the following process to estimate the width of a river. Standing directly across from the tree at point P, the scout walked 10 paces to X, and placed a stick at X, then walked 10 paces to Q. The scout then turned 90° and walked to point R, where the scout, stick, and tree were lined up.

   a. Which line segments have lengths equal to the width of the river?

   b. Why does this method work?

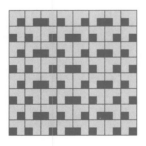

Objective J: *Draw tessellations of real objects.*
(Lesson 7-6)

49. The machine part at the right avoids waste when it is manufactured. Trace the shape and draw part of a tessellation, using it as a fundamental region.

50. The pattern below illustrates part of a tessellation. Trace a possible fundamental region.

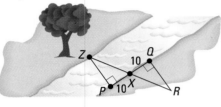

51. A section of a honeycomb in a beehive looks like a tessellation of a regular hexagon. Draw a section of a honeycomb.

52. A pattern for a button is made in the shape of a regular octagon. Can this shape tessellate the plane? Explain why or why not.

Objective K: *Use theorems about parallelograms to explain real situations.*
(Lessons 7-7, 7-8)

53. The sides of a corral $ABCD$ are $AB = 123'\ 4''$, $BC = 73'\ 6''$, $CD = 124'\ 1''$, and $AD = 73'\ 11''$. Then $ABCD$ is approximately in the shape of a(n) __?__.

54. Refer to the figure below. You have two equally long diagonal supports, $\overline{MN}$ and $\overline{PQ}$, for a bed. How should they be attached so that you can be certain $MPNQ$ is a parallelogram?

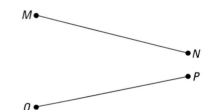

55. A piece of ribbon is cut at an angle so that $ABDC$ and $DCEF$ are parallelograms. Explain why $AB = EF$.

56. Two sidewalks meet at an angle forming quadrilateral $WXYZ$. Is $m\angle XWZ = m\angle ZYX$? Why or why not?

REPRESENTATIONS DEAL WITH PICTURES, GRAPHS, OR OBJECTS THAT ILLUSTRATE CONCEPTS.

There are no objectives for Representations in this chapter.

REFRESHER

Chapter 8, which discusses perimeter and area, assumes that you have mastered certain objectives in your previous study of algebra. Use these questions to check your mastery.

A. Multiply a monomial by a binomial.

1. $6(2 + x)$

2. $x(4 - y)$

3. $(h + \ell)a$

4. $\frac{1}{2}b(w + t)$

5. $(a - 2b)b$

6. $2xy(y + x)$

B. Factor.

7. $ax + 2x$

8. $9c^2 - 12c$

9. $4b - 20h$

10. $\pi r^2 + \pi h^2$

11. $\frac{1}{2}bh + \frac{1}{2}\ell h$

12. $12ar^2 + 6ar$

C. Multiply binomials.

13. $(y + 2)(y + 4)$

14. $(a - 7)(3b + 2)$

15. $(2r + t)(2r - t)$

16. $(s + 4)^2$

17. $(m + n)^2$

18. $\frac{1}{2}(e + f)(y + h)$

D. Estimate square roots to the nearest hundredth.

19. $\sqrt{2}$

20. $4\sqrt{3}$

21. $\sqrt{4^2 + 5^2}$

22. $\sqrt{\frac{17}{4}}$

E. Simplify square roots without using a calculator.

23. $\sqrt{100}$

24. $\sqrt{40}$

25. $\sqrt{18}$

26. $2\sqrt{45}$

F. Solve quadratic equations of the form $ax^2 = b$.

27. $x^2 = 25$

28. $y^2 + 9 = 25$

29. $40 = 2z^2$

30. $3 = \pi r^2$

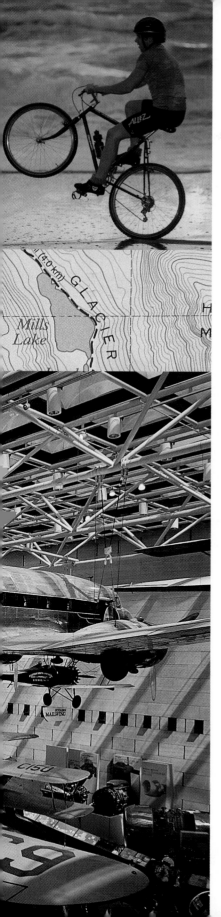

PERIMETERS AND AREAS

Pictured below is an aerial view of a museum surrounded by an elliptical walkway. The museum and the walkway both lie inside a large rectangle that is 2 blocks long and 1 block wide.

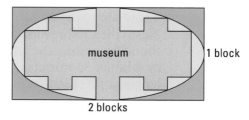

Area is a measure of the space occupied by a two-dimensional region. The area of the rectangle is 2 square blocks. So the area of the museum is less than 2 square blocks. The area of the elliptical region is larger than the area of the museum but smaller than the area of the rectangle. All of this can be seen by separating the three regions.

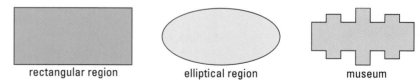

rectangular region   elliptical region   museum

Perimeter is different from area. The perimeter of a region is the length of its boundary. It tells you how far it would be to walk along the boundary. The perimeter of the rectangle is 6 blocks. It is harder to calculate the perimeter of the museum because the museum is a nonconvex 28-gon. But you can see that it would take longer to walk around all the walls of the museum than to walk around either the rectangle or the ellipse. The perimeter of the museum is larger than the perimeter of either the rectangle or the ellipse.

Thus the museum has the smallest area but the largest perimeter. So, although area and perimeter both measure how big something is, they are quite different.

In this chapter, you will learn formulas for the areas of many of the figures you have studied so far in this book. You will also learn how these formulas are related to each other, so that if you forget one of them, you may be able to derive it.

Time and Mileage Map

Below is a portion of a United States mileage and driving-time map. It is a network whose nodes are cities. The lengths of arcs in this network are given in miles and in hours and minutes. For instance, the length of the arc from Charleston to Lexington is 177 miles or 2 hours 48 minutes.

There are many paths on this network. For instance, there is a path from Nashville to Shreveport through Little Rock. The length of that path is found by adding the lengths of the individual segments. It is 562 miles or 9 hours 35 minutes. In general, the length of a path is the sum of the lengths of its segments.

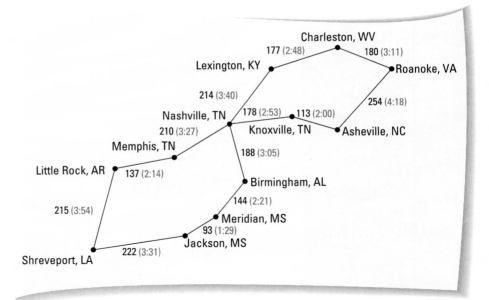

The network can be thought of as the union of two polygons, a heptagon and a hexagon, which share a vertex at Nashville. Think of traveling along the sides of one of these polygons, the hexagon *NLCRAK* (named using the first letters of the cities). You will have taken a tour through six cities (the vertices), ending where you started. According to this map, you will have traveled about 1116 miles and it would take you a little under 19 hours. In calculating the sum of the distances, you have calculated the *perimeter* of *NLCRAK*.

Definition
The perimeter of a polygon is the sum of the lengths of its sides.

The driving-time map illustrates that the length of a side of a polygon may be measured in various units. Usually the units are those of length in the metric system (meters, centimeters, and so on) or in the customary system (inches, miles, and so on). When calculating the perimeter of a polygon, it is important that the units for each side be the same.

If all the sides of a polygon have different lengths, there is no special formula for its perimeter. A formula for the perimeter p of a triangle with sides x, y, and z is just $p = x + y + z$.

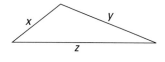

Biltmore House, Asheville, NC. *George Vanderbilt's summer house, constructed in 1895, contains art treasures, rare books, and more in its 250 rooms.*

Perimeters of Kites and Rectangles

When a polygon has some sides equal, then the calculations of the perimeter can be shortened. For instance, we know that a kite has two pairs of equal sides. If the lengths of these sides are a and b, then its perimeter p can be given by the formula

$$p = a + a + b + b$$

or, using repeated addition, $\qquad p = 2a + 2b$

or, factoring, using the Distributive Property,
$$p = 2(a + b).$$

Example 1

Kite *WXYZ* with ends *X* and *Z* has side lengths as shown at the right. Find its perimeter.

Solution 1

Use the definition of perimeter.
Perimeter of *WXYZ* = *WX* + *XY* + *YZ* + *ZW*.
Since *X* and *Z* are ends of the kite,
WX = *XY* and *YZ* = *ZW*. Substituting,
perimeter of WXYZ = 18 cm + 18 cm + 32 cm + 32 cm
WXYZ = 100 cm.

Solution 2

Use the formula $p = 2(a + b)$. If $a = WX$ and $b = ZW$, then
$$p = 2(a + b)$$
$$= 2(18 \text{ cm} + 32 \text{ cm})$$
$$= 2(50 \text{ cm})$$
$$= 100 \text{ cm}.$$

The lengths of two adjacent sides of a rectangle are the dimensions of the rectangle. Sometimes they are called the rectangle's length and width. Since the opposite sides of a rectangle are congruent, the perimeter of a rectangle with length ℓ and width w is $\ell + w + \ell + w$, or $2\ell + 2w$, or $2(\ell + w)$. So a formula for the perimeter of a rectangle is $p = 2(\ell + w)$. Formulas are nice to have because they allow you to apply what you know about algebra to geometry, and thus shorten calculations.

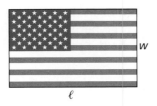

Example 2

Most rectangular flags are about 1.6 times as long as they are wide. If you have 10 meters of border material to strengthen the edges of a flag, about how large a flag can you make?

Solution

First, draw a picture. The edges are the length ℓ and width w of the rectangle. The perimeter p satisfies

$$p = 2(\ell + w).$$

Using $\ell = 1.6w$ and $p = 10$, substitute for p and ℓ.

$$10 = 2(1.6w + w)$$
$$10 = 2(2.6w)$$
$$10 = 5.2w$$

So $w = \frac{10}{5.2} \approx 1.923$ meters. Since $\ell = 1.6w$, $\ell \approx 3.077$ meters. From 10 meters of border material, you can make a flag about 1.92 meters wide and 3.08 meters long.

Equilateral Polygons

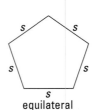

s

s

s

s

s

equilateral
pentagon

Rhombuses, squares, and regular polygons are all equilateral. A formula for the perimeter of an equilateral polygon follows directly from the definition of perimeter.

> **Equilateral Polygon Perimeter Formula**
> The perimeter p of an equilateral n-gon with sides of length s is given by the formula $p = ns$.

Since all regular polygons are equilateral, this formula applies to every regular polygon. For instance, in an equilateral triangle, $p = 3s$. In a square, $p = 4s$. In a regular octagon, $p = 8s$.

QUESTIONS

Covering the Reading

In 1–3, use the network pictured in this lesson.

1. What does the number 2:21 on the segment from Meridian, MS, to Birmingham, AL, mean?

2. A trucker with a perishable load should choose which route from Roanoke, VA, to Shreveport, LA?

3. Give the perimeter of the heptagon
   a. in miles.
   b. in hours.

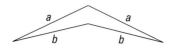

4. Give three different algebraic expressions for the perimeter of the polygon pictured at the left.

5. State the Distributive Property.

In 6–9, give the perimeter of each figure.

6. a rectangular-shaped piece of land $3\frac{1}{2}$ miles long and $\frac{1}{2}$ mile wide

7. an equilateral hexagon with one side of length 14 mm

8. a square with side t

9. a regular heptagon with side $(x + 1)$

10. The perimeter of a rectangle is 70. One side of the rectangle is 3 times the length of the other. What are the lengths of the sides?

11. A poster is to be 1.5 times as wide as it is high. If its edges total 3 meters in length, how wide will the poster be?

This poster shows a tapestry which hangs in the Parliament House in Canberra, Australia. Can you find a bird and Halley's comet woven into the eucalyptus forest?

Applying the Mathematics

12. The perimeter of a rhombus is 2 feet.
  a. Is this enough information to find the length of a side of the rhombus?
  b. If so, find that length. If not, why not?

13. A stop sign is a regular octagon. If the total length of its edges is 10′, what is the length of each side?

14. The perimeter of an equilateral triangle is p. What is the length of each side?

15. The boundary of the museum on page 435 has sides of three different lengths. Let the smallest sides have length s, the middle sides have length m, and the largest sides have length ℓ.
  a. What is a formula for the perimeter of the museum?
  b. If $s = 25$ meters, $m = 50$ meters, and $\ell = 100$ meters, what is the perimeter of the museum?

16. Suppose the dimensions of a rectangle are multiplied by 5. What happens to its perimeter?

17. A rectangle has perimeter 16.
  a. Graph all possible pairs of lengths x and widths y.
  b. Give an equation for the graph.
  c. Find one possible length and width where both are not integers.

18. One side of an equiangular pentagon has measure 7″. Can you compute its perimeter? Why or why not?

19. *Multiple choice.*   The measure of each angle of a regular *n*-gon is:

(a) $\frac{360}{n}$

(b) $\frac{n(n-3)}{2}$

(c) $(n-2) \cdot 180$

(d) $\frac{(n-2) \cdot 180}{n}$ *(Lesson 6-7)*

20. In the figure at the left, find m∠*BDC*.   *(Lesson 3-1)*

21. Is the network in the map on page 436 traversable?   *(Lesson 1-4)*

22. Fill in the blanks of these conversion formulas.   *(Previous course)*
a. 1 yard = __?__ feet
b. 1 kilometer = __?__ meters
c. 1 mile = __?__ feet

23. One inch is exactly 2.54 centimeters. About what part of an inch is two centimeters?   *(Previous course)*

24. Do the calculations, which are of a type often found when adding lengths.   *(Previous course)*
a. 3 feet 6 inches + 8 feet 11 inches
b. 8 · (2 feet 3 inches)
c. 2.4 meters + 62 centimeters

25. If $p = 2\ell + 2w$, $\ell = 11$, and $p = 25$, find *w*.   *(Previous course)*

26. Solve $x^2 = 200$. Give your answer to the nearest hundredth.
(Previous course)

27. Is $(x + 1)(2x - 3)$ the same as $2x^2 - 3$? Why or why not?
(Previous course)

28. *Multiple choice.* $(r + s)^2 =$ __?__
(a) $r^2 + s^2$

(b) $r^2 + rs + s^2$

(c) $r^2 + 2rs + s^2$

(d) none of these
(Previous course)

29. Refer to the map in this lesson.
a. What average speed did the map makers assume in the trip from Shreveport, LA, to Jackson, MS?
b. What average speed is assumed in the trip from Shreveport, LA, to Little Rock, AR?
c. What conditions could account for the different rates?
(Previous course)

Shown is the River Rose cruising down the Red River in Shreveport, Louisiana.

Exploration

30. If your school has one building, estimate the perimeter of that building. If your school has more than one building, estimate the perimeter of the largest building.

*Fundamental
Properties of
Area*

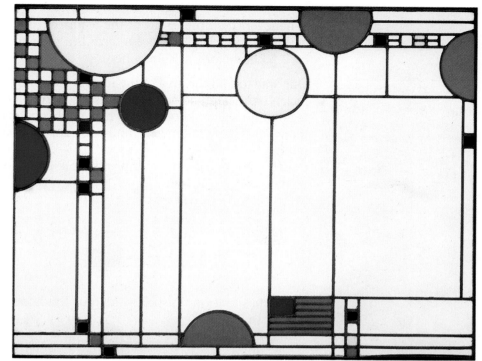

The Wright Way. *Frank Lloyd Wright (1867–1959) designed windows and furniture as well as buildings for which he is famous. He was conscious of area and lights as demonstrated by this center window of a stained-glass window triptych.*

What Is Area?

Area is a measure of the space covered by a two-dimensional region. The region may be small, like a microchip in a computer, or it may be large, like a country. The idea of area is the same. Tessellate the region with a fundamental region. Then count the number of copies of the fundamental region needed to cover the region. That count is the area.

Usually, the fundamental region is a square, and so we say that area is measured in **square units.** For instance, the rectangular region below at the left has dimensions 11 units and 8 units. Its area is 88 square units because 88 unit squares cover the region.

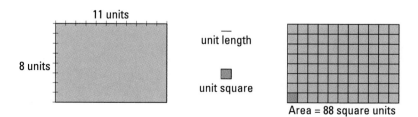

11 units

8 units

unit length

unit square

Area = 88 square units

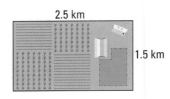
2.5 km

1.5 km

Fundamental Properties of Area

Whenever the dimensions of the rectangle are integers, the unit squares will fit exactly. But suppose a farm is a rectangular region, 1.5 kilometers by 2.5 kilometers, as shown here.

One way to find the area of the farm is shown by the figures below. First, pick a unit. Here the natural unit is 1 square kilometer. Then split the farm region into square kilometers. There are two whole-square kilometers, three half-square kilometers, and one quarter-square kilometer. Think of putting them end to end. The result is 3.75 square kilometers. This is exactly what you get by multiplying 1.5 kilometers by 2.5 kilometers.

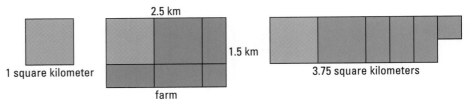

1 square kilometer

2.5 km

1.5 km

farm

3.75 square kilometers

This one situation illustrates the four fundamental properties of area which we assume.

Many plowed fields like these in the Midwest are rectangular.

Area Postulate

a. **Uniqueness Property**   Given a unit region, every polygonal region has a unique area.

b. **Rectangle Formula**   The area A of a rectangle with dimensions ℓ and w is ℓw. ($A = \ell w$)

c. **Congruence Property**   Congruent figures have the same area.

d. **Additive Property**   The area of the union of two nonoverlapping regions is the sum of the areas of the regions.

Activity

Measure to determine the area of rectangle *WXYZ*
a. in square centimeters.   b. in square millimeters.

W   X

Z   Y

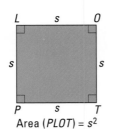

L   s   O

s   s

P   s   T

Area (*PLOT*) = s^2

From the Area Postulate, other properties of area can be deduced. For instance, a special case of the Rectangle Formula is that a formula for the area of any square with side s is $A = s^2$. This is pictured at the left.

The Uniqueness Property guarantees that a figure F can have only one area. Sometimes we write **Area(F)** for the area of the figure F. With this notation, the Congruence Property of Area becomes: If $F \cong G$, then Area(F) = Area(G). **Nonoverlapping** regions means regions that do not share interior points. They may share boundaries, as in the drawing below. The Additive Property of Area becomes: If F and G do not overlap, then Area $(F \cup G)$ = Area (F) + Area(G).

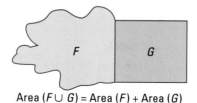

Area $(F \cup G)$ = Area (F) + Area (G)

Notice that the perimeter of $F \cup G$ does *not* equal the sum of the perimeters of F and G, where the common border is counted twice.

All the fundamental properties of area are used in Example 1.

Example 1

The floor plan of a ranch house is drawn on a coordinate system.
a. Find the dimensions of rooms I, II, and III if the unit is 1 foot.
b. Find the floor area of the house.

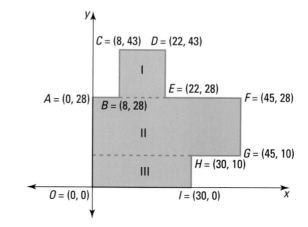

Solution
a. The Uniqueness and Congruence properties assure us that regions I, II, and III have areas that are unique and constant. I, II, and III are rectangles. The horizontal dimensions are found by subtracting appropriate pairs of the *x*-coordinates of the vertices. For example, $CD = |22 - 8| = 14$. The vertical dimensions are found by subtracting the *y*-coordinates of the vertices. For example,

$$BC = |28 - 43| = 15.$$

Dimensions of room I: CD = 14 ft and BC = 15 ft
Dimensions of room II: AF = 45 ft and FG = 18 ft
Dimensions of room III: OI = 30 ft and HI = 10 ft

b. By the Rectangle Formula:
$$Area(I) = 15 \text{ ft} \cdot 14 \text{ ft} = 210 \text{ sq ft}$$
$$Area(II) = 18 \text{ ft} \cdot 45 \text{ ft} = 810 \text{ sq ft}$$
$$Area(III) = 10 \text{ ft} \cdot 30 \text{ ft} = 300 \text{ sq ft}$$

By the Additive Property of Area, the area of the house is the sum of the areas of rooms I, II, and III.
$$Area(floor) = 210 + 810 + 300 \text{ sq ft}$$
$$= 1320 \text{ sq ft}$$

Check

One way to check is to consider the rectangle with vertices (0, 0), (45, 0), (45, 43), and (0, 43). This rectangle has area 43 · 45, or 1935 sq ft. It includes the entire floor plan plus rectangles in three corners. The sum of the areas of those corner rectangles subtracted from 1935 sq ft should give you the same answer as above.

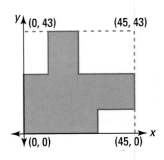

Example 2 describes a common problem involving area.

Example 2

A carpet dealer advertises a particular carpet for $18.95 a square yard. How much will it cost to carpet a rectangular room 9 feet wide and 12 feet long?

Solution

In doing calculations like these, it helps to keep track of the units. The unit "dollars per square yard" is written as $\frac{\$}{yd^2}$.

$$\text{Total cost} = \text{total area} \cdot \text{cost per unit area}$$
$$= (9 \text{ ft} \cdot 12 \text{ ft}) \cdot 18.95 \frac{\$}{yd^2}$$
$$= (3 \text{ yd} \cdot 4 \text{ yd}) \cdot 18.95 \frac{\$}{yd^2}$$
$$= 12 \text{ yd}^2 \cdot 18.95 \frac{\$}{yd^2}$$
$$= \$227.40$$

This cost does not include tax and any other charges such as padding or installation.

Example 2 could have been worked out in feet and square feet. Then you would need to realize that since 1 yard = 3 feet,

$$1 \text{ yard}^2 = (3 \text{ feet})^2 = 9 \text{ feet}^2.$$

That is, there are 9 square feet in 1 square yard, as pictured below.

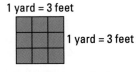

1 yard = 3 feet

1 square yard = 9 square feet

QUESTIONS

Covering the Reading

1. Rectangle $ABCD$ has dimensions 8.3 cm and 11.4 cm.
   a. What is an appropriate unit of area in this situation?
   b. Find Area($ABCD$).

2. In the figure at the left, all of the angles are right angles. Find the area of each region.
   a. $ABFG$ **b.** $CDEF$ **c.** $ABCDEG$

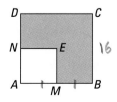

3. Which properties of area are used in answering Question 2?

4. **a.** The outline of a floor plan of a large house is given at the right. The unit of length is one meter. Find the area of the floor.
   b. In part **a**, which method seems easier to you: breaking the floor into smaller rectangles and adding their areas, or surrounding the floor with one rectangle and subtracting the areas of the corner rectangles? Explain the reasons for your choice.

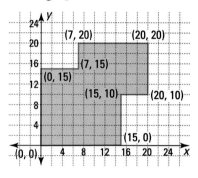

5. Give the answers to the Activity in this lesson.

6. Complete the Check in Example 1.

7. A carpet store is selling a particular carpet for $11.95 a square yard. How much will it cost to carpet a room that is 15 feet by 12 feet, before tax or any other charges?

8. How many square feet are in a square yard?

Applying the Mathematics

9. Find the area of the polygon with the given vertices.
   a. (0, 0), (0, 10), (10, 10), (10, 0)   **b.** (0, 0), (0, k), (k, k), (k, 0)

10. At the left, $ABCD$ and $AMEN$ are squares, M is the midpoint of $\overline{AB}$, and $BC = 16$. Find Area($BCDNEM$).

11. The area of a rectangle is 50 square yards. The length of the rectangle is 100 yards. What is the width of the rectangle?

12. To the nearest whole unit, find the side of a square with the given area.
    a. 49 square units   **b.** $\frac{3}{4}$ square unit
    c. 200 square units   **d.** 3141 square units

13. **a.** A desk is 24″ by 12″. What is its area in square inches?
    b. The same desk is 2′ by 1′. What is its area in square feet?
    c. How many square inches are in a square foot?

14. The sides of a square are tripled.
    a. What happens to its perimeter?
    b. What happens to its area?

15. In Olympic field hockey, the dimensions of the field are 100 yards by 60 yards. A piece of sod is 1.5 ft by 6 ft. Find the number of pieces of sod needed to cover an Olympic field-hockey field.

Review

16. If a regular pentagon has perimeter 13, what is the length of a side? *(Lesson 8-1)*

17. A parallelogram has perimeter 462 cm. One side is 185 cm. Find the lengths of the other three sides. *(Lesson 8-1)*

18. The length of one side of a rhombus is $3x$. Find its perimeter. *(Lesson 8-1)*

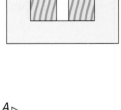

19. A pool is 50 m long and 25 m wide.
    a. A fence is built parallel to the sides of the pool and 10 m from each side. What is the perimeter of the fence?
    b. If a park district has budgeted money for 200 meters of fencing, how far from the pool can they put the fence? *(Lesson 8-1)*

20. The longest side of the triangle at the right is twice as long as the shortest side. The third side is $1\frac{1}{2}$ times as long as the shortest side. If the perimeter of the triangle is 45, how long are its three sides? *(Lesson 8-1)*

In 21 and 22, find all solutions. *(Previous course)*

21. $x^2 = 16$
22. $5y^2 = 240$

23. Trace right $\triangle ABC$ shown at the left. Draw $R(ABC) = A'B'C'$, where R is the rotation of 180° about the midpoint of the hypotenuse. What shape is $ABCB'$? *(Lesson 7-6)*

Exploration

24. Find a rectangular room in your home. Calculate its area to the nearest square foot.

25. In 1860, the total area of the United States was 3,021,295 square miles. By 1870, the total area had become 3,612,299 square miles. What caused such a large change?

Can't top it! Can't knock it! *Not quite mile-high, Mt. Katahdin (5258 ft) is the tallest peak in Maine and clearly visible from Millinocket Lake.*

An almanac gives the total land area of the 48 contiguous states of the United States as about 2,962,000 square miles. (The other two states, Alaska and Hawaii, add about 577,000 square miles to the area.) The contiguous United States is an irregular region about 3000 miles from east to west and 1600 miles from north to south. In this lesson, you will see how areas of irregular regions can be estimated.

A Typical Example

Here is a scale drawing of Millinocket Lake in northern Maine.

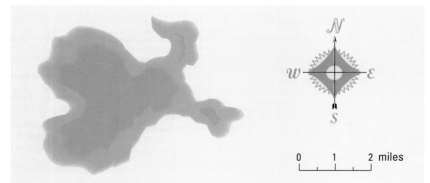

Like many shapes, the boundary of the lake is not the union of circular arcs and segments. Its shape is *irregular*. Still, it has an area; it takes up space. For all sorts of reasons, such as zoning, selling lakefront property, or tracking the fish in the lake, people might want to know its area.

To get a first approximation, you can draw a rectangle around the lake.

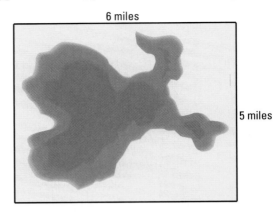

The area of the lake is less than the area of the rectangle, which is 6 miles · 5 miles. That is, the area of the lake is less than 30 square miles.

To get a better estimate, you might cover the lake with part of a tessellation of congruent squares. Here the squares are 1 mile on a side.

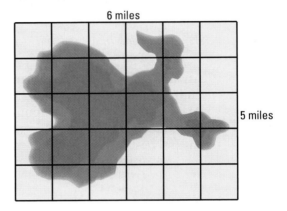

The squares above are too big to estimate the lake's area accurately. Within every square you would have to estimate how much of the square is covered by the lake. Smaller squares give a more accurate estimate. Below, the squares used are $\frac{1}{2}$ mile on a side.

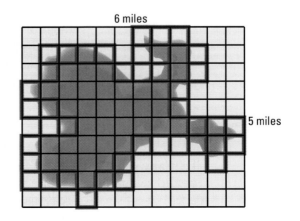

Computing the Area

The idea now is to count the number of small squares entirely inside the lake in the last figure on page 448. (There are 34.) Then count the number of squares partially covering the lake. (You should get 41). They are drawn with bold red edges. Instead of trying to estimate how much of each square is inside and then adding the fractions, it is easier to assume that, on average, half of each partly covered square is inside. So add *half* the second number to the first. $34 + \frac{41}{2} = 34 + 20.5 = 54.5$.

An estimate for the area of Millinocket Lake is 54.5 of these squares. Since each small square is $\frac{1}{2}$ mile on a side, the area of each small square is $\frac{1}{4}$ square mile, so an estimated area of the lake is $54.5 \cdot \frac{1}{4}$ square mile, or about 13.6 square miles.

In general, if I is the number of inside squares, B is the number of boundary squares, and U is the area of a single square, then an estimate for the total area is $(I + \frac{1}{2}B) \cdot U$.

Caution: When calculating area this way, do not spend too much time deciding whether a square is entirely inside or on the boundary. Your answer will always be an estimate, so do not search for an exact answer.

To get a closer estimate, use a grid with smaller squares. Then U is smaller. In the activity below, the sides of the squares are half the length of those above, so each side is $\frac{1}{4}$ mile and $U = \frac{1}{16}$ square mile. We have identified those squares that lie on the boundary.

Activity

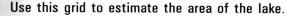

Use this grid to estimate the area of the lake.

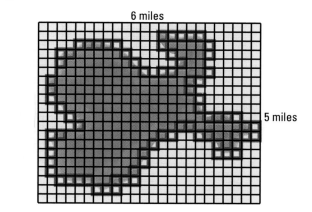

6 miles

5 miles

The above procedure can be continued using finer and finer grids. The estimates can be made to differ from the actual area by no more than 0.1 square mile, 0.01 square mile, or even less. When smaller and smaller squares are used, we say that the true area of the lake is the *limit* which the estimates approach. According to the Maine Department of Natural Resources, the official area of Millinocket Lake is 12.34 square miles.

The advantage of this method is that it works for any reasonably smooth curve. This same idea is used in calculus to calculate areas bounded by curves. Although it takes a long time to figure out areas in this way with pencil and paper, the method can be programmed on a computer. Also, when the curve can be described by an equation, there may exist a simple formula for its area.

QUESTIONS

Covering the Reading

1. Three tessellations of squares are used in this lesson to estimate the area of a lake. What is the area of an individual square
   a. in the first tessellation?
   b. in the second tessellation?
   c. in the finest tessellation?

2. Give three reasons people might have for estimating the area of a lake.

3. Suppose E is the number of squares entirely inside a region and P is the number of squares partially inside the region. If each square has area U square units, what formula gives an estimate, in square units, for the region's area?

4. a. Was your answer to the Activity greater than or less than the estimate of 13 square miles?
   b. Was it greater than or less than the official area of 12.34 square miles?

5. a. What is an advantage of the method of using grids to estimate area?
   b. What is a disadvantage?

6. The area of a region is the __?__ of the estimates made using finer and finer grids.

Applying the Mathematics

7. Recall that the resolution of TV screens or computer monitors of different sizes can be compared by calculating the number of dots (pixels) per square inch. Some Apple Macintosh™ computer screens measure about 10″ by 7.5″. There are 640 rows and 480 columns of dots. About how many dots per square inch is this?

8. The two figures below are congruent. The grid squares on the left are $\frac{1}{4}$ inch on a side, and the grid squares on the right are $\frac{1}{8}$ inch on a side. Estimate the area of the figure, using the given grid.

a.

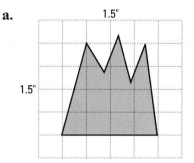

1.5"

1.5"

b.

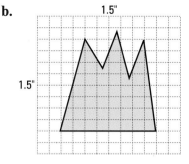

1.5"

1.5"

9. At the right is a scale drawing of Wilson Reservoir in Kansas on a grid of squares each 1 km on a side. Estimate the area of the reservoir.

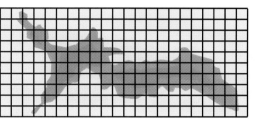

In 10–12, use this information. An *acre* is a unit of area commonly used to measure plots of land. Originally an acre was about the amount of land a farmer could plow in a day. Today an acre has an exact conversion equation: 640 acres = 1 square mile.

10. There are 5280 feet in a mile.
  a. How many square feet are in a square mile?
  b. How many square feet are in an acre?

11. Lake Dumont in the state of Michigan has an area of 215 acres. About what percent of a square mile is this?

12. A house is built on a rectangular half-acre lot. What might be the dimensions (in feet) of the rectangle?

13. A farm in Europe has an area of 30 square kilometers. A farm in the United States has an area of 20 square miles. If 1 mile ≈ 1.6 km, which farm is bigger?

Spanish spuds. *A farmer plows a potato field in Costa Blanca, Spain. Spain is one of the major potato-producing countries of the world.*

Review

14. At the right, a tilted square *EFGH* is placed inside a square *ABCD*. *(Lessons 8-1, 8-2)*
  a. If $AB = 7$ and $HE = 5$, what is the area of the shaded region?
  b. If $AB = x$ and $HE = 5$, what is the area of the shaded region?

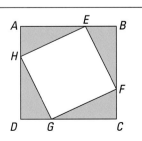

15. Suppose the sides of a rectangle are multiplied by 10. *(Lessons 8-1, 8-2)*
  a. What happens to its area?   **b.** What happens to its perimeter?

16. A person wishes to tile a kitchen floor with square tiles 8 inches on a side. If the kitchen measures 10 feet by 12 feet, how many tiles are needed? *(Lesson 8-2)*

17. A rectangle has area of 96 square units. If its width is 4 units, what is its perimeter? *(Lessons 8-1, 8-2)*

18. If the unit of the area of a figure is square kilometers, what is the natural unit for the perimeter of the figure? *(Lessons 8-1, 8-2)*

19. Use the figure at the right.
Given: $\angle PAQ \cong \angle DAR$,
$\qquad$ $\angle PQA \cong \angle DRA$, and
$\qquad$ $PQ = RD$.
To prove: $\triangle PAD$ is isosceles. *(Lesson 6-2)*

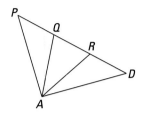

20. In $\odot O$ below, find the measures of $\overset{\frown}{AB}$, $\overset{\frown}{BC}$, and $\overset{\frown}{CA}$. *(Lesson 3-2)*

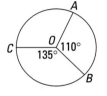

21. *True or false.* A solution to $2y^2 = 16$ is -2. *(Previous course)*

22. *Multiple choice.* If $\sqrt{48} = k\sqrt{3}$, what is k? *(Previous course)*
(a) 4 $\qquad\qquad\qquad\qquad$ (b) 16
(c) $\sqrt{45}$ $\qquad\qquad\qquad\qquad$ (d) cannot be determined

---

Exploration

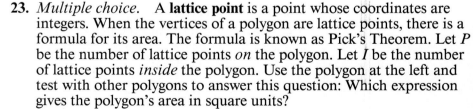

23. *Multiple choice.* A **lattice point** is a point whose coordinates are integers. When the vertices of a polygon are lattice points, there is a formula for its area. The formula is known as Pick's Theorem. Let P be the number of lattice points *on* the polygon. Let I be the number of lattice points *inside* the polygon. Use the polygon at the left and test with other polygons to answer this question: Which expression gives the polygon's area in square units?
(a) $\frac{1}{2}P + I - 1$ $\qquad\qquad\qquad$ (b) $\frac{1}{2}P + I$
(c) $\frac{1}{2}P + I + 1$ $\qquad\qquad\qquad$ (d) $\frac{1}{2}(P + I)$

Areas of Triangles

Most of the shapes you have studied so far have been special types of polygons, not the irregular shapes considered in the last lesson. Some polygonal shapes are so common that formulas have been developed to give their areas. All of these formulas can be derived from the Rectangle Formula.

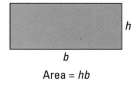

Area = *hb*

Area of a Right Triangle

It is easy to find the area of any right triangle ABC. Just rotate $\triangle ABC$ 180° about M, the midpoint of $\overline{AC}$, as you did in making a tessellation. The image is $\triangle CDA$. Since all four angles of $ABCD$ are 90° angles, $ABCD$ is a rectangle. By the Congruence and Additive Properties of the Area Postulate, the area of each triangle is half the rectangle. This argument proves the Right Triangle Area Formula.

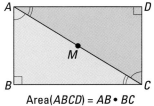

Area($ABCD$) = $AB \cdot BC$
Area($\triangle ABC$) = $\frac{1}{2}(AB \cdot BC)$

Right Triangle Area Formula
The area of a right triangle is half the product of the lengths of its legs.
$$A = \tfrac{1}{2}hb$$

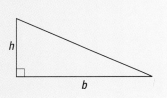

Example 1

Find the area of △PQR at the right.

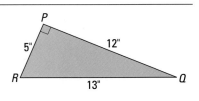

Solution

The legs of △PQR are 5″ and 12″.

So Area(△PQR) = $\frac{1}{2} \cdot 5 \cdot 12$

$= 30$ in^2.

Area of Any Triangle

From the area of a right triangle, a formula for the area of *any* triangle can be derived. The idea of *altitude* is needed. In a triangle, an **altitude** is the perpendicular segment from a vertex to the line containing the opposite side. In each drawing below, $\overline{AD}$ is the altitude to side $\overline{BC}$ of △ABC. The length of an altitude is called the **height** or the **altitude** of the triangle with that base.

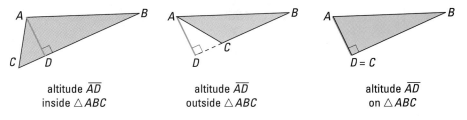

altitude $\overline{AD}$ inside △ABC

altitude $\overline{AD}$ outside △ABC

altitude $\overline{AD}$ on △ABC

In all cases, the same simple formula for the area of the triangle can be deduced.

Triangle Area Formula
The area of a triangle is half the product of a side (the base) and the altitude (height) to that side.

$$A = \tfrac{1}{2}hb$$

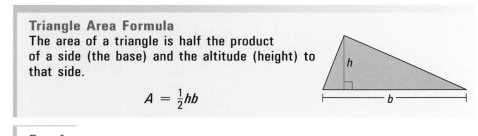

Proof:
Argument:
Case I: Altitude *inside* △ABC
The altitude splits △ABC into two right triangles. Let $BD = x$ and $DC = y$. Then $b = x + y$.

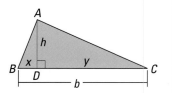

Area(△ABC) = Area(△ABD) + Area(△ADC)	Additive Property of Area
$= \frac{1}{2}hx + \frac{1}{2}hy$	Right Triangle Area Formula
$= \frac{1}{2}h(x + y)$	Distributive Property
$= \frac{1}{2}hb$	Substitution Property

Case II: Altitude *outside* △ABC
The area of △ABC can be found by
subtracting the areas of two right triangles.

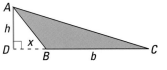

Area(△ABC) = Area(△ADC) − Area(△ADB)

$$= \tfrac{1}{2}h(x + b) - \tfrac{1}{2}hx$$

$$= \tfrac{1}{2}hx + \tfrac{1}{2}hb - \tfrac{1}{2}hx$$

$$= \tfrac{1}{2}hb$$

Case III: Altitude *on* △ABC
In this case, the triangle is a right triangle, so the formula works.

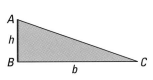

When a figure is given on the coordinate plane, there is a natural
unit square.

Example 2

Given coordinates as shown,
find the area of
a. △ABC. **b.** △ADE.
c. △ACE. **d.** △ABE.

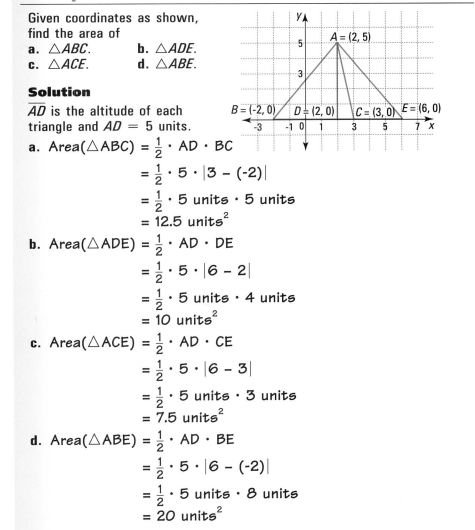

Solution

$\overline{AD}$ is the altitude of each
triangle and AD = 5 units.

a. Area(△ABC) = $\tfrac{1}{2} \cdot AD \cdot BC$

$$= \tfrac{1}{2} \cdot 5 \cdot |3 - (-2)|$$

$$= \tfrac{1}{2} \cdot 5 \text{ units} \cdot 5 \text{ units}$$

$$= 12.5 \text{ units}^2$$

b. Area(△ADE) = $\tfrac{1}{2} \cdot AD \cdot DE$

$$= \tfrac{1}{2} \cdot 5 \cdot |6 - 2|$$

$$= \tfrac{1}{2} \cdot 5 \text{ units} \cdot 4 \text{ units}$$

$$= 10 \text{ units}^2$$

c. Area(△ACE) = $\tfrac{1}{2} \cdot AD \cdot CE$

$$= \tfrac{1}{2} \cdot 5 \cdot |6 - 3|$$

$$= \tfrac{1}{2} \cdot 5 \text{ units} \cdot 3 \text{ units}$$

$$= 7.5 \text{ units}^2$$

d. Area(△ABE) = $\tfrac{1}{2} \cdot AD \cdot BE$

$$= \tfrac{1}{2} \cdot 5 \cdot |6 - (-2)|$$

$$= \tfrac{1}{2} \cdot 5 \text{ units} \cdot 8 \text{ units}$$

$$= 20 \text{ units}^2$$

QUESTIONS

Covering the Reading

In 1 and 2, give a formula for the area of the figure.

1. right triangle

2. triangle

3. Sketch a triangle with a *vertical* base and an altitude to that base
  a. with the altitude outside the triangle.
  b. with the altitude coinciding with a side.
  c. with the altitude interior to the triangle.

In 4 and 5, find the area of the triangle.

4.

5.

6. In the figure at the right, give the area of
  a. $\triangle EFH$.
  b. $\triangle FGH$.
  c. $\triangle EGH$.

7. a. How is Area($\triangle ABC$) related to the areas of the two right triangles in the figure at the right?
  b. If $BD = 7$ mm, $AC = 15$ mm, and $CD = 5$ mm, what is Area($\triangle ABC$)?

In 8 and 9, find the area of $\triangle XYZ$.

8.

9.

10. Refer to the figure below. Find the area of

  a. $\triangle PQR$.
  b. $\triangle PRS$.
  c. $\triangle PQS$.

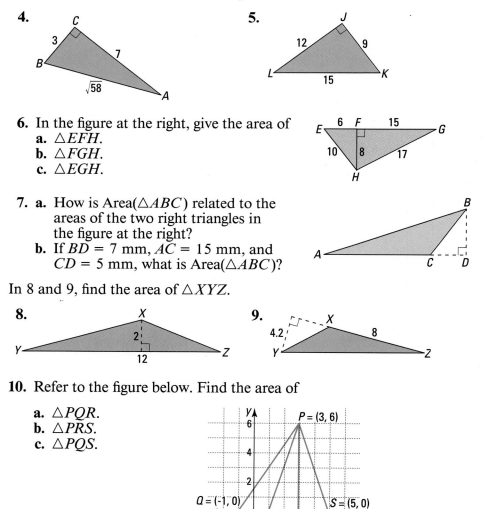

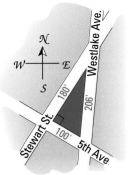

*Westlake Center in
Seattle, Washington.*

11. The Westlake Center in Seattle, Washington, is a casual place where you can get a meal, listen to a street band, and catch a ride on a monorail. It is next to some of the most exclusive shops in Seattle. It occupies a triangular lot, with approximate dimensions as shown at the left. What is the area of the lot?

12. Trace $\triangle XYZ$ at the right.
   a. Accurately draw one of its altitudes.
   b. Estimate its area in square millimeters by measuring a side, measuring the length of the altitude to that side, and using the Triangle Area Formula.
   c. Will you get the same area if you start with a different altitude? Why or why not?

13. Find the length of a side of a triangle with area 18 in^2 and altitude 12 in.

14. Suppose the sides of a right triangle are doubled. What happens to
   a. its perimeter?   **b.** its area?

15. Approximate dimensions (in meters) of a part of a roof ABC with $AC = BC$ are given at the right.
   a. What is the perimeter of this part of the roof?
   b. What is its area?

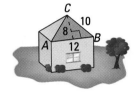

16. $\triangle ABC$ at the right has altitudes $\overline{AW}$ and $\overline{CF}$. If $AB = 8$, $CF = 6$, and $AW = 7$, find CB. (Hint: Compute Area($\triangle ABC$) in two different ways.)

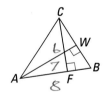

17. The grid at the right is a tessellation of unit squares. Find the exact area of quadrilateral $ABCD$.

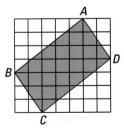

18. A quadrilateral $ABCD$ with perpendicular diagonals is drawn at the right.
   a. Prove that its area is half the product of the lengths of its diagonals:
   Area $ABCD = \frac{1}{2} AC \cdot BD$.
   b. To which of the following types of quadrilaterals does the result in part **a** apply: isosceles trapezoids, kites, parallelograms, rectangles, rhombuses, squares, or trapezoids?

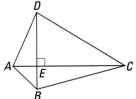

19. Each grid square below at the left is 70 miles on a side, and each grid square at the right is 35 miles on a side. Estimate the area of Texas using each grid. *(Lesson 8-3)*

a.

70 miles

70 miles {

b.

35 miles

35 miles ➤

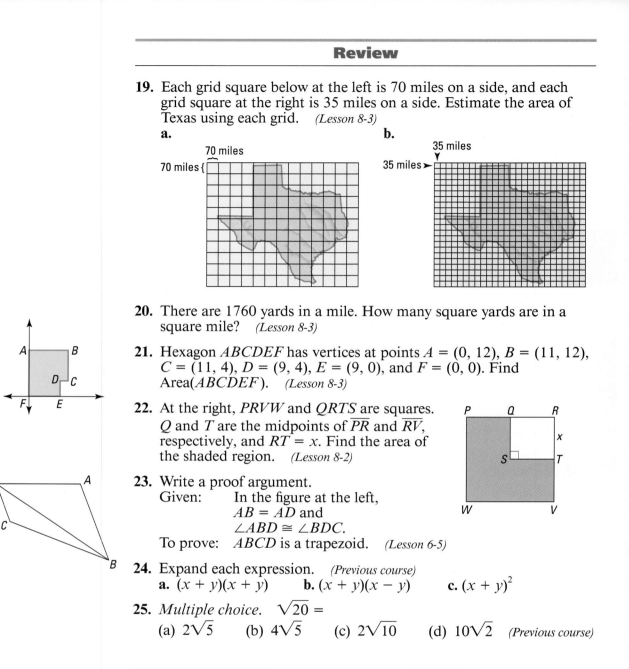

20. There are 1760 yards in a mile. How many square yards are in a square mile? *(Lesson 8-3)*

21. Hexagon $ABCDEF$ has vertices at points $A = (0, 12)$, $B = (11, 12)$, $C = (11, 4)$, $D = (9, 4)$, $E = (9, 0)$, and $F = (0, 0)$. Find Area($ABCDEF$). *(Lesson 8-3)*

22. At the right, $PRVW$ and $QRTS$ are squares. Q and T are the midpoints of $\overline{PR}$ and $\overline{RV}$, respectively, and $RT = x$. Find the area of the shaded region. *(Lesson 8-2)*

23. Write a proof argument.
Given: In the figure at the left,
             $AB = AD$ and
             $\angle ABD \cong \angle BDC$.
To prove: $ABCD$ is a trapezoid. *(Lesson 6-5)*

24. Expand each expression. *(Previous course)*
   a. $(x + y)(x + y)$ **b.** $(x + y)(x - y)$ **c.** $(x + y)^2$

25. *Multiple choice.* $\sqrt{20} =$
   (a) $2\sqrt{5}$ (b) $4\sqrt{5}$ (c) $2\sqrt{10}$ (d) $10\sqrt{2}$ *(Previous course)*

26. At the right, the named points are equally spaced along the rectangle $AEHL$. Each point A, B, C, D, and E is connected to each point H, I, J, K, and L. How many triangles in the drawing have the same area as $\triangle CLH$? Name them.

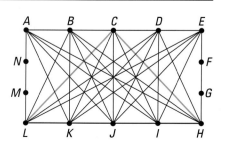

Areas of Trapezoids

Complex housing. *This modern housing development in the Ramat Polin section in Jerusalem utilizes pentagons and trapezoids in a most interesting way.*

Knowing a formula for the area of a triangle is useful because any polygon can be split into triangles. Recall that when this occurs, we say that the polygon has been *triangulated*. Pentagon *ABCDE* at the left below has been copied and triangulated at the right.

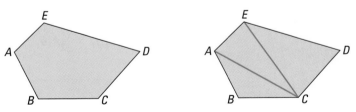

This idea provides an algorithm for finding the area of *any* polygon. Step 1: Triangulate the polygon. Step 2: Find the area of each triangle (by measuring lengths of sides and altitudes). Step 3: Add these areas to find the area of the polygon.

But an algorithm is not the same as a formula. There is no known general formula for the area of a polygon, even if you know the lengths of all its sides and the measures of all its angles. But if a polygon can be split into triangles with sides and their altitudes known, then there can be a formula. One kind of polygon that can be split in this way is the trapezoid.

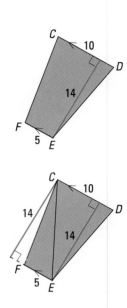

A Specific Example

The **height** or **altitude** of a trapezoid refers to any segment from one base perpendicular to the other, or to the length of that segment. This length is constant because the bases are parallel. In trapezoid *CDEF* at the left, the lengths of the bases (10 and 5) and its height (14) are given. This is enough to find the area.

First split the trapezoid into triangles *CEF* and *CDE*. The height of the trapezoid is the height of each triangle, and the bases of the trapezoid are the corresponding bases of the triangles. So the area of each triangle can be found.

$$\text{Area}(CDEF) = \text{Area}(\triangle CEF) + \text{Area}(\triangle CDE)$$
$$= \tfrac{1}{2}(5 \cdot 14) + \tfrac{1}{2}(10 \cdot 14)$$
$$= 35 + 70$$
$$= 105 \text{ square units}$$

A General Formula

The ideas used in the specific instance above can be applied to deduce a formula for the area of a trapezoid. A bonus is that a trapezoid area formula applies to all of the special kinds of quadrilaterals which are below the trapezoid in the hierarchy of quadrilaterals.

Trapezoid Area Formula
The area of a trapezoid equals half the product of its altitude and the sum of the lengths of its bases.
$$A = \tfrac{1}{2}h(b_1 + b_2)$$

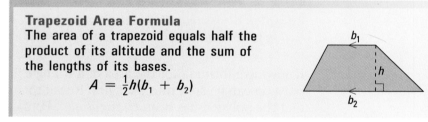

Proof:

Given: *ZOID* is a trapezoid with altitude h and bases b_1 and b_2.

To prove: $\text{Area}(ZOID) = \tfrac{1}{2}h(b_1 + b_2)$.

Drawing:

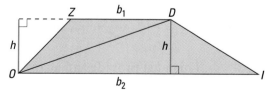

Argument:

$\text{Area}(ZOID) = \text{Area}(\triangle ZOD) + \text{Area}(\triangle DOI)$	Additive Property of Area
$= \tfrac{1}{2}h \cdot b_1 + \tfrac{1}{2}h \cdot b_2$	Triangle Area Formula
$= \tfrac{1}{2}h(b_1 + b_2)$	Distributive Property

Recall that $\frac{1}{2}(b_1 + b_2)$ is the mean or average of b_1 and b_2. So the area of a trapezoid equals the product of its altitude and the average of its bases.

Example 1

Compute the area of *ABCD* at the right.

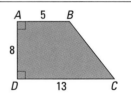

Solution

By the Two Perpendiculars Theorem, $\overline{AB} \parallel \overline{CD}$, so $\overline{AB}$ and $\overline{CD}$ are the bases of a trapezoid. Apply the Trapezoid Area Formula with $b_1 = 5$, $b_2 = 13$, and $h = 8$.

$$\text{Area}(ABCD) = \frac{1}{2}h(b_1 + b_2)$$
$$= \frac{1}{2} \cdot 8(5 + 13)$$
$$= 4 \cdot 18$$
$$= 72 \text{ units}^2$$

Check

The area of a trapezoid is the product of its altitude and the average of the bases. The average of 5 and 13 is $\frac{5 + 13}{2} = 9$. Since $8 \cdot 9 = 72$, the answer checks.

Areas of Parallelograms

Since every parallelogram is a trapezoid, the trapezoid formula applies to parallelograms as well. For example, *SPOT* is a parallelogram with altitude h. In a parallelogram, opposite sides are equal, so $b_1 = b_2 = b$.

$$\text{Area}(SPOT) = \frac{1}{2}h(b_1 + b_2)$$
$$= \frac{1}{2}h(b + b)$$
$$= \frac{1}{2}h(2b)$$
$$= hb$$

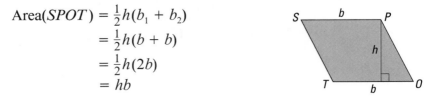

Parallelogram Area Formula
The area of a parallelogram is the product of one of its bases and the altitude to that base.
$$A = hb$$

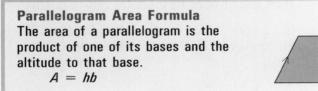

In Example 2, many lengths are given. You must be careful in selecting which lengths to use.

Example 2

Find the area of parallelogram *ABCD*.

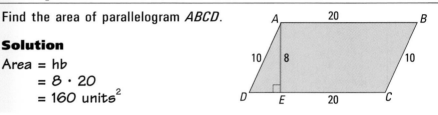

Solution

Area = hb

= 8 · 20

= 160 units^2

Notice that the sides with length 10 are not used.

QUESTIONS

Covering the Reading

1. Describe an algorithm for obtaining the area of any polygon.

2. Trace trapezoid *ABCD* at the left. Then triangulate it.

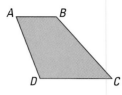

3. Give the area formula for any trapezoid.

4. $\frac{1}{2}(b_1 + b_2)$ is the __?__ or __?__ of b_1 and b_2.

In 5 and 6, use the figure at the right.

5. **a.** Name the bases and altitude of trapezoid *EFGH*.
   b. Find Area(*EFGH*).

6. Find Area(*EFIH*).

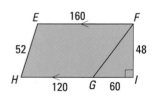

In 7–9, find the area of the largest trapezoid in the drawing.

7.

8.

9.

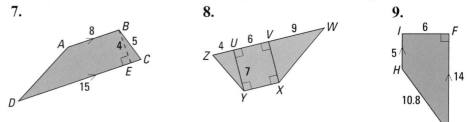

10. Give an area formula for any parallelogram.

In 11 and 12, **a.** Find the area of the parallelogram. **b.** List all lengths in the figure you did not need to answer part **a.**

11.

12.

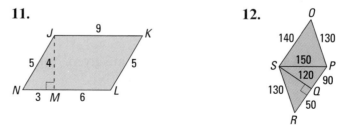

Applying the Mathematics

13. **a.** Trace the figure of Example 2.
    b. Translate $\triangle ADE$ by the vector $\overrightarrow{DC}$. Call E' the image of E.
    c. What is the shape of figure $ABE'E$?
    d. What is the area of figure $ABE'E$?
    e. Will this idea work for all parallelograms?

In 14 and 15, find the area of the trapezoid with the given vertices.

14.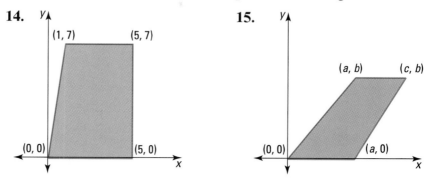

15.

16. A trapezoid has a pair of bases with lengths 20 and 15, and an area of 60. What is the altitude of the trapezoid?

17. At the right, $ABCF$ is an isosceles trapezoid with bases $\overline{AB}$ and $\overline{FC}$. $\overline{AE} \perp \overline{FC}$ and $\overline{BD} \perp \overline{FC}$. $FE = 2.5$, $ED = 8$, and $AE = 5$.
    a. Prove that $FE = DC$.
       (Hint: Use congruent triangles.)
    b. Find the area of $ABCF$.

Review

In 18 and 19, *multiple choice.* Use the figure below. *(Lesson 8-4)*

18. The area of $\triangle ABC$ is
    (a) $h + x + g$ units^2
    (b) hx units^2.
    (c) $\frac{1}{2} hxg$ units^2.
    (d) $\frac{1}{2} hx$ units^2.

19. The area of $\triangle ABE$ is
    (a) $h(x + y + z)$ units^2.
    (b) $\frac{1}{2} h(x + y + z)$ units^2.
    (c) $g + x + y + z + j$ units^2.
    (d) $\frac{1}{2} (g + j)(x + y + z)$ units^2.

20. Find, to the nearest tenth, the length of the altitude to $\overline{QR}$ in the figure at the right.
    (Lesson 8-4)

21. Paisley needs a piece of fabric cut in the shape at the right. Explain how she could estimate the amount of fabric needed for that piece. *(Lesson 8-3)*

22. How many trapezoids are in the figure for Question 8? *(Lesson 6-5)*

23. $\overline{AC}$ and $\overline{BD}$ are diameters of $\odot E$, as shown at the right. If $m\widehat{AB} = 72°$, what is $m\angle BEC$? *(Lesson 3-2)*

24. *Multiple choice.* $(a + 5)^2 =$
(a) $a^2 + 25$ (b) $a^2 + 5a + 25$
(c) $a^2 + 10a + 25$ (d) none of these
(Previous course)

25. Simplify $\sqrt{27}$. *(Previous course)*

Exploration

26. The state of Nevada is shaped roughly like a trapezoid.
   a. Use the dimensions given below to estimate the area of the trapezoid that includes Nevada.
   b. The Colorado River cuts off a region of southeast Nevada that is also roughly a trapezoid with north-south bases 80 and 120 miles long and an east-west length of 30 miles. Approximate the area of this region.
   c. From your calculations in parts **a** and **b**, estimate the area of Nevada.
   d. From a map or almanac, find the official area of Nevada in square miles. How close to the official area is your estimate from part **c**?

In the southeast region of Nevada is Lake Meade, which is formed by the Colorado River backing up at Hoover Dam shown here.

310 miles

210 miles

Nevada

520 miles

Creating Squares from Right Triangles

IN·CLASS

ACTIVITY

Work in a group of two or three. Your group will need a ruler, protractor, and scissors. You might want to use centimeter grid paper to draw your triangles.

This activity is preparation for a proof of the Pythagorean Theorem in Lesson 8-6.

1 Trace and cut out four copies of right triangle *ABC*.

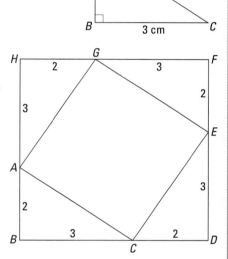

2 Put the copies together so that the legs of the triangles form the quadrilateral *BDFH*, as shown at the right.

3 **a.** Measure the sides and angles of *BDFH*.
   b. What type of figure is it?
   c. What is its area?

4 **a.** Measure the sides and angles of *ACEG*.
   b. What type of figure is it?

5 Compute the area of *ACEG* using the following relationship:
   Area(*ACEG*) = Area(*BDFH*) − Area(△*ABC*) − Area(△*CDE*) − Area(△*EFG*) − Area(△*GHA*).

LESSON 8-6

The Pythagorean Theorem

Harvest scene. *Egyptians used a rope with knots tied at equal intervals for measuring. These ropes were also used for constructing right triangles. See page 469.*

Square Roots

From the area of a square you can determine the length of any of its sides. If the area is A, the length of the side is $\sqrt{A}$. That is why $\sqrt{A}$ is called the **square root** of A. From its definition, $\sqrt{A} \cdot \sqrt{A} = A$.

area = A

side = $\sqrt{A}$

Example 1

The area of the square below is 400 cm^2. What is the length of a side?

400 cm^2

Solution

The length of a side is $\sqrt{400}$. Think: What number multiplied by itself is 400? That number is 20. **The length is 20 cm.**

Check

20 cm · 20 cm = 400 cm^2.

"The First Great Theorem"

The *Pythagorean Theorem* relates the lengths of the three sides of any right triangle.

> **Pythagorean Theorem**
> In any right triangle with legs of lengths a and b and hypotenuse of length c, $a^2 + b^2 = c^2$.

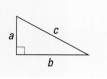

Ancient geometer.
Pythagoras had an acute interest in right triangles.

The Pythagorean Theorem received the name we give it because the Greek mathematician Pythagoras or one of his students proved it in the 6th century B.C. In the western world, this was the earliest proof known until recent times. Today we are aware of many proofs of this result from cultures throughout the world, and it is not clear where or when the first proof originated. The theorem which we call the "Pythagorean Theorem" was known to the Babylonians before 1650 B.C., and possibly known in India before 800 B.C. Howard Eves, a mathematical historian at the University of Maine, has appropriately called the Pythagorean Theorem "the first great theorem" in mathematics.

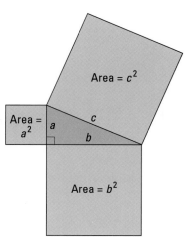

The Pythagorean Theorem can also be interpreted as a theorem about areas. It is called "The Theorem of Three Squares" in Japan.

> **Pythagorean Theorem (alternate statement)**
> In any right triangle, the sum of the areas of the squares on its legs equals the area of the square on its hypotenuse.

A collection of 370 different proofs of the Pythagorean Theorem was compiled by Elisha Loomis in 1940. It includes proofs by the 12th-century Hindu mathematician Bhaskara, by the 15th-century Italian Leonardo Da Vinci (better known as a painter, sculptor, architect, and engineer), by the 19th-century American James A. Garfield (better known as the 20th president of the United States), and many others. These and most early proofs of the Pythagorean Theorem deduced the theorem from the areas of triangles and rectangles. The proof we give in this lesson is of this type. It dates back at least to the year 1733, but is similar to early Chinese proofs. Howard Eves thinks it may have been the proof Pythagoras actually used. Later in this book you will see a completely different proof based on similar triangles.

A Proof of the Pythagorean Theorem

Look back at the In-class Activity on page 465. You should have found that $ACEG$ is a square with area 13. This means that the length of its side is $\sqrt{13}$. The sides of the four right triangles are thus 2, 3, and $\sqrt{13}$. Notice that $2^2 + 3^2 = (\sqrt{13})^2$. The general proof involves the same kind of figure as the In-class Activity, except that it uses a and b instead of 2 and 3.

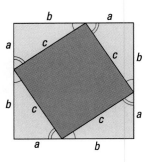

Proof:
Given: A right triangle with legs a and b and hypotenuse c.

To prove: $a^2 + b^2 = c^2$

Argument:
The right triangle and its congruent copies are shown at the left. The outer quadrilateral is a square because each side has length $(a + b)$ and it has four right angles. The purple quadrilateral also is a square because each side has length c and each angle is a right angle. (Do you see why?) Now we find the area of the purple square as you did in the Activity.

Area of purple square = Area(large square) − Area(4 right △s)

Since each side of the large square is $a + b$, its area is $(a + b)^2$.
Each of the four right triangles has area $\frac{1}{2}ab$.

So the purple square has area $(a + b)^2 - 4 \cdot \frac{1}{2}ab = (a^2 + 2ab + b^2) - 2ab$
$$= a^2 + b^2.$$

But the area of the purple square is c^2. So $c^2 = a^2 + b^2$.

The Pythagorean Theorem is useful in many kinds of problems.

Example 2

A conveyor belt carries bales of hay 35 feet up from ground level to a hayloft. The end of the belt is 42 feet from the barn. How far does a bale of hay travel on the conveyor belt?

Solution

The endpoints of the conveyor belt and the point directly below the hayloft determine a right triangle. Call x the length of the path.
$$x^2 = 35^2 + 42^2$$
$$= 1225 + 1764$$
$$= 2989$$
So $x = \sqrt{2989} \approx 54.7$ feet.

Given the lengths of the hypotenuse and a leg of a right triangle, you can use the Pythagorean Theorem to find the length of the other leg.

Example 3

A truck driver uses a portable conveyor belt 13 feet long to unload cartons to a loading dock 5 feet above the truck bed. How far does the conveyor belt extend into the truck?

Solution
From the Pythagorean Theorem,
$$5^2 + YZ^2 = 13^2.$$
$$25 + YZ^2 = 169$$
$$YZ^2 = 144$$
$$YZ = 12 \text{ feet}$$

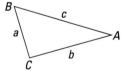

Check
Does $5^2 + 12^2 = 13^2$? Yes, $25 + 144 = 169$.

The Converse of the Pythagorean Theorem

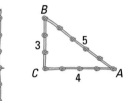

To make a right triangle, the ancient Egyptians took a rope with 12 equally spaced knots in it, and then bent it in two places to form a triangle with sides of lengths 3, 4, and 5. The angle formed by the sides of lengths 3 and 4 was used as a right angle.

Is the angle at C exactly 90°? Here we know that $3^2 + 4^2 = 5^2$, and wonder if the triangle is a right triangle. The converse of a theorem is not necessarily true, but the converse of the Pythagorean Theorem is true.

> **Pythagorean Converse Theorem**
> If a triangle has sides of lengths a, b, and c, and $a^2 + b^2 = c^2$, then the triangle is a right triangle.

Proof
Given: Triangle ABC with sides of lengths a, b, and c and with $c^2 = a^2 + b^2$.

To prove: $\triangle ABC$ is a right triangle.

Drawing:

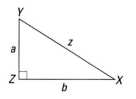

Argument:
Consider right $\triangle XYZ$ with legs of lengths a and b, the same as $\triangle ABC$, and hypotenuse z.

By the Pythagorean Theorem, in $\triangle XYZ$, $a^2 + b^2 = z^2.$
But it is given that $a^2 + b^2 = c^2.$
So, by substitution, $z^2 = c^2.$
Taking positive square roots of each side, $z = c.$

Thus the three sides of $\triangle ABC$ are congruent to the three sides of $\triangle XYZ$. So, by the SSS Congruence Theorem, $\triangle ABC \cong \triangle XYZ$. Since $\angle Z$ is a right angle, $\angle C$ is a right angle by the CPCF Theorem. Thus $\triangle ABC$ is a right triangle.

It is also true that if $a^2 + b^2 \neq c^2$, then a, b, and c cannot be lengths of sides of a right triangle. A proof is given in Chapter 11.

Example 4

A set of three numbers that can be the lengths of sides of a right triangle is called a **Pythagorean triple**. Determine if {11, 20, 23} is a Pythagorean triple.

Solution

Take the two smallest numbers, 11 and 20, and see if the sum of their squares equals the square of the largest number, 23.

$$\text{Does } 11^2 + 20^2 = 23^2?$$
$$\text{No, because } 11^2 + 20^2 = 521$$
$$\text{and } 23^2 = 529$$

The set {11, 20, 23} is not a Pythagorean triple.

QUESTIONS

Covering the Reading

1. If a square has area 225 square meters, how long is a side?

2. Babylonian manuscripts indicate knowledge of the Pythagorean Theorem a thousand years before Pythagoras. About how many years ago were the Babylonian manuscripts created?

3. Explain why the name "Theorem of Three Squares" is an appropriate name for what we call the Pythagorean Theorem.

4. Name two famous people who gave original proofs of the Pythagorean Theorem.

5. Use the figure at the right.
   a. What is the area of each triangle?
   b. What is the area of the large square?
   c. What is the area of the tilted square in terms of a and b?
   d. What is c in terms of a and b?

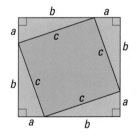

6. To find the length of the hypotenuse of a right triangle, which is quicker for you, the method of Question 5 or the Pythagorean Theorem?

In 7 and 8, find the length of the hypotenuse.

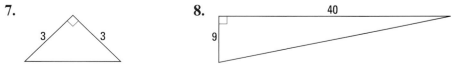

7.

8.

In 9 and 10, find the length of the missing leg.

9.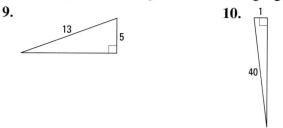

10.

11. Suppose that the hayloft in Example 2 was 20 feet high instead of 35 feet. To the nearest tenth of a foot, how far would the bale travel?

In 12–17, determine if the set is a Pythagorean triple.

12. {3, 4, 5}

13. {14, 8, 17}

14. {25, 24, 7}

15. {40, 9, 41}

16. {1.67, 2.76, 3.33}

17. {2, $2\frac{2}{3}$, $3\frac{1}{3}$}

Applying the Mathematics

18. If a square room has an area of 20 square feet, what is the length of a side of the room, to the nearest inch?

In 19–21, give answers to the nearest tenth. It helps to draw a picture.

19. Felicia walked from her home due north 10 miles, then due east 3 miles. How far was she from home?

20. Find the length of a diagonal of a rectangular field with sides 24 meters and 70 meters.

21. The base of an 8-foot ladder is placed 2 feet away from a wall. How high up the wall will the ladder reach?

22. How long would it take a person to walk around the trapezoidal field pictured here, at a rate of 90 meters per minute? (Hint: Draw a perpendicular line from Q to $\overline{TS}$.)

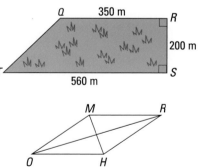

23. Find the perimeter of the rhombus shown at the right if $RO = 48$ and $MH = 14$.

24. One leg of a right triangle is twice the length of the other. How many times as long as the smaller leg is the hypotenuse? (That is, find the value of k below.)

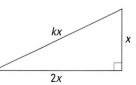

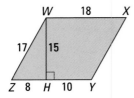

25. Find the area of the parallelogram at the left. *(Lesson 8-5)*

26. Surveyors were hired to find the area of the empty lot △*DNF* below. First, they marked off an east-west line $\overline{EW}$ through *A*. Then they measured the north-south lines and recorded:

segment	$\overline{DC}$	$\overline{NB}$	$\overline{FG}$	$\overline{CB}$	$\overline{BG}$
length (in feet)	80	200	160	66	250

What is the area of the lot? (Hint: Consider the areas of trapezoids in the figure.) *(Lesson 8-5)*

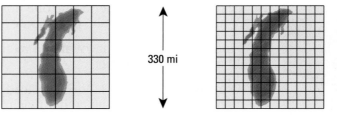

27. Find the area of the triangle of Question 7. *(Lesson 8-4)*

28. Estimate the area of Lake Michigan using the given grid. *(Lesson 8-3)*

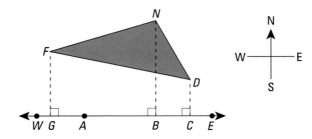

a.   330 mi   **b.**

29. Draw a circle. On the circle, identify an arc with each measure. *(Lesson 3-2)*

a. 90°   **b.** 80°   **c.** 220°

30. For any positive numbers *x* and *y* with $x > y$, the set of three numbers $\{x^2 - y^2, 2xy, x^2 + y^2\}$ is a Pythagorean triple. For instance, if $x = 3$ and $y = 2$, then $x^2 - y^2 = 5$, $2xy = 12$, and $x^2 + y^2 = 13$. Since $5^2 + 12^2 = 13^2$, {5, 12, 13} is a Pythagorean triple. Some Pythagorean triples appear in Questions 12–17. By substituting other values for *x* and *y*, find four other Pythagorean triples not listed there.

*Estimating
the
Circumference
of a Circle*

IN-CLASS

ACTIVITY

Work with another student. You will need a ruler, a compass, and some string.
You can use an automatic drawer if one is available.

In the figures below, regular polygons are inscribed in congruent circles
with radii of 1.5 cm.

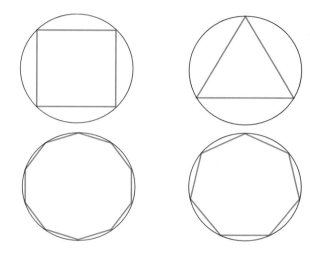

1 Measure the side of each polygon. (An automatic
drawer can also be used to create and measure the
figures above.)

2 Use the information from Step 1 to calculate the
perimeter of each polygon.

3 The perimeter of a circle is called its *circumference.*
Which of the four perimeters is the best approximation
to the circumference of the circle? Why do you think so?

4 Draw a polygon whose perimeter would give you a
better approximation of the circumference than these
four polygons.

5 Below right is a circle congruent
to the ones above. Use a string to
estimate the circumference of the
circle (in centimeters) by placing
the string on the circle and then
measuring its length.

6 Compute $\frac{\text{circumference}}{\text{length of diameter}}.$

Arc Length and Circumference

Driving on an arc. *The streets in Sun City, Arizona, are arranged in concentric circles.*

The word **circumference** is a synonym for *perimeter*. *Circumference* is used with closed curves like circles and ellipses, while *perimeter* is used for polygons and other figures.

When designing a ring for an individual client, a jeweler may measure the circumference of the client's ring finger just above a knuckle. This circumference helps the jeweler to determine the diameter of the finger. Then the jeweler can design and make a ring that will fit.

Relating the Circumference to the Diameter

A procedure like that in the first 4 steps of the In-class Activity on page 473 can be used to estimate the length of any arc of a circle. Recall that a chord is a segment whose endpoints are on a given circle.

At the right, the length of $\widehat{MN}$ is approximated by $MP + PQ + QR + RN$. By drawing more and more chords of increasingly shorter lengths, the total length of the chords approaches the length of the arc as a limit.

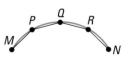

If this is done with an entire circle, the limit is the circumference of the circle. The ratio of the circumference C to the diameter d is equal in all circles. It is denoted by the famous number π, the Greek letter pi. You estimated pi in Step 6 of the In-class Activity.

Definition

$\pi = \frac{C}{d}$, where C is the circumference and d is the diameter of a circle.

The number π is irrational; π cannot be written either as a finite decimal or as a simple fraction. The decimal for π is infinite. Here are the first 50 decimal places.

3.14159 26535 89793 23846 26433 83279 50288 41971 69399 37510 . . .

Most scientific calculators have a key for π which gives the first 6 or 8 places in its decimal approximation. π is about 3.14159 or about $\frac{22}{7}$.

Solving the defining equation $\pi = \frac{C}{d}$ for C gives a formula for the circumference of any circle. Substituting $2r$ for d gives a formula for the circumference in terms of the circle's radius.

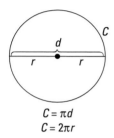

$C = \pi d$
$C = 2\pi r$

Circle Circumference Formula
If a circle has circumference C, diameter d, and radius r, then $C = \pi d$, or $C = 2\pi r$.

Substituting 3.14 for π in the circumference formula gives an estimate for C.

$$C \approx 3.14d$$

Exact and Approximate Answers

When you are asked to give an exact circumference, you should leave π in the answer; for example, in the circle below, $C = 7\pi$. If you are not asked for an exact answer, then you can use an approximation for π, such as 3.14 or the approximation given by your calculator. For the circle below, $C = 7\pi \approx 7 \cdot 3.14 = 21.98$.

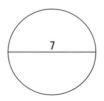

In real situations, the estimate you use for π depends on the accuracy of the data. In Example 1, the given information does not warrant a closer approximation than 3.14. You should not give the impression that your answer is more accurate than the data you are given.

Example 1

The larger bicycle wheel pictured at the left has a diameter of 22 inches. If a rider can get it to go 300 revolutions in a minute, how far will the bike have traveled in that time?

▶

Solution

One revolution moves the bike the length of the circumference of the wheel.

$$C \approx 3.14 \cdot 22 = 69.08 \text{ inches each revolution}$$

The value 69.08 is too accurate for the given information. It has four meaningful digits but the given information has at most two meaningful digits. So round to 69.
In 300 revolutions, the distance traveled is

$$300 \cdot 69 = 20{,}700 \text{ inches.}$$

Dividing by 12 gives the answer in feet. It travels about 1700 ft.

Check

We check by estimating. Since 22″ is a little less than 2 feet and π is a little more than 3, the circumference of the wheel is about 6 feet. In 300 revolutions, the bike should go about 1800 feet. The answer seems reasonable.

Computing Arc Length

The circumference of a circle can be thought of as the arc length of a 360° arc of the circle. Recall that the degree measure of the arc tells you how much of the circumference the arc covers. You can compute the length of an arc if you know the radius of its circle and the degree measure of the arc.

Example 2

In $\odot O$, $OB = 1.3$ cm and m$\angle AOB = 80$. Find the length of $\overset{\frown}{AB}$.

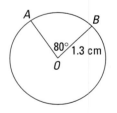

Solution

m$\angle AOB = 80$, so m$\overset{\frown}{AB} = 80°$. Thus $\overset{\frown}{AB}$ covers $\frac{80}{360}$ of the entire circumference of $\odot O$. So:

$$\text{length of } \overset{\frown}{AB} = \frac{80}{360} \cdot C$$
$$= \frac{80}{360} \cdot (2\pi r)$$
$$= \frac{80}{360} \cdot 2 \cdot \pi \cdot 1.3$$
$$= \frac{208\pi}{360} \text{ cm}$$
$$\approx 1.8 \text{ cm.}$$

QUESTIONS

Covering the Reading

1. *Circumference* is a synonym for __?__.

2. Refer to the In-class Activity.
   a. What was your answer in Step 3?
   b. What was your approximation for π in Step 6?

3. Define π.

4. π is often approximated by the fraction __?__ or the decimal __?__.

5. *Multiple choice.* The diameter of a circle is 15″. Which is the exact measure of its circumference?
   (a) $30\pi''$ (b) $15\pi''$ (c) 94.2″   (d) 47.1″

In 6 and 7, use $\odot F$ at the right.

6. **a.** Name all radii shown.
   b. Name all diameters shown.
   c. If $CF = 7$, then $FD = $ __?__.
   d. If $CA = 28$, then $FD = $ __?__.
   e. If $CA = 6x$, then $FC = $ __?__.

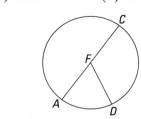

7. If $CA = 8$, find the circumference of the circle
   a. exactly.   **b.** to the nearest tenth.

8. In Example 1, about how far will the bike travel in five minutes if each wheel makes 210 revolutions per minute?

9. In $\odot O$ at the left, $m\overset{\frown}{AB} = 30°$. Find the length of $\overset{\frown}{AB}$
   a. exactly.   **b.** to the nearest hundredth.

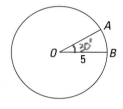

Applying the Mathematics

10. Write how you would explain the difference between arc length and arc measure to a younger person.

11. A circular pond is enclosed by a square wall ten meters on a side. What is the maximum possible distance around the edge of the pond?

12. On the Allen-Bradley Company building in Milwaukee, four clocks face in four different directions. The minute hand on each clock is 20′ long. How far does the tip of each minute hand travel in a day
    a. measured in degrees?   **b.** measured in feet?

13. Suppose it takes you 110 seconds to walk around a circular garden. At this rate, about how long would it take you to walk straight through the garden along a diameter?

14. Gail Ant is crawling along the crust of a large piece of 16″-diameter pizza. The piece has central angle 45°. How much further (to the nearest tenth of an inch) does she crawl than her cousin Kerr Ant who is on a medium piece of 12″-diameter pizza, central angle 45°?

Allen-Bradley clock, Milwaukee, Wisconsin

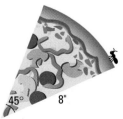

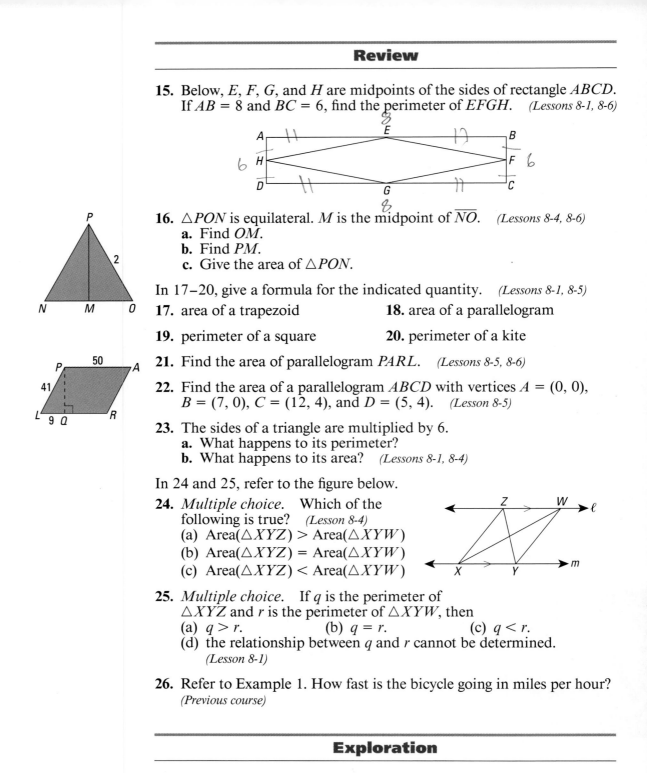

15. Below, E, F, G, and H are midpoints of the sides of rectangle $ABCD$. If $AB = 8$ and $BC = 6$, find the perimeter of $EFGH$. *(Lessons 8-1, 8-6)*

16. $\triangle PON$ is equilateral. M is the midpoint of $\overline{NO}$. *(Lessons 8-4, 8-6)*
  a. Find OM.
  b. Find PM.
  c. Give the area of $\triangle PON$.

In 17–20, give a formula for the indicated quantity. *(Lessons 8-1, 8-5)*

17. area of a trapezoid

18. area of a parallelogram

19. perimeter of a square

20. perimeter of a kite

21. Find the area of parallelogram $PARL$. *(Lessons 8-5, 8-6)*

22. Find the area of a parallelogram $ABCD$ with vertices $A = (0, 0)$, $B = (7, 0)$, $C = (12, 4)$, and $D = (5, 4)$. *(Lesson 8-5)*

23. The sides of a triangle are multiplied by 6.
  a. What happens to its perimeter?
  b. What happens to its area? *(Lessons 8-1, 8-4)*

In 24 and 25, refer to the figure below.

24. *Multiple choice.* Which of the following is true? *(Lesson 8-4)*
  (a) Area($\triangle XYZ$) > Area($\triangle XYW$)
  (b) Area($\triangle XYZ$) = Area($\triangle XYW$)
  (c) Area($\triangle XYZ$) < Area($\triangle XYW$)

25. *Multiple choice.* If q is the perimeter of $\triangle XYZ$ and r is the perimeter of $\triangle XYW$, then
  (a) $q > r$.   (b) $q = r$.   (c) $q < r$.
  (d) the relationship between q and r cannot be determined.
  (Lesson 8-1)

26. Refer to Example 1. How fast is the bicycle going in miles per hour? *(Previous course)*

Exploration

27. a. Using a tape measure or string, measure the circumference of your neck to the nearest half inch or centimeter.
  b. Assuming your neck is circular, use your measurement to estimate your neck's radius.
  c. What would be another way to determine its radius?

*Estimating
the Area of
a Circle*

IN-CLASS
ACTIVITY

Work on this Activity with a partner. You will need a ruler, compass, protractor, and scissors.

1 Trace the circle containing the decagon from the In-class Activity on page 473. Draw in the five symmetry lines that connect opposite vertices. Cut out the 10 wedges and rearrange them horizontally, as started below at the right.

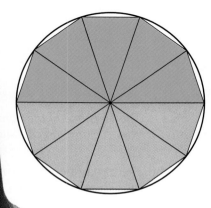

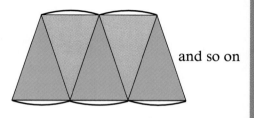

and so on

2 Trace a quadrilateral that approximates the shape you formed in Step 1. What type of quadrilateral is it?

3 Find the area of the quadrilateral you drew in Step 2. How did you calculate this area?

4 How does the area you found in Step 3 compare to the areas found by other students? If there are differences in your answers, what accounts for these differences? Do you think the actual area of the circle is greater than, equal to, or less than the area you calculated?

The Area of a Circle

Dome de dome dome. *Shown is one of the domes of Rowes Wharf, a shopping mall in Boston, as viewed from directly underneath its center.*

A circle is not a polygon. But it can be approximated as closely as you want by a polygon. We could try to get its area by finer and finer grids, using the method in Lesson 8-3. However, it is easier to use sectors, as in the In-class Activity on page 479. A **sector** is a region bounded by two radii and an arc of a circle. Dividing a circle into sectors is similar to triangulating a polygon.

The circle at the left below has radius r. It is split into 16 congruent sectors. Each sector resembles an isosceles triangle with altitude r, but with a curved base. At the right the sectors are rearranged as in the In-class Activity. Together they form a figure that is like a parallelogram with height r. Each base is a union of 8 arcs. So each base of the "parallelogram" has length half the circumference of the circle, or $\frac{1}{2}C$. Since $C = 2\pi r$, $\frac{1}{2}C = \pi r$.

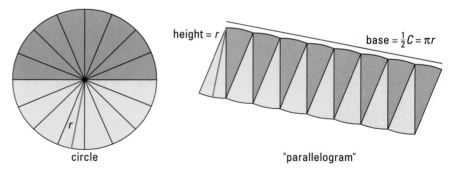

circle   "parallelogram"

As the number of sectors increases, the figure they form closely resembles a parallelogram with base πr and height r. That parallelogram would have area $bh = \pi r \cdot r = \pi r^2$. The areas of the "parallelograms" become better and better approximations to the area of the circle. *We say that the limit of the areas of the parallelograms is the area of the circle.* This gives a famous formula.

Circle Area Formula
The area A of a circle with radius r is πr^2.

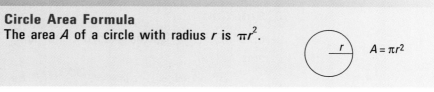

$A = \pi r^2$

Example 1

Find the area (to the nearest square inch) of the top of a manhole cover with diameter 22".

Solution
Since $d = 22"$, $r = 11"$.
Area(circle) $= \pi r^2$
$= 121\pi$
$\approx 121 \cdot 3.14$
≈ 380 square inches

Areas and Probability

The center of a target is called a *bull's-eye.* If an object randomly hits the target, then the probability that it will hit the bull's-eye is

$$\frac{\text{area of bull's-eye}}{\text{area of the target}}.$$

Example 2

In Olympic archery, the target's bull's-eye has a radius of 4 cm and the radius of the outside circle is 80 cm. What is the probability that an arrow shot hitting the target randomly hits the bull's-eye?

Solution

Probability of hitting a bull's-eye $= \dfrac{\text{Area of bull's-eye}}{\text{Area of the target}}$

$= \dfrac{\pi \cdot 4^2}{\pi \cdot 80^2}$

$= \dfrac{16\pi}{6400\pi}$

$= \dfrac{1}{400}.$

Areas of Sectors

Just as arc length is part of the circumference of a circle, the area of a sector is part of the area of a circle.

Example 3

In $\odot O$ at the right, the radius is 15. Find the area of the shaded sector.

Solution

$m\angle AOB = 30$, so the shaded sector is $\frac{30}{360}$ of the area of $\odot O$.

Area of shaded sector $= \frac{30}{360} \cdot$ (Area of $\odot O$)

$$= \frac{1}{12} \cdot \pi(15)^2$$

$$= \frac{225}{12}\pi$$

$$\approx 58.90 \text{ square units.}$$

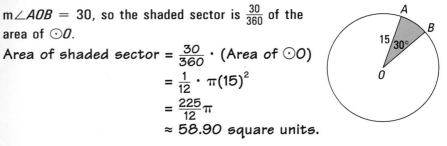

Check

The area of the circle is $\pi \cdot 15^2$, or 225π. $225\pi \approx 706.86$, which is about 12 times 58.90.

QUESTIONS

Covering the Reading

1. At the right, a circle with radius 10 is split into 16 congruent sectors. The 16 sectors can be put together (as in the lesson) into a "parallelogram."
   a. What is the height of the parallelogram?
   b. What is the base of the parallelogram?
   c. What is the area of the parallelogram?

2. Give the exact area of a circle with radius r.

3. Estimate the area to the nearest square inch of a circle with radius 70".

4. A penny has a diameter of approximately 18 mm. Find its area to the nearest square millimeter.

5. A dart board is shown at the right. If a dart randomly hits the target, what is the probability that it lands in the bull's-eye?

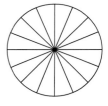

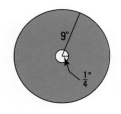

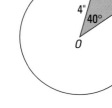

6. Find the area of the shaded sector at the left.

Applying the Mathematics

7. **a.** Give, to the nearest hundred square meters, the area that can be irrigated by a circular sprinkler 60 meters long rotating around a fixed point.
   b. Give the circumference of the irrigated region, to the nearest meter.

8. A circle has area 144π. Find
   a. its radius.　　　　**b.** its diameter.　　　　**c.** its circumference.

9. Refer to the cartoon.
   a. What question do you think is in the Geometry book?
   b. Answer that question.

FOX TROT

10. **a.** On a 10″ pizza, which measures 10″, the radius, diameter, or circumference?
    b. How many times as much of each ingredient is needed for an 18″ pizza than for a 10″ pizza with the same thickness?
    c. If an 18″ pizza is sliced into 8 congruent sectors, what is the area of a slice?

11. Use the concentric circles at the right. The radius of the larger circle is four times the radius of the smaller circle. If a dart lands randomly in the large circle, find the probability it will land
    a. in the inner circle.
    b. in the shaded region.

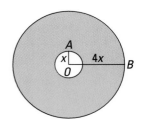

12. In the figure at the left, eight circular metal disks are to be cut out of a 12-cm by 24-cm piece of metal. The rest is not used.
    a. What is the area of the metal that is not used?
    b. What percent of the metal is not used?

12 cm

24 cm

These windmills in Zaanse Schans, Holland, near Amsterdam, are more decorative than functional.

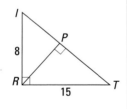

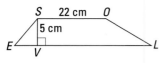

13. In $\odot O$ at the right, $OA = 15$ and $m\angle AOB = 72$.
   a. Find m $\widehat{AB}$.
   b. Find the length of $\widehat{AB}$.   *(Lessons 3-2, 8-7)*

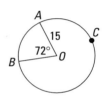

14. A windmill is turning in the wind. If the length of a blade is 12 meters and it makes 6 revolutions in a minute, how far will the tip of each blade travel in an hour?   *(Lesson 8-7)*

15. A car has tires with a diameter of 205 mm. How many revolutions does each tire make in traveling 300 meters?   *(Lesson 8-7)*

16. A rectangle is 16″ by 22″.
   a. Find the lengths of its diagonals.
   b. Find its perimeter.   *(Lessons 8-1, 8-6)*

17. The two legs of a right triangle have lengths $4x$ and $9x$.
   a. What is the area of the triangle?
   b. What is the perimeter of the triangle?   *(Lessons 8-1, 8-4, 8-6)*

18. Can 13, 84, and 85 be the lengths of sides of a right triangle?
   (Lesson 8-6)

19. In the triangle at the left, find RP to the nearest tenth.   *(Lessons 8-4, 8-6)*

20. Find the area of a kite with diagonals 42′ and 12′.   *(Lesson 8-4)*

21. Find the area of the polygon with vertices (4, 3), (4, -3), (-4, -3), and (-4, 3).   *(Lesson 8-2)*

22. Fill in the blanks in the following conversion formulas.   *(Previous course)*
   a. 1 centimeter = _?_ meter   **b.** 1 cm^2 = _?_ m^2
   c. 1 yard = _?_ in.   **d.** 1 yd^2 = _?_ in^2
   e. 1 mile = _?_ yards   **f.** 1 mi^2 = _?_ yd^2

23. *SOLE* is a trapezoid with bases $\overline{SO}$ and $\overline{EL}$. $\overline{SV} \perp \overline{EL}$. If SO = 22 cm, SV = 5 cm, and Area(*SOLE*) = 150 cm^2, find EL.   *(Lesson 8-5)*

24. Find a soft drink can or other can with a circular base.
   a. What is the capacity of the can (measured in ounces or milliliters)?
   b. Measure the diameter d as accurately as you can with a ruler or tape measure.
   c. Measure the circumference C with a tape measure or by rolling it on a ruler, again as accurately as you can.
   d. Calculate $\frac{C}{d}$ to the nearest hundredth.
   e. What number should $\frac{C}{d}$ approximate?
   f. Why isn't $\frac{C}{d}$ exactly that number?

A project presents an opportunity for you to extend your knowledge of a topic related to the material of this chapter. You should allow more time for a project than you do for typical homework questions.

PROJECTS

8

CHAPTER EIGHT

1 Perimeter of Your School

Measure the perimeter of your school. Write a report which includes:

a. a description of how you found the perimeter;

b. your actual measurements and calculations;

c. the perimeter of the school calculated from an actual floor plan; and

d. the actual error, percent of error, and the reasons for any differences between the answers in parts **b** and **c**.

2 Hero's Formula

Let s be half the perimeter of the triangle of sides a, b, and c. Then
$$\text{Area}(\triangle ABC) = \sqrt{s(s-a)(s-b)(s-c)}.$$
This formula was discovered by Archimedes, but it is known as **Hero's** or **Heron's Formula,** after the Greek mathematician and physicist Hero (or Heron) of Alexandria, who lived about 50 A.D.

a. Use Heron's formula to find the area of a triangle with sides 9, 12, and 15. Check by using an alternate method.

b. Find the area of a triangle with sides 10, 17, and 21.

c. Using an automatic drawer, draw three triangles, at least two of them scalene, and verify that the formula holds for these triangles.

3 Broadcast Areas

Radio stations have maps of the areas in which their signals are received. Contact four radio stations with various tower locations, broadcast frequencies, and transmission powers to find out their signal areas. Use the formulas and techniques of this chapter to calculate their broadcast areas. Do research to discover what affects the size and shape of these areas. Which factors are permanent? Which factors change as conditions change? Which other factors affect the size of a region in which a station can be received?

4 Quadrature of the Lune

In Question 18 of Lesson 2-8, you were asked to make a conjecture about the area of a figure like the one below, called a *lune*. The lune below is a region bounded by arcs $\overarc{PQR}$ and $\overarc{PR}$. In about 460 B.C., Hippocrates of Chios proved that the area of $\triangle POR$ was equal to the area of this lune.

a. Use a 90° central angle and form circles

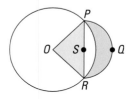

with different radii to calculate the area of the triangle and the corresponding lune. (You will need to find the area of $\triangle POR$ and of sector POR and subtract them to find the area X of the region bounded by $\overline{PR}$ and $\overarc{PR}$. Then find the area of the semicircle with diameter $\overline{PR}$ and subtract X from it to obtain the area of the lune.)

b. Generalize your findings for a circle of any radius.

c. Will the conjecture hold for angles of other sizes? Why or why not? Write a report with your drawings, calculations, and conclusions.

5 Floor Plans and Area

Find at least four advertisements that show a floor plan for apartments or houses. Calculate the area of the floor(s) and the area of each interior space, including hallways, closets, staircase wells, and so on. For each floor plan, show your calculations and write a brief description of how much *living space* (space which is used for sitting, eating, cooking, sleeping) the floor plan provides. Write an introductory paragraph explaining the source of your floor plans and how you determined living space and non-living space. What kind of space is most important in a home? Why do you think so? What percent of the space is used for non-living areas? What advantages and disadvantages do you see in the amount of living space in each floor plan?

6 Area of Puerto Rico

Make three copies of the map of Puerto Rico. Use the method of Lesson 8-3 to estimate its area by using finer and finer grids. Notice the scale in the lower left-hand corner in miles and kilometers.* Compare your answer to the area given in an atlas or almanac. (If a map of your city or town is available with scale of distance given, you may use that instead of the map below.) *You will need to calculate the area of the square of each grid in terms of the scale on the map.

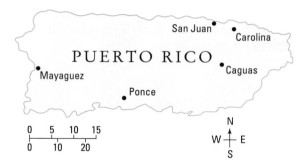

SUMMARY

This chapter is devoted to deriving and applying formulas for area, perimeter, and circumference. Perimeter, or circumference, measures the boundary of a figure. An equilateral n-gon with sides of length s has perimeter ns. A circle with diameter d has circumference πd. The length of an arc is a fraction of that circumference.

In contrast to perimeter, area measures the region enclosed by a figure. This region can be estimated by using congruent squares. With finer and finer grids, even the areas of irregular shapes can be estimated.

Mathematics captivates the imagination of many people because so many things can be derived from just a few simple statements. In this chapter, that idea is exemplified by the derivation of many area formulas from just a few basic properties.

A rectangle with dimensions h and b has area hb. Splitting it with a diagonal, two congruent right triangles are formed. Each has area $\frac{1}{2}hb$. By splitting it into right triangles, the area of *any* triangle can be shown to be $\frac{1}{2}hb$. Putting two triangles together, the area of any trapezoid is $\frac{1}{2}h(b_1 + b_2)$. A special case of a trapezoid is a parallelogram, whose area is hb. (See the List of Formulas at the back of the book for a summary of these and other formulas.)

This chapter contains some of the most important formulas in geometry. Areas of right triangles and squares help to develop the Pythagorean Theorem: In a right triangle with legs a and b and hypotenuse c, $c^2 = a^2 + b^2$. The areas of triangles can be put together to derive the formula $A = \pi r^2$ for the area of a circle.

VOCABULARY

Below are the most important terms and phrases for this chapter. For the starred (*) terms you should be able to give a definition of the term. For the other terms you should be able to give a general description and specific example of each.

Lesson 8-1
*perimeter of a polygon
dimensions of a rectangle
length and width of a
   rectangle
Equilateral Polygon Perimeter
   Formula

Lesson 8-2
square units
Area Postulate:
   Uniqueness Property
   Rectangle Formula
   Congruence Property
   Additive Property
Area(F)
nonoverlapping regions

Lesson 8-3
lattice point

Lesson 8-4
Right Triangle Area Formula
*altitude of a triangle
height of a triangle
Triangle Area Formula

Lesson 8-5
height or altitude of a
   trapezoid
Trapezoid Area Formula
Parallelogram Area Formula

Lesson 8-6
square root
Pythagorean Theorem
Pythagorean Converse
   Theorem
Pythagorean triple

Lesson 8-7
circumference
*π (pi)
Circle Circumference Formula
arc length

Lesson 8-8
sector
Circle Area Formula

PROGRESS SELF-TEST

Directions: Take this test as you would take a test in class. Then check your work with the solutions in the Selected Answers section in the back of this book. You will need graph paper and a calculator in addition to pencil and paper.

1. Give the perimeter of the parallelogram below.

In 2 and 3, give the area of the figure.

2.

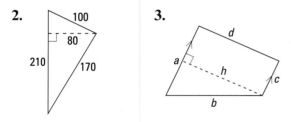

3.

4. If the perimeter of a regular hexagon is q, what is the length of a side?

5. A rectangle has area of 200 m^2 and length 25 m. What is its width?

6. A trapezoid has an area of 48 square units. Its height is 6 units and one base is 9 units. Find the length of the other base.

7. Give the circumference and area of a circle with diameter 12″

  a. exactly.

  b. estimated to the nearest inch and square inch.

8. In $\odot O$ at the right, $\overrightarrow{OC}$ bisects right angle DOB and $OB = 20$.

  a. What is the length of arc $\overarc{CD}$?

  b. What is the area of the shaded sector?

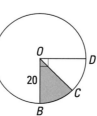

9. Find the length of $\overline{WX}$ in $\triangle WXY$ below.

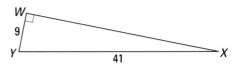

10. The two legs of a right triangle have lengths 20 and 21. What is the perimeter of the triangle?

11. a. Could 11, 60, and 61 be lengths of sides of a right triangle?

  b. Why or why not?

12. In the map and grid of the state of Rhode Island shown below, each small square has side length 5 miles. Use the method of this chapter to approximate the area of the land portion. Do not include the areas of the five small islands.

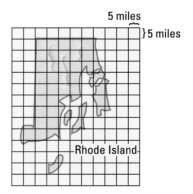

13. $\odot A$ with diameter 12 is contained in square $EFGH$ shown at the right. What is the area of the shaded region?

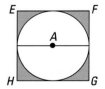

14. The sides of a rectangle are all multiplied by 4.

  a. What happens to its perimeter?

  b. What happens to its area?

PROGRESS SELF-TEST

15. Below, a frame 2″ wide is put around a rectangular painting that is 12″ by 17″. What is the outside perimeter of the frame?

16. A 5-meter ladder is resting on level ground against a wall. If its base is 1.8 meters away from the wall, how high up the wall will it reach? Answer to the nearest tenth of a meter.

17. In October, 1989, an earthquake measuring 6.9 on the Richter scale caused substantial damage in San Francisco, 80 miles from the epicenter of the quake in Santa Cruz, California. To the nearest hundred square miles, how much area was within 80 miles of the epicenter?

Quake's Cost. *The earthquake in San Francisco in October, 1989, caused billions of dollars of damage and destroyed over 100,000 buildings.*

18. A room measures 9 ft by 15 ft. How many square yards of carpeting are needed to cover the floor?

19. A park with perimeter 2640 ft is shaped like a square. What is its area?

20. A Japanese "star" archery target is pictured below. If an arrow hits the target at random, what is the probability it hits the bull's-eye?

21. A triangle on a coordinate grid has vertices (8, 0), (-1, 0), and (0, 1). Find its area.

22. Find the area of the octagon below.

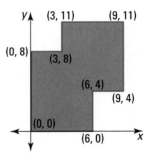

23. Name the cultures in which the Pythagorean Theorem was discovered.

CHAPTER REVIEW

Questions on SPUR Objectives

SPUR stands for **S**kills, **P**roperties, **U**ses, and **R**epresentations. The Chapter Review questions are grouped according to the SPUR Objectives for this chapter.

SKILLS DEAL WITH THE PROCEDURES USED TO GET ANSWERS.

Objective A: *Calculate perimeters of parallelograms, kites, and equilateral polygons given appropriate lengths, and vice versa.* (Lesson 8-1)

In 1–4, give the perimeter of the figure.

1. a kite in which one side has length 10 and another side has length 6

2. a rhombus in which one side has length *t*

3. a regular pentagon in which one side has length 47 meters

4. a square whose area is 324 square feet

5. The perimeter of a rectangle is 28 cm. One side has length 4 cm. What is the length of the other side?

6. If the perimeter of an equilateral triangle is *P*, what is the length of a side of the triangle?

7. *ABCD* is a parallelogram pictured at the right. If its perimeter is 75, what are the lengths of its sides?

8. An equilateral hexagon has perimeter 1. What is the length of each side?

Objective B: *Describe or apply a method for determining the area of an irregularly shaped region.* (Lesson 8-3)

In 9 and 10, each small square is 30 m on a side. Estimate the area of the island (in square meters).

9.   10.

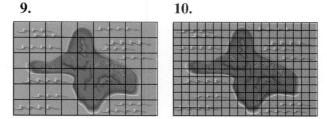

In 11 and 12, estimate the area of the mainland of South Carolina using the given grid.

11.   12.

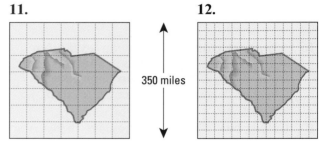

350 miles

Objective C: *Calculate areas of squares, rectangles, parallelograms, trapezoids, and triangles given relevant length of sides, and vice versa.* (Lessons 8-2, 8-4, 8-5)

In 13–16, calculate the area of the figure.

13. rectangle *MOST*

1.3 cm
3.5 cm

14. $\triangle EFG$ at the right in which $GE = 36x$

20x   16x

15. a trapezoid with bases 11 and 13 and altitude 6

16. the triangle with sides of lengths 13, 14, and 15 at the right

13   14
12
15

17. Find the area of the isosceles trapezoid below.

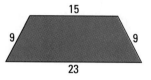

18. Find the area of a square with perimeter 100 feet.

19. A square has area 12.25. What is the length of a side of this square?

20. The bases of a trapezoid have lengths 20 feet and 30 feet and the trapezoid has area 800 square feet. What is the length of the altitude of the trapezoid?

21. $\triangle PQR$ at the right is a right triangle. $\overline{QA}$ is the altitude to $\overline{PR}$. Find QA.

22. Give dimensions of a rectangle with perimeter 200 ft and area less than 100 ft².

Objective D: *Apply the Pythagorean Theorem to calculate lengths and areas in right triangles and other figures.* *(Lesson 8-6)*

In 23 and 24, find the length of the missing side.

23.   **24.**

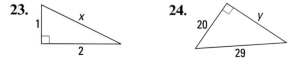

25. The two legs of a right triangle have lengths $6x$ and $7x$. What is its perimeter?

26. The hypotenuse of a right triangle is 50 and one leg is 40. What is the area of the triangle?

27. A rectangle has dimensions 60 cm and 45 cm. What is the length of a diagonal?

28. A square has area 16 square miles. What is the length of a diagonal?

Objective E: *Apply the Pythagorean Converse Theorem.* *(Lesson 8-6)*

In 29–32, could the numbers be the lengths of sides of a right triangle?

29. 8, 31, 32   **30.** 16, 30, 34

31. 1, 2, $\sqrt{3}$ **32.** $\sqrt{7}$, $\sqrt{11}$, 4

Objective F: *Calculate lengths and measures of arcs, the circumference, and the area of a circle given measures of relevant lengths and angles, and vice versa.* *(Lessons 8-7, 8-8)*

33. Give the circumference and area of a circle with radius 10
   a. exactly.
   b. estimated to the nearest hundredth.

34. A circle has area 144π. What is its diameter?

35. A circle has circumference $40x$ meters. What is its radius?

36. In $\odot Q$ at the right, $QY = 11$ cm and m$\angle YQZ = 50$.
   a. Find the length of $\overset{\frown}{YZ}$.
   b. Find the area of the shaded sector.

37. In $\odot O$ at the right, $\overline{BD}$ is a diameter, $OD = 15$, and m$\angle AOD = 20$.
   a. Find the length of $\overset{\frown}{AB}$ to the nearest tenth.
   b. Find the area of the sector bounded by A, O, and D.

PROPERTIES DEAL WITH THE PRINCIPLES BEHIND THE MATHEMATICS.

Objective G: *Relate various formulas for area.*
(Lessons 8-2, 8-4, 8-5, 8-8)

38. Explain how the area formula for a trapezoid is derived from the formula for the area of a triangle.

39. Explain how the area formula for a parallelogram is derived from the formula for the area of a trapezoid.

40. *Multiple choice.* In the figure at the right, Area $(\triangle ABC)$ __?__ Area $(\triangle ABD)$.
   (a) <   (b) =   (c) >

41. In the figure at the right, square *ABCD* has vertices on ⊙*O*. Find an exact value for the area of the shaded region.

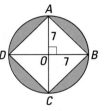

42. Refer to the figure below. Two metal disks are cut out of a 10-m by 20-m rectangle of metal. The leftover metal is recycled. What is the area of the metal that is recycled?

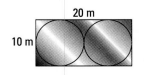

20 m
10 m

43. In the figure below, $\overline{AB}$ is a diameter of ⊙*O* and $\overline{AO}$ is a diameter of ⊙*P*. If *AP* = *x*, find an exact value for the area of the shaded region.

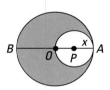

44. *WXYZ* below is a square with segments as marked. Find the area of the shaded region.

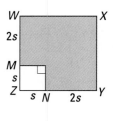

Objective H: *Apply the Pythagorean Theorem and perimeter formulas for parallelograms, kites, and equilateral polygons to real-world situations.* (*Lessons 8-1, 8-6*)

45. If a 20 foot ladder reaches $19\frac{1}{2}$ feet high on a wall, how far away from the wall is the bottom of the ladder?

46. A frame 3 cm wide is put around a rectangular-shaped painting that is 8 cm by 20 cm. What is the perimeter of the outside of the frame?

47. A stop sign is a regular octagon. For safety reasons, the manufacturer of the sign wishes to dull the edges by wrapping them with tape. If one edge of the sign has length *k*, what is the total length of tape needed?

48. A rectangular room is 9′ by 11′. Baseboard is to be put around the room in all places except the 3′ wide door. How many feet of baseboard are needed?

49. How long would it take for a person to walk around the trapezoidal field shown at the right at a rate of 300 feet per minute?

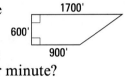

1700′
600′
900′

50. A school bus drops off Dennis and his little sister, Nati, at a corner near their house. To get to their house, Dennis walks north 68 yards along the sidewalk and then west 49 yards along their driveway. Nati cuts directly across the yard to get to the same place. How much farther does Dennis walk?

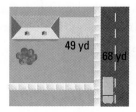

49 yd
68 yd

USES DEAL WITH APPLICATIONS OF MATHEMATICS IN REAL SITUATIONS.

Objective I: *Apply formulas for areas of squares, rectangles, parallelograms, trapezoids, and triangles to real-world situations.* (*Lessons 8-2, 8-4, 8-5*)

51. What is the area of a square park that is 210 meters on a side?

52. A person wishes to tile part of a bathroom floor with 1-inch-square tiles. How many tiles will be needed if the floor is a rectangular region 6′ long and 4′ wide?

53. What is the area of the field shown in Question 49?

54. A triangular piece of fabric is needed for a sail. If the sail is to be 14′ high and is 15′ long at the base, about how much fabric will be used?

Objective J: *Apply formulas for the area and circumference of a circle to real situations.* (Lessons 8-7, 8-8)

55. Due to a chemical spill, authorities evacuated all people within 3 km of the spill. To the nearest tenth of a square kilometer, what was the area of the evacuated region?

56. A car's tire has a radius of 1 foot. How many revolutions will the tire make if the car goes 1 mile? (1 mile = 5280 feet)

57. For his dog, Eugene wants to make a circular play area surrounded by a fence. He has 100 feet of fencing. What is the diameter of the largest play area he can make?

58. In the Olympic free pistol event, a target is 50 meters from the shooter. If a bullet lands at random in the target shown below, what is the probability it lands in the bull's-eye?

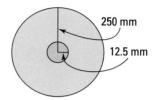

250 mm
12.5 mm

REPRESENTATIONS DEAL WITH PICTURES, GRAPHS OR OBJECTS THAT ILLUSTRATE CONCEPTS.

Objective K: *Determine the areas of polygons on a coordinate plane.* (Lessons 8-2, 8-4, 8-5)

59. Find the area of the decagon below.

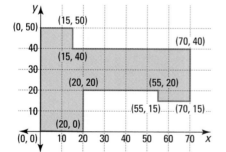

60. A triangle has vertices (7, -4), (-3, -4), and (-1, 11). Find its area.

61. The grid at the right is in unit squares. Find the area of quadrilateral *ABCD*.

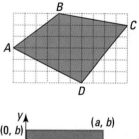

62. Find the area of the quadrilateral with the given vertices at the right. Assume *a*, *b*, and *c* are positive, with $b > c$.

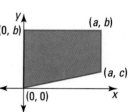

CULTURE DEALS WITH THE PEOPLES AND THE HISTORY RELATED TO THE DEVELOPMENT OF MATHEMATICS IDEALS.

Objective L: *Identify cultures in which the Pythagorean Theorem is known to have been studied.* (Lesson 8-6)

63. *Multiple choice.* In which of these cultures did the mathematician Pythagoras live?
(a) Greek   (b) Chinese
(c) Indian   (d) Babylonian

64. *True or false.* Over 200 different proofs of the Pythagorean Theorem are known.

65. *True or false.* In some cultures, the Pythagorean Theorem has a different name.

66. *Multiple choice.* When was the first known proof of the Pythagorean Theorem?
(a) over 2500 years ago
(b) 1500–2000 years ago
(c) about 1000 years ago
(d) less than 500 years ago

CHAPTER

9

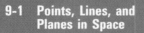

THREE-DIMENSIONAL FIGURES

Everything you touch or see, from paper to pencil, from house to car, from city to planet, does not lie in a single plane. All objects are 3-dimensional. The study of 3-dimensional figures is called *solid geometry*. However, mathematicians often analyze 3-dimensional shapes by looking at their parts in lower dimensions.

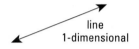

line
1-dimensional

plane
2-dimensional

Just as the 1-dimensional line has the plane as its 2-dimensional counterpart, every 2-dimensional figure has its counterpart in three dimensions.

2-dimensional figures	3-dimensional figures
Polygons	Polyhedral Surfaces
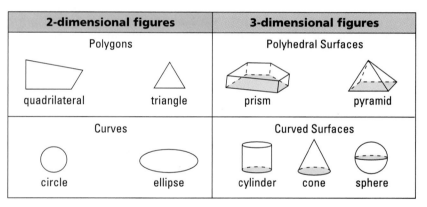	
quadrilateral triangle	prism pyramid
Curves	Curved Surfaces
circle ellipse	cylinder cone sphere

In this chapter, you will learn the important properties of the above basic 3-dimensional figures, how to draw them, and how they are used. You will also see how they are related to the 2-dimensional figures you already know.

One good turn deserves another. *When pushing a panel in a revolving door, you move parts of planes in space about their line of intersection.*

Just as you had ideas about what points and lines should be before we stated postulates about them, you may have ideas about planes. Most of the figures you have seen in the previous chapters lie in a single plane. Their points are coplanar. You can think of a plane as being flat and having no thickness, like a table top that goes on forever. In fact, mathematicians draw a plane on a page the same way as they would draw a tabletop. We name planes with a single capital letter, like X. We use a dashed line to represent parts of a figure "behind" the drawn plane.

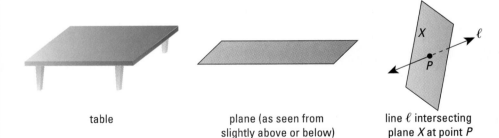

| table | plane (as seen from slightly above or below) | line ℓ intersecting plane X at point P |

Point-Line-Plane Postulate

Recall from Chapter 1 that the three terms *point*, *line*, and *plane* are undefined because we cannot define every word and because, in different geometries, they have different meanings. Our assumptions about points and lines were summarized in Lesson 1-7 in the Point-Line-Plane Postulate, which is repeated here.

Point-Line-Plane Postulate

a. Unique Line Assumption
Through any two points, there is exactly one line.

b. Number Line Assumption
Every line is a set of points that can be put into a one-to-one correspondence with the real numbers, with any point on it corresponding to 0 and any other point corresponding to 1.

c. Dimension Assumption
(1) Given a line in a plane, there is at least one point in the plane that is not on the line.
(2) Given a plane in space, there is at least one point in space that is not in the plane.

To convey the ideas about planes precisely, three new assumptions are added to the Point-Line-Plane Postulate. The remainder of this lesson discusses these assumptions and presents other information about planes.

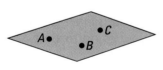

not a plane because of
Flat Plane Assumption

The Flat Plane Assumption conveys the idea that planes have no bumps and go on forever. It implies that segments connecting points do not jump out of planes.

d. Flat Plane Assumption
If two points lie in a plane, the line containing them lies in the plane.

Determining a Plane

The Unique Plane Assumption is sometimes stated: Three noncollinear points determine a plane.

e. Unique Plane Assumption
Through three noncollinear points, there is exactly one plane.

A three-legged stool works on the principle of the Unique Plane Assumption. It is stable on the plane containing the tips of its three legs. A chair with four legs, however, is often unstable. It is wobbly when the tips of its legs are not coplanar.

The stool is stable on the plane
containing A, B, and C.

The chair will wobble unless G lies in
the plane determined by E, F, and H.

Three noncollinear points A, B, and C also determine other figures. For example, they can determine a triangle ABC or $\angle ABC$; a line $\overleftrightarrow{AB}$ and a point C not on it; two intersecting lines $\overleftrightarrow{AC}$ and $\overleftrightarrow{BC}$; and so on. Through any of these figures, there is exactly one plane.

Intersecting Planes

There are, however, infinitely many planes through a single line. Think of an open door in various positions. At each position, a broad side of the door determines a plane through an imaginary line determined by the hinges. Two different planes that have a point in common are called intersecting planes.

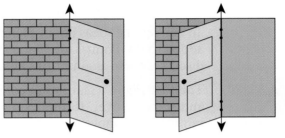

f. Intersecting Planes Assumption
If two different planes have a point in common, then their intersection is a line.

Because lines have no thickness, the Intersecting Planes Assumption implies that planes have no thickness. It also implies that planes go on forever. Even though planes cannot be drawn as if they go on forever, they do.

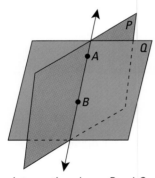

Intersecting planes P and Q
$P \cap Q = \overleftrightarrow{AB}$

QUESTIONS

Covering the Reading

In 1–4, name a 2- or 3-dimensional counterpart for the 1- or 2-dimensional idea.

1. line

2. polygon

3. collinear

4. perpendicular lines

5. Draw a plane as seen from slightly above.

6. Draw two intersecting planes.

7. **a.** When will a three-legged stool rest stably on a floor?
   b. When will a four-legged chair rest stably?
   c. To what part of the Point-Line-Plane Postulate are the answers to parts **a** and **b** related?

Applying the Mathematics

In 8–11, state whether the figure can be contained by exactly one plane.

8. a line and a point not on the line

9. two intersecting lines

10. three noncollinear points

11. a triangle

12. In pioneer days, homes often had dirt floors which were uneven. Would four-legged chairs have been practical? Why or why not?

13. Why is the surface of a table a better model for a plane than the entire tabletop?

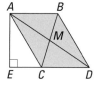

Review

14. In parallelogram $ABDC$ at the left, $AC = 8$, $ED = 10$, and $CD = 6$. Find Area($ABDC$) to the nearest tenth. *(Lessons 8-5, 8-6)*

15. In the figure at the left, D and B are nonoverlapping figures. *(Lesson 8-2)*
    a. *True or false.* Area($D \cup B$) = Area(D) + Area(B)
    b. *True or false.* The perimeter of $D \cup B$ = the perimeter of D + the perimeter of B.

16. Define *reflection-symmetric figure.* *(Lesson 6-1)*

In 17 and 18, give a precise definition. *(Lessons 2-4, 6-7)*

17. circle   18. regular polygon

19. Two lines intersect with angle measures as given in the figure at the left. Find x and y. *(Lesson 3-3)*

20. Three identical cubes used in the British game Crown and Anchor are pictured below. What symbol is on the face opposite the crown? *(Previous course)*

Exploration

21. Tripods are based on the Unique Plane Assumption. But tripods can tip over. When will this happen, and why?

Parallel and Perpendicular Lines and Planes

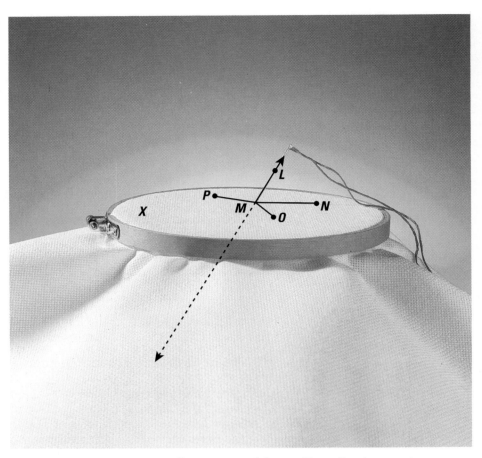

It's a stitch. *A sewing needle piercing a fabric is like a line intersecting a plane. Get the point?*

As stated earlier, virtually all ideas in two dimensions have counterparts in three dimensions. In this lesson, you will examine some ideas in three dimensions related to parallelism and perpendicularity. As in two dimensions, these ideas are related to measures of angles.

Angles Formed by a Line and a Plane

A line that is not in a plane can intersect the plane in at most one point. (If it intersects in two points, then by the Flat Plane Assumption, it lies in the plane.)

When a line intersects a plane, many angles can be formed with lines in the plane. In the picture above, $\overleftrightarrow{LM}$ intersects plane X at M. Three of the angles formed are $\angle LMO$, $\angle LMN$, and $\angle LMP$. The measure of the smallest of all the possible angles defines the angle measure between the line and the plane. This cannot be greater than 90°. When we represent a three-dimensional situation in a picture like the one above, it is difficult to tell if the smallest angle has been drawn.

Recall from Lesson 3-3 that the Leaning Tower of Pisa makes a smallest angle of 85° with the plane of the ground. But, as you could tell by walking around it, it is also possible to find angles from 85° to 95° between the ground and the edge of the tower. There are even lines on the ground that form a 90° angle with the tower, as shown at the left. But the Leaning Tower of Pisa is *not* perpendicular to the ground. We say that it makes an angle of 85° with the ground.

Definition:
If a line ℓ intersects a plane X at point P, then **line ℓ is perpendicular to plane X** ($\ell \perp X$) if and only if ℓ is perpendicular to every line in X that contains P.

Most flagpoles are perpendicular to the ground. This means that the pole is perpendicular to any line on the ground drawn through its base.

Based on the definition, it is very difficult to prove that a line is perpendicular to a plane. Infinitely many lines in the plane through the point of intersection would have to be considered! Fortunately, because of the following theorem, you only need to find two lines in the plane through the point of intersection that are perpendicular to the given line.

The proof shows that if two lines in a plane are perpendicular to a line t through their intersection P, then any other line in that plane through P is perpendicular to t. It involves an ingenious use of congruent triangles.

Line-Plane Perpendicularity Theorem
If a line is perpendicular to two different lines at their point of intersection, then it is perpendicular to the plane that contains those lines.

Proof
Given: Lines ℓ and m are in plane X and intersect at P; $t \perp \ell$; $t \perp m$.

To prove: $t \perp X$.

Drawing:

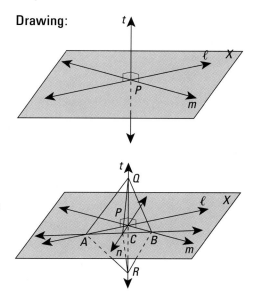

Argument:
Draw n as a line in X containing P. Draw an auxiliary line in plane X intersecting ℓ, m, and n in points A, B, and C, respectively. Choose points Q and R on line t so that they are on opposite sides of X and $PQ = PR$. Connect Q and R to A, B, and C.

▶

Now we proceed to prove four pairs of triangles congruent, one after the other. (You are asked in Question 6 to explain why the triangles are congruent.) Each pair of congruent triangles enables us to deduce that two segments or angles are congruent. Then applying the definitions of ⊥ bisector and of a line ⊥ to a plane finishes the proof.

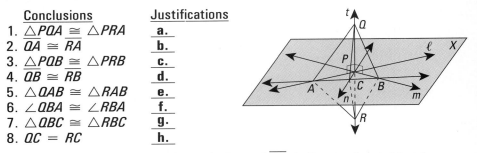

Conclusions	Justifications
1. △PQA ≅ △PRA	**a.**
2. $\overline{QA}$ ≅ $\overline{RA}$	**b.**
3. △PQB ≅ △PRB	**c.**
4. $\overline{QB}$ ≅ $\overline{RB}$	**d.**
5. △QAB ≅ △RAB	**e.**
6. ∠QBA ≅ ∠RBA	**f.**
7. △QBC ≅ △RBC	**g.**
8. QC = RC	**h.**

Since C is equidistant from the endpoints of $\overline{QR}$, it lies on the ⊥ bisector of $\overline{QR}$. P also lies on the ⊥ bisector of $\overline{QR}$. Thus $\overleftrightarrow{PC}$, which is line n, is the ⊥ bisector of $\overline{QR}$, which is on line t. Thus t ⊥ n. Since n can be any line through P in plane X, t is ⊥ to plane X.

In the above proof, we used theorems about figures in a plane to deduce theorems about figures in space. Because the proofs are sometimes quite long, we do not prove every property that we state for 3-dimensional figures. But every property could be proved from the postulates in this book.

One application of the Line-Plane Perpendicularity theorem is as follows. Suppose you wish to make certain that a post is perpendicular to the ground. Then you only have to draw two lines on the ground to the post and make certain that the post is perpendicular to each of those lines.

Parallel and Perpendicular Planes

The ideas of parallel and perpendicular lines have counterparts with planes. Two planes are **parallel planes** if and only if they have no points in common or they are identical. As with parallel lines, the **distance between parallel planes** is the length of a segment perpendicular to the planes with an endpoint in each plane. The **distance to a plane from a point** not on it is measured along the perpendicular segment to the plane from the point. A line is parallel to a plane if they do not intersect in a single point.

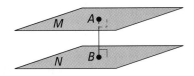

Planes *M* and *N* are parallel.
The distance between them is *AB*.

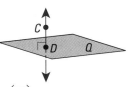

$\overleftrightarrow{CD}$ is ⊥ to plane *Q*.
The distance from
point *C* to plane *Q* is *CD*.

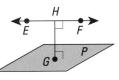

$\overleftrightarrow{EF}$ is // to plane *P*.
The distance from $\overleftrightarrow{EF}$
to plane *P* is *HG*.

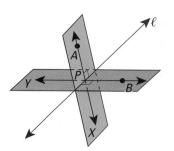

Dihedral Angles and Perpendicular Planes

When two planes intersect, four **dihedral angles** are formed. These dihedral angles are measured in the following way. Suppose the planes X and Y intersect at ℓ. Pick a point P on ℓ. Draw $\overleftrightarrow{PA} \perp \ell$ in X and $\overleftrightarrow{PB} \perp \ell$ in Y. The measures of the dihedral angles are the measures of the four angles formed by $\overleftrightarrow{PA}$ and $\overleftrightarrow{PB}$.

If $\overleftrightarrow{PA} \perp \overleftrightarrow{PB}$, then the planes are perpendicular.

QUESTIONS

Covering the Reading

In 1–3, draw the figure.

1. a line intersecting a plane at a 30° angle

2. a line perpendicular to a plane

3. two perpendicular planes

4. a. *True or false.* An edge of the Leaning Tower of Pisa is perpendicular to the plane of the ground.
  b. Why or why not?

5. When, by definition, is a line perpendicular to a plane?

6. Give the justifications for steps 1–8 in the proof on page 502.

7. The front and back walls of most classrooms are like parts of parallel planes. How is the distance between them measured?

8. Draw a dihedral angle with a measure of 120°.

9. How would you measure the distance from the upper right-hand corner of this page to a wall in front of you?

Applying the Mathematics

In 10–12, think of a classroom having west, east, north, and south walls; a floor; and a ceiling. What in the classroom illustrates each idea?

10. two perpendicular planes

11. three planes each perpendicular to the other two

12. a line perpendicular to a plane

13. Theorem: If two lines in a plane are perpendicular to the same line, then they are parallel. What is a three-dimensional counterpart to this theorem? Is it true?

14. A utility pole is knocked down in a storm. What is the approximate measure of the angle through which the pole falls?

Cast a different light.
Dramatic lighting emphasizes the dihedral angles formed by the ceilings and the walls.

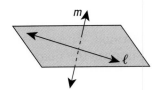

15. Lines in space that are not coplanar and do not intersect, like ℓ and m at the left, are called **skew lines.** Draw a cube and identify two skew lines on it.

16. You are replanting a young tree and you want to be sure its trunk is perpendicular to the ground. From at least how many locations do you need to look to check it? Explain how you will know it is perpendicular.

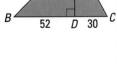

17. The *angle of inclination* of an object S in the sky is the measure of the angle formed by S, the observer O, and the plane of Earth. For instance, a star halfway to overhead has an angle of inclination of 45°.
  a. What is the largest possible angle of inclination?
  b. What is the angle of inclination of a star that is $\frac{1}{3}$ the way from the horizon to directly overhead?

Review

18. What assumed property of points, lines, and planes guarantees that a plane has no bumps? *(Lesson 9-1)*

19. Can two planes intersect in a segment? Why or why not? *(Lesson 9-1)*

20. Given $\odot Q$ at the right with $m\angle Q = 70$ and $QR = 30$ cm. *(Lessons 8-7, 8-8)*
  a. Find the length of $\overgroup{RTS}$.
  b. Find the area of the circle.
  c. To the nearest percent, how much of the interior of the circle lies in the shaded sector?

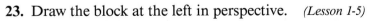

21. Given: $\overline{AD}$ is an altitude in $\triangle ABC$ at the left. *(Lessons 8-4, 8-6)*
  a. Find Area ($\triangle ABC$).
  b. Find AC.

22. Trace the regular hexagon at the right. Find the image of $ABCDEF$ under translation by vector $\vec{v}$. *(Lesson 4-6)*

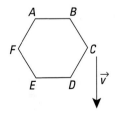

23. Draw the block at the left in perspective. *(Lesson 1-5)*

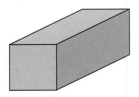

Exploration

24. Examine the legs of a chair at home or in school.
  a. Measure the angle between each leg and the plane of the floor.
  b. Are most of the chair legs you see perpendicular to the floor?

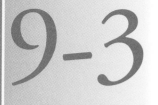
*Prisms and
Cylinders*

Recall that polygons and polygonal regions are different. A polygon is the boundary of a polygonal region. The region is the union of the boundary and its interior.

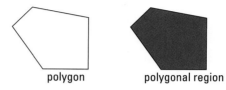

polygon   polygonal region

A similar distinction is made with 3-dimensional figures. In rough language—these are not good definitions—a **surface** is the boundary of a 3-dimensional figure. A **solid** is the union of the boundary and the region of space enclosed by the surface. Earth is a solid; a soap bubble is a surface. A brick is a solid; a carton is a surface. When drawing, you can distinguish a solid from a surface by shading and showing none of the hidden edges.

A carton or box is a surface.   A brick is a rectangular solid.

Boxes

The carton pictured above exemplifies the surface called a *box*. The union of a box and its interior is called a **rectangular solid.** A brick is usually a rectangular solid. Boxes are as important in three dimensions as rectangles are in two, so it is useful to have names for their parts. In the next two paragraphs, refer to the box drawn below.

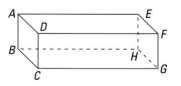

A **box** has six **faces,** each of which is a rectangular region. Some of the faces are drawn as parallelograms to give the appearance of three dimensions. When two of the faces are horizontal, all six faces can be identified by their locations: top ($AEFD$), bottom ($BHGC$), right ($EFGH$), left ($ABCD$), front ($DFGC$), and back ($AEHB$). The **opposite faces** lie in parallel planes. The plane of the front face is perpendicular to the planes

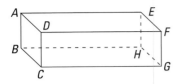

of the top, bottom, and sides. At each corner of the box, three planes meet and each of these planes is perpendicular to the other two. A **cube** is a box whose faces are all square.

The 12 segments, $\overline{AB}$, $\overline{AE}$, $\overline{AD}$, $\overline{BH}$, and so on, are the **edges** of the box. Each edge is perpendicular to two faces. For example, edge $\overline{CG}$ is perpendicular to the left face $ABCD$ and the right face $EFGH$. The endpoints of the edges—A, B, C, D, E, F, G, and H—are the 8 **vertices** of the box.

Most classrooms are like big boxes. The floor and two adjacent walls meet in a bottom corner of the room just as the bottom, front, and left sides of a box meet. The plane of the floor is perpendicular to the plane of each wall.

Classrooms and cartons exemplify the same kind of geometric figure. Even a piece of notebook paper can be thought of as a three-dimensional figure, for it has thickness (about 0.002 inch or 0.05 mm). Thus, the geometric figure that best describes notebook paper is a rectangular solid. Part of the power of geometry is that the same ideas can apply to things as small as parts of atoms or as large as galaxies, as wide as classrooms or as thin as paper.

Cylindric Solids and Surfaces

A rectangular solid is a special type of *cylindric solid*. In general, to form a cylindric solid, begin with a 2-dimensional region. Think of translating the region out of its plane into space in a fixed direction.

Definition
A **cylindric solid** is the set of points between a region and its translation image in space, including the region and its image.

Below, a circular region and a pentagonal region have been translated by the vector $\vec{v}$. Their translation images lie in a plane parallel to the original plane. Note that to appear three dimensional, circles are represented by ovals and the pentagons have also been distorted.

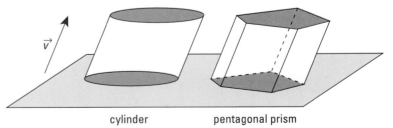

cylinder   pentagonal prism

The original region and its translation image are the **bases** of the cylindric solid. Bases are always congruent and always in parallel planes. The rest of the surface of the solid is the **lateral surface**. The union of the bases and the lateral surface is the **cylindric surface**. The **height** or **altitude** of the solid is the distance between the planes of the bases.

Two cylindric surfaces have special names.

> **Definitions**
> A **cylinder** is the surface of a cylindric solid whose base is a circle.
> A **prism** is the surface of a cylindric solid whose base is a polygon.

When the vector determining the translation is perpendicular to the planes of the bases, a **right cylinder** or **right prism** is formed. A nonright cylinder or prism is called **oblique.** The cylinder pictured on page 506 is oblique; a can is a right cylinder, like the one pictured below.

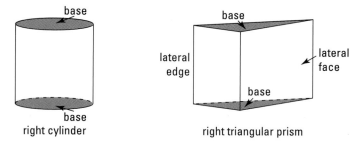

right cylinder right triangular prism

Prisms are also named by their bases. On page 506, an oblique pentagonal prism is shown. A triangular prism has triangles for its bases. The faces of the lateral surface of a prism are called **lateral faces** and they are always parallelograms. The intersection of two lateral faces is called a **lateral edge.**

Tin cans, ceramic tiles, hockey pucks, and new pencils (without erasers) are a few of the physical objects that illustrate cylindrical surfaces. You should be able to sketch any type of cylindric surface.

Activity

Sketch a right pentagonal prism using each of the following two algorithms, shown here for a hexagonal prism.

Algorithm 1

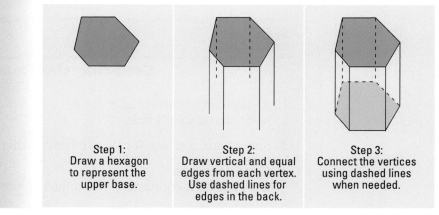

Step 1:
Draw a hexagon to represent the upper base.

Step 2:
Draw vertical and equal edges from each vertex. Use dashed lines for edges in the back.

Step 3:
Connect the vertices using dashed lines when needed.

Algorithm 2

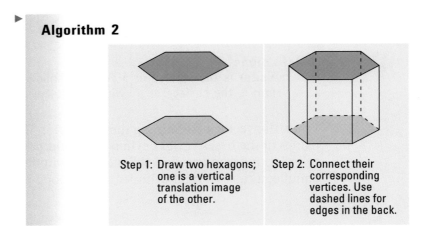

Step 1: Draw two hexagons; one is a vertical translation image of the other.

Step 2: Connect their corresponding vertices. Use dashed lines for edges in the back.

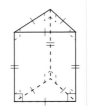

regular triangular prism

To use Algorithm 1 to sketch an oblique prism, you would draw parallel, equal, but nonvertical edges in Step 2. To use Algorithm 2, you would draw the second hexagon using a nonvertical translation vector.

A **regular prism** is a right prism whose base is a regular polygon. At the left is a sketch of a regular triangular prism. Notice that the bases are equilateral triangles, and all the lateral faces are congruent rectangles.

Hierarchy of Cylindric Surfaces

From their definitions, the various special types of cylindric surfaces fit nicely into a hierarchy.

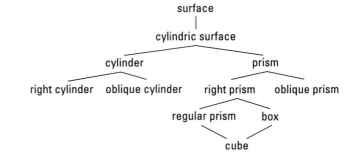

QUESTIONS

Covering the Reading

1. a. Sketch the surface of a brick.
 b. Sketch a solid brick.

2. Show your sketches for the Activity in this lesson.

3. Consider a box.
 a. A segment connecting two adjacent vertices is called a(n) __?__ of the box.
 b. How many lateral faces does the box have?
 c. How many faces does the box have?
 d. How many lateral edges does the box have?
 e. How many edges does the box have?

4. Refer to the box drawn at the right.
   a. Name faces in two parallel planes.
   b. If $AB = 7$, name all other edges that measure 7.

5. Give the special name for the cylindric surface
   a. whose base is a hexagon.
   b. whose base is a circle.

In 6–9, sketch a figure of the given type.

6. right triangular prism
        7. regular octagonal prism

8. right cylinder
        9. oblique cylindric solid

10. Refer to the hierarchy in this lesson. What types of figures is a tin can?

11. What type of figure is both a regular prism and a box?

Tottering twin towers.
The condemned Lindheimer Observatory at Northwestern University leaned dangerously during demolition.

Applying the Mathematics

In 12 and 13, tell which 3-dimensional figure most resembles the real-world object. Give as specific a name as you can, distinguishing solids from surfaces.

12. a CD (ignoring the hole in the middle)

13. an unsharpened pencil without an eraser

In 14 and 15, use the fact that the edges of an oblique prism are not perpendicular to the planes of the bases. Such prisms seem to lean. The amount of lean is measured from the perpendicular, as shown at the left.

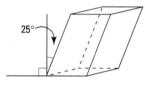

14. Sketch a triangular prism with a 30° lean.

15. Sketch a pentagonal prism with a 45° lean.

In 16 and 17, use the box drawn at the right. $BC = 4$, $AB = 12$, and $CG = 3$.

16. Find the area of
   a. *ABCD.*   **b.** *AEHD.*   **c.** *AEFB.*

17. a. Find the length of $\overline{BG}$.
    (Hint: There are right triangles in this drawing.)
   b. Find the length of $\overline{BH}$.
   c. Find the area of $\triangle BGH$.

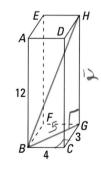

18. Refer to the oblique cylinder at the right.
   a. What is the height of the cylinder?
   b. What is the area of a base?

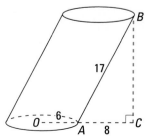

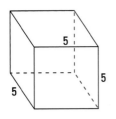

19. A cube is made of sticks, so it is hollow and you can stick your hand through it.
    a. If one edge has length 5, what is the total length of the sticks needed to make the cube?
    b. Generalize part **a** for a cube with edges of length *s*.

20. A computer artist wants to show a box on a computer screen. To do this, the artist has to think of the drawing as 2-dimensional even though it looks 3-dimensional. Given the coordinates of *A*, *B*, *C*, and *D* as shown below, what are the coordinates of *E*, *F*, *G*, and *H*?

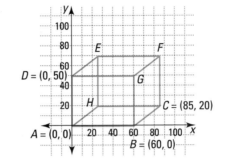

Review

21. Draw two perpendicular planes. *(Lesson 9-2)*

22. Name a 3-dimensional counterpart of intersecting lines. *(Lesson 9-1)*

23. How many planes contain two distinct points *A* and *B*? *(Lesson 9-1)*

24. Ropes are needed to stake a tent pole so that it remains perpendicular to the ground. Each of the ropes is staked into the ground 28 feet from the base of the pole and is 36 feet in length from the ground to where they meet the pole, one foot from the top. In the figure at the left, how tall is the tent pole? *(Lesson 8-6)*

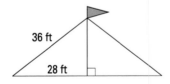

25. Draw a tessellation of the plane using equilateral triangles. *(Lesson 7-6)*

26. Define each term. *(Lesson 1-6)*
    a. *coplanar*
    b. *collinear*

Exploration

27. Some prisms have special properties relative to light. What are these properties?

28. The cells in honeycombs of bees are in the shapes of right hexagonal prisms. Why do bees use this shape?

29. Find two nonmathematical meanings of the word *lateral* in a dictionary. How is the mathematical meaning related to these meanings?

Lateral lateral. *In rugby, a player can pass the ball to a teammate on his or her side.*

Timeless shapes. *Pyramids have been prominent in architecture throughout time, from the pyramids at Giza, Egypt, to the Transamerica Building in San Francisco.*

Pyramids

Pyramids were built in many places in the ancient world as temples and burial sites. One of the wonders of the world is the collection of pyramids constructed by ancient Egyptians. The oldest of the Egyptian pyramids seems to have been designed by Imhotep for the Pharaoh Zoser around 2600 or 2800 B.C. The Transamerica Building in San Francisco, a more recently constructed building shaped like a pyramid, was built in 1972.

These structures are examples of *conic solids.* The boundary of a conic solid is a **conic surface.**

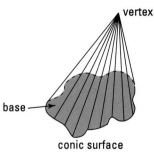

Definitions
Given a region (the **base**) and a point (the **vertex**) not in the plane of the base, a **conic solid** is the set of all points on segments joining the vertex and any point of the base.
A **pyramid** is the surface of a conic solid whose base is a polygon.
A **cone** is the surface of a conic solid whose base is a circle.

For example, a solid pyramid like the pyramids in Egypt is the set of points on segments joining a polygonal region (its base) and a point (its vertex) not in the plane of the region.

The parts of a pyramid are named in the figure at the right. The segments connecting the vertex of the pyramid to the vertices of the base are **lateral edges.** The other edges are called **base edges.** The polygonal regions formed by the edges are the **faces** of the pyramid. All faces other than the base are triangular regions. These triangular regions are the **lateral faces** of the pyramid.

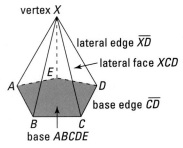

Pyramids are classified by their bases, just as prisms are. There are triangular pyramids, square pyramids, pentagonal pyramids, and so on. If the base of a pyramid is rotation-symmetric and the segment connecting the vertex to the center of symmetry of the polygonal base is perpendicular to the plane of the base, then the pyramid is a **right pyramid.** If a pyramid is not right, it is an **oblique pyramid.** A **regular pyramid** is a right pyramid whose base is a regular polygon. The Egyptian pyramids are regular square pyramids.

Activity

Sketch a right rectangular pyramid using the algorithm below, shown for a regular square pyramid.

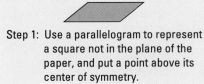

Step 1: Use a parallelogram to represent a square not in the plane of the paper, and put a point above its center of symmetry.

Step 2: Sketch the lateral edges, using dashed lines for the unseen edges.

Cones

From its definition, a cone is like a pyramid in that it has one base and a vertex. But the base of a cone is a circle. The line through the vertex and the center of the circle is the **axis** of the cone. When the axis is perpendicular to the plane of the circle, the cone is called a **right cone.** Otherwise the cone is an **oblique cone.** The surface of a cone other than the base is the **lateral surface** of the cone. A **lateral edge** of a cone is any segment whose endpoints are the vertex and a point on the circle.

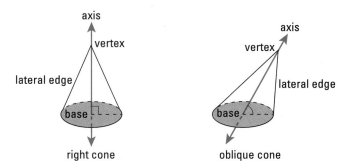

right cone   oblique cone

Buildings of the Muttart Conservatory in Edmonton, Alberta, Canada.

Heights of Pyramids and Cones

The **height** or **altitude** of a pyramid or cone is the distance from the vertex to the plane of the base. The height of a right pyramid or right cone is the length of the segment connecting the vertex to the base's center of symmetry.

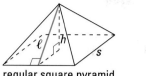

regular square pyramid

height *h*;
slant height ℓ

right cone

Each lateral face of a regular pyramid is an isosceles triangle congruent to all the other lateral faces. The altitude of any of the triangles forming one of the lateral faces of a regular pyramid, measured from the pyramid's vertex, is called the **slant height** of the pyramid (ℓ in the pyramid above). In a pyramid, the slant height is greater than the height but less than the length of a lateral edge. A right cone also has a **slant height**; it is the length of a lateral edge.

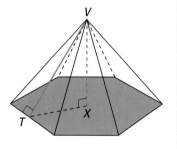

Example

The regular hexagonal pyramid shown at the left has height $VX = 12$. If $TX = 7$, find its slant height to the nearest tenth.

Solution

By the Pythagorean Theorem,
$$VT^2 = TX^2 + VX^2$$
$$VT^2 = 7^2 + 12^2$$
$$VT^2 = 193.$$
$$VT = \sqrt{193}$$
$$VT \approx 13.9 \text{ units}$$

Hierarchy of Conic Surfaces

Below is a hierarchy of conic surfaces. Compare this hierarchy with the hierarchy of cylindrical surfaces. They are very much alike.

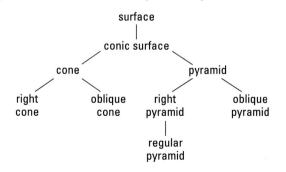

Notice the similarities between cones and pyramids, and between cylinders and prisms. In the next chapter, you will see similarities in the formulas for the volumes and surface areas of these figures.

QUESTIONS

Covering the Reading

1. Approximately how many centuries elapsed between the building of the oldest known Egyptian pyramid and construction of the Transamerica Building?

2. A regular square pyramid is sketched at the right. Identify
   a. its base.
   b. its vertex.
   c. a lateral edge.
   d. a lateral face.
   e. an edge of the base.

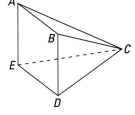

3. a. Show your answer to the Activity in this lesson.
   b. Draw an oblique triangular pyramid.

4. O is the center of the circle drawn at the right, and $\overline{MO}$ is perpendicular to the plane of the circle.
   a. What type of surface is shown?
   b. Name its axis.
   c. Name a lateral edge.
   d. Name its vertex.
   e. Name the base.
   f. What is its height?
   g. What is its slant height?

In 5–7, what type of surface is drawn?

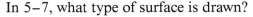

5.   6.   7.

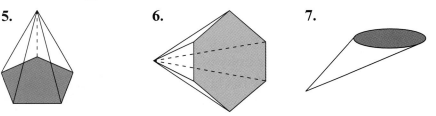

8. Sketch a cone with a base having radius 2 cm.

9. a. Sketch a regular octagonal pyramid.
   b. On your sketch, identify the height and the slant height.

10. In the regular pentagonal pyramid at the left, $RQ = 4$ and $PQ = 10$.
    a. Find its height.
    b. Find its slant height.

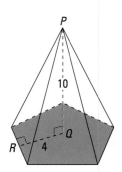

11. Given a right square pyramid, arrange each part in order from smallest to largest: the length of a lateral edge, the height of the pyramid, the slant height.

12. Suppose the base of a pyramid is a hexagon.
   a. How many lateral faces does the pyramid have?
   b. How many faces does the pyramid have?
   c. How many lateral edges does the pyramid have?
   d. How many edges does the pyramid have?

13. Repeat Question 12, but suppose the base is an *n*-gon.

14. In the early 20th century, Chinese officials wore conical hats as part of their summer uniform. Suppose the height of a conical hat is 17 cm and the radius of the base is 33 cm.
   a. Find the slant height.
   b. Find the circumference of its base.

In 15 and 16, a conic surface or conic solid has been cut by a plane parallel to its base. The part of the conic surface between and including the two parallel planes is called a **truncated surface.**

truncated cone

truncated pentagonal pyramid

15. Where in a circus might you find a truncated cone?

16. Draw a truncated hexagonal pyramid. (Hint: Think of the original vertex as a vanishing point.)

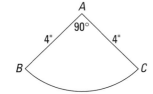

17. Cut a quarter disk out of a piece of paper with dimensions as shown at the left. Roll the paper so that $\overline{AB}$ and $\overline{AC}$ coincide and a cone is formed.
   a. What is the circumference of the base of this cone?
   b. What is the radius of the base of the cone (to the nearest tenth of an inch)?

18. Suppose you can see only the edges of a right square pyramid and they glow in the dark. Where are you in relation to the pyramid if you see it as shown?

   a.
   b.
   c.

19. Draw an oblique rectangular prism. *(Lesson 9-3)*

20. Pictured at the left is a prism. Explain why *ABED* is a parallelogram. *(Lesson 9-3)*

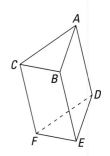

21. What is the difference between a right cylinder and an oblique cylinder? *(Lesson 9-3)*

In 22 and 23, refer to the cube at the right. *AT* = 5. *(Lessons 8-2, 8-6, 9-3)*

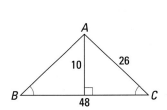

22. Find the indicated length.
   a. *AB*   **b.** *EA*   **c.** *ET*

23. What is the sum of the areas of all six faces?

24. How many different planes contain the indicated figure?
   a. three given noncollinear points
   b. a given △*ABC*
   c. a given line *(Lesson 9-1)*

25. Refer to △*ABC* at the right.
   a. Find the area of △*ABC*.
   b. Find the length of the altitude to $\overline{AC}$.
   (Lesson 8-4)

A triangular prism by Firth of Lorne near Oban, Argyll, Scotland.

26. Rectangle *F* has double the length and one-third the width of rectangle *G*. How do their areas compare? *(Lesson 8-2)*

Exploration

27. The pyramids of Egypt were called by ancient Greek writers one of the "seven wonders of the world."
   a. What were the other wonders of the world for those Greek writers?
   b. Identify a structure that has been called one of the wonders of today's world.

Planet Earth. *This famous photograph of Earth taken from the moon is dramatic confirmation that Earth is nearly spherical in shape.*

Ancient Greek astronomers deduced that Earth was a sphere based on the fact that Earth casts a consistent round shadow on the moon during eclipses of the moon. Nevertheless, many Europeans in the Middle Ages and Renaissance believed the world was flat. That belief persisted until 1522, when the voyage begun by Ferdinand Magellan, a Portuguese navigator, was completed, heralding the first circumnavigation of Earth.

The Sphere

A 3-dimensional counterpart of the circle is the *sphere*.

> **Definition**
> A **sphere** is the set of points in space at a certain distance (its radius) from a point (its center).

To draw a solid sphere (such as an orange or a baseball), shade the drawing as shown below at the left. To draw the surface (such as a beach ball or a bubble), draw only an outline, as shown below at the right, with arcs added to give the illusion of depth.

solid sphere

sphere

The terminology of circles extends to spheres. On page 517 the sphere at the right has center O and radius OA. The segments $\overline{OA}$ and $\overline{OB}$ are also called **radii.** Similarly, the **diameter** of a sphere is a number that is twice the radius, while any segment connecting two points of the sphere and containing the center of the sphere is a diameter.

Small Circles and Great Circles

Two types of intersections can occur with the intersection of a sphere and a plane (see the figures below). The intersection is a point if the plane just touches the sphere; otherwise it is a circle. (You are asked to prove that the intersection is a circle in Question 18.) If the plane contains the center of the sphere, the intersection is called a **great circle** of the sphere. Otherwise, the intersection is called a **small circle.**

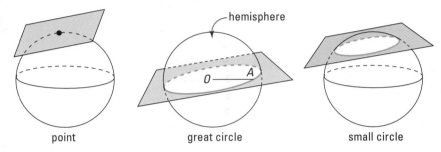

point   great circle   small circle

A great circle separates the sphere into two **hemispheres** which may or may not contain the great circle itself.

Earth as a Sphere

Earth is almost a solid sphere. One of its great circles is the equator, which separates the Northern Hemisphere from the Southern Hemisphere. Points on the equator are about 6378 km (or 3963 miles) from the center of Earth. However, Earth has been slightly flattened by its rotation. The North and South poles are about 6357 km (or 3950 miles) from the center of Earth.

Below are two sketches of Earth. The sketch at the left is of Earth as seen from slightly north of the equator. Notice how the oval representing the circle of the equator is widened to give the illusion of looking at it from above. In this sketch the South Pole cannot be seen. The sketch at the right shows Earth as seen from the plane of the equator.

Earth as seen from
above the equator and North Pole

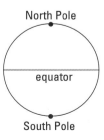

Earth as seen from
plane of the equator

Marble I *by Charles Bell is computer-generated art.*

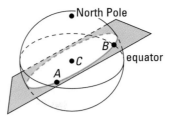

North Pole

Consider two points *A* and *B* on Earth. Since there is exactly one plane that contains three noncollinear points, if *A* and *B* are not endpoints of a diameter, there is a single plane containing *A*, *B*, and the sphere's center *C*. The intersection of that plane with the sphere is a great circle. Aircraft and ocean liners often travel along an arc of a great circle because it is the shortest route from *A* to *B*. The path they take is called a **great circle route** from *A* to *B*.

In about 230 B.C., the Greek mathematician Eratosthenes estimated the circumference of Earth in the following way. He noticed that at noon on the summer solstice (around June 21st), the Sun was directly overhead at Syene. He knew that Alexandria was about 5000 stades due north of Syene. (The stade was a Greek unit of length.) In another year, he calculated that at noon on the summer solstice at Alexandria, the Sun was 7.2° away from overhead.

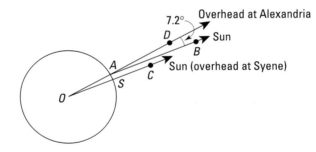

Today, Alexandria, Egypt is a city of over 3.3 million people.

Because Alexandria is due north of Syene, Eratosthenes could be certain that the distance between the cities is the great circle distance. He assumed that the Sun was far enough away that rays from anywhere on Earth to it would be parallel. Thus, using the above figure, $\overrightarrow{AB}$ // $\overrightarrow{OC}$. Consequently, due to the Corresponding Angles Postulate, m∠O = 7.2. Thus the distance from Alexandria to Syene is $\frac{7.2}{360}$ of the circumference of Earth. Let *C* be the circumference. Then

$$\frac{7.2}{360}C = 5000.$$
$$C = 250{,}000 \text{ stades}$$

We think one stade was equal to about 517 feet. This means that 250,000 stades equals about 24,500 miles. Today, with accurate readings taken from space, we know that the circumference of a north-south great circle is about 24,860 miles. The circumference at the equator is about 24,900 miles. The value obtained by Eratosthenes was within 2% of the actual value.

Plane Sections

A great circle is an example of a *plane section* of a figure.

> **Definition**
> A **plane section** of a three-dimensional figure is the intersection of that figure with a plane.

Biologists use plane sections of body tissue to study organs and blood. These sections are thin enough so that light from a microscope can shine through them. Doctors use special scanners to take x-rays of plane sections of the human body for diagnostic purposes. In geometry, two questions about plane sections arise naturally: What are the plane sections of a figure? How can plane sections be sketched?

We have already answered these questions for a sphere. For a solid prism or cylinder, there are different plane sections, depending on whether the intersecting plane

1. is parallel to the bases,
2. is not parallel to and does not intersect a base or bases, or
3. intersects one or both bases.

If the intersecting plane is parallel to the bases, then the section is a region congruent to the bases.

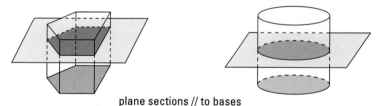

plane sections // to bases

Suppose the intersecting plane is not parallel to and does not intersect the bases. For a prism, the section will be a polygonal region of the same number of sides but not congruent to the base. For a cylinder, the section will be an *ellipse* and its interior.

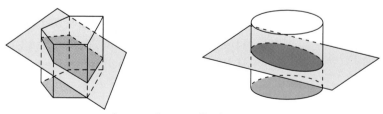

plane sections not // to bases

When a plane intersects a base as well as lateral faces, a variety of sections are possible. An example of a plane intersecting a solid pentagonal prism is shown below at the left. The section is a triangle. An example of a plane intersecting a solid cylinder is shown at the right. The section is a rectangle.

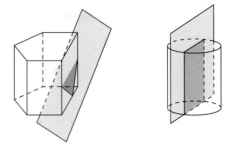

plane sections intersecting one or both bases

Activity 1

Without tracing, make drawings like the three plane sections of prisms shown above and on page 520, but with a *solid right quadrangular prism* (one with a quadrilateral base).

Activity 2

Without tracing, make drawings like the three plane sections of cylinders shown above and on page 520.

Plane Sections of Pyramids and Cones

For pyramids and cones, sections parallel to the base have shapes similar to the base, but they are smaller. You can sketch them by drawing segments or arcs parallel to the base.

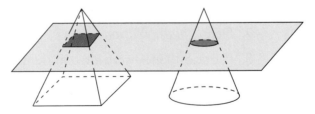

Conic Sections

Notice the figures at the top of page 522. At the far left are two right cones with the same axis, formed by rotating a line intersecting the axis about that axis. The plane sections formed are called the **conic sections.** The conic sections describe orbits of planets and paths of balls and rockets. Lenses and reflectors with these shapes have focusing properties used for long-range navigation (LORAN), telescopes, headlights, satellite dishes, flashlights, and whispering chambers. You are certain to study them in other mathematics courses.

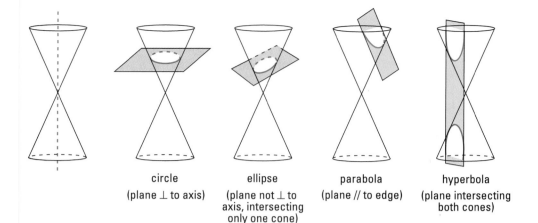

circle	ellipse	parabola	hyperbola
(plane ⊥ to axis)	(plane not ⊥ to axis, intersecting only one cone)	(plane // to edge)	(plane intersecting both cones)

QUESTIONS

Covering the Reading

1. Define *sphere*.

2. How does the definition of sphere differ from the definition of circle?

3. The intersection of a plane and a sphere is either a single point or a(n) __?__.

4. *Multiple choice.* If you think of Earth as a sphere, then the equator is
   (a) a diameter.   (b) a small circle.
   (c) a great circle.   (d) a chord.

5. **a.** What is the length of a diameter of Earth at the equator?
   b. What is the length of a diameter of Earth connecting the poles?
   c. Why are these different numbers?

6. What is meant by a *great circle route*?

7. Define *plane section*.

8. If Eratosthenes were alive in the United States today, he could repeat his calculation of the circumference of Earth. Wichita, Kansas, is 345 miles due north of Fort Worth, Texas. At a given moment, the sun is 5° lower in the sky in Wichita than Fort Worth. From this information, what estimate would Eratosthenes obtain for the Earth's circumference?

9. Show your drawings from Activity 1.

10. Show your drawings from Activity 2.

11. **a.** Draw a right square pyramid and a plane section parallel to its base.
    b. How do the section and the base compare?

12. **a.** Name the four types of conic sections.
    b. Name three situations where conic sections are used.

In 13–15, copy the figure shown.
a. Sketch a plane section parallel to the base(s).
b. Sketch a plane section not parallel to and not intersecting the base(s).
c. Name the shape of each plane section.

13.   **14.**   **15.**

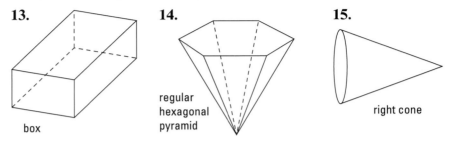

box   regular
hexagonal   right cone
pyramid

16. Slicing a grapefruit with a knife is a model for a plane intersecting a solid sphere. Describe the possible plane sections.

In 17 and 18, tell which 3-dimensional geometric figure most resembles each real-world object. Distinguish solids from surfaces.

17. table tennis ball   **18.** peach

19. Here is a proof that the intersection of a sphere and a plane not through its center is a circle. It is given that sphere O and plane M intersect in the curve as shown below. Fill in the missing justifications.

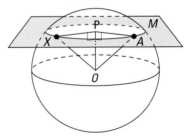

Argument:
Let P be the foot of the $\perp$ from point O to plane M. Let A and X be two points on the intersection. Then, by definition of a line perpendicular to a plane, $\overline{OP} \perp \overline{PA}$ and $\overline{OP} \perp \overline{PX}$. $\overline{OP} \cong \overline{OP}$ because of the **a.** and $\overline{OA} \cong \overline{OX}$ because **b.** . So $\triangle OPX \cong \triangle OPA$ by **c.** .

Thus, due to **d.** , $PX = PA$. Thus any point X on the intersection lies at the same distance from P as A does. So by the definition of circle, the intersection of sphere O and plane M is the circle with center P and radius PA.

20. A quadrangular pyramid has a quadrilateral as its base. Draw a quadrangular pyramid. *(Lesson 9-4)*

21. In the regular square pyramid at the right, $\overline{AX}$ is the height and F is the midpoint of $\overline{BC}$. If $AX = 15$ and $CD = 16$, find
a. XF.
b. Area($\triangle ABC$). *(Lessons 8-4, 9-4)*

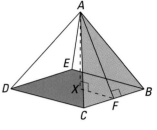

In 22 and 23, use the box at the right. $AB = 10$, $AE = 6$, and $AD = 8$.
(Lessons 8-2, 9-3)

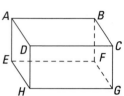

22. Find each value. (Some segments are not drawn.)
a. CD **b.** DE **c.** CE

23. Find the sum of the areas of the six faces.

24. Use the information in this lesson to give the circumference of the equator: **a.** to the nearest 100 miles **b.** to the nearest 100 kilometers. *(Lesson 8-7)*

25. Describe the location of the reflection image of a point P over a line ℓ when P is not on ℓ. *(Lesson 4-1)*

26. a. It is possible for a plane section of a cube to be a hexagon. Draw a cube or make a model to demonstrate how this is possible.
b. Is it possible for a plane section of a cube to be a pentagon?
c. Is it possible for a plane section of a cube to be a triangle?
d. Is it possible for a plane section of a cube to be a quadrilateral that is not a parallelogram?

27. On Earth, the union of the prime meridian and International Date Line approximates a great circle containing the North and South Poles.
a. Name a country through which the prime meridian passes.
b. Name the ocean that contains the International Date Line.
c. Into what hemispheres does this great circle separate Earth?
d. What is the purpose of this circle?

28. What point on Earth is opposite where you live?

*Reflection
Symmetry
in Space*

El Castillo de la Maya. *This great pyramid is part of the ruins of the ancient Mayan city Chichén Itzá located in present-day Yucatan, Mexico. This pyramid has four symmetry planes.*

Reflections in Space

The properties of reflections in two dimensions carry over to three dimensions. For example, when you look in a mirror, the mirror appears to lie halfway between you and your image. Also, an imaginary line from the tip of your nose to its image will always be perpendicular to the mirror. So, as with 2-dimensional figures, reflection symmetry in 3-dimensional figures is a powerful way of developing their properties.

In general, a plane M is the **perpendicular bisector** of a segment $\overline{AB}$ if and only if $M \perp \overline{AB}$ and M contains the midpoint of $\overline{AB}$. This enables 3-dimensional reflections (over planes) to have the same defining condition as their 2-dimensional counterparts (over lines).

Definition:
For a point P which is not on a plane M, the **reflection image of P over M** is the point Q if and only if M is the perpendicular bisector of $\overline{PQ}$. For a point P on a plane M, the **reflection image of P over M** is P itself.

Reflections over planes preserve the same properties as their 2-dimensional relatives: angle measure, betweenness, collinearity, and distance. So the definition of congruence in the plane can be extended to three dimensions.

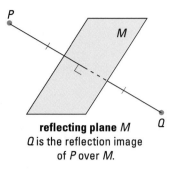

reflecting plane *M*
Q is the reflection image
of *P* over *M*.

Definition
Two three-dimensional figures *F* and *G* are **congruent figures** if and only if *G* is the image of *F* under a reflection or composite of reflections.

As in two dimensions, reflections switch orientation. Figures with the same orientation are **directly congruent.** Figures with different orientation are **oppositely congruent.** You are oppositely congruent to your image in a mirror.

Symmetry Planes

A 2-dimensional figure is reflection-symmetric if and only if it coincides with its image under some reflection. There is a corresponding definition for three dimensions.

Definition
A three-dimensional figure *F* is a **reflection-symmetric figure** if and only if there is a plane *M* (the **symmetry plane**) such that $r_M(F) = F$.

Both nonliving and living natural objects possess reflection symmetry. Symmetry allows rock collectors, chemists, and geologists to identify types of crystals. In animals, reflection symmetry is sometimes called **bilateral symmetry.**

crystal

beetle

The body surfaces of most people have approximate bilateral symmetry. The right side and left side are reflection images of each other over the plane that goes through the middle of the body from head to toe. Note that you cannot slide or turn a right hand onto a left hand, because they have different orientation.

Space figures may have any number of symmetry planes, from zero to infinitely many. The right cylinder below at the left has infinitely many vertical symmetry planes (any plane containing $\overline{PQ}$) and one horizontal symmetry plane. The regular triangular pyramid at the right has exactly three symmetry planes, one of which is drawn.

M is the ⊥ bisector of $\overline{PQ}$.   P is the ⊥ bisector of $\overline{AB}$.

Any right prism will have as many symmetry planes as the base has symmetry lines, plus an additional symmetry plane parallel to the bases. A right pyramid has as many symmetry planes as the base has symmetry lines. Regular prisms and regular pyramids have the most symmetry planes for their shapes.

Example

Determine the number of symmetry planes for a common cardboard box, like that shown here. Sketch one of the symmetry planes.

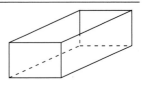

Solution

A box is a right rectangular prism, so the box has one symmetry plane for each symmetry line of its base, plus an additional plane parallel to the bases.

One symmetry plane is shown below.

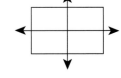

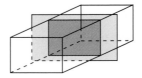

Activity

Trace the box above. Sketch the other two symmetry planes.

QUESTIONS

Covering the Reading

1. What properties of a figure are preserved by reflections in space?

2. *Multiple choice.*   If a human stands upright, the symmetry plane is:
   (a) // to the ground about waist high.
   (b) ⊥ to the ground, halfway between the front and back.
   (c) ⊥ to the ground, halfway between right and left sides.

3. How many symmetry planes does a box have?

In 4–7, a figure is given. **a.** Tell if the figure has reflection symmetry.
b. Indicate the number of symmetry planes.

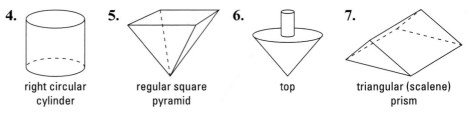

4. right circular cylinder   5. regular square pyramid   6. top   7. triangular (scalene) prism

8. Display your drawings from the Activity in this lesson.

9. How many symmetry planes does a cube have?

Applying the Mathematics

10. A symmetry plane of a regular triangular pyramid is drawn in this lesson.
    a. Draw this pyramid with another of its symmetry planes.
    b. Describe the plane section formed by your symmetry plane.

11. Draw the symmetry plane(s) of this oblique cylinder.

12. Is it possible for a prism to have only one plane of symmetry? Explain why or why not, and sketch one if it is possible.

13. *True or false.*   The cross section formed by any symmetry plane of a regular prism is a parallelogram. Explain your answer.

In 14 and 15, copy the figure shown.
a. Sketch a plane section parallel to the base.
b. Sketch a plane section not parallel to and not intersecting the base(s).
c. Name the shape of each section. *(Lesson 9-5)*

14.

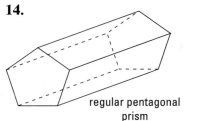

regular pentagonal
prism

15.

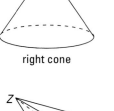

right cone

16. In the figure at the right, *FOUR* is a square. *Z* is not in the plane of *FOUR*.

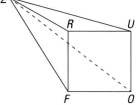

  a. Identify the figure.
  b. Name its base.
  c. Name its vertex.
  d. Name all the lateral edges.
  e. Name all of its faces.   *(Lesson 9-4)*

17. A regular square pyramid has a slant height of 13 and a base with side length 10.
  a. Draw and label a diagram of the pyramid.
  b. What is the height of the pyramid?   *(Lesson 9-4)*

In 18 and 19, tell which figure studied in this chapter most resembles the real-world object. Give as specific a name as you can, distinguishing solids from surfaces.   *(Lesson 9-3)*

18. a penny

19. a desk drawer

20. A cylinder has a base with area of 20.25π. Find the circumference of the base.   *(Lessons 8-7, 9-3)*

21. Find the area of $\triangle PQR$.   *(Lessons 8-4, 8-6)*

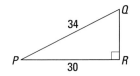

The cat's meow. *This pet carrier in the shape of a hexagonal prism provides creature comforts for the cat that loves to travel. How many symmetry planes does it have?*

22. Given that the figure at the right is a cube, write an argument that proves the following statements in order. *(Lessons 6-3, 7-8, 9-3)*

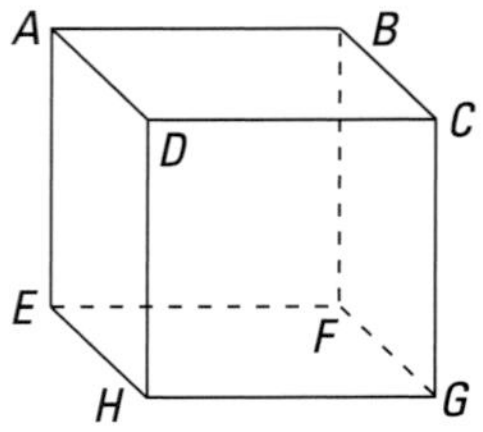

   a. *BE* = *CH*
   b. *BCHE* is a parallelogram.
   c. *EG* = *HF*
   d. *BH* = *CE*
   e. *BCHE* is a rectangle.

23. What qualities of a good definition are violated by the following definition of *Equator* from *Webster's II New Riverside Dictionary*? *(Lesson 2-4)*

   Equator: The great circle circumscribing the Earth's surface, the reckoning datum of latitudes and dividing boundary of Northern and Southern Hemispheres, formed by the intersection of a plane passing through the Earth's center perpendicular to its axis of rotation. *(Lesson 2-4)*

Exploration

---

24. Standing upright and still, you want to see the front of your entire body, head to toe, in a mirror. How far up and down the wall does the mirror have to go? (Hint: Try with an actual mirror. Block out the parts of the mirror that are not needed.)

25. Find a picture of your face, taken from the front. Except for scars, is the surface of your face *exactly* symmetric to a vertical plane down the middle?

26. a. What is *radial symmetry*?
    b. Name some living things that possess it.

Shingle Roof

Brick

Wood

Grade

Porch located on side

FRONT ELEVATION

Brick

Shingles

Grade

RIGHT SIDE ELEVATION

0   5   10   15   20
SCALE IN FEET

Views and Elevations

The house shown above was built in Highland, Indiana. Like many houses today, it was designed by an architect. Underneath the picture of the house are **views** or **elevations** from the front and right side. Architects draw views of a planned building from the top, sides, and other positions to help a client visualize the finished product. Computer-assisted design programs (CADs) also can provide many views or elevations of buildings. These elevations give accurate scale-model measurements; most photographs and perspective-type drawings do not.

In geometry, views of 3-dimensional figures are usually drawn without perspective as if the figures are solid but with all visible edges shown. (You were shown two views of Earth in Lesson 9-5.) From the views you can determine a possible shape of the original figure. The abbreviations L (left), R (right), F (front), and B (back) give guidance.

Example 1

Given the regular square pyramid at the right, draw views from the front, right side, and top.

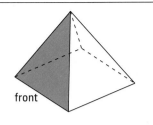
front

Solution

The bottoms of the front and side views are the same length as a side of the square base. The views are triangles, the same height as the height of the pyramid. The top view shows the base as a square and the vertex at the center of the square.

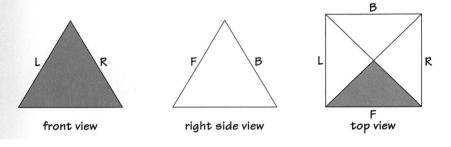

front view   right side view   top view

Activity

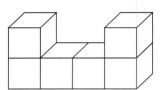

Make a shape out of six congruent cubes different from the one shown at the left. Draw views from the front, right side, and top.

Without experience, few people can visualize the shape of a structure by just looking at views. In this book, unless told otherwise, you should assume that the solids viewed are those we have already discussed—prisms, cylinders, pyramids, or cones—or combinations of them, as in Example 2.

Example 2

Here are three views of a prefabricated storage building made up of sections in the shape of congruent boxes.

Solution

a. How many stories tall is the building?
b. How many sections long is the building from front to back?
c. Where is the tallest part of the building located?

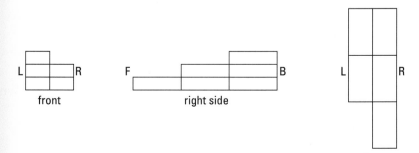

front   right side   top

▶

▶ **Solution**

a. The top view tells you nothing about the height of the building. The front or side view tells you **it is 3 stories high.**

b. The top view tells you **it is 3 sections long;** the right side view confirms this.

c. The front view tells you that the tallest part is somewhere on the left side. The right side view tells you that the tallest part is at the back. Combining these conclusions, you can conclude that **The tallest part of the building is at the back left.**

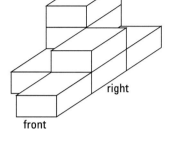

At the left is a drawing of a possible shape for the building whose views are given in Example 2. You are asked to draw another possible shape in Question 9.

It helps to make sketches to understand views. In making your sketches, you may want to put the hidden lines in at first and erase them as your sketch nears completion. Sometimes, in picturing these solids, the hidden lines are omitted so that the final shape is more easily viewed.

QUESTIONS

Covering the Reading

1. To an architect, what are *elevations*?

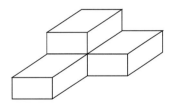

2. A building is sketched at the left.
   a. Draw a top view.
   b. Draw a view from the right side.
   c. Draw a front view.

In 3–6, sketch each shape as seen from each of the following positions: **a.** top, **b.** front, and **c.** right side.

3. a cube

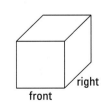

4. a right prism with an isosceles triangular base

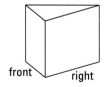

5. a table

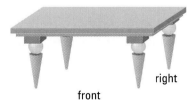

6. a block of buildings

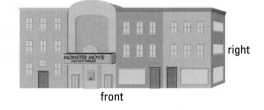

7. Give the three views you drew in the Activity of this lesson.

8. A building has the views shown below.
   a. How many stories tall is the building?
   b. How many sections long is the building from front to back?
   c. Where is the tallest part of the building located?

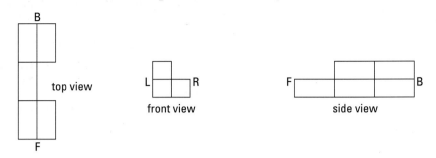

9. Sketch a possible shape, different from the solution given, for the building with the views given in Example 2.

Applying the Mathematics

In 10 and 11, refer to the elevations and scale at the beginning of this lesson.

10. About how wide (in feet) is the house?

11. About how high is it (in feet) from the ground to the highest point on the roof?

12. a. Which solid studied in this chapter has these views?

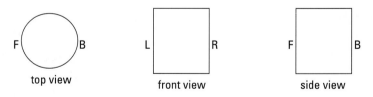

   b. Draw the solid.

13. At the right is a top view of a building (like the one in Example 2) made up of sections in the shape of congruent boxes. The number in each box tells how many are stacked in that space. Draw the building as seen from the front and from the right side.

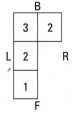

14. Draw the cup and saucer at the right as seen from overhead.

15. a. Give the number of symmetry planes of the prism shown in Question 4.
   b. Trace the drawing. Draw all symmetry planes and the cross sections they form.   *(Lessons 9-5, 9-6)*

In 16 and 17, use the right cone shown on the left.   *(Lessons 8-6, 9-4, 9-6)*

16. a. How many symmetry planes does the figure have?
   b. Sketch a plane section parallel to an edge not containing the vertex.
   c. Name the shape of the section in part **b.**

17. If the base of the cone has radius 8 and its slant height is 10, what is its height?

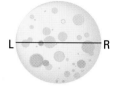

L————————R

18. At the left is a view of the moon and its equator as seen from Earth. The distance from *L* to *R* (through the center of the moon) is approximately 2160 miles. (Assume that the moon is a solid sphere.)   *(Lesson 9-5)*
   a. What is the approximate radius of the moon?
   b. How far is it from *L* to *R* along the surface of the moon?

19. How many great circles on Earth contain a given point in Baltimore and a given point in Tokyo?   *(Lesson 9-4)*

20. A regular heptagonal region is translated into a congruent region in a parallel plane. What solid figure is formed by all points on segments connecting the original region and its image?   *(Lesson 9-3)*

21. The area of Wyoming, the least populous state in the United States, is about 97,105 square miles. It is shaped like a trapezoid with east-west bases of length 350 miles and 370 miles. Estimate the north-south altitude of this trapezoidal state.

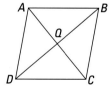

22. Given: *ABCD* at the left is a parallelogram with diagonals $\overline{AC}$ and $\overline{BD}$.
   To prove: $\triangle AQB \cong \triangle CQD$.   *(Lessons 7-2, 7-7)*

23. What is a transformation?   *(Lesson 4-1)*

Exploration

24. Choose an object in your home or classroom and draw it from four viewpoints: top, front, right, and left. For what kind of object will all the views differ?

25. a. Using a globe, draw Earth, including all visible land masses, as seen from below the South Pole.
   b. Is this view the same as a *bottom* view?

Building Cubes

IN-CLASS ACTIVITY

Work with a partner. You will need a ruler, scissors, and tape.

1 Trace the three figures below. Then cut out each of your drawings.

A.   B.

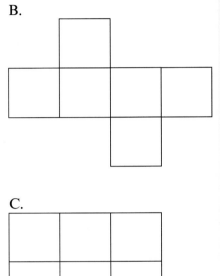

C.

2 Folding along the edges, try to make each
one into a cube.

3 a. Which figure(s) do not form a cube?
b. Why not?

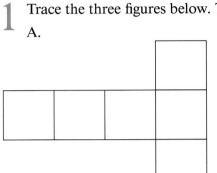

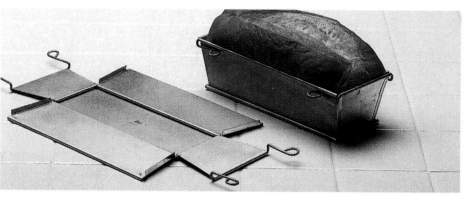

Dough net. *What advantages do you see in a baking pan which can be disassembled into a net?*

Polyhedra

Prisms and pyramids are special kinds of *polyhedra*. In general, a 3-dimensional surface which is the union of polygonal regions and which has no holes is called a **polyhedron.** The plural of *polyhedron* is either **polyhedrons** or **polyhedra.** Polyhedra can be classified by the number of their faces. Pictured below are a *tetrahedron* (4 faces) and a *hexahedron* (6 faces). The tetrahedron has 4 vertices and 6 edges. Both a box and a cube are special types of hexahedra.

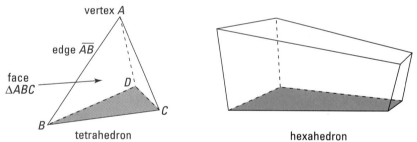

vertex *A*

edge $\overline{AB}$

face
$\triangle ABC$

tetrahedron

hexahedron

Regular polyhedra are the three-dimensional counterparts to regular polygons.

> **Definition**
> A regular polyhedron is a convex polyhedron in which all faces are congruent regular polygons and the same number of edges intersect at each of its vertices.

There are only five regular polyhedra; they are pictured here.

| regular tetrahedron (4 faces) | cube (6 faces) | regular octahedron (8 faces) | regular dodecahedron (12 faces) | regular icosahedron (20 faces) |

Nets for Polyhedra

In this book you have used perspective and nonperspective drawings, plane sections, and views to describe 3-dimensional figures in 2-dimensional drawings. You also can go the other way and use 2-dimensional drawings to make polyhedra and other 3-dimensional surfaces.

A **net** is a 2-dimensional figure that can be folded on its segments or curved on its boundaries into a 3-dimensional surface. To find the net for a particular surface, you can cut along the edges of the surface until it is flat. For instance, when cuts are made along some of the edges of a cube, then the cube can be flattened out.

Shown below is a net for a cube. In the In-class Activity, you folded a net like this into a cube. We have named the six faces by the first letters of the words up, down, left, right, back, and front. When you cut around the outside boundary, and folded along the common edges as shown, you formed a cube.

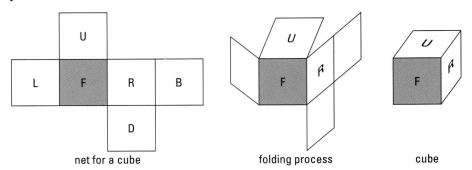

| net for a cube | folding process | cube |

Not all networks of six squares fold to make a cube. In the In-class Activity, you found that the 2-by-3 network of squares pictured below is not a net for a cube. It can be folded into the lateral surface of a prism, as shown at the right, but this is not a polyhedron since it has holes.

Prisms and pyramids often have relatively simple nets.

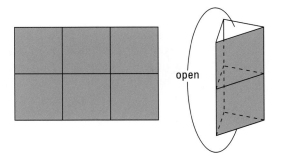

This piece of jewelry is in the shape of a regular icosahedron.

Cornered. *Many shapes have a function and purpose. This triangular prism is a practical space-saver.*

Example 1

Draw a net for the right triangular prism shown at the right.

Solution

The bases are congruent triangles. The lateral faces are rectangles with one pair of sides the same length as the corresponding sides of the base. A net is shown at the right. Notice how the tick marks would match up if this net were cut out and folded into a prism.

Activity

Consider a regular square pyramid like the one at the right. The base is a square, and the faces are isosceles triangles with their bases along the square's sides.

Step 1: To draw a net, first draw the square base.
Step 2: Draw four congruent isosceles triangles with each triangle having its base on an edge of the square.

The altitude of each triangular face must be greater than one-half the length of the side of the square. Why?

Nets for Cones and Cylinders

Cones and cylinders have simple nets.

Example 2

Draw a net for a cylinder *h* units high with base diameter *d*.

Solution

First, draw the circle that will be one base. The lateral surface of a cylinder is a rectangle. Since the diameter of the circle is *d*, the side of the rectangle that touches the circle has length πd, to match the circumference of the bases. The other side of the rectangle is the height *h* of the cylinder. Finally, draw a second circle, congruent to and across the lateral surface from the first. That will be the other base.

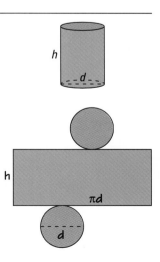

A net for the lateral surface of a cone is given in Question 10.

For greatest accuracy, you need to construct the bases and faces of a net using a straightedge and compass, or you should trace very carefully and measure the original and your copy with ruler and protractor. Printouts from an automatic drawer also can be useful.

QUESTIONS

Covering the Reading

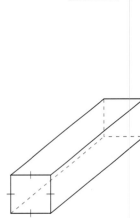

1. What is a *net*?

2. At the left is a net for a cube. Identified are the locations of three faces. Where will the other squares of the net end up on the cube?

3. Is a cone a polyhedron? Why or why not?

4. Match the name of each regular polyhedron with the shape of its faces.
   a. tetrahedron   (i) equilateral triangle
   b. cube   (ii) square
   c. regular octahedron   (iii) regular pentagon
   d. regular dodecahedron   (iv) regular hexagon
   e. regular icosahedron

5. Draw a net for the box shown at the left.

6. **a.** Show your net for the regular pyramid in this lesson's Activity.
   b. Trace your net in part **a**, cut it out, and fold it to make a pyramid.
   c. Why must the altitude of each triangular face be greater than half the length of a side of the square base?

7. Suppose you wanted to make the cylinder pictured at the right. Draw a net you could use. Make the net actual size. Trace your net and cut it out to make the cylinder.

8. Draw a net for a regular triangular prism with a base edge of 3″ and a height of 2″.

Applying the Mathematics

9. Connect the centers of the six faces of a cube. What polyhedron is formed?

10. **a.** A is the center of the partial disk at the right. If the shape is cut out and $\overline{AB}$ is moved to coincide with $\overline{AC}$, what 3-dimensional figure will be formed?
    b. Make an accurate partial disk with $AB = 4$ in. and m$\angle BAC = 100$. Cut out the shape and make the surface.

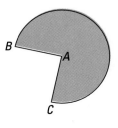

11. *Multiple choice.* Which cannot be a net for a cube?

(a) (b) (c) (d)

In 12 and 13, draw a net for the figure.

12.

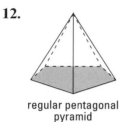

regular pentagonal
pyramid

13.

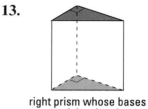

right prism whose bases
are right triangles

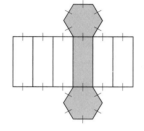

14. The net at the left is for what figure?

15. The can at the right has a paper label that exactly covers its lateral surface. What is the area of the label?

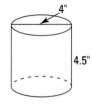

4″

4.5″

16. Pictured below are top, front, and left views of a solid. Draw a net for the surface of this solid.

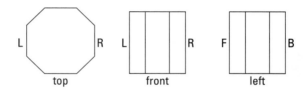

L   R   L   R   F   B

top   front   left

17. Draw a net for a cereal box.

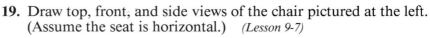

Review

18. Here are three views of a building. *(Lesson 9-7)*
  a. How many stories tall is the building?
  b. How many sections long is the building from front to back?
  c. Where is the tallest part of the building?

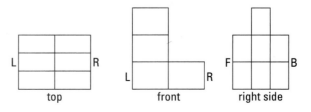

L   R   L   R   F   B

top   front   right side

19. Draw top, front, and side views of the chair pictured at the left. (Assume the seat is horizontal.) *(Lesson 9-7)*

20. What surface studied in this chapter has the following views? *(Lesson 9-7)*

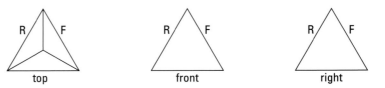

top   front   right

21. In parts **a–d**, give the number of vertices V, edges E, and faces F for each type of figure. *(Lessons 9-3, 9-4, 9-6)*
   a. any pyramid whose base is an octagon
   b. any prism whose bases are octagons
   c. any pyramid whose base is an n-gon
   d. any prism whose base is an n-gon
   e. *Multiple choice.* Descartes discovered and Euler proved a simple relationship between the numbers of vertices V, edges E, and faces F of any polyhedron. Which of the following is the relationship?
   (a) $V + E - F = 2$ (b) $F + E - V = 2$
   (c) $F + V - E = 2$ (d) $E + F - V = 2$
   f. Use algebra and your results in parts **c** and **d** to verify your choice in part **e**. (You will need to use two equations, one for pyramids and one for prisms.)

22. Refer to the right triangular prism in Example 1 in this lesson.
   a. Is the figure reflection-symmetric?
   b. If so, give the number of symmetry planes. *(Lesson 9-6)*

23. In the late fifteenth century, women's steeple caps were popular in France. It was customary for all hair to be concealed by this conical headdress. Find the slant height of a right conical steeple cap with height 24 inches and base area 16π square inches. *(Lessons 8-6, 9-4)*

24. Suppose, in a quadrilateral $ABCD$, that $AB = BC = CD = DA$. Why can you conclude that $ABCD$ is a rhombus? *(Lesson 6-3)*

25. Angles 1 and 2 are supplementary. Find t if $m\angle 1 = t$ and $m\angle 2 = t - 10$. *(Lesson 3-3)*

Exploration

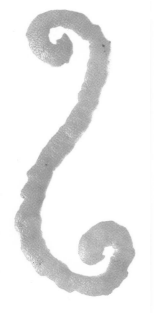

26. At the left is a net for a sphere made by peeling an orange in a single cut. Your teacher has a larger copy of this net which you should use for this question.
   a. Cut out the net and make a sphere from it.
   b. Find the radius of the sphere (in centimeters).

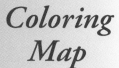

*Coloring
Map*

IN-CLASS
ACTIVITY

You will need colored pencils for this Activity. If colored pencils are not available, use a variety of patterns.

Below are maps of all the counties in the states of Vermont and Wyoming.

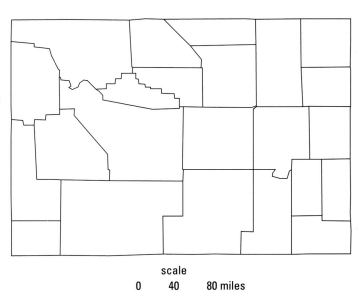

scale
0   20   40 miles

scale
0   40   80 miles

1 Trace or copy one of the maps (or both).

2 Color the counties using the following guidelines:
 (1) Each county has the same color throughout.
 (2) If two counties share a border, they must be
   different colors.
 (3) If two counties share only a corner, they can
   have the same color.
 (4) Try to use as few colors as possible.

Maps and the Four-Color Theorem

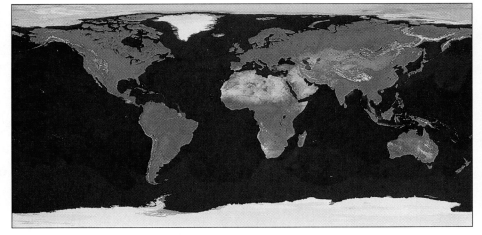

View from Tiros. *This picture of a cloudless world is a composite of the best images of each region as taken by the meteorological satellites. To see the distortion at the poles, compare Greenland (840,004 sq mi) to Australia (2,966,200 sq mi).*

Nets for a Sphere

A map of the world is a 2-dimensional approximation for 3-dimensional Earth. In Lesson 9-8, you learned how to make nets of prisms, cylinders, pyramids, and cones. The challenge that map makers have had to face is that it is not possible to make a flat net for a sphere. To create a globe, globe manufacturers take a net of the surface of Earth and attach it to a sphere. This net consists of a set of tapered sections in shapes called **gores.** The diagram below shows some of the gores in the net.

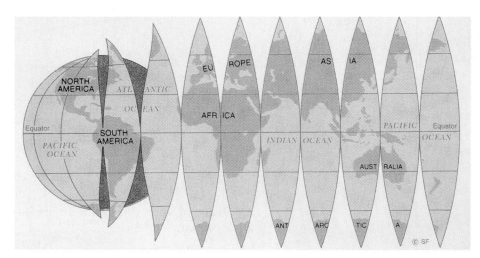

While gores give a fairly accurate view of the surface of Earth, they are very difficult to use as a map. One problem is that different parts of the same country, such as the United States, are on different gores, making it hard to see the country's actual shape. A second problem is more practical. The nonconvex shape of this map makes it easy to tear and inconvenient to use.

Above is a map of the world like those you see most often reproduced in newspapers and books. It is called a **Mercator projection,** named for Gerhardus Mercator, the Flemish cartographer who created it in 1569. While commonly used, this map is *not* an accurate picture of the world. Greenland looks almost as large as the continent of Africa. But the land area of Greenland is about 840,000 square miles, while the land area of Africa is about 11,700,000 square miles. Africa actually has about 14 times the area of Greenland.

A Mercator-projection map is made by projecting the surface of Earth onto the lateral face of a cylinder. The ray with endpoint at the center of Earth containing a point P on the surface of Earth also contains its image point P' on the cylinder. This is done with all the points on the surface of Earth for which rays containing them intersect the cylinder. The cylinder is then unrolled to give the map. So a Mercator projection is a net for a cylinder, not for a sphere.

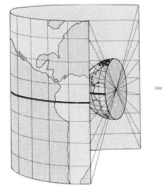
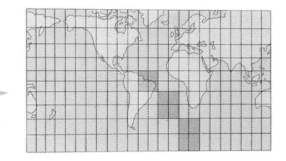

The process of creating a Mercator projection describes a transformation. Each point in the preimage (the surface of Earth) has a unique image point on the map. The preimage is not the entire surface of Earth, since the size of the cylinder will always prevent some points near the North and South poles from appearing on the map.

This transformation, however, is not an isometry because distance is not preserved. Thus distances near the poles are smaller than they appear on the map. Because area is based on distance, Greenland, near the North Pole, appears unusually large. However, the Mercator projection preserves betweenness and collinearity on the meridians (these great circles through the poles are projected onto vertical lines on the map) and on the curves of equal latitude (small circles on Earth are projected onto horizontal lines on the map). So the four directions—north, south, east, and west—are on perpendicular lines.

Coloring a Map

Here is a map of the 48 contiguous states of the United States (all but Alaska and Hawaii). To distinguish the states, each state has been colored under the guidelines of the In-class Activity on page 543. If two states share only a corner, as is the case with Colorado and Arizona, then they can have the same color. Notice that only four colors are needed for this map. Both of the maps in the In-class Activity required only four colors.

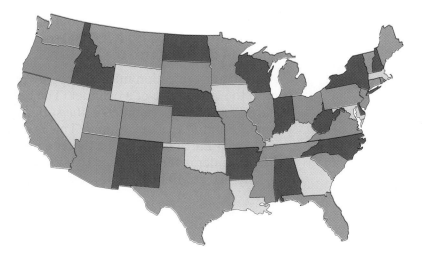

Four-Color Conjecture

In 1852, after studying many maps, Francis Guthrie, a British-born mathematician who later taught in South Africa, conjectured that *any* regions of a map on a sphere or a plane could be distinguished if four colors were used. For Guthrie, a region could have any shape as long as it remained connected. (The two unconnected parts of the state of Michigan were, for Guthrie, two separate regions.)

5 colors used

4 colors used

Guthrie's four-color conjecture states that every map *can* be colored using 4 colors or fewer. It does not say that a map *must* use four colors. For instance, the map at the top left has been colored using five different colors. However, it could have been colored using only four colors.

Example

Color this map using as few colors as possible.

Solution

This map could be colored with 4 colors, but since these regions share a corner, only 2 colors are needed.

Guthrie was not able to prove his conjecture. In fact, many mathematicians tried to prove the conjecture but failed. For almost 125 years, this conjecture, the "four-color problem", was one of the most famous undecided questions in mathematics. By 1975, mathematicians had proved that any map with fewer than 40 regions could be colored with four colors or fewer, but they had not proved that *any* map could be so colored.

By working on this problem, mathematicians discovered some new mathematics. Much of the mathematics of graph theory was discovered in the search for a proof to the four-color conjecture.

The Four-Color Theorem

In 1976, two mathematicians at the University of Illinois, Wolfgang Haken and Kenneth Appel, *proved* that Guthrie's four-color conjecture was correct. They could not prove this just by drawing maps and coloring, for there are infinitely many maps. First they showed that any map they needed to consider was one of 1,952 types of maps. Then they used a computer to help prove that for each type, no more than four colors would be needed to color it. Because of their proof, the four-color conjecture became the *Four-Color Theorem*.

The Four-Color Theorem
Suppose regions which share a border of some length must have different colors. Then any map of regions on a plane or a sphere can be colored in such a way that only four colors are needed.

Three-Color Maps

Work is still being done on this topic. In 1987, Elizabeth Wilmer, a student at Stuyvesant High School in New York City, won second prize in the nation in the Westinghouse Science Talent Search for her work on a three-color problem. She analyzed maps which could be colored with three colors. Here is an example of such a map with 19 regions.

QUESTIONS

Covering the Reading

1. How many years after Magellan's ships circumnavigated Earth did Mercator create his map of the world?

2. Globe makers use tapered plane sections called __?__.

3. A Mercator projection is a net for a __?__.

4. Of the gores, the globe, and the Mercator projection, which have the following flaws?
   a. The countries are disconnected.
   b. The areas of land masses are distorted.
   c. Only half the world is visible.

5. Why is the Mercator projection a transformation?

6. a. The total area of the contiguous United States is approximately 3,120,000 sq mi. The area of the U.S. is about how many times that of Greenland?
   b. In the Mercator projection, does Greenland look larger than, smaller than, or about equal to the mainland United States?

7. Who first stated the four-color conjecture?

8. Color the map at the left in as few colors as possible.

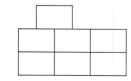

9. How many years elapsed between the statement and the proof of the four-color conjecture?

10. Who is Elizabeth Wilmer?

11.

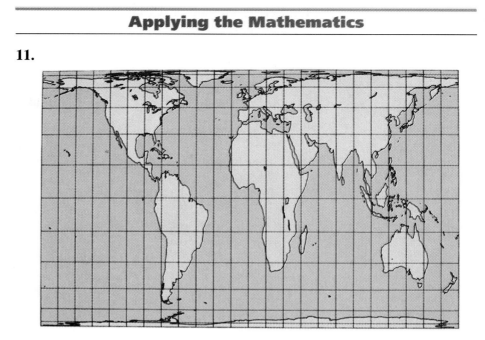

The Gall-Peters equal-area projection is a transformation that preserves relative areas of land masses. For instance, the ratio of the map area of Greenland to the map area of the United States equals the ratio of their actual areas, respectively. Examine the Gall-Peters projection shown above. Which of the A-B-C-D properties are *not* preserved by this transformation?

12. Oslo, Norway is due west of St. Petersburg, Russia. Also, Bi'r'Allāq, Libya is due west of Alexandria, Egypt.

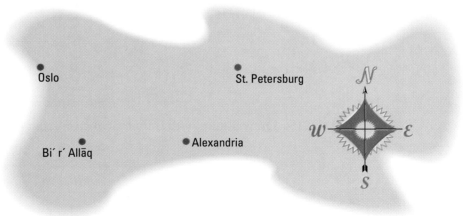

Oslo is west of a north-south line passing through Bi'r'Allāq, and St. Petersburg is east of a north-south line through Alexandria. Yet the distance from Oslo to St. Petersburg is less than the distance from Bi'r'Allāq to Alexandria. How can this be?

13. Below is a map of Africa. Borders of countries are drawn. Trace the map and color the mainland countries using the least number of colors. (Hint: Start where the countries are most densely packed, and finish with the outside countries.)

14. In the map below, the borders of countries are circular arcs. What is the least number of colors needed to color this map?

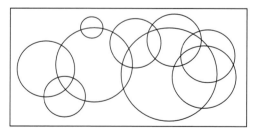

15. Draw a map with only four regions that needs four colors to color it.

Review

16. What shape can the net at the right form? *(Lesson 9-8)*

17. Draw a net for an open cylinder with one base. *(Lesson 9-8)*

18. Examine the apple juice can at the left. Sketch each view.
a. top   **b.** front   **c.** right   *(Lesson 9-7)*

apple juice

In 19–22, draw an example of each figure. *(Lessons 9-2, 9-3, 9-4, 9-5)*

19. an oblique cone **20.** a right prism with a nonconvex kite as a base

21. a solid sphere **22.** two intersecting planes

In 23 and 24, tell which geometric three-dimensional figure most resembles the real-world object. Give as specific a name as you can, distinguishing solids from surfaces. *(Lessons 9-3, 9-5)*

23. straw **24.** cantaloupe

25. *ABCDE* at the right is a regular pyramid. *AO* is the height. If *DE* = 12 and *AO* = 10, find each measure to the nearest tenth.
   a. slant height
   b. *BD*
   c. length of a lateral edge
   d. area of △*AOD* *(Lessons 8-4, 8-6, 9-4)*

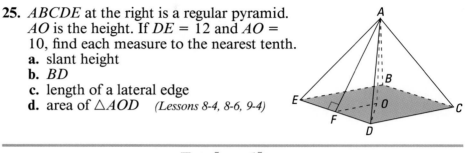

Exploration

26. Find a globe in a library, study, or store.
   a. How many colors are used to color the countries and the oceans?
   b. What conventions are used in coloring the regions?

27. Using the United States map in this lesson, name all the states that
   a. are shaded yellow.
   b. are green.
   c. are purple.
   d. are red.
   e. are not pictured.

28. An old riddle is thus: "An explorer walks one mile due south, turns and walks one mile due east, turns again and walks one mile due north. The explorer is now at the original starting point. The explorer sees a bear. What color is the bear?"
   a. Find all the points on Earth where an explorer could walk 1 mile south, 1 mile east and 1 mile north and end up at the original starting point.
   b. What kind of bears live near your answer to part **a**?

A project presents an opportunity for you to extend your knowledge of a topic related to the material of this chapter. You should allow more time for a project than you do for typical homework questions.

1 Nets for Packages

Many nets include flaps for adhesive. Undo three packages to determine the net from which the package was made. At least one package should be of a shape other than a box. Make a scale drawing of each net and package. Then design a net that would produce an appropriate package for a product of your choice.

2 Structures

Copy pictures or take photographs of buildings, dwellings, or other structures with a shape approximating each of the types of surfaces studied in this chapter: cylinders, prisms, cones, pyramids, other polyhedra, and spheres. (Do not use examples shown in this chapter.) Estimate the important dimensions of each structure. Arrange this information nicely on a poster or in a small booklet.

3 Architectural Views

Make a series of accurate drawings of the front, left side, right side, back, and top of the building in which you live. Include the scale you used. Then, do the same for a dream house of your design.

4 Geometry on a Sphere

Consider the following postulates and theorems from this book.

Point-Line-Plane Postulate
Corresponding Angles Postulate
Transitivity of Parallelism Theorem
Two Perpendiculars Theorem
Perpendicular to Parallels Theorem

a. Do these hold true if "points" are locations on a sphere, "lines" are great circles on the sphere, and a "plane" is the surface of the sphere? For each, explain why or why not using diagrams and words.

b. Do the five statements above hold true if great circles and small circles are considered to be the only lines in the "plane" of the sphere? Explain.

5 Regular Polyhedra

Use cardboard and tape to construct models of the regular polyhedra from the nets provided. The patterns shown below should be enlarged. Cut along solid lines, fold along dotted lines.

tetrahedron

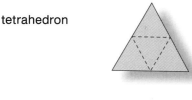

cube

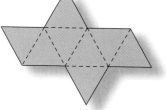

octahedron

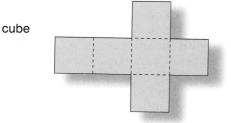

dodecahedron

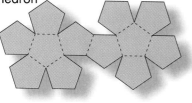

icosahedron

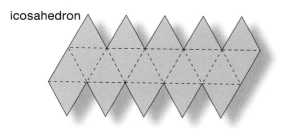

6 Continents and Their Countries

Create a map of the countries of South America and a map of the countries of Asia. Place the names of the countries on the maps. What is the least number of colors necessary to color each map? Color them, using that number of colors.

7 Plane Sections of a Cube

Make a series of large drawings of all the possible plane sections of a cube. How many are there? What are the minimum and maximum number of possible sides of a plane section of a cube? Make a conjecture about the minimum and maximum number of sides of the plane section of any regular prism with bases that are n-gons.

SUMMARY

The purpose of this chapter is to familiarize you with the common 3-dimensional figures, the surfaces and solids that are the figures of solid geometry. You should know their definitions and how they are related, be able to sketch them, identify plane sections, draw views from different positions, and be able to build the simpler surfaces from 2-dimensional nets. Below is a hierarchy relating many of these surfaces.

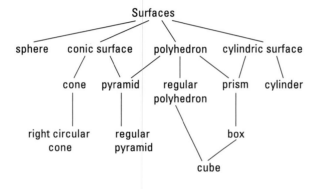

Many ideas from two dimensions extend to three. The basic properties of planes are given in the Point-Line-Plane Postulate. Intersecting planes form dihedral angles. If the measure of the dihedral angle is 90°, then the planes are perpendicular. Like lines, planes may be perpendicular or parallel. Spheres and circles have the same defining property except spheres are in three dimensions. Reflections and reflection symmetry are defined the same way in three dimensions as in two, except that the reflecting line is replaced by a reflecting plane.

Architects use views to describe 3-dimensional figures in two dimensions. Map makers have dealt differently with the impossibility of presenting the spherical Earth on a 2-dimensional map. Certain distortions have to be made in order for other properties to be preserved. The Four-Color Theorem applies to maps drawn on planes and spheres, whether or not shapes on the map are distorted.

VOCABULARY

Below are the most important terms and phrases for this chapter. For the starred (*) terms you should be able to give a definition of the term. For the other terms you should be able to give a general description and a specific example of each.

Lesson 9-1
Point-Line-Plane Postulate:
    Flat Plane Assumption
    Unique Plane Assumption
    Intersecting Planes
    Assumption
intersecting planes

Lesson 9-2
*line perpendicular to a plane
Line-Plane Perpendicularity
    Theorem
parallel planes
distance between parallel
    planes
distance to a plane from a
    point
dihedral angle
skew lines

Lesson 9-3
surface, solid
rectangular solid, box
faces of a box, opposite faces,
    cube
edges of a box
vertices of a box
*cylindric solid
base of a cylindric solid
lateral surface of a cylindric
    solid
cylindric surface
height or altitude of a
    cylindric solid
*cylinder
*prism
right prism, right cylinder
oblique prism, oblique
    cylinder
lateral faces of a prism
lateral edges of a prism
regular prism

Lesson 9-4
conic surface
*base of a conic solid
*vertex of a conic solid
*conic solid
*cone, *pyramid
lateral edges of a pyramid
base edges of a pyramid
faces of a pyramid
lateral faces of a pyramid
right pyramid, oblique
    pyramid
regular pyramid
axis of a cone
right cone, oblique cone
lateral surface of a cone
lateral edge of a cone
height or altitude of a
    pyramid
height or altitude of a cone
slant height of a pyramid
slant height of a cone
truncated surface

Lesson 9-5
*sphere
*the radius of a sphere
*center of a sphere
radii of a sphere
the diameter of a sphere
great circle
small circle
hemisphere
great circle route
*plane section
conic sections

Lesson 9-6
perpendicular bisector of a
    segment (in space)
*reflection image of a point
    over a plane
reflecting plane
*congruent figures (three-
    dimensional)
directly congruent figures in
    space
oppositely congruent figures
    in space
*reflection-symmetric figure
    in space
symmetry plane
bilateral symmetry

Lesson 9-7
views, elevations

Lesson 9-8
polyhedron, polyhedrons,
    polyhedra
tetrahedron, hexahedron
*regular polyhedron
octahedron, dodecahedron,
    icosohedron
net

Lesson 9-9
gores
Mercator projection
the Four-Color Theorem

PROGRESS SELF-TEST

Directions: Take this test as you would take a test in class. Then check your work with the solutions in the Selected Answers section in the back of the book. You will need graph paper and a calculator.

In 1–3, draw each figure.
1. two parallel planes
2. regular square prism
3. an oblique cone
4. Explain why a four-legged chair sometimes wobbles.

In 5–7, use the regular square pyramid below.

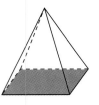

5. a. Trace the figure and sketch a plane section parallel to the base.
   b. Trace the figure and sketch a plane section not parallel to and not intersecting the base.
   c. Name the shape of each section.
6. How many symmetry planes does this figure have?
7. a. Draw a sphere and a plane section containing the center.
   b. Name the plane section. Be as specific as possible.
8. Sketch the top, front, and left views of this toy train car.

In 9 and 10, refer to the right cone below.
9. Draw each view.
   a. top
   b. front
   c. right

10. How many symmetry planes does it have?
11. If a great circle on a sphere has circumference 28π cm, what is the radius of the sphere?
12. Refer to the oblique cone with base $\odot O$, $OB = 10$ cm, $AO = 34$ cm, and $BD = 6$ cm.

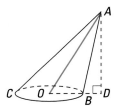

   a. What is the height of the cone?
   b. Find the area of the base of the cone.
13. In the right triangular prism below, $\overline{EF} \perp \overline{FG}$, $EF = 4''$, and $EG = 9''$. If the height of the prism is $22''$,
   a. find Area($EGHJ$).
   b. find Area($\triangle EFG$).

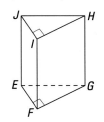

In 14 and 15, tell which three-dimensional figure most resembles the real-world object. Give as specific a name as you can, distinguishing whether it is a solid or a surface.

14. a stick of margarine

15. a blown-up balloon

16. Here are front, right, and top views of a building.

   a. How many stories tall is the building?

   b. How many sections long is the building from front to back?

   c. Where is the tallest part of the building located?

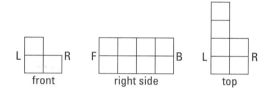

17. Give as specific a name as possible to the surface studied in this chapter with these views.

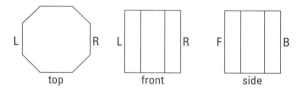

18. Draw a net for a regular pentagonal pyramid.

19. Name the figure that can be made from the net below.

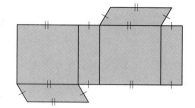

20. *Multiple choice.* In a Mercator projection, which of the following is preserved?
   (a) relative distance
   (b) relative areas of land mass
   (c) connectedness of land mass
   (d) betweenness

21. Draw a map with five regions that requires four colors to be colored.

CHAPTER REVIEW

Questions on SPUR Objectives

SPUR stands for **S**kills, **P**roperties, **U**ses, and **R**epresentations. The Chapter Review questions are grouped according to the SPUR Objectives for this chapter.

SKILLS DEAL WITH THE PROCEDURES USED TO GET ANSWERS.

Objective A: *Draw common 3-dimensional shapes.* *(Lessons 9-2, 9-3, 9-4, 9-5)*

In 1–8, draw each figure.

1. a line parallel to a plane
2. a line perpendicular to a plane
3. two perpendicular planes
4. a right cone
5. a right hexagonal prism
6. a sphere
7. a regular square pyramid
8. an oblique cylinder

Objective B: *Draw plane sections of common 3-dimensional shapes.* *(Lesson 9-5)*

In 9–11, use the figures below.

a. Trace the figure and sketch a plane section parallel to the base(s).

b. Trace the figure and sketch a plane section not parallel to and not intersecting the base(s).

c. Name the shape of each section.

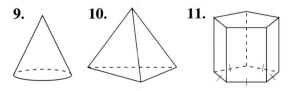

9. 10. 11.

12. **a.** Draw a sphere and a plane section not containing the center.

 b. Name the plane section. Be as specific as possible.

 c. How does this plane section differ from a great circle of the sphere?

Objective C: *Give views of a figure from the top, sides, or front.* *(Lesson 9-7)*

13. Give each view of the right circular cylinder at the right.

 a. top

 b. front

 c. right

14. Give each view of the regular square pyramid at the right.

 a. top

 b. front

 c. right

In 15 and 16, for each object sketch views of the **a.** top, **b.** front, and **c.** right.

15. a house   16. a cup

front

front

Objective D: *Given appropriate lengths, calculate areas and lengths in 3-dimensional figures.* *(Lessons 9-3, 9-4, 9-5)*

17. In the regular triangular pyramid shown at the right, $AC = 12$ and $AD = 10$.

 a. Find the perimeter of the base.

 b. Find the area of $\triangle ABD$.

18. Consider the oblique square prism shown below.

   a. Which segment's length is the height of the prism?

   b. If $GN = 15$ and $LN = 40$, what is the length of $\overline{KF}$?

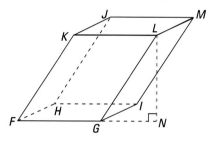

19. In the oblique cylinder at the right with base $\odot D$, $AC = 9$ cm, $BD = 10$ cm, and $CD = 3$ cm.

   a. Find the area of the base of the cylinder.

   b. Find the height of the cylinder.

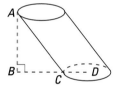

20. a. A sphere has radius 12″. What is the area of a great circle of the sphere?

   b. Find the circumference of a great circle of the sphere.

21. A right cone has a height of 24 mm and a slant height of 25 mm.

   a. Draw such a cone.

   b. Find the area of its base.

22. In the regular pentagonal pyramid shown at the right, $HO = 7$ and $HT = 16$. Find

   a. OT.

   b. Area($\triangle HOT$).

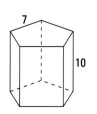

23. The regular pentagonal prism at the right has edges as marked.

   a. Find the perimeter of one lateral face.

   b. Find the sum of the areas of the five lateral faces.

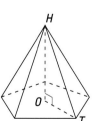

24. At the right is a right cylinder with parts marked.

   a. Find the area of a base.

   b. Find the area of the lateral surface.

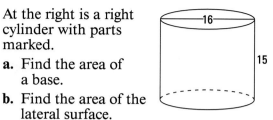

Objective E: *From 2-dimensional views of a figure, determine the 3-dimensional figure.* (Lesson 9-7)

In 25 and 26, use the given views of the buildings.

   a. How many stories tall is the building?

   b. How many sections long is the building from front to back?

   c. Where is the tallest part of the building located?

25.

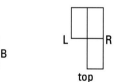

26.

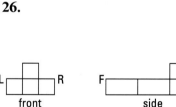

27. Which cylindric solid studied in this chapter has these views?

28. Give as specific a name as possible to the surface studied in this chapter with these views.

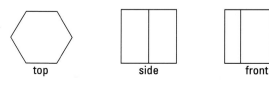

PROPERTIES DEAL WITH THE PRINCIPLES BEHIND THE MATHEMATICS.

Objective F: *Apply the properties of planes.*
(Lessons 9-1, 9-2)

29. Identify the figure that is the intersection of two planes.

30. Explain how the distance between parallel planes is determined.

In 31–33, use the figure below in which $\ell \perp m$ and $\ell \perp n$.

31. Is $\ell \perp R$? Explain your answer.

32. Is $\ell \perp p$? Explain your answer.

33. Is $m \perp n$? Explain your answer.

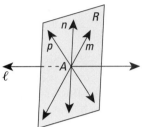

Objective G: *Determine symmetry planes in 3-dimensional figures.* *(Lesson 9-6)*

In 34 and 35, a figure is given.
a. Tell if the figure has bilateral symmetry.
b. Give the number of symmetry planes.

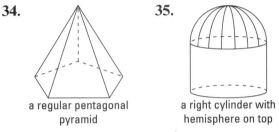

34. a regular pentagonal pyramid

35. a right cylinder with hemisphere on top

36. At least how many symmetry planes does a box have?

37. A regular prism which has an octagon for a base has __?__ symmetry planes.

USES DEAL WITH APPLICATIONS OF MATHEMATICS IN REAL SITUATIONS.

Objective H: *Recognize 3-dimensional figures in the real world.* *(Lessons 9-3, 9-4, 9-5)*

In 38–41, tell which three-dimensional figure most resembles the real-world object. Give as specific a name as you can, distinguishing solids from surfaces.

38. a birthday cake **39.** the planet Mars

40. a bubble **41.** a paperback book

Objective I: *Apply the Four-Color Theorem to maps.* *(Lesson 9-9)*

42. What is the Four-Color Theorem?

43. Draw a map with four regions that needs four colors in order to be colored.

44. Draw a map with nine regions that needs three colors in order to be colored.

45. Trace this map and color it with at most 4 colors.

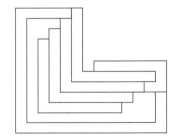

REPRESENTATIONS DEAL WITH PICTURES, GRAPHS, OR OBJECTS THAT ILLUSTRATE CONCEPTS.

Objective J: *Make a surface from a net, and vice versa.* *(Lesson 9-8)*

46. Tell whether each figure could be a net for a cube.

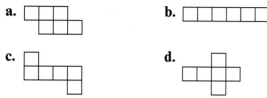

a.

b.

c.

d.

47. Draw a net for a square pyramid.

48. Draw a net for the lateral surface of a box with dimensions 4, 6, and 8.

49. Draw a net for a regular hexagonal prism with height 10 and a base with edge 7.

50. Name the figure that would be made from the net below.

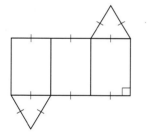

51. A cylinder has a height of 12 and a radius of 10. A net for this cylinder is given below.

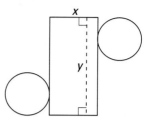

a. What is the value of x?

b. What is the value of y?

Objective K: *Interpret maps of the world.*
(Lesson 9-9)

In 52–54, each type of view of Earth has a flaw. Describe the problem with each.

52. globe

53. gores

54. Mercator projection

55. What preservation properties of isometries do not hold for a Mercator projection map?

CHAPTER

10

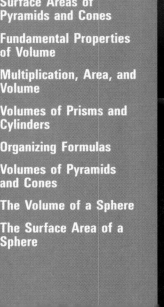

SURFACE AREAS AND VOLUMES

The two most important measures of 3-dimensional figures are *surface area* and *volume*. They are the counterparts of perimeter and area in 2-dimensional figures. Surface area, like perimeter, is a measure of a boundary, the surface of a 3-dimensional figure. Volume, like area, is a measure of the space enclosed by the figure.

Surface Area
helps in determining:

Volume
helps in determining:

how much paper is needed to make a box,

how much the box can hold,

how much land there is to explore on the moon,

how much material makes up the moon,

how much heat a bird loses through its skin,

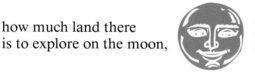

how much the bird weighs,

how much fabric is needed to cover a toy.

how much stuffing is needed to make the toy.

Surface area and volume are quite different, as the above examples show. In this chapter, you will learn how to calculate the surface area and volume of many of the 3-dimensional figures introduced in the last chapter. You also will learn how surface area and volume are related to each other, and you will explore the general properties underlying these measures.

It's all "hot air." *The amount of material used in a balloon is found by measuring the balloon's surface area when inflated. Plastic, nylon, or polyethylene are the materials used to make hot-air balloons.*

The cost of any container, from a suitcase to a new house, from a paper bag to a hot-air balloon, depends on the amount of material used to make it. The amount of material covering the container is its **surface area,** which we abbreviate **S.A.**

Consider a paper bag. It is approximately a box, a type of prism, with the upper base missing. Since each face of a box is a rectangular region, the surface area of the bag is the sum of the areas of five rectangles. The areas are easily seen by examining the net for the bag.

Example 1

A grocery bag has a base 7″ by 12″ and a height of 17″. What is its total surface area?

Solution
Draw the bag and a net for it.

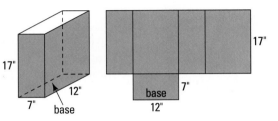

Surface area = area of base + sum of areas of lateral faces
$$= \quad 7 \cdot 12 \quad + (7 \cdot 17 + 12 \cdot 17 + 7 \cdot 17 + 12 \cdot 17)$$
$$= 730 \text{ square inches}$$

Lateral Area

The total area of the lateral surface of a solid is its **lateral area,** or **L.A.** In Example 1, the lateral area can be calculated by finding the sum of the areas of the four lateral faces.

$$
\begin{aligned}
\text{L.A.} &= 7 \cdot 17 + 12 \cdot 17 + 7 \cdot 17 + 12 \cdot 17 \\
&= (7 + 12 + 7 + 12) \cdot 17 \quad \text{Distributive Property} \\
&= 38 \cdot 17 \\
&= 646 \text{ in.}^2
\end{aligned}
$$

Since the base rectangle has perimeter $7'' + 12'' + 7'' + 12'' = 38''$ and the height of the bag is $17''$, the calculation can be written as

$$\text{L.A.} = 38 \cdot 17 = \text{perimeter of base} \cdot \text{height of prism.}$$

This idea works with any right prism (or cylinder) because the length of any lateral edge equals the height of the prism (or cylinder). It does not hold for oblique prisms and cylinders.

> **Right Prism-Cylinder Lateral Area Formula**
> The lateral area, L.A., of a right prism (or right cylinder) is the product of its height h and the perimeter (circumference) p of its base.
> $$\text{L.A.} = ph$$

Proof
Here are a representative right prism and right cylinder, each with height h. For the prism, the perimeter of either base is $a + b + c + d + e$. For the cylinder, the perimeter of either base is the circumference of the base, $2\pi r$.

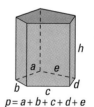

Here are their nets.

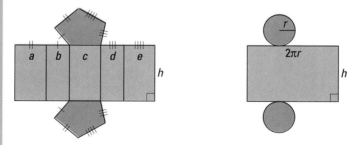

Now we compute the lateral areas. The lateral area is the area of the entire surface except the bases. The net for the lateral surface is a rectangle, and we use the formula for the area of a rectangle: base · height.

For the prism,
$$
\begin{aligned}
\text{L.A.} &= (a + b + c + d + e)h \\
&= ph.
\end{aligned}
$$

For the cylinder,
$$
\begin{aligned}
\text{L.A.} &= 2\pi r \cdot h \\
&= ph.
\end{aligned}
$$

Total Surface Area

The two congruent bases of every prism or cylinder have the same area. So the next theorem holds even for oblique prisms and cylinders.

> **Prism-Cylinder Surface Area Formula**
> The surface area, S.A., of any prism or cylinder is the sum of its lateral area L.A. and twice the area B of a base.
> $$S.A. = L.A. + 2B$$

In this chapter, the important formulas to learn are highlighted in boxes. Of course, you must know what each letter in the formula represents. As shown in Example 2, it is usually wise to begin a problem by writing an appropriate formula. Then substitute given values for the variables. Substitute for the area and perimeter only when needed.

Rolling a rectangular poster to fit into a cylinder allows it to be mailed without crushing.

Example 2

Consider the right cylinder shown.
a. Find its lateral area.
b. Find its surface area.

Solution

a. Begin by writing a formula for the lateral area of a right cylinder.
$$L.A. = ph$$
Since the base is a circle and $r = 3$,
$$p = 2\pi r = 2\pi \cdot 3.$$
Substituting,
$$L.A. = (2\pi \cdot 3)h$$
$$= 6\pi h.$$
Since $h = 5$, $\quad$ L.A. $= 30\pi$ square units.

b. In any cylinder, $\quad$ S.A. $= L.A. + 2B$.
Since the base is a circle, $B = \pi r^2$.
So, $\quad$ S.A. $= L.A. + 2(\pi r^2)$
$$= 30\pi + 2(\pi \cdot 3^2) \text{ since L.A.} = 30\pi \text{ and } r = 3$$
$$= 30\pi + 18\pi$$
$$= 48\pi \text{ square units.}$$

Notice that, since lateral area and surface area are areas, they are measured in square units.

QUESTIONS

Covering the Reading

1. Between surface area and volume, which is a measure of the boundary of a 3-dimensional figure and which is a measure of its interior?

In 2–5, is the quantity related more to surface area or to volume?

2. how much a railroad boxcar will hold

3. how much metal and wood are needed to build a boxcar

4. how much you weigh

5. how much you perspire after exercise

6. A small bag has dimensions shown on the left.
   a. Draw a net for the bag.   **b.** What is its surface area?

7. **a.** Define *lateral area*.
   b. Give a formula for the lateral area of a right prism.

8. The lateral area of a right cylinder is the product of its ＿?＿ and the ＿?＿ of the base.

9. In a prism or cylinder, S.A. = ＿?＿ + ＿?＿.

In 10 and 11, find the lateral area and surface area of each figure.

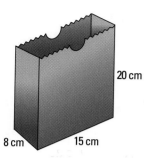

10.

2 cm

7 cm

right cylinder

11.

4.5"   6"

10"

right triangular prism

Applying the Mathematics

In 12 and 13, a net for a surface is given.
a. What is the surface?
b. Calculate its lateral area.
c. Calculate its surface area.

12.

5
5
5   5
5
5
5

13.

5 cm   4 cm
3 cm
9 cm

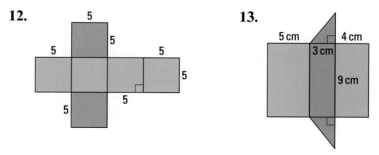

In 14–16, find the surface area of the given solid.

14. a box with dimensions 9 cm, 10 cm, and 11 cm

15. a box with dimensions L, W, and H

16. a cube with an edge of length s

In this Maltese version of bocce or boccie, weighted cylinders, instead of traditional spheres, are rolled with precision and accuracy.

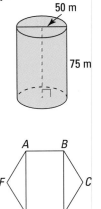

17. In one form of the game of bocce, cylindrical boules are thrown toward a cochonnet or target ball. A point is scored for each boule nearer to the cochonnet than the nearest opposing boule. How many square centimeters of metal are used to cover the lateral area of a cylindrical boule that is 10 cm high and has a base radius of 5 cm?

18. **a.** Draw an oblique cylinder.
    b. Explain whether the Right Prism-Cylinder Lateral Area Formula holds for your cylinder in part **a.**

19. Consider a box with dimensions 5, 6, and 7.
    a. Find its surface area. *10, 12, and 14*
    b. Double each dimension and find the new surface area.
    c. Divide your answer in **b** by the answer in **a.**
    d. Repeat parts **a–c** with a box of dimensions 2, 9, and 11. *4, 18, and 22*

20. A fuel storage tank has diameter 50 meters and height 75 meters. If a gallon of paint can cover about 45 square meters, about how many gallons are needed for two coats of paint on the exterior sides and top of the tank?

Review

21. Draw a net for a square pyramid and shade the regions that make up its lateral surface. *(Lesson 9-8)*

22. Write a proof argument.
    Given: Regular hexagon $ABCDEF$ at the left.
    To prove: $\triangle AEF \cong \triangle BDC$.
    (Lessons 6-7, 7-3)

23. Explain how to locate the center of a regular nonagon. *(Lesson 6-7)*

24. What is the cube of $\frac{2}{3}$? *(Previous course)*

Exploration

25. Find a paper bag in your house.
    a. What are its dimensions?
    b. What is its surface area?
    c. To make the bag, some of the paper has to overlap for gluing. Carefully undo the seams of the bag to make it lie flat. Draw its net, including its seams.
    d. How much more paper is used to make the bag than you calculated for your answer to part **b**?

26. Suppose you have 100 square inches of cardboard, scissors, and tape, and you wish to make a box with no top. What might be its dimensions?

Lateral Area of Regular Pyramids

IN-CLASS
ACTIVITY

Work on this Activity with another student.

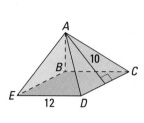

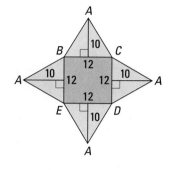

1 Shown above at the left is regular square pyramid *A-BCDE*. (Notice how the pyramid is named with the vertex, a dash, and then the name of the base.) Shown above at the right is a net for *A-BCDE*. The lateral area of *A-BCDE* is the sum of the areas of the 4 congruent triangles. Compute area($\triangle ABC$) + area($\triangle ACD$) + area($\triangle ADE$) + area($\triangle ABE$).

2 Shown below are regular hexagonal pyramid *K-LMNOPQ* and its net.
a. Compute the lateral area of *K-LMNOPQ*.
b. Let *p* represent the perimeter of *LMNOPQ*. Does your answer in part **a** equal $\frac{1}{2}\ell p$?

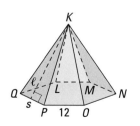

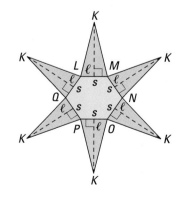

10-2

Surface Areas of Pyramids and Cones

Don't tread on me. *Highway cones are purposely brightly colored so drivers can see them and avoid them.*

Surface Area of Non-Regular Pyramids

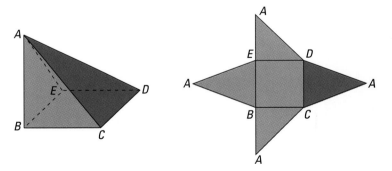

For a pyramid that is not regular, there is no simple formula for its lateral area. Above are oblique square pyramid *A-BCDE* and its net. Because its four faces are not congruent, you must calculate each of the four areas and add them.

However, all pyramids and cones have one base. So to calculate the total surface area, you can just add the area of the base to the lateral area.

> **Pyramid-Cone Surface Area Formula**
> The surface area, S.A., of any pyramid or cone is the sum of its lateral area L.A. and the area *B* of its base.
> $$\text{S.A.} = \text{L.A.} + B$$

Lateral Areas of Regular Pyramids

There is, however, a simple formula for the lateral area of a regular pyramid.

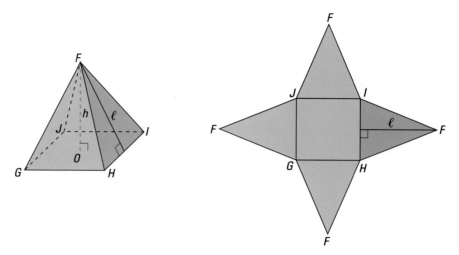

Shown above is a regular square pyramid F-$GHIJ$ like the one in the In-class Activity on page 569. The lateral area of F-$GHIJ$ is easy to calculate because its faces are congruent isosceles triangles. For each triangle the altitude from vertex F is the slant height ℓ of the pyramid.

$$\text{L.A. of } F\text{-}GHIJ = \text{Area}(\triangle FGH) + \text{Area}(\triangle FHI) + \text{Area}(\triangle FIJ) + \text{Area}(\triangle FJG)$$

$$= \tfrac{1}{2}\ell(GH) + \tfrac{1}{2}\ell(HI) + \tfrac{1}{2}\ell(IJ) + \tfrac{1}{2}\ell(GJ) \qquad \text{definition of lateral area}$$

$$= \tfrac{1}{2}\ell(GH + HI + IJ + GJ) \qquad \text{Distributive Property definition of perimeter}$$

$$= \tfrac{1}{2}\ell p,$$

where p is the perimeter of the base.

The formula L.A. $= \tfrac{1}{2}\ell p$ holds for any regular pyramid. Here is a proof argument.

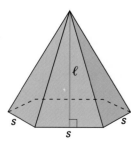

Let s be a side of the base (a regular n-gon) of a regular pyramid and ℓ be its slant height. Then the area of each lateral face is $\tfrac{1}{2}\ell s$. Since there are n lateral faces,

$$\text{L.A.} = n \cdot \tfrac{1}{2}\ell s$$
$$= \tfrac{1}{2}\ell \cdot ns.$$

But ns is the perimeter of the base. Substituting p for ns,

$$\text{L.A.} = \tfrac{1}{2}\ell p.$$

Sometimes, as in Example 1, you need to compute the slant height from other lengths.

Example 1

The pyramid of Khufu, a regular square pyramid which was originally 147 m tall and 231 m on a side of a base, is the largest of the Egyptian pyramids.
a. What is the approximate slant height of this pyramid?
b. What is its lateral area?

Solution

a. Draw a picture of the pyramid. We call it *V-ABCD*. *E* is the center of its base and *F* is the midpoint of an edge.

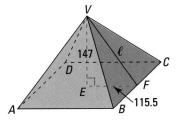

To compute the slant height ℓ, use $\triangle VEF$.
Since *V-ABCD* is a regular pyramid, $\triangle VEF$ is a right triangle.

E is the center of the square, so $EF = \frac{1}{2} \cdot 231 = 115.5$ meters.
VE is the height, 147 meters. By the Pythagorean Theorem,
$$\ell^2 = EF^2 + VE^2$$
$$= 115.5^2 + 147^2$$
$$\approx 34{,}950.$$
So, $\quad \ell \approx 187$ meters.

b. To find the lateral area, use the formula L.A. $= \frac{1}{2}\ell p$. p is the perimeter of the base, so $p = 4 \cdot 231 = 924$.
$$\text{L.A.} = \frac{1}{2}\ell p$$
$$\approx \frac{1}{2} \cdot 187 \cdot 924$$
$$\approx 86{,}400 \text{ square meters}$$

Lateral Areas of Right Cones

The formula L.A. $= \frac{1}{2}\ell p$ applies also to right cones. Consider the circle containing the vertices of the base of a regular pyramid. Imagine increasing *n*, the number of sides of the base, keeping the altitude *h* constant. A cone is the limit of the regular pyramid as *n* increases without bound. The slant height of the pyramids becomes the length of any edge of the cone. The perimeter of the *n*-gon becomes the circumference of the circular base of the cone.

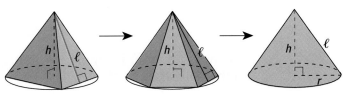

> **Regular Pyramid-Right Cone Lateral Area Formula**
> The lateral area, L.A., of a regular pyramid or right cone is half the product of its slant height ℓ and the perimeter (circumference) p of its base.
>
> $$\text{L.A.} = \tfrac{1}{2}\ell p$$

Example 2

Find the surface area of a right cone with slant height 13 and base of radius 10.

Solution

Use the formula S.A. = L.A. + B.

$$
\begin{aligned}
\text{S.A.} &= \text{L.A.} + B \\
&= \tfrac{1}{2}\ell p + B &&\text{Since L.A.} = \tfrac{1}{2}\ell p \\
&= \tfrac{1}{2}(\ell)(2\pi r) + \pi r^2 &&\text{Since } p = 2\pi r \text{ and } B = \pi r^2 \\
&= \tfrac{1}{2}(13)(2 \cdot \pi \cdot 10) + \pi(10)^2 &&\text{Substitute 13 for } \ell, \text{ and 10 for r.} \\
&= 130\pi + 100\pi \\
&= 230\pi
\end{aligned}
$$

Check

L.A. should be greater than B. (Do you see why?) $130p > 100p$, a rough check.

Cheers! *Cones are an effective means of amplifying voices because they force all the sound into a small space. The megaphones' surface usually displays school names and logos.*

QUESTIONS

Covering the Reading

In 1 and 2, refer to the In-class Activity on page 569.

1. Give your answer to Part 1.

2. Give your answers to Part 2.

3. *True or false.* All the lateral faces of a regular pyramid are congruent to each other.

In 4 and 5, consider the formula L.A. = $\tfrac{1}{2}\ell p$. For the figure given, what does each variable represent? Be as specific as possible.

4. regular pyramid

5. right cone

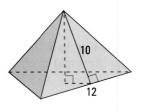

6. A regular triangular pyramid is shown at the left. Its slant height is 10 cm and a side of the base is 12 cm. Find its lateral area.

7. Khafre, another Egyptian pyramid, is a regular square pyramid originally 143 meters tall and 216 meters on a side of the base.
   a. Draw and label a diagram.
   b. What is its approximate slant height?
   c. What is its approximate lateral area?

8. A right cone is the limit of regular pyramids as the number of __?__ increases without bound.

9. A football field measures 120 yd by $53\frac{1}{3}$ yd. About how many times as large as the area of a football field is the lateral area of the pyramid of Khufu? (1 yard = 0.9144 meters)

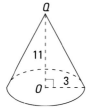

10. Use the right cone pictured at the left.
   a. Find its slant height.
   b. Find its lateral area.
   c. Find its surface area.

11. *Multiple choice.*   For a given cone, which is larger, its lateral area or the area of its base?
   a. lateral area
   b. area of base
   c. neither; they are equal
   d. The answer depends on the cone.

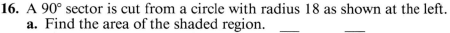

Applying the Mathematics

12. a. In a regular pyramid, which is greater, its height or its slant height?
   b. Explain your answer.

In 13–15, a surface is described. **a.** Draw a picture. **b.** Find its lateral area. **c.** Find its surface area.

13. a regular square pyramid with a base side of 10 and a slant height of 50

14. a regular triangular pyramid with slant height 7 and with a base whose side is 6 and whose area is $9\sqrt{3}$

15. a right cone with height 24 and base diameter 14

16. A 90° sector is cut from a circle with radius 18 as shown at the left.
   a. Find the area of the shaded region.
   b. If the shaded region is folded so that $\overline{AB}$ meets $\overline{CB}$, the lateral surface of a cone is formed. What is the lateral area of this cone?
   c. $\overparen{ATC}$ forms the perimeter of the base of the cone. What is the length of $\overparen{ATC}$?
   d. Use your answer to part **c** to find the area of the base of the cone.

17. Find the surface area of a box with dimensions 1′, 15″, and 18″. *(Lesson 10-1)*

18. Small cans of frozen juice are about 9.5 cm tall and 5.5 cm in diameter. The tops and bottoms are metal; the rest is cardboard.
a. About how much metal is used?
b. About how much cardboard is used? *(Lesson 10-1)*

19. Draw a net for a regular square pyramid with dimensions of your choosing. *(Lesson 9-8)*

20. Find the length of each lateral edge in the pyramid of Question 6. *(Lessons 8-6, 9-4)*

21. What are the four assumed properties of area? *(Lesson 8-2)*

22. Line t goes through the points (6, -4) and (8, 0).
a. Graph t.
b. Find the slope of t.
c. Find the slope of a line perpendicular to t. *(Lessons 3-6, 3-7)*

23. Solve each equation in your head. *(Previous course)*
a. $x + 3 = 64$
b. $y \cdot 3 = 64$
c. $z^3 = 64$

Exploration

24. Draw three circles of diameter 8″. Measure central angles of 45°, 60°, and 120°, and cut out the sectors with these measures as shown in the figures below. Make a cone from each shaded portion.

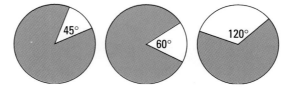

a. Which cone is the tallest?
b. Which cone has the greatest lateral area?
c. Which cone's base has the greatest area?
d. Finish this sentence: As the measure of the central angle increases and a cone is formed, the height of the cone __?__, its lateral area __?__, and its base area __?__.

These lily pads are shaped somewhat like the nets for cones found in Question 24.

These stylized cubes form building blocks the computer artist Brian Yen used to fill space efficiently.

The cube at the right has edges of length 1 unit, so the area of each face is 1 square unit. Since there are six faces, the surface area of this cube is 6 square units. Its *volume* is quite different from its surface area. The volume of this cube is 1 cubic unit or 1 unit^3. For this reason it is called the **unit cube.** Usually volume is measured in cubic units.

1 cubic unit

Volume is a measure of how much a figure will hold, its capacity.

Example 1

What is the volume of a paper bag with base 12″ by 7″ and height 17″?

Solution

Examine the drawing. There are 12 · 7 unit cubes in the layer along a base. There can be 17 layers. The volume is 12″ · 7″ · 17″, or 1428 cubic inches.

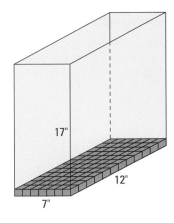

In Lesson 10-1 we found the surface area of this same bag to be 730 square inches. So volume and surface area are different quantities.

17"

12"

7"

Volume Postulate

Notice that the volume of the bag in Example 1 is the product of its three dimensions. This is one of the four fundamental properties of volume assumed in the Volume Postulate. Compare these assumptions with the fundamental properties of area assumed in the Area Postulate in Lesson 8-2.

> **Volume Postulate**
> a. **Uniqueness Property** Given a unit cube, every polyhedral region has a unique volume.
> b. **Box Volume Formula** The volume V of a box with dimensions ℓ, w, and h is found by the formula $V = \ell wh$.
> c. **Congruence Property** Congruent figures have the same volume.
> d. **Additive Property** The volume of the union of two nonoverlapping solids is the sum of the volumes of the solids.

In symbols, if Volume(S) is the volume of a solid S, and A and B are nonoverlapping solids, then the Additive Property of Volume states:

$$\text{Volume}(A \cup B) = \text{Volume}(A) + \text{Volume}(B).$$

Using this property, volumes of various figures can be calculated.

Example 2

Solids I and II are each the union of 5 unit cubes. Give the volume and surface area of each solid.

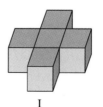 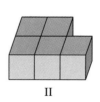

I   II

Solution
The volume for each solid is the same, since it is the sum of the volumes of the five cubes.

$$\text{Volume(I)} = \text{Volume(II)} = 5 \text{ cubic units}$$

In solid I, there are 12 lateral faces (3 on each of the four outside cubes), each with area 1 square unit, and each base has area 5 square units.

$$\text{solid I:} \quad \begin{aligned} \text{S.A.} &= \text{L.A.} + 2 \cdot \text{B} \\ &= 12 \cdot 1 + 2 \cdot 5 \\ &= 22 \text{ square units.} \end{aligned}$$

In solid II, there are 10 lateral faces. Each base has area 5 square units.

$$\text{solid II:} \quad \begin{aligned} \text{S.A.} &= \text{L.A.} + 2 \cdot \text{B} \\ &= 10 \cdot 1 + 2 \cdot 5 \\ &= 20 \text{ square units.} \end{aligned}$$

Cats Cubed.
This puzzle made of eight cubes connected at the edges can be rearranged to show different views. The puzzle's volume is the sum of the volumes of the eight cubes.

Contrasting Volume and Surface Area

Example 2 illustrates that solids may have equal volumes but unequal surface areas. However, many people judge the volume of a container by its surface area. They think, "if it looks bigger, then it holds more!"

Pictured at the left are a salad dressing bottle and a yogurt container. The two containers hold the same amount; they have the same volume. The salad dressing bottle has greater surface area, which may give the false impression that it has greater volume and holds more.

Liquid Volume

In both the metric system and the U.S. customary system, special units are used to measure the volumes of liquids. The basic units are the liter and the gallon. These liquid units can be converted to cubic inches or cubic centimeters using the conversion equations in the first row of the table below. Packaged liquids you might buy in a store, such as juices and yogurt, are often measured in smaller units, such as the milliliter, quart, or pint. Some conversion facts between liquid units are in the second row of the table.

Table of Liquid Measures	
U.S. customary system	**Metric system**
1 liquid gallon (gal) = 231 cubic inches	1 liter (L) = 1000 cubic centimeters
1 liquid quart (qt) = $\frac{1}{4}$ liquid gallon = 32 liquid ounces (oz)	1 milliliter (mL) = $\frac{1}{1000}$ liter
1 liquid pint (pt) = 16 liquid ounces	

These units are converted to each other using methods you may have learned in previous courses.

Example 3

A small carton contains 8 oz of milk. How many cubic inches is this?

Solution

32 oz = $\frac{1}{4}$ gallon, so 128 oz = 1 gallon = 231 cubic inches.
Now form a proportion.

$$\frac{128 \text{ oz}}{231 \text{ in}^3} = \frac{8 \text{ oz}}{x \text{ in}^3}$$

Solve this proportion.

$$128x = 8 \cdot 231$$
$$x = \frac{8 \cdot 231}{128}$$
$$= 14.4375$$

8 ounces of milk occupy 14.4375 cubic inches of volume.

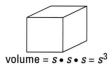

volume = $s \cdot s \cdot s = s^3$

The Volume of a Cube

The formula for the volume of a cube is a special case of the formula for the volume of a box, where each dimension is the same.

Cube Volume Formula
The volume V of a cube with edge s is s^3. $V = s^3$.

Because of the Cube Volume Formula, we call s^3 the "cube of s." If a cube has volume 8, its edge will satisfy $s^3 = 8$, so $s = 2$. We say that 2 is the cube root of 8, and write $2 = \sqrt[3]{8}$. In general, $\sqrt[3]{y}$ is the edge of a cube whose volume is y.

Definition
x is a **cube root of y,** written $x = \sqrt[3]{y}$, if and only if $x^3 = y$.

Calculators differ in the keys they employ to calculate a cube root. You should learn the sequence of keys you can use to find a cube root on your calculator.

Example 4

A cube has a volume of 50 cubic centimeters. What is the length of an edge, rounded to the tenths place?

Solution
Let s be the length of an edge. Since $s^3 = 50$, s is exactly $\sqrt[3]{50}$ cm. A calculator shows $\sqrt[3]{50} \approx 3.684\ldots$.
The length of an edge is approximately 3.7 cm.

Check
Is $(3.7)^3 \approx 50$? Yes, $3.7^3 = 50.653$, which is close enough for a check.

QUESTIONS

Covering the Reading

1. What is the volume of a box with dimensions 30 cm, 70 cm, and 84 cm?

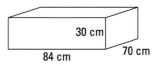

2. Two solids have the same surface area. Must they have the same volume?

3. Two cubes have the same surface area. Must they have the same volume?

4. Some sugar cubes have an edge length of approximately 0.6 in. What is the volume of one cube?

5. What is the volume, in cubic inches, of a pint of milk?

6. A two-liter container contains how many milliliters?

7. x is a cube root of y if and only if ___?___.

In 8–11, use a calculator to find each cube root. Round to the nearest tenth if necessary.

8. 3375   **9.** 2000.376

10. 110   **11.** 6.45

12. A cube has volume 40 cm³. What is the length of an edge
a. exactly?   **b.** to the nearest hundredth?

13. How can you check to see if you have found the correct cube root of a number?

Applying the Mathematics

14. A top is put on the paper bag of Example 1.
a. By how much does the top change the bag's surface area?
b. By how much does it change the bag's volume?

15. The volume of a box is 576 in³. If its base has area 48 in², what is its height?

16. Calculate $\sqrt[3]{25} + \sqrt[3]{100}$ to the nearest hundredth.

17. Some people use the formula $V = Bh$ for the volume of a box.
a. What is B in this formula?
b. Why does this formula work?

18. A cube has volume 29,791 cm³. What is its surface area?

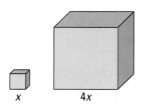

x $4x$

19. One cube has edges x centimeters long. Another cube has edges $4x$ centimeters long. Find the ratio of:
a. the total surface area of the smaller cube to that of the larger cube.
b. the volume of the smaller cube to that of the larger cube.

20. 1 in. = 2.54 cm.
a. What is the volume, in cubic centimeters, of a cube with edge length 1 in.?
b. 1 cubic inch ≈ ___?___ cubic centimeters

21. a. 1 yard = ___?___ feet
b. 1 square yard = ___?___ square feet
c. 1 cubic yard = ___?___ cubic feet

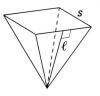

22. Find the surface area of the regular square pyramid pictured at the left. *(Lesson 10-2)*

23. a. In square inches, how much adhesive paper is needed to cover a cylinder 12″ in diameter and 7″ high?
  b. Suppose you can buy the adhesive paper only by the square foot. How many square feet will you need to buy to cover the cylinder of part **a**? *(Lesson 10-1)*

24. Determine the lateral area of the right cone at the left. *(Lesson 10-2)*

25. The tent at the right is in the shape of a right prism with isosceles triangle *GHI* as its base.
  a. If the top of the tent is 5 ft above the ground, find the area of the base of this prism.
  b. If the length of the tent is 7 feet, find the surface area of the tent (including the floor).
  (Lessons 7-7, 8-4, 9-3, 10-1)

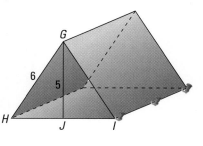

26. Containers holding small amounts can be made to appear to hold more than they do by making them long and thin. Give some examples of these kinds of containers.

27. Two polyhedra made up of 5 unit cubes are shown in Example 2 of this lesson.
  a. Draw some of the other possible polyhedra that are the union of 5 unit cubes.
  b. Which has the greatest surface area?
  c. Which has the least surface area?

Multiplication, Area, and Volume

Packing and stacking. *A common and practical way of transporting products cross-country or worldwide is to fill large, rectangular containers which can be stacked on boats, trains, planes, or trucks.*

Picturing Multiplication with Area

To obtain the area of a rectangle, you merely have to multiply its length by its width. Any positive numbers can be the length and the width. Thus the area of a rectangle is a *model* for the multiplication of two positive numbers. This enables multiplication to be pictured. For instance, the following picture shows that $(3.5)(2.3) = (3 + 0.5)(2 + 0.3) = 3 \cdot 2 + (3)(0.3) + (0.5)(2) + (0.5)(0.3)$.

The multiplication of two polynomials can also be pictured by area. For instance, in algebra you may have seen this picture of $(a + b)(c + d)$.

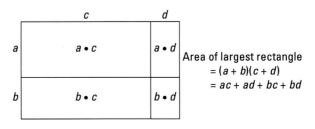

Area of largest rectangle
$= (a + b)(c + d)$
$= ac + ad + bc + bd$

Example 1

a. What binomial multiplication is pictured at the right?

b. Multiply the binomials.

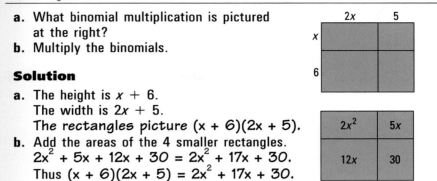

Solution

a. The height is $x + 6$.
The width is $2x + 5$.
The rectangles picture $(x + 6)(2x + 5)$.

b. Add the areas of the 4 smaller rectangles.
$2x^2 + 5x + 12x + 30 = 2x^2 + 17x + 30$.
Thus $(x + 6)(2x + 5) = 2x^2 + 17x + 30$.

The multiplication of two polynomials with more terms can also be pictured with an area diagram. Here is a picture of the multiplication of two trinomials $(a + b + c)(d + e + f)$.

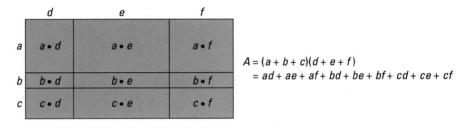

$$A = (a + b + c)(d + e + f)$$
$$= ad + ae + af + bd + be + bf + cd + ce + cf$$

Picturing Multiplication with Volume

The Box Volume Formula $V = \ell wh$ involves the multiplication of three numbers. So the volume of a box can model the multiplication of three polynomials. The biggest box below has dimensions $a + b$, $c + d$, and $e + f$. So its volume is the product of those three binomials. But its volume also is the sum of the volumes of the eight smaller boxes.

Notice that the product of the three binomials consists of all possible products in which one factor is taken from the first binomial, one from the second, and one from the third.

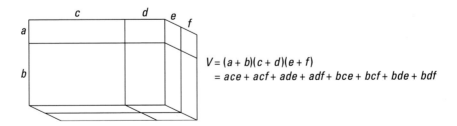

$$V = (a + b)(c + d)(e + f)$$
$$= ace + acf + ade + adf + bce + bcf + bde + bdf$$

Activity

Draw each of the 8 smaller boxes that make up the above figure, and label the length, width, and height of each.

Changing Dimensions of a Box

If you *multiply* each side of a box by a number, the change in volume is easy to find.

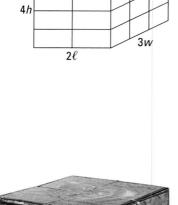

Example 2

A box has dimensions ℓ, w, and h. If the dimensions are multiplied by 2, 3, and 4, respectively, how is the volume of the box changed?

Solution

Draw a picture of the situation. The new volume is $(2\ell)(3w)(4h)$, which is $24\ell wh$, or 24 times the original volume.

In general, multiplying any one dimension of a box by a particular number (and leaving the others constant) multiplies the volume of the box by that number. For example, if the length of a box is doubled but nothing is done to the other dimensions, the volume is doubled. Adding to a dimension of a box does not affect the volume in such a simple way.

Example 3

The dimensions of a box are increased by 2, 3, and 4. What happens to the volume of the box?

Solution

Let the original dimensions of the box be x, y, and z. So the original volume is xyz. Let the new dimensions be ℓ, w, and h, where $\ell = z + 4$, $w = y + 3$, and $h = x + 2$.

$$V = \ell wh = (z + 4)(y + 3)(x + 2)$$

First, multiply two of the binomials.

$$(z + 4)(y + 3) = yz + 3z + 4y + 12$$

Then multiply $x + 2$ by your result.

$$(x + 2)(yz + 3z + 4y + 12) = xyz + 3xz + 4xy + 12x + 2yz + 6z + 8y + 24$$

The volume is increased by $3xz + 4xy + 12x + 2yz + 6z + 8y + 24$.

Check

Let $x = 6$, $y = 8$, $z = 10$. The volume of the original box was 480. The volume of the new box is $(6 + 2)(8 + 3)(10 + 4) = 8 \cdot 11 \cdot 14 = 1232$. Now substitute into the answer.

$$3 \cdot 6 \cdot 10 + 4 \cdot 6 \cdot 8 + 12 \cdot 6 + 2 \cdot 8 \cdot 10 + 6 \cdot 10 + 8 \cdot 8 + 24$$
$$= 180 + 192 + 72 + 160 + 60 + 64 + 24$$
$$= 752, \text{ which is } 1232 - 480. \text{ It checks.}$$

Another shape. Same volume. *This is another configuration of the puzzle shown on page 577.*

Example 3 points out how complicated the change in volume is if you *add* to the dimensions of a figure.

QUESTIONS

Covering the Reading

1. Use the area model for multiplication to write $5.6 \cdot 7.8$ as the sum of four products.

In 2 and 3, a binomial multiplication is pictured. **a.** What multiplication is pictured? **b.** Multiply the binomials.

2.

3.

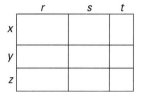

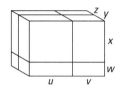

4. **a.** What trinomial multiplication is pictured at the left?
   b. Multiply the trinomials.

5. **a.** What multiplication of three binomials is pictured at the left?
   b. What is the sum of the volumes of the eight boxes at the left?
   c. Multiply the three binomials of part **a.** Does your product check with your answer to part **b**?

6. What is your answer to the Activity in this lesson?

In 7–10, perform the multiplication.

7. $(x + 9)(2x + 3)$

8. $(3a + 5)(7a + 6)$

9. $(a + 1)(a + 2)(a + 3)$

10. $(a + b + 2)(3a + 8b)$

In 11–13, a box has dimensions ℓ, w, and h. Describe how the volume is changed if the box's dimensions are changed in the indicated way.

11. Its length is multiplied by 5, and all other dimensions stay the same.

12. Its length, width, and height are all multiplied by 4.

13. Six is added to the length, and all other dimensions are kept the same.

Applying the Mathematics

14. In a store you see two bags X and Y of couscous ("koos'-koos"), coarse grains of wheat that are prepared by steaming. (Couscous is a common dish in North Africa.) Bag X is 1.4 times as high, 1.2 times as wide, and 1.2 times as deep as bag Y.
    a. How do the capacities of these bags compare?
    b. How should their prices compare?

15. The sum of the areas of the small rectangles at the left is $x^2 + 8x + 15$. If the length of the large rectangle is $x + 5$, what is its width?

16. Expand $(a + b)^3$
    a. by finding the volume of an appropriate cube and by finding the volume of the eight boxes that make up the cube.
    b. by multiplying $(a + b)(a + b)(a + b)$.

17. The makers of Bubblesuds laundry detergent increased the height of the container by 4 cm, keeping other dimensions the same. If this change increased the volume by 12%, what was the original height?

18. *Multiple choice.* A popcorn distributor wants to change the shape of popcorn boxes. The box is 6″ wide, 9″ high, and 2″ thick. Which change will keep the volume constant?
   (a) Add 1″ to the width, subtract 1″ from the height.
   (b) Increase the thickness by 25%, decrease the width by 25%.
   (c) Double the thickness, halve the height.

19. One dimension of a cube is increased by 5 units. Another dimension is decreased by 4 units. The third dimension is kept the same. Is it possible for the resulting prism to have the same volume as the original cube? Why or why not?

Review

20. Follow these steps to determine the number of gallons of liquid in a cubic foot. *(Lesson 10-3)*
   a. Finish the conversion equation: 1 gallon = ___ cubic inches.
   b. 1 foot = ___ inches, so 1 cubic foot = ___ cubic inches.
   c. Use the answers to parts **a** and **b** to complete the equation
      1 cubic foot = ___ gallons.

21. Find the cube root of π to the nearest hundredth. *(Lesson 10-3)*

22. A cube has volume 27 in^3. What is its surface area? *(Lessons 10-1, 10-3)*

23. A playing card is about 5.6 cm by 8.7 cm. If a stack of 72 cards is about 2.7 cm high, what is the volume of a single card? *(Lesson 10-3)*

24. Find the surface area of the right cone at the left. *(Lesson 10-2)*

25. Find the lateral area of the right triangular prism at the right. *(Lesson 10-1)*

26. Solve $\pi r^2 = 10$ for r. *(Previous course)*

Exploration

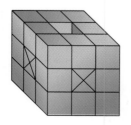

27. A hole is cut through the $3 \times 3 \times 3$ cube at the left, from top to bottom, as indicated. This leaves 24 cubes. Then the large cube with the hole is dipped in green paint.
   a. What is the surface area of the large cube with the hole? (Think of all faces with green paint.)
   How many of the smaller cubes that are left are green on
   b. 3 faces?   **c.** 2 faces?   **d.** 1 face?   **e.** no faces?

28. Repeat Question 27 if two more holes are cut through the cube, one at each place marked with an X.

Volumes of Prisms and Cylinders

Tanks a lot. *After oil is pumped from the ground, it is stored in tank farms like the one shown here. On the average, a tank farm can hold the total of one week of production from the wells in its cylindrical tanks.*

Oil, water, and other products are often stored in huge cylindrical tanks like those pictured above. It is natural to wonder how much water or oil is stored in each tank. The amount can be calculated if you know the dimensions of the tank and a formula for the volume of a cylinder. This formula, and other formulas for the volumes of other common 3-dimensional figures, can be derived from the Area and Volume Postulates you already have studied, using one other postulate given in this lesson.

Volumes of Right Prisms and Cylinders

Consider a prism or cylinder with a base of area B. Think of B unit squares covering the region. We think of this even if B is not an integer.

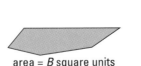

area = B square units

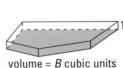

volume = B cubic units

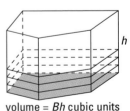

volume = Bh cubic units

If a prism with this base has height 1 unit, the prism contains B unit cubes, and so the volume of the 1-unit-high prism is B cubic units. This is pictured in the middle figure above. If you can stack h of these prisms, as in the figure above at the right, you form a prism with base B and height h. That prism has h times the volume of the middle prism, and so its volume is Bh. This argument shows that if a right prism or cylinder has a base with area B and height h, then a formula for its volume V is $V = Bh$.

A cubic foot of liquid contains about 7.48 gallons. (This is the correct answer for Question 20c of Lesson 10-4.) This information, combined with the volume formula derived on page 587, enables you to find the capacity (in gallons) of an oil storage tank.

Example 1

If a cylindrical oil tank has a diameter of 100 feet and is 70 feet high, how many gallons of oil can it hold? Make a guess before you go on.

Solution

It helps to draw a picture like the one at the right. We use the formula $V = Bh$. The radius of the base is 50', so
$B = \pi(50)^2 = 2500\pi \approx 7854$ sq ft.
Thus,
$$V = Bh$$
$$\approx 7854 \cdot 70$$
$$= 549,780 \text{ cubic feet}$$
$$\approx 550,000 \text{ cubic feet.}$$

We round because our calculations cannot be more accurate than the given data. Now we use the fact that each cubic foot contains about 7.48 gallons of fuel. So the capacity of the tank is

$$550,000 \text{ ft}^3 \cdot 7.48 \frac{gal}{ft^3} \approx 4,110,000 \text{ gallons.}$$

This is more than many people would estimate.

Volumes of Oblique Prisms and Cylinders

Now suppose you have an *oblique* prism or cylinder. Recall that in these figures, the lateral edges are not perpendicular to the planes of the bases. Pictured here are a right prism and an oblique prism with congruent bases and equal heights.

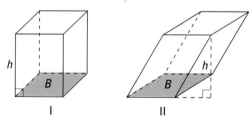

Imagine Prism I to be made up of a stack of thin slices like congruent sheets of paper. You can shift the slices until it takes the form of Prism II.

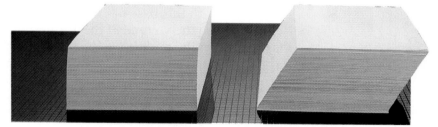

Note that the height, area of the base, and number of slices are the same in Prism I and Prism II. Consequently,

$$\text{Volume(Prism II)} = \text{Volume(Prism I)}.$$

Or, since the two prisms have equal heights and congruent bases,

$$\text{Volume(Prism II)} = Bh.$$

Cavalieri's Principle

The key ideas of this argument are: (1) the prisms have their bases in the same planes; (2) each slice is parallel to the bases; and (3) the slices in each prism have the same area. The conclusion is that these solids have the same volume. The first individuals to use these ideas to obtain volumes seem to have been the Chinese mathematician Zu Chongzhi (429–500) and his son Zu Geng. However, in the West, Francesco Bonaventura Cavalieri (1598–1647), an Italian mathematician, first realized the importance of this principle. So, in the West, the principle is named after him.

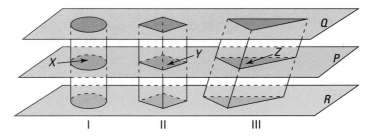

Volume Postulate

e. Cavalieri's Principle Let I and II be two solids included between parallel planes. If every plane *P* parallel to the given planes intersects I and II in sections with the same area, then Volume(I) = Volume(II).

Above, plane P is parallel to planes Q and R containing the bases, and all three solids have bases with area B. Since plane sections X, Y, and Z are translation images of the bases (this is how prisms and cylinders are defined), each also has area B. Thus the conditions for Cavalieri's Principle are satisfied. These solids have the same volume. But we know the volume of the box:

$$\begin{aligned}\text{Volume(II)} &= \ell \cdot w \cdot h \\ &= B \cdot h\end{aligned}$$

Thus, using Cavalieri's Principle,

$$\text{Volume(I)} = B \cdot h \text{ and}$$
$$\text{Volume(III)} = B \cdot h.$$

This proves the following theorem for all cylinders and prisms.

> **Prism-Cylinder Volume Formula**
> The volume V of any prism or cylinder is the product of its height h and the area B of its base. $V = Bh$

In an oblique prism or cylinder, the height sometimes can be determined using the Pythagorean Theorem.

Example 2

Find the volume of the oblique prism with parallelogram base pictured below.

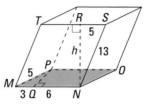

Solution

$V = Bh$. To find B, first look at the base parallelogram *MNOP*. Use the Pythagorean Theorem to find its height *PQ*.

$$MQ^2 + PQ^2 = MP^2$$
$$3^2 + PQ^2 = 5^2$$

So, $\qquad\qquad PQ = 4$

Thus, $\qquad B = MN \cdot PQ = 9 \cdot 4 = 36$ square units.

The height h of the prism also can be found by applying the Pythagorean Theorem.

$$RN^2 + RS^2 = NS^2$$
$$h^2 + 5^2 = 13^2$$

So, $\qquad\qquad h = 12$

Thus $\qquad V = Bh = 36 \cdot 12 = 432$ cubic units.

QUESTIONS

Covering the Reading

1. A cubic foot of liquid is how many gallons?

2. How many gallons of oil can a cylindrical tank with diameter 120 feet and height 60 feet hold?

3. *Multiple choice.* In this lesson, a stack of paper is used to illustrate all but which one of the following?
   (a) Cavalieri's Principle
   (b) An oblique prism and a right prism can have the same volume.
   (c) The volume of an oblique prism is Bh.
   (d) A cylinder and a prism have the same volume formula.

In 4–9, find the volume of each figure.

4. a square prism whose base has edge 5 meters, and whose height is 20 meters

5. a sewer pipe 200′ long with a radius of 2′

6. the oblique prism with parallelogram base drawn at the right

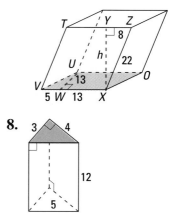

7.

8.

9. a right rectangular prism whose base is 3 feet by 7 feet and whose height is 10 feet

10. Cavalieri's Principle was discovered by mathematicians of what two nationalities?

11. State Cavalieri's Principle.

Applying the Mathematics

12. The cylindrical storage tank in Example 1 has a diameter of about 30 meters and a height of about 21 meters. About how many liters of oil does it hold?

13. If a cylinder has a height h and base with radius r, find a formula for its volume in terms of h and r.

14. Suppose eight pennies are stacked vertically as below at the left. Then they are moved to lean, as below at the right.

Which stack has the greater height, and why?

15. The volume of an oblique prism is 38 cubic meters. Its height is 4 meters. Find the area of each base.

16. An artist constructed a large cylindric solid as shown in the sketch at the left. What is its volume?

base area = 21 m²

17. Two cylindrical glasses have the same height, but the diameters of the glasses are 2.3″ and 3.3″.
    a. Can the second glass hold twice as much as the first?
    b. Why or why not?

18. A cylinder and cone have equal heights and bases of the same area. Why can't Cavalieri's Principle be applied in this situation?

19. Compare the volumes of two cylinders with the same height, but the radius of the second cylinder is half that of the first.

20. A milliliter of water has a mass of 1 gram and occupies 1 cm^3 of space. If a cylindrical can is 15 cm high, has radius 3 cm, and is filled with water, what is the mass of the water (to the nearest gram)?

Review

21. Perform the multiplication $(4x - 5y)(x + 2y + 7)$.   *(Lesson 10-4)*

22. The height of a box is multiplied by 4 and the length and width are each multiplied by 5. By how much is the volume of the box multiplied?   *(Lesson 10-4)*

23. A lead box for vials of plutonium measures 4 cm by 3.5 cm by 2.5 cm on the outside. The inside dimensions are 3 cm by 2.5 cm by 2 cm. How much lead was used to make the box?   *(Lesson 10-1)*

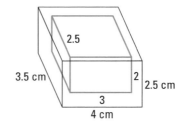

24. Find the surface area of the prism in Question 8.   *(Lesson 10-1)*

25. Arrange these quadrilateral area formulas from most general to most specific: $A = s^2$, $A = \frac{1}{2}h(b_1 + b_2)$, $A = \ell w$, $A = hb$.   *(Lesson 8-5)*

Exploration

26. Stack pennies obliquely as in Question 14. Try to get the top penny as far to the right as possible, as in the picture at the right. How long can you make the distance LR?

LESSON

10-6

Organizing Formulas

Now, why did I tie that string around my finger? *Many people have gimmicks for trying to remember names, birthdates, or other bits of information. The formulas for area and volume of various shapes are related. Use your hierarchy to make these connections, so you do not have to rely upon strict memorization.*

At this point, you have encountered formulas for perimeters and areas of many 2-dimensional regions and for lateral areas, surface areas, and volumes of some 3-dimensional figures. You will see more formulas later in this chapter. At this point, you may wonder which formulas to remember. The answer is simple: *Remember the formulas which apply to the most figures.*

Summarizing Surface Area and Volume Formulas

In this chapter there are eight—only eight—basic formulas. So far you have seen five of the eight. The chart below also includes the three other surface area and volume formulas you will see in the next three lessons. The chart is organized into columns based on the number of bases a figure has. Prisms and cylinders have two parallel bases. Cones and pyramids have one base. Spheres have no bases.

Surface Area and Volume Formulas of 3-Dimensional Figures		
Prisms/Cylinders (two parallel bases)	**Pyramids/Cones** (one base)	**Sphere** (no bases)
Lateral Area $\text{L.A.} = ph$	$\text{L.A.} = \frac{1}{2}\ell p$	
Surface Area $\text{S.A.} = \text{L.A.} + 2B$	$\text{S.A.} = \text{L.A.} + B$	$\text{S.A.} = 4\pi r^2$
Volume $V = Bh$	$V = \frac{1}{3}Bh$	$V = \frac{4}{3}\pi r^3$

If you organize the formulas in this way, you will find them easier to remember because the links between them are clearer. Of course, one thing is true for every formula: *To use a formula, you must know what each variable in the formula represents.*

Deriving Formulas

You may have seen some formulas that are not in the chart. For instance, the formula L.A. = $2\pi rh$ gives the lateral area of a right cylinder. This is not listed because a few general ideas help reduce the load of formulas to remember. *For special formulas for prisms and pyramids, substitute the formulas for the particular base area or perimeter. For special formulas for cones and cylinders, substitute the circle formulas πr^2 for B or $2\pi r$ for p.* The idea is simple: cones and cylinders are like pyramids and prisms, but with circular bases instead of polygonal bases.

For example, to obtain a specific formula for the lateral area of a right cylinder, use the formula L.A. = ph from the chart. Substitute $2\pi r$ for p to get L.A. = $(2\pi r)h = 2\pi rh$. You do not need to memorize a special formula for the lateral area of a right cylinder or a right cone.

In Lesson 10-3, you saw a formula for the volume of a box: $V = \ell wh$. You probably had seen this formula before. From this formula, all the other volume formulas can be deduced. The process we used—the process of proof—is the most powerful idea of all for remembering formulas. *If you cannot remember a formula, try to derive it from some simpler formulas you know to be true.* That is the way mathematicians recall many of the formulas they have to use. The difficulty with this advice is that finding a proof may take some time. Often there is not the time. So, if you do not want to spend your time proving formulas, you must either learn some of them by heart or have access to a list of the formulas.

Using General Formulas

Another way to avoid learning lots of formulas needlessly is to *use general formulas to obtain formulas for special types of figures.* These ideas are applied in the Example.

Secret formula. *Formulas in mathematics are usually not secret, but chemical formulas are often protected for safety. In the MGM film,* Dr. Jekyll and Mr. Hyde, *Spencer Tracy portrays the scientist who creates a potion that transforms his appearance and personality in a sinister way.*

Example

Find a formula for the surface area of a box in terms of its height h, length ℓ, and width w.

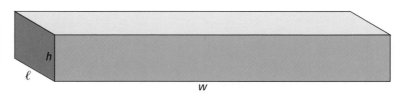

▶

▶ **Solution**

A box is a prism. Begin with the formula for the lateral area of a prism.

$$\text{L.A.} = ph$$

Since h is given, it need not be touched. But p is the perimeter of the base. Since the base is a rectangle with length ℓ and width w, its perimeter is $2\ell + 2w$. Substituting,

$$\text{L.A.} = (2\ell + 2w)h.$$

For the surface area, the areas of the two bases must be added to the lateral area.

$$\text{S.A.} = \text{L.A.} + 2B$$

Each base has area $B = \ell w$. Thus a formula is

$$\text{S.A.} = (2\ell + 2w)h + 2\ell w.$$

Using the Distributive Property, the formula can be rewritten as

$$\text{S.A.} = 2\ell h + 2wh + 2\ell w$$
$$= 2(\ell h + wh + \ell w).$$

That is to say, the surface area of the box pictured on page 594 is the sum of the areas of the faces visible in the drawing, times two.

QUESTIONS

Covering the Reading

1. What are the best formulas to remember?

In 2 and 3, choose from the following: boxes, cones, cylinders, prisms, pyramids, spheres.

2. In which figures does S.A. = L.A. + 2B?

3. In which figures does S.A. = L.A. + B?

4. **a.** List the five formulas in the chart which are in earlier lessons.
   b. List the three formulas in the chart that are in later lessons in this chapter.

5. In the formula L.A. = ph, what does each variable represent?

6. Describe the process for obtaining special formulas for cones and cylinders.

7. How do many mathematicians recall formulas?

In 8 and 9, consider a right cylinder with height h and base with radius r. From the general formulas,

8. deduce a specific formula for its lateral surface area.

9. deduce a specific formula for its volume.

Applying the Mathematics

10. Find a formula for the surface area of the right cylinder of Questions 8 and 9 in terms of its height h and the radius r of its base.

11. A right cone has slant height ℓ and its base has radius r.
    a. Find a formula for its L.A. in terms of ℓ and r.
    b. Find a formula for its S.A. in terms of ℓ and r.

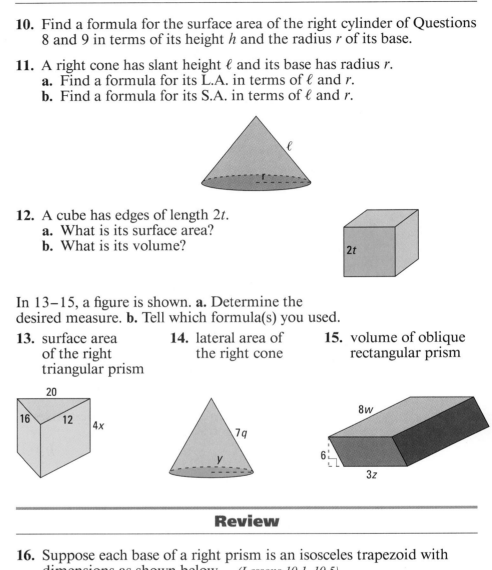

12. A cube has edges of length $2t$.
    a. What is its surface area?
    b. What is its volume?

In 13–15, a figure is shown. **a.** Determine the desired measure. **b.** Tell which formula(s) you used.

13. surface area of the right triangular prism

14. lateral area of the right cone

15. volume of oblique rectangular prism

Review

16. Suppose each base of a right prism is an isosceles trapezoid with dimensions as shown below.   *(Lessons 10-1, 10-5)*
    a. Find the volume of the prism.
    b. Find its lateral area.
    c. Find its surface area.

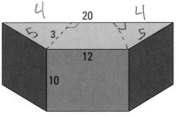

The Pyramids Office Building in Indianapolis, Indiana

17. Suppose the volume of a cylinder is $39,366\pi$ cm^3. If the radius of the base is 27 cm, what is the cylinder's height?   *(Lesson 10-5)*

18. Expand $(x + y)^2$.   *(Lesson 10-4)*

19. By how much is the volume of a cube of side 8 changed if k is added to the length, k is subtracted from the width, and the height remains 8? Draw a picture of the original cube and the resulting box. *(Lesson 10-4)*

20. Pyramids were built throughout the ancient world. Northeast of Mexico City is Teotihuacán's Pyramid of the Sun. This pyramid is presently 66 m high and has a rectangular base measuring 232 m by 200 m.   *(Lesson 10-2)*
   a. What is its lateral area? (Note: There are faces of two different sizes.)
   b. How does its lateral area compare with that of the pyramid of Khufu?

21. A ring between two concentric circles is shaded as shown at the left. If the radius of the small circle is r and the radius of the large circle is R, what is the area of the ring?   *(Lesson 8-8)*

22. In right triangle ABC at the right, find
   a. $m\angle C$.   **b.** BC.   **c.** AC.
   (Lessons 6-2, 8-6)

23. Find the slope of line ℓ below.   *(Lesson 3-6)*

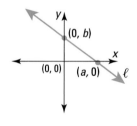

24. Graph $y = -\frac{1}{2}x + 3$.   *(Lesson 1-3)*

Exploration

25. a. Give a possible set of dimensions for a cylinder whose surface area is 200π units^2.
   b. Give dimensions for a second cylinder, not congruent to the first, whose surface area is 200π units^2.

Building a Prism from Pyramids

Work in small groups. You will need orange and blue pencils, scissors, and tape.

1 Trace two copies of the first net shown below. Label the first tracing "net I," and copy the first of each pair of vertex names onto net I. Label the second tracing "net II," and copy the second of each pair of vertex names onto your tracing. Copy the word *base* on the equilateral face of both nets. On net I color $\triangle ADG$ orange, and on net II color $\triangle FGD$ blue.

2 Trace one copy of net III and copy the vertex names onto it. Color $\triangle ADG$ orange and $\triangle FGD$ blue.

3 Make a pyramid from each net, keeping the tracing marks and the labels on the exterior of the pyramids.

4 By placing one base on top and one on bottom, and by matching the blue and orange triangles, form a regular triangular prism from the three pyramids.

5 Use these models as you read Lesson 10-7.

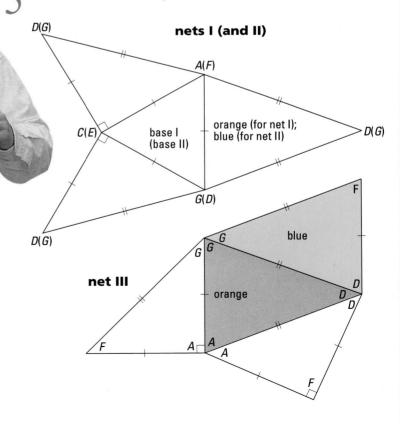

nets I (and II)

base I (base II)

orange (for net I); blue (for net II)

net III

blue

orange

Volumes of Pyramids and Cones

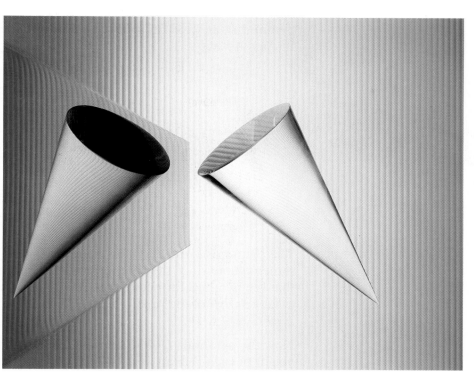

This computer-generated art shows two cones with the same volume.

In this lesson you will learn a simple formula for the volume of any pyramid or cone. The proof of that formula, however, is not so simple. Use the pyramids you constructed in the In-class Activity and examine the figures drawn as you read.

Creating a Prism from Pyramids

Consider the triangular pyramid *D-AGC* below at the left in which $\overline{DC}$ is perpendicular to the plane of the base $\triangle AGC$. (This is the pyramid from net I of the In-class Activity.) Let $B = \text{Area}(\triangle AGC)$ and $h = DC$. We want to compute the volume of *D-AGC*.

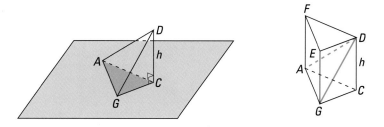

In the In-class Activity, you saw that three pyramids could be put together to form a triangular *prism* with congruent bases *AGC* and *FED* and height *DC*. This prism is pictured above at the right and has volume *Bh*.

We can show that the prism on page 599 has 3 times the volume of the pyramid D-AGC. First, split the prism into the three pyramids of the In-class Activity.

the three pyramids forming the prism

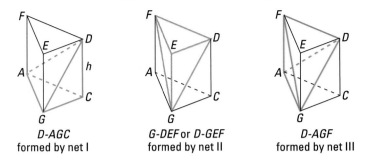

| D-AGC | G-DEF or D-GEF | D-AGF |
| formed by net I | formed by net II | formed by net III |

The pyramids formed by nets I and II, D-AGC and G-DEF, have the same volume since they are formed by the same net. Now we show that the pyramid formed by net III also has the same volume. Look at face AGEF of the prism. It is a rectangle with diagonal $\overline{FG}$, so $\triangle AFG \cong \triangle EGF$.

Now the pyramids D-AGF and D-GEF have the same volume since the height of each is from D to the plane of parallelogram AGEF, and the bases are congruent. But D-GEF and G-DEF are the same pyramid. So, by the Transitive Property of Equality, the three pyramids have the same volume. So each pyramid has a volume that is one third the volume of the prism.

Consequently, the volume of pyramid D-AGC is $\frac{1}{3}Bh$.

Volumes of Any Pyramids

By Cavalieri's Principle, the volume of any pyramid or cone is equal to the volume of a particular pyramid of the same height with a triangular base of equal area. For instance, if a cone has base area 6π units^2 and height h, a triangular pyramid with the same volume can be constructed if its base has area 6π units^2 and its height is h. Furthermore, these need not be regular pyramids or right cones.

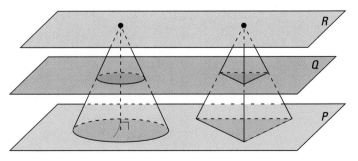

This argument proves the following theorem.

> **Pyramid-Cone Volume Formula**
> The volume V of any pyramid or cone equals $\frac{1}{3}$ the product of its height h and its base area B.
> $$V = \tfrac{1}{3}Bh$$

Thus, a pyramid or a cone which has the same base area and same height as a prism or cylinder has $\frac{1}{3}$ the volume of the prism or cylinder. Often the volume of a pyramid is easier to calculate than its surface area.

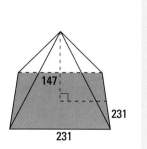

Example 1

Find the volume of the pyramid of Khufu. (Recall that this pyramid is a regular square pyramid 147 m high, and each side of its base is 231 m.)

Solution

Use the Pyramid-Cone Volume Formula
$$V = \tfrac{1}{3}Bh.$$
The base is a square, so $B = (231)^2 = 53{,}361$ square meters. From the given, $h = 147$ meters.
Thus,
$$V = \tfrac{1}{3} \cdot 53{,}361 \cdot 147$$
$$\approx 2{,}610{,}000 \text{ cubic meters.}$$
The volume is about 2,610,000 m³. This is over 5000 times the capacity of a middle-size house or large apartment.

If you know all but one quantity in a formula, you can solve an equation to determine the unknown.

Example 2

If a cone has a height of 6 in. and a volume of 40 in³, what is the radius of its base?

Solution

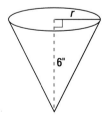

The relevant formula is $V = \tfrac{1}{3}Bh$.
Here $h = 6$ and $V = 40$.
Substituting,
$$40 = \tfrac{1}{3}B \cdot 6.$$
$$40 = 2B$$
So,
$$B = 20.$$
But $B = \pi r^2$. So, $\pi r^2 = 20$.
$$r^2 = \frac{20}{\pi}$$
So,
$$r = \sqrt{\frac{20}{\pi}} \text{ in. exactly,}$$
or
$$r \approx 2.52 \text{ in.}$$

QUESTIONS

Covering the Reading

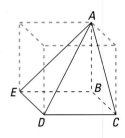

In 1–3, refer to the splitting of a prism into pyramids in this lesson.

1. Why do pyramids *D-AGC* and *G-DEF* have the same volume?

2. Why do pyramids *G-DEF* and *D-GEF* have the same volume?

3. Why do pyramids *D-GEF* and *D-AGF* have the same volume?

4. At the left, how does the volume of *A-BCDE* compare with the volume of the box?

5. Find the volume of a cone with height 8′ and base of radius 2′.

In 6–8, find the volume of the solid.

6. pyramid with right triangular base PCD and height AP

7. trapezoidal pyramid with height 12

8. right cone with height 27 and diameter 18

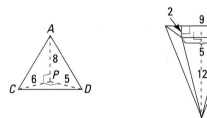
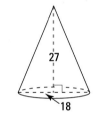

9. A cone has a volume of 40 cm³. Its height is 5 cm. What is the radius of the base?

10. A cone and a cylinder have identical bases and equal heights. If the volume of the cylinder is V, then the volume of the cone is ___?___.

Applying the Mathematics

11. Give the volume V of a cone in terms of its base radius r and height h.

12. The largest monument ever built is the Quetzalcóatl at Cholula de Rivadabia, a pyramid about 60 miles southeast of Mexico City. The Quetzalcóatl is 177 feet tall and its base covers 45 acres. (Note: 1 acre = 43,560 ft².) Determine the volume of the Quetzalcóatl to the nearest million cubic feet.

13. What happens to the volume of a pyramid if its base is kept the same but its height is multiplied by 31.8?

14. What happens to the volume of a cone if the height is kept the same but the radius of the base is multiplied by 7?

Largest monument. *The Quetzalcóatl pyramid is named for the Aztec god, Quetzalcóatl.*

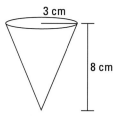
3 cm

8 cm

15. Consider the water cooler cup pictured at the left.
 a. How many cubic centimeters of liquid will it hold?
 b. How many times will it need to be used in order to fill a liter jug?
 c. How much paper is needed to make the cup?

16. A pyramid has a volume of 250 cubic centimeters. Its base is a square and its height is 7.5 centimeters. Find the length of a base edge.

Review

17. How much paper is needed to wrap a box that is 20 inches long, 13 inches wide, and 4.5 inches high? *(Lesson 10-6)*

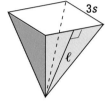
3s

ℓ

18. Give the lateral area of the regular square pyramid pictured at the left. *(Lesson 10-6)*

19. Find the volume and surface area of the right triangular prism pictured at the right. *(Lessons 10-1, 10-5)*

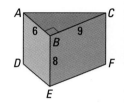

20. In parts **a** and **b**, express the volume of the largest rectangular prism pictured below.
 a. as a product of binomials
 b. as the sum of the volumes of eight smaller rectangular prisms *(Lesson 10-4)*

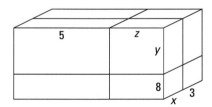

21. A two-by-four is a piece of wood that starts out measuring 2″ by 4″ by k' (it can have any length) but is planed to $1\frac{5}{8}″$ by $3\frac{5}{8}″$ by k'. What percent of the wood is lost in the planing? *(Lesson 10-3)*

Exploration

22. a. Make a cone using a net like that at the right, with $\ell \geq 3$ in.
 b. After the cone is made, make an open-top cylinder with the same height and base with the same radius as the cone.
 c. Fill the cone with dirt or sand. How many times must you empty the cone into the cylinder in order to fill up the cylinder?

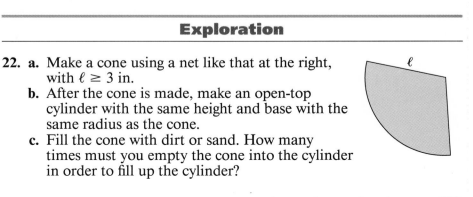
ℓ

Here is how the volume formulas of this chapter have been developed. We began with this postulate in Lesson 10-3.

$$V = \ell wh \quad \text{volume of a box}$$

Cavalieri's Principle was presented in Lesson 10-5, and the following formula was deduced.

$$V = Bh \quad \text{volume of a prism or cylinder}$$

In Lesson 10-7 we saw that a prism can be split into three pyramids with the same height and base. Using Cavalieri's Principle again, this led to another formula.

$$V = \tfrac{1}{3}Bh \quad \text{volume of a pyramid or cone}$$

In this lesson, still another application of Cavalieri's Principle results in a formula for the volume of a sphere. In this application, figures that look quite different will be shown to have the same volume.

Comparing a Sphere with Another Surface

Both the sphere and cylinder below have a height of $2r$. Each cone inside the cylinder has height r. An amazing result is that the volume of the sphere equals the volume between the cylinder and the two cones. The argument that follows uses plane sections and Cavalieri's Principle to prove this result.

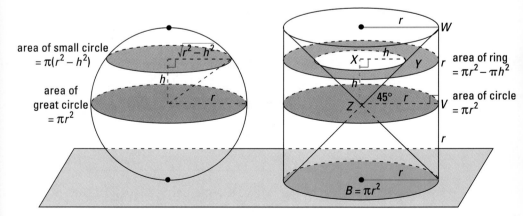

The purple sections in the figures on page 604 are the sections formed when a plane slices these figures in their middles. These sections are congruent circular regions with area πr^2. At h units above each horizontal section are two red sections. We now show that the areas of the red circle and the red ring are equal.

In the sphere, the red section is a small circle. By the Pythagorean Theorem, its radius is $\sqrt{r^2 - h^2}$. The area of the small circle is found using the familiar formula for the area of a circle.

$$\text{Area(small circle)} = \pi\left(\sqrt{r^2 - h^2}\right)^2 = \pi(r^2 - h^2)$$

For the region between the cylinder and the cones, the section is the red ring. The outside circle of the ring has radius r. The inside circle of the ring has radius XY. To find XY, examine $\triangle WVZ$. Since $VZ = WV = r$, $\triangle WVZ$ is an isosceles right triangle. Thus $m\angle VZW = 45$. This makes $m\angle XZY = 45$, so $\triangle XYZ$ must also be an isosceles right triangle. Thus $XY = XZ = h$.

The area of the ring now can be computed.

$$\text{Area(ring)} = \pi r^2 - \pi h^2 = \pi(r^2 - h^2)$$

Thus the red sections have equal area. Since the area of the small red circle equals the area of the ring no matter what height h is chosen, Cavalieri's Principle can be applied.

Finding the Formula

The volume of a sphere is thus the difference in the volume of the cylinder $(B \cdot 2r)$ and the volume of the two cones (each with volume $\frac{1}{3}B \cdot r$).

$$\begin{aligned}
\text{Volume of sphere} &= (B \cdot 2r) - 2 \cdot \left(\tfrac{1}{3}B \cdot r\right) \\
&= 2Br - \tfrac{2}{3}Br \\
&= \tfrac{4}{3}Br
\end{aligned}$$

But here the base of the cone and cylinder is a circle with radius r. So $B = \pi r^2$. Substituting,

$$\begin{aligned}
\text{Volume of sphere} &= \tfrac{4}{3} \cdot \pi r^2 \cdot r \\
&= \tfrac{4}{3}\pi r^3.
\end{aligned}$$

Sphere Volume Formula
The volume V of any sphere is $\frac{4}{3}\pi$ times the cube of its radius r.
$$V = \tfrac{4}{3}\pi r^3$$

The Sphere Volume Formula seems first to have been discovered by Archimedes (287–212 B.C.). Today, we apply his old formula using modern calculators.

Example 1

Find the volume of a sphere with radius 12.

Solution
Substitute into the volume formula.

$$V = \frac{4}{3}\pi r^3$$
$$= \frac{4}{3}\pi(12)^3$$
$$= 2304\pi \approx 7238 \text{ units}^3$$

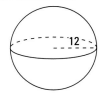

Example 2

A standard bowling ball cannot be more than 27 inches in circumference. What is the maximum volume of such a ball (to the nearest cubic inch) before the holes are drilled?

Solution
First find the radius of the ball.
Use the circumference formula.

$$C = 2\pi r$$
$$27 = 2\pi r$$

So,
$$r = \frac{27}{2\pi} \approx 4.3".$$

Now substitute into the volume formula $V = \frac{4}{3}\pi r^3$.

$$V = \frac{4}{3} \cdot \pi \cdot (4.3)^3$$
$$= \frac{4}{3} \cdot \pi \cdot 79.507$$
$$\approx 333 \text{ cubic inches}$$

Even after drilling the holes, there are more than 300 cubic inches of material in a standard bowling ball.

QUESTIONS

Covering the Reading

1. Cavalieri's Principle is used to derive three formulas in this chapter. Name them.

2. Who discovered the Sphere Volume Formula?

In 3 and 4, use the drawings below of a sphere and a cylinder with two cones inside it.

3. **a.** Give the area of the shaded cross section of the sphere.
   b. Give the area of the shaded ring between the cylinder and the cone.

4. Give the volume of
   a. the cylinder.
   b. the two cones.
   c. the solid region between the cylinder and the two cones.
   d. the sphere.

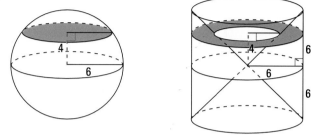

5. **a.** Draw a sphere with radius 3 cm.
   b. Find its volume.

6. *Multiple choice.* How much material is needed to make a standard bowling ball before the holes are drilled?
   (a) less than 100 cubic inches
   (b) between 100 and 200 cubic inches
   (c) between 200 and 300 cubic inches
   (d) between 300 and 400 cubic inches

Applying the Mathematics

7. The volume of a sphere is 268 cubic meters. Find its radius, to the nearest meter.

8. An inflated basketball has a diameter of about 9.4 inches. If it is shipped in a cube-shaped box with edge 9.4 inches, what percent of the box is filled by the basketball?

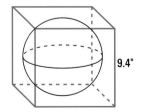

9. A filled ice cream cone has the shape of a **hemisphere** (half a sphere) atop a cone.
   a. If the cone has height 10 cm, and the radius of the hemisphere is 3 cm, how much ice cream is there?
   b. What percent of the ice cream have you eaten when only the ice cream in the cone is left?

10. A spherical water tank with diameter 16 meters supplies water to a small town. The town uses about 500 cubic meters of water per day. How long would a full tank last if
   a. 300 cubic meters were replaced each day?
   b. due to drought conditions no water were replaced?

11. Derive a formula for the volume of a sphere with diameter d.

Review

In 12 and 13, find the volume of the solid.   *(Lessons 10-3, 10-7)*

12.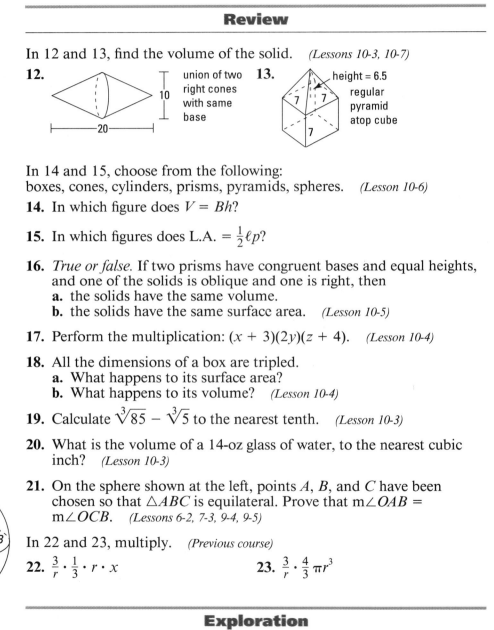

union of two right cones with same base

13. height = 6.5
regular pyramid atop cube

In 14 and 15, choose from the following:
boxes, cones, cylinders, prisms, pyramids, spheres.   *(Lesson 10-6)*

14. In which figure does $V = Bh$?

15. In which figures does L.A. $= \frac{1}{2}\ell p$?

16. *True or false.* If two prisms have congruent bases and equal heights, and one of the solids is oblique and one is right, then
   a. the solids have the same volume.
   b. the solids have the same surface area.   *(Lesson 10-5)*

17. Perform the multiplication: $(x + 3)(2y)(z + 4)$.   *(Lesson 10-4)*

18. All the dimensions of a box are tripled.
   a. What happens to its surface area?
   b. What happens to its volume?   *(Lesson 10-4)*

19. Calculate $\sqrt[3]{85} - \sqrt[3]{5}$ to the nearest tenth.   *(Lesson 10-3)*

20. What is the volume of a 14-oz glass of water, to the nearest cubic inch?   *(Lesson 10-3)*

21. On the sphere shown at the left, points A, B, and C have been chosen so that $\triangle ABC$ is equilateral. Prove that m$\angle OAB =$ m$\angle OCB$.   *(Lessons 6-2, 7-3, 9-4, 9-5)*

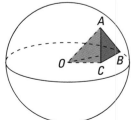

In 22 and 23, multiply.   *(Previous course)*

22. $\frac{3}{r} \cdot \frac{1}{3} \cdot r \cdot x$

23. $\frac{3}{r} \cdot \frac{4}{3} \pi r^3$

Exploration

24. Archimedes was an extraordinary scientist as well as a mathematician. Look in an encyclopedia or other reference and identify two scientific discoveries made by Archimedes.

LESSON

10-9

The Surface Area of a Sphere

A sphere of action. *A geodesic dome is a spherical shape supported by a lightweight frame of triangles. Shown here is B.C. Stadium in Vancouver.*

Shown below are a sphere and its net made out of gores. The surface area of the sphere is equal to the area of its net. But there is no easy way to compute the area of this net. Another strategy is needed.

Breaking a Sphere into Pyramids

Surprisingly, we derive the surface area formula from the volume formula. The idea is to consider a solid sphere as being made up of "almost pyramids" with vertices at the center of the sphere. One such "pyramid," with height h and base area B, is drawn in the sphere at the right. The solid is not exactly a pyramid because its base is not exactly a polygon. Even so, when the "almost pyramid" is small, its volume is close to that of a pyramid, namely $\frac{1}{3}Bh$. Since h equals r, the radius of the sphere, each "almost pyramid" has volume $\frac{1}{3}Br$.

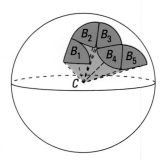

Now break up the entire sphere into "almost pyramids" with bases of areas B_1, B_2, B_3, and so on. The sum of all the B's is the surface area of the sphere.

The volume V of the sphere is the sum of the volumes of all the "almost pyramids" with base area B_1, B_2, B_3, . . .

$$V = \tfrac{1}{3}B_1 r + \tfrac{1}{3}B_2 r + \tfrac{1}{3}B_3 r + \tfrac{1}{3}B_4 r + \ldots$$
$$= \tfrac{1}{3}r(B_1 + B_2 + B_3 + B_4 + \ldots)$$
$$= \tfrac{1}{3} \cdot r \cdot \text{S.A.}$$

Finding the Formula

Now substitute $\tfrac{4}{3}\pi r^3$ for the volume V.

$$\tfrac{4}{3}\pi r^3 = \tfrac{1}{3} r \cdot \text{S.A.}$$

To solve for the surface area, multiply both sides by 3, then divide by r.

$$4\pi r^3 = r \cdot \text{S.A.}$$

Thus, $\qquad\qquad 4\pi r^2 = \text{S.A.}$

> **Sphere Surface Area Formula**
> The total surface area S.A. of a sphere with radius r is $4\pi r^2$.
> $$\text{S.A.} = 4\pi r^2$$

This formula indicates that the surface area of a sphere is equal to 4 times the area of a great circle of the sphere.

Example

Find the surface area of a beach ball with radius 20 cm.

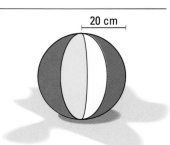

20 cm

Solution
S.A. $= 4\pi r^2 = 4 \cdot \pi \cdot 20^2 = 1600\pi$ exactly, or approximately 5027 square centimeters.

In the Example, the surface area measures how much plastic is needed to make the beach ball. To determine how much air is inside the inflated beach ball, volume is needed. The volume of the beach ball above is $\tfrac{4}{3}\pi r^3 = \tfrac{4}{3}\pi(20^3) \approx 33{,}500 \text{ cm}^3$.

Covering the Reading

1. Why is it hard to calculate the surface area of a sphere from its net?

2. A sphere can be imagined as the union of __?__, whose base areas add to the __?__ of the sphere.

3. State the Sphere Surface Area Formula.

4. The surface area of a sphere is __?__ times the area of one of its great circles.

In 5 and 6, a sphere is described. **a.** Find its exact surface area. **b.** Approximate its surface area to the nearest integer.

5. a sphere with radius 6

6. a sphere with diameter 100″

Applying the Mathematics

7. The area of the United States is about 3,600,000 square miles. What percent is this of the area of Earth, which is approximately a sphere with radius 3960 miles?

8. Derive a formula for the surface area of a sphere with diameter d.

9. B.C. Stadium in Vancouver, which was built as a pavilion at the 1986 Expo, has a geodesic dome with a diameter of about 215 meters. The geodesic dome can be approximated by a sphere. Estimate the cost of covering the dome if the gold foil that covered it cost $3.20 per square meter.

10. The moon has a diameter about $\frac{1}{4}$ that of Earth.
    a. How do their total surface areas compare?
    b. How do their volumes compare?
    c. How do your answers for parts **a** and **b** relate to the number $\frac{1}{4}$?

11. A sphere has volume 36π cubic meters. What is its surface area?

12. A sphere of radius r fits exactly into a cylinder, touching the cylinder at the top, bottom, and sides. How does the total surface area of the sphere compare to the lateral surface area of the cylinder?

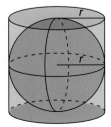

13. Two figure-8 pieces of cowhide are used to cover a baseball. The diameter of the baseball is about 2.9 inches.
    a. Find its surface area, to the nearest square inch.
    b. What must be the area of one cover?

Review

14. Refer to Question 7. Give the volume of Earth, to the nearest million cubic miles. *(Lesson 10-8)*

15. A right cone has a base radius of 15 and a volume of 1800π. Find its height. *(Lesson 10-7)*

16. Name the no-base, one-base, and two-base figures in the chart of 3-dimensional figures you have studied in this chapter. *(Lesson 10-6)*

17. Consider two cylindrical jars of jam. One jar of jam is twice as tall as the other, but only half as wide. Which jar holds more jam? *(Lessons 10-4, 10-5)*

18. You can make the lateral surface of two different cylinders by rolling an 8.5″ by 11″ piece of notebook paper along one or the other of its sides.
a. What is the lateral area of each cylinder?
b. Which has more volume? *(Lessons 10-1, 10-5)*

19. One cube has edges of length 13. A second cube has 125 times the volume of the first cube. To the nearest tenth, what is the length of an edge of the second cube? *(Lesson 10-4)*

20. Use the box at the left.
a. Write an expression for Area(*ABCD*).
b. Write an expression for the volume of the box.
c. Expand the expression in part **b.** *(Lessons 8-5, 10-4)*

21. Cheese is aged in large blocks. The block at the right is to be removed from the aging cellar, wrapped in foil, and shipped. *(Lesson 10-1)*
a. At a minimum, with no overlap, how many square centimeters of foil are needed?
b. A distributor cuts the block up into 15 cm × 7.5 cm × 3 cm slabs to sell to grocery stores. At a minimum, how much foil is needed to wrap all the slabs from the block?

15 cm
15 cm
30 cm

22. Write the statement *A trapezoid is a quadrilateral that has at least two parallel sides* as a conditional. *(Lesson 2-2)*

Exploration

In 23 and 24, use an almanac, atlas, dictionary, or other reference book.

23. What three countries each cover more of the surface of Earth than the United States covers? What percent (to the nearest tenth) of the surface does each cover?

24. What percent (to the nearest tenth) of the surface of Earth is covered by **a.** the Pacific Ocean? **b.** the Atlantic Ocean? **c.** all the oceans together?

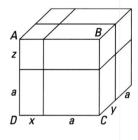

A project presents an opportunity for you to extend your knowledge of a topic related to the material of this chapter. You should allow more time for a project than you do for typical homework questions.

PROJECTS
10
CHAPTER TEN

1 Making Cones
In Question 24 of Lesson 10-2 you were asked to construct cones from sectors of circular disks of the same radius. Do this again and attempt to answer the question: Which central angle yields the cone with the most volume?

2 Sizes of Pyramids of Ancient Cultures
Choose one of the following:
a. the pyramids at Giza, Egypt
b. the pyramids near Mexico City
c. the step pyramids of ancient Babylon.
Look in an encyclopedia or other book for the dimensions of at least two pyramids from the location you have chosen. Draw an accurate picture of them. Calculate the lateral surface area and the volumes of these pyramids.

3 Prismoids
A prismoid (or prismatoid) is a polyhedron that has two parallel bases (that need not be congruent) and whose lateral faces are either quadrilaterals or triangles. The volume of a prismoid is given by the formula $V = \frac{1}{6} h(B_1 + 4B_m + B_2)$, where h is the height, B_1 and B_2 are the areas of the bases, and B_m is the area of a plane section midway between and parallel to the bases. Show that this formula works
a. for any prism or cylinder.
b. for any pyramid or cone (think of one base as being a point and use the fact that sides of the plane section are half the length of sides of the bases.).
c. for any sphere (think of both bases as being points).

5 Domes

Temples and other buildings are sometimes topped off with domes. Write an essay about at least two famous domes, including at least the following information:

a. their names and locations,
b. their shapes, dimensions, surface areas, and volumes, and
c. how and why they were built.

4 What Is Your Surface Area and Volume?

Do these three activities to estimate your surface area and volume.

a. Partially fill a bathtub with water. Then estimate your volume as accurately as you can by immersing yourself in the bathwater, determining the new level of the bathwater, and estimating how much volume of the bathtub you have occupied.

b. Estimate your surface area by covering all parts of your body with rectangular pieces of paper or cloth and adding up the areas.

c. Compare the two estimates in parts **a** and **b** with the surface area and volume of a cylinder that is your height and whose circumference is equal to the circumference of your waist. Do the surface area and volume of this cylinder provide a good approximation to your surface area and volume? Why or why not?

6 Prisms and Cylinders in Packaging

Imagine that you are asked to design containers for a new line of juices. You are to create a Family Size to hold 1200 mL in the shape of a cylinder and a Solo Size to hold 270 mL in the shape of a box. Design and construct two models for each of these sizes. Each model should have the required volume but different dimensions. Finally, write about the pros and cons of each model if used as a juice container.

SUMMARY

The lateral and total surface areas of a 3-dimensional figure are measures of its boundary, which is 2-dimensional. So these areas are given in square units. Volume is a measure of the space enclosed by a 3-dimensional figure, and is given in cubic units. Areas of rectangles can picture the product of two polynomials. Volumes of boxes can picture the product of three polynomials.

From the formula $V = \ell wh$ for the volume of a box, formulas for the volumes of other figures are developed. Pyramids and cones have $\frac{1}{3}$ the

volume of the prism or cylinder with congruent bases and the same height. The volume of a sphere has the same volume as a cylinder of the same height and radius, minus two cones.

Eight key formulas to remember are given in the table below. In all cases r represents radius, h represents height, ℓ represents slant height, p represents perimeter, and B represents area of a base. From these eight formulas you can derive other formulas. If you multiply any one dimension of a 3-dimensional figure by a particular number, the volume of the figure is multiplied by that number.

Surface Areas and Volumes of Common 3-Dimensional Figures		
	Prisms/Cylinders (two parallel bases)	Pyramids/Cones (one base)	Sphere (no bases)
Lateral Area	L.A. $= ph$	L.A. $= \frac{1}{2}\ell p$	
Surface Area	S.A. $=$ L.A. $+ 2B$	S.A. $=$ L.A. $+ B$	S.A. $= 4\pi r^2$
Volume	$V = Bh$	$V = \frac{1}{3}Bh$	$V = \frac{4}{3}\pi r^3$

VOCABULARY

Below are the most important terms and phrases for this chapter. For the starred (*) terms you should be able to give a definition of the term. For the other terms you should be able to give a general description and specific example of each.

Lesson 10-1
surface area, S.A.
lateral area, L.A.
*Right Prism-Cylinder Lateral
    Area Formula
*Prism-Cylinder Surface Area
    Formula

Lesson 10-2
*Pyramid-Cone Surface Area
    Formula
*Regular Pyramid-Right Cone
    Lateral Area Formula

Lesson 10-3
unit cube
volume, capacity
Volume Postulate:
    *Uniqueness Property
    *Box Volume Formula
    *Congruence Property
    *Additive Property
Cube Volume Formula
*cube root

Lesson 10-5
Volume Postulate:
    *Cavalieri's Principle
*Prism-Cylinder Volume Formula

Lesson 10-7
*Pyramid-Cone Volume Formula

Lesson 10-8
*Sphere Volume Formula
hemisphere

Lesson 10-9
*Sphere Surface Area Formula

PROGRESS SELF-TEST

Directions: Take this test as you would take a test in class. You will need a calculator. Then check your work with the solutions in the Selected Answers section in the back of the book.

1. In the right prism at the right, the bases are right triangles.
   a. Find its lateral area.
   b. Find its volume.

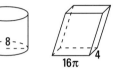

2. Find the height of a right cylinder whose base has radius 7″ and whose lateral area is 84π square inches.

3. A box has a volume of 400 cm³, a height of 10 cm, and a width of 5 cm. What is its length?

4. Refer to the regular square pyramid at the right that has base edges of length 20 and slant height of length 26.
   a. Find its lateral area.
   b. Find its volume.

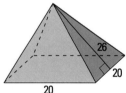

5. Find the volume of a right cone with height 15 and base radius 8.

6. If a pyramid has volume 96 cm³ and height 18 cm, what is the area of its base?

7. Give the surface area and volume of a sphere with radius 19 inches.

8. A sphere has a surface area of 100π units². What is its exact radius?

9. Give the cube root of 400 to the nearest whole number.

10. A right cone has slant height ℓ and base radius r. Find a formula for its lateral area in terms of ℓ and r.

11. A cube has edges of length $9t$.
    a. Find its surface area.
    b. Find its volume.

12. Suppose each edge of one box is 7 times the length of the corresponding edge of another box. How do their volumes compare?

13. Jupiter has a diameter about 11 times that of Earth. How do their total surface areas compare?

14. The two figures below have bases in parallel planes and the same height.
    a. Do they have the same volume?
    b. Why or why not?

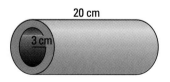

15. If a prism and a pyramid have congruent bases and their heights are equal, how do their volumes compare?

16. How many square meters of fabric do you need to cover a suitcase with dimensions 0.9 m by 0.7 m by 0.2 m?

17. How much paper covers the outside of a cone-shaped megaphone whose large end has radius 4″ and whose slant height is 18″? (Ignore the small open end.)

18. The largest asteroid, Ceres, has a diameter of about 485 miles. If Ceres were spherical in shape, what would be its volume?

19. Find the volume inside of a pipe 20 cm long with an inside radius of 3 cm, as drawn below.

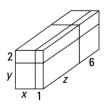

20. Multiply $(3x + 1)(2x + 8)$.

21. Give two expressions for the volume of the box pictured below.

CHAPTER REVIEW

Questions on SPUR Objectives

SPUR stands for **S**kills, **P**roperties, **U**ses, and **R**epresentations. The Chapter Review questions are grouped according to the SPUR Objectives for this chapter.

SKILLS DEAL WITH THE PROCEDURES USED TO GET ANSWERS.

Objective A: *Calculate lateral areas, surface areas, and volumes of cylinders and prisms from appropriate lengths, and vice versa.*
(Lessons 10-1, 10-3, 10-5)

1. Refer to the right cylinder at the right. Find its
   a. lateral area.
   b. surface area.
   c. volume.

2. Refer to the right square prism at the right. Find its
   a. volume.
   b. surface area.

3. The base of the prism at the right is a right triangle with legs of lengths 5 and 12. The distance between the bases of the prism is 24. Find the volume of the prism.

4. If a cylinder has a volume of 30π cubic units and a base with radius 3, what must its height be?

5. Find the surface area of a cube whose volume is 125 cubic units.

6. Find the volume of a right cylinder whose lateral area is 60π square centimeters and whose base has diameter 12 centimeters.

7. A cube has surface area $54e^2$. Find: **a.** the length of an edge. **b.** its volume.

8. Find the volume of a cylinder with diameter $8p$ and height equal to the radius of its base.

Objective B: *Calculate lateral areas, surface areas, and volumes of pyramids and cones from appropriate lengths, and vice versa.*
(Lessons 10-2, 10-7)

9. Find the surface area and volume of the right cone at the right.

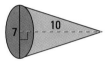

In 10 and 11, refer to the regular square pyramid at the right.

10. Find its volume.

11. Find its
   a. slant height.
   b. lateral area.
   c. surface area.

12. Find the volume of the cone at the right.

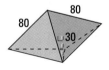

13. The slant height of a regular pentagonal pyramid is 20. The perimeter of the base also is 20. What is the lateral area of the pyramid?

14. If a pyramid has volume 75 cubic centimeters and its base has area 5 square centimeters, what is its height?

15. A cone has volume 900 ft^3. Its height is 20 ft. Find the radius of the base.

16. Find the volume of pyramid X-$ABCD$ below whose base is a rectangle with one side 11 and diagonal 61 and whose altitude is 20.

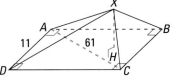

Objective C: *Calculate cube roots.* *(Lesson 10-3)*

17. Give the cube root of 27,000.

18. Approximate the cube root of 40 to the nearest tenth.

19. A cube has volume of 150. To the nearest hundredth, what is the length of an edge of the cube?

20. Calculate $\sqrt[3]{15} + \sqrt[3]{21}$ to the nearest hundredth.

Objective D: *Calculate the surface area and volume of a sphere from appropriate lengths, and vice versa.* *(Lessons 10-8, 10-9)*

21. Give the surface area and volume of a sphere with radius 72.

22. Give the surface area and volume of a sphere with diameter 3 mm.

23. A sphere has volume 288π. What is its radius?

24. A sphere has volume 40π cubic units. What is its surface area?

PROPERTIES DEAL WITH THE PRINCIPLES BEHIND THE MATHEMATICS.

Objective E: *Determine what happens to the surface area and volume of a figure when its dimensions are multiplied by some number(s).* *(Lessons 10-4, 10-8, 10-9)*

25. All of the dimensions of a box are tripled.

   a. What happens to the surface area?

   b. What happens to the volume?

26. The edges of a cube are multiplied by 9. What happens to its volume?

27. The diameter of a pizza is doubled. If its thickness remains the same, how do the volumes of the two pizzas compare?

28. The diameter of the Sun is about 109 times the diameter of Earth. How do their volumes compare?

Objective F: *Develop formulas for specific figures from more general formulas.* *(Lesson 10-6)*

29. At the right, the regular square pyramid has slant height ℓ and a base with side length s.

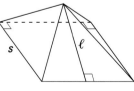

   a. Find a formula for the lateral area.

   b. Find a formula for the surface area.

30. A right cylinder has base radius r and height h.

   a. Find a formula for its lateral area.

   b. Find a formula for its volume.

31. A right cone has slant height ℓ and its base has radius r.

   a. Give a formula for its surface area.

   b. Give a formula for its volume.

32. A cube has edges of $2x$. Find

   a. its surface area.   **b.** its volume.

Objective G: *Know the conditions under which Cavalieri's Principle can be applied.* *(Lesson 10-5)*

33. A prism and pyramid have bases of the same area and equal heights. Can Cavalieri's Principle be applied in this situation? Why or why not?

34. Two cylindric solids have congruent bases and equal heights. One of the solids is oblique and one is right. *True or false.*

   a. The solids have the same volume.

   b. The solids have the same surface area.

In 35 and 36, the two figures have the same height. **a.** Do they have the same volume? **b.** Why or why not?

35.

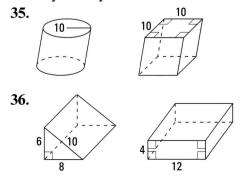

36.

USES DEAL WITH APPLICATIONS OF MATHEMATICS IN REAL SITUATIONS.

Objective H: *Apply formulas for lateral and surface area to real situations.*
(Lessons 10-1, 10-2, 10-9)

37. A jewelry box measures 12″ by 7″ by 4″. Find, to the nearest square inch, the amount of wrapping paper needed to exactly cover it.

38. How much paper is needed to make a cylindrical paper cup with a base diameter of 10 cm and a height of 12 cm?

39. The sides of the base of an ancient square pyramid are 100 cubits (an ancient unit). The pyramid is 50 cubits high. Find its lateral area.

40. A paper holder for an ice cream cone is in the shape of a right cone. If its slant height is 6 cm and its radius is 2 cm, what is its lateral area?

41. Venus is almost spherical with a radius of about 3000 km. To survey Venus completely, about how many square kilometers must be covered?

42. To the nearest square foot, how much leather is needed to make a basketball that is 9.4″ in diameter?

Objective I: *Apply formulas for volume to real situations.* *(Lessons 10-3, 10-5, 10-7, 10-8)*

43. How much wood is in a solid cube 2.5″ on a side?

44. Can a suitcase 3′ long, 1′ wide, and 2′ high hold one million one-dollar bills if a dollar bill is 6.125″ long, 2.562″ wide, and 0.004″ thick?

45. A silo is a cylinder whose base is a circle. One silo has base diameter 6 m and height 10 m. What is the volume of this silo?

46. How much can the paper cup of Question 38 hold?

47. Find the volume of the pyramid in Question 39.

48. An ice cream cone has diameter 2″ and a slant height of 3″. Find its volume to the nearest tenth of a cubic inch.

49. A chocolate golf ball has a radius of 2.6 cm. Ignoring the "dimples" on the ball, find its volume.

50. The Hagia Sophia in Istanbul was built in 532 A.D. Its great dome is a hemisphere with diameter 107′. Find the volume of the dome to the nearest thousand cubic feet.

REPRESENTATIONS DEAL WITH PICTURES, GRAPHS, OR OBJECTS THAT ILLUSTRATE CONCEPTS.

Objective J: *Represent products of two (or three) numbers or expressions as areas of rectangles (or volumes of boxes), and vice versa.* *(Lesson 10-4)*

In 51 and 52, perform the multiplication.

51. $(5x + 2)(4y + 3)$

52. $(a + 6)(2a + 1)(a + 8)$

In 53 and 54, give two expressions for the area of the rectangle.

53.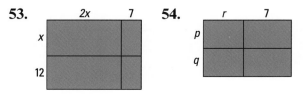

54.

In 55 and 56, give two expressions for the volume of the box.

55.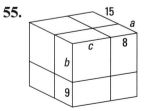

56.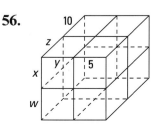

INDIRECT AND COORDINATE PROOFS

The 15 clues below are adapted from a famous puzzle called "Who Owns The Zebra?"

1. There are five houses in a row, each of a different color and inhabited by people of different nationalities, with different pets, drinks, and flowers.
2. The English person lives in the red house.
3. The Spaniard owns the dog.
4. Coffee is drunk in the green house.
5. The Ukrainian drinks tea.
6. The green house is immediately to the right (*your* right) of the ivory house.
7. The geranium grower owns snails.
8. Roses are in front of the yellow house.
9. Milk is drunk in the middle house.
10. The Norwegian lives in the first house on the left.
11. The person who grows marigolds lives in the house next to the person with the fox.
12. Roses are grown at the house next to the house where the horse is kept.
13. The person who grows lilies drinks orange juice.
14. The Japanese person grows gardenias.
15. The Norwegian lives next to the blue house.

Now, who drinks water? And who owns the zebra?

In this chapter, you will learn about the logic of indirect reasoning, and see and write some indirect proofs. Indirect reasoning is used to solve logic puzzles like "Who Owns the Zebra?" and also to prove many theorems. You may find that some indirect proofs are easier to write than direct proofs.

Recall that René Descartes, along with Pierre Fermat, invented the idea of coordinate graphs. Descartes was thrilled with his invention, which he called a *method*, for it used algebra to combine arithmetic and geometry, and so unified all the mathematics known up to that time. He used his method, which is now called **coordinate geometry** or **analytic geometry,** to solve many problems which were then considered to be very difficult or unsolvable. Descartes thought that mathematics and logic could provide the means to solve any problem in any field of endeavor. In this chapter, you will use coordinates the way Descartes used them, to *deduce* properties of figures.

The Logic of Making Conclusions

The "Who Owns the Zebra?" puzzle on page 621 is famous partially because nothing in the clues seems to have anything to do with water or zebras. Consequently, it doesn't seem as if there is enough information to figure out the answer to the question. However, by carefully using logic and ruling out possibilities, the owner of the zebra can be determined.

The logic used to solve this kind of puzzle is the same logic that is used in mathematics. Instead of reasoning from clues, mathematicians reason from postulates, definitions, and previously proved theorems. Instead of finding out who owns the zebra or who drinks water, mathematicians try to find out what is true about figures or numbers or other mathematical ideas.

Logic in Proofs

To begin the study of logic, look at the proof below. We have identified each statement with a letter so that the logic becomes clearer.

Given: $ABCD$ is a parallelogram. (p)
To prove: $m\angle 1 = m\angle 2$. (r)

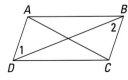

Argument:

Conclusions	Justifications
0. $ABCD$ is a parallelogram. (p)	Given
1. $\overline{AD} \parallel \overline{BC}$ (q)	definition of parallelogram ($p \Rightarrow q$)
2. $m\angle 1 = m\angle 2$ (r)	// Lines $\Rightarrow$ AIA $\cong$ Theorem ($q \Rightarrow r$)

The Law of Detachment

Recall that in a proof argument each justification for a conclusion is a true conditional, with the conclusion as the consequent. This idea is used in step 1 of the argument on page 622. The conditional $p \Rightarrow q$ is the justification for concluding q from the given. The general logical principle is called the *Law of Detachment* because q is "detached" from $p \Rightarrow q$. It is a generalization of the common sense idea that when $p \Rightarrow q$ is a true conditional and p is true, then q also must be true.

> **Law of Detachment**
> From a true conditional $p \Rightarrow q$ and a statement or given information p, you may conclude q.

The Law of Transitivity

Conclusions can be antecedents for making other conclusions. In step 2 of the proof argument on page 622, we use the conclusion q and a theorem $q \Rightarrow r$, to conclude r, giving the theorem as the justification. From steps 1 and 2, we can conclude $p \Rightarrow r$. This logic can be diagrammed as follows:

$$\text{If } p \Rightarrow q$$
$$\text{and } q \Rightarrow r,$$
$$\text{then } p \Rightarrow r.$$

This logical principle is called the *Law of Transitivity*. (It is sometimes called the *Transitive Property of Implication*.)

> **Law of Transitivity**
> If $p \Rightarrow q$ and $q \Rightarrow r$ are true, then $p \Rightarrow r$ is true.

Using Laws of Logic in Everyday Situations

Logical principles can be applied in everyday thinking.

Example 1

A commercial states:
> If you want to be popular, you must dress well.
> If you dress well, you must wear Brand X jeans.

What conclusion does the commercial want you to make?

Solution

Assign variables to the statements.
> p: You want to be popular.
> d: You dress well.
> x: You must wear Brand X jeans.

The commercial states: If p, then d. If d, then x.
Using the symbol $\Rightarrow$, you can write: $p \Rightarrow d$ and $d \Rightarrow x$. ▶

▶ By the Law of Transitivity, you can conclude $p \Rightarrow x$. In words:

    If you want to be popular, you must wear Brand X jeans.

Of course, the commercial wants you to believe that $p \Rightarrow d$ and $d \Rightarrow x$ are true conditionals. Then you would have to accept $p \Rightarrow x$. But many people would *not* accept $p \Rightarrow d$ and $d \Rightarrow x$. If either $p \Rightarrow d$ or $d \Rightarrow x$ is false, you have no reason to conclude $p \Rightarrow x$.

Using Laws of Logic in Geometry

In Example 2, the Law of Detachment and the Law of Transitivity are combined to make a conclusion about a geometric figure.

Example 2

What conclusion about *MBUS* can be made using all three statements below?
(1) Every rhombus is a kite.
(2) The diagonals of a kite are perpendicular.
(3) *MBUS* is a rhombus.

Solution

First, rewrite the first two statements as conditionals.
(1) If a figure is a rhombus, then it is a kite.
(2) If a figure is a kite, then its diagonals are perpendicular.
Next, represent each statement by means of appropriate letters.
(1) $r \Rightarrow k$
(2) $k \Rightarrow p$
(3) MBUS is an instance of r.

From (1) and (2), using the Law of Transitivity, you can conclude $r \Rightarrow p$:
    If a figure is a rhombus, then its diagonals are perpendicular.
From r and $r \Rightarrow p$, using the Law of Detachment, you can conclude p:
    The diagonals of MBUS are perpendicular.
Since the given statements (1)–(3) are true, the conclusion is also true.

Using Laws of Logic to Solve Puzzles

Lewis Carroll [1832–1898], the Englishman best known for writing *Alice's Adventures in Wonderland* and *Through the Looking Glass*, was a logician. That is, he studied the process of reasoning. Under his given name, Charles L. Dodgson, he was a professor at Oxford University and wrote books on logic. Here is a puzzle from one of his books.

Example 3

What can you conclude using all three statements?
(1) My gardener is well worth listening to on military subjects.
(2) No one can remember the battle of Waterloo, unless he is very old.
(3) Nobody is really worth listening to on military subjects, unless he can remember the battle of Waterloo.

The battle of Waterloo is one of the subjects in logic puzzles by Charles Dodgson. See Example 3 on page 624.

Solution

Assign a letter to each statement or clause in the given statements.

Let m: A person is worth listening to on military subjects.
w: A person can remember the battle of Waterloo.
o: A person is very old.

Now analyze the three numbered statements.

(1) is an instance of m (the gardener).
(2) is $w \Rightarrow o$.
(3) is $m \Rightarrow w$.

From (1) and (3), using the Law of Detachment, you can conclude *w*, for the gardener. Together with (2), using the Law of Detachment, you get an instance of *o*, again for the gardener. Lewis Carroll's desired conclusion is: **My gardener is very old.**

Conditionals with False Antecedents

Consider this statement:

If the moon is made of green cheese, then the sun is made of salsa.

The antecedent is "the moon is made of green cheese," a false statement. Is it possible to show an instance of the conditional? No, since an instance requires that the antecedent be true. Is it possible to state a counterexample? No, a counterexample also requires that the antecedent be true. Thus, if an antecedent is not true, you cannot trust any conclusions made from it. They could be true. They could be false.

QUESTIONS

Covering the Reading

1. What did Descartes think mathematics and logic could provide?

2. Which two logical principles are used in the proof on page 622?

3. If *p* is given and $p \Rightarrow q$ is true, then __?__ can be concluded.

In 4–6, what can be concluded from the pair of given statements?

4. (1) Every square is a rectangle.
(2) Every rectangle is a parallelogram.

5. (1) You will look better if you wear designer clothes.
(2) The better you look, the more popular you will be.

6. (1) If a figure is a prism, then the figure is a polyhedron.
(2) A figure is a polyhedron if it is a pyramid.

7. Name the law of logic used.
All cats like fish.
Garfield is a cat.
So Garfield likes fish.

8. Name the law of logic used in the proof argument below.
$s \Rightarrow t$ and $r \Rightarrow s$ are true.
So $r \Rightarrow t$ is true.

9. Explain why you cannot give an instance of the statement *If 2 = 3, then 100 = 200.*

Arthur Ashe (1943–1993)

Applying the Mathematics

In 10–13, assume this statement is true: *If a tennis player has won Wimbledon, then the player is world-class.* Using the Law of Detachment only, what (if anything) can you conclude if you also know the given statement is true?

10. Steffi Graf has won Wimbledon.

11. Pete Sampras has won Wimbledon.

12. Gabriela Sabatini has not won Wimbledon (as of 1995).

13. Arthur Ashe has won the U.S. Open Tennis Tournament.

14. What can you conclude from these two statements? Explain your answer in words.
(1) A triangle has two congruent angles if it is equilateral.
(2) $\triangle ABC$ has two congruent angles.

In 15–18, use this information: In some states if a person has a driver's license, then the person's age is greater than or equal to 16. Tell what you can conclude in those states if you also know the following:

15. Thayer has a driver's license.   **16.** Neesham drives a car legally.

17. Florence is 18 years old.   **18.** Isabel does not drive a car.

19. Lewis Carroll asked if this reasoning is correct. Is it?
(1) Dictionaries are useful.
(2) Useful books are valuable.
Conclusion: Dictionaries are valuable.

In 20 and 21, use the drawing at the right.

20. Let $HI = 15$, $EF = 12$, $EI = 16$, and $FG = 20$.
   a. Find FI.
   b. Find FH.
   c. Find GH.
   d. *Multiple choice.* Which, if any, of the following has not been used in finding GH?
     (i) Law of Detachment
     (ii) Law of Transitivity
     (iii) Pythagorean Theorem
     (iv) All of these are used.

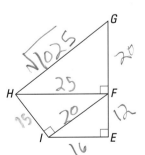

21. Explain why $GH^2 = FG^2 + HI^2 + IE^2 + EF^2$.

22. In quadrilateral *TANG*, angles have measures as marked. Is *TANG* a parallelogram? How do you know? *(Lesson 7-8)*

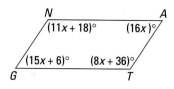

23. Draw a counterexample to this statement: *If a quadrilateral has two consecutive sides congruent, then it is a kite.* *(Lesson 6-4)*

24. a. Graph the lines with equations $y = 2x + 1$ and $y = 2x + 2$.
b. Are these lines parallel, intersecting, or coincident?
c. Are these lines horizontal, vertical, or oblique? *(Lessons 1-3, 3-6)*

25. Find the slope of the line through each pair of points. *(Lesson 3-6)*
a. $(7, 5)$ and $(-3, 1)$
b. (a, b) and $(2a, 2b)$ (Assume $a \neq 0$.)
c. (a, b) and (b, a) (Assume $a \neq b$.)

26. Fill in the blanks. $\sqrt{200} = \sqrt{\underline{\ ?\ }} \cdot \sqrt{2} = \underline{\ ?\ } \cdot \sqrt{2}$ *(Previous course)*

27. Let p be the statement *a figure is a rectangle.* State as many theorems as you can of the form $p \Rightarrow q$. For each theorem, tell what q means for rectangle *ABCD*.

28. Examine the clues in "Who Owns the Zebra?" Make a conclusion that is not given in the clues.

Who owns the snail?

Logicians . . . NOT! *Wayne and Garth are two gnarly negators who comically demonstrate how just to say "NOT!"*

In the movie *Wayne's World*, Wayne's friend Garth is unexpectedly left to host the Wayne's World television show alone. He says: "I'm having a good time. NOT!"

Writing Negations

The quote above and all indirect reasoning uses the idea of *negation*. The **negation** of a statement *p*, called **not-*p*,** is a statement that is true whenever statement *p* is false, and is false whenever statement *p* is true. We say the truth or falseness of a statement is its **truth value.** A statement and its negation always have opposite truth values.

Statement: You study for the next test.

Negation: You do not study for the next test.

Often you can write a negation by just inserting the word *not* in the statement. If the sentence already is negative, then the negation takes out the "not."

p: The quadrilateral *ABCD* is not a parallelogram.

not-p: The quadrilateral *ABCD* is a parallelogram.

Sometimes the word *not* can be avoided by considering all the alternatives.

p: $\triangle ABC$ is isosceles.

not-p: $\triangle ABC$ is scalene.

p:	Coplanar lines ℓ and m intersect at exactly one point.
correct *not-p*:	Coplanar lines ℓ and m are parallel.
incorrect *not-p*:	Coplanar lines ℓ and m have no points in common. This statement is not the negation of p because it does not include all alternatives to p (the lines may be identical).

The negation of a negation is the original statement. The statements p and *not-(not-p)* have the same truth value.

Inverses and Contrapositives of Conditionals

Recall that from any conditional $p \Rightarrow q$, you can form its converse $q \Rightarrow p$. Let p = a triangle is equilateral, and q = a triangle has three acute angles.

Original	$p \Rightarrow q$:	If a triangle is equilateral, then it has three acute angles.
Converse	$q \Rightarrow p$:	If a triangle has three acute angles, then it is equilateral.

Here the original conditional is true, and its converse is false.

Negating *both* the antecedent and the consequent of the original conditional gives a new conditional of the form **not-p $\Rightarrow$ not-q,** called the **inverse** of the original.

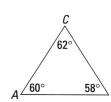

Inverse	*not-p $\Rightarrow$ not-q*:	If a triangle is not equilateral, then it does not have three acute angles.

This inverse is false. The triangle at the left is not equilateral, yet it has three acute angles.

If both parts of the original are negated and the antecedent and consequent are switched, the result is a statement of the form **not-q $\Rightarrow$ not-p.** This statement is called the **contrapositive** of the original.

Contrapositive	*not-q $\Rightarrow$ not-p*:	If a triangle does not have three acute angles, then it is not equilateral.

Notice that this statement is true, just as $p \Rightarrow q$ was true.

Example 1

Given the conditional *If you live in California, then you need a mountain bike*, write its converse, inverse, and contrapositive.

Solution

Converse ($q \Rightarrow p$):	If you need a mountain bike, then you live in California.
Inverse (not-p $\Rightarrow$ not-q):	If you *do not* live in California, then you *do not* need a mountain bike.

Contrapositive (not-q ⇒ not-p):   If you *do not need* a mountain bike, then you *do not live* in California.

The given and all three statements in the solution of Example 1 are false. Both Example 1 and the equilateral triangle example preceding it are instances of the following law of logic.

> **Law of the Contrapositive**
> A conditional ($p \Rightarrow q$) and its contrapositive (not-q ⇒ not-p) are either both true or both false.

That is, a statement and its contrapositive have the same truth value. To summarize: If a given conditional is true, you may conclude that its contrapositive is true, but its converse and inverse may be either true or false.

Making Conclusions Using the Law of the Contrapositive

The Law of the Contrapositive used along with the Law of Detachment and the Law of Transitivity allows you to make many more conclusions.

Example 2

Given are two statements.
(1) Every square is a kite.
(2) Quadrilateral *POTS* is not a kite.
What conclusion can be made using both statements?

Solution

First change statement (1) into a conditional: (1) If a figure is a square, then it is a kite. This is a true conditional. Thus, its contrapositive is also true: If a figure is not a kite, then it is not a square. Now, since (2) POTS is not a kite, you can use the Law of Detachment to conclude that POTS is not a square.

You now have studied enough laws of logic to figure out many logic puzzles. Here is a version of another one of Lewis Carroll's puzzles. Notice how the Law of the Contrapositive and the Law of Transitivity are both used.

Example 3

Given these statements, what can you conclude?
(1) Babies are illogical.
(2) Any person who can manage an alligator is not despised.
(3) Illogical persons are despised.

►

630

Solution

To avoid lots of writing, use variables to name the parts of the statements.

Let
$$B = \text{A person is a baby.}$$
$$D = \text{A person is despised.}$$
$$M = \text{A person can manage an alligator.}$$
$$I = \text{A person is illogical.}$$

Now (1) becomes $B \Rightarrow I$.

(2) becomes $M \Rightarrow \text{not-}D$.

(3) becomes $I \Rightarrow D$.

From (1) and (3) you can conclude (using transitivity) $B \Rightarrow D$.

The contrapositive of (2) is $D \Rightarrow \text{not-}M$. Using transitivity again, $B \Rightarrow \text{not-}M$: A baby cannot manage an alligator.

Why did the alligator cross the road? *Police officers in North Carolina protected this 12-foot long alligator for an hour while it soaked up the warmth of the asphalt in the early morning.*

QUESTIONS

Covering the Reading

In 1–4, write the negation of the statement.

1. $\angle A$ is acute.

2. The perimeter of an *n*-gon with side *s* is *ns*.

3. You were not late for school today.

4. $\triangle GHI$ is scalene.

5. The negation of a statement *p* is written __?__.

In 6–8, a statement is given.
a. Write its converse.
b. Write its contrapositive.
c. Write its inverse.
d. Tell which (if any) of these are true.

6. If $m\angle T = 45$, then $\angle T$ is acute.

7. $p \Rightarrow q$ (Assume $p \Rightarrow q$ is true.)

8. If an equation of a line is $y = k$, then the line is horizontal.

In 9–11, *multiple choice.* The four choices are
(a) negation
(b) converse
(c) inverse
(d) contrapositive.

9. If a statement is true, its __?__ must be true.

10. If a statement is true, its __?__ must be false.

11. If a statement is false, its __?__ must be false.

12. a. Give the contrapositive of this statement: *If △ABC is not a right triangle, then the Pythagorean Theorem does not hold for △ABC.*
  b. Is the contrapositive true?
  c. Is the original statement true?

In 13 and 14, make a conclusion using both of the statements.

13. (1) If a network has only even vertices, it is traversable.
  (2) The network below is not traversable.

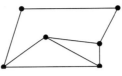

14. (1) If a line rises to the right, it has a positive slope.
  (2) The line $2x + 3y = 5$ has a slope of $-\frac{2}{3}$.

Applying the Mathematics

15. a. Make a conclusion using both these statements.
  (1) If $x = 3$, then $y = 4$.
  (2) $y = 5$
  b. Which two of the three laws—Detachment, Transitivity, and Contrapositive—are needed to make your conclusion?

16. Assuming these three statements are true, what final conclusion can you draw based on all three statements?
  (1) $p \Rightarrow q$
  (2) $q \Rightarrow r$
  (3) not-r

17. Joanne had a date, but her mother, whose word was law, told her, "If you don't apologize to your brother for the way you treated him, then you're not going out tonight." Joanne went on her date. Is it true that she apologized to her brother? Explain your reasoning.

18. José heard an ad that claimed "If you try our product, you won't be sorry." Later on, José was not sorry. Does that mean he tried the product?

In 19 and 20, make a final conclusion based on all the given statements.

19. (1) All equilateral triangles have three 60° angles.
  (2) m∠ABC = 59

20. (1) All unripe fruit is unwholesome.
  (2) All these apples are wholesome.
  (3) No fruit grown in the shade is ripe.
  (This question is adapted from a Lewis Carroll puzzle.)

21. Write the Law of Detachment.   *(Lesson 11-1)*

22. Make a conclusion about *ABCD* from all three statements.
(Lesson 11-1)
(1) *ABCD* is a rhombus.
(2) If a figure is a rhombus, then it is a parallelogram.
(3) Opposite sides of a parallelogram are parallel.

23. Find *PQ*.   *(Lesson 8-6)*

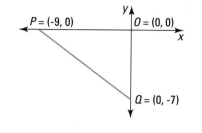

24. Write a proof argument using at least
one triangle congruence theorem.
Given: $\overline{AD} \parallel \overline{BC}$,
            $\overline{AB} \parallel \overline{CD}$.
To prove: $\overline{AB} \cong \overline{CD}$.   *(Lessons 7-2, 7-7)*

25. *Multiple choice.*   In Question 24, what have you proved?
(a) In a parallelogram, opposite sides are parallel.
(b) In a parallelogram, opposite sides are congruent.
(c) If opposite sides of a quadrilateral are congruent, the
quadrilateral is a parallelogram.
(d) If opposite sides of a quadrilateral are parallel, the quadrilateral
is a parallelogram.   *(Lessons 7-7, 7-8)*

In 26 and 27, solve.   *(Previous course)*

26. $6z^2 = 150$ **27.** $(w + 5)^2 = 289$

28. Give a counterexample to prove that $\sqrt{x^2 + y^2}$ is not always equal
to $x + y$.   *(Previous course)*

29. Fill in the blanks.   *(Previous course)*
$$\sqrt{4a^2 + 4b^2} = \sqrt{4(\underline{\;?\;})} = \sqrt{4}\,\sqrt{\underline{\;?\;}} = \underline{\;?\;}\,\sqrt{\underline{\;?\;}}$$

Exploration

30. Make up your own logic puzzle like the one in Example 3.

11-3

Ruling Out Possibilities

Process of elimination. *Toddlers' toys can help develop the ability to reason.*

The Law of Ruling Out Possibilities

Indirect reasoning is based on the idea of ruling out possibilities. The idea is simple, and is used even by animals and babies. If a child knows that a toy is either in a parent's right hand or in the parent's left hand, and the right hand is opened and found empty, the child will know to look in the left hand for the toy. This innate principle of reasoning is called the *Law of Ruling Out Possibilities*.

> **Law of Ruling Out Possibilities**
> When statement p or statement q is true, and q is not true, then p is true.

In mathematics, ruling out possibilities is often easy. For example, you know every angle in a triangle is either acute, right, or obtuse. If $\angle A$ in $\triangle ABC$ is not acute or right, then using the Law of Ruling Out Possibilities, you can conclude that $\angle A$ is obtuse. There is no other possibility.

In real life, if words are not carefully defined, you may not be able to rule out possibilities so easily. For example, if a person is not young, that does not necessarily mean the person is old.

Solving Logic Puzzles

In the questions for this lesson, you are asked to solve some logic puzzles. These puzzles use the idea of ruling out possibilities again and again. Notice how little information is given and how much you can deduce. The same thing happens in geometry.

Here are some hints for doing these puzzles: (1) Logic puzzles take a lot of time and analysis, so do not hurry. (2) Construct a grid and place an X in the square whenever something *cannot* occur. Place an O in the square when the situation *must* occur.

The famous composer and band conductor, John Philip Sousa (1854-1932) designed the sousaphone.

Example

Carol, Sue, Jill, Dave, and Jim each play a different instrument in the school band. The instruments they play are: clarinet, flute, saxophone, trombone, and sousaphone. From the clues below, determine which instrument each student plays.
(1) Carol plays either the clarinet, saxophone, or sousaphone.
(2) Sue does not play the flute.
(3) Dave does not play the trombone, saxophone, flute, or clarinet.
(4) Jim plays either the sousaphone or the saxophone.

Solution

The grid below can be used to solve this puzzle.

The first clue tells you that Carol does not play the flute and that Carol does not play the trombone. Two X_1s in the first row of the grid show this. (We call it X_1 so you can tell it comes directly from clue (1).)

The third clue tells you the four instruments Dave does not play. The four X_3s in the fourth row show this. Of course, this means that Dave must play the sousaphone. We show this with an O. Now we know that no one else plays the sousaphone, so we place four Xs in the column labeled "sousaphone."

	Clarinet	Flute	Saxophone	Trombone	Sousaphone
Carol		X_1		X_1	X
Sue					X
Jill					X
Dave	X_3	X_3	X_3	X_3	O
Jim					X

You are asked to complete this puzzle in Question 4.

Sometimes a single piece of given information may yield many conclusions, as in the following Activity.

a. In Question 11, clue (4) is "Neither Edgar nor Ms. Voila is the guard or the teller." Write five conclusions from this clue.

b. A grid is often very helpful for solving a logic problem like this one. The grid should have a space for each *last name-first name* pair, and also for each *first name-bank job* pair, and also each *bank job-last name* pair. Copy and mark the grid below with X₄s for the conclusions made in part **a.** Save your grid for Questions 5, 6, 7, and 11.

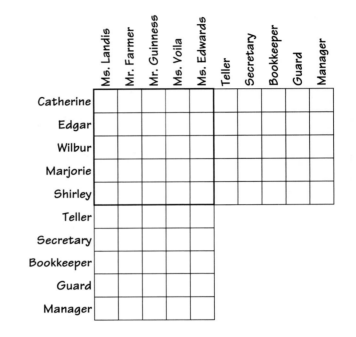

QUESTIONS

Covering the Reading

1. Marilyn tossed a coin. The face that showed was not "heads." You conclude that the face that showed was "tails." What principle of reasoning have you used?

2. If statement *m* or statement *n* is true, and *m* is not true, then ___?___ must be true.

3. **a.** What are the two possibilities for the intersection of two lines in a plane?
   b. If you know that two coplanar lines *s* and *t* are not parallel, what can you conclude?

4. Finish the Example of this lesson.

In 5–7, refer to Question 11 below. Record your responses in the grid you made for the Activity in this lesson.

5. From clue (3) alone, who cannot be the secretary?

6. Write at least two conclusions that follow from clue (1).

7. Write at least two conclusions that follow from clue (5).

Applying the Mathematics

In 8 and 9, make a conclusion from the given information.

8. Line m is not parallel to plane X and m is not in plane X.

9. $ABCD$ is a trapezoid with $\overline{AB} \parallel \overline{CD}$, but $ABCD$ is not a parallelogram.

10. The **Trichotomy Law** for real numbers is: Of two real numbers a and b, either $a < b$, $a = b$, or $a > b$, and no two of these can be true at the same time. Suppose you know that $\sqrt{2} \neq \frac{41}{29}$. What can you conclude by using the Trichotomy Law?

11. The Smalltown Bank has a teller, secretary, bookkeeper, guard, and manager named Mr. Farmer, Mr. Guinness, Ms. Landis, Ms. Voila, and Ms. Edwards, though not necessarily in that order. The two men are Edgar and Wilbur, while the three women are Catherine, Marjorie, and Shirley. From the clues below, determine the first and last names of each person and his or her position at the bank. (Use the grid you copied for the Activity in the lesson.)

    (1) Neither Catherine nor Marjorie is the teller, and neither is Ms. Edwards.
    (2) Shirley is not the guard.
    (3) The secretary is either Catherine or Ms. Landis.
    (4) Neither Edgar nor Ms. Voila is the guard or the teller.
    (5) Mr. Farmer, Edgar, and the bookkeeper have worked at the bank for more than five years.

12. Seven seniors, Joyce, Mike, Darlene, Gary, Wanda, Ken, and Brad, were asked about their career plans. These occupations were mentioned: lawyer, farmer, teacher, doctor, dentist, car dealer, and chemist. No occupation was selected by more than one student. Using the following clues, find out who mentioned which occupation.

    (1) Joyce doesn't want to be a doctor, car dealer, or chemist.
    (2) Mike doesn't want to be a doctor or car dealer either.
    (3) Gary wants to be either a teacher, dentist, or farmer.
    (4) Ken wants to be either a dentist or farmer.
    (5) Brad doesn't want to be a car dealer.
    (6) Darlene wants to be a dentist.

13. Consider this statement: *If my cousin is a good cook, I'll eat my hat!* Write its converse, inverse, and contrapositive. *(Lesson 11-2)*

In 14 and 15, make a conclusion using all three of these true statements. *(Lessons 11-1, 11-2)*

14. (1) Every integer is a real number.
(2) Every natural number is an integer.
(3) The complex number i is not a real number.

15. (1) Every square is a rhombus.
(2) Diagonals in a kite are perpendicular.
(3) A figure is a kite if it is a rhombus.

16. Write the negation of this statement in two different ways: $\triangle ABC$ is *isosceles.* *(Lesson 11-2)*

17. Write a proof argument.
Given: Diagonals of quadrilateral *ABED* intersect at *C*.
$\triangle ABC$ and $\triangle DCE$ are isosceles, both with vertex angle *C*.
To prove: $\triangle ACD \cong \triangle BCE$.
(Lesson 7-3)

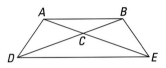

18. *Multiple choice.* In which geometry is there exactly one single line through two points? *(Lessons 1-1, 1-2, 1-4)*
(a) discrete geometry
(b) graph theory
(c) synthetic geometry

19. a. Write an equation for the horizontal line through the point (2, 11).
b. Write an equation for the vertical line through the point (*h*, *k*).
c. What is the intersection of the lines in parts **a** and **b**? *(Lesson 1-3)*

20. Rewrite without the radical sign $\sqrt{}$ or with a smaller number under the radical sign. *(Previous course)*
a. $\sqrt{50}$ **b.** $\sqrt{108}$ **c.** $\sqrt{\frac{1}{4}}$

21. If *z* is positive, then $\sqrt{16z^2}$ can be simplified to __?__. *(Previous course)*

22. Question 11 and the Example in the lesson are adapted from puzzles found in *Pencil Puzzle Treasury*, by Wayne Williams, Grosset and Dunlap, publishers, New York, 1978. Logic puzzles of this type can be found in *Games Magazine*, magazines by Dell Publishing Co., and other places. Find an example of a logic puzzle different from the ones given in this lesson, and solve it.

Reasoning it out. *In some trials, it is the lawyer's job to convince the jury by stating facts, showing contradictions, and building a sound argument.*

A lawyer, summing up a case for a jury, says, "The defendant, my client, is either guilty or not guilty. The prosecutors assume that my client committed the crime. In that case, my client must have been at the scene of the crime. But, as you remember, we brought in witnesses and telephone records that demonstrate my client was on a farm 15 miles away at the time the crime was committed. A person can't be in two places at one time! My client could not have been both on the farm and at the scene of the crime at the same time. So the prosecution's assumption that my client committed the crime cannot be true. Therefore it must be false. Ladies and gentlemen of the jury, the defendant is not guilty."

In this summing up, the lawyer has used *indirect reasoning.*

In **direct reasoning,** a person begins with given information known to be true. The Laws of Detachment and Transitivity are used to reason from that information to a conclusion. The proofs you have written so far in this book have been **direct proofs.**

In **indirect reasoning,** a person examines, and tries to rule out, all the possibilities other than the one thought to be true. This is exactly what you did in solving the logic puzzles of the previous lesson. You marked Xs in boxes to show which possibilities could not be true. When you had enough Xs, you knew that the only possibility left must be correct.

Contradictory Statements

You can rule out a possibility if you know it is false. But how can you tell if a statement is false? One way to tell that a statement is false is if you know its negation is true. For instance, suppose you know that $y = 5$ is true. Then $y \neq 5$ must be false. As another example, suppose you know $\triangle ABC$ is isosceles. Then the statement $\triangle ABC$ *is scalene* must be false.

You also know a statement is false if it contradicts another statement known to be true. For instance, if your friend Lillian is a senior, then she cannot be a junior. If you know 8 is a solution to an equation, and there is only one solution to that equation, then 3 cannot be a solution.

> **Definition**
> Two statements p and q are **contradictory** if and only if they cannot both be true at the same time.

Example 1

Let p be the statement $\angle V$ *is acute*. Let q be the statement $\angle V$ *is right*. Are p and q contradictory? Explain your answer.

Solution

Yes. An acute angle has a measure of less than 90. A right angle has measure 90. From the Angle Measure Postulate, an angle has only one measure, so it cannot be both acute and right at the same time.

Example 2

Let $p = ABCD$ is a rhombus. Let $q = ABCD$ is a rectangle. Are p and q contradictory? Explain your answer.

Solution

No. p and q can be true at the same time. For instance, square ABCD at the right is both a rhombus and a rectangle.

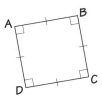

A **contradiction** is a situation in which two contradictory statements p and q are both asserted as true. Contradictions are false statements. In Example 1, the statement p *and* q ($\angle V$ is acute *and* $\angle V$ is right) is false, because p and q are contradictory. However, in Example 2, the statement p *and* q ($ABCD$ is a rhombus *and* $ABCD$ is a rectangle) is not a contradiction, because both p and q can be true at the same time.

Indirect Reasoning

Sometimes it is not easy to tell whether a statement is true or false. However, you may be able to employ the logic used by the lawyer in the situation described at the beginning of this lesson. This is an **indirect argument.**

Step 1: If you want to prove a statement to be false, start by reasoning from it. (The prosecutors thought the defendant was guilty. The lawyer reasoned from this.)

Step 2: Using valid logic, try to make the reasoning lead to a contradiction or other false statement. (The lawyer argued that the defendant would then have been in two places at the same time.)

Step 3: If the reasoning leads to a contradiction or other false statement, the assumed statement must be false. (The lawyer concluded that the defendant was not guilty.)

This logic exemplifies the *Law of Indirect Reasoning*. It is the fifth and last law of logic discussed in this book.

> **Law of Indirect Reasoning**
> If valid reasoning from a statement p leads to a false conclusion, then p is false.

Example 3

Show that the statement $3(4 + 2x) = 6(x + 1)$ is never true.

Solution

Step 1: Begin with the equation and reason from it as you would any normal equation to see what happens. Suppose there is a value of x with $3(4 + 2x) = 6(x + 1)$. Then, using the Distributive Property,
$$12 + 6x = 6x + 6.$$

Step 2: Adding $-6x$ to each side leads to the conclusion
$$12 = 6.$$

Step 3: $12 = 6$ is a false conclusion. So, by the Law of Indirect Reasoning, the original statement $3(4 + 2x) = 6(x + 1)$ is false for all values of x. So $3(4 + 2x) = 6(x + 1)$ is never true.

Indirect Proof

A proof whose argument uses the Law of the Contrapositive, the Law of Ruling Out Possibilities, or the Law of Indirect Reasoning is called an **indirect proof.** To write an argument for an indirect proof, you *suppose* that a statement is true, then reason from that statement, or *supposition.* Following are two examples of indirect proof.

Example 4

Write an indirect proof argument to show that $\sqrt{22,200} \neq 149$.
Given: The real numbers $\sqrt{22,200}$ and 149.
To prove: $\sqrt{22,200} \neq 149$.

Solution

No drawing is needed. We begin by analyzing the problem. There are only two possibilities here. To show that the first possibility is false, reason from it to produce a contradiction.
Argument:
1. Either $\sqrt{22,200} = 149$ or $\sqrt{22,200} \neq 149$. Suppose $\sqrt{22,200} = 149$.
2. Then, squaring both sides, $22,200 = 149^2$. However, by the meaning of power, $149^2 = 149 \cdot 149 = 22,201$.
3. The two statements in step 2 are contradictory. The supposition of step 1, $\sqrt{22,200} = 149$, has led to a false conclusion. By the Law of Indirect Reasoning, $\sqrt{22,200} = 149$ is false. Thus $\sqrt{22,200} \neq 149$.

Notice the steps in the argument. Start by stating all possibilities. Then pick one of the possibilities you think is *not* true (step 1 above) and make conclusions from it (step 2 above). Reason until you get a contradiction or a false statement. Then apply the Law of Indirect Reasoning to rule out the possibility. Rule out all other possibilities in the same way until the statement you wish to prove is the only one left (step 3 above).

Example 5

Write an indirect proof argument to show that no triangle has two obtuse angles.
Given: $\triangle ABC$
To prove: No two angles of $\triangle ABC$ are obtuse.

Solution

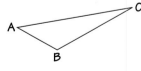

Argument:
1. It helps to draw a triangle $\triangle ABC$. First, list the possibilities. Either $\triangle ABC$ has two obtuse angles (say $\angle A$ and $\angle B$) or it does not. Suppose both $\angle A$ and $\angle B$ are obtuse.
2. Then, by the definition of obtuse, $m\angle A > 90$ and $m\angle B > 90$. By the Addition Property of Inequality, $m\angle A + m\angle B > 180$. Then, because $m\angle C > 0$ for any angle in a triangle, by the Angle Measure Postulate, $m\angle A + m\angle B + m\angle C > 180$. But the Triangle-Sum Theorem says that $m\angle A + m\angle B + m\angle C = 180$.
3. The last two statements in step 2 are contradictory. A false conclusion has been reached. By the Law of Indirect Reasoning, the supposition of step 1 is false. $\angle A$ and $\angle B$ cannot both be obtuse. Thus, no triangle has two obtuse angles.

In Example 5, the following statement was proved: (1) *If a figure is a triangle, then it does not have two obtuse angles.* Its contrapositive is (2) *If a figure has two obtuse angles, then it is not a triangle.* By the Law of the Contrapositive, statement (2) is also true.

QUESTIONS

Covering the Reading

1. What is *indirect reasoning*?

2. When are two statements contradictory?

In 3 and 4, you are given statements *p* and *q*.
a. Are *p* and *q* contradictory?   **b.** Explain your answer.

3. *p*: △*ABC* is isosceles.
   q: △*ABC* is equilateral.

4. *p*: ∠*U* is obtuse.
   q: ∠*U* is right.

5. State the Law of Indirect Reasoning.

6. Show that the statement $2 + 5x = 5x - 8$ is never true.

7. Write an indirect proof to show that $\sqrt{9800} \neq 99$.

8. Write an indirect proof to show that a quadrilateral cannot have four obtuse angles.

Applying the Mathematics

9. Write an indirect proof to prove that a triangle cannot have two right angles.

In 10 and 11, write an indirect proof argument.

10. Refer to △*GHI* at the right.
    Given: m∠*G* > m∠*H* > m∠*I*.
    To prove: △*GHI* is scalene.

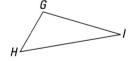

11. Given: In the figure at the right, *PX* > *QX*.
    To prove: *X* is not on the ⊥ bisector of $\overline{PQ}$.

12. The Law of the Contrapositive means that if a statement is true, then so is its contrapositive.
    a. State the contrapositive of the Pythagorean Theorem.
    b. If in △*ABC* at the left, $\overline{AC}$ is the longest side and $AB^2 + BC^2 \neq AC^2$, what can be concluded?

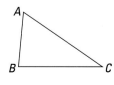

13. Refer to page 43 in Lesson 1-7. Write out the argument for the Line Intersection Theorem as an indirect proof.

14. Write an indirect argument to prove that there is no integer which is greater than all other integers.

15. (This puzzle is taken from *Quizzles*, by Wayne Williams.) Mary, Isobel, Marcia, Grace, and Ruth are on the Grand Avenue School basketball team. Each girl has a different hair color. The hair colors are blond, red, auburn, black, and brunette. As it happens, no two girls on the team are the same height; they are 5′11″, 5′10″, 5′8″, 5′7″, and 5′6″. From the clues given, try to determine the hair color and height of each girl on the team. (Hint: Use a grid like the ones shown in Lesson 11-3.)

   (1) Mary is taller than Ruth, who is two inches taller than the redhead.
   (2) The brunette is not 5′8″ tall.
   (3) Marcia and Mary are neither the tallest nor the shortest.
   (4) The girl with black hair is two inches taller than Ruth.
   (5) Isobel is taller than the blond, who is one inch taller than Grace.
   (Lesson 11-3)

16. Can you draw a logical conclusion that combines the following statements? Explain your reasoning. *(Lesson 11-3)*
   (1) Lines ℓ and m have no points in common.
   (2) Lines ℓ and m are not parallel.

17. Give the negation of this statement: $\triangle ABC \cong \triangle DEF$. *(Lesson 11-2)*

In 18 and 19, a statement is given. **a.** Write its contrapositive. **b.** Write its converse. **c.** Write its inverse. **d.** Indicate the true statement(s). *(Lesson 11-2)*

18. If $\triangle ABC \cong \triangle DEF$, then $\angle A \cong \angle D$.

19. Every prism is a box.

20. Can you draw a logical conclusion that combines the following statements? Explain your reasoning. *(Lesson 11-1)*
   (1) Every square is a parallelogram.
   (2) All rectangles are parallelograms.

21. *True or false.* If y_1 and y_2 are real numbers, then
$$(y_1 - y_2)^2 = (y_2 - y_1)^2. \quad \text{(Previous course)}$$

22. In everyday life, indirect reasoning is often used as follows. If you do *A*, then *B* will happen. But *B* is horrible (or dangerous or some other bad thing). Therefore, you should not do *A*. Give two examples of possible *A*s and *B*s.

LESSON

11-5

Proofs with Coordinates

So far, the proofs that you have seen have been in synthetic geometry, where points are locations. But figures in coordinate geometry have the same properties as those in synthetic geometry. So it is not surprising that there are proofs in coordinate geometry.

Proofs in coordinate geometry have two main differences from proofs in synthetic geometry. First, they involve ideas from algebra. Second, coordinate proofs are often nearly automatic, involving little more than a few calculations.

Recall that the slope m determined by two points (x_1, y_1) and (x_2, y_2) is defined as $m = \frac{y_2 - y_1}{x_2 - x_1}$. Recall also that two lines are parallel if and only if they have the same slope. These ideas are utilized in Example 1.

Example 1

Consider quadrilateral *ABCD* with vertices $A = (0, 0)$, $B = (8, 0)$, $C = (11, 12)$, and $D = (3, 12)$ as shown at the right. Prove or disprove that *ABCD* is a parallelogram.

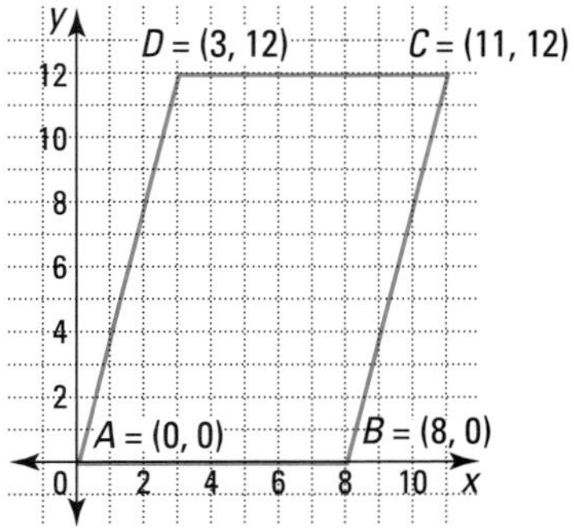

Solution

In the drawing, it appears that *ABCD* is a parallelogram. *ABCD* is a parallelogram if both pairs of opposite sides are parallel. So calculate the slopes of the sides of *ABCD*.

We write this argument in two-column form.

Conclusions	Justifications
1. slope of $\overline{AD} = \frac{12 - 0}{3 - 0} = 4$	definition of slope
slope of $\overline{BC} = \frac{12 - 0}{11 - 8} = 4$	
slope of $\overline{DC} = \frac{12 - 12}{11 - 3} = \frac{0}{8} = 0$	
slope of $\overline{AB} = \frac{0 - 0}{8 - 0} = \frac{0}{8} = 0$	
2. $\overline{AD} \parallel \overline{BC}$, $\overline{DC} \parallel \overline{AB}$	Parallel Lines and Slopes Theorem
3. ABCD is a parallelogram.	definition of parallelogram

The proof in Example 1 is about a single parallelogram. To prove a theorem that applies to many parallelograms, the coordinates of the vertices must be variables.

Example 2

Write a proof argument.

Given: Quadrilateral *PQRS* with $P = (a, b)$, $Q = (c, b)$, $R = (-a, -b)$, $S = (-c, -b)$, such that $a \neq c$ and $-a \neq c$.

To prove: *PQRS* is a parallelogram.

Solution

As in Example 1, calculate slopes.
Argument:

Slope of $\overline{PQ} = \dfrac{b - b}{c - a} = \dfrac{0}{c - a} = 0$

Slope of $\overline{RS} = \dfrac{-b - (-b)}{-a - (-c)} = \dfrac{0}{-a + c} = 0$

Slope of $\overline{PS} = \dfrac{b - (-b)}{a - (-c)} = \dfrac{2b}{c + a}$

Slope of $\overline{QR} = \dfrac{b - (-b)}{c - (-a)} = \dfrac{2b}{c + a}$

By the Parallel Lines and Slopes Theorem, $\overline{PQ} \parallel \overline{RS}$ and $\overline{PS} \parallel \overline{QR}$. So, by the definition of parallelogram, PQRS is a parallelogram.

Convenient Locations

The coordinates of the vertices in Example 2 were carefully chosen. Coordinate axes can always be positioned so that the vertices of a parallelogram have the locations of *P*, *Q*, *R*, and *S* above. Just put the origin at the intersection of its diagonals and make the x-axis parallel to one side. This is called a *convenient location* for the parallelogram. A **convenient location** for a figure is one in which its key points are described with the fewest possible number of variables. In Example 2, the vertices of the parallelogram are described with only the three variables *a*, *b*, and *c*.

The location in Example 2 turns out to be convenient because the parallelogram is rotation-symmetric with center of rotation (0, 0). When a polygon is reflection-symmetric, a convenient location can usually be found by locating the polygon so that it is symmetric to the *x*-axis or the *y*-axis. Otherwise, a convenient location can be found by placing the polygon with one vertex at (0, 0) and another vertex on one of the axes.

To remember convenient locations, recall how to find certain reflection and rotation images on a coordinate plane.

The reflection image of (*a*, *b*) over the *x*-axis is (*a*, -*b*).
The reflection image of (*a*, *b*) over the *y*-axis is (-*a*, *b*).
The image of (*a*, *b*) under a rotation of 180° about the origin is (-*a*, -*b*).

The point (a, b) and the images $(a, -b)$, $(-a, b)$, and $(-a, -b)$ are the vertices of a rectangle. This is one of the convenient locations for a rectangle. Here are convenient locations for some of the figures you have studied.

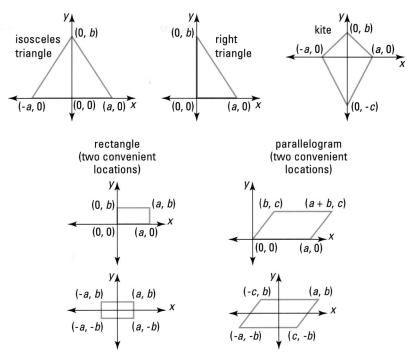

Indirect Coordinate Proofs

Coordinate proofs may be direct or indirect. Example 3 uses indirect reasoning to show that two lines are not perpendicular.

Example 3

If $T = (3, 6)$, $O = (-1, -2)$, and $W = (-3, 1)$, use an indirect argument to show that $\overleftrightarrow{WO}$ is *not* perpendicular to $\overleftrightarrow{WT}$.

Solution

A drawing shows that the lines look nearly perpendicular.

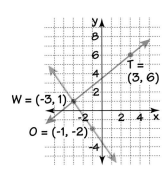

1. Either $\overleftrightarrow{WO} \perp \overleftrightarrow{WT}$ or they are not perpendicular. Suppose $\overleftrightarrow{WO} \perp \overleftrightarrow{WT}$. Then, by the Perpendicular Lines and Slopes Theorem, the product of their slopes is -1.

2. But by the definition of slope, Slope of $\overleftrightarrow{WT} = \frac{6 - 1}{3 - (-3)} = \frac{5}{6}$ and slope of $\overleftrightarrow{WO} = \frac{-2 - 1}{-1 - (-3)} = \frac{-3}{2}$.
The product of the slopes is $\left(\frac{5}{6}\right)\left(\frac{-3}{2}\right) = \frac{-15}{12}$.

3. The statements in steps 1 and 2 are contradictory. By the Law of Indirect Reasoning, the supposition of step 1 is false. $\overleftrightarrow{WO}$ is not perpendicular to $\overleftrightarrow{WT}$.

Notice how automatic coordinate proofs of parallelism and perpendicularity can be. Just calculate and compare slopes! This helps explain why Descartes was so optimistic about his method.

QUESTIONS

Covering the Reading

1. *True or false.* Most of the proofs in this text have been in synthetic geometry.

2. Given $G = (0, 0)$, $H = (6, 0)$, $I = (9, 1)$ and $J = (3, 1)$. Complete the proof argument below that $GHIJ$ is a parallelogram.

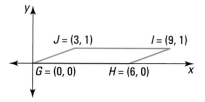

Argument:

Conclusions	Justifications
1. slope of $\overline{GJ} = \frac{1}{3}$	**d.**
slope of $\overline{HI} =$ **a.**	
slope of $\overline{GH} =$ **b.**	
slope of $\overline{IJ} =$ **c.**	
2. $\overline{GH} \mathbin{//} \overline{IJ}$ and $\overline{HI} \mathbin{//} \overline{GJ}$	**e.**
3. **f.**	definition of parallelogram

3. In quadrilateral $WXYZ$, $W = (0, 0)$, $X = (a, 0)$, $Y = (a + b, c)$, and $Z = (b, c)$. If $a \neq 0$ and $b \neq 0$, prove that $WXYZ$ is a parallelogram.

In 4–7, draw a figure of the indicated type in a convenient location on a coordinate system.

4. right triangle

5. rectangle

6. isosceles triangle

7. kite

8. If $X = (3, 7)$, $Y = (11, 3)$, and $Z = (4, 10)$, use an indirect proof to show that $\overleftrightarrow{XZ}$ is not perpendicular to $\overleftrightarrow{XY}$.

Applying the Mathematics

In 9–11, use quadrilateral *EFGH* shown at the right. Write an argument to prove each statement.

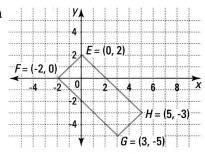

9. *EFGH* is a parallelogram.

10. *EFGH* is a rectangle.

11. $\overline{EG}$ and $\overline{FH}$ (not drawn) are not perpendicular.

12. Given $S = (1, 3)$, $P = (4, 4)$, $A = (3, 1)$, and $T = (0, 0)$.
   a. Draw *SPAT*.   **b.** Prove that $\overline{SA} \perp \overline{PT}$.

13. a. Draw a square in a convenient location on a coordinate system. (Hint: only one variable is needed.)
   b. Using this location, prove that its diagonals are perpendicular.

14. Full-grown zebras can range from 46 to 55 inches high at the shoulder and their weights can range from 550 to 650 pounds. Let h be these possible heights and w be these possible weights.
   a. Graph all possible ordered pairs (h, w).
   b. Describe the graph.

Review

15. Give an indirect proof to show that $\sqrt{39,600} \neq 199$. *(Lesson 11-4)*

16. a. Draw a quadrilateral with exactly
     (i) no right angles
     (ii) one right angle
     (iii) two right angles
     (iv) four right angles
   b. Use an indirect proof to show that no quadrilateral has exactly three right angles. *(Lesson 11-4)*

In 17 and 18, what (if anything) can you conclude using all of the statements?

17. (1) Either Julie walks to school or she rides her bicycle to school.
   (2) Julie's bicycle is being repaired. *(Lesson 11-3)*

18. (1) All people who grew up in Mississippi have a southern accent.
   (2) If you did not grow up in Mississippi, then you do not know the Ole Miss fight song.
   (3) Murray does not have a southern accent. *(Lessons 11-1, 11-2)*

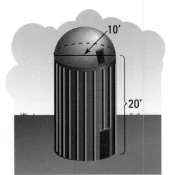

19. **a.** A silo is shaped with a hemisphere atop a cylinder. The cylinder is 10′ in diameter and 20′ high. If the outside is to be painted, what is the measure of the area to be painted?
    b. If the volume of a bushel of corn is about $\frac{1}{14}$ cubic feet, about how many bushels of corn could be stored in this silo?
    (Lessons 10-1, 10-5, 10-8, 10-9)

20. The orange scheelite crystal below is in the shape of a regular octahedron. Give the number of vertices, faces, and edges of this crystal. *(Lesson 10-7)*

21. **a.** What is the distance between two points on a number line with coordinates 50 and 500?
    b. What is the distance between (50, 100) and (500, 100)? *(Lesson 1-2)*

In 22 and 23, simplify if $a \geq 0$ and $b \geq 0$. *(Previous course)*

22. $\sqrt{a^2 b^2}$

23. $\sqrt{a^2 + b^2}$

Exploration

---

24. Three vertices of a parallelogram are (2, 6), (-1, 5), and (0, -4).
    a. Find at least two possible locations of the fourth vertex.
    b. Are there other possible locations for the fourth vertex?

Going the distance. *Running distances has been an event in the Special Olympics since they began in 1968.*

Distances on the Coordinate Plane

In the Example below, we find the distance between two points. The method is to draw a right triangle whose hypotenuse represents the distance, and then to apply the Pythagorean Theorem.

Example 1

Find the distance d between (-8, 50) and (30, -11).

Solution

Draw a rough graph. Then identify the coordinates of the third vertex of a right triangle, as shown here. The lengths of the legs are $30 - (-8)$, or 38, and $50 - (-11)$, or 61. Now use the Pythagorean Theorem.

$$d^2 = 38^2 + 61^2 = 5165.\ \text{So}$$
$$d = \sqrt{5165} \approx 71.87.$$

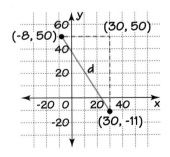

Activity

1. On a sheet of graph paper plot points $A = (3, 6)$ and $B = (-4, -2)$.
2. Draw a vertical line through A and a horizontal line through B. Let these lines intersect at C.
3. $\triangle ABC$ is a right triangle. Give the coordinates of C.
4. Find AC and BC.
5. Use the Pythagorean Theorem to find AB.

You could draw a right triangle every time you need to calculate distances on an oblique line, but it is easier to have a general formula. The process is the same as in Example 1 and the Activity.

Let $P = (x_1, y_1)$ be the first point and $R = (x_2, y_2)$ be the second point.

Find a third point Q so that $\overline{PR}$ is the hypotenuse of a right triangle $\triangle PQR$. Such a point is $Q = (x_2, y_1)$.

Using the Pythagorean Theorem,
$$PR^2 = PQ^2 + QR^2.$$
$$= (x_2 - x_1)^2 + (y_2 - y_1)^2$$

Taking the square roots of both sides,
$$PR = \sqrt{(x_2 - x_1)^2 + (y_2 - y_1)^2}.$$

This is a formula you should memorize.

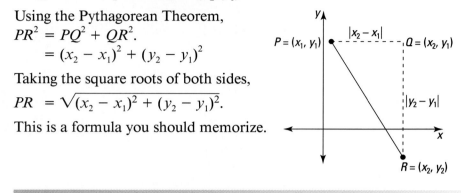

Distance Formula on the Coordinate Plane
The distance d between two points (x_1, y_1) and (x_2, y_2) in the coordinate plane is given by the formula

$$d = \sqrt{(x_2 - x_1)^2 + (y_2 - y_1)^2}.$$

With the Distance Formula, Example 1 can be solved without drawing a figure.

Let $(x_1, y_1) = (-8, 50)$ and let $(x_2, y_2) = (30, -11)$. If d is the distance between these points,

$$d = \sqrt{(x_2 - x_1)^2 + (y_2 - y_1)^2}.$$
$$= \sqrt{(30 - -8)^2 + (-11 - 50)^2}$$
$$= \sqrt{(38)^2 + (-61)^2}$$
$$= \sqrt{1444 + 3721}$$
$$= \sqrt{5165}$$
$$\approx 71.87$$

A grid can be put behind any drawing (blueprint, map, picture, and so on) to assign coordinates to points. With the Distance Formula, you can calculate distances between any two points on the drawing without having to measure.

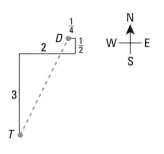

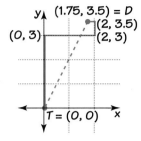

Example 2

Tom likes to bike from his apartment to Doris's house following the path shown at the left: 3 miles north, then 2 miles east, then $\frac{1}{2}$ mile north, then $\frac{1}{4}$ mile west. How far apart do they live?

Solution

Think of the path as being on a coordinate grid with Tom's apartment T at $(0, 0)$. Doris's house D is at the point $(1.75, 3.5)$. The distance between them is the length of $\overline{TD}$. Using the distance formula,

$$TD = \sqrt{(1.75 - 0)^2 + (3.5 - 0)^2}$$
$$= \sqrt{15.3125}$$
$$\approx 3.9.$$

They live about 3.9 miles apart.

Using the Distance Formula in Proofs

You can use the Distance Formula to prove that segments on the coordinate plane are congruent. Example 3 presents a coordinate proof of a theorem you have known for some time.

Example 3

Use coordinates to prove that the diagonals of a rectangle are congruent.

Solution

First, write the given and to prove in terms of a representative figure.

Given:　　RSTU is a rectangle.
To prove:　RT = US.

Drawing:
One of the convenient locations for any rectangle RSTU is with $R = (0, 0)$, $S = (a, 0)$, $T = (a, b)$, and $U = (0, b)$.

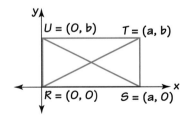

Argument:
Using the Distance Formula,

$$RT = \sqrt{(0 - a)^2 + (0 - b)^2} = \sqrt{a^2 + b^2}$$
$$\text{and } US = \sqrt{(a - 0)^2 + (0 - b)^2} = \sqrt{a^2 + b^2}.$$

By substitution, RT = US.

Again notice that the coordinate proof is simple and straightforward. However, it does require a knowledge of algebra and how to find square roots.

Covering the Reading

In 1–3, use the figure at the right to find the distance.

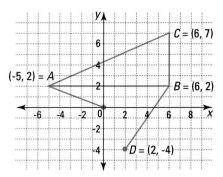

1. *AB*

2. *BC*

3. *AC*

4. Give your answers to the Activity on page 651.

5. Give the distance between (x_1, y_1) and (x_2, y_2).

In 6 and 7, find the distance between the points.

6. (14, -7), (28, 90)

7. (-10, 11), (-31, -25)

8. To get to a hospital from the middle of a nearby town, you can drive 8 miles east, turn right and go 4 miles south, and then turn right again and go 1 mile west. By helicopter, how far is it from the middle of the town to the hospital? (Ignore the altitude of the helicopter.)

9. A convenient location for an isosceles trapezoid is pictured at the right. Use this location to prove that the diagonals of an isosceles trapezoid are congruent.

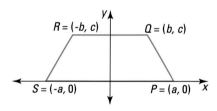

Applying the Mathematics

10. Let $J = (-5, 0)$, $K = (5, 8)$, and $L = (4, -1)$.
    a. Prove that $\triangle JKL$ is isosceles by using the Distance Formula.
    b. Is $\triangle JKL$ equilateral? Explain why or why not.

11. On a map, it can be seen that Charles lives one mile east and 1.5 miles south of school, while Cynthia lives 2 miles west and 0.8 mile south of the same school.
    a. Draw a graph with appropriate coordinates for the school, Charles's residence, and Cynthia's residence.
    b. How far do Charles and Cynthia live from each other?

12. Let $A = (-1, 3)$ and $B = (11, 2)$. Prove that the point $C = (3, -7)$ is on the circle with center B and radius BA. Include a picture.

13. Using coordinates, prove that each pair of opposite sides of a parallelogram are congruent.

14. Given: $A = (1, 0)$, $B = (4, 0)$, $C = (5, 1)$, and $D = (3, 3)$. Prove that $\overline{BC}$ is not parallel to $\overline{AD}$. *(Lesson 11-5)*

15. Prove that, when $z \neq 0$, the triangle with vertices $P = (3z, 4z)$, $Q = (7z, 2z)$, and $R = (2z, -8z)$ is a right triangle. *(Lesson 11-5)*

16. Show by indirect reasoning that a quadrilateral cannot have four acute angles. *(Lesson 11-4)*

In 17 and 18, what (if anything) can you conclude from the given statements, using the rules of logic? *(Lessons 11-1, 11-2, 11-3)*

17. (1) You are elected President of the U.S. if you win the majority of electoral votes.
(2) In 1960, Richard Nixon was not elected President.

18. (1) Last Saturday Notre Dame won the football game 6–3.
(2) You can score six points in football with three safeties, two field goals, or one touchdown.
(3) Notre Dame did not have any safeties or touchdowns.

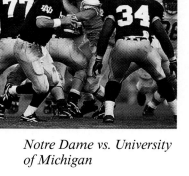

Notre Dame vs. University of Michigan

19. A parallelogram and a triangle have the same base and same altitude. How are their areas related? *(Lessons 8-5, 8-6)*

20. The measure of one acute angle of a right triangle is 45 more than the measure of the other.
a. Is this possible?
b. If so, find the measures. If not, tell why not. *(Lesson 3-3)*

21. Write $a - \dfrac{a + b}{2}$ as a single fraction. *(Previous course)*

22. Lynne has scored 92, 83, and 95 on her three tests so far this grading period. What is her average (or mean) score so far? *(Previous course)*

Exploration

23. The distance from point X to $(2, 8)$ is 17.
a. Show that X could be $(10, 23)$.
b. Name five other possible locations for point X.
(Hint: Draw a picture.)

11-7

Equations for Circles

Round and round they go. *As a carousel turns, the path a carved animal follows is a circle. The equation of each circle is determined by the distance each animal is from the center.*

Distance and Circles

A wonderful application of the Distance Formula is in finding an equation for a circle. At the right, a circle is drawn with center (3, 2) and radius 10. By adding or subtracting 10 from either coordinate of (3, 2), four points on the circle can be found. They are (13, 2), (3, 12), (-7, 2), and (3, -8). We seek an equation satisfied by the coordinates of these four points and all other points on the circle, and by no other points.

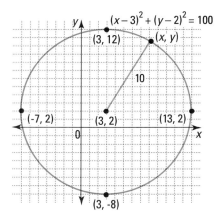

If a point (x, y) is on this circle, then it is 10 units away from the center. So, by the Distance Formula:

$$\sqrt{(x - 3)^2 + (y - 2)^2} = 10.$$

This is an equation for the circle.

However, most people prefer equations without square roots. Squaring both sides gives an equivalent equation.

$$(x - 3)^2 + (y - 2)^2 = 100$$

To check if this equation is correct, try the point (-7, 2), which is known to be on the circle. It should satisfy the equation. Substitute -7 for x and 2 for y.

$$\text{Does } (-7 - 3)^2 + (2 - 2)^2 = 100?$$
$$\text{Does } \quad (-10)^2 + \quad 0^2 \quad = 100? \text{ Yes.}$$

It is easy to generalize this example.

> **Equation for a Circle**
> The circle with center (h, k) and radius r is the set of points (x, y) satisfying
> $$(x - h)^2 + (y - k)^2 = r^2.$$

Drawing

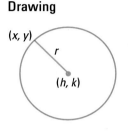

Proof

Given: The circle with center (h, k) and radius r, and a point (x, y) on the circle.

To prove: $(x - h)^2 + (y - k)^2 = r^2$.

Argument:
It is given that r is the radius. By the definition of circle, the distance from (h, k) to (x, y) is r. Express the distance from (h, k) to (x, y) using the Distance Formula.

$$\sqrt{(x - h)^2 + (y - k)^2} = r$$

Squaring both sides: $(x - h)^2 + (y - k)^2 = r^2$.

Example 1

Write an equation for the circle with center $(0, -4)$ and radius 7.

Solution

Here $r = 7$ and $(h, k) = (0, -4)$. So $h = 0$ and $k = -4$. Write the general equation for a circle, and substitute the values of h, k, and r into it.

$$(x - h)^2 + (y - k)^2 = r^2$$
$$(x - 0)^2 + (y - -4)^2 = 7^2$$

After simplifying, $x^2 + (y + 4)^2 = 49$ is the desired equation.

Check

Draw a picture and find the coordinates of a point on the circle. One point on the circle is $(7, -4)$. Does $(7, -4)$ satisfy the equation? Substitute 7 for x and -4 for y.

Does $7^2 + (-4 + 4)^2 = 49$?

Does $\quad\quad 7^2 + 0^2 = 49$? Yes.

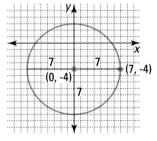

Other easily found points to check are $(0, 3)$, $(-7, -4)$, or $(0, -11)$. Remember that the center is not on the circle, so its coordinates will not satisfy the equation.

If you are given an equation for a circle, you can determine its center and radius.

Example 2

Find the center and radius of the circle with equation
$(x + 1)^2 + (y + 3)^2 = 25$.

Solution

Compare the general equation to the given equation.
$$(x - h)^2 + (y - k)^2 = r^2$$
$$(x + 1)^2 + (y + 3)^2 = 25$$
h must be -1 to get $(x + 1)$. k must be -3 to get $(y + 3)$. **The center is (-1, -3). We know $r^2 = 25$, so $r = 5$ or $r = -5$. Since r is a length, r cannot be negative. The radius is 5.**

Check

Draw a circle with center (-1, -3) and radius 5. One point on the circle is (-1, 2). Does it satisfy the equation?
Does $(-1 + 1)^2 + (2 + 3)^2 = 25$?
$$0^2 + 5^2 \quad = 25?\ \text{Yes.}$$

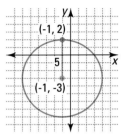

Circles Centered at the Origin

The circles with the simplest equations are those with center (0, 0), the origin. Then $h = 0$ and $k = 0$. So an equation is
$$(x - 0)^2 + (y - 0)^2 = r^2,$$
or just $\qquad\qquad\qquad x^2 + y^2 = r^2.$

Pictured below is the circle with center (0, 0) and radius 9. Its equation is $x^2 + y^2 = 81$.

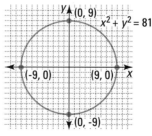

Four points on the circle are shown above. Others can be found by substituting values for x or y. For instance, consider the points of intersection of the circle with the line $x = 5$, as shown at the right. To find the coordinates of these points, substitute 5 for x in the equation. Then $5^2 + y^2 = 81$, making $y^2 = 56$, and so $y = \pm \sqrt{56}$. So the points $(5, \sqrt{56})$ and $(5, -\sqrt{56})$ are on this circle.

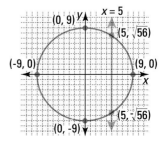

QUESTIONS

Covering the Reading

1. Is the point on the circle with center (3, 2) and radius 10?
   a. (13, 2)   **b.** (3, -8)
   c. (8, 7)   **d.** (9, -6)

2. Write the general form of the Equation for a Circle. Explain what each letter represents.

3. The equation for a circle is derived from what formula?

4. **a.** Write an equation for the circle with center (-3, 5) and radius 1.
   b. Give the coordinates of four points on this circle.

5. **a.** What is the distance between (x, y) and (7, 1)?
   b. Give an equation for the circle with center (7, 1) and radius 5.
   c. Graph this circle.
   d. Give the coordinates of four points on this circle.

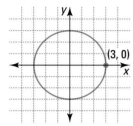

6. The circle at the left has center at the origin and contains (3, 0). What is an equation for this circle?

7. Consider the circle $x^2 + y^2 = 81$ graphed on page 658. Where does this circle intersect the line $x = 4$?

Applying the Mathematics

In 8–11, an equation for a circle is given. Determine:
a. its center.   **b.** its radius.   **c.** one point on the circle.

8. $(x - 5)^2 + (y - 11)^2 = 81$ 9. $(x + 1)^2 + y^2 = 2$

10. $x^2 + y^2 = \frac{4}{25}$ 11. $(x + 6)^2 + (y + 2)^2 = 1$

In 12 and 13, consider the circle $x^2 + y^2 = 121$.

12. Identify the coordinates of 10 points on this circle.

13. Find the circumference of this circle.

14. Give the coordinates of the 12 lattice points on the circle $x^2 + y^2 = 25$. (Recall that a lattice point is a point with integer coordinates.) Graph the points and the circle that contains them.

15. A circle is called **tangent** to a line if the circle and line have exactly one point in common. That point is called the **point of tangency.** A circle with center (2, -1) is tangent to the *x*-axis.
    a. Draw a picture.
    b. What are the coordinates of the point of tangency?
    c. Find an equation for the circle.
    d. Find the area of this circle.

In 16 and 17, calculate the distance between the points. Assume $x > 0$.
(Lesson 11-6)

16. (4, -7) and (-1, 5)   **17.** ($9x$, $-40x$) and the origin

18. Nancy lives 6 blocks west and 3 blocks north of the train station. Domaso lives 5 blocks south of the station. How many blocks is it from where Nancy lives to where Domaso lives?   *(Lesson 11-6)*

19. Prove that the quadrilateral $QRST$ with vertices $Q = (9a, 4b)$, $R = (6a, 2b)$, $S = (a, -7b)$, and $T = (-a, -14b)$ is a trapezoid. *(Lesson 11-5)*

20. In the box at the right, $CG = 8$, $HG = 12$ and $EH = 16$.
   a. Find EG.
   b. Find CE.   *(Lessons 8-6, 9-2)*

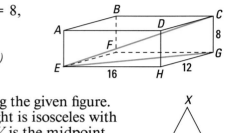

21. Write a proof argument using the given figure.
   Given: $\triangle XYZ$ at the right is isosceles with vertex angle X. V is the midpoint of $\overline{XY}$, and W is the midpoint of $\overline{XZ}$.
   To prove: $\triangle XVW$ is isosceles.   *(Lesson 6-2)*

22. Where can you aim to putt a golf ball G into the hole at H in one shot?   *(Lesson 4-3)*

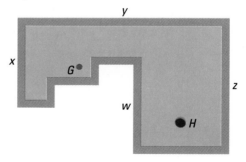

23. Suppose $\overleftrightarrow{AB}$ and $\overleftrightarrow{BC}$ have the same slope. What can you conclude about points A, B, and C?   *(Lessons 1-7, 3-6)*

24. Give an equation for a circle on which there are no lattice points and whose interior contains no lattice points. (See Question 14 if you have forgotten the meaning of *lattice point*.)

Connecting Midpoints in a Triangle

IN-CLASS

ACTIVITY

Work with a partner. You will need an automatic drawer for this activity.

1 Place a △*ABC* on the screen. Let *D* be the midpoint of $\overline{AB}$ and *E* be the midpoint of $\overline{AC}$. Draw $\overline{DE}$.

2 Place the following measurements on the screen: *DE*, *BC*, slope of $\overline{BC}$, slope of $\overline{DE}$, $\frac{BC}{DE}$. (Sample measurements are shown below.)

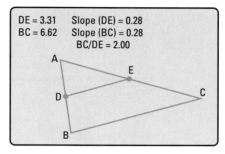

DE = 3.31   Slope (DE) = 0.28
BC = 6.62   Slope (BC) = 0.28
            BC/DE = 2.00

3 Select a vertex of △*ABC*, and distort the figure by moving that vertex. Record the screen for at least two other positions of △*ABC*.

4 Repeat Step 3, but move a different vertex.

5 What conjecture can you make about the slopes of $\overline{DE}$ and $\overline{BC}$?

6 What conjecture can you make about the lengths of $\overline{DE}$ and $\overline{BC}$?

11-8

Means and Midpoints

A Geometric Interpretation of the Mean of Two Numbers

Suppose you score 90 and 83 on two tests. Your average or **mean** score is the sum of these numbers divided by 2. It is

$$\frac{90 + 83}{2} = 86.5.$$

The mean has a physical interpretation. Think of a ruler with hooks for attaching weights (assume the ruler itself is weightless). If equal weights are hung from 90 and 83, the ruler will be horizontal if it is hung from a string at 86.5, or if it is placed to balance on a sharp object at 86.5.

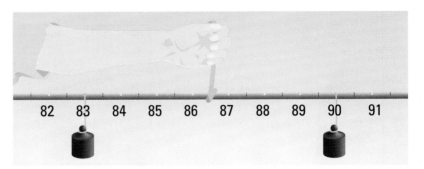

Notice that 86.5 is the *midpoint* of the segment connecting 90 and 83. It is 3.5 units away from both 90 and 83. With some algebra, the relationship between a mean and a midpoint can be proven for any segment on a number line.

Number Line Midpoint Formula
On a number line, the coordinate of the midpoint of the segment with endpoints *a* and *b* is $\frac{a + b}{2}$.

Proof:
Given: $\overline{PQ}$ with endpoint coordinates *a* and *b*.
To prove: The midpoint of $\overline{PQ}$ has coordinate $\frac{a + b}{2}$.
First, draw a picture. Let *M* be the point with coordinate $\frac{a + b}{2}$.

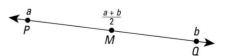

Argument: $MP = \left| \frac{a + b}{2} - a \right| = \left| \frac{a + b - 2a}{2} \right| = \left| \frac{b - a}{2} \right|$
$QM = \left| b - \frac{a + b}{2} \right| = \left| \frac{2b - (a + b)}{2} \right| = \left| \frac{b - a}{2} \right|$
So $MP = QM$, and so *M* is the midpoint of $\overline{PQ}$.

A Formula for the Midpoint in Two Dimensions

In two dimensions, the same idea holds. The coordinates of the midpoint of a segment are the averages of the coordinates of the endpoints. Notice that we state the theorem using the points (x_1, y_1) and (x_2, y_2) as endpoints. This makes the pattern easy to see. But in the proof we replace (x_1, y_1) by (a, b) and (x_2, y_2) by (c, d). This makes the algebra easier to follow.

> **Coordinate Plane Midpoint Formula**
> In the coordinate plane, the midpoint of the segment with endpoints (x_1, y_1) and (x_2, y_2) is $\left(\frac{x_1 + x_2}{2}, \frac{y_1 + y_2}{2}\right)$.

Drawing:

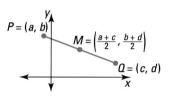

Proof

Given: $\overline{PQ}$ with $P = (a, b)$, $Q = (c, d)$, and $M = \left(\frac{a + c}{2}, \frac{b + d}{2}\right)$.

To prove: M is the midpoint of $\overline{PQ}$.

Argument:
Recall the definition of midpoint. To show that M is the midpoint of $\overline{PQ}$, we need to show that $PM = MQ$ and M is on $\overleftrightarrow{PQ}$.

The algebra that follows is cumbersome but straightforward. To show that $PM = MQ$, we calculate distances using the Distance Formula.

$$PM = \sqrt{\left(\frac{a + c}{2} - a\right)^2 + \left(\frac{b + d}{2} - b\right)^2}$$
$$= \sqrt{\left(\frac{a + c - 2a}{2}\right)^2 + \left(\frac{b + d - 2b}{2}\right)^2}$$
$$= \sqrt{\left(\frac{c - a}{2}\right)^2 + \left(\frac{d - b}{2}\right)^2}$$
$$MQ = \sqrt{\left(c - \frac{a + c}{2}\right)^2 + \left(d - \frac{b + d}{2}\right)^2}$$
$$= \sqrt{\left(\frac{2c - (a + c)}{2}\right)^2 + \left(\frac{2d - (b + d)}{2}\right)^2}$$
$$= \sqrt{\left(\frac{c - a}{2}\right)^2 + \left(\frac{d - b}{2}\right)^2}$$

Thus $PM = MQ$. To show that M is on $\overleftrightarrow{PQ}$, we calculate the slopes of $\overleftrightarrow{PM}$ and $\overleftrightarrow{MQ}$.

$$\text{slope of } \overleftrightarrow{PM} = \frac{\frac{b + d}{2} - b}{\frac{a + c}{2} - a} = \frac{b + d - 2b}{a + c - 2a} = \frac{d - b}{c - a}$$

$$\text{slope of } \overleftrightarrow{MQ} = \frac{d - \frac{b + d}{2}}{c - \frac{a + c}{2}} = \frac{2d - (b + d)}{2c - (a + c)} = \frac{d - b}{c - a}$$

The slopes are equal so $\overleftrightarrow{PM} \parallel \overleftrightarrow{MQ}$. Both lines contain point M, so $\overleftrightarrow{PM} = \overleftrightarrow{MQ}$ and M is on $\overleftrightarrow{PQ}$. So M is the midpoint of $\overline{PQ}$.

Fortunately, applying the Midpoint Formula is easier than proving it.

Example 1

If $P = (-10, 6)$ and $Q = (1, 8)$, find the midpoint of $\overline{PQ}$.

Solution

Call the midpoint M. Use the Midpoint Formula.

$$M = \left(\frac{-10 + 1}{2}, \frac{6 + 8}{2}\right) = \left(\frac{-9}{2}, \frac{14}{2}\right) = (-4.5, 7)$$

Check

Sketch a coordinate plane. $(-4.5, 7)$ looks halfway between $(-10, 6)$ and $(1, 8)$. This gives a rough check. You can calculate the slopes and lengths of $\overline{PM}$ and $\overline{MQ}$ for an exact check.

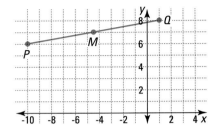

The Midpoint Connector Theorem

In the In-class Activity on page 661, you connected the midpoints of two sides of a triangle. You may have made a conjecture similar to the following theorem.

> **Midpoint Connector Theorem**
> The segment connecting the midpoints of two sides of a triangle is parallel to and half the length of the third side.

In this proof, we use $2a$, $2b$, and $2c$ as coordinates of vertices in the convenient location for the triangle. This avoids fractions when coordinates of midpoints are calculated.

Proof:

Given: $\triangle PQR$ with $Q = (0, 0)$, $R = (2a, 0)$, $P = (2b, 2c)$, M the midpoint of $\overline{PQ}$, and N the midpoint of $\overline{PR}$.

To prove: (1) $\overline{MN} \mathbin{/\mkern-5mu/} \overline{QR}$.
              (2) $MN = \frac{1}{2} QR$.

Argument:

First use the Coordinate Plane Midpoint Formula to find the coordinates of M and N.

$$M = \left(\frac{2b + 0}{2}, \frac{2c + 0}{2}\right) = (b, c)$$

$$N = \left(\frac{2b + 2a}{2}, \frac{2c + 0}{2}\right) = \left(\frac{2(a + b)}{2}, \frac{2c}{2}\right) = (a + b, c)$$

(1) Slopes help prove $\overline{MN} \mathbin{/\mkern-5mu/} \overline{QR}$.

    slope of $\overline{MN} = \dfrac{c - c}{(a + b) - b} = \dfrac{0}{a} = 0$

    slope of $\overline{QR} = \dfrac{0 - 0}{2a - 0} = \dfrac{0}{2a} = 0$

    Since the slope of $\overline{MN}$ equals the slope of $\overline{QR}$, $\overline{MN} \mathbin{/\mkern-5mu/} \overline{QR}$.

▶

▶ (2) The Distance Formula is used to prove $MN = \frac{1}{2}QR$.

$$MN = \sqrt{((b + a) - b)^2 + (c - c)^2} = \sqrt{a^2 + 0} = |a|$$
$$QR = \sqrt{(2a - 0)^2 + (0 - 0)^2} = \sqrt{4a^2} = 2|a|$$

Thus $QR = 2MN$.

So $MN = \frac{1}{2}QR$.

Example 2

$\triangle ABC$ has sides of lengths 15, 27, and 38. $\triangle LMN$ is formed by connecting the midpoints of $\triangle ABC$. What are the lengths of the sides of $\triangle LMN$?

Solution

Use the Midpoint Connector Theorem.

$LM = \frac{1}{2}AB = \frac{1}{2} \cdot 15 = 7.5$

$MN = \frac{1}{2}BC = \frac{1}{2} \cdot 38 = 19$

$LN = \frac{1}{2}AC = \frac{1}{2} \cdot 27 = 13.5$

Notice in Example 2 that the sides of $\triangle LMN$ are parallel to the sides of $\triangle ABC$. $\triangle LMN$ is called the *medial triangle* of $\triangle ABC$. You are asked to explore other properties of this medial triangle in the Questions.

QUESTIONS

Covering the Reading

In 1–3, use this information: The U.S. Weather Bureau calculates the mean temperature for a particular place on a particular day by averaging the high and low temperatures. Find the mean temperature if

1. the high is 20°C and the low is 12°C.

2. the high is 6°C and the low is -7°C.

3. the high is -1°C and the low is -9°C.

4. Suppose you score 95 and 87 on two tests. What is your mean score?

5. Give a number-line interpretation of the mean in Question 2.

6. What is the Number Line Midpoint Formula?

7. Give the coordinates of the midpoint of the segment connecting (x_1, y_1) to (x_2, y_2).

8. Find the midpoint of the segment with endpoints (-5, 11) and (13, 1).

Whether or not we have weather. *This temperature sign is on a building near the Shibuya Railway Station in Tokyo, Japan.*

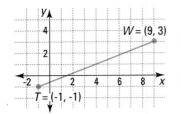

T = (-1, -1)   W = (9, 3)

9. Give the coordinates of the midpoint of $\overline{TW}$ at the left.

10. Give the coordinates of the midpoint of the segment with endpoints (a, b) and the origin.

11. At the right is a conveniently located right $\triangle PQR$. Let M be the midpoint of $\overline{PQ}$ and N the midpoint of $\overline{QR}$.
    a. Give the coordinates of M and N.
    b. Without using the Midpoint Connector Theorem, write a proof argument to show that $\overline{MN} \parallel \overline{PR}$ and $MN = \frac{1}{2} PR$.

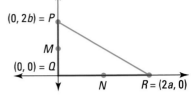

$(0, 2b) = P$
M
$(0, 0) = Q$
N
$R = (2a, 0)$

D
N
L
F
M
E

12. $\triangle LMN$ is the medial triangle of $\triangle DEF$ at the left. If $DE = 20$, $EF = 16$, and $DF = 24$, what are the lengths of the sides of $\triangle LMN$?

Applying the Mathematics

In 13–16, $\triangle ABC$ is the medial triangle of $\triangle LMN$ at the right.

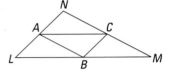

N
A
C
L
B
M

13. If $LN = 12$, what other length(s) can be found?

14. If $AB = 5.2$, what other length(s) can be found?

15. Explain why $ANCB$ is a parallelogram.

16. Write a proof argument to show that $\triangle ANC \cong \triangle BCM$.

17. At the right is a graph of the total number of African-American, Hispanic, and Asian families in the United States for 1980 and 1992.
    a. Find the midpoint of $\overline{PQ}$.
    b. What does the midpoint represent?

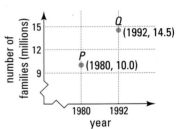

number of families (millions)
15
12
9
Q (1992, 14.5)
P (1980, 10.0)
1980   1992
year

18. Use quadrilateral $PQRS$ at the right.
    a. Show that the diagonals of $PQRS$ bisect each other.
    b. What type of quadrilateral is $PQRS$?

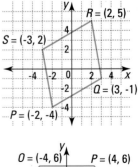

R = (2, 5)
S = (-3, 2)
Q = (3, -1)
P = (-2, -4)

19. Use quadrilateral $MNOP$ at the right. Use indirect reasoning to show that the diagonals of $MNOP$ do *not* bisect each other.

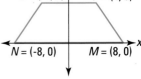

O = (-4, 6)   P = (4, 6)
N = (-8, 0)   M = (8, 0)

20. $\overline{UV}$ has midpoint M. The coordinates of U are (9, 11) and the coordinates of M are (-1, 7). Find the coordinates of V.

21. The endpoints of a diameter of a circle are (-3, 5) and (5, -10). Find
  a. The center of the circle.
  b. the radius of the circle.
  c. an equation for the circle.

Review

22. Consider the circle with equation $x^2 + y^2 = 75$.
  a. Find its center. **b.** Find its radius.
  c. Graph it. **d.** Find its area. *(Lesson 11-7)*

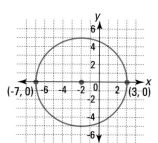

23. The circle at the left has center (-2, 0). What is an equation for it? *(Lesson 11-7)*

24. Write an indirect proof to show that $ABCD$ at the right is *not* a rhombus. *(Lessons 11-4, 11-6)*

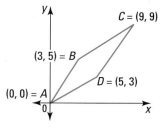

25. What can you conclude from all four statements (adapted from Lewis Carroll) using the laws of logic? *(Lessons 11-1, 11-2, 11-3)*
  (1) The only kinds of foods that my doctor allows me are kinds that are not very rich.
  (2) Nothing that agrees with me is unsuitable for dinner.
  (3) Birthday cake is always very rich.
  (4) My doctor allows me all kinds of food that are suitable for dinner.

26. The right square pyramid at the right has a height of 30 cm. The perimeter of the base is 64 cm.
  a. Find the volume of the pyramid.
  b. Find its surface area. *(Lessons 10-2, 10-7)*

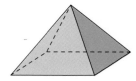

27. Draw a map with 6 regions that can be colored with 3 colors. *(Lesson 9-9)*

Exploration

28. The Midpoint Connector Theorem tells how the sides of a medial triangle and its original triangle are related.
  a. How are the angles related?
  b. How are the areas related?
  c. How are the perimeters related?

Off the wall. *Racquetball players bounce balls off planes which are perpendicular or parallel. Notice the three mutually perpendicular planes meeting at the corner.*

Coordinatizing Space

Points in space can be located using a **three-dimensional coordinate system.** For instance, you can locate points in a room by letting the origin be at a corner of the room where two walls and the floor intersect. Then with two coordinates, x and y, you can describe the location of any point on the floor. To describe an object in the room which is not on the floor (such as the bottom of a light hanging from the ceiling), you can use a third number to indicate the height from the floor. This is the **z-coordinate.** Thus, if the bottom of the light is 4 ft from the origin in the x-direction, 7 ft in the y-direction, and 8 ft in the z-direction (up), you could uniquely specify the position of the light by the **ordered triple** $(4, 7, 8)$. The x-coordinate is 4, the y-coordinate is 7, and the z-coordinate is 8. The three lines where the walls and floor meet are the **axes** of this 3-dimensional coordinate system.

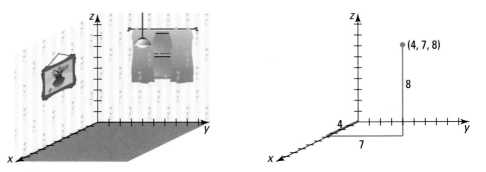

Now imagine extending each axis in its negative direction, as shown in the figure at the right. The three axes are called the **x-axis,** the **y-axis,** and the **z-axis.** The ordered triple (x, y, z) represents a point in *3-space.* The position of a point is given by its three distances from the origin $(0, 0, 0)$. The following example shows how to plot a point in three dimensions.

Example 1

Plot the point $R = (-7, 4, 3)$ on a 3-dimensional coordinate system.

Solution

Step 1: Draw or copy the axes from above.

Step 2: Move 7 units backward (in a negative direction) on the x-axis.

Step 3: From there move 4 units to the right, parallel to the y-axis.

Step 4: From there move 3 units up parallel to the z-axis.

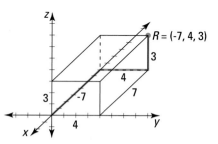

It helps to think of the point as the back upper right vertex of a box with base dimensions 7 and 4 and height 3.

Activity

List the x, y, z coordinates of each of the eight vertices of the box in Example 1. (Hint: Each vertex other than $(-7, 4, 3)$ has at least one coordinate that is 0.)

The Distance Formula in Space

Many of the formulas for 2-dimensional coordinates have counterparts in three dimensions. For instance, the Coordinate Plane Distance Formula states that the distance between (x_1, y_1) and (x_2, y_2) is $\sqrt{(x_2 - x_1)^2 + (y_2 - y_1)^2}$. Its counterpart is the 3-dimensional Distance Formula below.

Three-Dimension Distance Formula
The distance d between two points (x_1, y_1, z_1) and (x_2, y_2, z_2) is given by the formula
$$d = \sqrt{(x_2 - x_1)^2 + (y_2 - y_1)^2 + (z_2 - z_1)^2}.$$

The proof of this formula, which is given on the following page, is based on two successive applications of the Pythagorean Theorem.

Proof:

Given: $P = (x_1, y_1, z_1)$ and $Q = (x_2, y_2, z_2)$.

To prove: $PQ = \sqrt{(x_2 - x_1)^2 + (y_2 - y_1)^2 + (z_2 - z_1)^2}$.

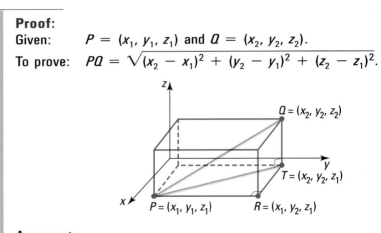

Argument:

In right $\triangle PRT$, $PT^2 = RT^2 + PR^2$.

In right $\triangle PTQ$, $PQ^2 = PT^2 + QT^2$

Substituting $RT^2 + PR^2$ for PT^2 gives
$$PQ^2 = RT^2 + PR^2 + QT^2.$$

So $\qquad PQ^2 = (x_2 - x_1)^2 + (y_2 - y_1)^2 + (z_2 - z_1)^2.$

Example 2

Find the distance between points P and Q, where: $P = (-5, 2, 1)$ and $Q = (4, 0, -3)$.

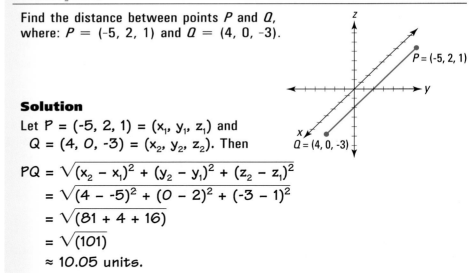

Solution

Let $P = (-5, 2, 1) = (x_1, y_1, z_1)$ and
$Q = (4, 0, -3) = (x_2, y_2, z_2)$. Then

$$
\begin{aligned}
PQ &= \sqrt{(x_2 - x_1)^2 + (y_2 - y_1)^2 + (z_2 - z_1)^2} \\
&= \sqrt{(4 - -5)^2 + (0 - 2)^2 + (-3 - 1)^2} \\
&= \sqrt{(81 + 4 + 16)} \\
&= \sqrt{(101)} \\
&\approx 10.05 \text{ units.}
\end{aligned}
$$

The Three-Dimension Distance Formula helps answer this question: What is the length of the diagonal of a box? A box with dimensions ℓ, w, and h is conveniently located below with one endpoint of the diagonal at the origin and the other at (ℓ, w, h).

The diagonal has length PQ.

$$
\begin{aligned}
PQ &= \sqrt{(\ell - 0)^2 + (w - 0)^2 + (h - 0)^2} \\
&= \sqrt{\ell^2 + w^2 + h^2}
\end{aligned}
$$

This argument proves the following formula.

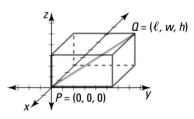

> **Box Diagonal Formula**
> In a box with dimensions ℓ, w, and h, the length d of the diagonal is given by the formula
> $$d = \sqrt{\ell^2 + w^2 + h^2}.$$

Spheres

In Lesson 11-7, an equation for the circle with center (h, k) and radius r was shown to be $(x - h)^2 + (y - k)^2 = r^2$. The 3-dimensional counterpart to the circle is the sphere. An equation for the sphere is analogous to the equation for a circle. Its proof uses the Three-Dimension Distance Formula.

> **Equation for a Sphere**
> The sphere with center (h, k, j) and radius r is the set of points (x, y, z) satisfying
> $$(x - h)^2 + (y - k)^2 + (z - j)^2 = r^2.$$

Proof

Given: The sphere with center (h, j, k) and radius r.

To prove: The sphere is the set of points (x, y, z) that satisfy $(x - h)^2 + (y - k)^2 + (z - j)^2 = r^2$.

Drawing:

Argument:
Since r is the radius, by the definition of sphere, the distance from (h, k, j) to (x, y, z) is r. That distance is given by the Three-Dimension Distance Formula:
$$\sqrt{(x - h)^2 + (y - k)^2 + (z - j)^2} = r.$$
Squaring both sides results in the theorem.
$$(x - h)^2 + (y - k)^2 + (z - j)^2 = r^2$$

Example 3

Find an equation for the sphere with center $(3, 0, -1)$ and radius 8.

Solution
Here $(h, k, j) = (3, 0, -1)$. So $h = 3$, $k = 0$, and $j = -1$. Also, $r = 8$. Substitute these values into the equation for a sphere.
$$(x - 3)^2 + (y - 0)^2 + (z - -1)^2 = 8^2$$
Simplify. $(x - 3)^2 + y^2 + (z + 1)^2 = 64$ is the desired equation.

The Midpoint Formula in Space

The Midpoint Formula from Lesson 11-8 also extends easily to three dimensions.

> **Three-Dimension Midpoint Formula**
> In space, the midpoint of the segment with endpoints (x_1, y_1, z_1) and (x_2, y_2, z_2) is
> $$\left(\frac{x_1 + x_2}{2}, \frac{y_1 + y_2}{2}, \frac{z_1 + z_2}{2}\right).$$

You can verify the Three-Dimension Midpoint Formula by calculating the distances between the endpoints and the midpoint, just as in two dimensions.

QUESTIONS

Covering the Reading

1. Any point in three dimensions can be located with an ordered __?__.

2. *True or false.* The intersection of two walls and the floor of a room can represent the origin of a coordinate system in three dimensions.

3. Draw a coordinate system and plot the points $A = (7, -1, -3)$ and $B = (0, -6, 1)$.

4. Give the list you wrote for the Activity in this lesson.

5. Find PQ when $P = (3, 7, -2)$ and $Q = (5, -11, 0)$.

6. In the box pictured below, calculate CF.

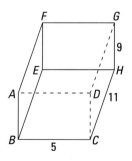

7. Find an equation for the sphere with center $(-5, 3, -10)$ and radius 13.

8. Refer to Question 5. Find the midpoint of $\overline{PQ}$.

9. Here is an equation for a sphere: $(x - 18)^2 + (y - 5)^2 + (z + 11)^2 = 36$.
   a. What point is the center of the sphere?
   b. What is the radius of the sphere?
   c. Give the coordinates of two points on the sphere.

10. A box has the following vertices: the origin, (0, 1, 0), (3, 0, 0), (3, 1, 0), (3, 1, 9), (3, 0, 9), (0, 0, 9), and (0, 1, 9).
  a. Draw the box on a 3-dimensional coordinate system.
  b. Determine its volume.

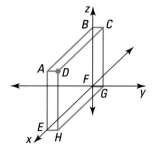

11. In the rectangular box at the left, $D = (10, 1, 6)$ and F is the origin.
  a. Find the coordinates of the points A, B, C, E, G, and H.
  b. Determine the volume of the box.
  c. Determine its surface area.

In 12 and 13, triangle ABC has vertices $A = (2, \text{-}1, 7)$, $B = (4, 0, \text{-}5)$, and $C = (\text{-}11, 8, 2)$.

12. Find the perimeter of $\triangle ABC$.

13. Find the midpoints of the three sides of $\triangle ABC$.

14. What is the length of the longest thin cylindrical tube that can be carried in a $16'' \times 18'' \times 5''$ carrying case?

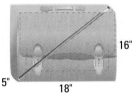

15. Find an equation for the sphere below with center (0, 0, 0).

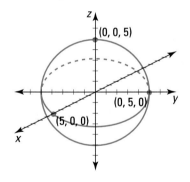

16. a. The figure below is a(n) __?__.
  b. Use this figure to complete the sentence. Then prove it. The segment joining the midpoints of the non-base sides of a(n) __?__ is parallel to the bases. *(Lessons 6-5, 11-8)*

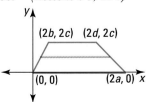

17. Below, $\overline{AD}$ is a median of $\triangle ABC$.
  a. Find the coordinates of D.
  b. Calculate the length of $\overline{AD}$.   *(Lesson 11-8)*

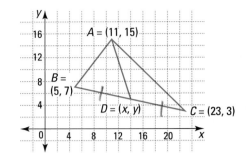

18. $\triangle TRI$ at the left is equilateral with each side having length $2a$. The smaller triangles are formed by joining the midpoints of the sides of the larger triangles. Find the perimeter of the smallest triangle shown. *(Lesson 11-8)*

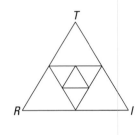

19. A circle has center $(3, -2)$ and passes through $P = (4, 1)$.
  a. What is its radius?
  b. Write an equation for it.   *(Lessons 11-6, 11-7)*

20. A right triangle has sides of lengths 12, 16, and 20. To study this triangle on a coordinate plane, what would be convenient coordinates for its vertices?   *(Lesson 11-5)*

21. Given: not-$m \Rightarrow n$,
         $m \Rightarrow p$,
         not-n.
  Draw a conclusion.   *(Lesson 11-1)*

22. The Pyramid Arena in Memphis, Tennessee, is a right square pyramid with a height of 321 feet and base area of 530,000 square feet. *(Lessons 10-2, 10-7)*
  a. What is the length of each edge of the base?
  b. What is the volume of this pyramid?
  c. A newspaper reported that this pyramid is two-thirds the size of the pyramid of Khufu. Refer back to page 572. Is the newspaper correct?

Pyramid Arena, Memphis, Tennessee

23. Solve each equation.   *(Previous course)*
  a. $\frac{x}{5} = \frac{11}{2}$ **b.** $\frac{2}{y} = \frac{4}{2.3}$ **c.** $\frac{3}{5} = \frac{2z}{13}$

Exploration

24. Measure the three dimensions of a rectangular bedroom or classroom. Set up a coordinate system for the room and determine the coordinates of the lights and windows. Include a drawing.

25. What is a *hypersphere*? What is a *hypercube*?

A project presents an opportunity for you to extend your knowledge of a topic related to the material of this chapter. You should allow more time for a project than you do for typical homework questions.

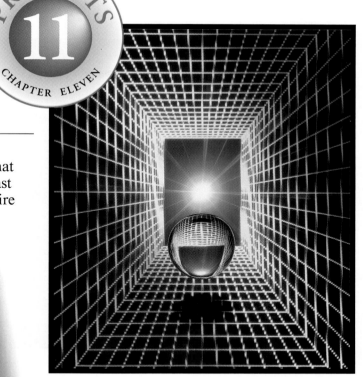

1 Logic Puzzle

Make up your own logic puzzle that has at least 3 categories and at least 8 necessary clues that would require a grid to solve. Make the grid and solve the puzzle.

3 3-Dimensional Locations

Pick a corner of your classroom or a room in your house and locate several objects using 3-dimensional coordinates. Make a poster showing the axes and the locations of the objects. Use the Distance Formula to calculate distances between the objects.

2 Lewis Carroll

Write a report on the life of Lewis Carroll (Charles Dodgson), including his mathematical and non-mathematical writings.

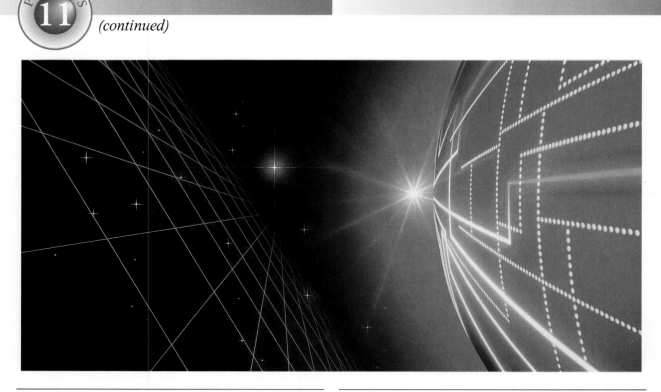

4 Varignon's Theorem

An amazing property of all quadrilaterals was discovered and proved by Pierre Varignon, a French mathematician (1654–1722): If the midpoints of consecutive sides of any quadrilateral are connected, then the quadrilateral so formed is a parallelogram.

a. Draw instances of Varignon's Theorem using quadrilaterals with different shapes, with at least one of them nonconvex.

b. Prove Varignon's Theorem for the plane using coordinate geometry or using synthetic geometry. (Hint: If you use coordinate geometry, begin with vertices $(0, 0)$, $(2a, 0)$, $(2b, 2c)$, and $(2d, 2e)$ to simplify the algebra. If you use synthetic geometry, use diagonals of the original quadrilateral as auxiliary lines.)

c. Consider a *space quadrilateral* as one that is formed by connecting four points in space that are not necessarily in the same plane. Does Varignon's Theorem hold for space quadrilaterals? Give evidence to support your answer.

5 Equations for Planes

A line in a 2-dimensional rectangular coordinate system is the set of points (x, y) that satisfy an equation of the form $ax + by = c$. Similarly, in a 3-dimensional rectangular coordinate system like the one in Lesson 11-9, a plane is the set of points (x, y, z) that satisfy an equation of the form $ax + by + cz = d$. Find an automatic drawer that graphs planes. Use it to graph planes that intersect all three axes, exactly two axes, and exactly one axis. Show these graphs.

6 Truth Tables

Examine a book on symbolic logic to find out what is meant by a *truth table*. Then show how each of the five logical principles in this chapter can be analyzed using truth tables.

SUMMARY

Every mathematical argument follows rules of logic. In this chapter, five rules are stated.

Law of Detachment: From a statement or given information p and a conditional $p \Rightarrow q$, you may conclude q.

Law of Transitivity: If $p \Rightarrow q$ and $q \Rightarrow r$, then $p \Rightarrow r$.

Law of the Contrapositive: A conditional ($p \Rightarrow q$) and its contrapositive (not-$q \Rightarrow$ not-p) are either both true or both false.

Law of Ruling Out Possibilities: When p or q is true and q is not true, then p is true.

Law of Indirect Reasoning: If reasoning from a statement p leads to a false conclusion, then p is false.

The last three of these laws comprise the basic logic used in indirect proofs. If you can prove that the contrapositive of a statement is true, then the statement is true. If you can rule out all possibilities but one, then the possibility left is true. If you reason from the negation of what you want to prove and arrive at a contradiction, then the negation is false, so what you want to prove must be true.

In coordinate geometry, figures are described by equations or by giving coordinates of key points. To deduce a general property of a polygon using coordinates, it is efficient to place the polygon in a convenient location on the coordinate plane. Either the figure is placed with one vertex at the origin and one or more sides on the axes, or the figure is placed so that it is symmetric to the x-axis or y-axis or origin.

Three formulas are involved in the coordinate proofs of this chapter. Let (x_1, y_1) and (x_2, y_2) be two points. The distance between them is $\sqrt{(x_2 - x_1)^2 + (y_2 - y_1)^2}$. This gives a way to tell whether segments are congruent. The midpoint of the segment joining the points is $\left(\frac{x_1 + x_2}{2}, \frac{y_1 + y_2}{2}\right)$. Many theorems which involve midpoints can be proved using the Midpoint Formula.

Just as there are equations for lines, there are equations for circles. An equation for the circle with center (h, k) and radius r is $(x - h)^2 + (y - k)^2 = r^2$.

A coordinate geometry of three dimensions can be built by extending the 2-dimensional coordinate system. An ordered pair becomes an ordered triple. The equation for a circle has an analogous equation for a sphere. The Distance Formula and the Midpoint Formula have their 3-dimensional counterparts. Many properties of 3-dimensional figures can be deduced using 3-dimensional coordinates. One example given in this chapter was the formula for the length of a diagonal of a box.

VOCABULARY

Below are the most important terms and phrases for the chapter. For the starred (*) terms you should be able to give a definition of the term. For the other terms you should be able to give a general description and a specific example of each.

Introduction
coordinate geometry
analytic geometry

Lesson 11-1
*Law of Detachment
*Law of Transitivity

Lesson 11-2
negation, not-p
truth value
not-$p \Rightarrow$ not-q, inverse
not-$q \Rightarrow$ not-p, contrapositive
*Law of the Contrapositive

Lesson 11-3
*Law of Ruling Out Possibilities
Trichotomy Law

Lesson 11-4
direct reasoning
direct proofs
indirect reasoning
*contradictory
*contradiction
indirect argument
*Law of Indirect Reasoning
*indirect proof

Lesson 11-5
convenient location for a figure

Lesson 11-6
*Distance Formula on the Coordinate Plane

Lesson 11-7
*Equation for a Circle

Lesson 11-8
mean
*Number Line Midpoint Formula
*Coordinate Plane Midpoint Formula
*Midpoint Connector Theorem
medial triangle

Lesson 11-9
three-dimensional coordinate system
z-coordinate
ordered triple
x-axis, y-axis, z-axis
*Three-Dimension Distance Formula
*Box Diagonal Formula
Equation for a Sphere
Three-Dimension Midpoint Formula

PROGRESS SELF-TEST

Directions: Take this test as you would take a test in class. Then check your work with the solutions in the Selected Answers section in the back of the book.

In 1 and 2, use points $P = (7, 15)$ and $Q = (30, -7)$.

1. Find PQ.

2. If M is the midpoint of $\overline{PQ}$, give the coordinates of M.

3. Let $R = (3, 4)$, $S = (8, 4)$, and $T = (11, 8)$. Find the perimeter of $\triangle RST$.

4. a. Draw a three-dimensional coordinate system and plot $A = (5, -6, 8)$.

   b. Find AO, where O is the origin.

   c. What are the coordinates of the midpoint of $\overline{AO}$?

In 5 and 6, consider the two statements (i) and (ii). a. What can you conclude using both statements? b. What law(s) of logic have you used to arrive at your answer to part a?

5. (i) $\angle A$ is an angle of $\triangle ABC$.
   (ii) $\angle A$ is neither acute nor obtuse.

6. (i) If a figure is a right cylinder, then its volume is the area of the base times its height.
   (ii) Figure Q is a right cylinder.

7. Consider this statement: *If a figure is a hexagon, then it is a polygon.* Write its

   a. converse.

   b. inverse.

   c. contrapositive.

   d. Tell which (if any) of these are true.

8. Write an argument to show why no triangle can have three angles all with measures less than 60.

9. Write an indirect proof to show that $\sqrt{80} \neq 40$.

10. $RHMB$ is located on the coordinate plane with coordinates $R = (16, 8)$, $B = (10, 0)$, $M = (0, 0)$, and $H = (6, 8)$. Write a proof argument to show $RHMB$ is a rhombus.

11. Write a proof argument to show that the diagonals of the quadrilateral at the right have the same midpoint.

$Z = (2b, 2c)$ $Y = (2a + 2b, 2c)$
$W = (0, 0)$ $X = (2a, 0)$

In 12 and 13, D, E, and F are midpoints of the sides of $\triangle ABC$, as shown below.

12. Explain why $BDFE$ is a parallelogram.

13. If $AB = 11$ and $BC = 22.3$, find as many other lengths as you can.

14. What (if anything) can you conclude using all the following statements?
    (1) All babies are happy.
    (2) If someone is teething, then that person is a baby.
    (3) Nate is sad.

15. Four cards—the Jack, Queen, King, and Ace—are from the seventeenth, eighteenth, nineteenth, and twentieth centuries, but not necessarily in that order. From the clues below, match each card to its century. (Note: the seventeenth century covers the years 1601–1700; the eighteenth: 1701–1800; and so on.)
    (1) The Queen is older than the King.
    (2) The Jack is exactly 100 years older than the Queen.
    (3) The Ace is older than the Jack.

16. Goodland, Kansas, is about 60 miles north and 12 miles west of Selkirk. Garden City is about 36 miles south and 34 miles east of Selkirk. What is the flying distance from Goodland to Garden City?

17. A grocery bag measures 7″ by 12″ by 17″. What is the length of the longest straw that will fit in the bag without sticking out of it?

18. Graph the circle with the equation $(x + 1)^2 + (y - 9)^2 = 25$.

19. What is an equation for a sphere with center $(0, -19, 4)$ and radius 6?

20. Give convenient coordinates for the vertices of an isosceles trapezoid.

CHAPTER REVIEW

Questions on SPUR Objectives

SPUR stands for **S**kills, **P**roperties, **U**ses, and **R**epresentations. The Chapter Review questions are grouped according to the SPUR Objectives for this chapter.

SKILLS DEAL WITH THE PROCEDURES USED TO GET ANSWERS.

Objective A: *Determine the length and the coordinates of the midpoint of a segment in the coordinate plane.* *(Lessons 11-6, 11-8)*

In 1 and 2, find the distance between the given points.

1. (3, 5) and (-7, -1)

2. (55, 90) and (64, 50)

In 3 and 4, give the coordinates of the midpoint of the segment with the given endpoints.

3. (3, -1) and (-7, -11)

4. (3, 2) and (6, -2)

In 5 and 6, a triangle has vertices (3, 2), (3, 7), and (6, 11).

5. What is its perimeter?

6. What are the coordinates of the midpoints of the sides?

In 7 and 8, refer to segment $\overline{AO}$ at the right.

7. Find AO.

8. If M is the midpoint of $\overline{AO}$, give the coordinates of M.

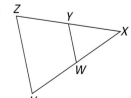

Objective B: *Apply the Midpoint Connector Theorem.* *(Lesson 11-8)*

In 9 and 10, W and Y are the midpoints of $\overline{XV}$ and $\overline{XZ}$ in the figure at the right.

9. If $YW = 40$, and $WV = 41$, determine all the other lengths that you can.

10. Write a proof argument to show that $m\angle XWY = m\angle XVZ$.

In 11 and 12, $\triangle EFG$ is the medial triangle of $\triangle BCD$, as shown below.

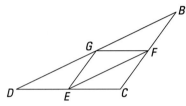

11. Explain why $BDEF$ is a trapezoid.

12. If $BD = 10$, $DC = 7$, and $BC = 6$, find the lengths of as many other segments as possible.

Objective C: *Plot points, find distances between them, and find coordinates of midpoints in 3-dimensional space.* *(Lesson 11-9)*

In 13 and 14, draw a coordinate system and plot the points.

13. (8, -2, 3)

14. (0, 12, -5)

In 15 and 16, two points are given. **a.** Find the coordinates of the midpoint of the segment joining them. **b.** Find the distance between them.

15. $P = (-2, 3, 6)$ and $Q = (0, -5, 1)$.

16. $R = (13, -5, 8)$ and $S = (0, 16, -3)$.

PROPERTIES DEAL WITH THE PRINCIPLES BEHIND THE MATHEMATICS.

Objective D: *Follow the basic laws of reasoning to make conclusions.* *(Lessons 11-1, 11-2, 11-3, 11-4)*

In 17–20, use all the statements. **a.** What (if anything) can you conclude? **b.** What laws of reasoning have you used?

17. (1) If a figure is a rectangle, then it is a trapezoid.
(2) *LOVE* is a rectangle.

18. (1) If corresponding angles formed by a transversal are congruent, then two lines are parallel.
(2) If alternate interior angles formed by a transversal are congruent, then so are corresponding angles.

19. (1) $x = 11$ or $y = 10$.
(2) $y = 7$

20. (1) If $\ell \perp m$, then m$\angle A = 90$.
(2) m$\angle A = 75$

21. *Multiple choice.*
a. If the conditional $p \Rightarrow q$ is true, which of the following must also be true?
(a) $p \Rightarrow$ not-q (b) $q \Rightarrow p$
(c) not-$p \Rightarrow$ not-q (d) not-$q \Rightarrow$ not-p
b. How is the correct answer to part **a** related to the given conditional?
(a) inverse   (b) contrapositive
(c) converse   (d) negation

22. Solving $(2x - 5)(3x + 4) = (x - 1)(6x - 1)$, Nella came up with the equation -20 = 1.
a. What should Nella conclude?
b. What law of logic is being used to make this conclusion?

Objective E: *Write the converse, inverse, or contrapositive of a conditional.* *(Lesson 11-2)*

In 23–26, write **a.** the converse, **b.** the inverse, and **c.** the contrapositive of the statement. **d.** Tell which (if any) of these are true.

23. If $x = 3$, then $x^2 = 9$.

24. If a figure is a rectangle, then it is a square.

25. All New Yorkers live in the U.S.

26. Use the figure at the right.
If m$\angle ABC = 40$, then
m$\angle DBC = 140$.

Objective F: *Write indirect proofs.*
(Lessons 11-4, 11-5, 11-6)

27. Explain why a quadrilateral cannot have four acute angles.

28. Give an indirect proof to show that $\sqrt{2400} \neq 49$.

29. Give an indirect proof to show that $\sqrt{2} \neq \frac{239}{169}$.

30. Use the figure below. Write an indirect proof to show that $\triangle ABC$ is not isosceles.

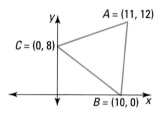

31. Write an indirect proof to show that *ABCD* below is not a trapezoid.

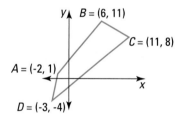

32. What is the difference between a direct proof and an indirect proof?

Objective G: *Use coordinate geometry to deduce properties of figures and prove theorems.*
(Lessons 11-5, 11-6, 11-8)

33. Prove that the triangle with vertices $A = (11, 2)$, $B = (23, 1)$, and $C = (2, 10)$ is isosceles.

34. Prove that the triangle with vertices $X = (q, 0)$, $Y = (0, q)$, and $Z = (2q, 3q)$ is a right triangle.

35. Prove that $X = (-1, 10)$, $Y = (6, 8)$, $Z = (3, -2)$, and $W = (-4, 0)$ are vertices of a parallelogram.

36. Show that the quadrilateral formed by joining midpoints of $ABCD$ shown below is a parallelogram.

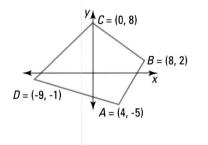

In 37 and 38, use the square below.

37. Prove that its diagonals are perpendicular to each other.

38. Prove that its diagonals have the same length.

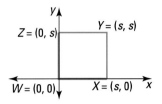

USES DEAL WITH APPLICATIONS OF MATHEMATICS IN REAL SITUATIONS.

Objective H: *Apply laws of reasoning in real situations.* *(Lessons 11-1, 11-2, 11-3, 11-4)*

In 39–41, given the statements, what (if anything) can you conclude using the rules of logic?

39. (1) I am allowed to watch TV at 8 P.M. if I have finished my homework.
(2) I am not allowed to watch TV at 8 P.M.

40. (1) Either Mary is too old for camp or she can go to camp.
(2) Mary cannot go to camp.

41. (1) All bats are mammals.
(2) No mammal can live without air.
(3) Air is not found on the moon.

42. Ted is looking for his lost homework paper. It isn't in his notebook. It isn't in his school locker. So he concludes that it is at home.
a. What reasoning law has he used?
b. What, if anything, is wrong with his reasoning?

In 43 and 44, the puzzles are from Lewis Carroll. What can you conclude?

43. (1) No name in this list is unsuitable for the hero of a romance.
(2) Names beginning with a vowel are always melodious.
(3) No name is suitable for the hero of a romance, if it begins with a consonant.

44. (1) Nobody who really appreciates Beethoven fails to keep silent while the Moonlight Sonata is being played.
(2) Guinea pigs are hopelessly ignorant of music.
(3) No one who is hopelessly ignorant of music ever keeps silent while the Moonlight Sonata is being played.

Ludwig van Beethoven (1770-1827)

Objective I: *Apply the Distance and Box Diagonal formulas in real situations.* *(Lessons 11-6, 11-9)*

45. A car drives 5 miles north, 2 miles east, then another 6 miles north, and another 3 miles east. By plane, how far is the car from its starting point?

46. A ship is located 2.3 km south and 1.4 km west of a lighthouse. Another ship is 1.6 km south and 0.8 km east of the lighthouse.
a. Draw a picture of this situation.
b. Calculate the distance between the ships.

47. A storage box is 8 inches by 12 inches by 24 inches. What is the longest dowel rod that can fit in it?

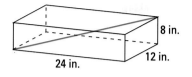

24 in. 12 in. 8 in.

48. A wooden crate is 45 mm by 30 mm by 80 mm. What is the length of the longest straw that will fit in it?

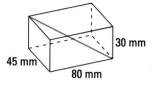

45 mm 30 mm 80 mm

REPRESENTATIONS DEAL WITH PICTURES, GRAPHS, OR OBJECTS THAT ILLUSTRATE CONCEPTS.

Objective J: *Graph and write an equation for a circle or a sphere given its center and radius, and vice versa.* *(Lessons 11-7, 11-9)*

49. What is an equation for the circle with center $(8, -1)$ and radius 15?

50. Write an equation for the circle below.

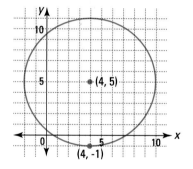

(4, 5)

(4, -1)

In 51 and 52, graph the given equation.

51. $(x + 1)^2 + (y - 4)^2 = 9$

52. $x^2 + y^2 = 36$

53. Find an equation for the sphere with center $(4, -3, 0)$ and radius 10.

54. Find the center and the radius of the sphere $(x - 1)^2 + (y + 2)^2 + (z - 5)^2 = 4$.

Objective K: *Give convenient locations for triangles and quadrilaterals in the coordinate plane.* *(Lesson 11-5)*

In 55–58, give convenient locations for the figure.

55. an isosceles triangle

56. a rectangle is symmetric to the x- and y-axes

57. a kite

58. a square

CHAPTER

12

SIMILARITY

Figures with different sizes but the same shape are found both in fun and serious pursuits. Models of planes, cars, trains, and ships, and dolls and dollhouses are scale models of real figures. Clothes designers, inventors, architects, and city planners use scale models to see how an object will look before the full-size version is made. Scientists magnify small things like insects or the atom, or make models of large objects, like Earth or our solar system, in order to study them.

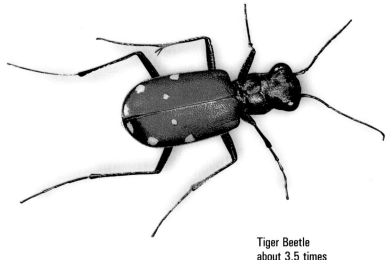

Tiger Beetle
about 3.5 times
actual size

Figures with the same shape (but not necessarily the same size) are *similar figures*. The concept of *similarity* is as important in analyzing figures as the concept of congruence. In this chapter, you will learn how to draw similar figures. Then you will study their basic properties and the transformations relating them, and see a few of their many applications.

12-1

The Transformation S_k

Relative size. *Many individual portraits are sized for the wallets, desktops, and photo albums of relatives and friends.*

A Specific Example

The transformations in this lesson are easy to do on a coordinate plane. You need only multiply the coordinates of points on the figure by a fixed number.

Example

In pentagon *ABCDE*, $A = (-9, 15)$, $B = (-15, -6)$, $C = (0, 3)$, $D = (12, 0)$, and $E = (3, 12)$. Describe the result when all the coordinates of the points of this figure are multiplied by $\frac{2}{3}$.

Solution

First draw *ABCDE*. Then multiply the coordinates of the vertices of *ABCDE* by $\frac{2}{3}$. The resulting five image points A', B', C', D', and E' are connected below.

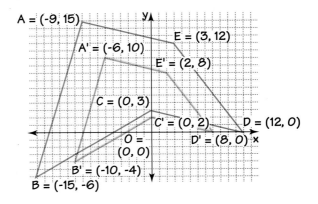

▶ To describe the result, examine the preimage and the image sides closely. You could write the following:

(1) Image and preimage sides seem to be parallel.

(2) Image sides appear to be $\frac{2}{3}$ as long as the preimage sides.

Closer examination gives something else.

(3) A point, its preimage, and the origin are collinear.

Do you see anything else?

The Example describes a transformation in which the image of (x, y) is $\left(\frac{2}{3}x, \frac{2}{3}y\right)$. This transformation is denoted $S_{\frac{2}{3}}$. To verify the first property of $S_{\frac{2}{3}}$ found in the Example, you can focus on a segment and its image, say $\overline{AE}$ and $\overline{A'E'}$. Using slopes, it is easy to show that $\overline{AE} \;/\!/\; \overline{A'E'}$.

$$\text{slope of } \overline{AE} \;=\; \frac{12-15}{3--9} = \frac{-3}{12} = -\frac{1}{4}$$
$$\text{slope of } \overline{A'E'} = \frac{8-10}{2--6} = \frac{-2}{8} = -\frac{1}{4}$$

Since the slopes are equal, $\overline{AE}$ and $\overline{A'E'}$ are parallel.

In the following Activity, the image of (x, y) is $(4x, 4y)$. The properties numbered (1) and (2) in the solution to the Example can be verified for this case.

Activity

On the coordinate plane below, $\overline{M'N'}$ is the image of $\overline{MN}$ under S_4.

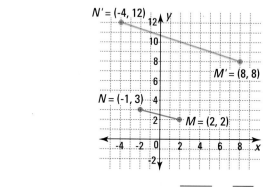

a. By calculating slopes, verify that $\overline{M'N'} \;/\!/\; \overline{MN}$.
b. By calculating distances, verify that $M'N' = 4 \cdot MN$.

Properties of the Transformation S_k

In general, the transformation which maps (x, y) onto (kx, ky) is denoted by the symbol S_k. The number k is called the **magnitude** of S_k. Any number but zero can be the magnitude, but in this book $k > 0$ unless otherwise stated. (In Question 24, you are asked to explore what happens when k is negative.) The transformation S_k has many important properties, all of which can be proved using coordinate geometry.

Properties of S_k Theorem

Let S_k be the transformation mapping (x, y) onto (kx, ky). Then under S_k,
(1) a line and its image are parallel, and
(2) the distance between two image points is k times the distance between their preimages.

Proof

Given: $P = (a, b)$, $Q = (c, d)$,
$S_k(P) = P' = (ka, kb)$, and $S_k(Q) = Q' = (kc, kd)$.

To prove: (1) $\overleftrightarrow{PQ} \,/\!/\, \overleftrightarrow{P'Q'}$.
(2) $P'Q' = k \cdot PQ$.

Drawing:

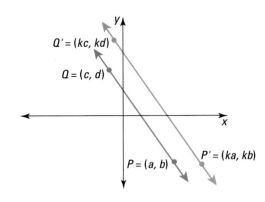

Argument:

(1) $\overleftrightarrow{P'Q'}$ is parallel to $\overleftrightarrow{PQ}$ if the slopes are the same.

$$\text{slope of } \overleftrightarrow{P'Q'} = \frac{kd - kb}{kc - ka} = \frac{k(d - b)}{k(c - a)} = \frac{d - b}{c - a}$$

$$\text{slope of } \overleftrightarrow{PQ} = \frac{d - b}{c - a}$$

Thus $\overleftrightarrow{PQ} \,/\!/\, \overleftrightarrow{P'Q'}$.

(2) The goal is to show that $P'Q' = k \cdot PQ$.
From the Distance Formula,

$$PQ = \sqrt{(c - a)^2 + (d - b)^2}.$$

Also from the Distance Formula,

$$
\begin{aligned}
P'Q' &= \sqrt{(kc - ka)^2 + (kd - kb)^2}. \\
&= \sqrt{(k(c - a))^2 + (k(d - b))^2} && \text{Distributive Property} \\
&= \sqrt{k^2(c - a)^2 + k^2(d - b)^2} && \text{Power of a Product} \\
&= \sqrt{k^2((c - a)^2 + (d - b)^2)} && \text{Distributive Property} \\
&= \sqrt{k^2}\sqrt{(c - a)^2 + (d - b)^2} && \text{Square Root of a Product} \\
&= k\sqrt{(c - a)^2 + (d - b)^2} && \text{Since } k > 0, \sqrt{k^2} = k. \\
&= k \cdot PQ && \text{Substitution}
\end{aligned}
$$

QUESTIONS

Covering the Reading

1. Name four occupations which use scale models.

In 2 and 3, refer to the Example.

2. Verify that $\overline{B'C'} \;//\; \overline{BC}$ by finding their slopes.

3. Verify that $B'C' = \frac{2}{3} BC$ by using the Distance Formula.

4. Show your answers to the Activity in this lesson.

5. Let $J = (5, -8)$, $K = (-6, 0)$, and $L = (-10, -4)$.
   a. Graph $\triangle JKL$ and its image under S_3.
   b. Describe what S_3 does to $\triangle JKL$.

6. a. S_k is the transformation which maps (x, y) onto __?__.
   b. The number k is the __?__ of S_k.

7. Name two properties of S_k.

8. A segment has endpoints (ka, kb) and (kc, kd).
   a. What is its slope?
   b. What is its length?

Applying the Mathematics

In 9 and 10, let O be the origin and A be the point $(-2, 7)$. Let $A' = S_3(A)$.

9. a. Give the coordinates of A'.
   b. By calculating distances, show that $OA + AA' = OA'$.
   c. *True or false.* A is between O and A'. Justify your answer.

10. Justify the conclusion that $OA' = 3 \cdot OA$.

11. An artist wished to draw the tiger beetle pictured on page 685 at double its actual size. So the artist drew an outline of the beetle at actual size onto graph paper as shown at the right. Then the coordinates of key points on the insect were multiplied by 2 and the image drawn. Repeat what the artist did, using your own paper.

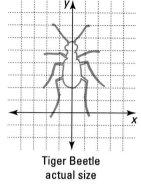

Tiger Beetle
actual size

12. a. Find $S_1(\triangle ABC)$, where $A = (4, -3)$, $B = (6, 1)$, and $C = (-2, 8)$.
    b. What is the image of (x, y) under S_1?
    c. Describe S_1.

13. At the right is a scale drawing of the side of a cabin. If each unit on the paper is $\frac{1}{8}$ inch and the drawing is $\frac{1}{96}$ actual size, what is the height of the actual cabin?

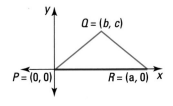

14. In three dimensions, S_k is the transformation mapping (x, y, z) onto (kx, ky, kz).
    a. Find the image of $P = (-3, 12, 4)$ under S_5. Call this point P'.
    b. Find the image of $Q = (2, -8, 0)$ under S_5. Call this point Q'.
    c. Using the Three-Dimension Distance Formula, verify that $Q'P' = 5 \cdot QP$.

15. At the right is a convenient location for any triangle $\triangle PQR$. Find $S_2(\triangle PQR)$.

Review

16. In the figure at the left, $\overleftrightarrow{KL} \parallel \overleftrightarrow{FG}$ and M is the midpoint of $\overline{LF}$.
    a. Write a proof argument which shows that $\triangle KLM \cong \triangle GFM$.
    b. Write a proof argument which shows that $MG = MK$.   *(Lesson 7-3)*

17. What are the *A-B-C-D* properties of isometries?   *(Lesson 4-7)*

18. State the Figure Reflection Theorem.   *(Lesson 4-2)*

19. The quadrilateral CEGB^b at the left joins the four notes which make up the *dominant-7th chord* of the key of F. Rotate the quadrilateral to find the notes which make up the dominant-7th chord of the key of A.   *(Lesson 3-2)*

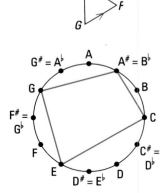

20. Draw this situation: B is between A and C and $AC = 5 \cdot AB$. *(Lesson 1-8)*

In 21–23, solve.   *(Previous course)*

21. $\frac{2}{9} = \frac{3}{x}$

22. $\frac{2y - 5}{5} = \frac{3y + 14}{8}$

23. $AB - 9 = \frac{3}{4} \cdot AB$ (Here, AB is the length of $\overline{AB}$.)

Exploration

24. Use $\triangle JKL$ of Question 5.
    a. Find $S_k(\triangle JKL)$ where $k = -3$.
    b. Describe what S_{-3} does to $\triangle JKL$.
    c. Repeat parts **a** and **b**, but let $k = -\frac{1}{2}$.
    d. Write some conjectures about S_k when k is negative.

*Drawing
Size-Change
Images*

IN·CLASS
ACTIVITY

**Work on this activity with another student. Choose different centers. Both of you
will need rulers.**

Here is a way to change the size of a figure without
using coordinates. At the right is a face made up of
segments, arcs, and dots. Suppose you wish to draw
a face like this one, but with corresponding lengths
3 times as long. Here is how to do this *size change
of magnitude* 3.

To begin, trace the above figure or draw another like it. Now choose
a point *O* in the plane to be the *center* of this size transformation.
O can be anywhere in the plane, but here, for clarity, we select *O*
above and to the left of the face. Then choose and label some key
points on the figure. Key points are endpoints of segments and
centers of arcs. We have chosen and labeled points *A* through *I*.

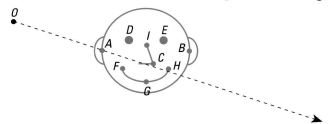

Find the image of each point. Here is how to find point *A*.

Step 1: Lightly draw ray $\overrightarrow{OA}$.
Step 2: Measure *OA*.
Step 3: Locate the point *A'* on $\overrightarrow{OA}$ so that $OA' = 3 \cdot OA$.
That is, *A'* is 3 times as far from the center as *A* is.
Point *A'* is the image of *A*.

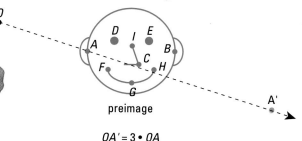

preimage

$$OA' = 3 \cdot OA$$

Repeat Steps 1–3 for the points *B* through *I*. Then
connect the image points in the same way the preimages
are connected.

691

Make a wish, any size. *Japanese Daruma dolls are made of papier-mâché and sold without eyes. When you make a wish, you draw one eye. If the wish comes true, you draw the other eye. The dolls shown are almost size-change images of each other.*

In the In-class Activity, you were shown how to draw a figure 3 times as large as a given figure. Below is the same process for a magnitude of 2.25. Because the magnitude is greater than one, the image is larger than the preimage.

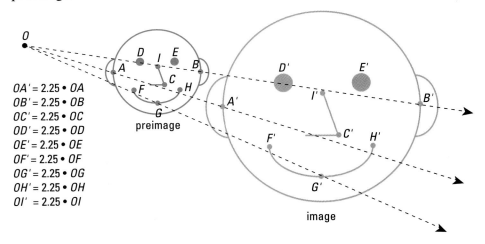

$OA' = 2.25 \cdot OA$
$OB' = 2.25 \cdot OB$
$OC' = 2.25 \cdot OC$
$OD' = 2.25 \cdot OD$
$OE' = 2.25 \cdot OE$
$OF' = 2.25 \cdot OF$
$OG' = 2.25 \cdot OG$
$OH' = 2.25 \cdot OH$
$OI' = 2.25 \cdot OI$

The same procedure can be used if the magnitude is less than or equal to one. Shown below is the face and its image when the magnitude is $\frac{1}{3}$ and the center is O. Notice OC'' is $\frac{1}{3}OC$. Now the image is smaller than the preimage.

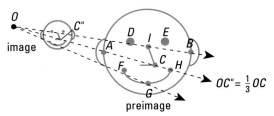

$OC'' = \frac{1}{3} OC$

The Definition of a Size Change

Because each preimage point has a unique image, this procedure defines a transformation. This transformation is called a *size transformation* or *size change*.

> **Definition**
> Let O be a point and k be a positive real number. For any point P, let $S(P) = P'$ be the point on $\overrightarrow{OP}$ with $OP' = k \cdot OP$. Then S is the **size change** or **size transformation** with **center** O and **magnitude** or **size-change factor** k.

When $k > 1$, S is called an **expansion.** When $0 < k < 1$, S is a **contraction.** When $k = 1$, S is called the **identity transformation** because each point coincides with its image and the figure keeps its identity. Pictured on the previous page are an expansion with magnitude 2.25 and a contraction with magnitude $\frac{1}{3}$.

When figures have the same shape and tilt, they are size-change images of each other. The center and magnitude of the size change can be found by drawing lines and measuring.

Example

EFGH is the image of ABCD under a size change. Find the center and magnitude of the size change.

Solution

Draw lines through the pairs of corresponding points. The lines will be concurrent at O, the center of the size change.

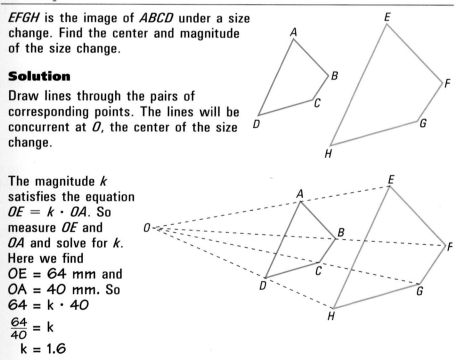

The magnitude k satisfies the equation $OE = k \cdot OA$. So measure OE and OA and solve for k. Here we find $OE = 64$ mm and $OA = 40$ mm. So

$$64 = k \cdot 40$$
$$\frac{64}{40} = k$$
$$k = 1.6$$

Check

Pick another point and its image. Does $OH = 1.6 \cdot OD$? We find $OH = 52$ mm and $OD = 32.5$ mm. So it checks.

Proving That S_k Is a Size-Change Transformation

The next theorem relates the size changes of this lesson to the transformation S_k that was defined in Lesson 12-1.

> **S_k Size-Change Theorem**
> When $k > 0$, the transformation S_k, where $S_k(x, y) = (kx, ky)$, is the size change with center $(0, 0)$ and magnitude k.

Proof
It must be shown that a point and its image under S_k satisfy the defining properties of a size change.

Given: $O = (0, 0)$, $P = (a, b)$, and   Drawing:
$P' = S_k(P) = (ka, kb)$.

To prove: (1) P' lies on $\overrightarrow{OP}$.
            (2) $OP' = k \cdot OP$.

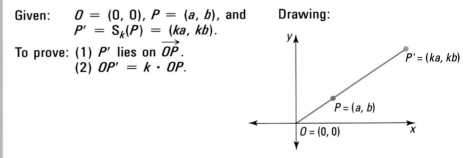

Argument:
First notice that $O' = S_k(O) = (k \cdot 0, k \cdot 0) = (0, 0) = O$.

(1) By part (a) of the Properties of S_k Theorem, $\overline{O'P'} \,/\!/\, \overline{OP}$.
     But there is only one line through O parallel to $\overrightarrow{OP}$.
     So, O, P, and P' must be on the same line. Thus P' lies on $\overrightarrow{OP}$.
(2) By part (b) of the Properties of S_k Theorem, $O'P' = k \cdot OP$.
     Since $O' = O$, then $O'P' = OP'$. By the Transitive Property of Equality, $OP' = k \cdot OP$.

The S_k Size-Change Theorem enables you to work with or without coordinates when doing size changes. In the next lesson, we apply the properties of S_k to deduce many properties of size changes.

QUESTIONS

Covering the Reading

1. Refer to the In-class Activity on page 691. Give the lengths of $\overline{OD}$ and $\overline{OD'}$.

2. For the expansion of the face in this lesson, what are the center and magnitude?

3. For the contraction of the face in this lesson, what are the center and magnitude?

In 4 and 5, trace the drawing below. Find the image of the flag under each transformation.

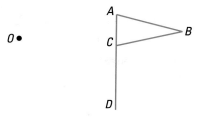

4. the size change with center O, magnitude 3

5. the size change with center P, magnitude $\frac{3}{4}$

6. Let S be a size transformation with size-change factor 6 and center O. Let A be any point. Fill in the blanks.
   a. S(A) is __?__ times as far from O as A is.
   b. Points A, O, and S(A) are __?__.

7. **a.** If k is the scale factor of an expansion, then __?__.
   b. If k is the scale factor of a contraction, then __?__.

8. If k is the scale factor of the identity transformation, then __?__.

9. Consider the proof that the transformation S_k is a size change of magnitude k with center O.
   a. What is given?
   b. What two statements were proved?

In 10–12, trace each figure. Use a ruler to determine the center and size-change factor k for the size transformation represented. (The figure is blue and the image is orange.)

10.

11.

12.

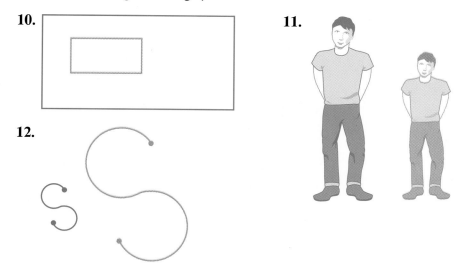

13. In the figure at the right, $A'B'C'D'$ is the image of $ABCD$ under the size change with center O.

    a. Is this size change an expansion or a contraction?

    b. If $OA = 10$ and $AA' = 4$, what is the magnitude of the size change?

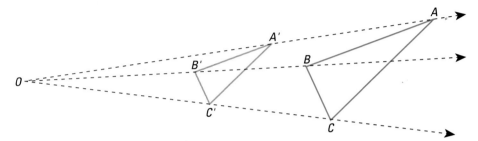

In 14–16, use the figure below. S is a size change with center O and $S(\triangle ABC) = \triangle A'B'C'$. The figure is a guide and not necessarily accurate.

14. If $k = \frac{2}{3}$ and $OA = 9$, then $OA' = \underline{\ ?\ }$ and $AA' = \underline{\ ?\ }$.

15. If $OB = 5$ and $OB' = 3$, then $k = \underline{\ ?\ }$.

16. If $\frac{OB'}{OB} = \frac{3}{4}$, then $k = \underline{\ ?\ }$.

17. In Question 11 of Lesson 12-1, you drew a size-change image of an insect. What are the center and magnitude of the size change?

18. Draw a figure whose linear dimensions are 1.5 times as large as those of the figure at the right. Use any method you wish.

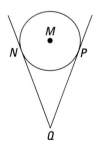

19. Trace $\triangle ABC$ below and draw its image under a size change with center A and magnitude $\frac{4}{5}$.

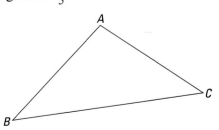

20. Let $A = (0, -4)$, $B = \left(-13, \frac{1}{3}\right)$, and $C = (7, 11)$. Give the coordinates of the vertices of the image of $\triangle ABC$ under S_3. *(Lesson 12-1)*

21. Given $P = (9, 2)$, $Q = (-4, 8)$, and $R = (35, -10)$.
    a. Show that P, Q, and R are collinear.
    b. Show that P is between Q and R.
    c. Find the images P', Q', and R' of P, Q, and R under S_4.
    d. Show that P', Q', and R' are collinear.
    e. Show that P' is between Q' and R'. *(Lesson 12-1)*

22. The rectangular field $ABCD$ pictured below is 300' by 400'. What is the perimeter of $\triangle ABC$? *(Lessons 8-1, 8-6)*

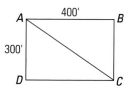

23. Is the figure of Question 22 traversable? If yes, give a path. If no, tell why not. *(Lesson 1-4)*

24. **a.** How many inches are in a mile?
    b. How many millimeters are in a kilometer? *(Previous course)*

25. Solve each equation. *(Previous course)*
    a. $\frac{z + 1}{2} = \frac{30}{40}$ **b.** $\frac{M}{5} = \frac{6}{M}$

26. On an automatic drawer, you can perform a size change on a figure by entering its center and magnitude. Draw a nonconvex pentagon using an automatic drawer. Show the image of this pentagon under a size change. (You pick the magnitude and the center.) Move the center so that it is inside, outside, and on the pentagon. How do the images compare?

12-3

Properties of Size Changes

You now have seen ways to find size changes with and without coordinates. You should be able to work without coordinates because figures are not always given on a coordinate plane. However, since coordinates enable some properties of size changes to be deduced rather easily, you should be able to work with coordinates. And, since any point can be the origin for a coordinate system, if a property of S_k can be deduced, then it holds for size changes with other centers.

Parallelism and Size Changes

For instance, the property that a line and its image under S_k are parallel can be used to draw images of polygons quickly.

Example

Let S be a size transformation with magnitude 0.6 and center O. Draw S($ABCD$).

$O \bullet$

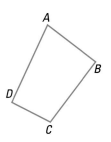

Solution

S(ABCD) is determined by the images of A, B, C, and D. Draw guide rays $\overrightarrow{OA}$, $\overrightarrow{OB}$, $\overrightarrow{OC}$, and $\overrightarrow{OD}$. Measuring, we find OA ≈ 56 mm. So OA' = 0.6 · OA ≈ 0.6 · 56 mm ≈ 34 mm. This locates A'. Since $\overline{A'B'}$ // $\overline{AB}$, we find B' by drawing a line through A' parallel to $\overleftrightarrow{AB}$. This line intersects $\overrightarrow{OB}$ at B'. Likewise, since $\overline{A'D'}$ // $\overline{AD}$, find D' by drawing a line parallel to $\overleftrightarrow{AD}$. This line intersects $\overrightarrow{OD}$ at D'. Continue this process until all the vertices of the image polygon are located.

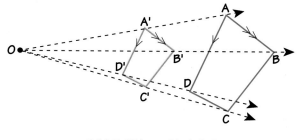

S(ABCD) = A'B'C'D'

How Do Size Changes Affect Distance?

In the Questions of the last two lessons, you have been asked to verify some of the properties of size changes. The property concerning distance is the most important property of size changes, and was proved in Lesson 12-1.

> **Size-Change Distance Theorem**
> Under a size change with magnitude $k > 0$, the distance between any two image points is k times the distance between their preimages.

Another way to think of this is that the *ratio* of the length of an image to the length of its preimage is k. That is, $A'B' = k \cdot AB$ implies that $\frac{A'B'}{AB} = k$.

Preservation Properties of Size Changes

Recall the A-B-C-D properties of reflections and other isometries. Size changes do not preserve distance, but they preserve the other three properties.

> **Size-Change Preservation Properties Theorem**
> Every size transformation preserves (1) angle measure, (2) betweenness, and (3) collinearity.

The proof of this theorem is given on the next page.

Proof

Given: S is a size transformation with center O.

                $S(A) = A'$, $S(B) = B'$, $S(C) = C'$.

To prove: (1) $m\angle ABC = m\angle A'B'C'$

            (2) If B is between A and C, then B' is between A' and C'.

            (3) If A, B and C are collinear, then A', B' and C' are collinear.

Drawing (1):

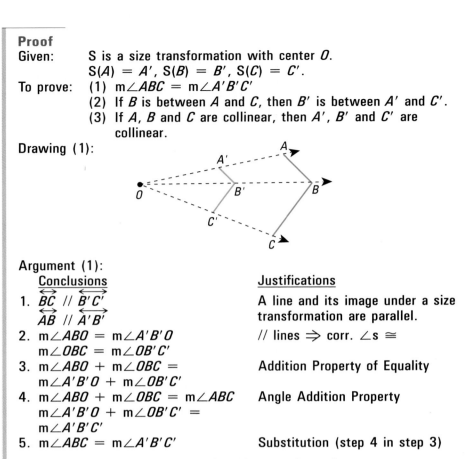

Argument (1):

Conclusions	Justifications
1. $\overleftrightarrow{BC}$ // $\overleftrightarrow{B'C'}$ $\overleftrightarrow{AB}$ // $\overleftrightarrow{A'B'}$	A line and its image under a size transformation are parallel.
2. $m\angle ABO = m\angle A'B'O$ $m\angle OBC = m\angle OB'C'$	// lines $\Rightarrow$ corr. $\angle$s $\cong$
3. $m\angle ABO + m\angle OBC =$ $m\angle A'B'O + m\angle OB'C'$	Addition Property of Equality
4. $m\angle ABO + m\angle OBC = m\angle ABC$ $m\angle A'B'O + m\angle OB'C' =$ $m\angle A'B'C'$	Angle Addition Property
5. $m\angle ABC = m\angle A'B'C'$	Substitution (step 4 in step 3)

The following argument proves that size transformations preserve betweenness.

Drawing (2):

Argument (2):

Conclusions	Justifications
0. B is between A and C.	Given
1. $AB + BC = AC$	definition of betweenness
2. $k \cdot (AB + BC) = k \cdot AC$	Multiplication Property of Equality
3. $k \cdot AB + k \cdot BC = k \cdot AC$	Distributive Property
4. $A'B' = k \cdot AB$ $B'C' = k \cdot BC$ $A'C' = k \cdot AC$	Size-Change Distance Theorem
5. $A'B' + B'C' = A'C'$	Substitution (step 4 into step 3)
6. B' is between A' and C'.	definition of betweenness

Argument (3):

Betweenness implies collinearity. So argument (2) also proves that collinearity is preserved.

The preservation of angle measure, betweenness, and collinearity implies that images of figures are determined by images of key points. Because size changes preserve collinearity, the image of a line is a line. Because size changes preserve betweenness, the image of a segment is a segment and the image of a ray is a ray. So the image of an angle is an angle, the image of a triangle is a triangle, and so on.

> **Figure Size-Change Theorem**
> If a figure is determined by certain points, then its size-change image is the corresponding figure determined by the size-change images of those points.

The Figure Size-Change Theorem was applied as early as Lesson 12-1 to find the image of a pentagon.

QUESTIONS

Covering the Reading

1. Trace the figure below. Let S be the size transformation with magnitude 1.4 and center *O*. Draw S(*ABCD*).

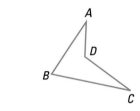

2. Suppose you know the lengths of a segment and its image under a size change. How can the magnitude of the size change be calculated?

3. *True or false.*   Size transformations preserve distance.

4. Suppose a photograph of an insect is k times its actual size. If a leg on the picture is 3 cm long, how long is the actual leg?

5. Size transformations preserve __?__, __?__, and __?__.

***Scare*ab.** *A rhinoceros beetle can grow to 6 cm in length.*

In 6–8, S is a size transformation of magnitude 200.

6. If S($\angle TJK$) = $\angle T'J'K'$ and m$\angle T'J'K'$ = 43, find m$\angle TJK$.

7. Suppose S($\overleftrightarrow{AB}$) = $\overleftrightarrow{CD}$. How are $\overleftrightarrow{AB}$ and $\overleftrightarrow{CD}$ related?

8. If S(*TINY*) = *HUGE*, then S(*N*) = __?__.

In 9–11, S is a size change with S(*A*) = *X*, S(*B*) = *T*, S(*C*) = *E*, and S(*D*) = *J*.

9. Draw a possible picture.   10. S($\angle BCD$) = __?__.

11. If m$\angle BAD$ = 73, then m$\angle$ __?__ = 73.

Applying the Mathematics

12. At the right, $\triangle DEF$ is a size-transformation image of $\triangle ABC$ with center O. $AB = 6$, $BC = 8$, $EF = 20$, and $DF = 30$. Find
  a. k, the magnitude of the size change.
  b. DE.
  c. AC.

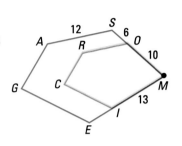

13. T is a size change with $T(MEGAS) = MICRO$ and lengths as indicated at the right.
  a. What is the magnitude of T?
  b. What is the center of T?
  c. Find OR.
  d. Find ME.

In 14–16, trace the figure. Use the Figure Size-Change Theorem to draw the image of the figure under the size change with center O and the given magnitude k.

14. $k = \frac{5}{6}$

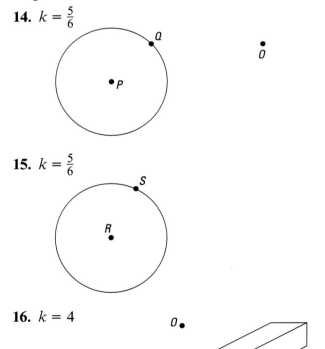

15. $k = \frac{5}{6}$

16. $k = 4$

17. A photograph that has width 5 cm and length 12 cm is enlarged. The new width is 7 cm.
  a. Find the size-change factor of the enlargement.
  b. Find the length of the enlargement.
  c. Find the areas of the photograph and its enlargement.
  d. The area of the enlargement is how many times the area of the original?
  e. The perimeter of the enlargement is how many times the perimeter of the original?

18. Use the figure at the right.
  Given: S($\triangle OAC$) = $\triangle OBD$.
  $OA = 6$, $AB = 1$, $BD = 4$,
  and $OC = 6.1$.
  Find the lengths of as many other segments as you can.

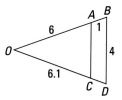

Review

19. Face A is a size-change image of face B. Trace the faces, and find the center and scale factor k.   *(Lesson 12-2)*

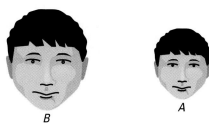

B   A

20. Many road maps can be thought of as contractions of a part of Earth. Suppose 1 cm on a map equals 1 km on Earth. What is the scale factor of the contraction for this map? (That magnitude is the *scale* of the map.)   *(Lesson 12-2)*

21. Suppose $P'Q'R'T'$ is the image of $PQRT$ under S_k with $P = (4, -6)$, $Q = (-9, 12)$, $R' = (5, 13)$, $T = (0, -8)$, and $P' = (2, -3)$. Find k and the coordinates of Q', R, and T'.   *(Lesson 12-1)*

22. a. The Hopi design shown below represents clouds with rain. Draw it under a size change with magnitude 2. Then draw an enlargement of the image, using a size change of magnitude 1.5. *(Lesson 12-2)*
  b. How does the final design compare to the original?

A Hopi hand-painted dance wand

23. Complete this hierarchy for isometries. *(Lesson 4-7)*

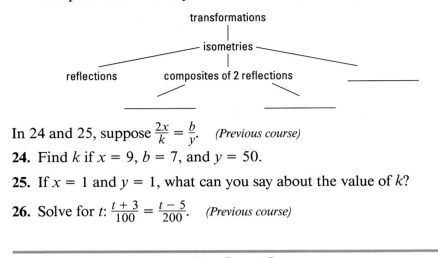

In 24 and 25, suppose $\frac{2x}{k} = \frac{b}{y}$. *(Previous course)*

24. Find k if $x = 9$, $b = 7$, and $y = 50$.

25. If $x = 1$ and $y = 1$, what can you say about the value of k?

26. Solve for t: $\frac{t + 3}{100} = \frac{t - 5}{200}$. *(Previous course)*

Exploration

27. a. Does the definition of size change given in Lesson 12-2 hold in three dimensions, or must it be modified? If so, how would you modify it?
  b. What properties mentioned in this lesson are preserved by size changes in three dimensions?

28. Consider this conjecture. If *MNOP* is a square and S is a size change, then S(*MNOP*) is a square.
  a. Draw an instance of this conjecture.
  b. Is this conjecture true? If so, prove it. If not, draw a counterexample.

Proportions

Anchors a*weigh*. *This stock anchor, one of four common types of anchors, is heavy enough to keep the vessel from drifting.*

What Is a Ratio?

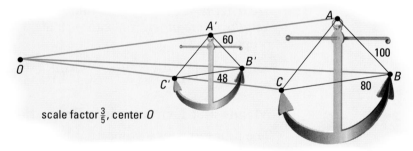

scale factor $\frac{3}{5}$, center O

A **ratio** is a quotient of two numbers, $\frac{m}{n}$, m/n, or $m : n$, where m and n are quantities of the same kind, such as lengths, populations, or areas. (If the quantities are of different kinds, $\frac{m}{n}$ is called a *rate*.)

Every size change involves equal ratios. For example, pictured above are a figure and its image under the size change with center O and magnitude $\frac{3}{5}$. Then, by the Size-Change Distance Theorem,

$$A'B' = \frac{3}{5} \cdot AB \quad \text{and} \quad B'C' = \frac{3}{5} \cdot BC.$$

Consequently, $\quad \frac{A'B'}{AB} = \frac{3}{5} \quad\quad$ and $\quad \frac{B'C'}{BC} = \frac{3}{5}.$

Thus in a size transformation, all ratios of image lengths to preimage lengths are equal. That is, $\frac{A'B'}{AB} = \frac{B'C'}{BC}$. This ratio is equal to the magnitude of the size change. The ratio is less than 1 if the size change is a contraction. It is greater than 1 if the size change is an expansion.

Proportions and Size Changes

A statement that two quotients are equal is called a **proportion.** Each equation below is a proportion.

$$\frac{B'C'}{BC} = \frac{3}{5} \qquad \frac{2}{7} = \frac{x}{9} \qquad \frac{y+3}{5} = \frac{7}{y} \qquad \frac{A'B'}{AB} = \frac{B'C'}{BC}$$

Four numbers that form a true proportion are called **proportional.** The numbers 5, 3, 10, and 6, in that order, are proportional because $\frac{5}{3} = \frac{10}{6}$. The numbers 1, 2, 3, and 4 are not proportional because $\frac{1}{2} \neq \frac{3}{4}$.

Proportional numbers always occur when there is a size change. For instance, when $\triangle ABC$ is a size-transformation image of $\triangle XYZ$, you can say, "The sides of the triangles are proportional." This means the three ratios of the corresponding sides are equal.

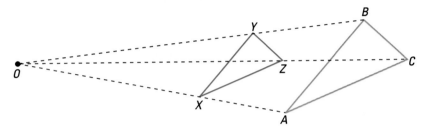

That is, $\frac{AB}{XY} = \frac{AC}{XZ} = \frac{BC}{YZ}$. Notice that in this proportion, *the numerators are lengths from one figure and the denominators are corresponding lengths in the other.* You can pick any two of these to form a true proportion.

$$\frac{AB}{XY} = \frac{AC}{XZ} \qquad \frac{AB}{XY} = \frac{BC}{YZ} \qquad \frac{AC}{XZ} = \frac{BC}{YZ}$$

Means and Extremes in a Proportion

The four terms in a proportion have two sets of names and are numbered in order. The 1st and 4th terms are the **extremes** and the 2nd and 3rd terms are the **means.**

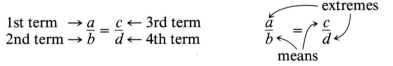

1st term $\rightarrow \dfrac{a}{b} = \dfrac{c}{d} \leftarrow$ 3rd term
2nd term $\rightarrow \phantom{\dfrac{a}{b}} \phantom{\dfrac{c}{d}} \leftarrow$ 4th term

In previous courses, you have learned that in any proportion, the product of the means equals the product of the extremes. This is the *Means-Extremes Property.*

> **Means-Extremes Property**
> If $\frac{a}{b} = \frac{c}{d}$, then $ad = bc$.

Proof

Given: $\frac{a}{b} = \frac{c}{d}$.

To prove: $ad = bc$.

Argument:

Multiply both sides of $\frac{a}{b} = \frac{c}{d}$ by bd.

$$bd \cdot \frac{a}{b} = bd \cdot \frac{c}{d}$$

$$b \cdot d \cdot a \cdot \frac{1}{b} = b \cdot d \cdot c \cdot \frac{1}{d}$$

Use the Commutative and Associative Properties of Multiplication.

$$\left(b \cdot \frac{1}{b}\right) \cdot a \cdot d = \left(d \cdot \frac{1}{d}\right) \cdot b \cdot c$$

Since the product of a number and its reciprocal is 1, these products can be simplified.

$$ad = bc$$

In geometry, *a*, *b*, *c*, and *d* are often lengths of segments.

Example 1

$\triangle QRS$ is the image of $\triangle TUV$ under a size change with center *O*. If $QR = 10$, $RS = 15$, and $TU = 25$, find *UV*.

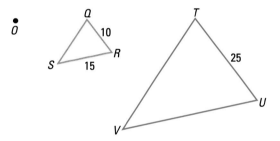

Solution 1

The sides of the triangles are proportional. Thus $\frac{TU}{QR} = \frac{UV}{RS}$.

Substituting,

$$\frac{25}{10} = \frac{UV}{15}.$$

Solve for *UV*, using the Means-Extremes Property:

$$25 \cdot 15 = 10 \cdot UV.$$

$$\frac{375}{10} = UV$$

$$UV = 37.5$$

Solution 2

The magnitude of the size change is $\frac{10}{25} = 0.4$.

So $0.4 \cdot UV = 15$, from which $UV = \frac{15}{0.4} = 37.5$.

Proportions in Scale Models

Proportions occur in scale models, maps, and wherever else there are size changes.

Dollhouse miniatures.
Accessories, such as the double boiler, the tea set, and the frying pan shown, are crafted to match the scale of the dollhouse.

Example 2

A dollhouse is $\frac{1}{12}$ actual size. If a table in the dollhouse is an oval $6\frac{1}{2}''$ long and $4\frac{3}{8}''$ wide, how long and wide is the actual table it models?

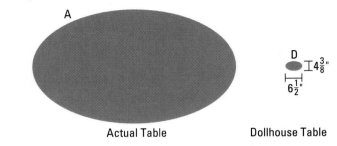

Actual Table   Dollhouse Table

Solution

Since corresponding lengths are proportional,

$$\frac{\text{dollhouse length}}{\text{actual length}} = \frac{1}{12} \quad \text{and} \quad \frac{\text{dollhouse width}}{\text{actual width}} = \frac{1}{12}.$$

Substitute for the dollhouse table dimensions and let L and W be the actual length and width.

$$\frac{6\frac{1}{2}''}{L} = \frac{1}{12} \qquad \qquad \frac{4\frac{3}{8}''}{W} = \frac{1}{12}$$

Use the Means-Extremes Property.

$$1 \cdot L = 12 \cdot 6\frac{1}{2}'' \qquad \qquad 1 \cdot W = 12 \cdot 4\frac{3}{8}''$$
$$L = 78'' \qquad \qquad W = 52\frac{1}{2}''$$

Check

$\frac{78''}{6.5''} = 12$ and $\frac{52.5''}{4.375''} = 12$, so the measurements of the actual table are 12 times the measurements of the model.

QUESTIONS

Covering the Reading

1. *Multiple choice.* Which is *not* a way of writing the ratio of 7 to 9?
   (a) $7:9$ (b) $7/9$
   (c) $\frac{7}{9}$ (d) 7.9

2. A ratio is a __?__ of two numbers.

3. A proportion is a statement that two __?__ are __?__.

4. Why are the numbers 2, 4, 6, and 8 *not* proportional?

5. Given $\frac{7}{x} = \frac{11}{y}$, name the
   a. extremes.
   b. means.
   c. first term.
   d. fourth term.
   e. third term.
   f. second term.

6. Suppose $\frac{r}{s} = \frac{t}{u}$. What conclusion is justified by the Means-Extremes Property?

7. In the figure at the right, $\triangle SMA$ is the image of $\triangle BIG$ under a size change. Write three equal ratios involving the sides of these triangles.

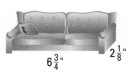

In 8 and 9, *ABCD* at the left is the image of *FGHE* under a size change with center *V*.

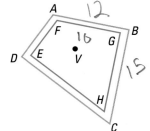

8. Fill in the missing length: $\frac{AB}{FG} = \frac{AD}{\square}$.

9. If $FG = 10$, $AB = 12$, and $BC = 15$, what is GH?

10. A dollhouse is $\frac{1}{12}$ actual size. If a couch in the dollhouse is a $6\frac{3}{4}''$ long and $2\frac{1}{8}''$ wide, how long and wide is the actual couch it models?

$6\frac{3}{4}''$ $2\frac{1}{8}''$

In 11 and 12, $\triangle GJK$ at the left is a size-change image of $\triangle GHI$.

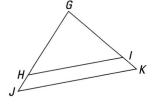

11. *True or false.*
   a. $\frac{GJ}{GH} = \frac{GK}{GI}$
   b. $\frac{GH}{GJ} = \frac{GI}{GK}$
   c. $\frac{JK}{HI} = \frac{GI}{GK}$

12. If $GH = 100$, $GJ = 130$, and $HI = 120$, then $JK =$ ___?___.

Applying the Mathematics

In 13 and 14, solve for *x*.

13. $\frac{a}{x} = \frac{b}{c}$

14. $\frac{x}{x+3} = \frac{9}{10}$

15. Weights and prices of fruit at a store usually are proportional. If 2.3 pounds of dates imported from Oman cost $2.39, what would 4 pounds of dates cost?

16. a. If you bike 13 miles in $1\frac{1}{2}$ hours, at that rate how many miles would you bike in $2\frac{1}{2}$ hours?
    b. If you drive *m* miles in *h* hours, at that rate how many miles can you drive in *r* hours?

17. A photograph measures 40 mm by 30 mm. It is enlarged so that the longer side measures 150 mm. What is the length of the shorter side of the enlargement?

18. Suppose S is a size change of magnitude 1.5, and S(*ABCDE*) = *UVWXY*. If m∠*BDE* = 47 and *AE* = 30, find two other measures. *(Lessons 12-2, 12-3)*

19. Let S be a size transformation of magnitude $\frac{1}{2}$ centered at the origin. If *A* = (16, -6) and *B* = (10, 8), verify that the distance between S(*A*) and S(*B*) is half the distance between *A* and *B*. *(Lessons 12-1, 12-3)*

20. Trace the figure below. Draw the image of the figure under the size change with center *H* and magnitude $\frac{3}{5}$. *(Lesson 12-2)*

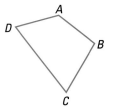

21. The distance between the lens and the negative is 2 inches. How far away from the lens must Ann place the photo paper if the developed picture is to be 5 times as large as the negative? *(Lesson 12-2)*

A photo enlarger projects an image from a negative onto photographic paper.

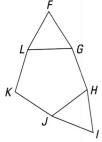

22. Show that the transformation T with equation T(*x*, *y*) = (2*x*, 3*y*) does not preserve distance. *(Lesson 12-1)*

23. Given: *FGL*, *GHJKL*, and *HIJ* at the left are regular polygons.
a. Write an argument to prove that △*FGL* ≅ △*HIJ*.
b. What is m∠*GHI*? *(Lessons 6-7, 7-2)*

24. Finish this definition. Two figures α (alpha) and β (beta) are congruent, written α ≅ β, if and only if __?__. *(Lesson 5-1)*

25. Draw a quadrilateral with sides of four different lengths. Next, draw a second quadrilateral with sides that are twice as long as those of the first quadrilateral. Must the corresponding angles of the two quadrilaterals be congruent? Explain why or why not.

Model Plans. *Model enthusiasts often use detailed plans for constructing replicas of real-life objects.*

The word "congruent" refers to figures with the same size and shape. In Lesson 4-8, three terms relating to this idea were precisely defined:

> congruent figures,
> ≅ (is congruent to), and
> congruence transformation.

These precise definitions enable an in-depth study of figures with the same size and shape.

In this lesson, we define three corresponding terms for figures that have the same shape but not necessarily the same size:

> similar figures,
> ~ (is similar to), and
> similarity transformation.

A Precise Definition for Similar Figures

You have seen that under a size change, figures and their images have the same shape. So under any definition, these figures should be similar. But also, performing an isometry (reflection, rotation, translation, or glide reflection) on a figure does not change its shape. The definition of *similar figures* encompasses all of these possibilities.

> **Definition**
> Two figures *F* and *G* are **similar**, written *F* ~ *G,* if and only if there is a composite of size changes and reflections mapping one onto the other.

Triangles *ABC*, *PQR*, *SQT*, and *XYZ* below are similar.

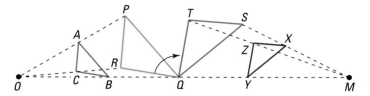

ΔPQR is a size-change image of ΔABC with center O and magnitude 1.8.
ΔSQT is a rotation image of ΔPQR with center Q and magnitude -90°.
ΔXYZ is a size-change image of ΔSQT with center M and magnitude $\frac{2}{3}$.

The symbol "~" is read "is similar to." Thus, in the figure below, you can write "*WXYZ ~ ABCD*" and say "Quadrilateral *WXYZ* is similar to quadrilateral *ABCD*." As with congruence, corresponding vertices are written in corresponding order.

$S_{1.6}(WXYZ) = ABCD$

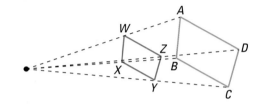

Which Transformations Are Similarity Transformations?

Transformations that give rise to similar figures are called *similarity transformations*.

Definition
A transformation is a **similarity transformation** if and only if it is the composite of size changes and reflections.

Recall that the size change S_1 with magnitude 1 is the identity transformation. It does nothing to a figure. Thus, all congruent figures are images of each other under a composite of reflections and S_1. So all congruent figures are similar figures.

Here is a hierarchy of transformations you have studied, with similarity transformations, size changes, and the identity transformation included.

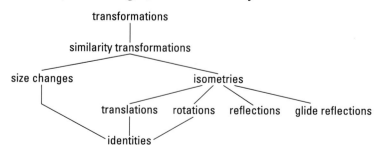

Properties of Similar Figures

The basic properties of similar figures come from preservation properties of similarity transformations. Compare these lists.

Preserved under reflections	Preserved under size transformations
Angle measure	Angle measure
Betweenness	Betweenness
Collinearity	Collinearity
Distance	

The properties common to both columns are preserved by similarity transformations. Thus similarity transformations have the A-B-C preservation properties, while distance is not preserved. Still, because size transformations are involved, similarity transformations multiply distance by a constant amount. Thus the ratios of image lengths to preimage lengths are equal.

Similar Figures Theorem
If two figures are similar, then
(1) corresponding angles are congruent, and
(2) corresponding lengths are proportional.

The Similar Figures Theorem allows lengths and angle measures in similar figures to be found.

Example 1

In the figure below, $r_\ell \circ S_{2.5}(ABCD) = WXYZ$, and the center of $S_{2.5}$ is O.
a. If $m\angle B = 85$, what angle in $WXYZ$ has measure 85?
b. If $CD = 12$, what length in $WXYZ$ can be determined, and what is it?

Solution
$ABCD \sim WXYZ$ because $r_\ell \circ S_{2.5}$ is a composite of a reflection and a size change.

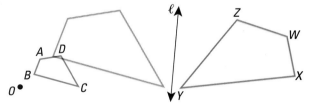

a. By the Similar Figures Theorem, corresponding angles are congruent. $\angle A \cong \angle W$, $\angle B \cong \angle X$, $\angle C \cong \angle Y$, and $\angle D \cong \angle Z$. Since $m\angle B = 85$, so also $m\angle X = 85$.
b. By the Similar Figures Theorem, corresponding sides are proportional. These ratios equal 2.5, the magnitude of the size change. $\frac{WX}{AB} = \frac{XY}{BC} = \frac{YZ}{CD} = \frac{ZW}{DA} = 2.5$. Since $CD = 12$, the length of the corresponding side YZ can be determined.
$$\frac{YZ}{12} = 2.5$$
$$YZ = 12 \cdot 2.5 = 30$$

The Ratio of Similitude

The ratio of a length on an image to the corresponding length on a similar preimage is called the **ratio of similitude**. Unless otherwise specified, $F \sim G$, with *ratio of similitude* k, means that lengths in G divided by lengths in F equal k. The ratio of similitude is the product of the size-change factors of all the size transformations used in the similarity transformation. In Example 1 on page 713, the ratio of similitude is 2.5.

When examining similar figures, always look first for corresponding vertices. These give the pairs of congruent angles. Look next at corresponding sides.

Example 2

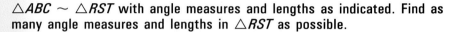

$\triangle ABC \sim \triangle RST$ with angle measures and lengths as indicated. Find as many angle measures and lengths in $\triangle RST$ as possible.

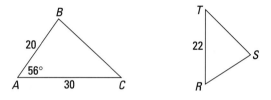

Solution

Angle measures: $\angle R$ corresponds to $\angle A$. In similar figures, corresponding angles are congruent. So $m\angle R = m\angle A = 56$.

Lengths: Since corresponding sides are proportional, $\frac{TR}{CA} = \frac{SR}{BA} = \frac{ST}{BC}$.

Any one of these ratios equals the ratio of similitude. Now substitute the three known lengths.

$$\frac{22}{30} = \frac{SR}{20} = \frac{ST}{BC}$$

Use the equality of the left and middle ratios.

$$\frac{22}{30} = \frac{SR}{20}$$

By the Means-Extremes Property,

$$30 \cdot SR = 440.$$

$$SR = \frac{440}{30} = 14\frac{2}{3} \approx 14.67$$

Neither BC nor ST nor the measures of the other four angles can be found using the Similar Figures Theorem.

In Example 2, the SAS condition is satisfied, so enough information is given to determine the measures of all sides and angle measures in each triangle. But to do this, a theorem from trigonometry (called the Law of Cosines) is needed. We do not cover that theorem in this book.

Covering the Reading

In 1–3, use the sketch below. Figure III is the image of Figure I under $S \circ r_\ell$, where S is the size transformation with center C, magnitude 2.3.

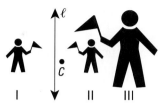

1. $S \circ r_\ell$ is what kind of transformation?

2. Figures I and III are __?__.

3. Figures I and II are both __?__ and __?__.

4. Define *similar figures*.

5. The symbol "~" is read __?__.

6. *True or false.*   If two figures are congruent, then they are similar.

7. *True or false.*   If two figures are similar, then they are congruent.

8. Tell whether the property is preserved by every similarity transformation.
   a. angle measure   **b.** betweenness
   c. distance   **d.** orientation

In 9–11, is the given transformation a similarity transformation?

9. a reflection   10. a size change with magnitude $\frac{1}{3}$

11. the composite of a rotation and a size change

In 12–14, $ABCDE \sim FGHIJ$. The ratio of similitude is $\frac{4}{7}$.

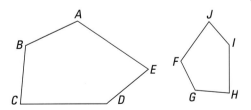

12. *True or false.* $m\angle I = \frac{4}{7} \cdot m\angle D$.

13. Suppose $FJ = 10$.
    a. Which other segment length can be determined?
    b. What is that length?

14. If $DE = x$, then $IJ = $ __?__.

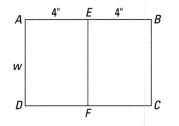

Applying the Mathematics

15. If $\triangle ABC \sim \triangle DEF$, $BC = 12$, $DE = 8$, and $EF = 16$, find AB.

16. Presto Printing makes folded cards which, when opened, have an outside boundary similar to that when they are folded. That is, in the figure at the left, $ABCD \sim ADFE$. Find the width w of the cards.

17. A model plane is similar to a real plane with ratio of similitude of $\frac{1}{45}$. If the wingspan of the model is $18''$, what is the wingspan of the actual plane?

18. Let $P = (4, 7)$ and $Q = (-4, -8)$. Suppose $\overline{P'Q'} = S_6 \circ S_8(\overline{PQ})$. Find the length of $\overline{P'Q'}$.

19. Let S be a similarity transformation and let ℓ and m be lines. Tell whether the statement is true or false and justify your answer.
a. $S(\ell) \parallel \ell$ 　　　　　　　　**b.** If $\ell \parallel m$, then $S(\ell) \parallel S(m)$.

Review

20. If 3 cans of tuna cost \$2.00, how much are 5 cans? *(Lesson 12-4)*

21. Solve $\frac{5}{x} = \frac{11}{x + 3}$ for x. *(Lesson 12-4, Previous course)*

22. In the figure at the right, $\triangle ABC$ is the image of $\triangle DBE$ under a size change with center B.
a. *True or false.* $\overleftrightarrow{DE} \parallel \overleftrightarrow{AC}$.
b. *True or false.* $\frac{BD}{AB} = \frac{ED}{AC}$.
c. Find x.　*(Lessons 12-2, 12-3, 12-4)*

23. A spherical rubber ball has an outside diameter of 10 cm. The ball is hollow and the rubber is 1 cm thick. How much rubber is used in making the ball? *(Lesson 10-8)*

24. In triangles XYZ and APD, $\angle X \cong \angle A$, $\angle Y \cong \angle P$, and $\angle Z \cong \angle D$.
a. Is $\triangle XYZ$ necessarily congruent to $\triangle APD$?
b. Justify your answer to part **a.** *(Lesson 7-1)*

Exploration

25. Scale models of objects are similar to objects that are larger or smaller than the models. Find a scale model.
a. For your scale model, what is the ratio of similitude?
b. A length of $1''$ on the model corresponds to what length on the object it models?　(Note: Save your scale model to do Question 22 of Lesson 12-6.)

26. In Example 2, we noted that BC can be found by using a theorem called the Law of Cosines. Look in some other book to find out what this "law" is.

The Fundamental Theorem of Similarity

Packaging nautical history. *Gil Charbonneau, a former sailor, constructs models of sailing ships for museums and collectors. See Example 3.*

Similar figures may be 2-dimensional or 3-dimensional because a similarity transformation can be done in three dimensions just as easily as in two. Any point can be the center of the size change and any real number, the scale factor. The Similar Figures Theorem holds for space figures—corresponding distances are multiplied by the scale factor, and corresponding angles are congruent.

A 3-Dimensional Example

In similar figures, perimeters, areas, and volumes satisfy a very important relationship, which is the subject of this lesson.

Activity

Below, Box I ~ Box II, with ratio of similitude 5. Find
a. the perimeter of the largest face of each box.
b. the surface area of each box.
c. the volume of each box.

3
6   10

Box I

Box II

In the Activity, you should have noticed that the face of Box II has 5 times the perimeter of the face of Box I, and that Box II has 25 times the surface area and 125 times the volume of Box I.

These answers could have been predicted without calculating any perimeters, surface areas, or volumes. The ratio of similitude of the boxes is 5. The ratio of corresponding lengths is 5, or 5^1. The ratio of corresponding surface areas is 25, or 5^2, because area involves the multiplication of 2 dimensions. The ratio of corresponding volumes is 125, or 5^3, because volume involves the multiplication of 3 dimensions.

A 2-Dimensional Example

Example 1

$\triangle ABC \sim \triangle DEF$. Give the ratios of the perimeters and areas of these triangles.

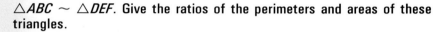
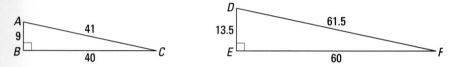

Solution
The perimeters and areas can be calculated directly.

For $\triangle ABC$:   For $\triangle DEF$:

Perimeter = 9 + 40 + 41   Perimeter = 13.5 + 60 + 61.5

           = 90 units   = 135 units

Area = $\frac{1}{2} \cdot 40 \cdot 9$ Area = $\frac{1}{2} \cdot 13.5 \cdot 60$

       = 180 square units   = 405 square units

Thus, the ratio of perimeters = $\frac{135}{90}$ = 1.5

and   the ratio of areas = $\frac{405}{180}$ = 2.25.

The ratio of similitude is 1.5. The ratio of perimeters is 1.5^1, while the ratio of areas is 1.5^2.

The General Theorem

These examples are instances of the following theorem, which applies to all similar figures in 2 or 3 dimensions.

> **Theorem**
> If $G \sim G'$ and k is the ratio of similitude, then
> (1) Perimeter(G') = $k \cdot$ Perimeter(G)   or $\frac{\text{Perimeter}(G')}{\text{Perimeter}(G)} = k$,
>
> (2)   Area(G') = $k^2 \cdot$ Area(G)   or $\frac{\text{Area}(G')}{\text{Area}(G)} = k^2$, and
>
> (3)   Volume(G') = $k^3 \cdot$ Volume(G)   or $\frac{\text{Volume}(G')}{\text{Volume}(G)} = k^3$.

Proof

Argument (1):

Suppose lengths $a, b, c, d, e, \ldots$ make up the perimeter of G. Then lengths $ka, kb, kc, kd, ke, \ldots$ make up the perimeter of G'.

$$\begin{aligned} \text{Perimeter}(G') &= ka + kb + kc + \ldots \\ &= k(a + b + c + \ldots) \\ &= k \cdot \text{Perimeter}(G) \end{aligned}$$

Argument (2):

Let $A = \text{Area}(G)$. Then you could think of the area of G as the sum of the areas of A squares with sides of length 1 (unit squares). Then the area of G' is the sum of areas of A squares with sides of length k units. Since each square in G' has area k^2,

$$\text{Area}(G') = A \cdot k^2 = k^2 \cdot \text{Area}(G).$$

Argument (3):

The argument is much like that for the area. Let $V = \text{Volume}(G)$. Then the volume of G equals that of V cubes with edges of length 1 (unit cubes). The volume of G' is the sum of the volumes of V cubes each with edges of length k. Since each cube in G' has volume k^3,

$$\text{Volume}(G') = V \cdot k^3 = k^3 \cdot \text{Volume}(G).$$

Part (2) of the theorem can be seen in area formulas, such as $A = \pi r^2$ (circles), $A = \ell w$ (rectangles), $A = \frac{1}{2} h(b_1 + b_2)$ (trapezoids), L.A. $= ph$ (cylinders or prisms). Notice that each area formula involves the product of two lengths. (That's why the result is measured in square units.) So if each length is multiplied by k, the area is multiplied by k^2.

Volume formulas illustrate part (3) of the theorem. Consider some volume formulas: $V = \ell wh$ (boxes), $V = \frac{4}{3} \pi r^3$ (spheres), $V = \frac{1}{3} Bh$ (pyramids or cones). In each, there are three lengths multiplied. (For pyramids or cones, it looks as if there are only two quantities multiplied, but B is an area, so for B two lengths are multiplied.) Since each length is multiplied by k, the volume is multiplied by k^3.

The previous theorem and the Similar Figures Theorem combine to produce the most important theorem relating measures in similar figures.

Fundamental Theorem of Similarity

If two figures are similar with ratio of similitude k, then
(a) corresponding angle measures are equal;
(b) corresponding lengths and perimeters are in the ratio k;
(c) corresponding areas and surface areas are in the ratio k^2; and
(d) corresponding volumes are in the ratio k^3.

Example 2

Pentagons *ABCDE* and *FGHIJ* are similar with ratio of similitude $\frac{2}{3}$. If *FGHIJ* has area 50 square units, what is the area of *ABCDE*?

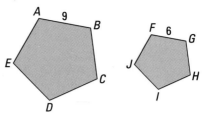

Solution

The ratio of areas is the square of the ratio of similitude.

$$\frac{\text{Area}(FGHIJ)}{\text{Area}(ABCDE)} = \left(\frac{2}{3}\right)^2 = \frac{4}{9}$$

Substituting,

$$\frac{50}{\text{Area}(ABCDE)} = \frac{4}{9}, \text{ from which}$$

$$\text{Area}(ABCDE) = \frac{450}{4} = 112.5 \text{ square units.}$$

Example 3

Two models of the same schooner (a boat with front and rear masts) are shown below. Each model is similar to the original, so they are similar to each other. The larger model is 120 cm long; the smaller, 30 cm long.

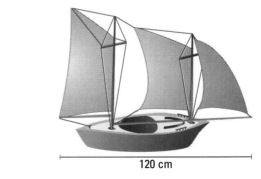

30 cm   120 cm

a. How do their heights compare?
b. How do the areas of their sails compare?
c. How do the volumes of their hulls (the bodies of the boats) compare?

Solution

First compute the ratio of similitude.

$$k = \frac{\text{length of large schooner}}{\text{length of small schooner}} = \frac{120 \text{ cm}}{30 \text{ cm}} = 4$$

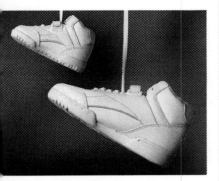

Similarity check. *Can you find similar pentagons on these two shoes?*

a. The height is a length, so the ratio of the heights is *k*.

$$\frac{\text{height of large schooner}}{\text{height of small schooner}} = k = 4.$$

So the height of the large schooner is 4 times that of the small schooner.

b. The ratio of the areas of the sails is k^2.

$$\frac{\text{area of large sail}}{\text{area of small sail}} = k^2 = 16$$

The areas of sails in the large schooner are 16 times those of the small schooner.

c. To compare the volumes of their hulls, use k^3.

$$\frac{\text{volume of larger hull}}{\text{volume of smaller hull}} = k^3 = 64$$

The volume of the large hull is 64 times the volume of the small hull.

QUESTIONS

Covering the Reading

1. Can a similarity transformation be done in three dimensions?

2. **a.** Give your results for the Activity in this lesson.
   b. Give the ratios of the perimeters, surface areas, and volumes.

3. At the right, $\triangle ABC \sim \triangle DEF$.
   a. Give the ratio of similitude.
   b. Give the ratio of the perimeters.
   c. Give the ratio of the areas.

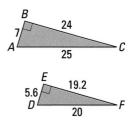

4. Area is the product of __?__ lengths.

5. Volume is the product of __?__ lengths.

6. The ratio of similitude of two figures is $\frac{5}{3}$. Find the ratio of their
   a. perimeters.   **b.** areas.
   c. volumes.   **d.** corresponding sides.

7. Two squares have sides 5 inches and 13 inches. Find the ratio of their perimeters.

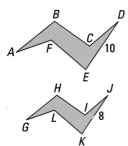

8. Hexagons *ABCDEF* and *GHIJKL* at the left are similar with ratio of similitude $\frac{4}{5}$. If *GHIJKL* has area 72 square meters, what is the area of *ABCDEF*?

9. Two models of the same airplane are similar to each other. The larger model is 90 cm long, and the smaller is 30 cm long.

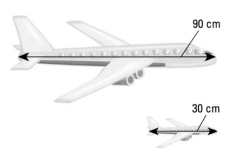

90 cm

30 cm

a. How do their wingspans compare?
b. How do the areas of their tailpieces compare?
c. How do the volumes of their bodies compare?

Applying the Mathematics

25

11

cylinder *Q*

10. $R \sim R'$, $k = 4$, and Volume(R) = 34 cubic units. Find Volume(R').

11. Right cylinder Q has height 25 and the radius of its base is 11.
a. Find its volume and surface area.
b. Let cylinder Z be the image of cylinder Q under a size change with magnitude 3. Cylinder Z will have _?_ times the volume and _?_ times the surface area of cylinder Q.
c. Find the volume and the surface area of cylinder Z.

12. Two similar nonregular 15-gons have perimeters of 20 ft and 28 ft. What is the ratio of the length of a side of the smaller to the length of the corresponding side of the larger?

13. Corresponding edges of two similar pyramids are 8 and 12 inches.
a. Find the volume of the larger pyramid if the smaller has volume 100 cubic inches.
b. Find the volume of the smaller pyramid if the larger has volume 100 cubic inches.

14. On two similar solid brass statues of Martin Luther King, Jr., the lengths of the left ear are 3 cm and 5 cm.
a. If the base area of the larger statue is 50 cm², find the base area of the smaller statue.
b. The volume of brass in the smaller statue is 216 cm³. What volume of brass is in the larger statue?

15. The volumes of two spheres are 288π mm³ and 7776π mm³.
a. What is a ratio of similitude for these spheres?
b. What is a ratio of their surface areas?

16. *Multiple choice.* The bases of two quadrangular prisms are similar with ratio of similitude 1.5, but the prisms have the same height. What is the ratio of their volumes?

Martin Luther King, Jr. received the Nobel prize for peace in 1964.

(a) 1.5   (b) 2.25   (c) 3   (d) 4.5
(e) cannot be determined from the given information

17. Below, Figure $Q \sim$ Figure R. Determine the ratio of similitude.
(Lesson 12-5)

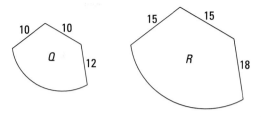

18. $\triangle ABC \sim \triangle DEF$ with ratio of similitude 2.5. If $AB = 3$, $BC = 5$, and m$\angle B = 135$, find as many lengths and angle measures in $\triangle DEF$ as you can. *(Lesson 12-5)*

19. A comic strip measures 6″ by 2″. The smaller sides of a reduction measure $\frac{1''}{2}$. What are the lengths of the larger sides of the reduction?
(Lesson 12-4)

20. A goal post and its shadow are shown below. The length of the shadow of each vertical post is 75 ft. The crossbar is 10 ft high and its shadow falls 24 ft from the base of the goal post. What is the height h of the goal post? *(Lesson 12-4)*

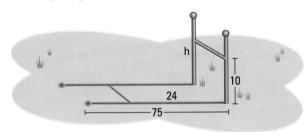

21. A can of tennis balls usually holds three balls, as pictured at the left. What percent of the volume of the can is not used by the balls? (Hint: Let r be the radius of the can and proceed from there.)
(Lessons 10-5, 10-8)

22. In Question 25 of Lesson 12-5, you were asked to find a scale model of an actual object. Weigh the scale model you found. If the actual object was made from the same materials, what would be its weight?

Can There Be Giants?

Giants are common characters in children's stories; an example is the giant from *Jack and the Beanstalk*. Saturday morning cartoons often have giant creatures with human shapes. The tallest man on record was Robert Wadlow from Alton, Illinois. On June 27, 1940, at age 22, he was measured at 8 ft, 11.1 in. Wadlow was about 1.5 times the height of a typical male. Can humans be much taller? The Fundamental Theorem of Similarity provides the answer.

Giants in *Gulliver's Travels*

Let's look at what the theorem reveals, using an example from a famous novel. In *Gulliver's Travels*, Jonathan Swift writes about a man named Gulliver who visits the land of Brobdingnag, where the Brobdingnagians are similar to us but 12 times as tall. Their volume, and thus their

Measuring-up. *Tailor Sol Winkelman fits a suit for Robert Wadlow in 1936. At this time Wadlow was 8′5″ tall.*

weight, would be 12^3 times ours. Since $12^3 = 1728$, they would weigh about 1728 times what we weigh. If you weigh 140 pounds, a similar Brobdingnagian would weigh 241,920 pounds! He or she would support this weight on feet covering a region whose area is only 12^2, or 144 times what you stand on. So each bone of a Brobdingnagian would have to carry 12 times as much weight as yours.

Even champion weight lifters seldom lift more than twice their body weight—and when they do, it is only for a few seconds. Imagine what lifting 12 times your weight would do. It would quickly break your bones!

You might think that a giant body would find some way of dealing with the extra weight. But it can't. Wadlow had to wear a leg brace to support his weight. One day while getting out of a car (which was hard for him to do), the brace cut a deep wound in his leg, which became infected. Eighteen days after his height was measured, he died in Manistee, Michigan.

The Fundamental Theorem of Similarity in Nature

Animals have developed within the constraints imposed by the Fundamental Theorem of Similarity. Elephants have legs with large horizontal cross-sectional areas to support their great weight. Thoroughbred race horses have thin legs which enable them to run fast, but their legs are small for their bodies, and when a thoroughbred falls, its legs often break. Draft horses which pull wagons have thicker, stronger legs, but these horses are slow. A mosquito can walk on the surface of water without sinking. It is so light that it will not break the surface tension of the water. It also has thin legs which support its light body. But that body has a relatively large surface area. When a raindrop forces a mosquito's body into the water, the surface tension acts like glue on the body's surface, and the thin legs cannot pull the mosquito from the water.

Food and Clothing Needs

In general, the amount of food needed by an animal is proportional to its volume. The Brobdingnagians would need to consume 1728 times the food needed by Gulliver. A person Gulliver's size needs about 2500 calories a day to maintain body weight (about 16 calories per pound). A Brobdingnagian would require 1728 times 2500 calories a day. That's a lot of food.

Gulliver also visited the land of Lilliput, where people were $\frac{1}{12}$ his height. For Gulliver, as for us, a new shirt would require about two square yards of material. Clothing is related to surface area, so we multiply by the square of the ratio of similitude. The Lilliputians, being $\frac{1}{12}$ Gulliver's height, would require only $\left(\frac{1}{12}\right)^2$ or $\frac{1}{144}$ times the two square yards needed by Gulliver. The Brobdingnagians would require 12^2 or 144 times as much material. Thus, geometry answers questions about clothing and food needs as well as the properties of giants.

Giants by Gully! *This sketch from a 1795 French edition of* Gulliver's Travels *shows Gulliver in the corn fields of Brobdingnag.*

Example 1

A pizzeria sells 12″-diameter pizzas with cheese and one topping for $9.99 (plus tax). Suppose the price of a pizza is based on the amount of ingredients used to make the pizza, and that the pizzas have the same thickness. What would be the charge for a 14″-diameter pizza with cheese and one topping?

▶

▶ **Solution**

Since the tops of both pizzas are circles, they are similar, with $k = \frac{14}{12} = \frac{7}{6}$.
An immediate response is to charge $\frac{7}{6} \cdot \$9.99$. However, this is not correct because the crust and toppings are proportional to the *area*, not the linear dimensions. So we must multiply by $\left(\frac{7}{6}\right)^2$.
The price, based only on amount of ingredients, would be
$$\$9.99 \cdot \left(\frac{7}{6}\right)^2 = \$9.99 \cdot \frac{49}{36} \approx \$13.60.$$

Usually the prices of pizzas are not based only on ingredients. Other factors, such as salaries, equipment, competition, and overhead must be taken into account.

Example 2

A solid clay figurine weighs 5 kilograms. A similar one, twice as tall, stands next to it. Could a 4-year-old child pick up the taller figurine?

Solution

Weight is proportional to volume. Thus the multiplying factor for weight is the cube of the ratio of similitude, in this case 2^3. **The taller figurine weighs $2^3 \cdot 5 = 40$ kg (about 90 pounds).**
A child of 4 would not be able to pick up the figurine, and many adults would be surprised at the weight of the figurine.

This Mayan clay figurine was found in Campeche, Mexico, and dates from 700–1000 A.D.

The Fundamental Theorem of Similarity was known to Euclid, but the structural applications were not recognized until over 1800 years later by the Italian scientist Galileo. He considered this discovery as important as his more famous discovery that when heavier-than-air objects of different weights are dropped from the same height, they fall to the ground at the same rate.

QUESTIONS

Covering the Reading

In 1–3, according to *Gulliver's Travels* and the Fundamental Theorem of Similarity,

1. Brobdingnagians are __?__ times the height of Gulliver, so they weigh __?__ times as much.

2. Lilliputians are __?__ times the height of Gulliver, so they weigh __?__ times as much.

3. Brobdingnagians are __?__ times the height of Lilliputians so they weigh __?__ times as much.

4. Who was Robert Wadlow?

Baby giant. *Only in pictures will you see a toddler this size. This scene is from* Honey, I Blew Up the Kid.

In 5–8, consider an imaginary female giant 27 feet tall, which is about 5 times the height of an average woman. If the giant and an average-height woman have similar shapes, how would the following quantities compare?

5. weights

6. nose lengths

7. areas of footprints

8. wrist circumferences

9. *True or false.* Champion weight lifters often lift weights five times their own weights.

10. *True or false.* Prices of pizza are proportional to their diameters.

11. Why does an elephant need thicker legs for its height than a mosquito needs for its height?

12. A scale model is $\frac{1}{15}$ actual size. If it is made from the same materials as the original object, then its weight will be __?__ times the weight of the original. The amount of paint needed to cover the exterior will be __?__ times the amount of paint used to cover the original.

Applying the Mathematics

13. Two similar solid clay figurines are 40 cm and 50 cm tall. If the shorter one weighs 8 kg, how much will the taller one weigh?

14. A pizza store manager calculates that the ingredients in a 16″ pizza cost the store $1.50. At this rate, what do the ingredients cost in a 12″ pizza with the same thickness?

15. A 90-cm-tall statue weighs 120 kg. If a similar statue weighs 40 kg, how tall (to the nearest centimeter) is it?

16. The surface area of Earth is about 13 times that of the moon.
   a. What is the ratio of their radii, considering them both to be spheres?
   b. What is the ratio of their volumes?

17. When his height was measured at 8′11.1″, Robert Wadlow weighed 439 lbs. How much would a 5′10″ man weigh if he were similar in shape to Wadlow?

18. Suppose two boxes have congruent bases, but one box is twice the height of the other.
   a. Are the boxes similar?
   b. How do their volumes compare?
   c. How do their surface areas compare?

Review

19. A hexagon has area 70 units^2. What is the area of its image under a size change of magnitude $\frac{2}{5}$? *(Lesson 12-6)*

20. At the right, △*PQR* ~ △*MST*. Find as many missing lengths and angle measures as possible.
(*Lessons 12-4, 12-5*)

21. Define similar figures.   (*Lesson 12-5*)

22. Trace △*ABC* and point *P* at the right. Then draw the image of △*ABC* under a size change with center *P*, magnitude 4.
(*Lesson 12-2*)

23. Write a proof argument using the given figure.
  Given: *K* is the midpoint of $\overline{FJ}$. The segments $\overline{KI}$ and $\overline{KG}$ are drawn parallel to $\overline{FH}$ and $\overline{HJ}$.
  To prove: △*FGK* ≅ △*KIJ*.   (*Lesson 7-3*)

Pictured is 1992 gold medal winner Naim Suleymanoglu from Turkey.

Exploration

24. Below are the winning total amounts lifted in the men's weightlifting competition in the 1992 Olympic Games. "Weight Class" refers to the lifter's maximum allowable weight.
  a. Calculate the ratios of weight lifted to weight lifter's maximum weight.
  b. What trends do you see?
  c. Give an explanation for any trends you find.

Weight Class	Winner	Team	Total (2 lifts)
52.0 kg	Ivan Ivanov	Bulgaria	265.0 kg
56.0 kg	Chun Byung Kwan	S. Korea	287.5 kg
60.0 kg	Naim Suleymanoglu	Turkey	320.0 kg
67.5 kg	Israel Militossian	Unified Team	337.5 kg
75.0 kg	Fedo Kassapu	Unified Team	357.5 kg
82.5 kg	Pyrros Dimas	Greece	370.0 kg
90.0 kg	Kakhi Kakhiaehvili	Unified Team	412.5 kg
100.0 kg	Victor Tregoubov	Unified Team	410.0 kg
110.0 kg	Ronny Weller	Germany	432.5 kg
Over 110.0 kg	Aleksandr Kourlovitch	Unified Team	450.0 kg

25. Find the costs of two different-size pizzas (with the same ingredients) at a local pizza parlor.
  a. What is the ratio of the diameters of the pizzas?
  b. What is the ratio of the costs of the pizzas?
  c. Are costs based on areas?

A project presents an opportunity for you to extend your knowledge of a topic related to the material of this chapter. You should allow more time for a project than you do for typical homework questions.

PROJECTS 12 CHAPTER TWELVE

1 Finish This Story

A pharaoh in ancient Egypt asked a goldsmith to make two statues, identical except for scale, of the pharaoh and his son. The statue of the pharaoh was to be six times the height of the statue of his son. The statue of the pharaoh required 60 pieces of gold, which the pharaoh supplied to the goldsmith. After that statue was completed, the goldsmith asked for 10 pieces of gold to make the statue for the son. The pharaoh gave the goldsmith the 10 pieces, the goldsmith made the statue, and the pharaoh was so pleased with the result that he rewarded the goldsmith with 5 gold pieces. Some time later, one of the pharaoh's advisors told him that the goldsmith had cheated him.

Write a conclusion to this story, which will answer the following questions: Was the pharaoh really cheated? If so, by how much? If not, why is the advisor wrong? If the goldsmith hired a lawyer to defend him, what might the lawyer say? If the advisor hired a lawyer to defend him, what would that lawyer say?

2 Golden Ratio

A rectangle known as the golden rectangle has the following property: If the square on its short side is cut off, then the remaining rectangle is similar to the original, as shown here.

The ratio of the long side to the short side of the golden rectangle is called the golden ratio, and is often represented by the Greek letter ϕ (phi). (This is the same letter used to describe the empty set, but it is not related at all to that use.)

a. Research the golden rectangle and golden ratio and write a report on them. What are some of their properties and applications?

b. Use an automatic drawer or straightedge and compass to construct a golden rectangle. Write an explanation of how you did this, complete with sketches.

3 Models of Everyday Objects

Make a 3-dimensional scale model of a real object. Calculate the surface areas and volumes of the original object and your scale model. Write a report on how you built your model, including calculations.

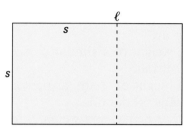

▶

4 **Similar People**
a. Record your height and weight, and use the method of Lesson 12-7 to determine:
(i) the weight of a similar person a foot (.3048 meter) taller than you are.
(ii) the height of a similar person 20 pounds (9.07 kg) lighter than you are.
(iii) the weight of a similar person 7 feet tall.
(iv) the height of a similar person who weighs 50 pounds.
b. Using your height and weight, create a chart to show the weights of people with similar shape, but taller or shorter, in two-inch intervals from 4'6" to 6'6".
c. Consult a chart which gives average male and female heights and weights. Do the average weights in this chart assume people are similar? Why or why not?

5 **Pantograph**
A pantograph is a mechanical device used for drawing similar figures. In the pantograph shown below, P is fixed. When Q traces a figure, S draws the image of the figure under a size change with magnitude $\frac{PC}{PA}$. Write a report on how a pantograph works, and how drawings can be enlarged or shrunk with it.

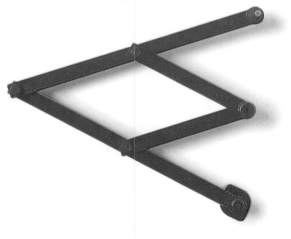

6 **Prices of Fruit Drinks**
Find a place where you can purchase fruit drinks in at least three sizes. For example, some places have small, medium, and large drinks. Some have extra-large drinks.

a. Are the containers similar? Support your answer by referring to the dimensions of the containers.
b. Calculate the volume of the container for each size drink.
c. Divide each volume by the cost of the drink for that container.
d. Which size has the lowest price per unit volume? Which size do you feel is the best buy, and why?

SUMMARY

In the coordinate plane, a size change centered at the origin can be achieved by multiplying coordinates of points by a given scale factor. Since a coordinate system can be created with any point as the origin, size transformations can be centered anywhere. Size transformations can occur in two or three dimensions.

Two figures are similar if and only if one can be mapped onto the other by a composite of reflections and size transformations. The ability to draw or construct similar figures is necessary in the making of scale drawings, toys or scale models, maps, blueprints, and other diagrams. In similar figures, angles and their images are congruent. Lengths of image segments are

k times the lengths of preimage segments, where k is a positive number called the ratio of similitude. Areas of images are k^2 times the areas of their preimages. Volumes of images are k^3 times the volumes of their preimages. These relationships between two similar figures help explain why large animals need relatively thicker legs than small animals, and why there cannot be giants.

When one quantity is k times another, then the ratio of the quantities equals k. An equality of two ratios is called a proportion. Whenever there are similar figures, corresponding lengths are proportional. Solving proportions can help you determine unknown measurements.

VOCABULARY

Below are the most important terms and phrases for this chapter. For the starred (*) terms you should be able to give a definition of the term. For the other terms you should be able to give a general description and a specific example of each.

Lesson 12-1
S_k, magnitude
Properties of S_k Theorem

Lesson 12-2
*size change
*size transformation
*center of size transformation
*magnitude of size transformation
*size-change factor k
expansion
contraction
identity transformation
S_k Size-Change Theorem

Lesson 12-3
Size-Change Distance Theorem
Size-Change Preservation Properties Theorem
Figure Size-Change Theorem

Lesson 12-4
ratio
proportion, proportional
extremes, means
Means-Extremes Property

Lesson 12-5
*similar figures, $F \sim G$
*similarity transformation
Similar Figures Theorem
ratio of similitude

Lesson 12-6
Fundamental Theorem of Similarity

PROGRESS SELF-TEST

Take this test as you would take a test in class. You will need a ruler and compass. Check your work with the solutions in the Selected Answers section in the back of the book.

In 1 and 2, trace the figure. Draw the image of △ABC under each size change.

1. center O, magnitude $\frac{3}{4}$.

2. center A, magnitude 2.

3. Trace the figure at the right. Draw the image of $\odot E$ under a size change with center F, magnitude 1.5.

4. Trace the figure below in which the preimage is blue and the image is orange. Use a ruler to determine the center and size-change factor k for the size change represented.

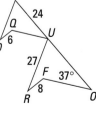

5. $QUAD \sim FOUR$ with side and angle measures as indicated at the right. Find as many missing lengths and angle measures as possible.

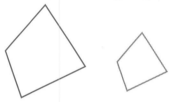

In 6 and 7, S(△VXZ) = △WXY in the figure.

6. $VZ = 12$, $WY = 4$, and $WX = 6$. What is XV?

7. **a.** Is this S an expansion or a contraction?
   b. What is the center of the size change?

8. *Multiple choice.* Size changes preserve
   (a) angle measure.   (b) distance.
   (c) area.   (d) volume.

9. Define similar figures.

10. △$P'Q'R'$ is a size-change image of △PQR with a magnitude of 5. How do the areas of △$P'Q'R'$ and △PQR compare?

11. A pyramid has volume 729 cm^3. What is the volume of a similar pyramid $\frac{1}{3}$ as high?

12. At the right, △$QRS \sim$ △VTU.
    a. Find QS.
    b. Find TU.

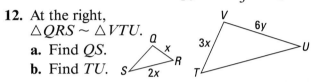

13. A photo slide measures 5 cm by 3 cm. If the shorter dimension of a similar print is 25 cm, what is its longer dimension?

14. A car can go 41 km in 28 minutes. At that rate, how long will it take to travel 100 km?

15. A solid figurine is 4″ tall and weighs 5 pounds. What will a similar solid figurine of the same material weigh if it is 12″ tall?

16. Suppose a person 2 meters tall has feet 33 cm long. If a person with similar physique were 0.4 meter tall, how long would this person's feet be?

17. If there were a person 6 times as tall as you with a similar physique, the person would weigh __?__ times as much. This weight would be supported by about __?__ times the area.

18. Graph △ABC and S$_{\frac{1}{3}}$(△ABC) if $A = (6, -6)$, $B = (9, 0)$, and $C = (-6, 10)$.

19. If $P = (5, -1)$ and $Q = (1, 2)$, verify that the distance between S$_8(P)$ and S$_8(Q)$ is $8 \cdot PQ$.

20. At the right, kite $KITE$ is conveniently located on a coordinate plane. Find the coordinates of the vertices of S$_4(KITE)$.

CHAPTER REVIEW

Questions on SPUR Objectives

SPUR stands for **S**kills, **P**roperties, **U**ses, and **R**epresentations. The Chapter Review questions are grouped according to the SPUR Objectives for this chapter.

SKILLS DEAL WITH THE PROCEDURES USED TO GET ANSWERS.

Objective A: *Draw size-transformation images of figures.* *(Lessons 12-2, 12-3)*

In 1–6, first trace the figure.

1. Draw the image of △*ABC* under a size change with center *O*, magnitude 1.5.

2. Draw the image of △*ABC* above under a size transformation with center *A*, magnitude 0.8.

3. Draw the image of *DEFG* below under a size transformation with center *I*, scale factor $\frac{2}{3}$.

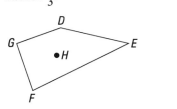

4. Draw the image of *DEFG* above under a size change with scale factor $\frac{5}{3}$, center *H*.

5. Draw the image of *ABCDE* under a size transformation with center *O*, magnitude 3.2.

6. Draw *A′B′C′D′E′*, the image of *ABCDE*, under a size change with center *E*, magnitude 1.

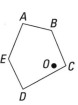

Objective B: *Use proportions to find missing parts in similar figures.* *(Lessons 12-4, 12-5)*

7. Below, *HOURS* is the image of *PENTA* under a size change. Some side and angle measures are indicated. Find as many missing lengths and angle measures as possible. (Be careful about which vertices correspond.)

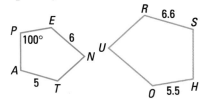

8. Let S be a size transformation with S(△*JKL*) = △*JMN*, as pictured below.
 a. Find *JN*.
 b. Find *LN*.
 c. Find *MN*.

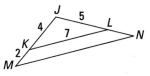

In 9 and 10, △*TUV* ~ △*WXV*.

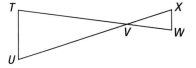

9. If *TV* = 10, *VW* = 5, and *VX* = 6, find *VU*.
10. *TU* = 70, *XW* = 42, and *TV* = 112.5.
 a. What other length can be found?
 b. Find it.

11. $\triangle ABC \sim \triangle DEF$ in the figure at the right.
   a. Find DF.
   b. Find BC.

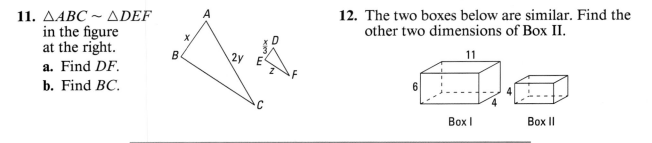

12. The two boxes below are similar. Find the other two dimensions of Box II.

Box I   Box II

PROPERTIES DEAL WITH THE PRINCIPLES BEHIND THE MATHEMATICS.

Objective C: *Recognize and apply properties of size transformations.* *(Lessons 12-2, 12-3)*

13. Suppose S is a size transformation of magnitude 4 and $S(\triangle CAT) = \triangle FUN$.
   a. $S(T) = \underline{\ ?\ }$.
   b. If $m\angle NUF = 71$, find $m\angle CAT$.
   c. If $AC = 11$, find UF.

14. *Multiple choice.* A size change of magnitude .5 does *not* preserve
   (a) angle measure.   (b) betweenness.
   (c) collinearity.   (d) distance.

In 15 and 16, the preimage is blue and the image is orange. Trace the figure and then use a ruler to determine the center and size-change factor k for the size change represented.

15.

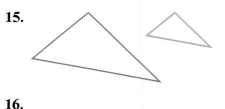

16.

In 17 and 18, $\triangle RST$ is the image of $\triangle OQP$ under a size change.

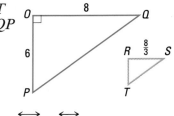

17. a. Is this size change an expansion or contraction?
   b. *True or false.* $\overleftrightarrow{PQ} \parallel \overleftrightarrow{ST}$

18. a. Trace the figure and locate the center of the size change.
   b. What is the magnitude of the size change?
   c. Find the lengths of $\overline{PQ}$, $\overline{RT}$, and $\overline{ST}$.

Objective D: *Use the Fundamental Theorem of Similarity to find lengths, perimeters, areas, and volumes in similar figures.* *(Lesson 12-6)*

19. If $\triangle A'B'C'$ is the image of $\triangle ABC$ under a size change of magnitude 11, how do the areas of $\triangle A'B'C'$ and $\triangle ABC$ compare?

20. How are the volumes of similar figures related to the ratio of similitude?

21. A hexagon has area 20 cm^2 and a shortest side with length 6 cm. A similar hexagon has a shortest side of length 8 cm. What is the area of this similar hexagon?

22. A prism has volume 64 cubic meters. What is the volume of a similar prism $\frac{3}{4}$ as high?

USES DEAL WITH APPLICATIONS OF MATHEMATICS IN REAL SITUATIONS.

Objective E: *Identify and determine proportional lengths and distances in real situations.* *(Lessons 12-4, 12-5)*

23. A photograph is 5″ by 8″. If a similar photograph is 10″ in its longer dimension, what is its shorter dimension?

24. A recipe for *sevyan*, a noodle pudding served on the holiday *Id Alfitr* by Indian and Pakistani Muslims, calls for seven ounces of sevyan noodles for five servings. How many ounces of sevyan noodles are needed for two servings?

25. TV screens are nearly all similar. If a 9″ screen (measured along a diagonal) is 7″ wide, how wide is a 26″ screen?

26. If a marathon runner covers 15 miles in 1 hour 45 minutes, at that rate how many miles can the runner cover in $2\frac{1}{2}$ hours?

27. A poster of a certain landscape is 15″ × 20″. If a similar photograph of the same landscape is 6″ × 8″, how tall will an 8″ spire in the poster appear in the photograph?

28. A highway from Chicago to Milwaukee is 90 miles long and 6″ long on a map. If the actual highway from Chicago to St. Louis is 300 miles long, how long is this highway on the map?

Objective F: *Apply the Fundamental Theorem of Similarity in real situations.*
(Lessons 12-6, 12-7)

29. A solid figurine is 20 cm tall and weighs 3 kg. How much will a similar solid figurine of the same material weigh if it is 32 cm tall?

30. If a 10-inch diameter pizza costs $5.89, and if cost is proportional to the quantity of ingredients used to make it, what should a 16-inch diameter pizza cost if it has the same kinds of ingredients and the same thickness?

31. Dolls are often $\frac{1}{12}$ actual size. The same cloth used for a real coat can be used to make how many doll coats?

32. An elephant 16 feet high can weigh 7 tons. (A ton is 2000 pounds.) If a similar elephant were 1 foot high, how many pounds would it weigh?

33. If there were a person 8 times as tall as you and with your physique, the person would weigh __?__ times as much as you do. This weight would be supported by __?__ times the area that supports you.

34. *True or false.* Larger animals need thicker legs to support their weight.

REPRESENTATIONS DEAL WITH PICTURES, GRAPHS, OR OBJECTS THAT ILLUSTRATE CONCEPTS.

Objective G: *Perform and analyze size transformations on figures in the coordinate plane.* *(Lesson 12-1)*

In 35–37, $A = (6, -5)$, $B = (-10, 0)$, $C = (-3, 8)$, $D = (12, 20)$.

35. Graph *ABCD* and its image under $S_{\frac{1}{2}}$.

36. Graph *ABCD* and its image under $S_{2.5}$.

37. List the coordinates of the image of *ABCD* under S_k.

38. Let S be the size transformation of magnitude 5, center (0, 0).

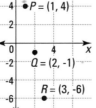

   a. Using the graph at the right, find S(*P*), S(*Q*), and S(*R*).

   b. Fill in the blank and then prove the statement. The distance between S(*P*) and S(*Q*) is __?__ times the distance between *P* and *Q*.

   c. Verify that the slope of $\overline{PQ}$ equals the slope of the line through S(*P*) and S(*Q*).

   d. Verify that S(*P*), S(*Q*), and S(*R*) are collinear.

   e. Verify that S(*Q*) is between S(*P*) and S(*R*).

In 39 and 40, refer to the figure *HEXAGO* below. Let S be the size transformation of magnitude $\frac{1}{2}$, center (0, 0).

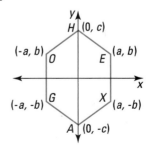

39. What are the coordinates of S(*E*), S(*H*), and S(*G*)?

40. Let S(*HEXAGO*) = (*BCDFIJ*). What size transformation T could be applied to *BCDFIJ* so that T(*BCDFIJ*) = (*HEXAGO*)?

SIMILAR TRIANGLES AND TRIGONOMETRY

As you saw in Chapter 12, similarity is a powerful idea with many applications. One group of important applications not discussed in that chapter is the use of similarity to obtain the measures of lengths that are inaccessible or that are too large or too small to be measured directly.

Surveyors, pilots, navigators, and many other people estimate lengths or distances as part of their job. For this they often use trigonometry, which literally means "triangle measure." Trigonometry is based on simple properties of similar right triangles, yet it allows us to estimate the heights of buildings or mountains, the distances across lakes or rivers, the depths of craters on the moon, and even distances to stars.

In this chapter, you will first study the conditions under which triangles are similar. Then, after investigating properties of similar right triangles, you will learn about the sine, cosine, and tangent of an angle, which are the beginnings of trigonometry.

The SSS Similarity Theorem

Model team. *Computer assisted design (CAD) and teamwork combine to produce a detailed prototype car of the future.*

Imagine that you are the head of a design team. You ask all the team members to construct triangles with sides of lengths 2, 4, and 5. How will these triangles compare?

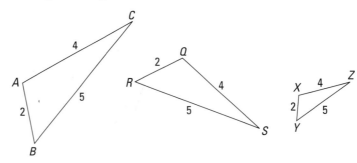

Three triangles that the team members might make are drawn above. Notice that because no unit was named, the triangles may not be congruent. Some people might use centimeters as a unit, others could use inches, and others could just use some arbitrary unit. However, the triangles do look similar.

Below, △QRS is measured using the same unit as △ABC. Note that the ratios of the corresponding sides are equal. From this information, it can be proved that the triangles are similar.

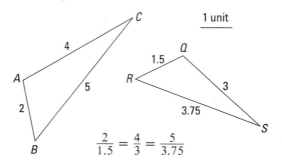

$$\frac{2}{1.5} = \frac{4}{3} = \frac{5}{3.75}$$

Just apply a size change of magnitude $\frac{4}{3}$ to $\triangle QRS$. The resulting image $\triangle Q'R'S'$ (not drawn), has sides of the same length as $\triangle ABC$. So $\triangle Q'R'S' \cong \triangle ABC$ by the SSS Congruence Theorem. Since $\triangle ABC$ is the image of $\triangle QRS$ under a similarity transformation, $\triangle ABC \sim \triangle QRS$. This idea can be applied to any two triangles whose three pairs of sides are proportional. The result is called the SSS Similarity Theorem.

SSS Similarity Theorem
If three sides of one triangle are proportional to three sides of a second triangle, then the triangles are similar.

Proof

Given: $\frac{XY}{AB} = \frac{YZ}{BC} = \frac{XZ}{AC}$.

To prove: $\triangle ABC \sim \triangle XYZ$.

Drawing:

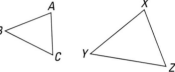

Argument:

Let $k = \frac{XY}{AB}$. Then, by the Transitive Property of Equality, $k = \frac{YZ}{BC}$ and $k = \frac{XZ}{AC}$. Apply any size transformation with magnitude k to $\triangle ABC$.

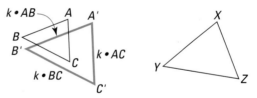

In the image $\triangle A'B'C'$, $A'B' = k \cdot AB$, $B'C' = k \cdot BC$, and $A'C' = k \cdot AC$.

But
$$k \cdot AB = \frac{XY}{AB} \cdot AB = XY$$
$$k \cdot BC = \frac{YZ}{BC} \cdot BC = YZ$$
$$k \cdot AC = \frac{XZ}{AC} \cdot AC = XZ.$$

Thus the three sides of $\triangle A'B'C'$ have the same lengths as the three sides of $\triangle XYZ$. So, by the SSS Congruence Theorem, $\triangle A'B'C' \cong \triangle XYZ$. The definition of congruence tells us there is an isometry mapping $\triangle A'B'C'$ onto $\triangle XYZ$. So there is a composite of a size change (the one we started with) and an isometry mapping $\triangle ABC$ onto $\triangle XYZ$. So, by the definition of similarity, $\triangle ABC \sim \triangle XYZ$.

Applying the SSS Similarity Theorem

One way to tell whether two triangles are similar is to order the sides of each triangle by their lengths. Then compare the ratios formed by corresponding lengths.

Example 1

True or false.   A triangle with sides 3, 4, and 6 is similar to a triangle with sides 8, 6, and 12.

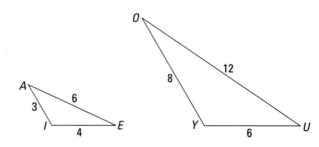

Solution

For each triangle put the sides in order from shortest to longest. Then form three ratios.

$$\frac{6}{3} \qquad \frac{8}{4} \qquad \frac{12}{6}$$

Since all the ratios are equal, the triangles are similar due to the SSS Similarity Theorem. The answer is **true**.

In Example 1, the ratio of similitude is either 2 or $\frac{1}{2}$, depending on which triangle is first. The corresponding sides tell you which vertices correspond. One way to write the similarity is $\triangle IEA \sim \triangle YOU$. This tells which corresponding angles are congruent. For instance, $\angle E \cong \angle O$.

Example 2

Given $\triangle ABC$ and $\triangle XYZ$ below with sides and approximate angle measures as indicated.

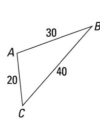
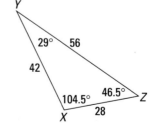

a. Ratios of which sides are equal?
b. Are the triangles similar?
c. $\triangle BAC \sim$ __?__
d. Find the measure of each angle of $\triangle ABC$.

▶ **Solution**

a. The shortest side of △*ABC* must correspond to the shortest side of △*XYZ*. In general, the sides correspond in the order of their lengths. Write the extended proportion.

$$\text{Is} \quad \frac{28}{20} = \frac{42}{30} = \frac{56}{40}?$$

Yes, since $\frac{28}{20} = 1.4$, $\frac{42}{30} = 1.4$, and $\frac{56}{40} = 1.4$.

Thus, $\frac{XZ}{AC} = \frac{XY}{AB} = \frac{YZ}{BC}$.

b. Since all three ratios of sides are equal, the triangles are similar due to the SSS Similarity Theorem.

c. The vertices of the congruent angles correspond, so △BAC ~ △YXZ.

d. The congruent angles are opposite the corresponding sides.
$\overline{BC}$ corresponds to $\overline{YZ}$, so ∠A ≅ ∠X. m∠A = 104.5.
$\overline{AC}$ corresponds to $\overline{XZ}$, so ∠B ≅ ∠Y. m∠B = 29.
$\overline{AB}$ corresponds to $\overline{XY}$, so ∠C ≅ ∠Z. m∠C = 46.5.

QUESTIONS

Covering the Reading

1. What is the literal meaning of *trigonometry*?

2. Imagine that you are a teacher. You ask each of your students to draw a triangle with sides of lengths 7, 9, and 12.
   a. Must all the triangles drawn be congruent? Why or why not?
   b. Must all the triangles drawn be similar? Why or why not?

3. One way to prove △*ABC* ~ △*DEF* is to find a size-change image of △*ABC* that is __?__ to △*DEF*.

4. State the SSS Similarity Theorem.

5. If △*RST* ~ △*UVW*, name two ratios equal to $\frac{RS}{UV}$.

6. *True or false.* A right triangle with sides 5, 12, and 13 is similar to a right triangle with sides 60, 65, and 25. Justify your answer.

7. *True or false.* A right triangle with sides 20, 21, and 29 is similar to a right triangle with sides 41, 9, and 40. Justify your answer.

8. Examine triangles *ABC* and *DEF* at the right.
   a. Ratios of which sides are equal?
   b. Are the triangles similar?
   c. Find the measure of each angle of △*DEF*.
   d. △*DEF* ~ __?__

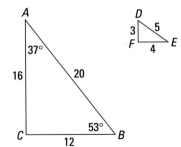

In 9 and 10, the triangles are similar with corresponding sides parallel.
a. List the corresponding vertices.
b. Find a ratio of similitude.
c. Determine as many missing side lengths as possible.

9.

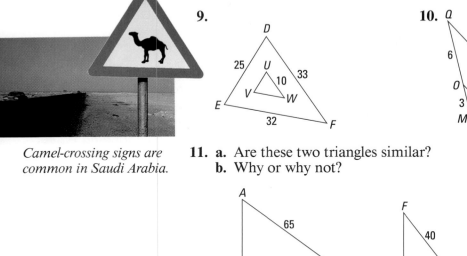

10.

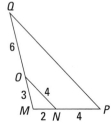

Camel-crossing signs are common in Saudi Arabia.

11. a. Are these two triangles similar?
b. Why or why not?

12. Are the two triangles at the right similar? If so, write the similarity with the vertices in the proper order. If not, explain why not.

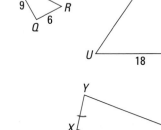

13. Use the figure at the right to write a proof argument.
Given: X is the midpoint of $\overline{WY}$.
V is the midpoint of $\overline{WZ}$.
To prove: $\triangle WXV \sim \triangle WYZ$.

14. In Lesson 2-8, the following conjecture was made. If D, E, and F are midpoints of the sides of $\triangle ABC$, then $\dfrac{\text{area}(\triangle ABC)}{\text{area}(\triangle DEF)} = 4$. Explain why this conjecture is true.

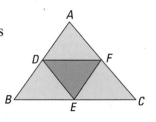

15. Suppose that $\triangle DEF$ is the image of $\triangle ABC$ under a size change of magnitude 5. Then $\triangle ABC$ is the image of $\triangle DEF$ under a size change of magnitude __?__. *(Lesson 12-2)*

In 16–20, the given formula finds what quantity in what figure? *(Lessons 10-5, 10-6, 10-7)*

16. $V = \frac{1}{3} Bh$

17. S.A. $= 2\ell w + 2wh + 2\ell h$

18. $p = a + b + c$

19. $V = \pi r^2 h$

20. L.A. $= \pi r \ell$

In 21 and 22, refer to the figure below in which $HG = 9$, $GI = 40$ and $HI = 41$

21. **a.** Find the area of $\triangle GHI$.
    b. Find GJ, the length of the altitude to the hypotenuse. (Hint: use the area formula)
    (Lesson 8-4)

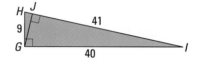

22. If $m\angle I = 13$, find as many other angle measures as you can. *(Lesson 5-7)*

23. *Multiple choice.* $\frac{2x + 10}{2} = $ __?__ *(Previous course)*
    (a) $\frac{x + 5}{2}$ (b) $x + 5$ (c) $x + 10$ (d) $2x + 5$

24. *Multiple choice.* $\frac{a + b}{b} = $ __?__ *(Previous course)*
    (a) $\frac{a}{b} + 1$ (b) $\frac{a}{b} + b$ (c) $a + 1$ (d) a

25. Solve for x.
    $\frac{10}{x} = \frac{x}{20}$ *(Previous course)*

26. Is there an SSSSS Similarity Theorem for pentagons? If so, how do you know? If not, draw a counterexample.

LESSON

13-2

The AA
and SAS
Similarity
Theorems

Shaping music. *Many of the instruments used by the Orchestra of Russian Folk Music are similar in shape. See Question 8.*

Similarity Counterparts of the Triangle Congruence Theorems

For each triangle congruence theorem there is a counterpart triangle similarity theorem. In the triangle similarity theorems, the letter "A" still denotes a pair of congruent angles but "S" denotes a *ratio* of corresponding sides.

Three triangle similarity theorems are used more often than the others.

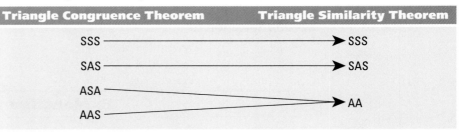

Triangle Congruence Theorem	Triangle Similarity Theorem
SSS ⟶	SSS
SAS ⟶	SAS
ASA	
AAS	AA

The strategy used in proving the AA and SAS Similarity Theorems is the same as that used in proving the SSS Similarity Theorem. A size change is applied to one triangle so that its image is congruent to the other triangle. The key decision in each proof is the choice of the magnitude k of the size change. Once k is chosen, the only other thing to do is identify the triangle congruence theorem to use. That theorem always turns out to be the corresponding triangle congruence theorem.

AA Similarity

Consider the two triangles at the top of the next page. Two angles in each are congruent. The lengths of the included sides are also shown.

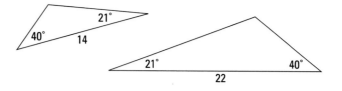

Under a size change of magnitude $\frac{22}{14}$, the image of the smaller triangle is a triangle with two angles of 40° and 21° and an included side of length $\frac{22}{14} \cdot 14$, or 22. So the image is congruent to the larger triangle by the ASA Congruence Theorem. This argument holds when any two angles of one triangle are congruent to two angles of another.

AA Similarity Theorem
If two angles of one triangle are congruent to two angles of another, then the triangles are similar.

Proof
Given: Triangles ABC and XYZ with $\angle A \cong \angle X$ and $\angle B \cong \angle Y$.
To prove: $\triangle ABC \sim \triangle XYZ$.

Drawing: The congruent angles signal the corresponding vertices. This indicates the corresponding sides and enables a picture to be drawn and marked.

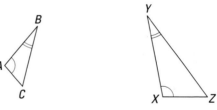

Argument:
X and A are corresponding vertices, as are Y and B. So $\overline{XY}$ and $\overline{AB}$ are corresponding sides, let $k = \frac{XY}{AB}$ be the magnitude of a size transformation applied to $\triangle ABC$.

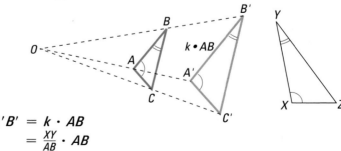

Then $A'B' = k \cdot AB$
$= \frac{XY}{AB} \cdot AB$
$= XY$.

Also, since size transformations preserve angle measure, $\angle A \cong \angle A'$ and $\angle B \cong \angle B'$. By the Transitive Property of Congruence, $\angle A' \cong \angle X$ and $\angle B' \cong \angle Y$. So $\triangle A'B'C' \cong \triangle XYZ$ by the ASA Congruence Theorem. Thus $\triangle ABC$ can be mapped onto $\triangle XYZ$ by a composite of size changes and reflections. So $\triangle ABC \sim \triangle XYZ$.

For instance, the triangles at the right have congruent angles as indicated. So $\triangle ABC \sim \triangle XYZ$ by the AA Similarity Theorem. One ratio of similitude is any of the equal ratios $\frac{XY}{AB} = \frac{YZ}{BC} = \frac{XZ}{AC}$. Substituting the given lengths into these ratios, $\frac{XY}{AB} = \frac{70}{20} = \frac{50}{AC}$. From this the ratio of similitude is $\frac{70}{20}$. Solving the proportion $\frac{70}{20} = \frac{50}{AC}$ gives $AC = 14\frac{2}{7}$.

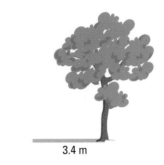

The AA Similarity Theorem justifies a method for determining the height of an object.

Example 1

A meter stick casts a shadow 70 cm long at the same time that a tree casts a shadow 3.4 m long. How tall is the tree?

1 m

70 cm   3.4 m

Solution

The key to the solution is that the sun is so far away that its rays can be considered to be parallel. Consequently, right triangles are formed with congruent acute angles.

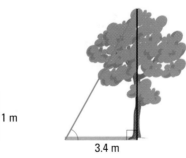

1 m

70 cm   3.4 m

By the AA Similarity Theorem, the triangles are similar. So corresponding sides are proportional.

$$\frac{1 \text{ m}}{70 \text{ cm}} = \frac{h}{3.4 \text{ m}}$$

Convert to the same units. We choose to convert to centimeters.

$$\frac{100 \text{ cm}}{70 \text{ cm}} = \frac{h}{340 \text{ cm}}$$

Solve the proportion.
$$70h = 34,000$$
$$h \approx 486 \text{ cm}$$

The tree is about 490 cm, or 4.9 m tall.

SAS Similarity

The third important triangle similarity theorem is the *SAS Similarity Theorem*. You are asked to prove it in Question 16.

> **SAS Similarity Theorem**
> If, in two triangles, the ratios of two pairs of corresponding sides are equal and the included angles are congruent, then the triangles are similar.

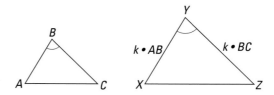

Specifically, if in two triangles ABC and XYZ, $\angle B \cong \angle Y$, and $\frac{AB}{XY} = \frac{BC}{YZ}$, then $\triangle ABC \sim \triangle XYZ$.

Example 2

In the figure at the right, T is the midpoint of $\overline{PS}$, and Q is the midpoint of $\overline{PR}$. Write an argument which proves that $\triangle PTQ \sim \triangle PSR$.

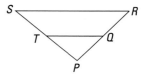

Solution 1

$\frac{PT}{PS} = \frac{PQ}{PR} = \frac{1}{2}$. $\angle P$, included by these sides, is in both triangles. So, by the SAS Similarity Theorem, $\triangle PTQ \sim \triangle PSR$.

Solution 2

By the Midpoint Connector Theorem, $\overline{QT} \parallel \overline{RS}$. Thus $\angle TQP \cong \angle R$ and $\angle PTQ \cong \angle S$, because $\parallel$ lines $\Rightarrow$ corr. $\angle$s $\cong$. Then $\triangle PTQ \sim \triangle PSR$ by the AA Similarity Theorem.

QUESTIONS

Covering the Reading

1. For each triangle similarity theorem, give the corresponding triangle congruence theorem(s).
   a. SSS Similarity Theorem
   b. AA Similarity Theorem
   c. SAS Similarity Theorem

2. Describe the general strategy used to prove all three triangle similarity theorems.

3. Given: △*ABC* and △*DEF* below. What is the magnitude of a size change that can be applied to △*ABC* to produce an image congruent to △*DEF*?

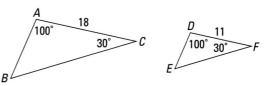

4. State the AA Similarity Theorem.

5. What triangle congruence theorem is used in the proof of the AA Similarity Theorem?

6. Suppose a person 5′6″ tall casts a shadow 3′ long at the same time that a flagpole casts a shadow 13′4″ long. To the nearest foot, how tall is the flagpole?

7. A flat-roofed garage casts a shadow 5 meters long. At the same time, a meter stick casts a shadow 1.2 meters long. Determine the height of the garage.

8. Russian *balalaika* orchestras consist of as many as six different sized balalaikas, long-necked lutes with triangular-faced soundboxes. The shapes of two balalaika soundboxes are shown below.
   a. What is the magnitude of a size change applied to △*PQR* which would produce an image congruent to △*XYZ*?
   b. Which triangle congruence theorem is the justification that the image of the smaller triangle is congruent to the larger triangle?

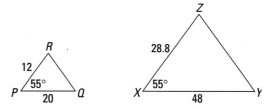

In 9 and 10, two triangles are given.
a. Prove that the triangles are similar.
b. Find a ratio of similitude.
c. Determine as many missing angle measures or side lengths as possible using theorems you have had.

9.

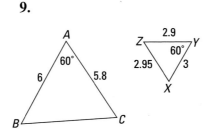

10.

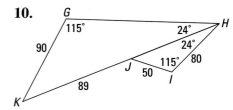

11. Use the figure at the right. Complete the proof.

Given: $WY = 3 \cdot VY$,
$XY = 3 \cdot YZ$.
To prove: $\triangle WXY \sim \triangle VZY$.

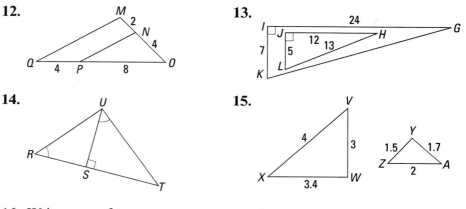

Applying the Mathematics

In 12–15, each figure contains at least two triangles.
a. Are two triangles similar?
b. If so, what triangle similarity theorem guarantees their similarity? If not, explain why not.
c. If the two triangles are similar, write the similarity with vertices in correct order.

12.

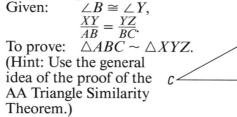

13.

14.

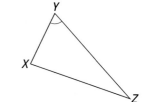

15.

16. Write a proof argument to prove the SAS Similarity Theorem using the given figure.

Given: $\angle B \cong \angle Y$,
$\dfrac{XY}{AB} = \dfrac{YZ}{BC}$.
To prove: $\triangle ABC \sim \triangle XYZ$.
(Hint: Use the general idea of the proof of the AA Triangle Similarity Theorem.)

17. In the director's chair pictured below, $\overline{AE}$ and $\overline{BD}$ intersect at C, and $\overleftrightarrow{AB} \parallel \overleftrightarrow{DE}$. Prove that $\triangle ABC$ and $\triangle EDC$ are similar.

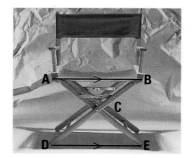

18. a. Are triangles *LMN* and *PQR*, pictured below, similar?
  b. If so, why? If not, why not?   *(Lesson 13-1)*

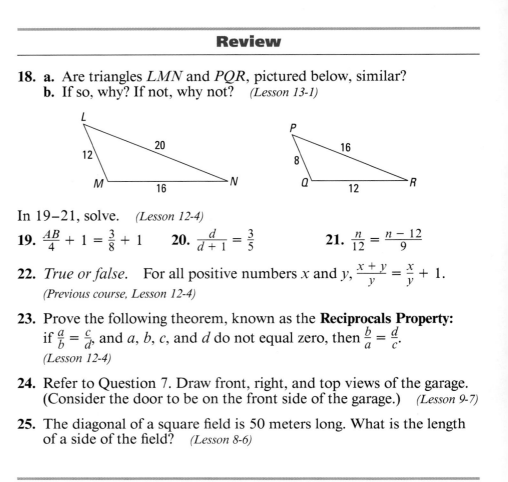

In 19–21, solve.   *(Lesson 12-4)*

19. $\frac{AB}{4} + 1 = \frac{3}{8} + 1$ **20.** $\frac{d}{d+1} = \frac{3}{5}$ **21.** $\frac{n}{12} = \frac{n-12}{9}$

22. *True or false.*   For all positive numbers x and y, $\frac{x+y}{y} = \frac{x}{y} + 1$.
  (Previous course, Lesson 12-4)

23. Prove the following theorem, known as the **Reciprocals Property:**
  if $\frac{a}{b} = \frac{c}{d}$, and a, b, c, and d do not equal zero, then $\frac{b}{a} = \frac{d}{c}$.
  (Lesson 12-4)

24. Refer to Question 7. Draw front, right, and top views of the garage.
  (Consider the door to be on the front side of the garage.)   *(Lesson 9-7)*

25. The diagonal of a square field is 50 meters long. What is the length
  of a side of the field?   *(Lesson 8-6)*

26. Is there an HL Similarity Theorem and an SsA Similarity Theorem
  for triangles? If so, how do you know? If not, draw a
  counterexample.

The Side-Splitting Theorem

A Proportion Within a Triangle

Pictured below is an asymmetric roof. Still, the parallel beams split the sides of the roof, forming similar triangles.

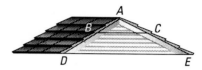

It turns out that the lengths AB, BD, AC, and CE are proportional. We call this result the *Side-Splitting Theorem*. Its proof depends on the AA Similarity Theorem and properties of algebra.

> **Side-Splitting Theorem**
> If a line is parallel to a side of a triangle and intersects the other two sides in distinct points, it splits these sides into proportional segments.

Proof:
Given: $\overleftrightarrow{PQ} \parallel \overleftrightarrow{BC}$.
To prove: $\frac{AP}{PB} = \frac{AQ}{QC}$.

Drawing:

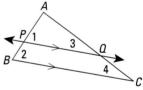

Argument:
$\angle 1 \cong \angle 2$ and $\angle 3 \cong \angle 4$ since $\parallel$ lines $\Rightarrow$ corr. $\angle$s $\cong$. Thus, by the AA Similarity Theorem, $\triangle APQ \sim \triangle ABC$. Now, in these triangles, corresponding sides are proportional.

$$\frac{AB}{AP} = \frac{AC}{AQ}$$

Now we split AB and AC into two parts using the Betweenness Theorem.

$$\frac{AP + PB}{AP} = \frac{AQ + QC}{AQ}$$

Write each side of the equation as the sum of two fractions.

$$\frac{AP}{AP} + \frac{PB}{AP} = \frac{AQ}{AQ} + \frac{QC}{AQ}$$
$$1 + \frac{PB}{AP} = 1 + \frac{QC}{AQ}$$

Subtract 1 from both sides. $\quad \frac{PB}{AP} = \frac{QC}{AQ}$

Use the Reciprocals Property. $\quad \frac{AP}{PB} = \frac{AQ}{QC}$

Measure *AB*, *BC*, *AD*, and *DE*. Form a true (or nearly true) proportion with these measurements.

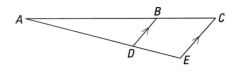

Example 1

Suppose beams $\overline{MN}$ and $\overline{ST}$ are parallel and split the sides of the roof into the lengths (in inches) as indicated.
a. Find *RN*.
b. Find *ST*.

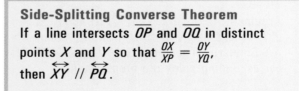

Solution

a. From the Side-Splitting Theorem,

$$\frac{RM}{MS} = \frac{RN}{NT}.$$

Substitute. $\frac{140}{80} = \frac{RN}{120}$

Solve for *RN*. $80 \cdot RN = 120 \cdot 140$
$RN = 210$ inches

b. $\triangle RMN \sim \triangle RST$. Thus $\frac{RM}{RS} = \frac{MN}{ST}$. Substituting, $\frac{140}{220} = \frac{168}{ST}$.
Thus $140 \cdot ST = 168 \cdot 220$ and $ST = 264$ inches.

The Converse of the Side-Splitting Theorem

The converse of the Side-Splitting Theorem is also true: If a line intersects two sides of a triangle and forms segments whose lengths are proportional, then the line is parallel to the third side.

Side-Splitting Converse Theorem
If a line intersects $\overline{OP}$ and $\overline{OQ}$ in distinct points *X* and *Y* so that $\frac{OX}{XP} = \frac{OY}{YQ}$, then $\overleftrightarrow{XY} \parallel \overleftrightarrow{PQ}$.

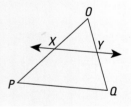

Proof:

Given: $\frac{OX}{XP} = \frac{OY}{YQ}$ in the figure above.

To prove: $\overleftrightarrow{XY} \parallel \overleftrightarrow{PQ}$.

Argument:
First we prove that the triangles are similar. Then corresponding angles are used to get the parallel lines. Using the Reciprocals Property,

$$\frac{XP}{OX} = \frac{YQ}{OY}.$$

Add 1 to both sides (in the form of $\frac{OX}{OX}$ on the left side, and $\frac{OY}{OY}$ on the right).

$$\frac{OX}{OX} + \frac{XP}{OX} = \frac{OY}{OY} + \frac{YQ}{OY}$$

Combine the fractions. $\frac{OX + XP}{OX} = \frac{OY + YQ}{OY}$

$OX + XP = OP$ and $OY + YQ = OQ$ by the Betweenness Theorem.
Substituting, $\frac{OP}{OX} = \frac{OQ}{OY}.$

Thus, two pairs of sides are proportional. By the Reflexive Property, $\angle XOY \cong \angle POQ$. So $\triangle OPQ \sim \triangle OXY$ by the SAS Similarity Theorem. The corresponding angles in the similar triangles are congruent, so $\angle OPQ \cong \angle OXY$. These are corresponding angles for $\overleftrightarrow{XY}$ and $\overleftrightarrow{PQ}$ with transversal $\overleftrightarrow{OP}$. Since corr. $\angle$s $\cong \Rightarrow$ // lines, $\overleftrightarrow{XY}$ // $\overleftrightarrow{PQ}$.

You may use the Side-Splitting Converse Theorem to conclude that lines are parallel.

Example 2

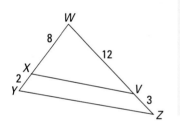

Refer to the figure at the left. Is $\overline{XV}$ // $\overline{YZ}$? Why or why not?

Solution

Does $\frac{WX}{XY} = \frac{WV}{VZ}$? Yes, $\frac{8}{2} = \frac{12}{3}$.

So, by the Side-Splitting Converse Theorem, $\overline{XV}$ // $\overline{YZ}$.

An alternate way to solve Example 2 is to prove $\triangle WXV \sim \triangle WYZ$ by SAS Similarity and show $\overline{XV}$ // $\overline{YZ}$ using corresponding angles.

Example 3

The flag of Lesotho, a country in southern Africa, has a blue stripe with one edge of the stripe forming the diagonal $\overline{TE}$. The other edge $\overline{AB}$ of the blue stripe is parallel to $\overline{TE}$. If the dimensions of the flag are 40 cm by 60 cm, and AC is 42 cm, where is B located?

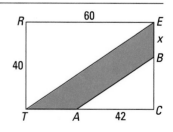

Solution

From the Side-Splitting Theorem, since $\overline{AB}$ is parallel to $\overline{TE}$,

$$\frac{CA}{AT} = \frac{CB}{BE}.$$

Let $BE = x$. Then
$$\frac{42}{18} = \frac{40 - x}{x}$$
$$42x = 720 - 18x$$
$$60x = 720$$
$$x = 12$$

B should be located 12 cm from E.

Lesotho is located in the east central part of South Africa. Its economy depends upon agriculture and production from diamond mines.

QUESTIONS

Covering the Reading

1. Given m // $\overleftrightarrow{XY}$, complete the proportions.

   a. $\frac{AC}{AX} = \underline{\quad?\quad}$

   b. $\frac{AB}{BY} = \underline{\quad?\quad}$

2. Write the results you found from the Activity in this lesson.

3. In the triangle below at the left, if $\overleftrightarrow{BC}$ // $\overleftrightarrow{DE}$, then $CE = \underline{\quad?\quad}$.

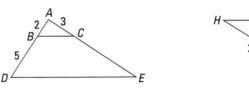

4. In the triangle above at the right, if $\overleftrightarrow{GI}$ // $\overleftrightarrow{HJ}$, then $FJ = \underline{\quad?\quad}$.

5. State the Side-Splitting Converse Theorem.

In 6 and 7, use the figures below. Is $\overline{XV}$ // $\overline{YZ}$? Why or why not?

6.

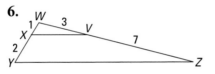

7.

8. At the right, the beam $\overline{BC}$ splits the sides of this asymmetric roof. If $AB = 20$, $BD = 30$, $AE = 60$, and $\overline{BC}$ // $\overline{DE}$, find AC and CE.

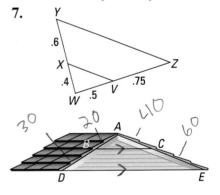

9. In the drawing at the right, $\overleftrightarrow{MN}$ // $\overleftrightarrow{PQ}$, $MP = 6$, $NQ = 5$, $ON = 30$, and $MN = 14$. Find each number.

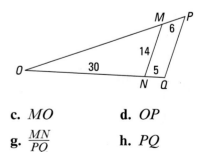

   a. $\frac{NQ}{ON}$ b. $\frac{MP}{MO}$ c. MO d. OP

   e. $\frac{ON}{OQ}$ f. $\frac{OM}{OP}$ g. $\frac{MN}{PQ}$ h. PQ

10. In Question 9, what is the magnitude of the size change S with center O and $S(\triangle MNO) = \triangle PQO$?

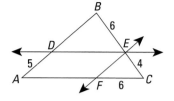

11. Given: $\overleftrightarrow{DE}$ // $\overleftrightarrow{AC}$, $\overleftrightarrow{EF}$ // $\overleftrightarrow{AB}$, and lengths as indicated in the figure at the left.
    a. Find DB.
    b. Find FA.
    c. $\angle A$ and $\angle AFE$ are __?__ angles.
    d. Write a proportion using BD, DA, CF, and FA.

12. Fill in the justifications for the proof argument of the following generalization of the Side-Splitting Theorem. If m // n // p, then $\frac{a}{b} = \frac{c}{d}$.

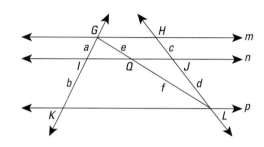

Argument:

Conclusions	Justifications
1. $\frac{a}{b} = \frac{e}{f}$	a.
2. $\frac{e}{f} = \frac{c}{d}$	b.
3. $\frac{a}{b} = \frac{c}{d}$	c.

In 13 and 14, use the map shown below. 8th, 9th, and 10th Streets are parallel. The distances along Rasci Road are as given.

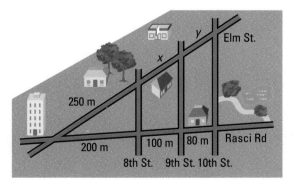

13. Find x and y, two distances along Elm Street.

14. The city wants to replace curbs on 8th, 9th, and 10th Streets between Elm and Rasci. If it is 120 meters from Elm St. to Rasci Rd. along 8th St., how many meters of curbs will be needed for 9th and 10th Streets? (Remember that curbs go on both sides of the street.)

In 15–17, determine from the markings and other information whether the two triangles are similar. If so, indicate the corresponding vertices and state the theorem or definition that justifies your conclusion. *(Lessons 13-1, 13-2)*

15.

16.

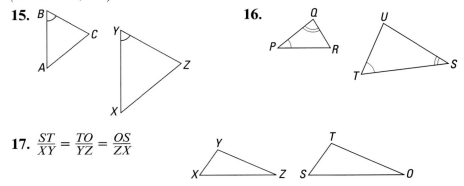

17. $\dfrac{ST}{XY} = \dfrac{TO}{YZ} = \dfrac{OS}{ZX}$

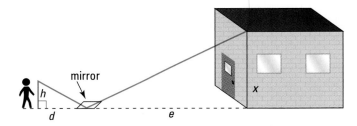

18. In Lesson 13-2, you saw one way to use similar triangles to find the height of a structure. Here is another. Place a mirror on the ground between you and the building. Move until you can see the top of the building in the mirror. Measure the distances d and e and the height h of your eyes from the ground.

mirror

x

h

d

e

   a. Explain why $\dfrac{x}{h} = \dfrac{e}{d}$.
   b. Henry's eyes are 5′8″ off the ground when he stands 4′3″ from a mirror and sees the top of the building in it. If the mirror is 15′6″ from the building, how tall is the building? *(Lesson 13-2)*

19. Refer to the figure at the left to complete the proof.

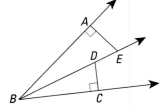

   Given: $\overrightarrow{BE}$ bisects $\angle ABC$; $\overline{AE} \perp \overline{AB}$, $\overline{DC} \perp \overline{CB}$.
   To prove: $\triangle ABE \sim \triangle CBD$.   *(Lesson 13-2)*

20. The fraction $\dfrac{4}{\sqrt{3}}$ equals $\dfrac{4}{\sqrt{3}} \cdot \dfrac{\sqrt{3}}{\sqrt{3}}$, or $\dfrac{4\sqrt{3}}{3}$. This last form is easier to add to other multiples of $\sqrt{3}$, and is sometimes thought to be simpler. In this way, simplify the following. *(Previous course)*
   a. $\dfrac{5}{\sqrt{3}}$ **b.** $\dfrac{1}{\sqrt{2}}$ **c.** $\dfrac{3}{\sqrt{6}}$

21. *Multiple choice.* $\sqrt{75} = \underline{\ ?\ }$ *(Previous course)*
   (a) $5\sqrt{3}$ (b) $25\sqrt{3}$ (c) $3\sqrt{5}$ (d) $3\sqrt{25}$

22. *Multiple choice.* $\sqrt{16 + 16} = \underline{\ ?\ }$ *(Previous course)*
   (a) $2\sqrt{16}$ (b) $16\sqrt{2}$ (c) $4\sqrt{2}$ (d) 8

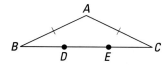

23. In the figure at the left, $AB = AC$ and D and E split $\overline{BC}$ into three congruent segments. Carefully trace the figure (or use an automatic drawer) and draw $\overline{AD}$ and $\overline{AE}$. Do they trisect $\angle BAC$? That is, do they split $\angle BAC$ into three angles of equal measure? Justify your answer.

Exploration

---

24. Here is an algorithm to divide a line segment $\overline{AB}$ into n congruent parts. (The figures show the case $n = 5$.)

Step 1: Draw any ray $\overrightarrow{AC_1}$, that is not collinear with $\overrightarrow{AB}$. For convenience, make C_1 close to A

Step 2: On $\overrightarrow{AC_1}$ mark off n segments of lengths AC_1, so that $AC_1 = C_1C_2 = C_2C_3 = C_3C_4 = C_4C_5 \ldots$

Step 3: Draw $\overline{BC_n}$ (here $n = 5$, so $\overline{BC_5}$ is drawn.)

Step 4: Draw parallels to $\overline{BC_n}$ through $C_1, C_2, C_3, \ldots$ Let the parallels intersect $\overline{AB}$ at $B_1, B_2, B_3, \ldots B_{n-1}$ thus $AB_1 = B_1B_2 = B_2B_3 = \ldots = B_{n-1}B$. (In the figure below, $\overline{AB}$ has been divided into five congruent parts.)

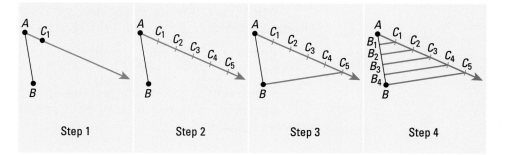

| Step 1 | Step 2 | Step 3 | Step 4 |

a. Trace $\overline{AB}$ below on a separate sheet of paper. Follow the algorithm to divide $\overline{AB}$ into 7 congruent parts.

b. How do you know that each part of $\overline{AB}$ is $\frac{1}{7}$ of $\overline{AB}$?

Similarity in Right Triangles

IN·CLASS
ACTIVITY

Work with a partner. For this activity, both of you will need a ruler, protractor, and compass.

1 Trace right $\triangle ABC$ at the right or draw one like it. Measure $\angle A$ and $\angle B$.

2 Draw $\overline{CD}$, the altitude to hypotenuse $\overline{AB}$.

3 Use a protractor to determine these four angle measures.
    a. $m\angle CDB$ **b.** $m\angle CDA$
    c. $m\angle DCB$ **d.** $m\angle DCA$

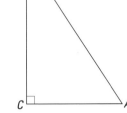

4 There are three triangles in your figure. Which pairs of triangles are similar? For each pair write the similarity correspondences with the vertices in correct order.

5 For each pair of similar triangles, write the three equal ratios of the sides.

6 In Question 5, you should find that in each set of ratios, one side appears twice. For each set of ratios, write the proportion that includes that side.

LESSON

13-4

Geometric Means in Right Triangles

Right triangles mean stronger buildings. *These prefabricated house frames reduce the length of time it takes to build a house.*

You may recall that the average of two numbers is also called their **arithmetic mean.** It is the coordinate of the midpoint of two points on a number line. In this lesson, you will learn about another kind of mean, the *geometric mean.* The geometric mean appears in surprising places.

What Is the Geometric Mean?

It is possible for the two means in a proportion to be equal, as in

$$\frac{2}{10} = \frac{10}{50}.$$

When this happens, the number that appears twice is called a geometric mean of the other two numbers. In the proportion above, 10 is a geometric mean of 2 and 50. In this course, we are interested only in positive geometric means of positive numbers.

Definition
Let *a*, *b*, and *g* be positive numbers. *g* is the **geometric mean** of *a* and *b* if and only if
$$\frac{a}{g} = \frac{g}{b}.$$

If *g* is the geometric mean of *a* and *b*, then $\frac{a}{g} = \frac{g}{b}$. Then by the Means-Extremes Property $g^2 = ab$. So $g = \pm \sqrt{ab}$. Thus $\sqrt{ab}$ is the positive geometric mean. This is a proof argument for the following theorem.

Geometric Mean Theorem
The positive geometric mean of the positive numbers *a* and *b* is $\sqrt{ab}$.

Example 1

Find the geometric mean of 7 and 12.

Solution

Let g be the geometric mean. Using the Geometric Mean Theorem,

$$g = \sqrt{7 \cdot 12}$$
$$= \sqrt{84}$$

The geometric mean of 7 and 12 is $\sqrt{84}$.

Check

Use the definition of geometric mean. $\sqrt{84}$ is about 9.17. Is $\frac{7}{9.17} \approx \frac{9.17}{12}$? Yes, $0.7634 \approx 0.7642$.

The arithmetic mean of a and b is exactly halfway between them. The geometric mean, $\sqrt{ab}$, is also always between a and b. But it is equal to the arithmetic mean of a and b only if $a = b$; otherwise, it is smaller than the arithmetic mean. Both arithmetic and geometric means have applications in arithmetic, algebra, and geometry.

Similar Triangles Within a Right Triangle

The name "geometric mean" comes from relationships among lengths in any right triangle. These relationships were discovered by the ancient Greeks and are described in Euclid's *Elements*. The proofs of these relationships depend mainly on the AA Similarity Theorem.

$\triangle ABC$ below at the left is split into two triangles, CBD and ACD, by drawing $\overline{CD}$, the altitude to its hypotenuse, as in the In-class Activity. We use the small letters a and b for the lengths of the legs of right $\triangle ABC$, and c for its hypotenuse. The letter h is the altitude to the hypotenuse. The point where the altitude intersects the hypotenuse splits it into two lengths, x and y, so $x + y = c$.

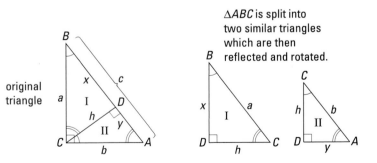

$\triangle ABC$ is split into two similar triangles which are then reflected and rotated.

$m\angle A = m\angle BCD$, since both angles are complements of $\angle B$. So $\triangle\text{I}$, $\triangle\text{II}$, and $\triangle ABC$ are all right triangles, and each includes an angle equal in measure to $\angle A$. Thus, by the AA Similarity Theorem, all three triangles are similar: $\triangle ABC \sim \triangle CBD \sim \triangle ACD$. Since the triangles are similar, corresponding sides are proportional.

Now look for proportions in which the same quantity appears twice. Consider △I and △II. Both have the side $\overline{CD}$ with length h.

$$\frac{BD}{CD} = \frac{CD}{AD}, \text{ or } \frac{x}{h} = \frac{h}{y}.$$

For △I and △ABC, $\overline{BC}$ is a side in both. $\frac{BD}{BC} = \frac{BC}{AB}$. For △I and △ABC, $\overline{AC}$ is a side in both. $\frac{AD}{AC} = \frac{AC}{AB}$. Thus,

$$\frac{x}{a} = \frac{a}{c} \quad \text{and} \quad \frac{y}{b} = \frac{b}{c}.$$

Thus in △ABC, the altitude h and the legs a and b are geometric means of other lengths. This argument proves the following theorem:

Right-Triangle Altitude Theorem
In a right triangle,
(1) the altitude to the hypotenuse is the geometric mean of the segments into which it divides the hypotenuse, and
(2) each leg is the geometric mean of the hypotenuse and the segment of the hypotenuse adjacent to the leg.

Using the Geometric Mean Theorem, $h = \sqrt{xy}$, $a = \sqrt{cx}$, and $b = \sqrt{cy}$.

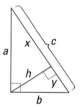

Notice the patterns of letters in these equations and in the solution to Example 2.

Example 2

$\overline{CD}$ is the altitude to the hypotenuse of right triangle *ABC*, as shown below. If *AD* = 3 and *BD* = 12, find *CD*, *CA*, and *CB*.

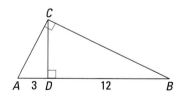

Solution
CD is the geometric mean of *AD* and *BD*.
So CD = $\sqrt{AD \cdot BD}$ = $\sqrt{3 \cdot 12}$ = $\sqrt{36}$ = 6.
CA is the geometric mean of *DA* and *BA*.
So CA = $\sqrt{DA \cdot BA}$ = $\sqrt{3 \cdot 15}$ = $\sqrt{45}$ = 3$\sqrt{5}$.
CB is the geometric mean of *DB* and *AB*.
So CB = $\sqrt{DB \cdot AB}$ = $\sqrt{12 \cdot 15}$ = $\sqrt{180}$ = 6$\sqrt{5}$.

QUESTIONS

Covering the Reading

1. Give your answer to Parts 3–6 of the In-class Activity on page 758.

In 2 and 3, find the geometric mean of the given numbers to the nearest hundredth.

2. 2 and 50

3. 9 and 12

4. *True or false.* If g is the geometric mean of a and b, and $a < b$, then g is closer to a than to b.

In 5–7, use the figure at the right.

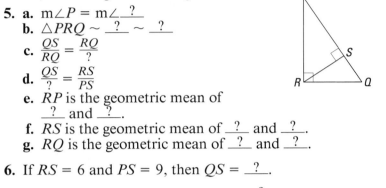

5. a. $m\angle P = m\angle \underline{\ ?\ }$
   b. $\triangle PRQ \sim \underline{\ ?\ } \sim \underline{\ ?\ }$
   c. $\dfrac{QS}{RQ} = \dfrac{RQ}{?}$
   d. $\dfrac{QS}{?} = \dfrac{RS}{PS}$
   e. RP is the geometric mean of $\underline{\ ?\ }$ and $\underline{\ ?\ }$.
   f. RS is the geometric mean of $\underline{\ ?\ }$ and $\underline{\ ?\ }$.
   g. RQ is the geometric mean of $\underline{\ ?\ }$ and $\underline{\ ?\ }$.

6. If $RS = 6$ and $PS = 9$, then $QS = \underline{\ ?\ }$.

7. If $RS = 6$ and $SQ = 4$, then $QR = \underline{\ ?\ }$.

8. Using the diagram at the right find each length.
   a. NC
   b. CE
   c. IE

Applying the Mathematics

9. At the left is the famous 3-4-5 right triangle. The altitude to the hypotenuse has been drawn. Find the lengths of x, y, and h.

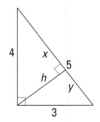

10. *True or false.* Refer to the figure at the right.
    a. $\triangle QRT \sim \triangle RST$
    b. $m\angle QRS = 90$

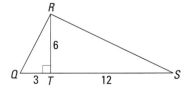

11. Nancy taught the members of Girl Scout Troop 36 that they could use a notebook to estimate distances. To show them how, she used the lifeguard tower at Henson Beach. She held the corner of her notebook near her eye (5 feet off the ground) and moved back from the tower until she could sight both the top and bottom of the tower along with two adjacent edges of the notebook. Then she asked her friend Amy to measure her (Nancy's) distance from the tower by pacing. Amy measured the distance as 8 feet.
  a. Which part of the Right-Triangle Altitude Theorem could Nancy now use to estimate the height of the lifeguard tower?
  b. How tall is the tower?

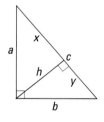

12. Provide the missing justifications in this argument proving the Pythagorean Theorem.
  Given: The right triangle at the left with lengths as indicated.
  To prove: $a^2 + b^2 = c^2$.
  Argument:

Conclusions	Justifications
1. a is the geometric mean of c and x b is the geometric mean of c and y.	**a.**
2. $a = \sqrt{cx}$, $b = \sqrt{cy}$	**b.**
3. $a^2 = cx$, $b^2 = cy$	Multiplication Property of Equality
4. $a^2 + b^2 = cx + cy$	**c.**
5. $a^2 + b^2 = c(x + y)$	**d.**
6. $x + y = c$	Betweenness Theorem
7. $a^2 + b^2 = c^2$	**e.**

13. The altitude $\overline{CD}$ of right triangle ABC splits the hypotenuse into segments of lengths 6 and 9. Find the lengths of the altitude and the two legs.

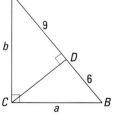

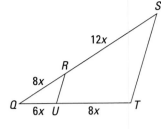

14. Refer to the figure at the left. Is $\overline{RU} \parallel \overline{ST}$? Explain your answer. *(Lesson 13-3)*

15. Write the proof argument.

Given: $WXYZ$ at the right is a trapezoid with bases $\overline{WX}$ and $\overline{YZ}$.

To prove: $\triangle WXU \sim \triangle YZU$.   *(Lesson 13-2)*

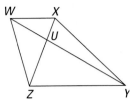

In 16–18, each figure contains two triangles.
a. Are the triangles similar?
b. If so, which similarity theorem justifies their similarity?
c. Write the similarity with vertices in correct order.   *(Lessons 13-1, 13-2)*

16.   **17.**   **18.**

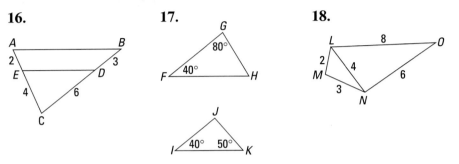

19. In the box at the left, suppose all the vertices are connected.
a. How many other segments have the same length as $\overline{AC}$?
b. How many other segments have the same length as $\overline{AG}$?
c. A 75-cm rod just fits in a carton whose inner dimensions are 60 cm by 40 cm by 24 cm. In how many positions will the rod fit?   *(Lessons 9-2, 11-9)*

20. A prism has the same base as a pyramid but twice the height. How do their volumes compare?   *(Lessons 10-5, 10-7)*

21. In Question 12, why is it inappropriate to use the Pythagorean Theorem as the justification for Conclusion 7?   *(Lesson 5-2)*

22. a. Write a fraction equal to $\dfrac{1}{\sqrt{3}}$ with no radical sign in its denominator.

b. Round $\dfrac{1}{\sqrt{3}}$ to the nearest millionth.   *(Previous course)*

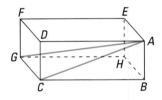

Two power. *Technicians in this "clean-room" dress as surgeons to prevent environmental contamination of computer chips. A computer's capacity for memory is expressed in powers of 2.*

Exploration

23. List the first ten positive integer powers of 2.
a. Which of these are geometric means of other powers in your list.
b. Find a pattern which predicts when one power of a number is a geometric mean of other powers of that number.

24. Given: scalene $\triangle ABC$ such that C is *not* a right angle. Is it possible to find a point D on $\overline{AB}$ so that $\overline{CD}$ splits $\triangle ABC$ into 2 smaller similar triangles? Why or why not?

13-5

Special Right Triangles

Baseball Square. *Shown is Wrigley Field, home of the Chicago Cubs. See Example 1*

Certain right triangles have such relationships among their sides and angles that they are considered special.

Properties of Isosceles Right Triangles

By drawing the diagonals of a square, eight isosceles right triangles are formed. (Can you find all eight?) Notice that all isosceles right triangles are similar, because they all have the same angles: 45°, 45°, and 90°. Sometimes they are called **45-45-90 triangles.** In a 45-45-90 triangle, if you know the length of one side, you can find the lengths of the others. Suppose the congruent sides of one of these triangles have length x and the hypotenuse has length c, as in the triangle at the right below.

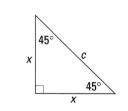

By the Pythagorean theorem $\qquad c^2 = x^2 + x^2.$
So $\qquad\qquad\qquad\qquad\qquad c^2 = 2x^2.$
Taking the positive square root, $\quad c = x \cdot \sqrt{2}.$

The result is a relationship among the sides of any isosceles right triangle.

Isosceles Right Triangle Theorem
In an isosceles right triangle, if a leg is x, then the hypotenuse is $x\sqrt{2}$.

Example 1

A major league baseball diamond is a 90-ft square. How far is it from home plate to second base, to the nearest tenth of a foot?

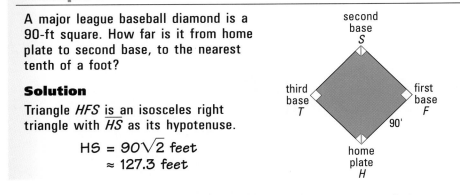

Solution

Triangle *HFS* is an isosceles right triangle with $\overline{HS}$ as its hypotenuse.

$$HS = 90\sqrt{2} \text{ feet}$$
$$\approx 127.3 \text{ feet}$$

In Example 1, a leg was known. In Example 2, a hypotenuse is known.

Example 2

At the right, a square *ABCD* is inscribed in a circle with diameter 12. What is an exact value for the length of a side of the square?

Solution

The word **inscribed** means that all vertices of the figure lie on the circle. Let *BC* = *x*. Then, by the Isosceles Right Triangle Theorem,

$$AC = x\sqrt{2}.$$
So $\qquad\qquad 12 = x\sqrt{2}.$

So $\qquad\qquad x = \dfrac{12}{\sqrt{2}}.$

This is a perfectly good exact answer, but many people prefer fractions to have rational denominators. To rationalize the denominator, multiply both numerator and denominator by $\sqrt{2}$.

$$\frac{12}{\sqrt{2}} = \frac{12 \cdot \sqrt{2}}{\sqrt{2} \cdot \sqrt{2}} = \frac{12\sqrt{2}}{2} = 6\sqrt{2}$$

Check

You should check that $6\sqrt{2} = \dfrac{12}{\sqrt{2}}$ by finding decimal approximations on your calculator. Also check that the lengths satisfy the Pythagorean Theorem:
$$(6\sqrt{2})^2 + (6\sqrt{2})^2 = 12^2.$$

Properties of 30-60-90 Triangles

Another special right triangle is the **30-60-90 triangle.** All 30-60-90 triangles are similar. (Do you see why?) One way to form a 30-60-90 triangle is to draw any altitude of an equilateral triangle, as shown at the right.

Again the lengths of all sides can be determined if one side length is known.

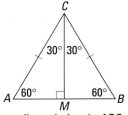

equilateral triangle *ABC*
30-60-90 triangles *CAM* and *CBM*

Proof

Given: $\triangle ABC$ with $m\angle A = 30$, $m\angle B = 60$,
$m\angle C = 90$. The shorter leg is opposite
the smaller acute angle, so let $BC = x$.

To prove: (1) $AB = 2x$;
            (2) $AC = x\sqrt{3}$.

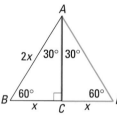

Argument:
The idea is to think of $\triangle ABC$ as half an equilateral
triangle, and use the Pythagorean Theorem.

(1) Reflect $\triangle ABC$ over $\overleftrightarrow{AC}$. Let $D = r_{\overleftrightarrow{AC}}(B)$. By definition of congruence,
$\triangle ABC \cong \triangle ADC$. Thus, $\triangle ADC$ is a 30-60-90 right triangle, with
$CD = x$. Also, $m\angle BCD = 180$ so C lies on $\overleftrightarrow{BD}$ and ABD is a
triangle. $\triangle ABD$ has three 60° angles, making it equilateral. Since
$BD = 2x$, then $AB = 2x$.

(2) Now apply the Pythagorean Theorem to $\triangle ABC$ to get AC.

$$AC^2 + BC^2 = AB^2$$
$$AC^2 + x^2 = (2x)^2$$
$$AC^2 + x^2 = 4x^2$$
$$AC^2 = 3x^2$$

Taking the positive square roots of each side:
$$AC = x\sqrt{3}.$$

Example 3

A natural beehive is a tessellation of regular hexagons. An apiarist
measured a side of a large cell as 8.0 mm but forgot to measure the
height h of each cell. Approximate the height to the nearest 0.1 mm.

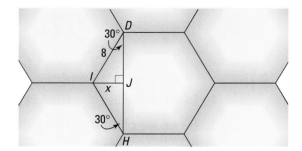

Bee careful. *An apiary is a
man-made home for bees
consisting of thousands of
hexagonal cells.*

Solution

The figure on page 767 isolates one cell. *DH* is the desired height. From properties of a regular hexagon, m∠*DIH* = 120 and *DI* = *IH*. From the Isosceles △ Coincidence Theorem, altitude *IJ* bisects ∠*DIH* and so △*DIJ* is a 30-60-90 right triangle with hypotenuse 8.0 mm. Let *IJ*, the length of the shortest side, be *x*. Using the 30-60-90 Triangle Theorem, the hypotenuse *DI* = 2x = 8.0 mm, so x = 4.0 mm. The longer leg *DJ* = x√3 = 4√3.

Now,
$$DH = 2 \cdot 4\sqrt{3}$$
$$= 8\sqrt{3}$$
$$\approx 13.9 \text{ mm.}$$

Applications to Other Triangles

With these theorems, lengths can be found in some non-right triangles that contain a 30°, 45°, or 60° angle.

Example 4

In △*XYZ* at the right, m∠*Z* = 45, m∠*Y* = 30, and *XY* = 8. Find *YZ* and *XZ*.

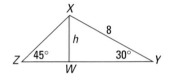

Solution

Right triangles are needed, so An auxiliary line is drawn, the altitude $\overline{XW}$ from X to $\overline{YZ}$. Let *XW* = h. This forms a 45-45-90 right triangle *XZW* and a 30-60-90 right triangle *XYW*. Now apply the theorems of this lesson. The side $\overline{XW}$, opposite the 30° angle in △XYW, is half of 8, so h = 4. Also, YW = h√3 = 4√3. Since △XZW is isosceles, ZW = h = 4. Adding, YZ = ZW + YW = 4 + 4√3 ≈ 10.93. XZ = h · √2 = 4√2 ≈ 5.66.

Check

An accurate picture verifies these lengths.

Of course, you could always measure to get approximate lengths and angle measures in a triangle. One advantage of having the theorems of this lesson is that for these common triangles you do not have to measure. Another advantage is that they give exact values.

QUESTIONS

Covering the Reading

1. In an isosceles right triangle, the acute angles each measure ___?___.

2. If one leg of an isosceles right triangle has length 10 cm, the hypotenuse has length ___?___.

3. On a Major League baseball field, how far is it from first base to third base?

4. *True or false.*
   a. All right triangles are similar.
   b. All right triangles with a 60° angle are similar.
   c. All isosceles right triangles are similar.

5. In a right triangle with a 30° angle, the __?__ has double the length of the __?__ leg.

In 6–9, find each missing length.

6.

7.

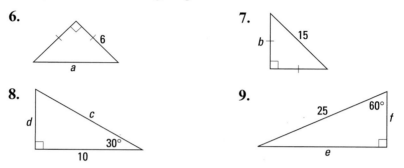

8.

9.

10. If the shortest side of a 30-60-90 triangle is 6 cm, what are the lengths of the other two sides (to the nearest 0.1 cm)?

11. Refer to Example 3. Suppose the cell of a natural beehive is a regular hexagon with a side of length 5 mm. What is the height of each cell?

12. In Example 4, what is m∠*ZXY*?

13. In Example 4, what is the perimeter of △*XYZ*?

Applying the Mathematics

14. A square has side *s*. What is the length of its diagonals?

15. What is the length of one of the altitudes of an equilateral triangle with a side of length 14?

16. Use the figure below. If *OL* = *h*, find the perimeter of △*BLT* in terms of *h*.

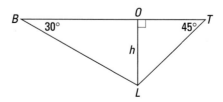

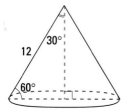

17. How many 30-60-90 triangles are formed when all the diagonals of a regular hexagon are drawn?

18. Find the total surface area and volume to the nearest tenth of the right cone at the left.

In 19 and 20, use the drawing at the right. *(Lesson 13-4)*

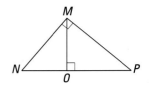

19. If $NO = 2$ and $MO = 3$, find OP.

20. If $NO = 5$ and $OP = 10$, find MO.

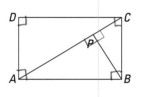

21. Find x and y in the figure at the left. *(Lesson 13-3)*

22. At the right $\overleftrightarrow{AB}$ // $\overleftrightarrow{CD}$ // $\overleftrightarrow{EF}$, $AX = 5$, $XD = 7$, and other lengths are as indicated. Find XB, BD, and CE. *(Lesson 13-3)*

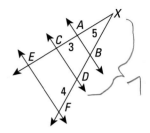

23. $\overline{BP} \perp \overline{AC}$ in rectangle $ABCD$ at the left. Name all triangles in the figure which are similar to $\triangle ADC$. Be sure to list them with their vertices in the correct order. *(Lessons 13-1, 13-2)*

24. A square and a circle each have area 400 square meters. Which has the larger perimeter? *(Lessons 8-1, 8-2, 8-7, 8-8)*

25. There are many triangles that could be considered special. One such triangle is formed in a regular heptagon by a side, a shorter diagonal, and a longer diagonal.

    a. Use an accurate drawing tool to draw a regular heptagon. (One way is to draw seven equally spaced points on a circle, as has been done at the right.)

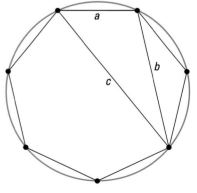

    b. Measure a, b, and c. Calculate $\left(\frac{c}{a}\right)^2 + \left(\frac{a}{b}\right)^2 + \left(\frac{b}{c}\right)^2$. Your answer should be very close to a whole number. Which whole number?

    c. Measure the three angles of this triangle. How do they seem to be related?

Ratios of Legs in Right Triangles

I N - C L A S S

A C T I V I T Y

Work with a partner. You will need a ruler and protractor, or an automatic drawer.

In right triangle ABC below, m$\angle A = 26$ and $\overline{B_1C_1} \parallel \overline{B_2C_2} \parallel \overline{B_3C_3} \parallel \overline{BC}$.

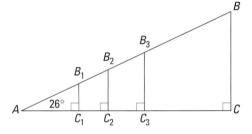

Trace the figure above, or draw one of your own by hand or on an automatic drawer.

1 Measure B_1C_1 and AC_1 to the nearest millimeter and estimate $\frac{B_1C_1}{AC_1}$ to two decimal places.

2 Do the same for B_2C_2 and AC_2, B_3C_3 and AC_3, and BC and AC.

3 How close to each other are the ratios $\frac{BC}{AC}$, $\frac{B_1C_1}{AC_1}$, $\frac{B_2C_2}{AC_2}$, and $\frac{B_3C_3}{AC_3}$?

4 Calculate the average of the four ratios in Step 3.

5 Each of the ratios in Step 3 is an estimate of the *tangent* of 26°. To get a calculator estimate of this value, make certain your calculator is in degree mode. Then press tan 26 **ENTER** if your calculator is a graphing calculator, 26 tan if you are using a scientific calculator.

6 How close to each other are your values from Steps 4 and 5?

7 Draw another right triangle ABC with a 26° angle at A and a right angle at C, where $\overline{AC}$ and $\overline{BC}$ are neither horizontal nor vertical. Calculate $\frac{BC}{AC}$ for this triangle.

8 How does your value from Step 7 compare with your values from Steps 4 and 5?

If you know two sides of a right triangle, you can always find the third side by using the Pythagorean Theorem. In Lesson 13-5, you learned that if you know one side and one acute angle of a 30-60-90 or 45-45-90 triangle, you can find the other two sides. But most triangles do not contain these angles. In this lesson and the next lesson, you will learn how you can use one side and one acute angle of *any* right triangle to find the lengths of the other two sides.

Ratios of Legs in Right Triangles

Consider two right triangles *ABC* and *XYZ* with congruent acute angles. The triangles might be formed by figures and shadows at the same time of day. The triangles are similar because of the AA Similarity Theorem.

The drawing at the right follows custom by using *a* to represent the side opposite angle *A*, *b* to represent the side opposite angle *B*, and so on. (You should be careful to write small letters so they are different from capital letters.) Since corresponding sides are proportional,

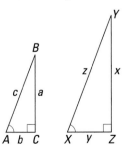

$$\frac{a}{x} = \frac{b}{y}.$$

Multiplying both sides by $\frac{x}{b}$, yields a ratio of legs in each triangle.

$$\frac{a}{x} \cdot \frac{x}{b} = \frac{b}{y} \cdot \frac{x}{b}$$
$$\frac{a}{b} = \frac{x}{y}$$

The ratios of the legs are equal.

The Tangent Ratio

The legs a and x are **opposite** the congruent angles A and X. The legs b and y are **adjacent to** angles A and X. This argument proves that in a right triangle, the ratio

$$\frac{\text{leg opposite angle } A}{\text{leg adjacent to angle } A}$$

is the same for any angle congruent to A. This ratio is called the *tangent of angle A*.

Definition

In right triangle *ABC* with right angle *C*, **the tangent of** ∠*A*, written **tan *A*,** is

$$\frac{\text{leg opposite } \angle A}{\text{leg adjacent to } \angle A}.$$

In the In-class Activity on page 771, you were asked to calculate ratios for tan 26°.

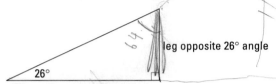

leg opposite 26° angle

26°

leg adjacent to 26° angle

$$\tan 26° = \frac{\text{leg opposite 26° angle}}{\text{leg adjacent to 26° angle}}$$

But estimating using a drawing is tedious and not very precise. Today, most people use calculators to approximate values of tangents. Every scientific or graphics calculator contains a [tan] key. To use a [tan] key, make sure your calculator is measuring angles in degree mode (there are other modes, but we do not discuss them in this book). Your calculator may display 0.4877325886. That means tan 26° ≈ 0.4877325886.

Using Tangents to Find Lengths

When the measure of an angle and one leg in a right triangle are known, the tangent can help to find the length of the other leg.

Example 1

At a location 50 m from the base of a tree, the *angle of elevation* of the tree top is 33°. Determine the height of the tree to the nearest meter.

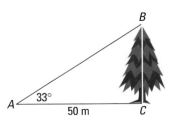

B

A

33°

50 m

C

Solution

The angle of elevation is $\angle A$ in the figure on page 773.

$$\tan 33° = \frac{\text{leg opposite } \angle A}{\text{leg adjacent to } \angle A}$$

$$\tan 33° = \frac{BC}{50}$$

Multiplying both sides by 50,

$$BC = 50 \cdot \tan 33°$$
$$= 50 \boxed{\times} 33 \boxed{\tan} \boxed{=} \text{ or } 50 \boxed{\times} \boxed{\tan} 33 \boxed{\text{ENTER}}$$
$$\approx 32.47038$$
$$\approx 32 \text{ meters.}$$

The exact value of a tangent of an acute angle can be determined if you know the lengths of the two legs in a right triangle with that angle. This value can then be used to determine the measure of the angle.

Example 2

In the right triangle at the left, find $m\angle D$.

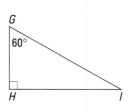

Solution

$$\tan D = \frac{\text{leg opposite } \angle D}{\text{leg adjacent } \angle D} = \frac{EF}{ED} = \frac{5}{12}$$

To find $m\angle D$, use the $\tan^{-1}$ key, the inverse tangent function, on your calculator. $\tan^{-1}x$ is the angle whose tangent is x. Key in

$\boxed{\tan^{-1}}$ $\boxed{(}$ 5 $\boxed{\div}$ 12 $\boxed{)}$ $\boxed{\text{ENTER}}$ or 5 $\boxed{\div}$ 12 $\boxed{=}$ $\boxed{\tan^{-1}}$. Our calculator shows 22.61986495, which means $m\angle D \approx 22.62$.

Exact Values of the Tangent Ratio

You can use special triangles to find exact values of some tangents.

Example 3

Give an exact value for $\tan 60°$.

Solution

Draw a 30-60-90 triangle. In $\triangle GHI$, $\tan 60° = \tan G = \frac{\text{leg opposite } \angle G}{\text{leg adjacent } \angle G} = \frac{HI}{GH}$. $\overline{GH}$ is the shorter leg. Call its length x. Then $HI = x\sqrt{3}$. Substituting, $\tan 60° = \frac{x\sqrt{3}}{x} = \sqrt{3}$.

Check

On your calculator, find $\tan 60° \approx 1.732050808 \approx \sqrt{3}$.

The tangent of an angle is an example of a **trigonometric ratio.** You will study two other trigonometric ratios in the next lesson. With these ratios, you can find all sides and angles of a right triangle whenever you have enough information to satisfy any one of the triangle congruence conditions.

QUESTIONS

Covering the Reading

1. Refer to the In-class Activity on page 771.
   a. What average value did you get in Step 4?
   b. How close was this value to a calculator value for tan 26°?

2. Draw a right triangle with a 40° angle. Measure the sides, and use those measurements to estimate tan 40°.

3. Use a calculator to estimate tan 73° to the nearest thousandth.

4. When the sun is 32° up from the horizon, the wall of a store casts a shadow 25 meters long. How high is the wall?

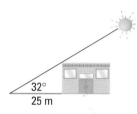

In 5 and 6, use $\triangle DEF$ at the right.
Give the exact value of

5. tan E.   6. tan F.

7. tan 30°.   8. tan 45°.

9. Draw a right triangle with legs of 4 units and 6 units. Use your triangle to estimate answers to the following questions.
   a. What is m$\angle A$ if tan $A = \frac{2}{3}$?
   b. What is m$\angle A$ if tan $A = \frac{3}{2}$?
   c. Check your answers to parts **a** and **b** with a calculator.

Applying the Mathematics

10. Suppose a person whose eye level is 5′ off the ground and who is 20′ away from a flagpole has to look up at a 40° angle to see the top of the pole. How high is the pole?

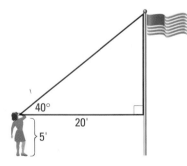

11. A size change of magnitude 3.5 is applied to $\triangle DEF$ at the left, resulting in $\triangle D'E'F'$ (not drawn). What is tan E'?

12. Refer to $\triangle ABC$ and $\triangle XYZ$ at the beginning of this lesson. Prove that $\frac{a}{c} = \frac{x}{z}$.

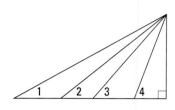

13. Use the figure at the left. Consider angles 1, 2, 3, and 4.
   a. Which has the largest tangent?
   b. Which has the smallest tangent?

14. Without calculating, explain why tan 75° > tan 74°.

Review

15. An altitude of an equilateral triangle has length 100. **a.** What is the perimeter of the triangle? **b.** What is the triangle's area?
(Lessons 8-4, 13-5)

16. Find the area of trapezoid *ZONK* if m∠*Z* = m∠*O* = 45, *ZO* = 24, and *NK* = 9. *(Lessons 8-5, 13-5)*

In 17 and 18, use the figure at the right.

17. Given $\overline{AE}$ // $\overline{BD}$, *BC* = 180, *AC* = 200, and *CE* = 240. Find the length of $\overline{DE}$. *(Lesson 13-3)*

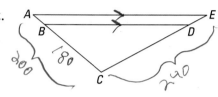

18. Suppose *BC* = 13, *AB* = 0.8, *ED* = 0.9, and *CD* = 14. Is $\overline{AE}$ // $\overline{BD}$? Why or why not? *(Lesson 13-3)*

19. The diagonals of any trapezoid split the trapezoid into four nonoverlapping triangles.
   a. Must two of these triangles be congruent? Justify your answer.
   b. Must two of these triangles be similar? Justify your answer.
   (Lesson 13-2)

20. About how long is the longest straw that can fit into a box 3″ by 4″ by 8″ in the way shown at the left. *(Lesson 11-9)*

In 21 and 22, trace the figure at the right, in which ℓ // *m*.

21. Translate △*ADB* by the translation vector $\overrightarrow{AD}$.
(Lesson 4-6)

22. Draw $r_\ell \circ r_m (DAB)$. Describe this transformation. *(Lesson 4-4)*

Exploration

23. Choose three angle measures (other than 90) whose sum is 180. (For example, you could choose 27, 97, and 56.)
   a. Use a calculator to find the sum of the tangents of the numbers you have chosen.
   b. Calculate the product of the tangents of the numbers you have chosen.
   c. Repeat parts **a** and **b** with a different set of three angle measures.
   d. Make a conjecture based on your results.

LESSON

13-7

The Sine and Cosine Ratios

Seaing the way. *Before radio communication, sailors needed to use a sextant to determine their position at sea. Sextants measure the angle between the horizon and a celestial body such as a star or the sun.*

The tangent is the ratio of the lengths of two legs in a right triangle. When a leg is compared to the hypotenuse, the *sine* or *cosine* ratio results.

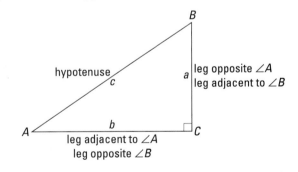

Definition

In right triangle *ABC* with right angle *C*,

the **sine of** ∠*A*, written **sin A**, is $\dfrac{\text{leg opposite } \angle A}{\text{hypotenuse}}$;

the **cosine of** ∠*A*, written **cos A**, is $\dfrac{\text{leg adjacent to } \angle A}{\text{hypotenuse}}$.

In the above triangle, $\sin A = \frac{a}{c}$ and $\cos A = \frac{b}{c}$. In any other right triangle with an acute angle congruent to ∠*A*, the sine and cosine of that angle will have these same values because of the AA Similarity Theorem.

Example 1

Right triangle *ABC* has side lengths as indicated below. Find each value.
a. sin *A*
b. cos *A*
c. sin *B*
d. cos *B*

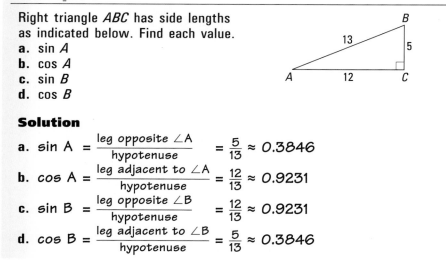

Solution

a. $\sin A = \dfrac{\text{leg opposite } \angle A}{\text{hypotenuse}} = \dfrac{5}{13} \approx 0.3846$

b. $\cos A = \dfrac{\text{leg adjacent to } \angle A}{\text{hypotenuse}} = \dfrac{12}{13} \approx 0.9231$

c. $\sin B = \dfrac{\text{leg opposite } \angle B}{\text{hypotenuse}} = \dfrac{12}{13} \approx 0.9231$

d. $\cos B = \dfrac{\text{leg adjacent to } \angle B}{\text{hypotenuse}} = \dfrac{5}{13} \approx 0.3846$

In Example 1, the reason that sin *A* = cos *B* is that each is equal to $\frac{BC}{AB}$. Also sin *B* = cos *A* since both equal $\frac{AC}{AB}$. This is because the leg opposite one acute angle is the leg adjacent to the other. Also recall that angles *A* and *B* are complementary. This is the origin of the term "cosine"; *cosine* is short for *complement's sine*.

Estimating Sines and Cosines

As with tangents, you can estimate values of sines and cosines with drawings or a calculator. To approximate sin 54° or cos 54°, you can draw a right triangle with a 54° angle, as shown at the left. Measuring the sides of this triangle, we find lengths (to the nearest mm) of 21 mm, 29 mm, and 36 mm.

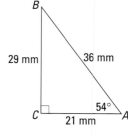

From these measurements, $\sin 54° = \frac{BC}{AB} \approx \frac{29}{36} \approx 0.81$ and $\cos 54° = \frac{AC}{AB} \approx \frac{21}{36} \approx 0.58$. A calculator gives greater accuracy. For sin 54°, press

54 [sin] or [sin] 54 [ENTER].

An 8-digit display will show 0.80901699 so sin 54° ≈ 0.8090. For cos 54°, press

54 [cos] or [cos] 54 [ENTER].

The display may show 0.58778525 so cos 54° ≈ 0.5878.

Exact Values of Sines and Cosines

Drawings and calculators usually give approximations to sines and cosines. When the lengths of sides are known exactly, as in special triangles, you can find exact values for the sine and cosine ratios.

Example 2

Find exact values of sin 30° and cos 30°.

Solution

Sketch a 30-60-90 triangle as shown at the right. The legs have lengths x and $x\sqrt{3}$, and the hypotenuse has length $2x$.

$$\sin 30° = \frac{x}{2x} = \frac{1}{2}$$

$$\cos 30° = \frac{x\sqrt{3}}{2x} = \frac{\sqrt{3}}{2}$$

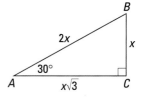

Check

A calculator shows sin 30° = .5 and cos 30° ≈ .8660254. These decimal values agree with the exact values found in the solution.

Applications of Sines and Cosines

Sines and cosines have a great number and variety of applications, and a long history. The ancient Babylonians and Greeks measured the sides and angles of triangles carefully, needing such measurements for navigation, surveying, and astronomy. The first table of trigonometric values was constructed by Claudius Ptolemy in the second century A.D. Values like our present-day sine, cosine, and tangent values were first obtained by the German astronomer Regiomontanus (1436–1476). The abbreviations *sin*, *cos*, and *tan* were first used by Euler.

Example 3

A particular 30-foot extension ladder is safe if the angle it makes with the ground is between 65° and 80°.
a. What is the farthest up on a vertical wall that a 30-foot ladder of this type can reach and still be safe?
b. How far, at a minimum, should the foot of the ladder be placed from the base of the wall?

Solution

First draw a picture. To reach the maximum height, use m∠A = 80. In the figure drawn, *BC* is the height for part **a** and *AC* is the horizontal distance for part **b**.

An artist's rendering of Ptolemy.

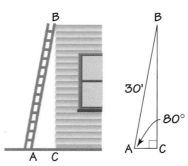

a. $\overline{BC}$ is the leg opposite the 80° angle. Since the hypotenuse is known, the sine should be used.

$$\sin 80° = \frac{\text{leg opposite } 80° \text{ angle}}{\text{hypotenuse}}$$

A calculator shows $\sin 80° \approx 0.985$. Substituting,

$$0.985 \approx \frac{BC}{30}.$$

Solve for BC.　　$BC \approx 30 \cdot 0.985 \approx 29.6$ feet
The ladder can safely reach about 29.6 feet high on the wall.

b. $\overline{AC}$ is the leg adjacent to the 80° angle. This suggests using the cosine.

$$\cos 80° = \frac{\text{leg adjacent to } 80° \text{ angle}}{\text{hypotenuse}}$$

A calculator shows $\cos 80° \approx 0.174$. Substituting,

$$0.174 \approx \frac{AC}{30}.$$

Solve for AC.　　　　$AC \approx 30 \cdot 0.174 \approx 5.2$ feet.
5.2 feet = 5 feet, 2.4 inches. If the ladder is placed less than 5.2 feet away it is too steep to be used safely.

Check

The Pythagorean Theorem can check both answers at once.
Does　　　　　　　　　　　$AC^2 + BC^2 = AB^2$?
$$(5.2)^2 + (29.6)^2 = 30^2$$
The left side is 903.2, the right side is 900. This is close enough, given the approximations used for $\sin 80°$ and $\cos 80°$ and the rounding done to get 5.2 and 29.6.

The measure of an acute angle can be found if its sine or cosine is known, as the following example shows.

Example 4

Suppose the bottom of the ladder in Example 3 is placed 6 feet from the wall.
a. What angle does the ladder make with the ground?
b. Explain whether this is a safe position for the ladder.

Solution

a. Draw a picture, as at the right. The side adjacent to $\angle A$ and the hypotenuse are known. So the cosine should be used.

$$\cos A = \frac{6}{30} = 0.2$$

From this, $m\angle A = \boxed{\cos^{-1}} \ .2 \approx 78.5°.$

b. 78.5° is between 65° and 80°, so it is a safe position for the ladder.

QUESTIONS

Covering the Reading

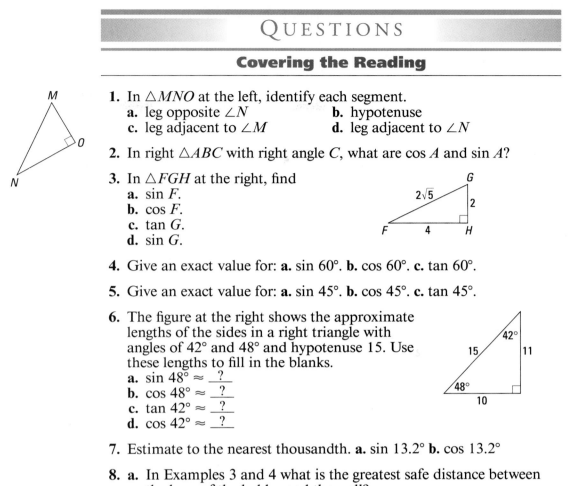

1. In △*MNO* at the left, identify each segment.
   a. leg opposite ∠*N*   b. hypotenuse
   c. leg adjacent to ∠*M*   d. leg adjacent to ∠*N*

2. In right △*ABC* with right angle *C*, what are cos *A* and sin *A*?

3. In △*FGH* at the right, find
   a. sin *F*.
   b. cos *F*.
   c. tan *G*.
   d. sin *G*.

4. Give an exact value for: **a.** sin 60°. **b.** cos 60°. **c.** tan 60°.

5. Give an exact value for: **a.** sin 45°. **b.** cos 45°. **c.** tan 45°.

6. The figure at the right shows the approximate
   lengths of the sides in a right triangle with
   angles of 42° and 48° and hypotenuse 15. Use
   these lengths to fill in the blanks.
   a. sin 48° ≈ __?__
   b. cos 48° ≈ __?__
   c. tan 42° ≈ __?__
   d. cos 42° ≈ __?__

7. Estimate to the nearest thousandth. **a.** sin 13.2° **b.** cos 13.2°

8. a. In Examples 3 and 4 what is the greatest safe distance between
      the base of the ladder and the wall?
   b. How high will it reach at that distance?

In 9–11, consider a 20′ ladder with the same safety restrictions as in
Example 3.

9. How far up on a vertical wall can the ladder safely reach?

10. How far, at minimum, should the bottom of the ladder be from
    the wall?

11. Is it safe to place the base of this ladder 6 feet from the wall?

Applying the Mathematics

12. a. Use your calculator to estimate sin 89° to the nearest
       ten-thousandth.
    b. Use a triangle to explain why the value is so near 1.

13. In the triangles at the right, ∠*B* ≅ ∠*B*′.
    a. By measuring lengths of sides, estimate
       sin *B* and sin *B*′.
    b. Do your results agree with what you
       expected?

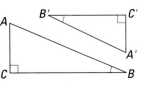

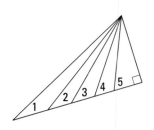

14. a. Of angles 1, 2, 3, 4 and 5 pictured at the left, which angle has the largest sine?
   b. Which angle has the largest cosine?

15. What length wire is needed as a brace from the top of a 20-foot pole, if the brace is to make an angle of 85° with the ground?

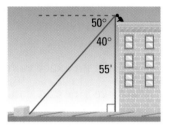

16. From the top of a building you look down at an object on the ground. If your eyes are 55 feet above the ground, and the angle of sight, called the *angle of depression*, is 50° below the horizontal, how far is the object from you?

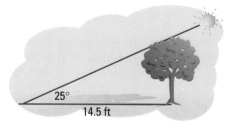

Review

17. When the sun is 25° up from the horizon, a tree casts a shadow 14.5 feet long. How tall is the tree?   *(Lesson 13-6)*

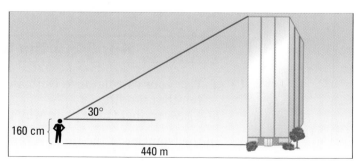

18. From eye level, 160 cm off the ground, about 440 meters away from a skyscraper, a person has to look up at an angle of 30° to see the top of the building. About how high (in meters) is the building? *(Lesson 13-6)*

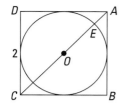

19. At the left, circle O is inscribed in the square. Find AE. *(Lesson 13-5)*

20. Are the two triangles in the figure at the right similar? Explain why or why not. *(Lesson 13-2)*

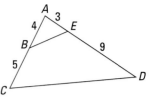

21. a. Write the converse, the inverse, and the contrapositive of the following statement: If $\triangle ABC \sim \triangle DEF$, then $m\angle ABC = m\angle DEF$.
b. Which of the statements you wrote in part **a** are true? *(Lessons 11-2, 13-2)*

22. Isosceles $\triangle ABC$ has vertex angle A. If its sides are 7, 7, and 12, find the length of the altitude from A. *(Lessons 6-2, 8-6)*

23. Trace the figure at the right. Then draw the image of $ABCD$ under the translation described by the vector $\overrightarrow{EF}$. *(Lesson 4-6)*

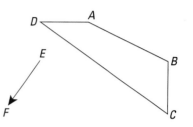

Exploration

24. a. Use a calculator to fill in this table of sine and cosine values.

x	$\sin x$	$\cos x$	x	$\sin x$	$\cos x$
5			50		
10			55		
15			60		
20			65		
25			70		
30			75		
35			80		
40			85		
45					

b. For which of these values of x does $(\sin x)^2 + (\cos x)^2 = 1$?
c. Generalize part **b.**

13-8

More Work with Vectors and Area

Go the distance. *The Panama Canal lies on a vector about 40° north of west. Its use shortens a ship's journey between the east and west coasts of the United States by 8000 nautical miles.*

A few lessons are not enough to show all the applications of sines and cosines. Sometimes entire books are devoted to trigonometry. In this lesson, we give applications that result directly from the definitions of the sine and cosine. The first application is to vectors, the second to area.

Finding Components of a Vector

Recall from Lesson 4-6 that a vector is a quantity that can be characterized by its direction and magnitude. Consider a ship that is moving at 25 miles per hour in a direction 30° south of east. This speed and direction is called its **velocity**. The velocity can be represented by the vector $\overrightarrow{OA}$ below. The magnitude of $\overrightarrow{OA}$ is 25 and the direction of $\overrightarrow{OA}$ is 30° clockwise from the positive x-axis. The velocity can be broken down into its eastern component and its southern component using sines and cosines.

Example 1

Find the eastern and southern components of the velocity of a ship moving at 25 miles per hour in a direction 30° south of east.

Solution

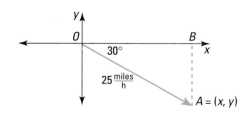

► The eastern component is $\vec{OB}$. The southern component is $\vec{BA}$.

Since
$$\frac{OB}{OA} = \cos 30°,$$
$$OB = OA \cos 30°$$
$$= 25 \cos 30°$$
$$\approx 25 \cdot 0.866 = 21.65 \text{ miles/hr.}$$

Similarly, because
$$\frac{BA}{OA} = \sin 30°,$$
$$BA = OA \sin 30°$$
$$= 25 \sin 30°$$
$$= 25 \cdot 0.5 = 12.5 \text{ miles/hr.}$$

So the position of the ship is changing by about 21.65 miles/hr to the east and 12.5 miles/hr to the south.

In Example 1, one component was found using the sine of the given angle. The other component was found using the cosine. However, you do not have to memorize which is which. It is usually safer to draw a picture each time and use the definitions of the sine and cosine.

Finding the Direction of a Vector From Its Components

To find the direction of a vector from its components, the tangent is required.

Example 2

An airplane has to fly to a location 60 km east and 100 km north of its present location.
a. In what direction should it fly?
b. How far will it travel?

Solution

Again, draw a picture. A sample is shown at the right.
a. The direction can be found from m∠*FOE*. We know that

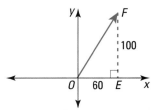

$$\tan \angle FOE = \frac{100}{60}.$$

So m∠*FOE* = $\tan^{-1} \frac{100}{60}$. You can key in

100 ÷ 60 = INV tan or INV tan (100 ÷ 60) ENTER.

You should see 59.0362 . . . displayed. This means that m∠*FOE* ≈ 59°.
The direction the plane should fly is about 59° north of east.
You could also say that the plane should fly about 31° east of north.
b. *OF* can be found using the Pythagorean Theorem.
$$OF = \sqrt{60^2 + 100^2} = \sqrt{13600} \approx 116.6$$
The plane will travel about 117 km.

Another Triangle Area Formula

The same idea that was applied in Example 1 can be used to find the area of a triangle if you know two sides and the included angle.

Example 3

In $\triangle PQR$, $PQ = 12$, $PR = 30$, and $m\angle P = 41°$. What is the area of $\triangle PQR$?

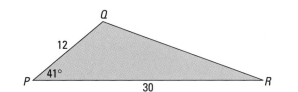

Solution

The area is given by the formula $A = \frac{1}{2}hb$, where h is the altitude to a base b. We let the longer side $\overline{PR}$ be the base, and draw the altitude from Q to that side, intersecting $\overline{PR}$ at T. $\triangle PQT$ is a right triangle. We need to find h.

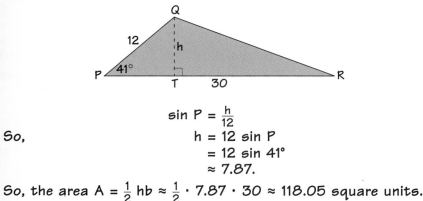

$$\sin P = \frac{h}{12}$$

So,
$$h = 12 \sin P$$
$$= 12 \sin 41°$$
$$\approx 7.87.$$

So, the area $A = \frac{1}{2} hb \approx \frac{1}{2} \cdot 7.87 \cdot 30 \approx 118.05$ square units.

The process that was used in Example 3 can be generalized to give the following elegant formula. The proof is left for you as Question 10.

SAS Triangle Area Formula
In any $\triangle ABC$, area($\triangle ABC$) = $\frac{1}{2} ab \sin C$.

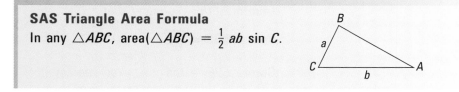

QUESTIONS

Covering the Reading

1. Consider a plane traveling at 340 mph in a direction 15° north of west. Picture this with a vector.

2. In $\triangle XYZ$ at the left, explain why $XY = YZ \cdot \sin Z$.

In 3–5, find the indicated side of the right triangle.

3.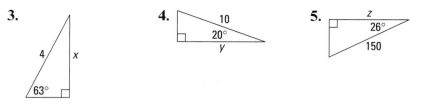

4.

5.

6. Give the northern and eastern components of the velocity of a tornado moving at 20 mph in a direction 15° north of east.

7. An airplane is to fly 200 km south and 40 km west of its present location.
  a. In what direction should it fly?
  b. How far will it travel?

In 8 and 9, give the area of the triangle.

8.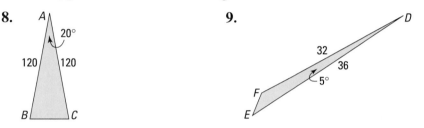

9.

Applying the Mathematics

10. Prove the SAS Triangle Area Formula, Area $\triangle ABC = \frac{1}{2} ab \sin C$, when C is an acute angle. (Hint: Use the figure on page 786. Draw in the altitude from B to side $\overline{AC}$).

11. A regular pentagon is cut from a disk of metal of radius 5 cm.
  a. Find the area of the pentagon.
  b. How much material is cut off, to the nearest cm^2?

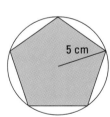

12. What is the area of $\triangle STU$ below?

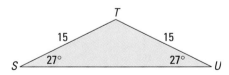

13. A trade triangle was formed between Johor, Malaysia; Singapore; and Batam Island, Indonesia in the 1960s to promote economic growth. Suppose a ship traveled 20 km on a trade route from Johor on a course 9° east of south to Singapore. Then it traveled 50 km on a course 24° south of east to Batam Island. How far (to the nearest km) is Johor from Batam Island?

Funnel cloud. *Paths of tornadoes average 16 miles long and can be several hundred yards wide. They travel from 30 to 40 mph with winds as high as 500 mph.*

In 14 and 15, consider $\triangle ABC$ at the right.
(Lessons 13-6, 13-7)

14. Give exact values of sin A, cos A, and tan A.

15. Give exact values of sin B, cos B, and tan B.

16. In regular hexagon *GHIDEF*, each side has length 10. Find the length of $\overline{GI}$. (Hint: Find the measures of the angles of $\triangle GHI$.) *(Lessons 6-7, 13-5)*

17. In a 5-12-13 right triangle, find the length of the altitude to the hypotenuse and the lengths of the segments into which the altitude splits the hypotenuse. *(Lesson 13-4)*

18. A triangle has sides of 12, 13, and 14. Tell whether or not it is similar to each triangle. Explain why or why not.
a. a triangle with sides of 13, 14, and 12
b. a triangle with sides of 120, 130, 14
c. a triangle with sides of 14, 15, and 16 *(Lesson 13-1)*

19. Make a conclusion using all of the following statements.
a. No x is a y.
b. If an animal is not a y, then the animal is not a z.
c. Rudolph is a z. *(Lesson 11-4)*

20. a. Refer to the figures at the right. Describe the figure formed by rotating circle C in space about line ℓ.
b. Describe the figure formed by rotating circle A in space about line ℓ. *(Lesson 9-5)*

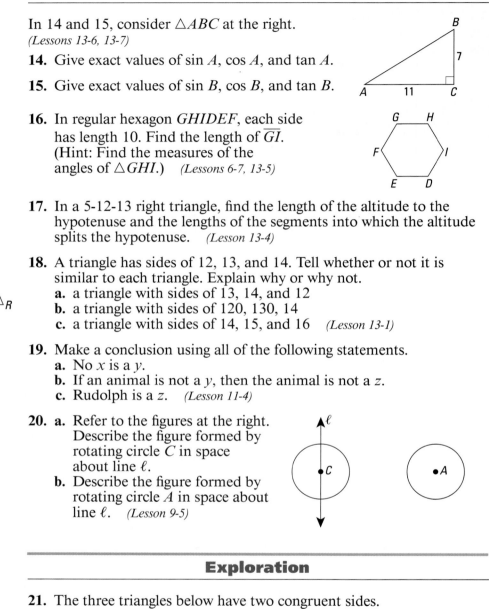

21. The three triangles below have two congruent sides.

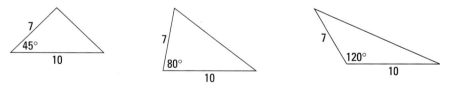

Explore what happens to the area of these triangles as the angle between the two sides increases from 0° to 180°. Write your explanation.

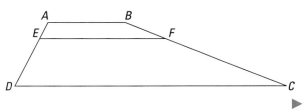

A project presents an opportunity for you to extend your knowledge of a topic related to the material of this chapter. You should allow more time for a project than you do for typical homework questions.

1 Measuring with Shadows

Four methods are given in this chapter for estimating heights of tall objects. They are found in Example 1 of Lesson 13-2, in Question 18 of Lesson 13-3, in Question 11 of Lesson 13-4, and in Example 1 of Lesson 13-6. Use at least three of these methods to estimate the height of at least three tall objects. Describe the methods you used and what you found.

2 Splitting Sides of a Trapezoid

Suppose you know the lengths of the four sides of a trapezoid $ABCD$ with bases $\overline{AB}$ and $\overline{CD}$ as pictured here. Further, suppose that you split the trapezoid by a segment $\overline{EF}$ parallel to the bases. Explain how, if you know the length of $\overline{AE}$, you can find

a. the lengths of $\overline{BF}$ and $\overline{FC}$, and

b. the length of $\overline{EF}$. If you cannot give a general process, answer the question for a number of special cases.

3 Combining Forces

By adding components of vectors, you can determine the result of two forces acting on an object. For instance, suppose a plane is flying at a speed and direction which, in still air, would be 550 mph, 25° north of west. Also, suppose there is a wind of 80 mph, 10° south of east. What is the resulting ground speed and direction of the plane in this wind condition? Answer this question and make up and answer at least two other questions using similar ideas.

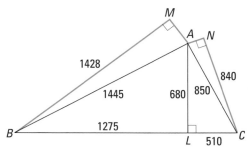

4 Regular Polygons and Unit Circles

Each circle below has radius 1 unit. The regular polygon at the left is said to be circumscribed about the circle. The circle is inscribed in the polygon. The regular polygon below it is said to be inscribed in the circle. The circle is circumscribed about the polygon. There are formulas for the area and perimeter of these polygons in terms of the tangent and sine.

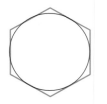

Perimeter =
$2n \cdot \tan \left(\frac{180°}{n}\right)$ units

Area =
$n \cdot \tan \left(\frac{180°}{n}\right)$ square units

circumscribed regular hexagon

Perimeter =
$2n \cdot \sin \left(\frac{180°}{n}\right)$ units

Area =
$n \cdot \sin \left(\frac{180°}{n}\right)$
$\cdot \cos \left(\frac{180°}{n}\right)$ square units

inscribed regular octagon

a. Show that these formulas work when the regular n-gon is a square.

b. Prove any two of the formulas.

5 Oblique Triangles with Integer Sides and Integer Altitudes

The special triangles of this chapter do not have sides that are all integers. But you have seen many right triangles with sides that are all integers. For example, there are the 3-4-5 and 8-15-17 right triangles. We took two triangles of these types and put them together. $\triangle ALC$ below is a 3-4-5 triangle. $\triangle ABL$ is an 8-15-17 triangle. By expanding the triangles, we formed an oblique triangle whose sides are all integers and whose altitudes are also integers. Form two other oblique triangles whose three sides and three altitudes are all integers. Describe how you formed the triangles and how you determined the lengths of the altitudes.

$\triangle ABC$ with sides $AB = 1445$, $AC = 850$, and $BC = 1785$, and altitudes $AL = 680$, $BM = 1428$, and $CN = 840$

SUMMARY

When triangles are similar, corresponding angles are congruent and corresponding sides form equal ratios. For triangles, the conditions guaranteeing similarity correspond to those for congruence. There are SSS, AA, and SAS Similarity Theorems, where A indicates corresponding congruent angles and S indicates equal ratios of corresponding sides.

Any line parallel to one side of a triangle and intersecting the other two sides forms similar triangles, and so splits these sides into proportional segments. The converse of this Side-Splitting Theorem is also true. When the altitude to the hypotenuse of a right triangle is drawn, two triangles are formed, each similar to the original. Three of the resulting proportions involve a geometric mean, as one length appears as both means in the proportion. The altitude itself is the geometric mean of the segments into which it divides the hypotenuse. Either leg is the geometric mean of the hypotenuse and the segment of the hypotenuse closest to it.

Right triangles are important because any triangle and thus any polygon can be split up into them, and because from knowing only one acute angle and one side, all lengths can be determined. Certain right triangles occur so often they are special. Any square can be split into two isosceles right triangles, with angles of 45°, 45°, and 90° and sides of lengths x, x, and $x\sqrt{2}$. Any equilateral triangle can be split into two right triangles with angles of 30°, 60°, and 90° and sides of lengths x, $x\sqrt{3}$, and $2x$.

When right triangles have congruent angles, the triangles are similar, and the ratios of corresponding sides within the triangles are equal. These are the trigonometric ratios. Three of the ratios are the tangent, sine, and cosine. The trigonometric ratios can be estimated using an accurate drawing or a scientific calculator, or calculated exactly if you are given lengths of sides in the triangle or are dealing with one of the special triangles.

Lengths in right triangles can be described in terms of sines and cosines. These descriptions give rise to the formula Area $= \frac{1}{2} ab \sin C$ for the area of any triangle. Sines and cosines also enable horizontal and vertical components of vectors to be described.

VOCABULARY

Below are the most important terms and phrases for this chapter. For the starred (*) terms you should be able to give a definition of the term. For the other terms you should be able to give a general description and a specific example of each.

Lesson 13-1
SSS Similarity Theorem

Lesson 13-2
AA Similarity Theorem
SAS Similarity Theorem
Reciprocals Property

Lesson 13-3
Side-Splitting Theorem
Side-Splitting Converse Theorem

Lesson 13-4
arithmetic mean
*geometric mean
Geometric Mean Theorem
Right-Triangle Altitude Theorem

Lesson 13-5
45-45-90 triangle
Isosceles Right Triangle Theorem
inscribed figure
30-60-90 triangle
30-60-90 Triangle Theorem

Lesson 13-6
leg opposite an angle
leg adjacent to an angle
*tangent of an angle A,
*tan A
$\tan^{-1} x$
trigonometric ratio

Lesson 13-7
*sine of an angle A, *sin A
*cosine of an angle A, *cos A

Lesson 13-8
velocity
SAS Triangle Area Formula

Projects
Polygon circumscribed about a circle
Circle inscribed in a polygon
Polygon inscribed in a circle
Circle circumscribed about a polygon

PROGRESS SELF-TEST

Directions: Take this test as you would take a test in class. You will need a calculator with trigonometric functions. Check your work with the solutions in the Selected Answers section in the back of the book.

1. Use the figure at the right. If $BC = 6$ and $BD = 2$, find AB.

2. Refer to the figure at the right. If $XY = 60$ and $YZ = 75$, find WZ.

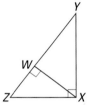

In 3 and 4, use the figure at the right, in which $\overleftrightarrow{WX} \,/\!/\, \overleftrightarrow{YZ}$.

3. If $VW = 11$, $WY = 13$, and $VZ = 30$, what is VX?

4. If $WX = 8$, $YZ = 20$, and $WY = 15$, find VW.

5. Use the figure at the right. If $AB = 7$, find each length.
   a. BC
   b. AC

6. An equilateral triangle has sides of length 30. Find the length of an altitude.

7. Refer to the triangles below. Are the triangles similar? Why or why not?

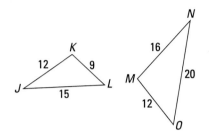

8. In $\triangle TRI$, $TR = 36$, $RI = 11$, and $m\angle TRI = 55$. What is the area of the triangle?

9. Write a proof argument.
   Given: $\overleftrightarrow{AB} \,/\!/\, \overleftrightarrow{DE}$ and $\overline{AE}$ intersects $\overline{BD}$ at C in the figure below.
   To prove: $\triangle ABC \sim \triangle EDC$.

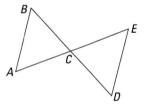

10. Estimate cos 31.2° to the nearest thousandth.

11. Give the exact value of sin 60°.

12. Define tan B for $\triangle ABC$ with right angle C.

13. Refer to the figure below. Of the numbered angles, which has the smallest tangent?

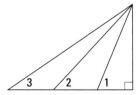

PROGRESS SELF-TEST

14. In right triangle ABC below, if $\sin B = \frac{9}{11}$, find $\cos A$.

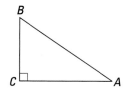

15. In triangle DEF below, find $\sin E$.

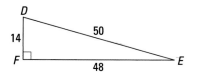

16. Washington, Adams, and Jefferson streets are parallel. Given the distances as indicated below, find x the length on Abigail Avenue between Washington and Adams.

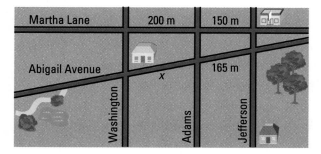

17. How far up on a vertical wall can a 15-foot ladder reach if the angle it makes with the ground is 80°?

18. As of 1995, Nation's Bank Tower in Atlanta was the tallest building in the southern United States. At noon on some days, it casts a shadow 67 m long at the same time a vertical meter stick casts a shadow 21 cm long. About how tall, in meters, is Nation's Bank Tower?

19. A boat has to travel to a spot 6.2 km west and 4.8 km north of its present location.
  a. In what direction should it go?
  b. How far will it travel?

20. From eye level 5 ft above the ground, a person has to look up at an angle of 35° to see the top of a tree 40 ft away. How tall is the tree?

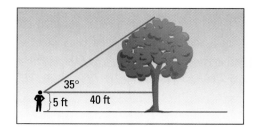

CHAPTER REVIEW

Questions on SPUR Objectives

SPUR stands for **S**kills, **P**roperties, **U**ses, and **R**epresentations. The Chapter Review questions are grouped according to the SPUR Objectives for this chapter.

SKILLS DEAL WITH THE PROCEDURES USED TO GET ANSWERS.

Objective A: *Find lengths in figures by applying the Side-Splitting Theorem and the Side-Splitting Converse Theorem.*   *(Lesson 13-3)*

1. In $\triangle JMN$ below, $\overline{KL} \parallel \overline{MN}$. Find each length.
   a. MN
   b. LN

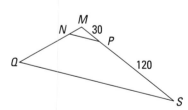

2. In the figure below, $\overleftrightarrow{NP} \parallel \overleftrightarrow{QS}$, $MP = 30$, and $PS = 120$. If $MQ = 90$, then $MN = \underline{\ ?\ }$.

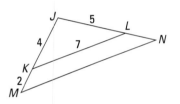

In 3 and 4, use the figures below. Is $\overline{UV} \parallel \overline{YZ}$? Why or why not?

3.

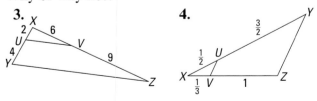

4.

5. If $\overline{AB} \parallel \overline{CD}$ in the figure below, find DE.

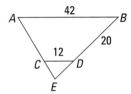

6. If $\overleftrightarrow{WV} \parallel \overleftrightarrow{UT}$ in the figure below, what is SW?

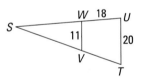

Objective B: *Calculate lengths using the Right-Triangle Altitude Theorem.*   *(Lesson 13-4)*

In 7–9, refer to the figure below.

7. If $AD = 9$ and $DB = 4$, then $CD = \underline{\ ?\ }$.
8. If $AC = 7$ and $AB = 12$, then $AD = \underline{\ ?\ }$.
9. If $DC = 12$ and $AD = 18$, then $BC = \underline{\ ?\ }$.

10. In $\triangle RQS$ below, $\overline{QR} \perp \overline{RS}$ and $\overline{RT} \perp \overline{QS}$. If $RT = 4$ and $TS = 8$, find each length.

    a. QT **b.** QR **c.** RS **d.** QS

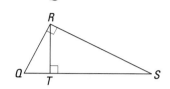

Objective C: *Calculate lengths of sides in isosceles right triangles and in 30-60-90 triangles.* (*Lesson 13-5*)

11. In $\triangle ABC$ below, find AC and BC.

12. In $\triangle DEF$ below, find DE and EF.

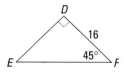

13. An equilateral triangle has sides of length 12. Find the length of an altitude.

14. A square has sides of length q. What is the length of a diagonal?

In 15 and 16, use the figure below.

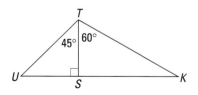

15. If $ST = 7$, find each length.

    a. TU **b.** US **c.** SK **d.** TK

16. If $SK = 13$, find each length.

    a. ST **b.** SU **c.** TK

Objective D: *Determine sines, cosines, and tangents of angles and use the SAS Triangle Area Formula.* (*Lessons 13-6, 13-7, 13-8*)

17. Refer to $\triangle ABC$ below. By measuring, determine $m\angle A$ (to the nearest degree), and $\tan A$, $\sin A$, and $\cos A$ (to the nearest hundredth).

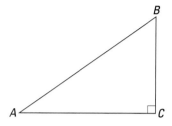

In 18 and 19, use the figure below. Choose from the numbered angles.

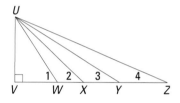

18. Which angle has the largest tangent?

19. Which angle has the largest sine?

In 20–22, use $\triangle ABC$ below. Give exact values for each.

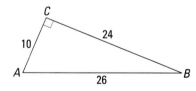

20. $\sin A$

21. $\cos B$

22. $\tan B$

In 23 and 24, determine the area of the triangle.

23.   **24.**

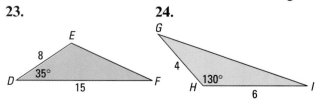

Objective E: *Estimate or determine exact values of the trigonometric ratios.* *(Lessons 13-6, 13-7)*

In 25 and 26, estimate to the nearest thousandth.

25. sin 57.5°

26. tan 22.1°

In 27–30, give exact values.

27. sin 30°

28. tan 60°

29. tan 45°

30. cos 45°

PROPERTIES DEAL WITH THE PRINCIPLES BEHIND THE MATHEMATICS.

Objective F: *Determine whether or not triangles are similar using the AA, SAS, or SSS Similarity Theorems.* *(Lessons 13-1, 13-2)*

31. One triangle has sides 40, 45, and 50.

  a. Is this triangle similar to a second triangle with sides 10, 9, and 8?

  b. If so, why? If not, why not?

32. *True or false.* All equilateral triangles are similar. Explain your answer.

In 33 and 34, are the triangles similar? If so, why? If not, why not?

33.

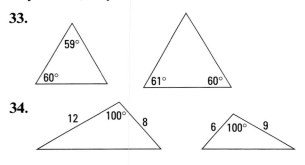

34.

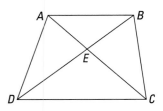

35. Write a proof argument.

  Given: $ABCD$ is a trapezoid with $\overleftrightarrow{AB} \parallel \overleftrightarrow{CD}$ and diagonals intersecting at E.

  To prove: $\triangle ABE \sim \triangle CDE$.

36. Write a proof argument.

  Given: $\overline{AC} \perp \overline{CD}$, $BC = x$, $AC = 2x$, and $DC = 4x$.

  To prove: $\triangle ABC \sim \triangle DAC$.

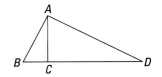

Objective G: *Know the definitions of sine, cosine, and tangent.* *(Lessons 13-6, 13-7)*

In 37–39, define for $\triangle ABC$ with right angle C.

37. cos A **38.** sin A **39.** tan A

In 40–42, use the figure below.

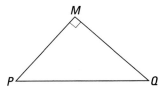

40. $\dfrac{MP}{MQ}$ is the tangent of which angle?

41. $\dfrac{MQ}{PQ}$ is the __?__ of angle P.

42. Give the cosine of angle Q.

In 43 and 44, use right triangle $\triangle ABC$ below.

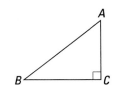

43. Write a ratio for tan B.

44. If cos $A = x$, find sin B.

USES DEAL WITH APPLICATIONS OF MATHEMATICS IN REAL SITUATIONS.

Objective H: *Use the Triangle Similarity and Side-Splitting Theorems to find lengths and distances in real situations.* *(Lessons 13-2, 13-3)*

45. A tree casts a shadow 9 meters long. At the same time, a vertical meter stick casts a shadow 60 cm long. How tall is the tree?

46. The height of a ramp at a point 1.5 meters from its bottom is 1.3 m. If the ramp runs for 4 m along the ground, what is its height at its highest point? (Round your answer to the nearest 0.1 m.)

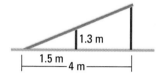

47. The beam $\overline{BC}$ is parallel to $\overline{DE}$ and splits the sides of the asymmetric roof. In the drawing below, if $AD = 50$, $AE = 70$, $BC = 25$, and $DE = 75$, find

 a. AB. **b.** CE.

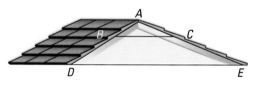

48. First Street runs north and south. Elm, Maple, and Pine run east and west. Distances between them are as indicated below. There is 2000 ft of frontage on Slant Street between Elm and Maple. How much frontage is there on Slant Street between Maple and Pine?

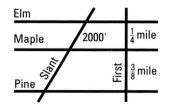

Objective I: *Use sines, cosines, and tangents to determine unknown lengths in real situations.* *(Lessons 13-6, 13-7)*

49. The tallest tree in the world (as of 1991) is a sequoia in Humboldt Redwoods State Park, California. When the sun is 57° up from the horizon, the tree casts a shadow 237 feet long. About how tall is this tree?

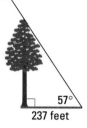

50. How far up a vertical wall can a 4-meter ladder reach if the angle it makes with the ground is 75°?

51. From eye level 2 meters off the ground and 25 meters from a sculpture, a sightseer on a bus has to look up at an angle of 20° to see the top of the sculpture. How high is the sculpture?

52. From the window of a building you look down at a clown in a parade. If your eyes are 60 feet above the clown, and the angle of sight, measured from the building, is 25°, how far is the clown from you?

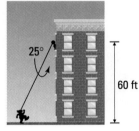

Objective J: *Determine components of vectors in real situations.* *(Lesson 13-8)*

53. A boat is moving at 60 knots on a course 72° north of east. Find the northern and eastern components of its velocity.

54. A helicopter has to fly to a place 15 km west and 7 km south of its present location.

 a. In what direction should it fly?

 b. How far will it travel?

REPRESENTATIONS DEAL WITH PICTURES, GRAPHS, OR OBJECTS THAT ILLUSTRATE CONCEPTS.

There are no representation objectives for this chapter.

14

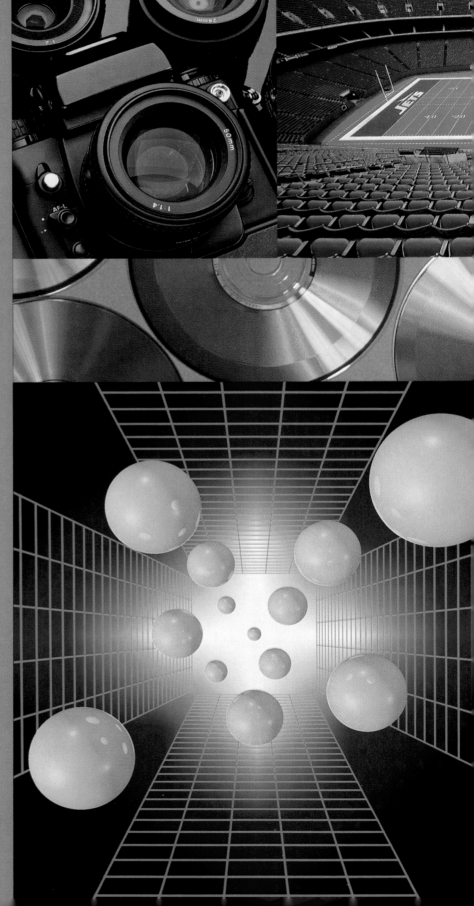

FURTHER WORK WITH CIRCLES

The great Italian scientist Galileo Galilei wrote in 1623:

. . . the universe . . . is written in the language of mathematics, and its characters are triangles, circles, and other geometrical figures, without which it is humanly impossible to understand a single word of it; without these, one is wandering about in a dark labyrinth.

Galileo's views have been mirrored in cultures around the world. The ancient Greeks believed that the sun, planets, and other celestial objects went around Earth in circles. The ancient Chinese believed that two forces, yin and yang, represented in the drawing below at the left, worked to create all that is. Today's Olympic Games symbol consists of five interlocking circles.

Circles involve ideas from many of the previous chapters of this book, so the study of circles is especially appropriate as a last chapter. You have seen circles arise from a variety of sources.

as cross-sections of cylinders, cones, and other 3-dimensional surfaces;

as limits of regular polygons;

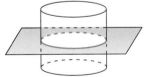

$n = 4$

$n = 8$

$n = 12$

$n = 16$

as shortest paths on the surface of a sphere;

as a set of points at a fixed distance from a center;

and as the graph of the solutions (x, y) to $(x - h)^2 + (y - k)^2 = r^2$.

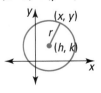

This chapter examines arcs of circles, and segments and angles associated with circles. The results are fascinating and often quite surprising; they are not obvious, yet not difficult to prove.

LESSON

14-1

*Chord
Length and
Arc
Measure*

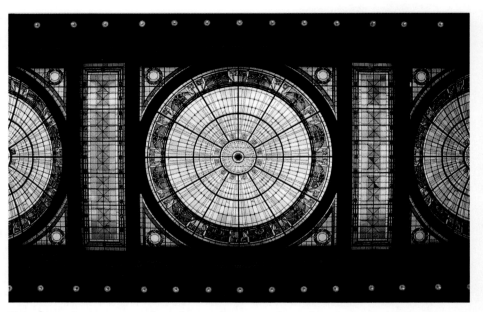

Heads up! *Arcs, central angles, and sectors abound in this stained-glass panel in the roof of Penn Station in Baltimore, Maryland.*

Reviewing Central Angles, Chords, and Arcs

Recall some names of figures related to circles. An angle with its vertex at the center of a circle is a *central angle* of the circle. The arc of the circle in the interior of the angle is said to be **intercepted** by the angle. The *measure of the intercepted arc* is defined as the measure of its central angle.

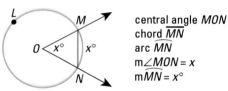

central angle MON
chord $\overline{MN}$
arc $\overset{\frown}{MN}$
$m\angle MON = x$
$m\overset{\frown}{MN} = x°$

In the drawing above, $m\angle MON$ seems to be about 45, thus arc $\overset{\frown}{MN}$ also measures 45°. $\overset{\frown}{MN}$ is a minor arc because its measure is less than 180°. The *major arc* $\overset{\frown}{MLN}$ has measure 360° − 45°, or 315°. An arc with measure 180° is a *semicircle*. If $\overset{\frown}{MN}$ is an arc of a circle, the segment $\overline{MN}$ is called the *chord of arc* $\overset{\frown}{MN}$.

Congruent Circles and Congruent Arcs

If two circles $\odot X$ and $\odot Y$
have equal radii, then
$\odot X$ can be mapped onto $\odot Y$
by the translation vector $\overrightarrow{XY}$.
So $\odot X$ and $\odot Y$ are congruent.

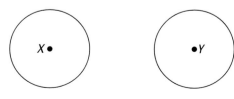

If the two circles do not have equal radii, no isometry will map one onto the other, since isometries preserve distance. These arguments prove that *circles are congruent if and only if they have equal radii.*

Suppose two arcs $\overset{\frown}{AB}$ and $\overset{\frown}{CD}$ have the same measure, as in the circle at the right. Then you can rotate $\overset{\frown}{AB}$ about O by the measure of $\angle AOC$ to the position of $\overset{\frown}{CD}$. Then the chord $\overline{AB}$ rotates to $\overline{CD}$, and $\overline{AB} \cong \overline{CD}$. Thus, in a circle, arcs of the same measure are congruent and have congruent chords. This proves part (1) of the next theorem for arcs in the same circle. The proof of part (2) is left for you as Question 15.

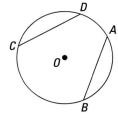

Arc-Chord Congruence Theorem
In a circle or in congruent circles,
(1) if two arcs have the same measure, they are congruent and their chords are congruent.
(2) if two chords have the same length, their minor arcs have the same measure.

In circles with *different* radii, however, arcs of the same measure are not congruent, nor are their chords congruent. This is pictured below. Arcs with the same measure are similar, though, because one can be mapped onto the other by a composite of a translation, a size change, and a rotation.

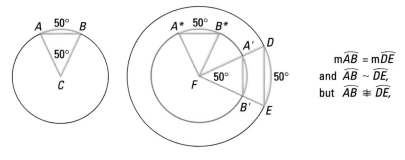

$$m\overset{\frown}{AB} = m\overset{\frown}{DE}$$
and $\overset{\frown}{AB} \sim \overset{\frown}{DE}$,
but $\overset{\frown}{AB} \not\cong \overset{\frown}{DE}$,

Curlers on ice. *Four concentric circles form the target in the sport of curling.*

Activity

Refer to the figure above. Describe a translation, rotation, and size change whose composite maps $\overset{\frown}{AB}$ onto $\overset{\frown}{DE}$.

Properties of Chords

As shown in $\odot O$ at the left, when $\overset{\frown}{AB}$ is not a semicircle, $\triangle ABO$ is isosceles. You have learned that in an isosceles triangle, the bisector of the vertex angle, the perpendicular bisector of the base, the altitude from the vertex, and the median from the vertex all lie on the same line. (Recall the Isosceles Triangle Coincidence Theorem in Lesson 6-2.) In the language of circles and chords, this leads to the following theorem.

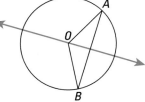

Proof
Each part is only a restatement of a property of isosceles triangles. You are asked in Question 10 to fill in the blanks in the proof below.
(1) This says the altitude of an isosceles triangle is also the median.
(2) This says the _?_ of an isosceles triangle is also a(n) _?_.
(3) This says the _?_ of an isosceles triangle is also a(n) _?_.
(4) This says the _?_ of an isosceles triangle is also a(n) _?_.

How to Find the Length of a Chord

Given the measure of an arc, can the length of its chord be found? This can always be done using trigonometry if the radius of the circle is known. In fact, this was the kind of problem that first led to trigonometry.

Example 1

Find the length of a chord of a 103° arc in a circle of radius 20 cm.

Solution

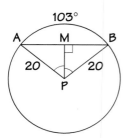

A picture is drawn at the left. The length of $\overline{AB}$ is desired. $\triangle PAB$ is isosceles with vertex angle P. Let M be the midpoint of $\overline{AB}$. Then $\overline{PM}$ bisects $\angle APB$ and $\overline{PM} \perp \overline{AB}$. Thus $\triangle APM$ is a right triangle and $m\angle APM = \frac{103}{2} = 51.5$.

To find AM, use trigonometry.
$$\sin 51.5° = \frac{AM}{20}$$
$$AM = 20 \cdot \sin 51.5° \approx 15.65$$
Since M is a midpoint, $AB = 2 \cdot AM \approx 31.30$ cm.

If an arc in a circle has a measure of 60°, 90°, or 120°, its length can be found without trigonometry.

Example 2

A circle has radius 10″. Find the length of a chord of
a. a 60° arc.　　　**b.** a 90° arc.　　　**c.** a 120° arc.

▶

Solution

For each situation, draw a picture.

a. △AOB is an isosceles triangle with a 60° vertex angle, so it is equilateral. AB = 10".

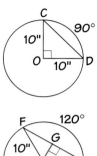

b. Since m∠COD = 90, △COD is an isosceles right triangle. So CD = $\sqrt{2} \cdot 10"$ ≈ 14.14".

c. Since m∠FOE = 120, m∠F = 30 and m∠E = 30. Draw altitude $\overline{OG}$ to the base of △OEF. The two triangles formed are 30-60-90 triangles.
OE = 2 · OG, so OG = 5".
GE = $\sqrt{3} \cdot OG$, so GE = $5\sqrt{3}$". Therefore, FE = 2 · GE = $10\sqrt{3}$" ≈ 17.32".

Recall that a polygon whose vertices lie on a given circle is called an inscribed polygon. So Example 2 provides a way of finding the length of a side of an inscribed regular hexagon, square, or equilateral triangle if you know the radius of the circle. The center of the inscribed regular polygon is the center of the circle.

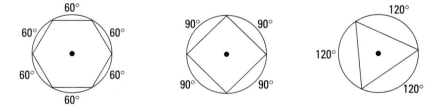

QUESTIONS

Covering the Reading

1. The measure of a minor arc of a circle is between __?__ and __?__.

2. The measure of a major arc of a circle is between __?__ and __?__.

3. *Multiple choice.* Two circles are congruent if and only if their radii are
   (a) parallel.
   (b) perpendicular.
   (c) equal in length.
   (d) none of these.

4. *True or false.* In $\odot Z$ and $\odot A$ below, $m\angle Z = m\angle A$ and $ZY > AC$.

    a. $m\widehat{XY} = m\widehat{BC}$
    b. $\widehat{XY} \sim \widehat{BC}$
    c. $\widehat{XY} \cong \widehat{BC}$

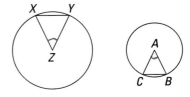

5. Give the results you found in the Activity in this lesson.

In 6–9, use $\odot O$ pictured at the right.

6. If $m\angle POQ = 98$, then $m\widehat{PQ} = \underline{\ ?\ }$.

7. Explain why $\triangle OPQ$ is isosceles.

8. If $\overrightarrow{OM} \perp \overline{PQ}$ then $\overrightarrow{OM} \underline{\ ?\ } \overline{PQ}$.

9. If $\overrightarrow{OM}$ bisects $\angle POQ$, then $\overrightarrow{OM} \underline{\ ?\ } \overline{PQ}$.

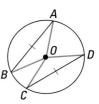

10. Fill in the blanks in the proof of the Chord-Center Theorem.

In 11–14, use $\odot O$ with radius 25 m shown at the left. Find the length of a chord of

11. a 53° arc.   **12.** a 60° arc.   **13.** a 90° arc.   **14.** a 120° arc.

Applying the Mathematics

15. Use the figure at the right to complete this proof of part (2) of the Arc-Chord Congruence Theorem.
Given: $AB = CD$.
To prove: $m\widehat{AB} = m\widehat{CD}$.
(Hint: The measure of an arc equals the measure of its central angle.)

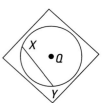

16. The circular hole with center Q shown at the right has radius 3 feet. A 5-foot board $\overline{XY}$ is to be wedged into the hole. What will be the distance from Q to $\overline{XY}$? (Hint: Draw some auxiliary radii and a perpendicular.)

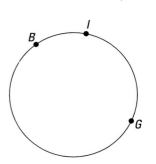

17. Trace the circle through points B, I, and G, or draw one like it by hand or with an automatic drawer.
    a. Draw the $\perp$ bisector of $\overline{BI}$.
    b. Draw the $\perp$ bisector of $\overline{IG}$.
    c. At what point do the lines of parts **a** and **b** intersect?
    d. Which part of which theorem justifies your answer to part **c**?

Notice the circular windows with square frames on the Lampoon Building similar to the situation in Question 16.

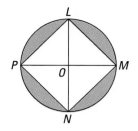

18. At the left, *LMNP* is a square inscribed in ⊙*O*. Suppose the radius of the circle is 1.
  a. Find the length of a side of the square.
  b. Find the area of the shaded region.

19. a. A regular hexagon is inscribed in a circle of radius 1. What is its perimeter?
  b. A regular hexagon is inscribed in a circle of radius *x*. What is its perimeter?

Review

20. The diagonals of trapezoid *ABCD* intersect at point *E*, as pictured below. Is △*ABE* ~ △*CDE*? Explain why or why not.   *(Lesson 13-2)*

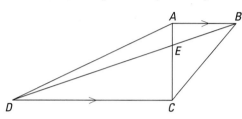

21. What is the volume of a square pyramid with base sides 2 m and height 1.4 m?   *(Lesson 10-7)*

22. In the figure at the left, find m∠*A*.   *(Lesson 5-7)*

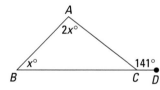

In 23 and 24, use ⊙*Q* at the right.   *(Lesson 3-2)*
23. If m$\widehat{ABC}$ = 178° and m$\widehat{BC}$ = 94°, find m$\widehat{AB}$.

24. m$\widehat{BA}$ + m$\widehat{BC}$ + m$\widehat{CD}$ + m$\widehat{AD}$ = __?__

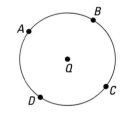

Exploration

25. This question extends Question 18 to polygons of more than 4 sides.

  a. Suppose a regular octagon is inscribed in a circle of radius 1. What is the area of the region between the circle and the octagon? (Hint: use the SAS Area Theorem.)
  b. Select any regular polygon with more than 8 sides and inscribe it in a circle of radius 1. Find the area between the regular polygon and the circle.

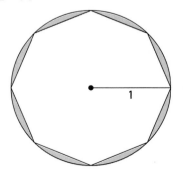

14-2

The Inscribed Angle Theorem

Picture perfect. *Sometimes in order to fit an entire building in a picture, the photographer must be located in a hard-to-reach place.*

A Situation Leading to Inscribed Angles

The **picture angle** of a camera lens is a measure indicating how wide a field of vision can be captured in one photo. A normal camera lens in a 35-mm camera has a picture angle of 62°. A wide-angle lens may have a picture angle as large as 118°. A telephoto lens has a smaller picture angle, perhaps 18°.

Here is a situation in photography. You want to take a picture of a building, and you want to get the entire front of the building in your picture. (Assume the height of the building is not a problem.) Suppose your camera has only one normal lens with a 62° field. The diagram at the right pictures the situation as seen from above.

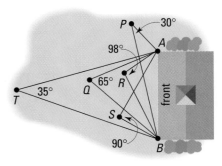

Let A and B be the endpoints of the building. If you stand at Q, m$\angle AQB$ is larger than 62, so the picture cannot include the entire front of the building. The same is true for points R and S. But at points P and T, the whole building easily fits in the picture. There is a natural question: where are all the points that are the vertex of a 62° angle through points A and B? The answer is surprising. It is found by considering *inscribed angles* in a circle.

What Are Inscribed Angles?

Definition
An angle is an **inscribed angle** in a circle if and only if
a. the vertex of the angle is on the circle, and
b. each side of the angle intersects the circle
at a point other than the vertex.

inscribed angle *ABC*

Activity

a. Below, inscribed angles *A*, *B*, and *C* intercept the same arc $\overset{\frown}{MN}$. Use a
protractor to measure these angles. (You will need to extend sides.)
Then measure $\angle MON$ to determine m$\overset{\frown}{MN}$.

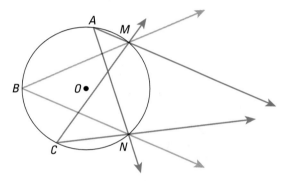

b. If you have access to dynamic software, create points *A* and *B* on
circle *O*. Choose a point *C* on circle *O* but not on minor arc $\overset{\frown}{MN}$.
Measure $\angle MCN$ and minor arc $\overset{\frown}{MN}$. Move point *C* around the major
arc $\overset{\frown}{MN}$. What happens to m$\angle MCN$?

In the Activity above, you should find that all the inscribed angles have
the same measure, and that the intercepted arc has twice that measure.
This is true of all inscribed angles and their intercepted arcs.

Inscribed Angle Theorem
In a circle, the measure of an inscribed angle is one-half the measure
of its intercepted arc.

Proof:
The steps of the argument depend on the position of the circle's center
O relative to the inscribed $\angle ABC$. There are three possibilities. They
are referred to as Case I, Case II, and Case III.

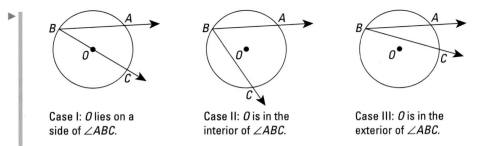

Case I: O lies on a side of $\angle ABC$.

Case II: O is in the interior of $\angle ABC$.

Case III: O is in the exterior of $\angle ABC$.

For all three cases what is given and what is to prove are the same.

Given: $\angle ABC$ inscribed in $\odot O$.

To prove: $m\angle ABC = \frac{1}{2} \cdot m\widehat{AC}$.

Argument:

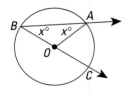

Case I: Draw the auxiliary segment $\overline{OA}$. Since $\triangle AOB$ is isosceles, $m\angle B = m\angle A$. Call this measure x. By the Exterior Angle Theorem, $m\angle AOC = 2x$.

Because the measure of an arc equals the measure of its central angle, $m\widehat{AC} = 2x = 2 \cdot m\angle B$.

Solving for $m\angle B$, $m\angle B = \frac{1}{2}m\widehat{AC}$.

Case I proves that $m\angle B = \frac{1}{2}m\widehat{AC}$ when one side of $\angle B$ contains the center of the circle. This fact is used in the proofs of the other cases.

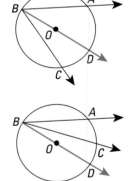

Case II: The auxiliary ray $\overrightarrow{BO}$ is needed.

$$
\begin{aligned}
m\angle ABC &= m\angle ABD + m\angle DBC & \text{Angle Addition Property} \\
&= \tfrac{1}{2}m\widehat{AD} + \tfrac{1}{2}m\widehat{DC} & \text{by the result in Case I} \\
&= \tfrac{1}{2}(m\widehat{AD} + m\widehat{DC}) & \text{Distributive Property} \\
&= \tfrac{1}{2}m\widehat{AC} & \text{Arc Addition}
\end{aligned}
$$

Case III: The proof is like the proof of Case II. Here $m\angle ABC = m\angle ABD - m\angle CBD$.

You are asked to complete the proof in Question 7.

Using the Inscribed Angle Theorem

Example 1

Four points A, B, C, and D split a circle into arcs with measures as shown. Find the measures of the angles of $ABCD$.

Solution

$m\angle A = \frac{1}{2}m\widehat{BCD} = \frac{1}{2}(100° + 148°) = 124°$

$m\angle B = \frac{1}{2}m\widehat{ADC} = \frac{1}{2}(62° + 148°) = 105°$

$m\angle C = \frac{1}{2}m\widehat{DAB} = \frac{1}{2}(62° + 50°) = 56°$

$m\angle D = \frac{1}{2}m\widehat{ABC} = \frac{1}{2}(50° + 100°) = 75°$

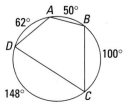

▶ **Check**
The four angle measures add to 360° as they should. Furthermore, the measures look correct.

The next example demonstrates a surprising and useful consequence of the Inscribed Angle Theorem.

Example 2

Let $\overset{\frown}{PQR}$ be a semicircle. Find m$\angle PQR$.

Solution
Complete the circle.
From the Inscribed Angle Theorem,
$\angle PQR = \frac{1}{2}m\overset{\frown}{PSR}$. Since $\overset{\frown}{PSR}$ is a
semicircle, m$\overset{\frown}{PSR} = 180°$. Thus
m$\angle PQR = \frac{1}{2} \cdot 180 = 90$.

Example 2 proves the following.

Theorem
An angle inscribed in a semicircle is a right angle.

What about the camera problem stated at the beginning of this lesson? The answer is given in Example 2 of the next lesson.

How many semicircles can you find in this Boston scene?

QUESTIONS

Covering the Reading

1. What is the *picture angle* of a camera lens?

2. Use the diagram at the right. A person stands at point *P* to take a picture of the house. Will the entire front of the house be in the picture
   a. if the person uses a normal camera lens?
   b. if the person uses a telephoto lens?
   c. if the person uses a wide-angle lens?

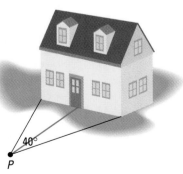

3. Use the circle below at the left. What is m∠ABC?

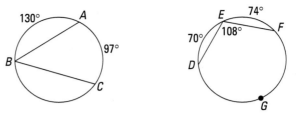

4. In the circle above at the right, what is m$\widehat{DGF}$?

5. Give the measurements of the angles and arcs you found in the Activity of this lesson.

6. The proof of the Inscribed Angle Theorem has three cases. How do the cases differ?

7. Finish this proof of Case III of the Inscribed Angle Theorem.

Conclusions	Justifications
m∠ABC = m∠ABD − m∠CBD	a.
$= \frac{1}{2}$m$\widehat{AD}$ − $\frac{1}{2}$m$\widehat{CD}$	b.
= c.	Distributive Property
$= \frac{1}{2}$m$\widehat{AC}$	Arc Addition

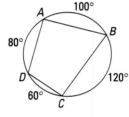

8. Find the measures of the four angles of the quadrilateral pictured at the left.

9. An angle inscribed in a semicircle has what measure?

In 10 and 11, $\overline{MN}$ is a diameter of ⊙O, as pictured below.

10. a. m$\widehat{ML}$ = _?_
b. m$\widehat{MPN}$ = _?_

11. a. m∠M = _?_
b. m∠N = _?_
c. m∠L = _?_
d. △LMN is a(n) _?_ triangle.

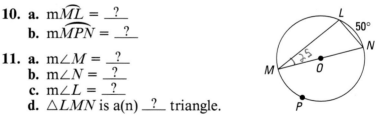

Applying the Mathematics

12. Use the figure at the right.
a. △TUV is _?_ in the circle.
b. m∠U = _?_

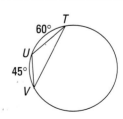

13. In the figure below, $\overline{YW}$ and $\overline{XZ}$ are diameters.
   a. m∠WOZ = __?__
   b. m∠Y = __?__

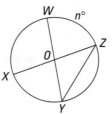

14. $\overline{AC}$ contains the center of the circle below. Calculate the measures of as many angles as you can.

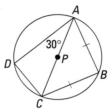

15. In ⊙O below, m∠M = 35 and m∠P = 92. If $MP = NQ$, find m $\widehat{MP}$.

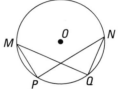

16. Fill in the justifications in this proof of the following theorem.

Theorem
In a circle, if two inscribed angles intercept the same arc, then they have the same measure.

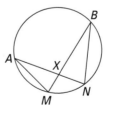

   Given: Inscribed angles AMB and ANB.
   To prove: m∠AMB = m∠ANB.

   Argument:

	Conclusions	Justifications
1.	m∠AMB = $\frac{1}{2}$m$\widehat{AB}$	a.
2.	m∠ANB = $\frac{1}{2}$m$\widehat{AB}$	b.
3.	m∠AMB = m∠ANB	c.

17. Use the given of Question 16 to prove that △AXM ~ △BXN.

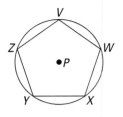

18. Find the perimeter of an equilateral triangle inscribed in a circle of radius 24. *(Lesson 14-1)*

19. Regular pentagon $VWXYZ$ is inscribed in $\odot P$ at the left.
a. What is m$\widehat{WX}$?
b. If $PW = 50$, find the perimeter of the pentagon. *(Lesson 14-1)*

20. Willow, Lake, and Central are east-west streets. Greenwood and Landwehr are north-south streets. Milwaukee is oblique to east-west streets. The distance on Milwaukee from Willow to Milwaukee and Central is 3 miles. What is the distance on Milwaukee from Lake to Willow? *(Lesson 13-3)*

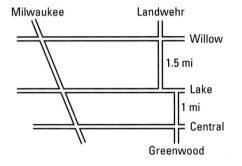

21. The length of a ballroom is twice its width, and its area is 280 square yards. How long is it from one corner of the ballroom to the opposite corner? *(Lessons 8-2, 8-6)*

22. Refer to the figure at the right. $LOGARITH\underline{M}$ is a regular nonagon. L is the midpoint of $\overline{NO}$. What is m$\angle MLN$? *(Lesson 6-7)*

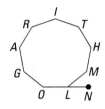

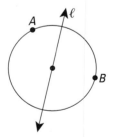

23. Trace the drawing at the left. Then draw line m so that $r_m \circ r_\ell(A) = B$. *(Lesson 4-5)*

24. The circle with center C and radius r is defined to be the set of all points whose distance from C is equal to r. Give a good definition of the *interior* of this circle. *(Lesson 2-4)*

25. Solve for x: $100n - 200 + x = 100n$. *(Previous course)*

26. Examine a camera in your household or in a store. What is its normal picture angle? (You may need to refer to the instruction manual.)

LESSON

14-3

Locating the Center of a Circle

Suppose you bought a circular coffee-table top and wished to place it on a revolving pedestal base so that it rotates evenly. To do this you need to match the center of the base and the center of the table. In Lesson 3-8 you learned how to draw perpendicular bisectors of segments. If two segments are chords of the same circle, and are not parallel, then the intersection of their perpendicular bisectors is the center of the circle. That method for locating the center of a circle is called the *perpendicular bisector method*.

The Right-Angle Method

A second method relies on a theorem proved in the previous lesson: If an inscribed angle is a right angle, then its intercepted arc is a semicircle. So the segment connecting the endpoints of the intercepted arc is a diameter. Draw two diameters and you have the center. We call this the *right-angle method* for locating the center of the circle.

Example 1

Locate the center of the circle drawn at the right.

Solution

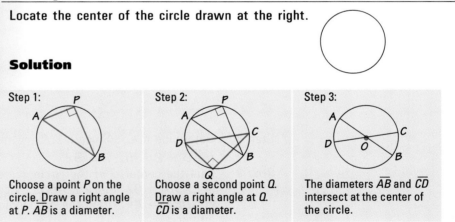

Step 1: Choose a point *P* on the circle. Draw a right angle at *P*. $\overline{AB}$ is a diameter.

Step 2: Choose a second point *Q*. Draw a right angle at *Q*. $\overline{CD}$ is a diameter.

Step 3: The diameters $\overline{AB}$ and $\overline{CD}$ intersect at the center of the circle.

This method is often used by people in drafting; the right angles are drawn with stiff metal instruments called *T-squares* or *ells*. You can use the corner of an index card or a piece of duplicating paper.

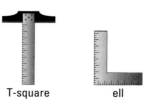

T-square   ell

Using the Inscribed Angle Theorem

A third method also uses the Inscribed Angle Theorem. This method is shown in Example 2.

Example 2

A camera has a 46° field of vision. Where are all the places a person could stand so the front of the building just fills the picture?

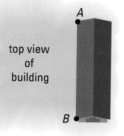

top view of building

Solution

Consider a point P so that $m\angle APB = 46$. The front of the building will just fit in the picture if you stand at point P. There is one circle that contains points P, A, and B. $\overline{AB}$ is a chord in that circle.

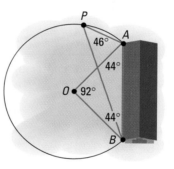

Step 1: Determine $m\widehat{AB}$. Since $m\angle P = \frac{1}{2}m\widehat{AB}$,
$m\widehat{AB} = 2 \cdot m\angle P = 92°$.

Step 2: Find the center O of the circle containing $\widehat{AB}$ such that $m\angle AOB = 92$. Since $\triangle AOB$ is isosceles, $m\angle OAB = m\angle OBA = 44$. Draw the 44° angles at A and B. Their sides intersect at O.

Step 3: Draw $\odot O$ with radius OA. Any point P on the major arc $\widehat{APB}$ of the circle will satisfy $m\angle APB = 46$.

From anywhere on major arc $\widehat{APB}$, the front of the building will just fit the picture. Inside the circle you will get only part of the building in your picture. Outside the circle you will see more than the building in your picture.

You can use trigonometry to determine the distance from certain positions of point P to the building.

Eagle eye. *Photographer Margaret Bourke White on the Chrysler Building, New York.*

Example 3

A photographer wishes to photograph a building 130' long with a lens that has an 80° field of vision. The picture is to be taken directly in front of the middle of the building. At least how far from the building will the photographer need to stand?

Solution

Let $\overline{QS}$ be the front of the building, as shown at the right. R is the midpoint of $\overline{QS}$. With the 80° lens the photographer needs to stand at a point P on the perpendicular bisector of $\overline{QS}$.

$QR = \frac{1}{2}QS = 65'$ and $PQ = PS$.

$\triangle PQS$ is isosceles and $\overrightarrow{PR}$ bisects $\angle QPS$. So $m\angle RPQ = 40$.

Now use trigonometry with right $\triangle PQR$.

$$\frac{65}{PR} = \tan 40°$$
$$65 = PR \cdot \tan 40°$$
$$\frac{65}{\tan 40°} = PR$$
$$77.5' \approx PR$$

If the photographer stands on $\overrightarrow{RP}$ 77.5' or more from the building, the entire front of the building will be in the picture.

Check

We know from right $\triangle PQR$ that $m\angle PQR = 50°$. From the solution, $\tan \angle PQR = \frac{PR}{RQ} \approx \frac{77.5}{65} \approx 1.19$. From a calculator, $\tan 50° \approx 1.19$. So the value of PR checks.

QUESTIONS

Covering the Reading

1. Trace the circle at the left below. Find its center using the right-angle method.

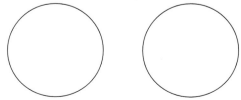

2. Trace the circle at the right above. Construct its center using the perpendicular bisector method.

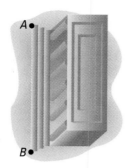

In 3 and 4, suppose you have a camera with a lens having a 56° field of vision.

3. Trace the monument at <u>the</u> left and diagram all the places where you could stand so that $\overline{AB}$, the monument's front, just fits into your picture.

4. Assume $AB = 60$ yards. If you wanted to stand in front of the middle of the monument and fit the entire monument into your picture, at least how far from the monument would you need to stand?

5. Each year, the classes at Emmy Noether High School take class pictures on the steps to the main entrance. The steps are 50 meters across.
   a. A photographer has an 88° wide-angle lens and wants to stand in front of the middle of the steps, as close to the students as possible. Where should the photographer stand?
   b. At the same time, parents of class members want to take pictures of the class. If they all have 88° wide-angle lenses and want the steps to fill the picture, where can they stand?

Applying the Mathematics

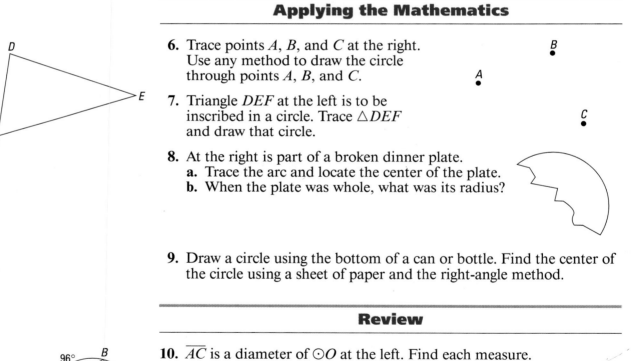

6. Trace points A, B, and C at the right. Use any method to draw the circle through points A, B, and C.

7. Triangle DEF at the left is to be inscribed in a circle. Trace $\triangle DEF$ and draw that circle.

8. At the right is part of a broken dinner plate.
   a. Trace the arc and locate the center of the plate.
   b. When the plate was whole, what was its radius?

9. Draw a circle using the bottom of a can or bottle. Find the center of the circle using a sheet of paper and the right-angle method.

Review

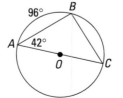

10. $\overline{AC}$ is a diameter of $\odot O$ at the left. Find each measure.
    a. m∠C　　　　b. m$\overarc{BC}$　　　　c. m$\overarc{AC}$　　(Lesson 14-2)

11. $BEHIVS$ is a regular hexagon. The diagonals from V are drawn.
    a. Find the measures of the numbered angles.
    b. *True or false.* $\triangle BEV$ is a right triangle.
    (Lesson 14-2)

12. *IJKL* is a quadrilateral inscribed in circle *P*. m∠*ILK* = *x*° and
m∠*IJK* = *y*°.
a. Which arc measures 2*x*°?
b. Which arc measures 2*y*°?
c. What number does 2*x* + 2*y* equal?
d. What is the sum of the measures of
angles *ILK* and *IJK*?
e. What is the sum of the measures of
angles *LIJ* and *LKJ*?
f. What does this prove about opposite angles
in an inscribed quadrilateral?   *(Lesson 14-2)*

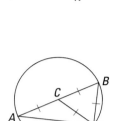

13. In the figure at the right, $\overline{AB}$ is a diameter of
⊙*C*, and △*BCD* is equilateral.
a. Find the measures of as many angles as
you can.
b. If *BD* = *x*, then what is the length
of $\overline{AC}$?
c. If *BD* = 7, what is the length of $\overline{AD}$?
(Lessons 5-7, 6-8, 13-5, 14-2)

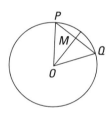

In 14 and 15, $\overline{PQ}$ is a chord of ⊙*O*.

14. a. If *M* is the midpoint of $\overline{PQ}$, then __?__.
b. If $\overline{OM} \perp \overline{PQ}$, then __?__.
c. If $\overline{OM}$ bisects ∠*POQ*, then __?__.
(Lesson 14-1)

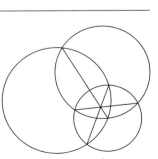

15. If *PQ* = 12 and *PQ* is 8 units from
the center, what is *OQ*?   *(Lesson 14-1)*

16. a. Give the contrapositive of this statement: *If a figure is a
rectangle, then its diagonals are congruent.*
b. Is the contrapositive true?   *(Lesson 11-2)*

17. Sod is to be put on a circular golf-course green, 50' in diameter.
How much sod is needed?   *(Lesson 8-8)*

18. If 10 people are at a party, and each person shakes hands once with
every other person there, how many handshakes are there?
(Lesson 6-8)

*Saint-Faustin golf course
in Laurentides, Quebec,
Canada.*

Exploration

---

19. Each of the three circles at the right
overlaps the other two. The three chords
common to each pair of circles are drawn.
They seem to have a point in common.
Using a compass and ruler or an automatic
drawer show other examples of this
situation and decide whether the chords
will always have a point in common.

Capital circularity. *Circular intersections as well as cross streets and diagonal avenues were part of the plans of Pierre-Charles L'Enfant, designer of Washington, D.C.*

Each side of a central angle or an inscribed angle intersects the circle in which it is found. In this lesson and the next one, other angles for which both sides intersect a circle will be discussed. The Inscribed Angle Theorem enables measures of these angles to be determined.

Angles Formed by Chords

First, consider an angle whose vertex is in the interior of a circle. For instance, consider the situation pictured at the right, where two chords $\overline{AB}$ and $\overline{CD}$ intersect at E. Given $m\widehat{AD} = x°$ and $m\widehat{BC} = y°$, what is $m\angle CEB$?

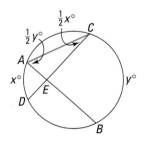

To find $m\angle CEB$, draw the auxiliary segment $\overline{AC}$. Now $\angle BAC$ and $\angle ACD$ are inscribed angles whose measures are half their intercepted arcs. So $m\angle BAC = \frac{1}{2}y$ and $m\angle ACD = \frac{1}{2}x$. $\angle CEB$ is an exterior angle of $\triangle ACE$. Its measure is the sum of the measures of $\angle BAC$ and $\angle ACD$.

Thus, $\qquad\qquad m\angle CEB = \frac{1}{2}x + \frac{1}{2}y$, or $\frac{1}{2}(x + y)$.

This argument proves the following startling theorem.

Angle-Chord Theorem
The measure of an angle formed by two intersecting chords is one-half the sum of the measures of the arcs intercepted by it and its vertical angle.

Example 1

In $\odot O$ below, $m\overset{\frown}{XRY} = 200°$ and $m\overset{\frown}{VW} = 60°$. Find $m\angle XQY$.

Solution

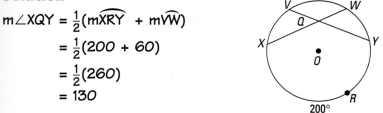

$m\angle XQY = \frac{1}{2}(m\overset{\frown}{XRY} + m\overset{\frown}{VW})$

$\qquad = \frac{1}{2}(200 + 60)$

$\qquad = \frac{1}{2}(260)$

$\qquad = 130$

Activity

In Example 1, explain why the given information does not uniquely determine $m\overset{\frown}{WY}$.

Angles Formed by Secants

The measure of an angle between chords can be found even when the lines containing the chords intersect in the exterior of the circle. Such lines are called *secants*.

Definition
A secant to a circle is a line that intersects the circle in two points.

Here is one way to find the measure of an angle formed by secants to the same circle. Below at the left, $\angle AEC$ is formed by two secants and intercepts the two arcs $\overset{\frown}{AC}$ and $\overset{\frown}{BD}$ with measures 84° and 26°. To find $m\angle E$, draw $\overline{AD}$, as pictured below at the right.

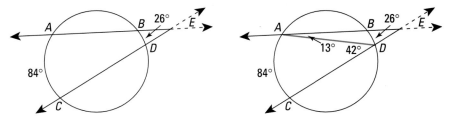

Again there are two inscribed angles. One of them is $\angle ADC$, an exterior angle to $\triangle ADE$. By the Exterior Angle Theorem,
$$m\angle DAE + m\angle E = m\angle ADC.$$
Solve for $m\angle E$. $\qquad\qquad m\angle E = m\angle ADC - m\angle DAE$

$\qquad\qquad\qquad\qquad m\angle E = \frac{1}{2}m\overset{\frown}{AC} - \frac{1}{2}m\overset{\frown}{BD}$

Substitute the given measures. $\qquad = \frac{1}{2}\cdot 84 - \frac{1}{2}\cdot 26$

$\qquad\qquad\qquad\qquad\qquad = 42 - 13$

So $m\angle E = 29$.

In solving for $m\angle E$, we have proved the following theorem.

Example 2

In $\odot R$ below, $\overset{\frown}{mKTM} = 195°$ and $\overset{\frown}{mJL} = 51°$. Find $m\angle P$.

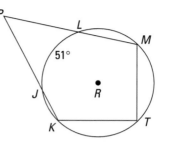

Solution
First, sort out which arcs are needed. $\angle P$ intercepts arcs $\overset{\frown}{KTM}$ and $\overset{\frown}{JL}$.
From the Angle-Secant Theorem,

$$m\angle P = \tfrac{1}{2}(\overset{\frown}{mKTM} - \overset{\frown}{mJL})$$
$$= \tfrac{1}{2}(195 - 51)$$
$$= \tfrac{1}{2} \cdot 144$$
$$= 72.$$

In finding measures of angles formed by chords or secants you have a
choice between drawing an auxiliary segment and calculating, or
applying a theorem. You should be able to do both.

QUESTIONS

Covering the Reading

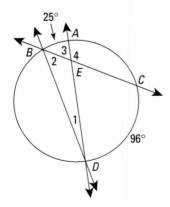

1. Use the figure at the left. Find each measure.
   a. $m\angle 1$
   b. $m\angle 2$
   c. $m\angle 3$
   d. $m\angle 4$

2. Use the figure at the right. Give the
   measure of the angle in terms of the
   measures of the arcs.
   a. $m\angle 5$
   b. $m\angle 6$
   c. $m\angle 7$
   d. $m\angle 8$

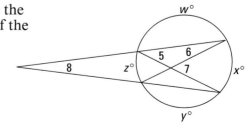

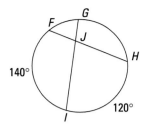

In 3 and 4, use the figure at the left.

3. Which additional arc measure(s) do you need in order to find m∠*FJI*?

4. Suppose $\overline{GI}$ is a diameter. Find m∠*FJI*.

5. Answer the question in the Activity of this Lesson.

6. In the figure at the right, m⌢*LP* = 80° and m⌢*MO* = 50°. Find each measure.
   a. m∠*MQO*
   b. m∠*N*

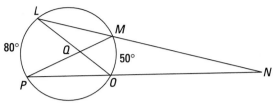

7. Define *secant*.

Applying the Mathematics

8. A point *P* is outside ⊙*O*. How many secants of ⊙*O* contain *P*?

9. In the circle at the left, m⌢*RU* = 101° and $\overline{RS} \perp \overline{TU}$. Find m⌢*ST*.

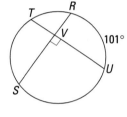

10. Rectangle *VWXY* is inscribed in the circle at the right, and m⌢*WX* = 40°.
   a. What is the measure of each acute angle formed by the intersection of the diagonals?
   b. What is m∠*VYW*?

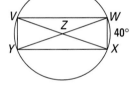

11. Use the figure at the right. If m⌢*BD* = 53° and m∠*C* = 45, what is m⌢*AE*?

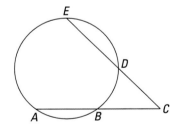

Review

12. Trace or copy the circular arc at the right. Then find its center to complete the whole circle. *(Lesson 14-3)*

13. Suppose you want to photograph the entire front of the house at the right with a camera that has a picture angle of 118°. At least how far in front of the middle of the house shown would you need to stand? *(Lesson 14-2)*

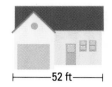
|—52 ft—|

14. Find the measure of each angle of pentagon *ABCDE* at the left. *(Lesson 14-2)*

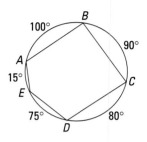

15. A chord is 7 inches away from the center of a circle whose radius is $12\frac{1}{2}$ inches. Find the length of the chord. *(Lesson 14-1)*

16. Use indirect reasoning to explain why no triangle in Euclidean geometry can have two 95° angles. *(Lesson 11-4)*

17. What is the value of *y* in the figure at the right? *(Lesson 5-7)*

18. a. State the Quadratic Formula.
b. Use the Quadratic Formula to solve $2x^2 + 5x - 1 = 0$. *(Previous course)*

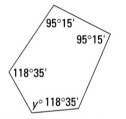

Exploration

19. At the right the sides of an inscribed pentagon *ABCDE* are extended to form a five-pointed star.
a. What is the sum of the measures of angles *F, G, H, I,* and *J* if *ABCDE* is regular?
b. What is the largest and smallest this sum can be if *ABCDE* is not regular?

20. Repeat Question 19 for a five-pointed star inscribed in a circle.

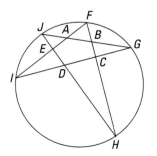

How far can you see? *Shown are Barbra Streisand and Jack Nicholson in a scene from the movie* On A Clear Day You Can See Forever, *but even on a clear day you cannot see forever because Earth is round. See the Example.*

What Is a Tangent to a Circle?

The word *tangent* comes from the Latin word meaning "touching." It has two meanings in geometry. One meaning is a ratio of sides in a right triangle. A second meaning is more directly related to touching. Think of a wheel (circle) as tangent to a ramp (a line) as it rolls up or down the ramp. If the wheel and the ramp are very hard, they are thought to have only one point in common in a given plane.

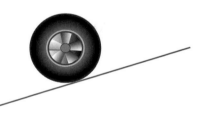

> **Definition**
> A **tangent to a circle** is a line in the plane of the circle which intersects the circle in exactly one point (the **point of tangency**).

Tangents to circles can be constructed easily, due to the following theorem.

> **Theorem**
> If a line is perpendicular to a radius of a circle at the radius's endpoint on the circle, then the line is tangent to the circle.

Proof

Drawing: Here is the theorem restated in terms of a figure.

Given: ⊙O with point P on it, $\overline{OP} \perp \ell$.

To prove: ℓ is tangent to the circle.

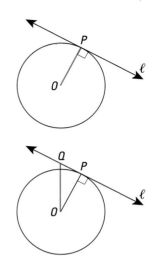

Argument:
Either ℓ is tangent (it intersects the circle at P only), or ℓ also intersects the circle at another point Q. Suppose Q is another point on ℓ and on the circle. Since Q is on ℓ, △OPQ is a right triangle with hypotenuse OQ. So $OQ > OP$. But since Q is on the circle, $OQ = OP$. The statements $OQ > OP$ and $OQ = OP$ are contradictory. By the Law of Indirect Reasoning, the supposition must be false. Thus, ℓ is tangent to the circle.

The converse of this theorem is true, but its proof is longer. Again we use an indirect argument.

Theorem

If a line is tangent to a circle, then it is perpendicular to the radius containing the point of tangency.

Proof

Drawing: What is given and what is to be proved are stated in terms of the figure at the right.

Given: m is tangent to ⊙O at point P.

To prove: $\overline{OP} \perp m$.

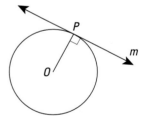

Argument:
Suppose $\overline{OP}$ is not perpendicular to m. Then there is a different segment $\overline{OQ}$, with $\overline{OQ} \perp m$ and Q on m. Locate R on m so that Q is between R and P and $QR = QP$. Then △OQR ≅ △OQP because of the SAS Congruence Theorem. So $OR = OP$ by the CPCF Theorem. This means R is on ⊙O (it is the same distance from O as P is). So m contains two points on the circle. Thus m is not a tangent, which contradicts the given. So, by the Law of Indirect Reasoning, $\overline{OP} \perp m$.

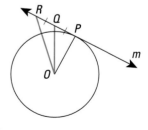

The two theorems of this lesson can be combined into one biconditional.

Radius-Tangent Theorem

A line is tangent to a circle if and only if it is perpendicular to the radius at the radius's endpoint on the circle.

Activity

Trace or copy this circle. Draw the lines tangent to the circle at points *A*, *B*, and *C*.

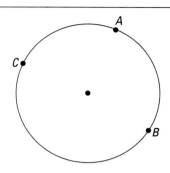

Tangents to Spheres

As you know, many properties of 2-dimensional figures extend to 3-dimensional figures. The idea of tangency extends very easily to spheres. A **tangent to a sphere** is a line or a plane which intersects the sphere in exactly one point. A common example of a plane tangent to a sphere is a ball in one of its positions as it rolls down a ramp.

You can apply the Radius-Tangent Theorem to calculate how far you can see from the top of a building or hill (assuming nothing blocks your view).

Example

How far is it to the horizon from a point *P* that is 200 feet above the ground?

Solution

We assume Earth is a sphere with center *C* and radius 3960 miles. Any tangent line from *P* to Earth intersects Earth at what we call the horizon. Let *T* be a point on the horizon and *X* be the point on Earth directly below *P*. We wish to calculate *PT*.

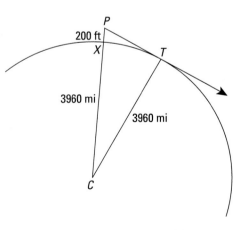

P, *C*, and *T* determine a plane. Drawn here is part of the cross-section of Earth determined by that plane. Notice that the figure drawn here is *very* distorted to fit everything into it.

Because of the Radius-Tangent Theorem, $\angle CTP$ is a right angle. So *PT* can be found by using the Pythagorean Theorem. To find *CP*, start by converting the 200 ft to miles.

$$200 \text{ ft} = 200 \text{ ft} \cdot \frac{1 \text{ mi}}{5280 \text{ ft}} = \frac{200}{5280} \text{ mi} \approx 0.038 \text{ mi}$$

▶

▶ This means that $CP \approx 3960.038$ miles. The little amount .038 is important. Don't round here!

From the Pythagorean Theorem,
$$PT^2 + 3960^2 \approx 3960.038^2.$$
$$PT^2 \approx 3960.038^2 - 3960^2$$
$$\approx 301$$
So
$$PT \approx \sqrt{301} \approx 17.3.$$
It is approximately 17 miles to the horizon from a point 200 feet high.

The numbers in the Example are difficult to deal with. It would be nice to have a formula that gives the distance to the horizon from a point *h* units above the surface of Earth. For these reasons, it is desirable to use variables to obtain a general formula. But then, why work only with Earth? Letting *r* be the radius of any sphere and *h* be the height above the sphere (in some units), the distance to the horizon can be shown to be $\sqrt{2rh + h^2}$. (In the Example, $r = 3960$ mi and $h = 0.038$ mi.) You are asked to derive this formula in Question 15.

QUESTIONS

Covering the Reading

1. **a.** According to the definition, when is a line tangent to a circle?
   b. According to the Radius-Tangent Theorem, when is a line tangent to a circle?

2. At the left, ℓ is tangent to $\odot O$ at P. Must ℓ be perpendicular to $\overline{OP}$?

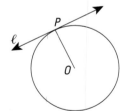

3. In the proof of the first theorem of this lesson, what contradiction is reached?

4. *Multiple choice.* The first two theorems of this lesson are
   (a) converses of each other. (b) inverses of each other.
   (c) contrapositives of each other. (d) negations of each other.

5. To prove the theorems in this lesson, what logical principle was applied?

6. Refer to the Example.
   a. Why is $\overline{PT} \perp \overline{TC}$?
   b. Where does the distance 3960 miles come from?

7. If Earth were flat, how far could you see from a point 2 meters above ground level, assuming nothing was in the way?

8. How far is the horizon from the observation deck of the Society Center in Cleveland, Ohio, which is 888 feet above ground level?

In 9 and 10, give a real-world example of the given mathematical idea.

9. a line tangent to a circle   10. a plane tangent to a sphere

Applying the Mathematics

11. **a.** Show your drawing from the Activity in this lesson.
    b. Label the triangle formed by the tangent lines $\triangle DEF$. Then draw the angle bisectors of $\triangle DEF$. What is their point of concurrency?

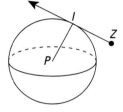

12. In the figure at the left, $\overrightarrow{ZI}$ is tangent to sphere P at point I. How many other rays from point Z are tangent to sphere P?

13. **a.** Extend the Radius-Tangent Theorem of this lesson to apply to spheres.
    b. Is the extension true?

14. Determine how far it is to the horizon from a point 100 meters above the surface of the moon. (The radius of the moon is about 1750 kilometers.)

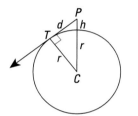

15. Use the figure at the left to show that if $\overline{PT}$ is tangent to circle C at point T, then $d = \sqrt{2rh + h^2}$.

16. Refer to the figure at the right and write a proof argument.
    Given: Point P outside $\odot N$; $\overline{PX}$ and $\overline{PY}$ are tangent to $\odot N$ at points X and Y.
    To prove: $PXNY$ is a kite. (This will prove that the two tangents to circle N from any point P have the same length. This is a result you should remember.)

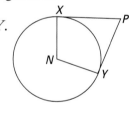

17. Copy the figure below and draw the four tangents that circles O and P have in common. These are called **common tangents** to the circles.

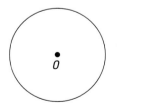

18. At the right, $\overline{PT}$ is tangent to $\odot O$ at T. $PT = 12$ and $PO = 15$.
    a. What is the area of the circle?
    b. What is the distance from P to N, the nearest point on the circle?

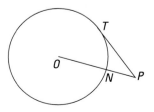

19. Use the figure below. Find the measure of the indicated angle.
(Lessons 14-2, 14-4)

   a. ∠1   **b.** ∠2   **c.** ∠3   **d.** ∠E

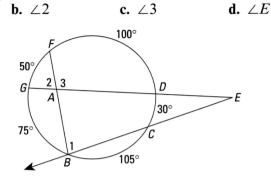

20. An 8″ chord is 3″ from the center of the circle as shown at the left. What is the area of the circle? *(Lessons 8-8, 14-1)*

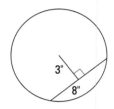

21. Below, $ABCD \sim JKHG$. Find as many missing lengths and angle measures as you can. *(Lesson 12-5)*

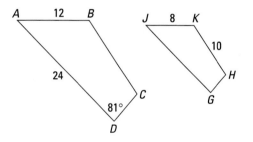

22. The measures of the exterior angles of a quadrilateral are $4x$, $8x - 23$, $9x + 7$, and $5x - 40$.
   a. What is x?
   b. What is the measure of each interior angle? *(Lesson 7-9)*

23. A polygon is circumscribed about a circle if each of its sides is tangent to the circle. At the left, a quadrilateral has been circumscribed about circle O.

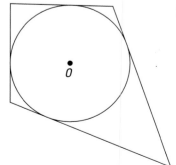

   a. Experiment with various circles and polygons. Tell whether you think these conjectures are *true or false.*
     i. If a quadrilateral is circumscribed about a circle, then at least two of its sides are congruent.
     ii. If a parallelogram is circumscribed about a circle, then it is a square.
     iii. If an isosceles trapezoid is circumscribed about a circle, then it is a rectangle.
     iv. A polygon with an odd number of sides greater than 3 cannot be circumscribed about a circle.
   b. Prove at least one of the conclusions you make.

14-6

Angles Formed by Tangents

Big wheels. *On most bicycle wheels that have ever been built, spokes are tangent to a circular axle in the middle of the wheel.*

Angles Formed by a Tangent and a Chord

In Lesson 14-4, angles were formed by chords or secants. In this lesson, angles which have at least one side tangent to a circle are explored. Consider first the angle formed by a tangent and a chord. Suppose $\overleftrightarrow{BC}$ is tangent to $\odot O$ at B and $m\overarc{AB} = 75°$, as pictured here.

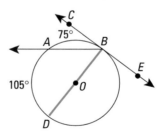

To find $m\angle ABC$, draw the diameter containing B and O. The semicircle $\overarc{BAD}$ has measure 180°, so $m\overarc{AD} = 105°$. Thus $m\angle ABD = \frac{1}{2} \cdot 105 = 52.5$. Now, since $\overleftrightarrow{CB} \perp \overline{BD}$, $\angle CBA$ is complementary to $\angle ABD$. So $m\angle CBA = 37.5$. In general,

$$\begin{aligned} m\angle ABC &= 90 - m\angle ABD \\ &= \tfrac{1}{2} \cdot 180 - \tfrac{1}{2} \cdot m\overarc{AD} \\ &= \tfrac{1}{2}(180 - m\overarc{AD}) \\ &= \tfrac{1}{2}m\overarc{AB}. \end{aligned}$$

This argument proves the following theorem.

Example 1

In the figure at the right, $\overleftrightarrow{AB}$ is tangent to circle O at A. If $m\angle BAC = 80$, find $m\widehat{AC}$.

Solution

From the Tangent-Chord Theorem,

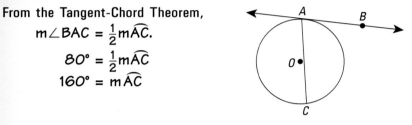

$$m\angle BAC = \tfrac{1}{2}m\widehat{AC}.$$
$$80° = \tfrac{1}{2}m\widehat{AC}$$
$$160° = m\widehat{AC}$$

Angles Formed by Tangents and Secants

Measures of angles between tangents and secants are calculated the same way we calculate measures of angles between secants.

Proof

Here is a proof for an angle between a secant and tangent. The proof for an angle between two tangents is similar. It is left to you as Question 10.

Drawing: Secant $\overleftrightarrow{EB}$ and tangent $\overleftrightarrow{EC}$ form $\angle E$, as shown below.

Given: $m\widehat{AC} = x°$ and $m\widehat{BC} = y°$.

To prove: $m\angle E = \tfrac{1}{2}(x - y)$.

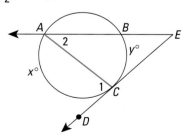

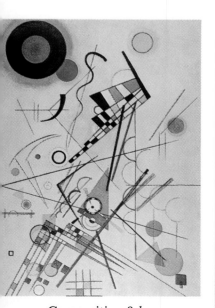

Composition 8 *by Russian artist Wassily Kandinsky (1866-1944)*

Argument:

Conclusions	Justifications
1. Draw $\overline{AC}$.	Point-Line-Plane Postulate
2. $m\angle 1 = \frac{1}{2}x$	a
3. $m\angle 2 = \frac{1}{2}y$	b
4. $m\angle 1 = m\angle 2 + m\angle E$	c
5. $m\angle E = m\angle 1 - m\angle 2$	d
6. $m\angle E = \frac{1}{2}x - \frac{1}{2}y$	e
7. $m\angle E = \frac{1}{2}(x - y)$	f

Question 8 asks you for the justifications for steps 2–7.

Example 2

Refer to the figure below. $\overrightarrow{AB}$ is tangent to the circle at B. Find $m\angle A$.

Solution

The circle measures 360°, so

$$m\widehat{BD} = 360° - 160° - 60°$$
$$= 140°.$$

$$m\angle A = \frac{1}{2}(m\widehat{BD} - m\widehat{BC})$$
$$= \frac{1}{2}(140° - 60°) = 40°$$

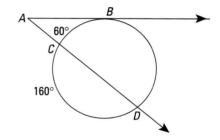

Angles Formed by Two Tangents

The two tangents to a circle from a point determine the measure of a smaller intercepted arc, a larger intercepted arc, and the angle formed by the tangents. If you know any one of these, you can find the other two.

Example 3

In the figure at the right, $m\angle P = 30$. What is $m\widehat{QSR}$?

Solution

Let $m\widehat{QSR} = x°$. Then $m\widehat{QTR} = (360 - x)°$.
By the Tangent-Secant Theorem,

$$m\angle P = \frac{1}{2}(m\widehat{QTR} - m\widehat{QSR})$$
$$= \frac{1}{2}((360 - x) - x)$$

Substitute for $m\angle P$ and solve for x.

$$30 = \frac{1}{2}(360 - 2x)$$
$$30 = 180 - x$$

So $\qquad x = 150°.$

Check

When $x = 150$, then $360 - x = 210$. Does $m\angle P$ equal half the difference of the two arcs? Yes, 30 is half of $210 - 150$.

You now have studied all the measures of angles made by lines that intersect a circle. You are asked to summarize the relevant theorems in Question 12.

QUESTIONS

Covering the Reading

In 1 and 2, use the diagram at the right in which $\overleftrightarrow{BC}$ is tangent to $\odot A$ at B, and m$\overparen{BD} = 110°$.

1. Find m$\angle ABC$.

2. Find m$\angle DBC$.

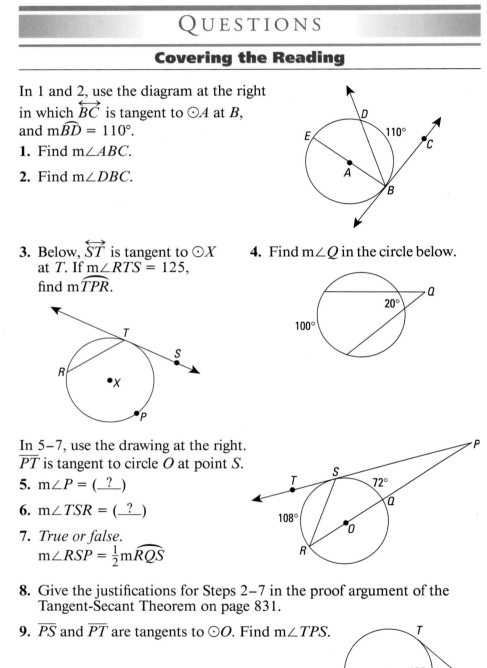

3. Below, $\overleftrightarrow{ST}$ is tangent to $\odot X$ at T. If m$\angle RTS = 125$, find m$\overparen{TPR}$.

4. Find m$\angle Q$ in the circle below.

In 5–7, use the drawing at the right. $\overline{PT}$ is tangent to circle O at point S.

5. m$\angle P = (\underline{})$

6. m$\angle TSR = (\underline{})$

7. *True or false.*
m$\angle RSP = \frac{1}{2}$m$\overparen{RQS}$

8. Give the justifications for Steps 2–7 in the proof argument of the Tangent-Secant Theorem on page 831.

9. $\overline{PS}$ and $\overline{PT}$ are tangents to $\odot O$. Find m$\angle TPS$.

In 10 and 11, use the drawing below in which $\overrightarrow{PQ}$ and $\overrightarrow{PR}$ are tangents to $\odot O$ at points Q and R.

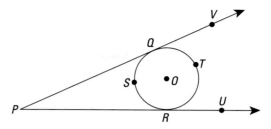

10. Prove the second part of the Tangent-Secant Theorem. That is, prove that $m\angle P = \frac{1}{2}(m\overset{\frown}{QTR} - m\overset{\frown}{QSR})$.
    (Hint: Draw $\overline{QR}$ and use an exterior angle of $\triangle PQR$.)

11. If $m\angle P = 25$, find the measures of arcs $\overset{\frown}{QSR}$ and $\overset{\frown}{QTR}$.

Applying the Mathematics

12. Match each angle at the left with a way to compute its measure at the right.
    a. angle between two chords
    b. angle between two secants
    c. angle between two tangents
    d. angle between secant and tangent
    e. angle between chord and tangent
    f. inscribed angle
    g. central angle

    (i) the intercepted arc
    (ii) $\frac{1}{2}$ the intercepted arc
    (iii) $\frac{1}{2}$ the sum of the intercepted arcs
    (iv) $\frac{1}{2}$ the difference of the intercepted arcs

13. In the figure at the right, $\overrightarrow{PR}$ is tangent to $\odot O$ at R. $\overrightarrow{PQ}$ contains O. If $m\angle P = 41$, what is $m\overset{\frown}{QR}$?

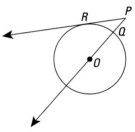

14. At the right, $\overline{AB}$ and $\overline{AC}$ are tangents to the circle and arc measurements are as marked. Find the measures of the indicated angles.
    a. $\angle ABD$
    b. $\angle CAB$
    c. $\angle ACD$

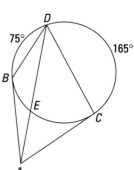

15. Write a proof argument.

Given: $\overrightarrow{CB}$ and $\overrightarrow{CD}$ are tangent to $\odot A$, as shown below.
To prove: $m\widehat{BD} + m\angle C = 180$.

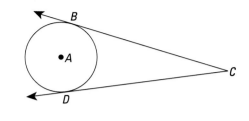

Review

16. You are in a plane at 35,000 feet on a cloudless day. How far is it from the plane to a point on the horizon? *(Lesson 14-5)*

17. Prove that the tangents to a circle at the endpoints of a diameter are parallel. *(Lesson 14-5)*

18. Suppose you have a camera with a picture angle of 38°. Trace the building below. Then show all the places where you could stand so that the building's front just fits into your picture. *(Lesson 14-2)*

19. In right $\triangle ABC$ at the right, $\overline{BD}$ is the altitude to the hypotenuse. If $AD = 3$ and $AC = 15$, what is BD? *(Lesson 13-4)*

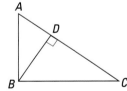

20. a. What is the mean of 10 and 20?
b. What is the geometric mean of 10 and 20?
(Lesson 11-8)

21. If $\frac{4}{x} = \frac{q}{10}$, write three other true proportions involving 4, q, x, and 10. *(Lesson 12-4)*

22. Find all solutions to the equation $4x^2 + 10x = 6$. *(Previous course.)*

Exploration

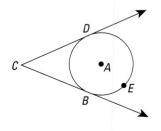

23. Refer to the figure at the left. $\overrightarrow{CD}$ and $\overrightarrow{CB}$ are tangent to $\odot A$ at D and B, respectively.
a. On a number line, graph the possible measures of angle C.
b. Prove your result using the result in Question 15.

*Investigating
Lengths of
Chords,
Secants, and
Tangents*

IN-CLASS

ACTIVITY

Work on this activity with a partner. You will need either an automatic drawer or a ruler and compass.

Chords, secants, and tangent segments to a circle have lengths that are related in simple ways. This activity explores those lengths.

1 Draw a circle with a radius of at least 4 cm. Identify four points A, B, C, and D in order on it so that $ABCD$ is not a trapezoid. Draw the quadrilateral and the diagonals $\overline{AC}$ and $\overline{BD}$. Call their point of intersection E. You should have a figure that looks something like this.

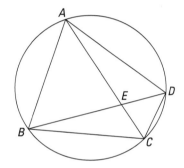

2 Measure AE, BE, CE, and DE. Multiply and divide pairs of these measures. Do you get any equal products or equal quotients? If so, write them down.

3 Extend the sides $\overline{AD}$ and $\overline{BC}$ of the quadrilateral. Call their point of intersection F. You should have a figure that looks something like this.

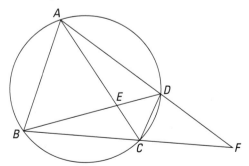

4 Measure AF, BF, CF, and DF. Again multiply and divide the measures. Do you get any equal products or equal quotients this time?

5 Move A, B, C, and D to different positions on the circle and repeat Steps 1-4.

Lengths of Chords, Secants, and Tangents

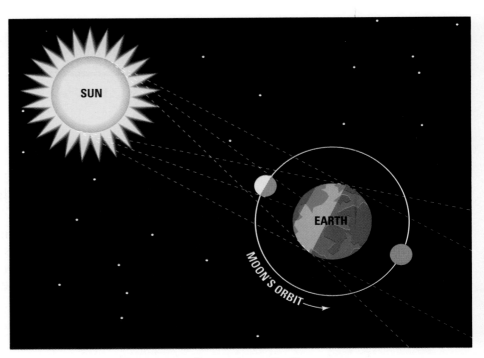

In the shadow. *Tangents outline regions of eclipses.*

A Simple Relationship Among Lengths of Segments in a Circle

The theorem relating the lengths of the segments of the In-class Activity is simple and amazing, and was known to Euclid. We still use his proof. The surprise is that the same proof works, letter for letter, for two quite different figures.

Secant Length Theorem
Suppose one secant intersects a circle at A and B, and a second secant intersects the circle at C and D. If the secants intersect at P, then
$$AP \cdot BP = CP \cdot DP.$$

Proof
Given: $\odot O$; secants $\overleftrightarrow{AB}$ and $\overleftrightarrow{CD}$ intersect at P.
To prove: $AP \cdot BP = CP \cdot DP$.
Drawing: There are two figures, depending on whether P is inside or outside the circle.

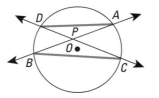

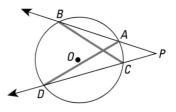

Argument:

$AP \cdot BP$ will equal $CP \cdot DP$ if it can be proved that $\frac{AP}{CP} = \frac{DP}{BP}$. This suggests forming triangles and trying to prove them similar. Follow the conclusions and justifications first for the figure at the left, then for the figure at the right replacing $\angle PAD \cong \angle PCB$ in Step 2 with $\angle P \cong \angle P$ (Reflexive Property).

Conclusions	Justifications
1. Draw $\overline{DA}$ and $\overline{BC}$.	Two points determine a line.
2. $\angle PAD \cong \angle PCB$, $\angle ADP \cong \angle CBP$	In a circle, inscribed angles intercepting the same arc are congruent.
3. $\triangle DPA \sim \triangle BPC$	AA $\sim$ Theorem (Step 2)
4. $\frac{AP}{CP} = \frac{DP}{BP}$	Corresponding sides of similar figures are proportional.
5. $AP \cdot BP = CP \cdot DP$	Means-Extremes Property

Example 1

Given chords $\overline{AB}$ and $\overline{CD}$ intersecting at P, with lengths as shown, find PB.

Solution

In general, $PA \cdot PB = PC \cdot PD$.

Substitute, $3 \cdot PB = 5 \cdot 6$.

$3 \cdot PB = 30$

$PB = 10$

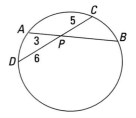

Example 2

Given secants $\overleftrightarrow{A_1B_1}$ and $\overleftrightarrow{A_2B_2}$ intersecting at E, with lengths as shown, find A_2B_2.

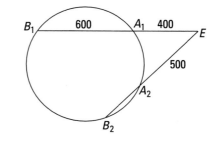

Solution

A_2B_2 is found by first getting EB_2 and then subtracting EA_2.

$$EA_1 \cdot EB_1 = EA_2 \cdot EB_2$$

Substituting, $400 \cdot 1000 = 500 \cdot EB_2$.

$800 = EB_2$

Since $A_2B_2 = EB_2 - EA_2$,

$A_2B_2 = 800 - 500 = 300$.

The Secant Length Theorem has a surprising application. Examine the drawing below.

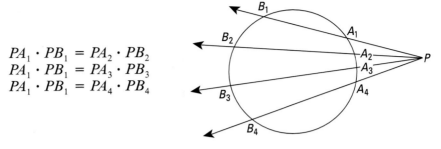

$$PA_1 \cdot PB_1 = PA_2 \cdot PB_2$$
$$PA_1 \cdot PB_1 = PA_3 \cdot PB_3$$
$$PA_1 \cdot PB_1 = PA_4 \cdot PB_4$$

Suppose a circle and a point P are given. For any secant through P intersecting the circle in two points A and B, there is a product $PA \cdot PB$. This product is the *same* number for every secant through P. In Example 2, it is the number 400,000. In Example 1, it is the number 30.

The Swiss geometer Jacob Steiner (1796–1863) called the product the **power of the point P for the circle O.**

The Length of a Tangent from a Point to a Circle

The power of a point external to a circle is easily calculated if you know the length of a tangent from P to $\odot O$.

Tangent Square Theorem
The power of point P for $\odot O$, is the square of the length of a segment tangent to $\odot O$ from P.

Proof

Given: Point P outside $\odot O$, and $\overleftrightarrow{PX}$ tangent to $\odot O$ at T.

To prove: The power of point P for $\odot O$ is PT^2.

Drawing: A figure is shown at the right.

Argument:
Draw $\overrightarrow{TO}$ which intersects $\odot O$ at B. Let $\overline{PB}$ intersect $\odot O$ at A and B. Since $\overline{PT} \perp \overline{TB}$ and since $\angle TAB$ is inscribed in a semicircle, $\triangle PTB$ is a right triangle with altitude $\overline{TA}$. Thus $PT^2 = PA \cdot PB$ by the Right Triangle Altitude Theorem. Thus the power of point P for $\odot O$ is PT^2.

Since $PT^2 = PA \cdot PB$, $PT = \sqrt{PA \cdot PB}$. Thus the length of a tangent $\overline{PT}$ to the circle from a point is the geometric mean of the lengths of segments of any secant drawn from that point.

Example 3

At the right, $\overleftrightarrow{AR}$ is tangent to the circle at R.
If $AP = 3$ and $AR = 6$, find AQ and PQ.

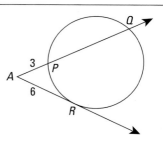

Solution

Use the Tangent Square Theorem:
$AR^2 = AP \cdot AQ$.
Substituting, $\qquad 6^2 = 3 \cdot AQ$
$\qquad\qquad\qquad 36 = 3 \cdot AQ$
So $\qquad\qquad AQ = 12$.
Then $\qquad\quad PQ = AQ - AP$
$\qquad\qquad\qquad\quad = 12 - 3 = 9$.

Some problems involving the power of a point require the solving of a quadratic equation.

Example 4

In the figure below, $\overleftrightarrow{BT}$ is tangent to the circle at T, $BT = 12$,
and $ER = 32$. Find BE.

Solution

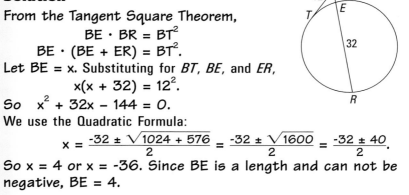

From the Tangent Square Theorem,
$$BE \cdot BR = BT^2$$
$$BE \cdot (BE + ER) = BT^2.$$
Let $BE = x$. Substituting for BT, BE, and ER,
$$x(x + 32) = 12^2.$$
So $\quad x^2 + 32x - 144 = 0$.
We use the Quadratic Formula:
$$x = \frac{-32 \pm \sqrt{1024 + 576}}{2} = \frac{-32 \pm \sqrt{1600}}{2} = \frac{-32 \pm 40}{2}.$$
So $x = 4$ or $x = -36$. Since BE is a length and can not be negative, $BE = 4$.

QUESTIONS

Covering the Reading

In 1 and 2, refer to the figure at the right.

1. $AT \cdot TQ = \underline{\quad?\quad}$.

2. If $AT = 6$, $TQ = 4$ and $TR = 3$, then $TP = \underline{\quad?\quad}$.

3. In the figure at the left, $DP \cdot DT = \underline{\quad?\quad}$.

4. Refer to the figure at the right. Let $TW = 3$,
$WX = 3$ and $TU = 2$.
  a. Calculate TV.
  b. What is the power of point T for this circle?
  c. Calculate UV.

5. In the In-class Activity, what were your equal products and quotients
   a. in Step 2?   **b.** in Step 4?

6. Refer to the figure at the left, with two intersecting chords.
   a. $x =$ __?__
   b. The power of P in this circle is __?__.

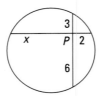

In 7 and 8, refer to the figure at the right.

7. a. If $JX = 2$ and $XY = 6$, then $JQ =$ __?__.
    b. What is the power of point J?

8. JQ is the geometric mean of __?__ and __?__.

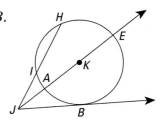

9. In the figure at the left, $\overleftrightarrow{BA}$ is tangent to $\odot O$ at A. If $AB = 15$ and $CD = 45$, find BC.

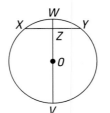

Applying the Mathematics

10. In $\odot K$ at the right, $\overleftrightarrow{JB}$ is tangent to $\odot K$ at B.
Suppose $JI = 10$, $JA = 8$, and $AE = 12$.
   a. Find HI.
   b. Find JB.
   c. What is the power of point J?

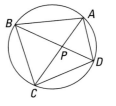

11. In $\odot O$ at the left, diameter $WV = 16$. If $\overline{XY} \perp \overline{WV}$ and $XY = 10$, find WZ.

12. Use the figure at the right in which
$\overline{BD}$ and $\overline{AC}$ intersect at P.
Prove or disprove:
If $PA = PD$, then $PB = PC$.

13. Is the Secant Length Theorem true if the
word "circle" in it is replaced by "sphere"?
That is, if A, B, C, and D are points on a
sphere, does $PA \cdot PB = PC \cdot PD$? Explain
your answer.

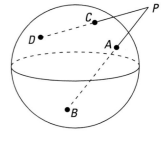

Review

14. Find the measures of the angles of $\triangle BDC$ at the left.
   (Lessons 14-2, 14-6)

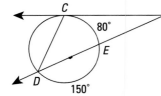

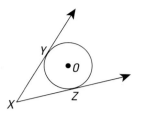

15. At the left, $\overrightarrow{XY}$ and $\overrightarrow{XZ}$ are tangent to $\odot O$ at the left. If $m\angle X = 45$, find $m\widehat{YZ}$.   *(Lesson 14-6)*

16. From a point P above Earth, the horizon is 43 miles away. How high above Earth is P?   *(Lesson 14-5)*

17. Find the measure of $\angle A$ below.   *(Lesson 14-4)*

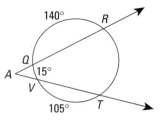

In 18 and 19, suppose you have a camera with a 64° picture angle. A top view of a building is shown below. Trace the diagram.

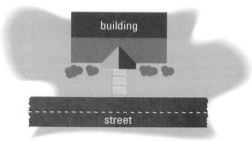

18. Draw the set of locations where you can stand so that the entire front of the building just fills your picture.   *(Lesson 14-2)*

19. If the building is 48 meters across, at least how far in front of the middle of the building would you need to stand so the building just fills your picture?   *(Lesson 14-2)*

20. In right triangle ABC at the left, find
   a. AC.   **b.** $\tan B$.
   c. $\sin A$.   **d.** $\cos B$.   *(Lessons 8-6, 13-6, 13-7)*

Exploration

21. a. Is the Secant Length Theorem true if the secants intersect on the circle?
   b. If true, prove it. If not true, show a counterexample.

22. Two chords intersect, as shown at the left. Is it possible for the lengths AP, BP, CP, and DP to be four consecutive integers?

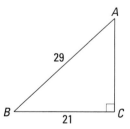

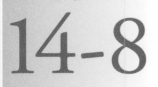

LESSON

14-8

The Isoperimetric Inequality

Can a corner spectator see center court? *The best view of a sporting event is not from the corner of an arena. See page 845.*

What Figure Has the Maximum Area for a Given Perimeter?

A basketball court is rectangular in shape, and it is easier to put benches in straight rows than in circles. So why are large basketball arenas often circular? The reason is due to an important property of circles involving areas and perimeters.

Activity

Here are 4 figures, each with a perimeter of 12 units. Calculate the area of each. (Hint: Divide the regular hexagon into six equilateral triangles.)

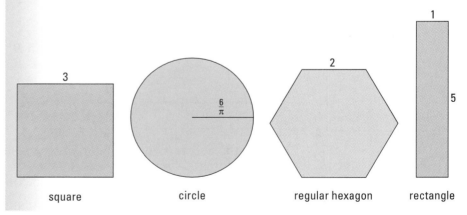

These figures are drawn to the same scale, so you can compare their areas by sight. It looks as if the circle has the maximum area. Your calculations should bear this out. That the circle has the most area for a particular perimeter is known as the Isoperimetric Theorem. *Isoperimetric* means "having equal perimeters."

> **Isoperimetric Theorem**
> Of all plane figures with the same perimeter, the circle has the maximum area.

The proof of this theorem requires advanced calculus, a subject usually not studied until college. The reason the proof is difficult is that it requires discussing all sorts of curves.

Pictured below are two such curves. At the left is an ellipse which is close to circular and thus encloses a substantial area for its perimeter. At the right is a non-convex curve with the same perimeter as the ellipse. As you can see, it encloses very little area for its perimeter.

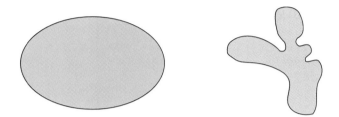

When limited to polygons with a given number of sides, regular polygons have the maximum area for their perimeter. As the number of sides increases, these regular polygons look more and more like circles.

How Large Can the Area Be?

Using the Isoperimetric Theorem, the maximum area enclosed by any perimeter p can be found. The result is known as the *Isoperimetric Inequality*.

> **Isoperimetric Inequality**
> If a plane figure has area A and perimeter p, then $A \leq \frac{p^2}{4\pi}$.

> **Proof**
> Given: A plane figure with area A and perimeter p.
>
> To prove: $A \leq \frac{p^2}{4\pi}$.
>
> Argument:
> Consider a circle with area A and perimeter (circumference) p.
> Then $p = 2\pi r$. So its radius equals $\frac{p}{2\pi}$ and its area
> $$A = \pi r^2 = \pi\left(\frac{p}{2\pi}\right)^2 = \pi \cdot \frac{p^2}{4\pi^2} = \frac{p^2}{4\pi}.$$ So for a circle,
> $A = \frac{p^2}{4\pi}$. By the Isoperimetric Theorem, the area of any other plane
> figure must be less than $\frac{p^2}{4\pi}$, so for any plane figure, $A \leq \frac{p^2}{4\pi}$.

Playing with a full deck.
In India, playing cards called ganjifa *are traditionally circular. These ganjifa are handpainted.*

Example 1

Suppose a figure has perimeter 30 cm.
a. What is its maximum possible area?
b. What is its minimum possible area?

Solution

a. The maximum possible area is given by the Isoperimetric Inequality with $p = 30$.

$$A \leq \frac{30^2}{4\pi}$$

$$A \leq \frac{900}{4\pi}$$

$$A \leq 71.6 \text{ sq cm (approximately)}$$

The maximum possible area is about 71.6 sq cm, when the figure is a circle.

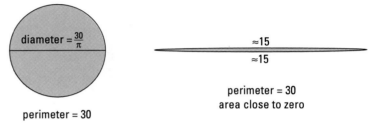

b. The area can be as small as you want, but not zero. **The minimum is as close to zero as you like.**

What Is the Least Perimeter Possible for a Given Area?

The inequality $A \leq \frac{p^2}{4\pi}$ gives the range of possible areas A for a fixed perimeter p. By solving the inequality for p, we obtain the range of possible perimeters when the area is fixed. Here is how:

Multiply both sides by 4π. $\qquad\qquad\qquad 4\pi A \leq p^2$
Rewrite the inequality to put p^2 on the left. $\qquad p^2 \geq 4\pi A$
Take the positive square root of each side. $\qquad p \geq \sqrt{4\pi A}$

Thus of all plane figures with a given area A, the perimeter p is at least $\sqrt{4\pi A}$. This result is another way of stating the Isoperimetric Theorem.

> **Isoperimetric Theorem (alternate statement)**
> Of all plane figures with the same area, the circle has the least perimeter.

If the area of a circle is A, the perimeter (circumference) of the circle is exactly $\sqrt{4\pi A}$. Any other figure with this area has a greater perimeter.

Example 2

Suppose a square and a circle both have an area of 25 sq ft. Verify that the perimeter (circumference) of the circle is less than the perimeter of the square.

Solution

The square's area is 25 sq ft, so a side is 5 ft, and its perimeter is 20 ft. For the circle, if $A = 25$, then $p = \sqrt{4\pi(25)} = \sqrt{100\pi}$, or about 17.72 ft. Since $17.72 < 20$, the circle needs less perimeter to enclose the same area.

The restatement of the Isoperimetric Theorem explains why the circle is a popular shape for arenas. In an arena, people like to sit close to the action. Since each person takes up about the same amount of space, each person can be thought of as one unit of area. To minimize the farthest distance from seats to the center of the court, the perimeter enclosing them should be as small as possible. The smallest perimeter is given by a circle. Another way of stating this is: Of all figures with a given area, the circle has the smallest width.

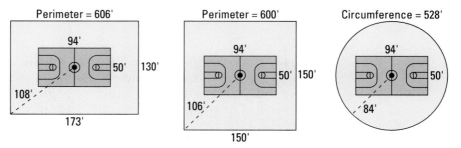

Each of these arenas has an area of about 22,500 ft^2.

QUESTIONS

Covering the Reading

1. Refer to the Activity in this lesson. Give the area of the figure.
   a. square
   b. circle
   c. regular hexagon
   d. rectangle

2. Of all figures with perimeter 12, which has the maximum area?

3. Of all figures with perimeter 12, which has the least area?

4. Of all triangles with perimeter 12, which has the maximum area?

5. Draw a non-polygonal figure whose area is small for its perimeter.

6. Draw a polygon whose area is small for its perimeter.

7. Consider all figures with area 600 square meters. Which has the least perimeter?

8. A circle has area 9π sq cm. What is its circumference?

9. A square has area 9π sq cm. What is its perimeter?

10. Which answer should be larger, that for Question 8 or that for Question 9?

11. Complete the statements.
    a. Of all figures with the same area, the __?__ has the __?__ perimeter.
    b. Of all figures with the same perimeter, the __?__ has the __?__ area.

12. What is an advantage of making a large basketball arena circular rather than rectangular?

Applying the Mathematics

13. a. A fence encloses pentagonal region *ABCDE* at the right. Find the area of this region.
    b. Find the area of the largest region that could be enclosed by this amount of fencing.

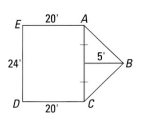

14. If a rectangle has perimeter 4*s*, then its sides can be called $s - t$, $s + t$, $s - t$, and $s + t$.
    a. What is the area of this rectangle?
    b. For what value of *t* is the area the greatest?

15. Many nomads of Mongolia today live in yurts, cylindrical tents that are made of animal skins and are easy to put up or take down. Some Native Americans used to live in tepees, cone-shaped tents with the same properties as the yurts. Explain why the circular base is a wise shape.

16. In ⊙O below, suppose $XY = 8$, $YZ = 20$, and $XW = 10$.
   a. Find WV.
   b. What is the power of point X for ⊙O?   *(Lesson 14-7)*

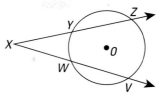

17. In the circle at the left, $AE = 16$, $BE = 14$, and $CE = 18$.
   Find DE.   *(Lesson 14-7)*

In 18 and 19, refer to the figure at the
right. $\overrightarrow{AB}$ is tangent to ⊙O at point B and
m∠$A = 40$.   *(Lesson 14-6)*

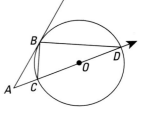

18. Find each measure.
   a. m$\widehat{BD}$
   b. m$\widehat{BC}$

19. Find m∠CBD.

20. In the figure below, find x.   *(Lesson 14-4)*

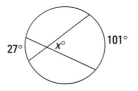

21. a. What is the volume of a right square pyramid with base area of
   80 sq ft and height 16 ft?
   b. If the pyramid in part **a** undergoes a transformation by $S_{\frac{2}{5}}$, what
   is the volume of its image?   *(Lessons 10-7, 12-6)*

22. State **a.** the converse, **b.** the inverse, and **c.** the contrapositive of the
   following statement: *If we finish the next lesson, our class will have
   done every lesson in the book.*   *(Lessons 2-2, 11-2)*

23. a. What is the volume of a sphere with radius 7 cm?
   b. What is the surface area of this sphere?   *(Lessons 10-8, 10-9)*

24. Estimate $\sqrt[3]{\pi}$ to the nearest hundredth.   *(Lesson 10-3)*

Exploration

25. Give the dimensions and a diagram of a polygon whose perimeter is
   100 feet and whose area is greater than 625 square feet.

Ice house but warm! *Inuits who build igloos capitalize on the fact that the sphere is an efficient figure.*

The Shape of Igloos

Some Inuits of the northern islands of Canada live in igloos in the winter. Ice for the igloo is always available, so an igloo can be built anywhere. Also, a small igloo can be built in about an hour. (How long do you think it took to build the building you live in?) A large igloo, which takes only a little longer to build, usually has an entranceway leading to a smaller part and then to a main part of the igloo. Usually the bases of the smaller and main parts are circles. As you have learned, this shape gives the maximum floor area for a given perimeter.

Both the smaller and main parts of the igloo have the shape of a hemisphere. By using a hemisphere, Inuits can enclose more space with the same amount of ice than with a cylinder, cone, or polyhedron that covers the same base. The sphere plays the same role in 3-dimensional relationships between surface area and volume as the circle does in 2-dimensional relationships between perimeter and area. To emphasize the sameness of the roles, mathematicians use the same term, "isoperimetric," to indicate "same boundary."

> **Isoperimetric Theorem (3-dimensional version)**
> Of all solids with the same surface area, the sphere has the largest volume.

Verifying that the Sphere Has the Maximum Volume

Like its counterpart in two dimensions, a proof of the space version of the Isoperimetric Theorem requires advanced mathematics. But the theorem can be verified in different ways. First is with an example. The sphere and the cylinder below have the same surface area, 144π square units.

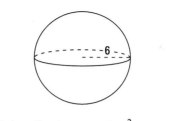

S.A. of sphere = $4\pi r^2$
= 144π sq units

S.A. of cylinder = L.A. + $2B$
= $2\pi rh + 2\pi r^2$
= 144π sq units

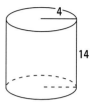

Fore! Water. *This 25,000-gallon water tank is spherical because that shape holds the most water for its surface area.*

Which has the greater volume? Their volumes can be easily found from formulas.

volume of sphere = $\frac{4}{3}\pi r^3$
= 288π units^3

volume of cylinder = Bh
= $\pi r^2 h$
= 224π units^3

The sphere has 64π units^3, almost 30% more volume than the cylinder. In fact, the sphere is the most efficient container. You would see spherical containers everywhere if they didn't roll!

A common experience provides a second verification. Suppose you blow air into a plastic bag. The bag, being unable to stretch to change its surface area, will tend to assume a shape as close to a sphere as it can. If you blow in more air than the sphere can hold, the bag will burst.

What Figure Has the Minimum Surface Area for a Given Volume?

Now, rather than a constant boundary, keep the capacity of the figure constant. To do this, consider shapes with the same volume. Soap bubbles consist of some soapy water and a fixed volume of air trapped inside. Because of surface tension, the bubble takes a shape to minimize the surface area surrounding the trapped air. That shape is a sphere.

> **Isoperimetric Theorem (alternate 3-dimensional version)**
> Of all solids with the same volume, the sphere has the least surface area.

For instance, a plastic container that would hold a gallon of milk, but use the least amount of plastic, would be shaped like a sphere. The Example verifies the alternate statement of the theorem numerically.

Example

A cube and a sphere each have volume 1000 cubic meters. Calculate the surface area for each figure.

Solution

First draw figures and write down relevant formulas.

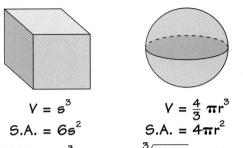

$$V = s^3$$
$$S.A. = 6s^2$$

$$V = \frac{4}{3}\pi r^3$$
$$S.A. = 4\pi r^2$$

For the cube, $1000 = s^3$, so $s = \sqrt[3]{1000} = 10$.
The surface area of the cube is $6s^2 = 600$ square meters.

For the sphere,
$$1000 = \frac{4}{3}\pi r^3$$
$$\frac{750}{\pi} = r^3$$
$$238.7 \approx r^3$$

Take the cube root to find $r \approx 6.2$.
The surface area of the sphere is $4\pi r^2 \approx 4 \cdot \pi \cdot 6.2^2 \approx 483$ square meters. The sphere has considerably less surface area than the cube.

The cube "wastes" surface near its edges and corners.

Figures with Large Surface Areas for a Given Volume

Just as a 2-dimensional figure can have a large perimeter and a small area, so a 3-dimensional figure can have a large surface area and a small volume. Think of sponges like the natural and artificial ones pictured at the left. The artificial sponge has the shape of a rectangular solid with volume 12 cubic inches, whereas the natural sponge has a more irregular shape. Both sponges are nonconvex curved spaces whose volume is quite small in comparison to their surface areas. The holes give the sponges their large surface areas. This large irregular surface enables the sponges to hold a lot of water. Some water is held because in almost any position of the sponge some of the sponge is under the water, cupping it. Other water clings to the large surface just the way the inside of a glass remains wet after you pour out its contents.

Concluding Remarks

The Isoperimetric Theorems involve square and cube roots, π, polygons, circles, polyhedra, and spheres. They explain properties of arenas, fences, soap bubbles, and sponges. They demonstrate the broad applicability of geometry and the unity of mathematics.

Many people enjoy the way mathematics connects diverse topics. Others like mathematics for its uses. Still others like the logical way mathematics fits together and grows. The Isoperimetric Theorems exemplify all of these properties of mathematics. We hope that you have found it to be an enjoyable way to end your study of geometry using this book.

QUESTIONS

Covering the Reading

1. Why is an igloo an efficient dwelling?

2. Of all solids with the same surface area, the __?__ has the maximum __?__.

3. The surface area of a solid is 600 square meters. To the nearest 100 cubic meters, what is the maximum possible volume of the solid?

4. Of all solids with the same volume, the __?__ has the least __?__.

5. Which statement, that of Question 2 or Question 4, explains the shape of a soap bubble?

6. In Question 3, what is the least possible volume of the solid?

7. A cube has volume 8 cubic units. What is its surface area?

8. Explain why sponges are able to hold so much water.

9. List three reasons given in this lesson why people enjoy mathematics.

10. The plastic milk container that would have the least material for a given amount of milk would be shaped like a sphere. Why are milk containers *not* shaped like spheres?

Applying the Mathematics

11. a. A sphere has volume 36π cubic meters. What is its surface area?
    b. Give possible dimensions for a cylinder with volume 36π cubic meters. What is the surface area of the cylinder you identified?
    c. Give dimensions for a right cone with volume 36π cubic meters. What is its surface area?
    d. According to the Isoperimetric Inequality, the surface area in part a is __?__ than the surface areas in part b or c.

12. A sink-side water purifier depends on a cylindrical charcoal filter of diameter 3 inches and height 8.25 inches. The manufacturer claims the filter has over 100 acres of surface area.
    a. Can this claim possibly be true?
    b. If so, why would anyone want to have so much surface area? If not, why can't the claim be true?

13. A sphere has surface area πx. Find its volume in terms of x.

Review

14. Consider all figures with area 12 square meters.
  a. Which has the least perimeter?
  b. What is the perimeter?   *(Lesson 14-8)*

15. A circle and a square both have perimeters 96 inches.
  a. Calculate their areas.
  b. Which has the smaller area?   *(Lesson 14-8)*

16. $\overleftrightarrow{QM}$ is tangent to $\odot X$ at L below. If $NQ = 6$ and $PN = 12$, find QL.   *(Lesson 14-7)*

17. $\triangle ABC$ is inscribed in the circle below. What are the measures of its angles?   *(Lesson 14-2)*

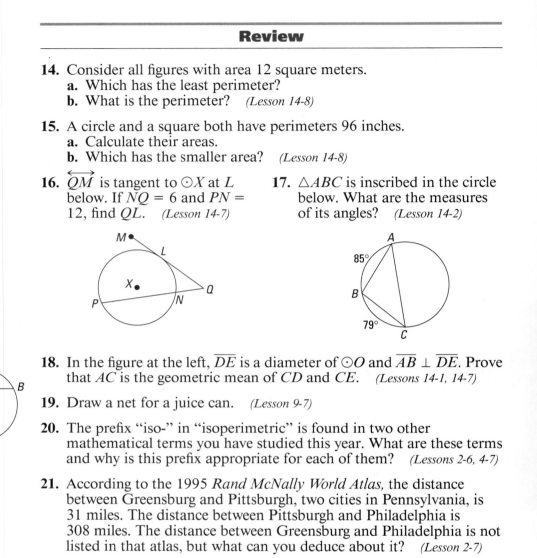

18. In the figure at the left, $\overline{DE}$ is a diameter of $\odot O$ and $\overline{AB} \perp \overline{DE}$. Prove that AC is the geometric mean of CD and CE.   *(Lessons 14-1, 14-7)*

19. Draw a net for a juice can.   *(Lesson 9-7)*

20. The prefix "iso-" in "isoperimetric" is found in two other mathematical terms you have studied this year. What are these terms and why is this prefix appropriate for each of them?   *(Lessons 2-6, 4-7)*

21. According to the 1995 *Rand McNally World Atlas,* the distance between Greensburg and Pittsburgh, two cities in Pennsylvania, is 31 miles. The distance between Pittsburgh and Philadelphia is 308 miles. The distance between Greensburg and Philadelphia is not listed in that atlas, but what can you deduce about it?   *(Lesson 2-7)*

Exploration

22. According to legend, the first person to use the Isoperimetric Inequality was Dido, the queen of Carthage.
  a. Where is, or was, Carthage?
  b. How did Dido use this inequality? (Hint: Look in an encyclopedia or dictionary under Dido.)

23. Develop an Isoperimetric Inequality for three dimensions. That is, find a relationship between the volume V and surface area S.A. of any 3-dimensional figure.

Dido and Aeneas.

A project presents an opportunity for you to extend your knowledge of a topic related to the material of this chapter. You should allow more time for a project than you do for typical homework questions.

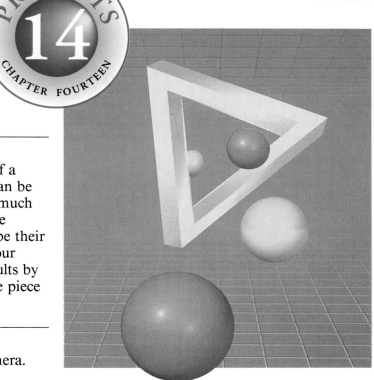

1 Cutting Down on Waste

You wish to cut five discs out of a square piece of tin. The discs can be any size. If you want to use as much tin as possible, where should the discs be located in the square, and what should be their radii? Explain how you have come to your conclusion. Then, demonstrate your results by cutting comparable disks out of a square piece of cardboard or wood.

2 Picture Angles

For this project you need a camera. Choose a large object such as a building. Measure the width of the building and the distance to a point where you can stand so as to fit just the building in a picture. Determine the picture angle of the camera. Draw a diagram indicating where you will stand. Then take shots from five different locations so that the building just fits into your picture, and display them on a poster.

3 Circles Associated with Triangles

Mathematicians have studied properties of many circles associated with triangles. For instance, the *circumcircle* of a triangle is the circle that passes through its vertices. Every triangle also has an *incircle,* three *excircles,* and a *nine-point circle.* Find out where these circles are located for a given triangle, and draw all of them for at least two different scalene triangles.

4 The Universe

Earlier in this chapter, you read the quote of Galileo: ". . . the universe . . . is written in the language of mathematics, and its characters are triangles, circles, and other geometrical figures, without which it is humanly impossible to understand a single word of it; without these, one is wandering about in a dark labyrinth." Write an essay supporting or criticizing this claim of Galileo. Illustrate your essay with examples.

▶

5 **Japanese Temple Geometry Theorems**
In the Edo period of the history of Japan (1603–1867), when Japan isolated itself from the rest of the world, there was a tradition of posing geometry problems

on tablets hung under the roofs of shrines and temples. Many of the theorems involve tangent lines and circles and were unknown outside of Japan until recently, and many are astounding. Here are four of these theorems, with the dates of their appearance given in parentheses. In these problems the radius of circle O_1 is r_1, of circle O_2 is r_2, and so on. Pick at least two of them and draw two different accurate drawings verifying the relations in the theorem.

a. (1824) If circle O_1, is tangent to circle O_2, and if circle O_3 is tangent to the two circles O_2 and O_1 and to their common tangent $\overleftrightarrow{AB}$, then

$$\frac{1}{\sqrt{r_3}} = \frac{1}{\sqrt{r_1}} + \frac{1}{\sqrt{r_2}}.$$

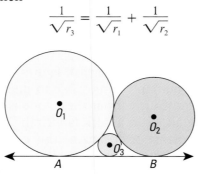

b. (1842) If $\overline{AB}$ is a chord of circle O with radius r, the five circles O_1 to O_5 are each tangent to each other, to the chord, and to circle O, and if $r_1 = r_5$, $r_2 = r_4$, and $4r_3 = r$, then

$$r_1 = \frac{4}{9}r_3.$$

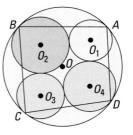

c. (1782) If circles O_1 to O_4 (of different sizes) are tangent to each other and to circle O at points A, B, C, and D, then

$$AB \cdot CD = AD \cdot BC.$$

d. (1791) Let $\overline{AC}$ and $\overline{BD}$ be any two chords of a circle. Let r_1, r_2, r_3, and r_4 be the radii of the largest circles that can be drawn in the 4 regions determined by $\overline{AC}$ and $\overline{BD}$, as shown below. (The circles are tangent to the chords, the blue circle, and the pink circle.) Then

$$\frac{1}{r_1} + \frac{1}{r_3} = \frac{1}{r_2} + \frac{1}{r_4}.$$

SUMMARY

The theorems and applications of this chapter are related to many of the ideas in earlier chapters.

Perpendicular lines are important in circles. A line perpendicular to a chord bisects the chord if and only if it contains the center of the circle. If the sides of an inscribed angle are perpendicular, then the angle intercepts a semicircle. These theorems give ways of finding the center of a circle. A line perpendicular to a radius at its endpoint on a circle is tangent to the circle. All other lines through that point are secants to the circle.

The chapter also applies congruence and similarity. In a circle or in congruent circles, arcs of the same measure are congruent if and only if they have congruent chords. Inscribed angles which intercept the same arc are congruent. If two chords $\overline{AB}$ and $\overline{CD}$ intersect at point E, $\triangle EAC$ and $\triangle EDB$ are similar. As a result, $AE \cdot BE = CE \cdot DE$. This theorem holds if the word "secant" is substituted for "chord" and A, B, C, and D are the points at which two secants intersect the circle. $AE \cdot BE$ or $CE \cdot DE$ is the power of point E with respect to the circle.

Relationships between measures of angles and arcs are all derived from the definition that the measure of an arc is equal to the measure of its central angle. The measure of an inscribed angle is half the intercepted arc. From this property, you can find all the places to take a photo so that an object just fills the photo. The measure of an angle between two chords is half the sum of the intercepted arcs. The angle between secants or tangents is half the difference of the intercepted arcs.

The Isoperimetric Theorems relate perimeters and areas, or surface areas and volumes. In two dimensions: (1) of all figures with the same perimeter, the circle has the maximum area; (2) of all figures with the same area, the circle has the least perimeter. In three dimensions: (1) of all figures with the same surface area, the sphere has the maximum volume; (2) of all figures with the same volume, the sphere has the least surface area. Many properties of real objects can be explained by these inequalities.

VOCABULARY

Below are the most important terms and phrases for this chapter. For the starred (*) terms you should be able to give a definition of the term. For the other terms you should be able to give a general description and a specific example of each.

Lesson 14-1
intercepted arc
Arc-Chord ≅ Theorem
Chord-Center Theorem

Lesson 14-2
picture angle of a camera lens
*inscribed angle
Inscribed Angle Theorem

Lesson 14-3
right-angle method to find
   the center of a circle

Lesson 14-4
Angle-Chord Theorem
*secant to a circle
Angle-Secant Theorem

Lesson 14-5
*tangent to a circle
point of tangency
Radius-Tangent Theorem
tangent to a sphere
common tangents

Lesson 14-6
Tangent-Chord Theorem
Tangent-Secant Theorem

Lesson 14-7
Secant Length Theorem
power of a point P for a circle
Tangent Square Theorem

Lesson 14-8
Isoperimetric Theorem
Isoperimetric Inequality
Isoperimetric Theorem
   (alternate statement)

Lesson 14-9
Isoperimetric Theorem
   (3-dimensional version)
Isoperimetric Theorem
   (alternate 3-dimensional
   version)

PROGRESS SELF-TEST

Directions: Take this test as you would take a test in class. Then check your work with the solutions in the Selected Answers section in the back of the book. You will need a straightedge, compass, and protractor.

1. In a circle with radius 210 mm, find the length of a chord of a 120° arc.

2. Trace the figure below. Then find the center of the circle that contains points C, D, and E.

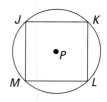

3. Square $JKLM$ is inscribed in $\odot P$ below. If the radius of $\odot P$ is 30, find the perimeter of $JKLM$.

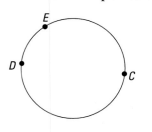

In 4 and 5, use the figure below, in which $m\widehat{DC} = 80°$ and $m\angle DEC = 100$.

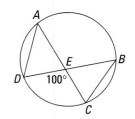

4. Find $m\angle B$.
5. Find $m\widehat{AB}$.

6. In $\odot Z$ below, $m\widehat{US} = 30°$, $m\widehat{UV} = 80°$, and $m\widehat{ST} = 140°$. Find $m\angle R$.

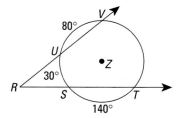

7. In circle O below, if $AQ = 19$, $BQ = 40$, and $CQ = 38$, find QD.

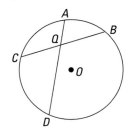

8. In $\odot G$ below, if $WX = 12$, $XY = 16$, and $WZ = 10$, find ZV.

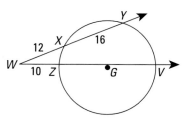

9. In the figure below, ℓ is tangent to $\odot O$ at Q. If $\overline{OQ}$ intersects chord $\overline{XY}$ at the midpoint M of $\overline{XY}$, prove that $\ell \parallel \overline{XY}$.

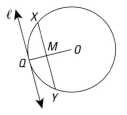

PROGRESS SELF-TEST

In 10 and 11, use the figure below. $\overrightarrow{PT}$ and $\overrightarrow{PU}$ are tangents to $\odot O$ at T and U.

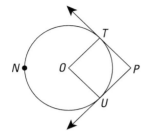

10. **a.** *True or false.* If $m\widehat{UT} = 90°$, then *PUOT* is a square.

   b. Explain your answer to part **a.**

11. If $m\angle P = 92$, find $m\widehat{TNU}$.

12. Write a proof argument using the given figure.
   Given: $\overline{BC} \parallel \overline{AD}$.
   To prove: $m\widehat{AB} = m\widehat{CD}$.

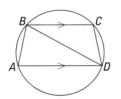

13. A clerk has a 30-cm strip of wire screen to fence in some supplies on a desk.

   a. How should the screen be shaped to fence in the most area?

   b. What is this area?

14. A sphere and a cube each have volume 240 cubic feet. Which has the larger surface area and why?

15. Refer to the stage below which is 60 feet wide. You have a camera lens with a 64° picture angle.

   a. Draw a picture indicating where you can stand so that the entire stage will just be seen in your picture.

   b. If you are standing in front of the center of the stage, how near to the center of the stage can you be and still photograph the entire stage?

16. How far can you see from a 100-foot high tower if there are no obstructions?

CHAPTER REVIEW

Questions on SPUR Objectives

SPUR stands for **S**kills, **P**roperties, **U**ses, and **R**epresentations. The Chapter Review questions are grouped according to the SPUR Objectives for this chapter.

SKILLS DEAL WITH THE PROCEDURES USED TO GET ANSWERS.

Objective A: *Calculate lengths of chords and arcs.* *(Lesson 14-1)*

In 1 and 2, a circle has a radius of 55 mm.

1. Find the length of a 60° arc.
2. Find the length of a chord of a 144° arc.
3. *ABCD* is a square inscribed in $\odot O$ below and *OA* = 12.

  a. Find *AB*.
  b. Find the area of the shaded region.

4. In $\odot P$, *RT* = 18, and $\overline{RT}$ is 7 units away from the center. What is *PT*?

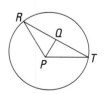

5. Regular octagon *STOPZIGN* is inscribed in $\odot Q$ at the right.

  a. What is $m\widehat{IZ}$?

  b. If *QT* = 15, find the perimeter of the octagon.

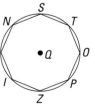

6. A regular hexagon is inscribed in $\odot O$ at the right. If the radius of $\odot O$ is 12, what is the length of each side of the hexagon?

Objective B: *Calculate measures of inscribed angles from measures of intercepted arcs, and vice versa.* *(Lesson 14-2)*

In 7 and 8, use the circle at the right.

7. Find $m\widehat{CB}$.
8. Find $m\widehat{ABC}$.

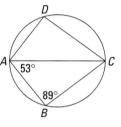

In 9 and 10, $\triangle PQR$ is inscribed in the circle at the right.

9. Find $m\angle Q$.
10. Find $m\angle P$.

Objective C: *Calculate measures of angles between chords, secants, or tangents from measures of intercepted arcs, and vice versa.* *(Lessons 14-2, 14-4, 14-6)*

In 11 and 12, use circle *Z* at the right.

11. If $m\widehat{DG} = 100°$ and $m\widehat{EF} = 140°$, what is $m\angle EHF$?

12. If $m\angle EHD = 51$ and $m\widehat{GF} = 37°$, what other arc measure can be found, and what is that measure?

13. In the figure below, m$\widehat{BC}$ = 30°, and m$\widehat{DE}$ = 125°. Find the measures of as many angles in the figure as you can.

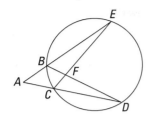

In 14–16, use ⊙L with measures as marked.

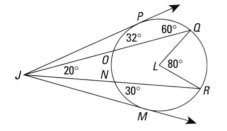

14. Find m∠*PJQ*.
15. Find m$\widehat{ON}$.
16. Find m∠*PJM*.

Objective D: *Locate the center of a circle given sufficient information.*   *(Lesson 14-3)*

17. Trace the figure below. Then find the center of the circle using the right angle method.

18. Trace the circular arc below. Find the center of the circle containing the circular arc. Then draw the entire circle.

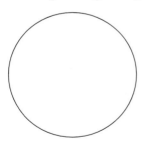

19. Trace the three points below at the left. Draw the circle through points *A*, *B*, and *C*.

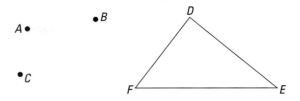

20. Trace △*DEF* above at the right. Draw the circle containing all the vertices of △*DEF*.

Objective E: *Apply the Secant Length Theorem and the Tangent Square Theorem.*   *(Lesson 14-7)*

21. In the figure below, *A*, *B*, *C*, and *D* all lie on ⊙*O*. If *AX* = 12, *XB* = 40, and *DX* = 48, find *CX*.

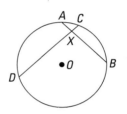

22. In the figure below, $\overleftrightarrow{QR}$ is tangent to ⊙*Z* at *R*. If *QR* = 8 and *QX* = 4, find *YX*.

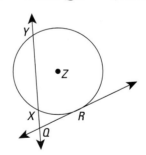

In 23 and 24, refer to ⊙*O* below.

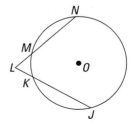

23. If *LN* = 20, *MN* = 15, and *LJ* = 25, find *KJ*.
24. If *LM* = 6, *MN* = 7, and *KJ* = 8, find *LK*.

PROPERTIES DEAL WITH THE PRINCIPLES BEHIND THE MATHEMATICS.

Objective F: *Make deductions from properties of radii, chords, and tangents, and know sufficient conditions for radii to be perpendicular to them.*
(Lessons 14-1, 14-5, 14-6)

In 25 and 26, refer to $\odot O$ below. $\overline{AB}$ is a diameter, $\overline{AB} \perp \overline{CD}$ and ℓ is tangent to $\odot O$ at point A.

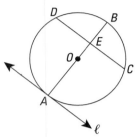

25. Justify: $DE = EC$.

26. Write an argument that proves $\ell \,/\!/\, \overline{CD}$.

In 27 and 28, write a proof argument using the given figure.

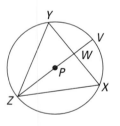

27. Given: $m\widehat{YZ} = m\widehat{XY} = m\widehat{XZ}$.
     To prove: $\triangle XYZ$ is equilateral.

28. Given: W is the midpoint of $\overline{XY}$.
     To prove: $\triangle ZYX$ is isosceles.

In 29 and 30, $\overline{AB}$ and $\overline{AC}$ are tangent to $\odot D$ at points B and C in the figure below.

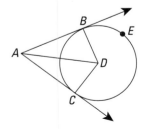

29. Find all angle measures that can be found, if $m\widehat{BEC} = 210°$.

30. Write an argument to prove $\triangle ABD \cong \triangle ACD$.

Objective G: *Make deductions from properties of angles formed by chords, tangents, or secants.*
(Lessons 14-2, 14-4, 14-6)

In 31 and 32, $ABDC$ at the right is a rectangle.

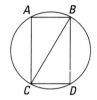

31. Explain why $\overline{BC}$ is a diameter.

32. Suppose $m\widehat{AB} = x$. What is $m\angle ACB$?

In 33 and 34, write a proof argument using the given figure.

33. Given: $\overleftrightarrow{AB}$ is tangent to circle O at B.
     To prove: $m\angle ABD = m\angle C$.

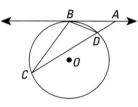

34. Given: $\overrightarrow{XW}$ and $\overrightarrow{XY}$ are tangent to $\odot P$ below at W and Y, respectively.
     To prove: $m\angle X = 180 - m\widehat{WY}$.

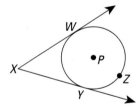

Objective H: *Apply the Isoperimetric Theorems and the Isoperimetric Inequality to determine which figures have the greatest or least area, perimeter, or volume.* *(Lessons 14-8, 14-9)*

35. Consider all figures with area 800 square feet.
   a. Which has the least perimeter?
   b. What is the perimeter of the figure in part **a**?

36. a. Of all figures with perimeter 2000 centimeters, which has the most area?
   b. What is that area?

37. The surface area of a solid is 10,000 square meters. What figure has the largest volume for this surface area?

38. A circle and a square both have perimeters of 32 inches.

   a. Calculate their areas.

   b. Which has the larger area?

39. a. Of all solids with surface area 48 square feet, which has the maximum volume?

   b. What is that volume?

40. A sphere and a cylinder both have volume 20,000 cubic meters.

   a. Can you tell which has the larger surface area?

   b. Why or why not?

USES DEAL WITH APPLICATIONS OF MATHEMATICS IN REAL SITUATIONS.

Objective I: *Given the angle width of a lens and the width of an object, determine the set of points from which the object just fits in the picture.* *(Lesson 14-3)*

In 41 and 42, a photographer wants to take a picture showing the entire length of a football field (from *A* to *B*). A scale drawing is shown. The 50-yard line bisects the segment connecting the goal posts at *C*, and *AB* = 120 yards.

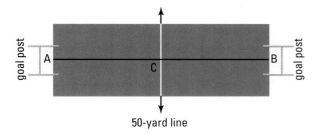

50-yard line

41. a. Locate all points where the photographer could stand to exactly fit $\overline{AB}$ if the camera lens has a picture angle of 62°.

   b. At least how far from *C* will the photographer be if he or she stands on the extended 50-yard line?

42. a. Locate all points where the photographer could stand to exactly fit $\overline{AB}$ if the camera lens has a picture angle of 84°.

   b. At least how far from *C* will the photographer be standing on the extended 50-yard line?

Objective J: *Determine the maximum distance that can be seen from a particular elevation.* *(Lesson 14-5)*

In 43–46, assume the radius of Earth is 3960 miles, or 6373 kilometers, and assume there are no hills or obstructions.

43. How far can you see from the top of the Eiffel Tower in Paris, 300 meters up?

44. How far can you see from the top of the CN Tower in Toronto, the world's tallest self-supporting structure, 555 meters up?

45. How far can you see if you are standing on the ground and your eyes are 5 feet above the ground?

46. How far can you see from a plane 4 miles high?

Objective K: *Apply the Isoperimetric Theorems and the Isoperimetric Inequality in real situations.* *(Lessons 14-8, 14-9)*

47. a. Of all containers that can hold a liter of orange juice, what is the shape of the container that has the least surface area?

   b. Why is this shape seldom used?

48. A farmer is making a pigpen with 60 feet of fencing. What is the area of the shape that gives the most room for the pigs?

49. Draw a figure with a large perimeter for its area.

50. Identify a figure with a large surface area for its volume.

REPRESENTATIONS DEAL WITH PICTURES, GRAPHS, OR OBJECTS THAT ILLUSTRATE CONCEPTS.

There are no Representation Objectives for this chapter.

Theorems are statements that have been proved, or can be proved, from the postulates. They are given here in order of appearance.

Theorems

Chapter 1

Line Intersection Theorem: Two different lines intersect in at most one point. *(Lesson 1-7, p. 43)*

Chapter 3

Linear Pair Theorem: If two angles form a linear pair, then they are supplementary. *(Lesson 3-3, p. 140)*

Vertical Angles Theorem: If two angles are vertical angles, then they have equal measures. *(Lesson 3-3, p. 141)*

Parallel Lines and Slopes Theorem: Two nonvertical lines are parallel if and only if they have the same slope. *(Lesson 3-6, p. 158)*

Transitivity of Parallelism Theorem: In a plane, if ℓ // m and m // n, then ℓ // n. *(Lesson 3-6, p. 158)*

Two Perpendiculars Theorem: If two coplanar lines ℓ and m are each perpendicular to the same line, then they are parallel to each other. *(Lesson 3-7, p. 162)*

Perpendicular to Parallels Theorem: In a plane, if a line is perpendicular to one of two parallel lines, then it is also perpendicular to the other. *(Lesson 3-7, p. 162)*

Perpendicular Lines and Slopes Theorem: Two nonvertical lines are perpendicular if and only if the product of their slopes is -1. *(Lesson 3-7, p. 162)*

Chapter 4

Figure Reflection Theorem: If a figure is determined by certain points, then its reflection image is the corresponding figure determined by the reflection images of those points. *(Lesson 4-2, p. 192)*

Two Reflection Theorem for Translations: If m // ℓ, the translation $r_m \circ r_\ell$ has magnitude two times the distance between ℓ and m, in the direction from ℓ perpendicular to m. *(Lesson 4-4, p. 206)*

Two Reflection Theorem for Rotations: If m intersects ℓ, the rotation $r_m \circ r_\ell$ has center at the point of intersection of m and ℓ and has magnitude twice the measure of the non-obtuse angle formed by these lines, in the direction from ℓ to m. *(Lesson 4-5, p. 211)*

Chapter 5

Corresponding Parts of Congruent Figures (CPCF) Theorem: If two figures are congruent, then any pair of corresponding parts is congruent. *(Lesson 5-1, p. 245)*

A-B-C-D Theorem: Every isometry preserves Angle measure, Betweenness, Collinearity (lines), and Distance (lengths of segments). *(Lesson 5-1, p. 246)*

Equivalence Properties of ≅ Theorem: For any figures F, G, and H:

Reflexive Property of Congruence: $F \cong F$.

Symmetric Property of Congruence: If $F \cong G$, then $G \cong F$.

Transitive Property of Congruence: If $F \cong G$ and $G \cong H$, then $F \cong H$. *(Lesson 5-2, p. 251)*

Segment Congruence Theorem: Two segments are congruent if and only if they have the same length. *(Lesson 5-2, p. 252)*

Angle Congruence Theorem: Two angles are congruent if and only if they have the same measure. *(Lesson 5-2, p. 252)*

Euclid's First Theorem: If circle A contains point B and circle B contains point A and the circles intersect at C, then $\triangle ABC$ is equilateral. *(Lesson 5-4, p. 264)*

// Lines ⇒ AIA ≅ Theorem: If two parallel lines are cut by a transversal, then alternate interior angles are congruent. *(Lesson 5-4, p. 265)*

AIA ≅ ⇒ // Lines Theorem: If two lines are cut by a transversal and form congruent alternate interior angles, then the lines are parallel. *(Lesson 5-4, p. 266)*

AEA ≅ ⇒ // Lines Theorem: If two lines are cut by a transversal and form congruent alternate exterior angles, then the lines are parallel. *(Lesson 5-4, p. 267)*

Perpendicular Bisector Theorem: If a point is on the perpendicular bisector of a segment, then it is equidistant from the endpoints of the segment. *(Lesson 5-5, p. 271)*

Uniqueness of Parallels Theorem (Playfair's Parallel Postulate): Through a point not on a line, there is exactly one line parallel to the given line. *(Lesson 5-6, p. 277)*

Triangle-Sum Theorem: The sum of the measures of the angles of a triangle is 180°. *(Lesson 5-7, p. 282)*

Quadrilateral-Sum Theorem: The sum of the measures of the angles of a convex quadrilateral is 360°. *(Lesson 5-7, p. 283)*

Polygon-Sum Theorem: The sum of the measures of the angles of a convex n-gon is $(n - 2) \cdot 180°$. *(Lesson 5-7, p. 284)*

Chapter 6

Flip-Flop Theorem:
(1) If F and G are points and $r_\ell(F) = G$, then $r_\ell(G) = F$.
(2) If F and G are figures and $r_\ell(F) = G$, then $r_\ell(G) = F$. *(Lesson 6-1, p. 301)*

Segment Symmetry Theorem: Every segment has exactly two symmetry lines: (1) its perpendicular bisector, and (2) the line containing the segment. *(Lesson 6-1, p. 302)*

Side-Switching Theorem: If one side of an angle is reflected over the line containing the angle bisector, its image is the other side of the angle. *(Lesson 6-1, p. 303)*

Angle Symmetry Theorem: The line containing the bisector of an angle is a symmetry line of the angle. *(Lesson 6-1, p. 303)*

Circle Symmetry Theorem: A circle is reflection-symmetric to any line through its center. *(Lesson 6-1, p. 303)*

Symmetric Figures Theorem: If a figure is symmetric, then any pair of corresponding parts under the symmetry is congruent. *(Lesson 6-1, p. 304)*

Isosceles Triangle Symmetry Theorem: The line containing the bisector of the vertex angle of an isosceles triangle is a symmetry line for the triangle. *(Lesson 6-2, p. 310)*

Isosceles Triangle Coincidence Theorem: In an isosceles triangle, the bisector of the vertex angle, the perpendicular bisector of the base, and the median to the base determine the same line. *(Lesson 6-2, p. 310)*

Isosceles Triangle Base Angles Theorem: If a triangle has two congruent sides, then the angles opposite them are congruent. *(Lesson 6-2, p. 310)*

Equilateral Triangle Symmetry Theorem: Every equilateral triangle has three symmetry lines, which are the bisectors of its angles (or equivalently, the perpendicular bisectors of its sides). *(Lesson 6-2, p. 312)*

Equilateral Triangle Angle Theorem: If a triangle is equilateral, then it is equiangular. *(Lesson 6-2, p. 312)*

Corollary: Each angle of an equilateral triangle has measure 60°. *(Lesson 6-2, p. 312)*

Quadrilateral Hierarchy Theorem: Every property true of all figures of one type on the hierarchy is also true of all figures of all the types below it to which the first type is connected. *(Lesson 6-3, p. 319)*

Kite Symmetry Theorem: The line containing the ends of a kite is a symmetry line for the kite. *(Lesson 6-4, p. 324)*

Kite Diagonal Theorem: The symmetry diagonal of a kite is the perpendicular bisector of the other diagonal and bisects the two angles at the ends of the kite. *(Lesson 6-4, p. 325)*

Rhombus Diagonal Theorem: Each diagonal of a rhombus is the perpendicular bisector of the other diagonal. *(Lesson 6-4, p. 325)*

Theorem: If a quadrilateral is a rhombus, then it is a parallelogram. *(Lesson 6-4, p. 326)*

Trapezoid Angle Theorem: In a trapezoid, consecutive angles between a pair of parallel sides are supplementary. *(Lesson 6-5, p. 330)*

Isosceles Trapezoid Symmetry Theorem: The perpendicular bisector of one base of an isosceles trapezoid is the perpendicular bisector of the other base and a symmetry line for the trapezoid. *(Lesson 6-5, p. 330)*

Isosceles Trapezoid Theorem: In an isosceles trapezoid, the non-base sides are congruent. *(Lesson 6-5, p. 331)*

Rectangle Symmetry Theorem: The perpendicular bisectors of the sides of a rectangle are symmetry lines for the rectangle. *(Lesson 6-5, p. 331)*

Theorem: If a figure possesses two lines of symmetry intersecting at a point P, then it is rotation-symmetric with a center of symmetry at P. *(Lesson 6-6, p. 337)*

Center of a Regular Polygon Theorem: In any regular polygon there is a point (its center) which is equidistant from all of its vertices. *(Lesson 6-7, p. 343)*

Regular Polygon Symmetry Theorem: Every regular n-gon possesses (1) n symmetry lines, which are the perpendicular bisectors of each of its sides and the bisectors of each of its angles, (2) n-fold rotation symmetry. *(Lesson 6-7, p. 344)*

Chapter 7

Theorem: If two angles in one triangle are congruent to two angles in another triangle, then the third angles are congruent. *(Lesson 7-1, p. 366)*

SSS Congruence Theorem: If, in two triangles, three sides of one are congruent to three sides of the other, then the triangles are congruent. *(Lesson 7-2, p. 370)*

SAS Congruence Theorem: If, in two triangles, two sides and the included angle of one are congruent to two sides and the included angle of the other, then the triangles are congruent. *(Lesson 7-2, p. 372)*

ASA Congruence Theorem: If, in two triangles, two angles and the included side of one are congruent to two angles and the included side of the other, then the two triangles are congruent. *(Lesson 7-2, p. 373)*

AAS Congruence Theorem: If, in two triangles, two angles and a non-included side of one are congruent respectively to two angles and the *corresponding* non-included side of the other, then the triangles are congruent. *(Lesson 7-2, p. 374)*

Isosceles Triangle Base Angles Converse Theorem: If two angles of a triangle are congruent, then the sides opposite them are congruent. *(Lesson 7-3, p. 380)*

HL Congruence Theorem: If, in two right triangles, the hypotenuse and a leg of one are congruent to the hypotenuse and a leg of the other, then the two triangles are congruent. *(Lesson 7-5, p. 390)*

SsA Congruence Theorem: If two sides and the angle opposite the longer of the two sides in one triangle are congruent, respectively, to two sides and the corresponding angle in another triangle, then the triangles are congruent. *(Lesson 7-5, p. 392)*

Properties of a Parallelogram Theorem: In any parallelogram, (a) opposite sides are congruent; (b) opposite angles are congruent; (c) the diagonals intersect at their midpoints. *(Lesson 7-7, p. 405)*

Theorem: The distance between two given parallel lines is constant. *(Lesson 7-7, p. 406)*

Parallelogram Symmetry Theorem: Every parallelogram has 2-fold rotation symmetry about the intersection of its diagonals. *(Lesson 7-7, p. 407)*

Sufficient Conditions for a Parallelogram Theorem: If, in a quadrilateral, (a) one pair of sides is both parallel and congruent, or (b) both pairs of opposite sides are congruent, or (c) the diagonals bisect each other, or (d) both pairs of opposite angles are congruent, then the quadrilateral is a parallelogram. *(Lesson 7-8, p. 413)*

Exterior Angle Theorem: In a triangle, the measure of an exterior angle is equal to the sum of the measures of the interior angles at the other two vertices of the triangle. *(Lesson 7-9, p. 417)*

Exterior Angle Inequality: In a triangle, the measure of an exterior angle is greater than the measure of the interior angle at each of the other two vertices. *(Lesson 7-9, p. 417)*

Unequal Sides Theorem: If two sides of a triangle are not congruent, then the angles opposite them are not congruent, and the larger angle is opposite the longer side. *(Lesson 7-9, p. 419)*

Unequal Angles Theorem: If two angles of a triangle are not congruent, then the sides opposite them are not congruent, and the longer side is opposite the larger angle. *(Lesson 7-9, p. 419)*

Chapter 8

Equilateral Polygon Perimeter Formula: The perimeter p of an equilateral n-gon with sides of length s is given by the formula $p = ns$. *(Lesson 8-1, p. 438)*

Right Triangle Area Formula: The area of a right triangle is half the product of the lengths of its legs. *(Lesson 8-4, p. 453)*

Triangle Area Formula: The area of a triangle is half the product of a side (the base) and the altitude (height) to that side. *(Lesson 8-4, p. 454)*

Trapezoid Area Formula: The area of a trapezoid equals half the product of its altitude and the sum of the lengths of its bases. *(Lesson 8-5, p. 460)*

Parallelogram Area Formula: The area of a parallelogram is the product of one of its bases and the altitude to that base. *(Lesson 8-5, p. 461)*

Pythagorean Theorem: In any right triangle with legs of lengths a and b and hypotenuse of length c, $a^2 + b^2 = c^2$. *(Lesson 8-6, p. 467)*

Pythagorean Theorem (alternate statement): In any right triangle, the sum of the areas of the squares on its legs equals the area of the square on its hypotenuse. *(Lesson 8-6, p. 467)*

Pythagorean Converse Theorem: If a triangle has sides of lengths a, b, and c, and $a^2 + b^2 = c^2$, then the triangle is a right triangle. *(Lesson 8-6, p. 469)*

Circle Circumference Formula: If a circle has circumference C, diameter d, and radius r, then $C = \pi d$, or $C = 2\pi r$. *(Lesson 8-7, p. 475)*

Circle Area Formula: The area A of a circle with radius r is πr^2. *(Lesson 8-8, p. 481)*

Chapter 9

Line-Plane Perpendicularity Theorem: If a line is perpendicular to two different lines at their point of intersection, then it is perpendicular to the plane that contains those lines. *(Lesson 9-2, p. 501)*

The Four-Color Theorem: Suppose regions which share a border of some length must have different colors. Then any map of regions on a plane or a sphere can be colored in such a way that only four colors are needed. *(Lesson 9-9, p. 547)*

Chapter 10

Right Prism-Cylinder Lateral Area Formula: The lateral area, L.A., of a right prism (or right cylinder) is the product of its height h and the perimeter (circumference) p of its base. *(Lesson 10-1, p. 565)*

Prism-Cylinder Surface Area Formula: The surface area, S.A., of any prism or cylinder is the sum of its lateral area L.A. and twice the area B of a base. *(Lesson 10-1, p. 566)*

Pyramid-Cone Surface Area Formula: The surface area, S.A., of any pyramid or cone is the sum of its lateral area L.A. and the area B of its base. *(Lesson 10-2, p. 570)*

Regular Pyramid-Right Cone Lateral Area Formula: The lateral area, L.A., of a regular pyramid or right cone is half the product of its slant height ℓ and the perimeter (circumference) p of its base. *(Lesson 10-2, p. 573)*

Cube Volume Formula: The volume V of a cube with edge s is s^3. *(Lesson 10-3, p. 579)*

Prism-Cylinder Volume Formula: The volume V of any prism or cylinder is the product of its height h and the area B of its base. *(Lesson 10-5, p. 590)*

Pyramid-Cone Volume Formula: The volume V of any pyramid or cone equals $\frac{1}{3}$ the product of its height h and its base area B. *(Lesson 10-7, p. 601)*

Sphere Volume Formula: The volume V of any sphere is $\frac{4}{3}\pi$ times the cube of its radius r. *(Lesson 10-8, p. 605)*

Sphere Surface Area Formula: The total surface area S.A. of a sphere with radius r is $4\pi r^2$. *(Lesson 10-9, p. 610)*

Chapter 11

Distance Formula on the Coordinate Plane: The distance d between two points (x_1, y_1) and (x_2, y_2) in the coordinate plane is given by the formula $d = \sqrt{(x_2 - x_1)^2 + (y_2 - y_1)^2}$. *(Lesson 11-6, p. 652)*

Equation for a Circle: The circle with center (h, k) and radius r is the set of points (x, y) satisfying $(x - h)^2 + (y - k)^2 = r^2$. *(Lesson 11-7, p. 657)*

Number Line Midpoint Formula: On a number line, the coordinate of the midpoint of the segment with endpoints a and b is $\frac{a + b}{2}$. *(Lesson 11-8, p. 662)*

Coordinate Plane Midpoint Formula: In the coordinate plane, the midpoint of the segment with endpoints (x_1, y_1) and (x_2, y_2) is $\left(\frac{x_1 + x_2}{2}, \frac{y_1 + y_2}{2}\right)$. *(Lesson 11-8, p. 663)*

Midpoint Connector Theorem: The segment connecting the midpoints of two sides of a triangle is parallel to and half the length of the third side. *(Lesson 11-8, p. 664)*

Three-Dimension Distance Formula: The distance d between two points (x_1, y_1, z_1) and (x_2, y_2, z_2) is given by the formula $d = \sqrt{(x_2 - x_1)^2 + (y_2 - y_1)^2 + (z_2 - z_1)^2}$. *(Lesson 11-9, p. 669)*

Box Diagonal Formula: In a box with dimensions ℓ, w, and h, the length d of the diagonal is given by the formula $d = \sqrt{\ell^2 + w^2 + h^2}$. *(Lesson 11-9, p. 671)*

Equation for a Sphere: The sphere with center (h, k, j) and radius r is the set of points (x, y, z) satisfying $(x - h)^2 + (y - k)^2 + (z - j)^2 = r^2$. *(Lesson 11-9, p. 671)*

Three-Dimension Midpoint Formula: In space, the midpoint of the segment with endpoints (x_1, y_1, z_1) and (x_2, y_2, z_2) is $\left(\frac{x_1 + x_2}{2}, \frac{y_1 + y_2}{2}, \frac{z_1 + z_2}{2}\right)$. *(Lesson 11-9, p. 672)*

Varignon's Theorem: If the midpoints of consecutive sides of any quadrilateral are connected, then the quadrilateral so formed is a parallelogram. *(Chapter 11 Projects, p. 676)*

Chapter 12

Properties of S_k Theorem: Let S_k be the transformation mapping (x, y) onto (kx, ky). Then, under S_k,
(1) a line and its image are parallel, and
(2) the distance between two image points is k times the distance between their preimages. *(Lesson 12-1, p. 688)*

S_k Size-Change Theorem: When $k > 0$, the transformation S_k, where $S_k(x, y) = (kx, ky)$, is the size change with center $(0, 0)$ and magnitude k. *(Lesson 12-2, p. 694)*

Size-Change Distance Theorem: Under a size change with magnitude $k > 0$, the distance between any two image points is k times the distance between their preimages. *(Lesson 12-3, p. 699)*

Size-Change Preservation Properties Theorem: Every size transformation preserves (1) angle measure, (2) betweenness, and (3) collinearity. *(Lesson 12-3, p. 699)*

Figure Size-Change Theorem: If a figure is determined by certain points, then its size-change image is the corresponding figure determined by the size-change images of those points. *(Lesson 12-3, p. 701)*

Means-Extremes Property: If $\frac{a}{b} = \frac{c}{d}$ then $ad = bc$. *(Lesson 12-4, p. 706)*

Similar Figures Theorem: If two figures are similar, then: (1) corresponding angles are congruent, and (2) corresponding lengths are proportional. *(Lesson 12-5, p. 713)*

Theorem: If $G \sim G'$ and k is the ratio of similitude, then
(1) Perimeter$(G') = k \cdot$ Perimeter(G)
   or $\frac{\text{Perimeter}(G')}{\text{Perimeter}(G)} = k$,
(2)   Area$(G') = k^2 \cdot$ Area(G)
   or $\frac{\text{Area}(G')}{\text{Area}(G)} = k^2$, and
(3)   Volume$(G') = k^3 \cdot$ Volume(G)
   or $\frac{\text{Volume}(G')}{\text{Volume}(G)} = k^3$.
(Lesson 12-6, p. 718)

Fundamental Theorem of Similarity:
If two figures are similar with ratio of similitude k, then: (a) corresponding angle measures are equal; (b) corresponding lengths and perimeters are in the ratio k; (c) corresponding areas and surface areas are in the ratio k^2; and (d) corresponding volumes are in the ratio k^3. *(Lesson 12-6, p. 719)*

Chapter 13

SSS Similarity Theorem: If the three sides of one triangle are proportional to three sides of a second triangle, then the triangles are similar. *(Lesson 13-1, p. 739)*

AA Similarity Theorem: If two angles of one triangle are congruent to two angles of another, then the triangles are similar. *(Lesson 13-2, p. 745)*

SAS Similarity Theorem: If, in two triangles, the ratios of two pairs of corresponding sides are equal and the included angles are congruent, then the triangles are similar. *(Lesson 13-2, p. 747)*

Side-Splitting Theorem: If a line is parallel to a side of a triangle and intersects the other two sides in distinct points, it splits these sides into proportional segments. *(Lesson 13-3, p. 751)*

Side-Splitting Converse Theorem: If a line intersects $\overline{OP}$ and $\overline{OQ}$ in distinct points X and Y so that $\frac{OX}{XP} = \frac{OY}{YQ}$ then $\overleftrightarrow{XY} \; /\!/ \; \overleftrightarrow{PQ}$. *(Lesson 13-3, p. 752)*

Geometric Mean Theorem: The positive geometric mean of the positive numbers a and b is $\sqrt{ab}$. *(Lesson 13-4, p. 759)*

Right Triangle Altitude Theorem: In a right triangle,
(1) the altitude to the hypotenuse is the geometric mean of the segments into which it divides the hypotenuse, and
(2) each leg is the geometric mean of the hypotenuse and the segment of the hypotenuse adjacent to the leg. *(Lesson 13-4, p. 761)*

Isosceles Right Triangle Theorem: In an isosceles right triangle, if a leg is x, then the hypotenuse is $x\sqrt{2}$. *(Lesson 13-5, p. 765)*

30-60-90 Triangle Theorem: In a 30-60-90 (right) triangle, if the short leg is x, then the longer leg is $x\sqrt{3}$, and the hypotenuse is $2x$. *(Lesson 13-5, p. 767)*

SAS Triangle Area Formula: In any $\triangle ABC$, the area of $\triangle ABC = \frac{1}{2} ab \sin C$. *(Lesson 13-8, p. 786)*

Chapter 14

Arc-Chord Congruence Theorem: In a circle or in congruent circles:
(1) If two arcs have the same measure, they are congruent and their chords are congruent.
(2) If two chords have the same length, their minor arcs have the same measure.
(Lesson 14-1, p. 801)

Chord-Center Theorem:
(1) The line that contains the center of a circle and is perpendicular to a chord bisects the chord.
(2) The line that contains the center of a circle and the midpoint of a chord bisects the central angle of the chord.
(3) The bisector of the central angle of a chord is the perpendicular bisector of the chord.
(4) The perpendicular bisector of a chord of a circle contains the center of the circle.
(Lesson 14-1, p. 802)

Inscribed Angle Theorem: In a circle, the measure of an inscribed angle is one-half the measure of its intercepted arc. *(Lesson 14-2, p. 807)*

Theorem: An angle inscribed in a semicircle is a right angle. *(Lesson 14-2, p. 809)*

Theorem: In a circle, if two inscribed angles intercept the same arc, then they have the same measure. *(Lesson 14-2, p. 811)*

Angle-Chord Theorem: The measure of an angle formed by two intersecting chords is one-half the sum of the measures of the arcs intercepted by it and its vertical angle. *(Lesson 14-4, p. 818)*

Angle-Secant Theorem: The measure of an angle formed by two secants intersecting outside a circle is half the difference of the arcs intercepted by the angle. *(Lesson 14-4, p. 820)*

Theorem: If a line is perpendicular to a radius of a circle at the radius's endpoint on the circle, then the line is tangent to the circle. *(Lesson 14-5, p. 823)*

Theorem: If a line is tangent to a circle, then it is perpendicular to the radius containing the point of tangency. *(Lesson 14-5, p. 824)*

Radius-Tangent Theorem: A line is tangent to a circle if and only if it is perpendicular to the radius at the radius's endpoint on the circle. *(Lesson 14-5, p. 825)*

Tangent-Chord Theorem: The measure of an angle formed by a tangent and a chord is half the measure of its intercepted arc. *(Lesson 14-6, p. 830)*

Tangent-Secant Theorem: The measure of the angle between two tangents, or between a tangent and a secant, is half the difference of the intercepted arcs. *(Lesson 14-6, p. 830)*

Secant Length Theorem: Suppose one secant intersects a circle at A and B, and a second secant intersects the circle at C and D. If the secants intersect at P, then $AP \cdot BP = CP \cdot DP$. *(Lesson 14-7, p. 836)*

Tangent Square Theorem: The power of point P for $\odot O$ is the square of the length of a segment tangent to $\odot O$ from P. *(Lesson 14-7, p. 838)*

Isoperimetric Theorem: Of all plane figures with the same perimeter, the circle has the maximum area. *(Lesson 14-8, p. 843)*

Isoperimetric Inequality: If a plane figure has area A and perimeter p, then $A \leq \frac{p^2}{4\pi}$, and $A = \frac{p^2}{4\pi}$ only when the figure is a circle. *(Lesson 14-8, p. 843)*

Isoperimetric Theorem (alternate statement): Of all plane figures with the same area, the circle has the least perimeter. *(Lesson 14-8, p. 844)*

Isoperimetric Theorem (3-dimensional version): Of all solids with the same surface area, the sphere has the largest volume. *(Lesson 14-9, p. 848)*

Isoperimetric Theorem (alternate 3-dimensional version): Of all solids with the same volume, the sphere has the least surface area. *(Lesson 14-9, p. 849)*

Isoperimetric Inequality for Three-Dimensional Figures: If a solid has volume V and surface area s, then $V \leq \sqrt{\frac{s^3}{36\pi}}$ and $V = \sqrt{\frac{s^3}{36\pi}}$ only when the solid is a sphere. *(Lesson 14-9, p. 852)*

Postulates

Postulates are statements that are assumed true. The postulates listed below may be different from those found in other geometry books.

Postulates of Euclidean Geometry

Point-Line-Plane Postulate

a. Unique Line Assumption: Through any two points, there is exactly one line.
(Lesson 1-7, p. 41; Lesson 9-1, p. 497)

b. Number Line Assumption: Every line is a set of points that can be put into a one-to-one correspondence with the real numbers, with any point on it corresponding to 0 and any other point corresponding to 1.
(Lesson 1-7, p. 42; Lesson 9-1, p. 497)

c. Dimension Assumption: (1) Given a line in a plane, there is at least one point in the plane that is not on the line. (2) Given a plane in space, there is at least one point in space that is not in the plane.
(Lesson 1-7, p. 42; Lesson 9-1, p. 497)

d. Flat Plane Assumption: If two points lie in a plane, the line containing them lies in the plane.
(Lesson 9-1, p. 497)

e. Unique Plane Assumption: Through three noncollinear points, there is exactly one plane.
(Lesson 9-1, p. 497)

f. Intersecting Planes Assumption: If two different planes have a point in common, then their intersection is a line. *(Lesson 9-1, p. 498)*

Some Postulates from Arithmetic and Geometry

Distance Postulate

a. Uniqueness Property: On a line, there is a unique distance between two points. *(Lesson 1-8, p. 48)*

b. Distance Formula: If the two points on a line have coordinates x and y, the distance between them is $|x - y|$. *(Lesson 1-8, p. 48)*

c. Additive Property: If B is on $\overline{AC}$, then $AB + BC = AC$. *(Lesson 1-8, p. 49)*

Triangle Inequality Postulate

The sum of the lengths of any two sides of a triangle is greater than the length of the third side.
(Lesson 2-7, p. 103)

Angle Measure Postulate

a. Unique Measure Assumption: Every angle has a unique measure from 0° to 180°.

b. Unique Angle Assumption: Given any ray $\overrightarrow{VA}$ and any real number r between 0 and 180, there is a unique angle BVA in each half-plane of $\overleftrightarrow{VA}$ such that $m\angle BVA = r$.

c. Zero Angle Assumption: If $\overrightarrow{VA}$ and $\overrightarrow{VB}$ are the same ray, then $m\angle AVB = 0$.

d. Straight Angle Assumption: If $\overrightarrow{VA}$ and $\overrightarrow{VB}$ are opposite rays, then $m\angle AVB = 180$.

e. Angle Addition Property: If $\overrightarrow{VC}$ (except for point V) is in the interior of $\angle AVB$, then $m\angle AVC + m\angle CVB = m\angle AVB$. *(Lesson 3-1, p. 126)*

Postulates of Equality

For any real numbers a, b, and c:

Reflexive Property of Equality: $a = a$.
Symmetric Property of Equality: If $a = b$, then $b = a$.
Transitive Property of Equality: If $a = b$ and $b = c$, then $a = c$. *(Lesson 3-4, p. 145)*

Postulates of Equality and Operations

For any real numbers a, b, and c:

Addition Property of Equality:
If $a = b$, then $a + c = b + c$.

Multiplication Property of Equality:
If $a = b$, then $ac = bc$.

(Lesson 3-4, p. 145)

Postulates of Inequality and Operations

For any real numbers a, b, and c:

Transitive Property of Inequality:
If $a < b$ and $b < c$, then $a < c$.

Addition Property of Inequality:
If $a < b$, then $a + c < b + c$.

Multiplication Properties of Inequality:
If $a < b$ and $c > 0$, then $ac < bc$.
If $a < b$ and $c < 0$, then $ac > bc$.

(Lesson 3-4, p. 146)

Postulates of Equality and Inequality

For any real numbers a, b, and c:

Equation to Inequality Property: If a and b are positive numbers and $a + b = c$, then $c > a$ and $c > b$.

Substitution Property: If $a = b$, then a may be substituted for b in any expression.

(Lesson 3-4, p. 147)

Corresponding Angles Postulate

Suppose two coplanar lines are cut by a transversal.
a. If two corresponding angles have the same measure, then the lines are parallel.
b. If the lines are parallel, then corresponding angles have the same measure.

(Lesson 3-6, p. 156)

Reflection Postulate

Under a reflection:
a. There is a 1-1 correspondence between points and their images.
b. Collinearity is preserved. If three points A, B, and C lie on a line, then their images A', B', and C' are collinear.
c. Betweenness is preserved. If B is between A and C, then the image B' is between the images A' and C'.
d. Distance is preserved. If $\overline{A'B'}$ is the image of $\overline{AB}$, then $A'B' = AB$.
e. Angle measure is preserved. If $\angle A'C'E'$ is the image of $\angle ACE$, then $m\angle A'C'E' = m\angle ACE$.
f. Orientation is reversed. A polygon and its image, with vertices taken in corresponding order, have opposite orientations. *(Lesson 4-2, pp. 191, 193)*

Area Postulate

a. **Uniqueness Property:** Given a unit region, every polygonal region has a unique area.
b. **Rectangle Formula:** The area A of a rectangle with dimensions ℓ and w is ℓw.
c. **Congruence Property:** Congruent figures have the same area.
d. **Additive Property:** The area of the union of two nonoverlapping regions is the sum of the areas of the regions.

(Lesson 8-2, p. 442)

Volume Postulate

a. **Uniqueness Property:** Given a unit cube, every polyhedral region has a unique volume.
b. **Box Volume Formula:** The volume V of a box with dimensions ℓ, w, and h is found by the formula $V = \ell wh$.
c. **Congruence Property:** Congruent figures have the same volume.
d. **Additive Property:** The volume of the union of two nonoverlapping solids is the sum of the volumes of the solids.
(Lesson 10-3, p. 577)
e. **Cavalieri's Principle:** Let I and II be two solids included between parallel planes. If every plane P parallel to the given planes intersects I and II in sections with the same area, then Volume(I) = Volume(II). *(Lesson 10-5, p. 589)*

Trichotomy Law: Of two real numbers a and b, either $a < b$, $a = b$, or $a > b$, and no two of these can be true at the same time. *(Lesson 11-3, p. 637)*

Postulates of Logic

Law of Detachment: From a true conditional $p \Rightarrow q$ and a statement or given information p, you may conclude q. *(Lesson 11-1, p. 623)*
Law of Transitivity: If $p \Rightarrow q$ and $q \Rightarrow r$ are true, then $p \Rightarrow r$ is true. (also known as Transitivity Property of Implication) *(Lesson 11-1, p. 623)*
Law of the Contrapositive: A conditional ($p \Rightarrow q$) and its contrapositive (*not-q* $\Rightarrow$ *not-p*) are either both true or both false. *(Lesson 11-2, p. 630)*
Law of Ruling Out Possibilities: When p or q is true and q is not true, then p is true.
(Lesson 11-3, p. 634)
Law of Indirect Reasoning: If valid reasoning from a statement p leads to a false conclusion, then p is false. *(Lesson 11-4, p. 641)*

Formulas

In this book, we use many measurement formulas. The following symbols are used: a, b, c = sides; A = area; B = area of base; b_1, b_2 = bases; C = circumference; d = diameter; d_1, d_2 = diagonals; e = edge; h = height; ℓ = slant height (in conics); ℓ = length; L.A. = lateral area; n = number of sides; p = perimeter; r = radius; s = side; S.A. = surface area; V = volume; w = width; θ = measure of angle.

Two-Dimensional Figures

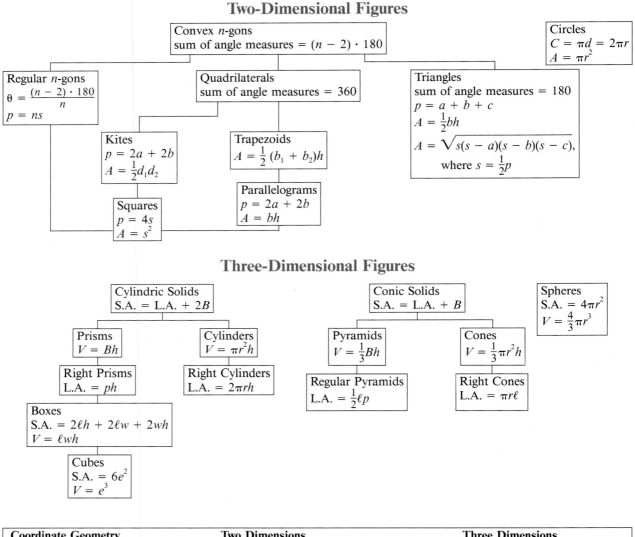

Convex *n*-gons
sum of angle measures = $(n - 2) \cdot 180$

Circles
$C = \pi d = 2\pi r$
$A = \pi r^2$

Regular *n*-gons
$\theta = \dfrac{(n - 2) \cdot 180}{n}$
$p = ns$

Quadrilaterals
sum of angle measures = 360

Triangles
sum of angle measures = 180
$p = a + b + c$
$A = \frac{1}{2}bh$
$A = \sqrt{s(s - a)(s - b)(s - c)}$,
where $s = \frac{1}{2}p$

Kites
$p = 2a + 2b$
$A = \frac{1}{2}d_1 d_2$

Trapezoids
$A = \frac{1}{2}(b_1 + b_2)h$

Parallelograms
$p = 2a + 2b$
$A = bh$

Squares
$p = 4s$
$A = s^2$

Three-Dimensional Figures

Cylindric Solids
S.A. = L.A. + 2B

Conic Solids
S.A. = L.A. + B

Spheres
S.A. = $4\pi r^2$
$V = \frac{4}{3}\pi r^3$

Prisms
$V = Bh$

Cylinders
$V = \pi r^2 h$

Pyramids
$V = \frac{1}{3}Bh$

Cones
$V = \frac{1}{3}\pi r^2 h$

Right Prisms
L.A. = ph

Right Cylinders
L.A. = $2\pi rh$

Regular Pyramids
L.A. = $\frac{1}{2}\ell p$

Right Cones
L.A. = $\pi r \ell$

Boxes
S.A. = $2\ell h + 2\ell w + 2wh$
$V = \ell wh$

Cubes
S.A. = $6e^2$
$V = e^3$

Coordinate Geometry	Two Dimensions	Three Dimensions
Distance	$d = \sqrt{(x_2 - x_1)^2 + (y_2 - y_1)^2}$	$d = \sqrt{(x_2 - x_1)^2 + (y_2 - y_1)^2 + (z_2 - z_1)^2}$
Midpoint of a segment	$M = \left(\dfrac{x_1 + x_2}{2}, \dfrac{y_1 + y_2}{2}\right)$	$M = \left(\dfrac{x_1 + x_2}{2}, \dfrac{y_1 + y_2}{2}, \dfrac{z_1 + z_2}{2}\right)$
Slope of a line	$m = \dfrac{y_2 - y_1}{x_2 - x_1}$	

Conversion Formulas

	Customary	**Metric**
Length	ft = foot, in. = inch, yd = yard, mi = mile	cm = centimeter, mm = millimeter, m = meter, km = kilometer

1 ft = 12 in.   1 cm = 10 mm
1 yd = 3 ft   1 m = 100 cm
1 mi = 5280 ft   1 km = 1000 m

1 in. = 2.54 cm
1 ft = 0.3048 m
1 yd = 0.9144 m
1 mi $\approx$ 1.609 km

Area

1 sq ft = 144 sq in.   1 sq cm = 100 sq mm
1 sq yd = 9 sq ft   1 sq m = 10,000 sq cm

1 sq in. = 6.4516 sq cm
1 sq yd $\approx$ 0.836 sq m
1 sq mi $\approx$ 2.5889 sq km

Land Area

1 sq mi = 640 acres   1 hectare = 10,000 sq m

1 hectare $\approx$ 2.471 acres

Volume

$1 \text{ ft}^3 = 1728 \text{ in}^3$ $1 \text{ cm}^3 = 1000 \text{ mm}^3$
$1 \text{ yd}^3 = 27 \text{ ft}^3$ $1 \text{ m}^3 = 1,000,000 \text{ cm}^3$

$1 \text{ in}^3 \approx 16.387 \text{ cm}^3$
$1 \text{ yd}^3 \approx 0.765 \text{ m}^3$

Liquid Volume

pt = pint, qt = quart, gal = gallon, oz = ounce   mL = milliliter, L = liter

$1 \text{ pt} = 28.875 \text{ in}^3 = 16 \text{ oz}$ $1 \text{ mL} = 1000 \text{ mm}^3$
$1 \text{ qt} = 57.75 \text{ in}^3 = 32 \text{ oz}$ $1 \text{ L} = 1000 \text{ cm}^3$
$1 \text{ gal} = 231 \text{ in}^3 = 128 \text{ oz}$

1 oz $\approx$ 29.574 mL
1 qt $\approx$ 0.946 L
1.057 qt $\approx$ 1 L

Weight (Mass)

lb = pound, oz = ounce   g = gram, kg = kilogram, mg = milligram

1 lb = 16 oz   1 g = 1000 mg
1 ton = 2000 lb   1 kg = 1000 g
       1 metric ton = 1000 kg

1 oz $\approx$ 28.350 g
1 lb $\approx$ 0.4536 kg
2.2 lb $\approx$ 1 kg
1 ton $\approx$ 1.016 metric ton

LESSON 1-1 (pp. 6–10)

13. a. 786,432 **b.** 307,200
15. a. black squares
b. by clustering the squares
17. about 60 square feet
19. See right.

19.

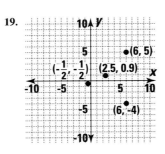

LESSON 1-2 (pp. 11–16)

13. The road distance is measured along roads, which do not always go directly from one point to another. **15. a.** $AB = 150$, $BC = 225$, $AC = 375$ **b.** True **17. a.** rows of columns **b.** walls and steps **c.** Answers may vary. **d.** Sample: a landscape plan showing fences and trees or bushes **19.** Sample: **See below.** **21. See below.**

19. **21.**

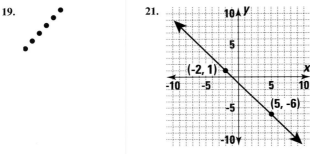

LESSON 1-3 (pp. 17–21)

13. $y = -5$ **15. a, b.** See below. **c.** $(3, -2)$ **17. a.** Sample: (10, 95), (16, 80), (20, 70), (30, 45), (40, 20) **b.** Samples: **See below.** **c.** Yes **19.** about 1.8 cm **21. a.** 94 miles **b.** about 1.7 hours (1 hour 42 minutes)

15. a, b. **17. b.**

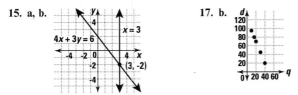

LESSON 1-4 (pp. 22–28)

15. a. 0 even, 4 odd **b.** No **17.** No; there are more than 2 odd nodes. **19. a.** cities **b.** flight routes **c.** No **21.** Yes, **See below.** **23. See below.** **25.** 5 **27.** False

21. **23.**

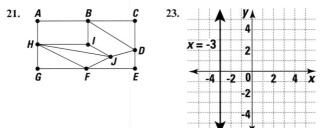

LESSON 1-5 (pp. 30–34)

9. See below. 11. Answers may vary. **13.** No. There are more than 2 odd nodes. **15. a. See below. b.** No **17.** (b) **19. a. See below. b.** Answers may vary; the actual point of intersection is (2, 4). **21. See below.**

9. **15. a.**

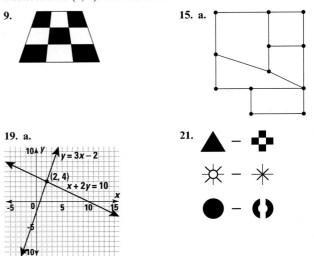

19. a. **21.**

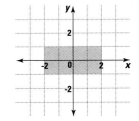

LESSON 1-6 (pp. 35–40)

11. a. C, B, and D **b.** Sample: A, E, and F **13.** Sample: 3; concord, harmony, agreement, concord **15.** Sample: 2; satire, irony, satire **17.** Sample: **See below. 19.** No **21. See below. 23.** (b) **25.** 5.3 **27. See below.**

17.

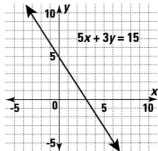

21. **27.**

LESSON 1-7 (pp. 41–45)
11. a. True **b.** Number Line Assumption **13.** True
15. a. See below. **b.** Yes **17. a.** Sample: See right. **b.** Sample:
See right. **19. a.** 6 **b.** 12 **c.** No, there are more than two odd
nodes. **21. a.** horizontal **b.** oblique **23.** $^-16$ or 8

15. a.

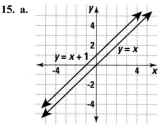

17. a.

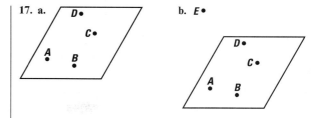

b. E•

LESSON 1-8 (pp. 46–51)
13. Sample: $^-1.8$ **15.** 60 **17.** The graph of $x > 1$ does not contain
an endpoint. **19. a.** Yes. Sample: $AB + BE = AE$ **b.** No. Sample:
$BI + IO \neq BO$ **21.** Unique Line Assumption, Dimension
Assumption **23.** point, line, plane **25.** U opposite M, G opposite
S, P opposite C

CHAPTER 1 PROGRESS SELF-TEST (pp. 56–57)
1. $|^-31 - 22| = |^-53| = 53$ **2.** See right. **3.** See right.
4. $14.4 - 5.3 = 9.1$ **5.** two **6.** circularity **7.** $|4 - 32| = |^-28| = 28''$
8. The air distance is measured directly from airport to airport; the
road distance is probably measured from two other points and is
not a straight path. **9.** Sample: See right. **10.** Sample: There can
be more than one arc connecting two nodes. See right. **11.** False
12. (a) **13.** See right. **14.** vertical **15.** oblique **16. a.** Yes
b. Sample: V to Z to T to V to W to Z to X to W **17.** ray
18. Sample: $\overleftrightarrow{AD}$ **19.** $11 + BC + 18 = 41; BC = 12$ **20.** No, the
rays have different endpoints. **21. a.** numbers which indicate hours
b. the border of the clock face **22.** $(81 + 26) - 96 = 11$ miles
23. a. See right. **b.** The hallways do not form a traversable network
because they have more than two odd nodes.

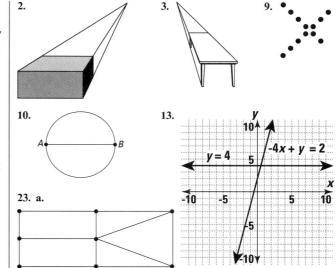

The chart below keys the **Progress Self-Test** questions to the objectives in the **Chapter Review** on pages 58–61 or to the **Vocabulary** (Voc.) on page 55. This will enable you to locate those **Chapter Review** questions that correspond to questions missed on the **Progress Self-Test.** The lesson where the material is covered is also indicated on the chart.

Question	1	2	3	4	5	6	7	8	9	10
Objective	K	C	C	H	E	G	K	I	A	F
Lesson	1-2	1-5	1-5	1-8	1-6	1-6	1-2	1-2	1-1	1-4
Question	**11**	**12**	**13**	**14**	**15**	**16**	**17**	**18**	**19**	**20**
Objective	G	F	L	L	L	B	H	D	H	D
Lesson	1-6	1-4	1-3	1-3	1-3	1-4	1-8	1-8	1-8	1-8
Question	**21**	**22**	**23**							
Objective	J	I	J							
Lesson	1-1	1-2	1-4							

CHAPTER 1 REVIEW (pp. 58–61)
1. Sample: See page 874. **3.** Sample: See page 874. **5. a.** 5 **b.** 2
7. a. No **b.** It has more than two odd nodes. **9.** Sample:
See page 874. **11. a.** not in perspective **13.** $\overrightarrow{UT}$ **15.** Yes, a segment
is named by its endpoints in either order. **17.** 2 **19.** 1 **21.** (a)
23. True **25.** There can be more than 1 overlapping line that
contains the same 2 points. **27.** circularity **29.** to explain
undefined terms and to serve as a starting point for logically
deducing other statements **31.** (c) **33.** False **35.** $^-17 < x < 14$

37. a. See right. **b.** ray **39.** 16.8 **41.** 8° **43.** 96 miles
45. a. Sample: See right. **b.** Sample: B to C to D to G to C to H to
B to A to H to F to A to E to F to G to E to D. **47. a.** tables
b. chairs **49.** 2.5 **51.** 78 **53.** $|x - y|$ or $|y - x|$ **55.** $\frac{1}{4}$ or $\frac{3}{4}$
57. See right. **59.** See right. **61.** oblique **63.** horizontal

1. **3.** **9.**

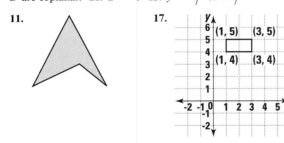

37. a.

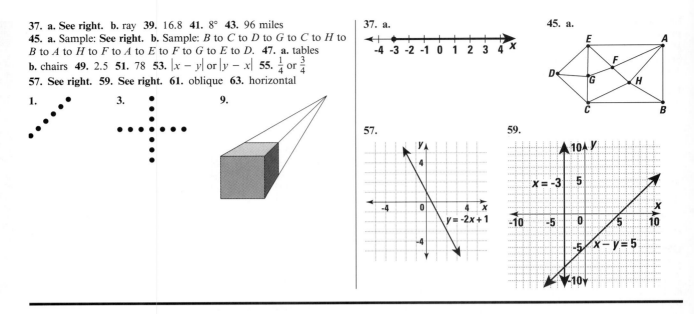

45. a.

57.

59.

LESSON 2-1 (pp. 64–68)
11. Sample: See below. **13.** True **15. a.** Answers may vary. Figure
I will fit our definition. **b.** Answers may vary. **17.** Answers may
vary. The figure will fit our definition. See below. **19.** Answers may
vary for each answer; it will fit our definition because A, B, C, and
D are coplanar. **21.** $a = -6$ **23.** $y = \frac{540}{7}$ or $77\frac{1}{7}$

11.

17.

LESSON 2-2 (pp. 69–75)
15. a. You are in Springfield, Massachusetts. **b.** Sample: You are in
Springfield, Illinois. **c.** false **17.** Sample: If $x \geq 9$, then $x > 8$.
19. She will wear Nyebox shoes. **21. a.** Sample: If a person is 14
years old, then that person is a teenager. **b.** Sample: If a person is
14 years old, then that person is a girl. **23. a.** nonconvex
b. nonconvex **c.** convex **25.** $x = 6$

LESSON 2-3 (pp. 76–80)
9. a. If a polygon has four sides, then it is a quadrilateral. **b.** If a
polygon is a quadrilateral, then it has four sides. **11.** Sample:
A $4' \times 5'$ rectangle. **13. a.** Sample: robin **b.** Sample: ostrich
15. When $T = 10$, the antecedent is true. It is an instance of the
conditional. **17.** Sample: to avoid circularity **19.** 56.6 miles
21. $W = \frac{P - 2L}{2}$

LESSON 2-4 (pp. 81–86)
9. a. contains more information than is necessary **b.** uses terms
not yet defined **c.** inaccurate **11.** characteristics $\Rightarrow$ term **13.** A set
is convex if and only if all segments connecting points of the set lie
entirely in the set. **15.** Sample: A bisector of a segment is a line,
ray, or segment that intersects the segment at its midpoint.
17. a. A bird is a dodo if and only if it is extinct, was flightless, and

was once found on the island of Mauritius. **b.** *Raphus cucullatus*:
it would not be informative, and would lead directly to circularity.
c. Answers may vary. Samples: No, because it does not accurately
describe the bird defined and there may have been other extinct
flightless birds on the island of Mauritius. Yes; because the
dictionary provides information, not just a definition. **19. a.** If A is
between B and C, then $\overrightarrow{AB}$ and $\overrightarrow{AC}$ are opposite rays.
b. Yes **21.** See below. **23.** $161 > q > 71$ or $71 < q < 161$

21.

LESSON 2-5 (pp. 87–93)
13. $G \cup H$ = all residents of Indonesia; $G \cap H$ = all residents of
Bali, Indonesia. **15.** $G \cup H$ = students in all geometry classes;
$G \cap H = \varnothing$ **17.** See below. **19.** (b) **21.** If P is on circle O with
radius r, then $PO = r$. If $PO = r$, then P is on circle O with
radius r. **23. a.** If $AC = AB - BC$, then B is on $\overrightarrow{AC}$, but not
between A and C. **b.** Yes **c.** Yes **25.** Sample: See below.
27. $x = 120$; check: Does $120 = 2(180 - 120)$? Yes, $120 = 2(60)$.

17. **25.**

LESSON 2-6 (pp. 95–100)
15. Sample: See below. **17.** See below. **19.** octagon **21.** First
figure: It is not a union of segments. **23.** Second figure: Some
segments intersect more than two others. **25. a.** $\{-5, -3, 0, 3, 5, 10\}$
b. $\{0\}$ **27. a.** 10 **b.** -2 **c.** $\frac{x + y}{2}$ **29. a.** $\overleftrightarrow{XY}$ **b.** $\overline{XY}$ **c.** XY

15. **17.**

figure
|
polygon
|
triangle   quadrilateral
|
isosceles triangle

874

LESSON 2-7 (pp. 101–106)

13. when $AC = CB$ **15. a.** for B on $\overline{AC}$ **b.** for B not on $\overline{AC}$
c. never true **17.** 15, 55 **19.** Sample: See below. **21. a.** $\overline{TO}$
b. Quadrilateral $OMTS \cup \overline{TO}$ **c.** $\overline{MT} \cup \overline{MO}$ **23.** $y = -\frac{1}{2}x + 2$

19.

LESSON 2-8 (pp. 107–112)

11. a. (i) **b.** (iii) **c.** (ii) **13. a.** (iii) **b.** (iii) **c.** (i) **15.** Answers will
vary. **17.** Answers will vary. (The statement is not true.) See below.
19. Yes **21. a.** convex **b.** nonconvex **23. a.** If you are looking for
the "ultimate" new baby gift, then call 1-800-555-BABY for details.
b. If you call 1-800-555-BABY, then you are looking for the
"ultimate" new baby gift.

17.

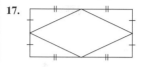

CHAPTER 2 PROGRESS SELF-TEST (pp. 116–117)

1. It includes too much information (violates property III).
2. a. (iii) **b.** (i) **c.** (iv) **3.** <u>Two angles have equal measure</u> if
<u>they are vertical angles</u>. **4.** If a figure is a trapezoid, then it is a
quadrilateral. **5. a.** $ABCD \cup \overline{AC}$ **b.** $\overline{AD} \cup \overline{DC}$ **6.** C **7.** The shelf
falls if and only if there are over 50 books on that shelf.
8. Sample: One might receive a passing grade as a result of earning
high test scores, even though one hadn't done their homework
every night. **9.** nonconvex **10.** Sample: See right. **11.** If a figure is
an angle, then it is the union of two rays with a common endpoint.
If a figure is the union of two rays with a common endpoint, then
it is an angle. **12.** Let $T = y$. By the definition of midpoint $BO =$
OT. $BO = 2 - (-10) = 12$. So $BT = 12$.
$BT = y - 2 = 12$. Solve $y - 2 = 12$ for y. $y = 14$ **13. a.** Sample:
$x = 10$ **b.** Sample: $x = 8$ **14.** The Triangle Inequality Postulate
says that the sum of any two sides of a triangle is greater than the
third side. $11 + 11 > 12$, $11 + 12 > 11$, $12 + 11 > 11$. Yes.

15. Use the Triangle Inequality Postulate. The distance is either
148 or 428 miles. **16.** Two of the segments do not intersect two
others. **17.** See below. **18.** By the definition of midpoint,
$PL = PT$. $3(7 - x) = 2x - 5$. So $x = \frac{26}{5} = 5.2$. $TL = 3(7 - x) +$
$2x - 5 = 10.8$ **19.** 18-gon **20. a.** R or S **b.** False **c.** $\overline{OC}$, $\overline{OI}$,
$\overline{OR}$, $\overline{OT}$ **21.** If a figure is an equilateral triangle, then it is an
isosceles triangle. **22. a.** If a figure is an isosceles triangle, then it is
an equilateral triangle. **b.** No. An isosceles triangle might have only
two sides congruent. **23. a.** (i) **b.** (ii) **c.** (iii)

10.

17.

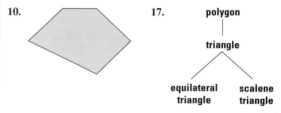

The chart below keys the **Progress Self-Test** questions to the objectives in the **Chapter Review** on pages 118–120 or to the **Vocabulary**
(Voc.) on page 115. This will enable you to locate those **Chapter Review** questions that correspond to questions missed on the **Progress Self-Test**. The lesson where the material is covered is also indicated on the chart.

Question	1	2	3	4	5	6	7	8
Objective	F	B	G	G	I	I	C	K
Lesson	2-4	2-6	2-2	2-2	2-5	2-5	2-4	2-3

Question	9	10	11	12	13	14	15	16
Objective	A	A	G	D	H	J	M	B
Lesson	2-1	2-1	2-4	2-4	2-2	2-7	2-7	2-6

Question	17	18	19	20	21	22	23
Objective	N	D	L	B	G	E, H	H
Lesson	2-6	2-4	2-6	2-6	2-2	2-3, 2-2	2-2, 2-8

CHAPTER 2 REVIEW (pp. 118–120)

1. nonconvex **3.** nonconvex **5.** Sample: See page 876. **7.** Sample:
See page 876. **a.** Sample: A, B **b.** Sample: $\overline{AB}$, $\overline{CD}$ **c.** Sample: $\overline{AC}$
9. a. (iv) **b.** (ii) **c.** (i) **11.** Not all segments intersect two other
segments. **13.** If $\triangle ABC$ has three 60° angles, then $\triangle ABC$ is
equilateral. **15.** $r \Rightarrow p$ **17.** 4.5 **19.** 36 **21. a.** If the temperature of
water is less than 32°F, then it freezes. **b.** True **23.** $t \Rightarrow r$
25. It does not accurately describe the idea being defined. (violates

property II) **27.** circle, line, intersects, point **29.** If a figure is a
radius, then it is a segment. **31.** <u>A figure is a rectangle</u> if
<u>it is a square</u>. **33.** You are at 0° latitude if and only if you are at
the equator. **35. a.** (i) **b.** (iii) **37. a.** Sample: See page 876.
b. Sample: See page 876. **39.** {5, 9} **41.** Sample: See page 876.
43. No **45.** between 3 cm and 11 cm **47.** If Bill is able to watch
his favorite TV program, then he will hurry home. **49.** If a person
wears Nyebox shoes, then that person can play like the pros.
51. pentagon **53.** hexagon **55.** from 9 to 15 blocks **57.** See
page 876.

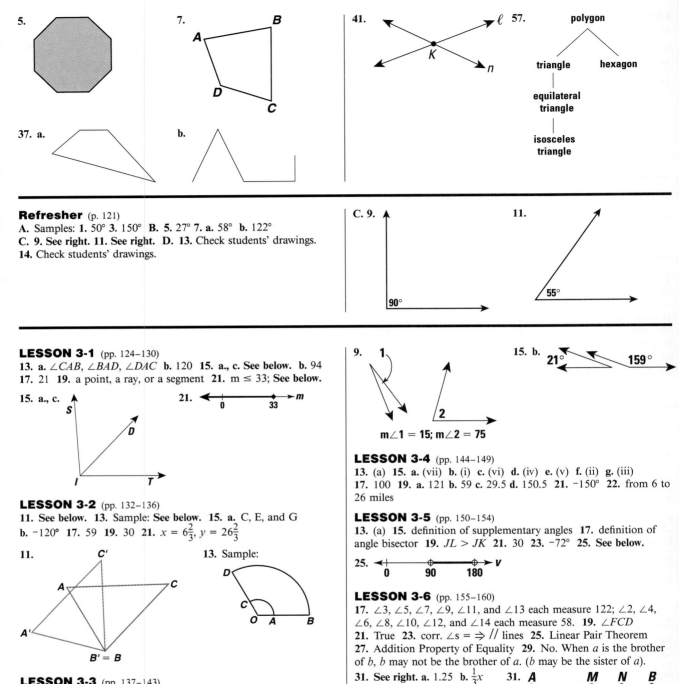

5. **7.**

A
B
D
C

41. ℓ **57.** polygon

K
n

triangle   hexagon

equilateral
triangle

isosceles
triangle

37. a.   **b.**

Refresher (p. 121)
A. Samples: **1.** 50° **3.** 150° **B. 5.** 27° **7. a.** 58° **b.** 122°
C. 9. See right. **11.** See right. **D. 13.** Check students' drawings.
14. Check students' drawings.

C. 9.   **11.**

90°

55°

LESSON 3-1 (pp. 124–130)
13. a. ∠CAB, ∠BAD, ∠DAC **b.** 120 **15. a., c. See below. b.** 94
17. 21 **19.** a point, a ray, or a segment **21.** m ≤ 33; **See below.**

15. a., c.   **21.**

S
D
I   T

0   33   m

9. 1

2

m∠1 = 15; m∠2 = 75

15. b. 21° 159°

LESSON 3-2 (pp. 132–136)
11. See below. 13. Sample: **See below. 15. a.** C, E, and G
b. −120° **17.** 59 **19.** 30 **21.** $x = 6\frac{2}{3}$, $y = 26\frac{2}{3}$

11.   **13.** Sample:

C'
A   C
A'
B' = B

D
C
O A   B

LESSON 3-3 (pp. 137–143)
9. Sample: **See right. 11.** 49 and 41 **13.** Points C, P, and B do
not seem to be collinear. Thus ∠APB and ∠BPD, and ∠DPC and
∠CPA are linear pairs. ∠APB and ∠BPD, ∠DPC and ∠CPA,
∠CPA and ∠APB are adjacent angles. There are no vertical angles.
15. a. If two angles are supplementary, then they form a linear
pair. **b.** Sample: the two angles may not be adjacent. **See right.**
17. 0 < t < 45 **19.** −120° or 240° **21. a.** 20° **b.** 180° **c.** 200°
23. They are all equal.

LESSON 3-4 (pp. 144–149)
13. (a) **15. a.** (vii) **b.** (i) **c.** (vi) **d.** (iv) **e.** (v) **f.** (ii) **g.** (iii)
17. 100 **19. a.** 121 **b.** 59 **c.** 29.5 **d.** 150.5 **21.** −150° **22.** from 6 to
26 miles

LESSON 3-5 (pp. 150–154)
13. (a) **15.** definition of supplementary angles **17.** definition of
angle bisector **19.** JL > JK **21.** 30 **23.** −72° **25. See below.**

25.
0   90   180   V

LESSON 3-6 (pp. 155–160)
17. ∠3, ∠5, ∠7, ∠9, ∠11, and ∠13 each measure 122; ∠2, ∠4,
∠6, ∠8, ∠10, ∠12, and ∠14 each measure 58. **19.** ∠FCD
21. True **23.** corr. ∠s = ⇒ // lines **25.** Linear Pair Theorem
27. Addition Property of Equality **29.** No. When a is the brother
of b, b may not be the brother of a. (b may be the sister of a).
31. See right. a. 1.25 **b.** $\frac{1}{3}x$ **31.** A   M   N   B

LESSON 3-7 (pp. 161–166)
9. If two nonvertical lines are perpendicular, then the product of
their slopes is −1. If the product of the slopes of two nonvertical
lines is −1, then they are perpendicular. **11. a.** 90 **b.** 50
13. a. // lines ⇒ corresponding ∠s = **b.** ℓ ⊥ m and
m // n ⇒ ℓ ⊥ n **c.** // lines ⇒ corr. ∠s = **d.** definition of
perpendicular lines **15.** $\frac{9}{2}$ **17.** m∠2 = 70, m∠3 = 110,
m∠4 = 70, m∠5 = 110,
m∠6 = 110, m∠7 = 70,   **23.**
m∠8 = 110
19. corr. ∠s = ⇒ // lines
21. m∠6 ≈ 100 **23. See right.**

24°

LESSON 3-8 (pp. 167–171)

11. a. See right. **b.** Sample: protractor and straightedge **c.** Sample: Place the straight edge of the protractor on the point for the clubhouse. Rotate the protractor until the 90° mark falls on the "road." Draw the line along the edge of the protractor from the clubhouse to the road. **13. a., b.** Sample: See right. **c.** Answers may vary. **d.** $m\angle DCA = m\angle BAC$; $m\angle ECB = m\angle ABC$ **15. a.** 360 **b.** The sum is 360. **17. a.** $\ell \parallel n$ **b.** $\ell \perp n$ **c.** $\ell \perp n$ **d.** $\ell \parallel n$ **19.** $m\angle 1 = 84$ **21. a.** 144 **b.** no pair **c.** $\angle EPD$ and $\angle DPC$, $\angle EPF$ and $\angle FPC$

11. a.

13. a., b.

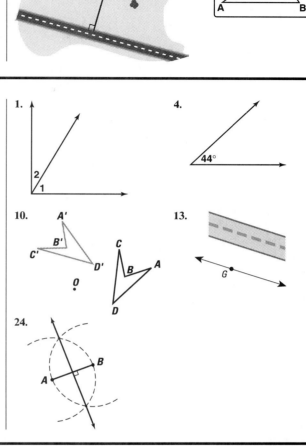

Chapter 3 Progress Self-Test (pp. 176–177)

1. Sample: See right. **2.** 77 **3.** $180 - 2x$. **4.** See right.
5. $90 - 15 = 75$ **6. a.** $m\angle 1 + m\angle 2 = 90$; $5x - 25 + 4x + 16 = 90$; $9x - 9 = 90$; $9x = 99$; $x = 11$ **b.** $m\angle 1 = 5 \cdot 11 - 25 = 30$
7. $105 = m\angle 2 + m\angle 1 = 6t + 9t = 15t$, so $t = 7$. **8.** (d)
9. $m\angle ABC = m\angle CBD$; $14y - 4 = 37 - y$; $15y = 41$; $y = \frac{41}{15} = 2\frac{11}{15}$ **10.** A rotation of 80° about point O means image point A' is found by drawing an arc $AA' = 80°$ from ray OA in a counterclockwise rotation. Locate points B', C', and D' in the same manner. See right. **11.** Each section contains $\frac{360}{5}$ or 72°, so the rotation from A to B is $-72°$ in a clockwise rotation or 288° in a counterclockwise rotation. **12. a.** $180° - 76° = 104°$ because arc PRT is a semicircle. **b.** 104° because an arc has the same measure as its central angle. **c.** $360° - 76° = 284°$ RTP is a major arc and a circle contains 360°. **13.** See right. **14.** (c) **15.** $\frac{1-0}{4-0} = \frac{1}{4}$
16. $\frac{4}{5}$, $-\frac{5}{4}$ **17.** 3 **18.** $y = 2x - 6$; $2 \cdot m = -1$; $m = -\frac{1}{2}$
19. $m\angle 6 = m\angle 8$ by Corresponding Angles Postulate, so $5x - 12 = 3x + 14$; $2x = 26$; $x = 13$ so $m\angle 6 = 5 \cdot 13 - 12 = 53$
20. $n \perp m$ by Perpendicular to Parallels Theorem, so $m\angle 5 = 90°$. Then $11x + 5 = 90$; $11x = 85$; $x = \frac{85}{11} = 7\frac{8}{11}$ **21.** definition of midpoint **22.** $a = a$ **23.** $GP + RM = 14$ **24.** See right.
25. $x + 9x = 180$; $10x = 180$; $x = 18$; So $9x = 9 \cdot 18 = 162$ The measure of the angle is 162.

1.

4.
44°

10.
A'
B'
C'
O
C
D'
B
A
D

13.
G

24.
B
A

The chart below keys the **Progress Self-Test** questions to the objectives in the **Chapter Review** on pages 178–181 or to the **Vocabulary** (Voc.) on page 175. This will enable you to locate those **Chapter Review** questions that correspond to questions missed on the **Progress Self-Test.** The lesson where the material is covered is also indicated on the chart.

Question	1	2	3	4	5	6	7	8	9	10
Objective	A	A	A	A	I	B	B	H	B	E
Lesson	3-3	3-3	3-3	3-1	3-3	3-3	3-1, 3-3	3-1	3-1	3-2
Question	11	12	13	14	15	16	17	18	19	20
Objective	I	F	J	H	K	L	K	L	C	C
Lesson	3-2	3-2	3-8	3-6	3-6	3-6, 3-7	3-6	3-7	3-6	3-7
Question	21	22	23	24	25					
Objective	H	G	G	D	B					
Lesson	3-5	3-4	3-4	3-8	3-3					

CHAPTER 3 REVIEW (pp. 178–181)

1. a. Sample: $\angle OJM$ **b.** Sample: $\angle NJN$ **c.** Sample: $\angle NJO$ and $\angle NJM$ **d.** $\angle LJM$ and $\angle MJL$ **3.** Sample: See page 878. **5.** 102, 102, 78 **7.** See page 878. **9.** $t = 9$ **11.** $m\angle 1 = 92$, $m\angle 2 = 88$ **13.** $x = 2\frac{13}{17}$ **15.** 36 **17.** $m\angle 3 = m\angle 6 = m\angle 8 = 83$, $m\angle 2 = m\angle 4 = m\angle 5 = m\angle 7 = 97$ **19.** $x = 8$ **21.** See page 878. **23.** Sample: See page 878.

25. See page 878. **27.** about 100° **29. a.** 80° **b.** 100° **c.** 50° **d.** 230° **31.** If $a = b$ and $b = c$, then $a = c$. **33.** $MS = RP$ **35.** (c) **37.** (a) **39.** Linear Pair Theorem; definition of supplementary angles **41.** 78 **43.** See page 878. **45.** 45° **47.** $-270°$ or 90° **49.** See page 878. **51.** See page 878. **53.** $\frac{3}{11}$ **55.** $-\frac{3}{5}$ **57.** (d) **59.** $10, -\frac{1}{10}$ **61.** $\frac{7}{4}$

3. Sample:

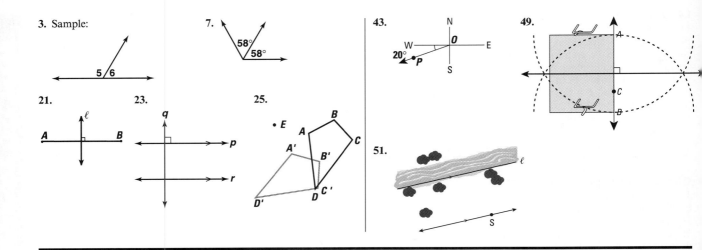

7.

43.

49.

21.

23.

25.

51.

LESSON 4-1 (pp. 184–189)
13. a. B and ℓ, C and m, D and n **b.** $r_\ell(A) = B$; $r_m(A) = C$; $r_n(A) = D$ **15. See below. 17. a.** HELP! I'M TRAPPED INSIDE THIS PAGE! **b.** N **19. a.** Yes, both lines have slope $\frac{1}{4}$. **b.** No; the slope of n is -3 and $\left(\frac{1}{4}\right) \cdot (-3) \neq -1$. **21. a.** Linear Pair Theorem; def. of supplementary angles **b.** corr. $\angle s = \Rightarrow /\!/$ lines **23.** Sample: **See below.** This network is traversable yet it has no odd nodes.

15.

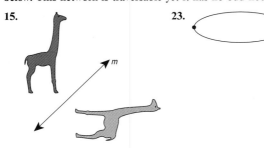

23.

7.

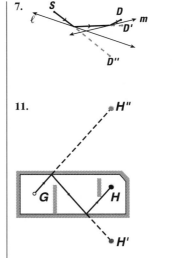

9.

11.

LESSON 4-2 (pp. 191–196)
15. (d) **17.** (e) **19. See below. 21. See below. 23.** *DCBAZTXD*
25. a. True **b.** $\overline{EF}$ and $\overline{CD}$ are each perpendicular to line m, so they are parallel by the Two Perpendiculars Theorem. **27.** $-3, \frac{1}{3}$
29. Nothing can be concluded about its truth.

LESSON 4-4 (pp. 203–209)
11. $\overline{AB}$ **13. a., b. See below. c.** Lengths of sides and angle measures remain the same. **15. See below. 17. a.** $(8, 5)$ **b.** $(-8, -5)$
19. Vertical Angles Theorem

19.

21.

13a., b.

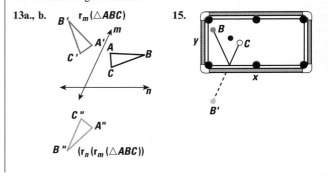

15.

LESSON 4-3 (pp. 197–202)
7. See right. 9. See right. 11. See right. 13. a. angle measure, betweenness, collinearity, distance **b.** orientation
15. counterclockwise **17.** $P' = (2, 0)$, $Q' = (4, 3)$, $R' = (-2, 7)$
19. a. (iii) **b.** (iii) **c.** (i) **d.** (ii) **21.** Let $x = 45$

LESSON 4-5 (pp. 210–215)
9. a., b. See below. c. 180° **11.** Reflections preserve distance.
13. a. $^-128°$ **b.** 128° **c.** 64° **15. See below. 17.** $y = -x - 1$
19. Western Ave. is parallel to Kelley Ave. **21. a.** 24 **b.** 2

9a, b.

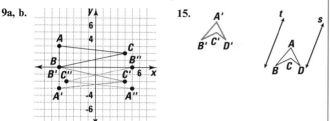

15.

LESSON 4-6 (pp. 216–221)
11. Answers will vary. **13.** Answers will vary. **15. See below.**
17. See below. 19. See below. 21. 0, since P lies on the y-axis.
23. A circle is the set of points in a plane at a certain distance from a certain point.

15. **17.**

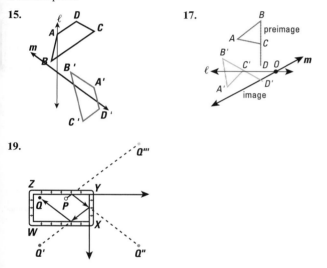

19.

LESSON 4-7 (pp. 222–227)
11. See below. 13. the stem **15. a.** Sample: translation and
rotation **b.** Sample: The composite must preserve orientation, so it
can only be a translation or rotation. **17. See below. 19. a.** angle
measure, betweenness, collinearity, distance, orientation **b.** the same
as rotations **21. a. See below. b.** Q, $^-180°$

11. **17.**

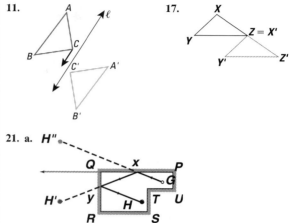

21. a.

LESSON 4-8 (pp. 228–232)
11. False. Any two circles whose radii are not equal cannot be
congruent because congruence transformations preserve distance.
See below. 13. One position is a rotation image of another.
15. Yes, it is a translation. **17.** Transitive Property of Equality
19. Reflexive Property of Equality, Symmetric Property of
Equality, Transitive Property of Equality

11.

CHAPTER 4 PROGRESS SELF-TEST (pp. 236–237)
1. a., b. See right. 2. See right. 3. See right. 4. Reflections
preserve distance. **5.** definition of reflection **6.** If the three
reflections are over three coplanar lines which intersect in zero or
one point, the composite is a reflection; otherwise, it is a glide
reflection. So the answers are (a) and (d). **7.** (7, $^-12$) because a
vector (a, b) translates a point (x, y) to image point $(x + a, y + b)$.
8. angle measure, betweenness, collinearity, distance, orientation
9. $M' = (^-2, 0)$, $N' = (5, 1)$, $P' = (^-3, ^-4)$ **10.** $\angle L$ **11. See page
880. 12. See page 880.** $r_t \circ r_s$ rotates $\triangle PQR$ from s to t, clockwise
about the point of intersection of the lines with a magnitude equal
to twice the non-obtuse angle between s and t. **13. a. See page 880.**
b. The final image is a translation perpendicular to ℓ and m in the
direction from m to ℓ with a magnitude twice the distance between
ℓ and m. **14.** 25°, Q **15.** translation ten units in a direction from
m to ℓ, parallel to line n **16.** 11 **17.** counterclockwise **18.** The
vector $\overrightarrow{EC}$ gives the direction and magnitude for the translation of
the six vertices of the preimage $ABCDEF$. **See page 880. 19.** Work
backwards. Find $r_y(H) = H'$; then find $r_x(H') = H''$. Draw the
path from G toward H'', then the path toward H', and finally the
path to H. **See page 880. 20.** The orientation from I to II is
reversed, so the isometry is either a single reflection or a glide
reflection. Since no single line of reflection is evident, it must be a
glide reflection. **21.** glide reflection **22. a.** D and L must be on the
reflecting line. **See page 880. b.** $\overleftrightarrow{DL}$ **23.** (b)

1. a., b. 2. 3.

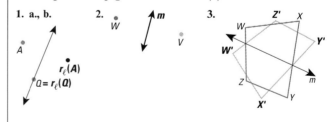

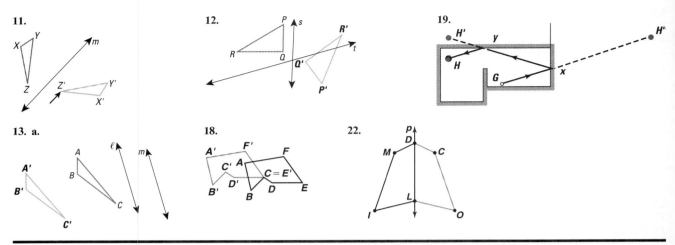

11.

12.

19.

13. a.

18.

22.

The chart below keys the **Progress Self-Test** questions to the objectives in the **Chapter Review** on pages 238–241. This will enable you to locate those **Chapter Review** questions that correspond to questions missed on the **Progress Self-Test**. The lesson where the material is covered is also indicated on the chart.

Question	1	2	3	4	5	6	7	8	9	10
Objective	A	A	B	E	E	F	K	F	K	F
Lesson	4–1	4–1	4–2	4–2	4–1	4–7	4–6	4–5	4–2	4–5
Question	11	12	13	14	15	16	17	18	19	20
Objective	C	D	D	G	G	E	E	C	I	H
Lesson	4–7	4–5	4–4	4–5	4–4	4–1	4–2	4–6	4–3	4–7
Question	21	22	23							
Objective	H	A	J							
Lesson	4–7	4–2	4–8							

CHAPTER 4 REVIEW (pp. 238–241)

1. a., b. See below. **3.** See below. **5.** Sample: See right.
7. See right. **9.** See right. **11.** See right. **13.** See right.
15. a counterclockwise rotation about Z of $\approx 104°$ See right.
17. a. See page 881. **b.** translation $\perp$ to m and n, in the direction from m to n with a magnitude twice the distance from m to n
19. ℓ is the $\perp$ bisector of AB. **21.** Reflections preserve angle measure. **23.** 84 **25.** counterclockwise **27.** (d) **29.** False **31.** $\overline{GI}$
33. a translation 8 inches to the left in the direction of $\overrightarrow{QP}$)
35. vertical, 3″ **37.** rotation **39.** glide reflection **41.** translation
43. See page 881. **45.** See page 881. **47.** (c) and (a) **49.** (b)
51. $(3, -7)$ **53.** $(2, 15)$ **55.** $A' = (-3, 7)$, $B' = (-3, 1)$, $C' = (0, -5)$, $D' = (2, 8)$, See page 881.

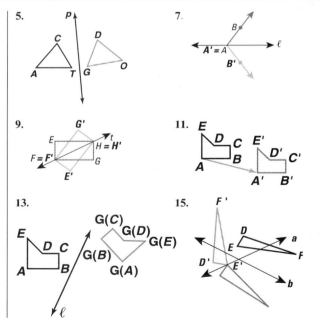

1. a., b.

3.

5.

7.

9.

11.

13.

15.

17. a.

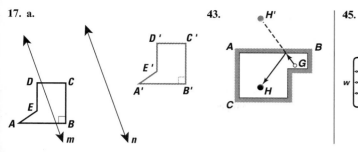

43.

45.

55.

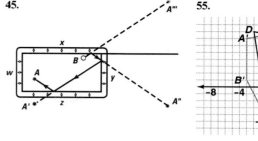

LESSON 5-1 (pp. 244–249)

11. a. $\triangle DPF \approx \triangle LPG$ **b.** directly congruent **c.** $\overline{DP} \approx \overline{LP}$, $\overline{PF} \approx \overline{PG}$, $\overline{DF} \approx \overline{LG}$; $\angle D \approx \angle L$, $\angle DPF \approx \angle LPG$, $\angle F \approx \angle G$
13. See below. **15. a.** (d) **b.** The picture is smaller than the front of the building. **17.** Sample: **See below.** $\triangle ABC$ and $\triangle A'B'C'$ are oppositely congruent. **19. a., b. See below. c.** R is a rotation of $-180°$ or $180°$. $CDAB$ is rotated about the intersection of ℓ and m.

13.

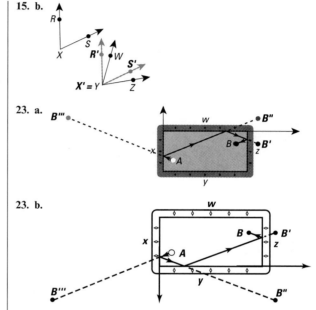

17.

19. a., b.

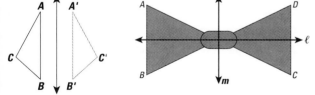

LESSON 5-2 (pp. 250–255)

13. An isosceles triangle has two congruent sides. **15. a.** $\overrightarrow{XY}$
b. See right. c. about $-20°$ **d.** Yes, the two angles are the same by the Angle Congruence Theorem. **17. a.** m$\angle X$ **b.** QZ **19.** when one is the image of the other under an isometry **21.** $\angle FIG$
23. a. See right. b. Sample: **See right.** **25.** definition of circle

15. b.

23. a.

23. b.

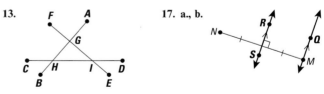

LESSON 5-3 (pp. 256–261)

9. Sample: $\angle C \approx \angle E$ (CPCF Theorem) and $\overline{BC} \approx \overline{DE}$ (CPCF Theorem) **11.** Sample: $\overline{AB} \approx \overline{BC}$, $\overline{BC} \approx \overline{CD}$ (Segment Congruence Theorem) and $\overline{AB} \approx \overline{CD}$ (Transitive Property of Congruence)
13. Sample: **See below.** $\angle AGE \approx \angle FGB$, $\angle DHB \approx \angle AHC$, $\angle CIF \approx \angle EID$ **15. a.** Sample: Diane attends a public school and it is a holiday, so she is not in the school. **b.** Sample: Suzanne: David will go to New Jersey next week. Cui: How do you know that? Suzanne: He told me he would go there. **17. a., b. See below. c.** $-\frac{1}{n}$
19. 15

13.

17. a., b.

LESSON 5-4 (pp. 263–268)

11. Sample: Since the lines are parallel, $\angle 2$ is congruent to the alternate interior angle to its right. This angle is congruent to the corresponding angle to its right. By transitivity of congruence, $\angle 2$ is also congruent to it. This reasoning can be repeated until $\angle 1$ is reached. **13.** m$\angle 1 = 38$, m$\angle 2 = 57$

15.

Conclusions	Justifications
$\overline{AB} \approx \overline{BC}$	definition of midpoint
$\overline{BC} \approx \overline{CD}$	definition of midpoint
$\overline{AB} \approx \overline{CD}$	Transitive Property of Congruence

17. Yes, each reflection image is congruent to the original R; they are congruent to each other by the Transitive Property of Congruence. **19. a.** Yes **b.** No **21.** True **23. a.** Samples: A, H, I, M, N, W **b.** Samples: E, F

LESSON 5-5 (pp. 270–275)

7. a. See below. b. argument:

Conclusions	Justifications
0. $\overleftrightarrow{WY}$ is the $\perp$ bisector of $\overline{XY}$	given
Let $\ell = \overleftrightarrow{WY}$	
1. $r_\ell(X) = Z$, $r_\ell(W) = W$, $r_\ell(Y) = Y$	definition of reflection
2. $r_\ell(\angle WXY) = \angle WZY$	Figure Reflection Theorem
3. $\angle WXY \approx \angle WZY$	definition of congruence

9. argument:

Conclusions	Justifications
0. m is the $\perp$ bisector of $\overline{XY}$ n is the $\perp$ bisector of $\overline{YZ}$ O is on m, O is on n	given
1. $OX = OY$, $OY = OZ$	Perpendicular Bisector Theorem
2. $OX = OY = OZ$	Transitive Property of Equality
3. X, Y, Z lie on $\odot O$ with radius OX	definition of circle

11. argument:

Conclusions	Justifications
0. ℓ is the $\perp$ bisector of $\overline{CD}$ $r_\ell(E) = F$	given
1. $r_\ell(C) = D$, $r_\ell(A) = A$, $r_\ell(B) = B$	definition of reflection
2. r_ℓ(Quadrilateral $ABCE$) = Quadrilateral $ABDF$	Figure Reflection Theorem
3. Quadrilateral $ABCE$ $\approx$ Quadrilateral $ABDF$	definition of congruence

13. a. definition of circle **b.** definition of midpoint **c.** Transitive Property of Equality **15.** $\angle Q$ and $\angle K$, $\angle Z$ and $\angle R$, $\angle P$ and $\angle A$, $\overline{QZ}$ and $\overline{KR}$, $\overline{ZP}$ and $\overline{RA}$, $\overline{QP}$ and $\overline{KA}$ **17.** translation **19.** between 53 cm and 129 cm

7. a.

LESSON 5-6 (pp. 276–280)

13. Not uniquely determined; there are infinitely many points between A and B. **15.** not always possible; The uniquely determined perpendicular bisectors of $\overline{AB}$ and $\overline{CD}$ are not necessarily the same. **17.** (c)
19. Sample:

Conclusions	Justifications
0. m is the $\perp$ bisector of $\overline{AB}$ and $\overline{CD}$	given
1. $r_m(A) = B$, $r_m(C) = D$, $r_m(D) = C$	definition of reflection
2. $r_m(\triangle ACD) = \triangle BDC$	Figure Reflection Theorem.
3. $\triangle ACD \approx \triangle BDC$	definition of congruence

21. $m\angle 1 = m\angle 4 = m\angle 8 = m\angle 7 = 100$, $m\angle 2 = m\angle 5 = m\angle 3 = m\angle 6 = 80$ **23. a.** 60° or −300° **b.** 240° or −120° **25.** Answers may vary. Sample conjecture and proof: Because of // lines ⇒ AIA ≈ Theorem and the Corresponding Angles Postulate, it can be proved that the angles of the new triangle are congruent respectively to the angles of $\triangle ABC$.

LESSON 5-7 (pp. 281–287)

15. 82°26′ **17. a.** $x = 97$ **b.** $m\angle P = 97$, $m\angle Q = 117$, $m\angle R = 47$, $m\angle S = 99$ **19. See below.** not uniquely determined **21. See below.** not uniquely determined

23.

Conclusions	Justifications
0. $\overrightarrow{OA}$ bisects $\angle BOF$, $\overrightarrow{OB}$ bisects $\angle AOC$	given
1. $m\angle BOA = m\angle AOF$ $m\angle BOC = m\angle BOA$	definition of angle bisector
2. $m\angle BOC = m\angle AOF$	Transitive Property of Equality
3. $m\angle EOF = m\angle BOC$	Vertical Angles Theorem
4. $m\angle AOF = m\angle EOF$	Transitive Property of Equality

25. R: 15°E of S or 75°S of E; Q: 30°S of W or 60°W of S

19.

21.

CHAPTER 5 PROGRESS SELF-TEST (p. 291)

1. $m\angle RCS = 47$ (CPCF Theorem) **2.** $\overline{OR}$ (CPCF Theorem) **3.** Step 1: $\odot Q$ with radius QP. Step 2: $\odot P$ with radius PQ. Step 3: Label the points of intersection of $\odot A$ and $\odot B$ as M and N. Step 4: $\overline{QM}$ and $\overline{PM}$ to form $\triangle QPM$. $\triangle QPM$ is equilateral. **See page 883. 4. a.** $m\angle 5 + m\angle 3 = 180$; $9z - 52 + 2z + 45 = 180$; $11z - 7 = 180$; $11z = 187$; $z = 17$ **b.** $m\angle 3 = 2 \cdot 17 + 45 = 34 + 45 = 79$ **5.** Segment Congruence Theorem: If two segments have the same length, then they are congruent. **6. a.** Given **b.** definition of reflection **c.** definition of reflection **d.** Figure Reflection Theorem **e.** definition of congruence

7. argument:

Conclusions	Justifications
0. M is the midpoint of $\overline{AB}$ $\overline{AM} \approx \overline{BC}$	given
1. $AM = MB$	definition of midpoint
2. $\overline{AM} \approx \overline{MB}$	Segment Congruence Theorem

(Step 1 could be omitted; then the justification should be "definition of midpoint.")

3. $\overline{MB} \approx \overline{BC}$	Transitive Property of Congruence

8. argument:

Conclusions	Justifications
0. $\overleftrightarrow{ES}$ is the $\perp$ bisector of $\overline{JT}$, E is on m, S is on m	given
1. $r_m(J) = T$, $r_m(E) = E$, $r_m(S) = S$,	definition of reflection
2. $r_m(\triangle JES) = \triangle TES$	Figure Reflection Theorem
3. $\triangle JES \approx \triangle TES$	definition of congruence

9. $2x + 3x + 4x = 180$; $9x = 180$; $x = 20$; $2x = 40$ (the measure of the smallest angle); 40, 60, 80 **10.** $115 + 60 + 3x + 2x + 5 = 360$; $5x + 180 = 360$; $5x = 180$; $x = 36$; $m\angle Q = 2x + 5 = 2 \times 36 + 5 = 77$ **11.** Substitute 30 for n in the formula $S = (n - 2) \cdot 180$. $S = (30 - 2) \cdot 180 = 5040$ **12.** Point Z is on $\overline{VZ}$, the $\perp$ bisector of $\overline{WY}$; therefore, by Perpendicular Bisector Theorem, $\overline{WZ} \approx \overline{YZ}$ **13. a. See right.** uniquely determined **b. See right.** not uniquely determined

14. Step 1: Construct n, the $\perp$ bisector of $\overline{MN}$. Step 2: Construct m, the $\perp$ bisector of $\overline{NO}$. Step 3: Label the intersection of m and n as C. Step 4: Construct the circle with center C, and radius $\overline{OM}$. The circle passes through the vertices of $\triangle MNO$. **See right.**
15. definitely; assumed

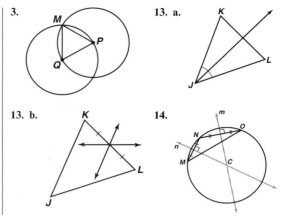

3. **13. a.** **13. b.** **14.**

The chart below keys the **Progress Self-Test** questions to the objectives in the **Chapter Review** on pages 292–297 or to the **Vocabulary** (Voc.) on page 290. This will enable you to locate those **Chapter Review** questions that correspond to questions you missed on the **Progress Self-Test**. The lesson where the material is covered is also indicated on the chart.

Question	1	2	3	4	5	6	7	8	9	10
Objective	A	A, E	B	C	E	G	F	G	D	D
Lesson	5-1	5-1	5-4	5-4	5-2	5-5	5-4	5-5	5-7	5-7

Question	11	12	13	14	15
Objective	D	I	H	B	J
Lesson	5-7	5-5	5-6	5-4	5-6

CHAPTER 5 REVIEW (pp. 292–297)
1. a. Sample: **See right. b.** $m\angle PTO$ **c.** TH **3.** $m\angle E$ **5.** $PL = 16$ cm, $RN = NT = PM = ML = 8$ cm **7. a.** 38 **b.** 76 **c.** 114 **9. See right.**
11. See page 884. 13. 12 **15. a.** 40 **b.** 57 **c.** $\angle 4$ and $\angle 7$; $\angle 3$ and $\angle 1$; $\angle 1$ and $\angle 6$; $\angle 1$ and $\angle 5$; $\angle 3$ and $\angle 5$; $\angle 3$ and $\angle 6$ **17.** 86.4
19. a. 58 **b.** $m\angle D = 59$, $m\angle E = 60$, and $m\angle F = 61$ **21.** $m\angle R = 58$, $m\angle S = 174$, $m\angle STU = 40$, $m\angle TUR = 88$ **23.** $\overline{PA} \approx \overline{LO}$, $\overline{AT} \approx \overline{OG}$, $\overline{PT} \approx \overline{LG}$, $\angle P \approx \angle L$, $\angle A \approx \angle O$, $\angle T \approx \angle G$
25. $EHGF$, definition of congruence **27.** Reflections preserve distance. **29.** Transitive Property of Congruence **31.** Angle Congruence Theorem **33.** 0. Given 1. Transitive Property of Congruence 2. AIA $\approx \Rightarrow$ // Lines Theorem
35. argument:

Conclusions	Justifications
0. $\angle 2 \approx \angle 3$	Given
1. $\angle 2 \approx \angle 1$	Vertical Angles Theorem
2. $\angle 1 \approx \angle 3$	Transitive Property of Congruence

37. argument:

Conclusions	Justifications
0. $\overrightarrow{OB}$ bisects $\angle AOC$	Given
$\overrightarrow{OC}$ bisects $\angle BOD$	
1. $m\angle AOB = m\angle BOC$	definition of $\angle$ bisector
$m\angle BOC = m\angle COD$	
2. $m\angle AOB = m\angle COD$	Transitive Property of Equality

39. a. Given **b.** definition of reflection **c.** definition of reflection **d.** definition of reflection **e.** Figure Reflection Theorem **f.** definition of congruence

41. argument:

Conclusions	Justifications
0. $\overleftrightarrow{AJ}$ is the $\perp$ bisector of $\overline{RM}$	Given
1. $r(A) = A$, $r(J) = J$, $r(M) = R$	definition of reflection
2. $r(\angle JAM) = \angle JAR$	Figure Reflection Theorem
3. $\angle JAM \approx \angle JAR$	definition of congruence

43. argument:

Conclusions	Justifications
0. $r_p(A) = C$	Given
p is the $\perp$ bisector of $\overline{EB}$	
D is on p	
1. $r_p(D) = D$, $r_p(B) = E$	definition of reflection
2. $r_p(\angle ABD) = \angle CED$	Figure Reflection Theorem
3. $\angle ABD \approx \angle CED$	definition of congruence

45. Yes. **See page 884. 47.** No, the diagonal is not necessarily the bisector of $\angle REC$. **See page 884. 49.** (b) **51.** // lines $\Rightarrow$ AIA $\approx$ Theorem **53.** True **55.** Postulates were viewed as statements assumed to be true instead of statements definitely true.

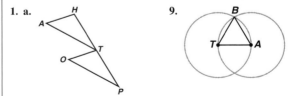

1. a. **9.**

11.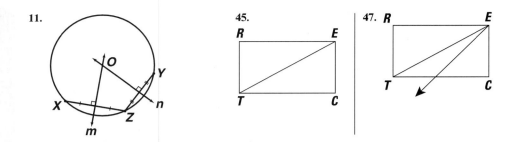

45.

47.

LESSON 6-1 (pp. 300–307)

13. a. See below. **b.** none See below. **c.** See below. **15.** See below.
17. a. 8 **b.** △*UWV* **c.** No **19.** (a)
21.

Conclusions	Justifications
0. $r_m(E) = F$; $r_m(G) = H$	Given
1. *m* is the ⊥ bisector of $\overline{GH}$	definition of reflection
2. $r_m(H) = G$	definition of reflection
3. $EH = FG$	Reflections preserve distance

23. See below. **25.** 2

13. a.

13. b.

13. c.

15.

23.

LESSON 6-2 (pp. 309–315)

11. (d) **13.** 41.5 **15. a.** *AFEDCB* **b.** $\overline{AB}$ and $\overline{AF}$, $\overline{BC}$ and $\overline{FE}$, $\overline{CD}$ and $\overline{ED}$ **c.** m∠*F* and m∠*B*, m∠*E* and m∠*C*, m∠*FAD* and m∠*BAD*, m∠*ADE* and m∠*ADC* **17.** Yes, the angle measures are 36, 72, 108, and 144, respectively. Sample: **See below.** **19.** $(^-x, y)$
21. It is either 27 or 3.

17.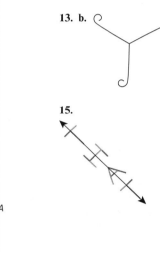

LESSON 6-3 (pp. 316–321)

17. a. $\overline{ZW} \perp \overline{WX}$; $\overline{WX} \perp \overline{XY}$; $\overline{XY} \perp \overline{ZY}$; $\overline{YZ} \perp \overline{ZW}$ **b.** rectangle
19. a. $\overline{FG} \perp \overline{GH}$; $\overline{GH} \perp \overline{HE}$; $\overline{HE} \perp \overline{EF}$; $\overline{EF} \perp \overline{FG}$; $FG = GH = HE = EF$ **b.** square **21. a.** definition of circle **b.** definition of circle **c.** definition of kite **23.** trapezoid, isosceles trapezoid, parallelogram, rectangle **25.** trapezoid, isosceles trapezoid
27. 84 and 12; 48 and 48. **29.** $\overline{AB} \approx \overline{CB}$, $\overline{AF} \approx \overline{CD}$, $\overline{FE} \approx \overline{DE}$. ∠*A* ≈ ∠*C*, ∠*F* ≈ ∠*D*, ∠*ABE* ≈ ∠*CBE*, ∠*FEB* ≈ ∠*DEB*.

LESSON 6-4 (pp. 323–328)

9. $E' = r_{\overleftrightarrow{CD}}(E)$, so $\overleftrightarrow{CD}$ is the ⊥ bisector of $\overline{EE'}$. *C and D* are on line $\overleftrightarrow{CD}$. So $DE = DE'$, $CE = CE'$ since reflections preserve distance. From the definition of kite, $ECE'D$ is a kite. **11. a.** 10″ **b.** 10″ **c.** 23″ **d.** 90 **13.** False **15.** False **17.** A trapezoid is isosceles if and only if it has a pair of base angles equal in measure.
19. always **21.** sometimes **23. a.** 21 **b.** No. Since the angles are either 157° or 123°, the transversal should slant more in a clockwise direction.

LESSON 6-5 (pp. 329–334)

7. a. Isosceles Triangle Base Angles Theorem **b.** definition of trapezoid **c.** definition of isosceles trapezoid
9. a.

Conclusions	Justifications
1. *ABCD* has a line of symmetry (call it *m*), which is the ⊥ bisector of $\overline{AB}$ and $\overline{DC}$.	Isosceles Trapezoid Symmetry Theorem
2. $r_m(A) = B$, $r_m(C) = D$	definition of reflection
3. $AC = BD$	Reflections preserve distance.

b. The diagonals of an isosceles trapezoid are equal in measure.
11. a. $180 - x$ **b.** $180 - 2y$ **13.** $\overleftrightarrow{AC}$, $\overleftrightarrow{BD}$, the ⊥ bisector of $\overline{AB}$, the ⊥ bisector of $\overline{AD}$ **15.** $RT = 3$; $IT = 4$; $KE = 8$; $KT = 6$; m∠*KIE* = 49; m∠*KIT* = 98; m∠*KEI* = m∠*TEI* = 20.5; m∠*ITK* = m∠*IKT* = 41; m∠*TKE* = m∠*KTE* = 69.5; m∠*IKE* = m∠*ITE* = 110.5; m∠*TRE* = m∠*TRI* = m∠*KRE* = m∠*KRI* = 90 **17. a.** *B* and *D* **b.** $\overleftrightarrow{BD}$ **19. a.** See below. **b.** See below.
c. See page 885. **21.** See page 885.

19. a. **19. b.**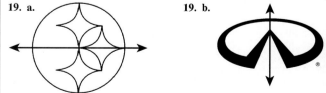

19. c.

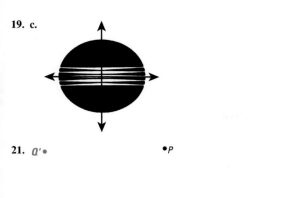

21. $Q' \bullet$　　　　　　$\bullet P$

$\bullet Q$

LESSON 6-6 (pp. 335–339)

9. 60 **11.** Sample: **See below.** **13.** 2-fold rotation symmetry
15. See below. 17. See below. $ABCD$ is a trapezoid because slope
of $\overline{DC}$ = slope of $\overline{AB}$ = 0, so $\overline{DC} \parallel \overline{AB}$. **19.** isosceles right
triangle **21.** m∠4 = 135; m∠7 = 140; m∠2 = 95; m∠3 = 40;
m∠1 = 45 **23.** Sample: **See below.**

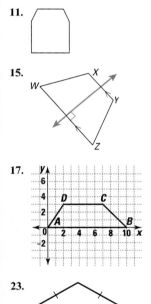

11.

15.

17.

23.

LESSON 6-7 (pp. 341–346)

15. Sample: The reflection symmetry enables a wrench to be placed
on the nut either to tighten it or to loosen it; the rotation symmetry
enables the wrench to be placed in a number of different directions
on the nut. **17. See below. a.** regular hexagon **b.** All the sides are
congruent since they are formed by equilateral triangles with
common sides, and all the vertex angles are congruent since each is
120. **19. a.** False **b.** True **21. a. See below. b.** $n = 3$
23. m∠P = 71, m∠R = 148 **25. a.** 70 **b.** 90 **c.** 20 **d.** 40

17.

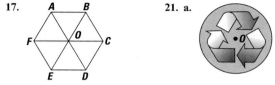

21. a.

LESSON 6-8 (pp. 347–351)

11. a. Let the 9 teams be vertices of a regular 9-gon. **b.** Sample:
wk 1: 1-2, 9-3, 8-4, 7-5, 6 bye; wk 2: 2-3, 1-4, 9-5, 8-6, 7 bye; **c.** wk
3: 3-4, 2-5, 1-6, 9-7, 8 bye; wk 4: 4-5, 3-6, 2-7, 1-8, 9 bye; wk 5:
5-6, 4-7, 3-8, 2-9, 1 bye; wk 6: 6-7, 5-8, 4-9, 3-1, 2 bye; wk 7: 7-8,
6-9, 5-1, 4-2, 3 bye; wk 8: 8-9, 7-1, 6-2, 5-3, 4 bye; wk 9: 9-1, 8-2,
7-3, 6-4, 5 bye; **d.** 27 **e.** Replace each bye with team 10.
13. Sample: wk 1: 1-2, 5-3, 4-6; wk 2: 2-3, 1-4, 5-6; wk 3: 3-4, 2-5,
1-6; wk 4: 4-5, 3-1, 2-6; wk 5: 5-1, 4-2, 3-6 **15.** 7 **17. a.** 6 **b.** It is
6-fold rotation symmetry. **19. See below. 21.** Locate m so that the
angle between ℓ and $m = \frac{1}{2}$m∠AOB. **See below. 23.** $\frac{360 + s}{180}$

19.

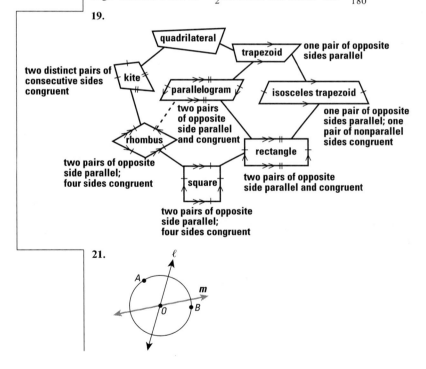

21.

CHAPTER 6 PROGRESS SELF-TEST (pages 356–357)

1. See below. **2.** See below. **3.** See right. **4.** Sample: See right.
5. Sample: See right. **6.** Since the vertex is $\angle CBA$, so $m\angle A = m\angle C$. $6x = 50 - 2x$, $x = \frac{25}{4}$; $m\angle B = 180 - m\angle A - m\angle C = 180 - 6x - 50 + 2x = 180 - 50 - 4x = 105$ **7.** $m\angle H = \frac{(5-2)\cdot 180}{5} = 108$; $\triangle HDG$ is an isosceles triangle with $HD = HG$, so $m\angle HDG = m\angle HGD$. $m\angle HDG = \frac{180 - 108}{2} = 36$ **8.** $m\angle W = m\angle X = 100$; $m\angle Y = m\angle Z = 180 - 100 = 80$ **9. a.** $SR = SP = 23$, $QR = QP = 9.1$ **10. a.** Since each rhombus is a parallelogram, $\overleftrightarrow{MR} \parallel \overleftrightarrow{HO}$, so $m\angle 2 = m\angle 1 = 78$. ($\parallel$ Lines $\Rightarrow$ AIA = Theorem)
b. From Rhombus Diagonal Theorem, $m\angle 1 + m\angle 3 = 90$. So, $m\angle 3 = 90 - 78 = 12$ **11.** $BADC$ **12.** $\angle CAD$ **13.** $\overline{IH}$ and $\overline{KJ}$
14. (c) A scalene $\triangle$ has 0 symmetry lines; an isosceles $\triangle$ has 1; an equilateral $\triangle$, 3. **15.** (b) **16.** rectangle **17.** True **18.** False.
Sample: See right.

19.

Conclusions	Justifications
0. $AB = BC$	Given
1. $m\angle A = m\angle BCA$	Isosceles $\triangle$ Base Angles Theorem
2. $m\angle BCA = m\angle ECD$	Vertical Angles Theorem
3. $m\angle A = m\angle ECD$	Transitive Property of Equality

20.

Conclusions	Justifications
1. $CY = CT = CI$	definition of circle
2. $TY = TC = TI$	definition of circle
3. $CY = CI = TY = TI$	Transitive Property of Equality
4. $CITY$ is a rhombus	definition of rhombus

21. a., b. See right. **22.** See right. **23.** Sample: 1-2 3-5 4 bye; 2-3 4-1 5 bye; 3-4 5-2 1 bye; 4-5 1-3 2 bye; 5-1 2-4 3 bye

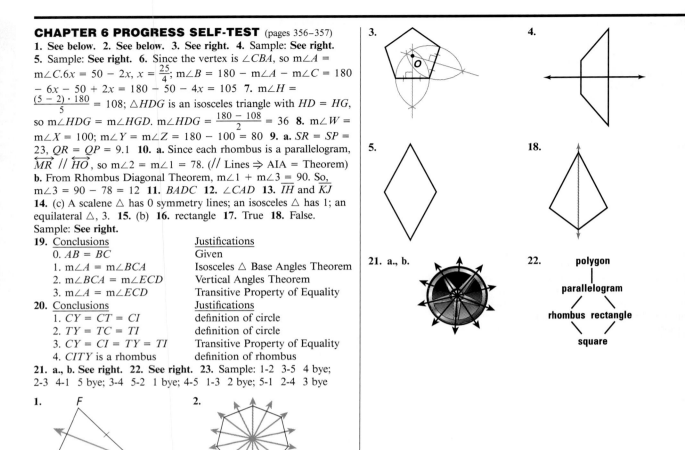

3.

4.

5.

18.

21. a., b.

22.
polygon
parallelogram
rhombus rectangle
square

1.
F
G   H

2.

The chart below keys the **Progress Self-Test** questions to the objectives in the **Chapter Review** on pages 358–361 or to the **Vocabulary** (Voc.) on page 355. This will enable you to locate those **Chapter Review** questions that correspond to questions you missed on the **Progress Self-Test.** The lesson where the material is covered is also indicated on the chart.

Question	1	2	3	4	5	6	7	8	9	10
Objective	A	A	A	A	B	C	C	D	D	D
Lesson	6-2	6-7	6-6	6-5	6-3, 6-7	6-2	6-2, 6-7	6-3, 6-5	6-3, 6-4	6-4

Question	11	12	13	14	15	16	17	18	19	20
Objective	E	E	E	F	F	G	G	G	H	H
Lesson	6-1	6-1	6-6	6-2	6-7	6-3	6-3	6-4	6-2	6-3

Question	21	22	23
Objective	I	K	J
Lesson	6-1, 6-6	6-3	6-8

CHAPTER 6 REVIEW (pp. 358–361)

1. See page 887. **3.** See page 887. **5.** See page 887. **7.** See page 887. **9.** Sample: See page 887. **11.** Sample: See page 887.
13. Sample: See page 887. **15.** $m\angle H = m\angle G = 73$ **17.** 34
19. 30 **21.** $m\angle Q = 43$, $m\angle P = m\angle R = 137$ **23.** $m\angle 2 = 90$, $m\angle 3 = m\angle 5 = 27$, $m\angle 4 = m\angle 6 = 63$ **25.** 60 **27.** 165 **29.** $\overline{XY}$ and $\overline{XW}$, $\overline{YZ}$ and $\overline{ZW}$ **31.** $\angle EBA$ **33.** True **35.** parallelogram, rhombus, rectangle, square **37.** $\angle J$ **39.** (c) **41.** True **43.** isosceles trapezoid **45.** True **47.** Yes, $\overleftrightarrow{DB}$ **49.** All sides have equal lengths. **51.** False **53.** True **55.** Given; definition of isosceles $\triangle$; Transitive Property of Equality

57.

Conclusions	Justifications
0. $m\angle F = m\angle FHI$	Given
1. $\overline{EF} \;/\!/\; \overline{GH}$	AIA $= \Rightarrow /\!/$ Lines Theorem
2. $EFHG$ is a trapezoid	definition of trapezoid

59.

Conclusions	Justifications
1. $OQ = OR$	definition of circle
2. $PQ = PR$	definition of circle
3. $OQPR$ is a kite	definition of kite

61. a. reflection-symmetric over 1 line **b.** no rotation symmetry
63. a. reflection-symmetric over 4 lines **b.** has rotation-symmetry
with $n = 4$ and center point O **65.** Sample: **See right.**
67. Sample:

A-B	K-C	J-D	I-E	H-F	G-L
B-C	A-D	K-E	J-F	I-G	H-L
C-D	B-E	A-F	K-G	J-H	I-L
D-E	C-F	B-G	A-H	K-I	J-L
E-F	D-G	C-H	B-I	A-J	K-L
F-G	E-H	D-I	C-J	B-K	A-L
G-H	F-I	E-J	D-K	C-A	B-L
H-I	G-J	F-K	E-A	D-B	C-L
I-J	H-K	G-A	F-B	E-C	D-L
J-K	I-A	H-B	G-C	F-D	E-L
K-A	J-B	I-C	H-D	G-E	F-L

69. See right. 71. scalene triangle, kite, hexagon, heptagon,
octagon

1.
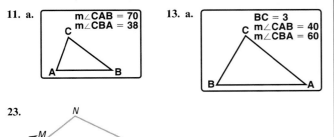

3.
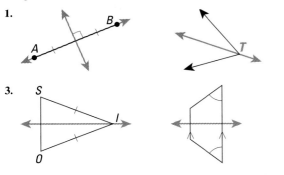

5.
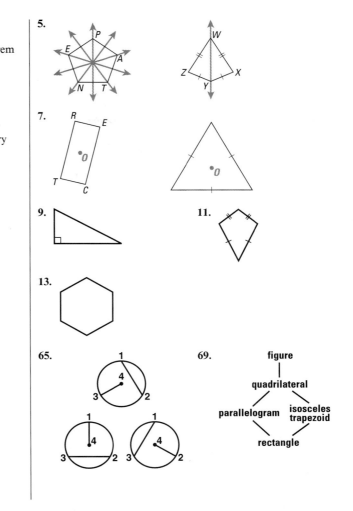

7.

9.

11.

13.

65.

69.
figure
|
quadrilateral
/   \
parallelogram   isosceles
trapezoid
|
rectangle

LESSON 7-1 (pp. 364–369)
11. a. Sample: **See below. b.** Any opinion is allowed. The answer
will be found to be no. **13. a. See below. b.** Any opinion is
allowed. The answer will be found to be yes. **15. a.** No unique
triangle can be drawn. **b.** The answer is no. **17.** 135 **19.** isosceles
trapezoid **21.** $m\angle DCA = 40$, $m\angle CBP = 40$ $m\angle DAC = 50$,
$m\angle PBA = 50$, $m\angle PCB = 50$ $m\angle CDA = m\angle BCD = m\angle ABC =$
$m\angle BAD = 90$, $m\angle CPB = 90$, $m\angle APB = 90$ **23. See below.**

11. a.
m∠CAB = 70
m∠CBA = 38

13. a.
BC = 3
m∠CAB = 40
m∠CBA = 60

23.

LESSON 7-2 (pp. 370–377)
15. a. $\triangle JKL$ **b.** LJ **c.** L **d.** $m\angle K$ **17.** $\triangle ABD \cong \triangle CBD$ by the
AAS Congruence Theorem.
19.

Conclusions	Justifications
0. $m\angle FDE = m\angle C'DE$	
G, D, and E are collinear	Given
1. $m\angle FDG = 180 - m\angle FDE$	Linear Pair Theorem
$m\angle C'DG = 180 - m\angle C'DE$	(and algebra)
2. $m\angle FDG = m\angle C'DG$	Transitive Property of Equality
	(and substitution)
3. $\overleftrightarrow{EG}$ bisects $\angle FDC'$	definition of angle bisector

21. a. 5 **b. See below. 23.** 84 **24.** Sample: $\angle O \cong \angle N$; $\angle L \cong \angle E$;
$\angle D \cong \angle W$; $OL \cong NE$ (CPCF Theorem) **25. See page 888.**

21. b.
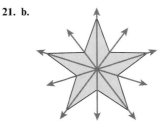

25.

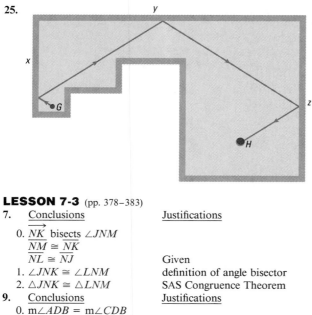

LESSON 7-3 (pp. 378–383)

7.

Conclusions	Justifications
0. $\overrightarrow{NK}$ bisects $\angle JNM$	
$\overline{NM} \cong \overline{NK}$	
$\overline{NL} \cong \overline{NJ}$	Given
1. $\angle JNK \cong \angle LNM$	definition of angle bisector
2. $\triangle JNK \cong \triangle LNM$	SAS Congruence Theorem

9.

Conclusions	Justifications
0. $m\angle ADB = m\angle CDB$	
$m\angle ABD = m\angle CBD$	Given
1. $BD = BD$	Reflexive Property of Equality
2. $\triangle ABD \cong \triangle CBD$	ASA Congruence Theorem
3. $AB = CB$	CPCF Theorem

11. a. SAS Congruence Theorem **b.** CPCF Theorem **13.** $\triangle EFG \cong \triangle IHG$; AAS Congruence Theorem **15. a.** Yes, assuming units of length measurements are the same. **b.** SAS Congruence Theorem **17. See below. 19.** Sample: **See below.**

17.

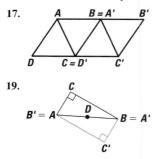

19.

LESSON 7-4 (pp. 384–388)

5.

Conclusions	Justifications
0. $AD = AE$,	
$m\angle D = m\angle E$	Given
1. $\angle A \cong \angle A$	Reflexive Property of Congruence
2. $\triangle AEB \cong \triangle ADC$	ASA Congruence Theorem
3. $EB = CD$	CPCF Theorem

7. Since ℓ, the $\perp$ bisector of $\overline{AB}$ and $\overline{ED}$, is a symmetry line of $ABCDEF$ by the Regular Polygon Symmetry Theorem, $r_\ell(A) = B$, $r_\ell(C) = F$. Therefore, $\overline{AC} \cong \overline{BF}$ because reflections preserve distance.

9. Sample Argument:

Conclusions	Justifications
0. $\overline{QR} \parallel \overline{TU}$; S is the midpoint of $\overline{QU}$.	Given
1. $QS = SU$	definition of midpoint
2. $m\angle QSR = m\angle UST$	Vertical Angles Theorem
3. $m\angle SQR = m\angle SUT$	$\parallel$ Lines $\Rightarrow$ AIA $\cong$ Theorem
4. $\triangle QSR \cong \triangle UST$	ASA Congruence Theorem
5. $ST = SR$	CPCF Theorem
6. S is the midpoint of $\overline{RT}$	definition of midpoint

11. a. Vertical Angles Theorem **b.** SAS Congruence Theorem (step 1 and given) **c.** CPCF Theorem **d.** Vertical Angles Theorem **e.** Transitive Property of Congruence **13.** 108 **15.** True **17.** $\overline{AB}, \overline{AC}$, and point F

LESSON 7-5 (pp. 389–396)

9. a-c. See below. d. No, there are two possible triangles XZW.

11.

Conclusions	Justifications
1. $\overline{AB} \cong \overline{AB}$	Reflexive Property of Congruence
2. $\triangle PBA \cong \triangle RBA$	HL Congruence Theorem (step 1 and given)
3. $\overline{PB} \cong \overline{RB}$	CPCF Theorem
4. $PBRA$ is a kite	definition of kite

13. Let T be the top of the maypole, M the point on the maypole that is the same height as June and April's hands, J the position of June's hand, and A the position of April's hand. $\overline{JA}$ is parallel to the ground, since J and A are equal in height (given). So, $\overline{JA} \perp \overline{TM}$ by the Perpendicular to Parallels Theorem. Since $\overline{TM} \cong \overline{TM}$ and $\overline{TJ} \cong \overline{TA}$, $\triangle TJM \cong \triangle TAM$ by the HL Congruence Theorem. Thus, $\overline{JM} \cong \overline{AM}$ by the CPCF Theorem. **15. f.** Isosceles Trapezoid Symmetry Theorem **g.** reflections preserve distance **17. a.** (Art is reduced in size) **See below. b.** Yes by the AAS Congruence Theorem. **19.** $\triangle AGY, \triangle DEX, \triangle DEY$

9. a-c.

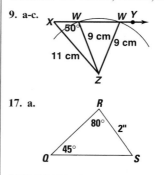

17. a.

LESSON 7-6 (pp. 397–403)

13. Sample: **See page 889. 15.** $\triangle ABC \cong \triangle CDA$, so $\angle BAC \cong \angle DCA$, $\angle BCA \cong \angle DAC$ (CPCF Theorem). So $\overline{AB} \parallel \overline{CD}$, $\overline{BC} \parallel \overline{AD}$ (AIA $\cong \Rightarrow \parallel$ lines.) $ABCD$ is a parallelogram. (definition of parallelogram).

17.

Conclusions	Justifications
0. $\overline{AD} = \overline{BC}$	
$\overline{AD} \perp \overline{AB}$	
$\overline{BC} \perp \overline{DC}$	Given
1. $BD = BD$	Reflexive Property of Equality
2. $\triangle ABD \cong \triangle CDB$	HL Congruence Theorem
3. $\angle ABD \cong \angle CDB$	CPCF Theorem
4. $\overline{AB} \parallel \overline{DC}$	AIA $\cong \Rightarrow \parallel$ lines

19.

Conclusions	Justifications
1. $\overline{BD} \cong \overline{BD}$	Reflexive Property of Congruence
2. $\triangle BAD \cong \triangle DCB$	ASA Congruence Theorem (step 1 and given)
3. $AB = CD$	CPCF Theorem

21. a. No **b.** There is no such segment unless $\triangle PQR$ is an isosceles triangle with $PQ = PR$.

13.

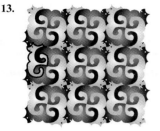

LESSON 7-7 (pp. 405–410)

13. The distance between two given parallel lines is constant.
15. From Properties of a Parallelogram Theorem, $AB = CD$, $CD = EF$, $EF = GH$. From Transitive Property of Equality, $AB = GH$. **17.** Sample: **See below. 19.** No; the congruent sides are not corresponding sides. **21.** It cannot be justified.
23. Transitive Property of Congruence

17.

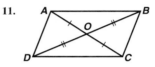

LESSON 7-8 (pp. 411–415)

11. See below. Sample argument: $\triangle ABO \cong \triangle CDO$ by SAS, so $\angle ABD \cong \angle CDB$ by CPCF, and so $\overline{AB} \, // \, \overline{CD}$ by the AIA $\cong \Rightarrow //$ lines Theorem. Similarly, using the triangles AOD and COB, $\overline{AD} \, // \, \overline{CB}$. Thus $ABCD$ is a parallelogram by the definition of parallelogram. **13.** the second scout; part c of the Sufficient Conditions for a Parallelogram Theorem
15. m$\angle BCD = 70$, m$\angle ADC =$ m$\angle ABC = 110$ **17.** $PQRS$ is a parallelogram, so $PQ = RS$ (Properties of a Parallelogram Theorem)
19. a.

Conclusions	Justifications
0. $\triangle PTS$ is isosceles: m$\angle PTQ =$ m$\angle STR$	Given
1. $PT = ST$	definition of isosceles $\triangle$
2. $\angle P \cong \angle S$	Isosceles $\triangle$ Base Angles Theorem
3. $\triangle TPQ \cong \triangle TSR$	ASA Congruence Theorem

b.

Conclusions	Justifications
0. $\triangle TPQ \cong \triangle TSR$	From a.
1. $TQ = TR$	CPCF Theorem
2. $\triangle TQR$ is isosceles	definition of isosceles $\triangle$

11.

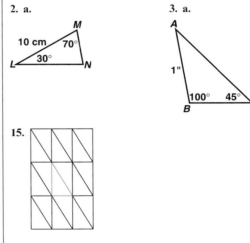

LESSON 7-9 (pp. 416–422)

11. True **13.** $\overline{GJ}$ **15. a.** x **b.** $x + y$ **c.** $180 - y - 2x$
17. a parallelogram: Sufficient Conditions for a Parallelogram Theorem part (c) **19.** No **21.** Sample: **a.** $AB = DE$
b. HL Congruence Theorem

CHAPTER 7 PROGRESS SELF-TEST (pp. 426–427)

1. a. m$\angle CBD = 180 -$ m$\angle ABD$ (linear pair) $= 180 - 60 = 120$
b. m$\angle C < 60$ (by Exterior Angle Inequality) **c.** m$\angle D < 60$
2. a. (Art is reduced in size.) **See right. b.** Yes **c.** by the ASA Congruence Theorem **3. a.** (Art is reduced in size.) **See right.**
b. Yes **c.** by the AAS Congruence Theorem **4. a.** No **b.** The given information is SSA condition but not SsA Congruence, so it does not yield a unique triangle. **5. a.** Yes **b.** SSS Congruence Theorem
6. If two angles and a non-included side in one triangle are congruent to two angles and the corresponding non-included side in a second triangle, then the triangles are congruent.
7. a. $\triangle ABC \cong \triangle CDA$ by the ASA Congruence Theorem
b. $\triangle WVU \cong \triangle ZXY$ by the HL Congruence Theorem.
8. a. $\triangle ABC \cong \triangle CDA$ **b.** ASA Congruence Theorem
9. Sample argument:

Conclusions	Justifications
0. M is the midpoint of $\overline{AC}$, $\overline{AB} \, // \, \overline{CD}$	Given
1. $\overline{AM} \cong \overline{CM}$	definition of midpoint
2. $\angle AMB \cong \angle CMD$	Vertical Angles Theorem
3. $\angle BAM \cong \angle DCM$	// lines $\Rightarrow$ AIA $\cong$
4. $\triangle ABM \cong \triangle CDM$	ASA Congruence Theorem

10. Argument:

Conclusions	Justifications
0. $WX = WY$ $\angle WUY \cong \angle WVX$	Given
1. $\angle W \cong \angle W$	Reflexive Property of Congruence
2. $\triangle UWY \cong \triangle VWX$	AAS Congruence Theorem
3. $\overline{WU} \cong \overline{WV}$	CPCF Theorem
4. $\triangle WUV$ is isosceles	definition of isosceles $\triangle$

11. Since $AB = AC$, $AD = AD$, and $\angle ADB$ and $\angle ADC$ are both right angles, $\triangle ADB \cong \triangle ADC$ by the HL Congruence Theorem. Thus, $BD = CD$ by the CPCF Theorem. **12.** Because one pair of opposite sides is parallel and congruent. **13.** $OR = x$, $LR = y$, $PR = 2x$ **14.** Yes, because the diagonals have the same midpoint.
15. Sample: **See below. 16.** (f) by the Unequal Angles Theorem

2. a.

3. a.

15.

The chart below keys the **Progress Self-Test** questions to the objectives in the **Chapter Review** on pages 428-432 or to the **Vocabulary** (Voc.) on page 425. This will enable you to locate those **Chapter Review** questions that correspond to questions you missed on the **Progress Self-Test**. The lesson where the material is covered is also indicated on the chart.

Question	1	2	3	4	5	6	7	8	9	10
Objective	B	A	A	A	A	Voc.	C	C	D	E
Lesson	7-9	7-5	7-2	7-5	7-2	7-2	7-5	7-2	7-3	7-3, 7-4

Question	11	12	13	14	15	16
Objective	I	K	F	G	J	H
Lesson	7-5	7-8	7-7	7-8	7-6	7-9

CHAPTER 7 REVIEW (pp. 428–432)

1. a. No **b.** There are many noncongruent triangles that fit the given information. **3. a.** Yes **b.** ASA Congruence Theorem **5. a.** See right. **b.** No. There are many noncongruent triangles that fit the given information. **7. a.** See right. **b.** Yes. SsA Congruence Theorem **9. a.** $= 48$ **b.** < 132 **c.** $= 132$ **11.** $\angle 1$. $\angle 1$ is an exterior angle of $\triangle TWX$, TWY, and TWZ. Therefore m$\angle 1 >$ m$\angle 2$, m$\angle 1 >$ m$\angle 3$ and m$\angle 1 >$ m$\angle 4$. **13. A.** AAS Congruence Theorem $\triangle MOP \cong \triangle MNP$ **15.** HL Congruence Theorem, $\triangle KLM \cong \triangle KJM$ **17.** AAS $\cong$ Theorem $\triangle GFE \cong \triangle GIH$ **19.** Sample: $\overline{AC} \cong \overline{DF}$ SSS Congruence Theorem

21.

Conclusions	Justifications
1. $\overline{AC} \cong \overline{AC}$	Reflexive Property of Congruence
2. $\triangle ADC \cong \triangle ABC$	ASA Congruence Theorem (step 1 and given)

23.

Conclusions	Justifications
1. $QA = QB$; $PA = PB$	definition of circle
2. $\overline{PQ} \cong \overline{PQ}$	Reflexive Property of Congruence
3. $\triangle APQ \cong \triangle BPQ$	SSS Congruence Theorem (steps 1 and 2)

25. Cannot be proved. This is the SSA condition, but the longer $\cong$ sides are not opposite the congruent angles. So SsA Congruence Theorem does not apply.

27.

Conclusions	Justifications
1. $\overline{AD} \cong \overline{AD}$	Reflexive Property of Congruence
2. $\triangle ABD \cong \triangle ACD$	SSS Congruence Theorem (step 1 and given)
3. $\angle BAD \cong \angle CAD$	CPCF Theorem

29.

Conclusions	Justifications
1. $\angle KJM \cong \angle KJL$	definition of angle bisector
2. $\overline{JK} \cong \overline{JK}$	Reflexive Property of Congruence
3. $\triangle KJM \cong \triangle KJL$	SAS Congruence Theorem (steps 1, 2, and given)
4. $\angle M \cong \angle L$	CPCF Theorem

31. Sample argument:

Conclusions	Justifications
1. $\overline{ON} \cong \overline{NE}$	definition of midpoint
2. $\angle NUO \cong \angle NAE$; $\angle NOU \cong \angle NEA$	// Lines $\Rightarrow$ AIA $\cong$ Theorem
3. $\triangle NUO \cong \triangle NAE$	AAS Congruence Theorem (steps 1 and 2)
4. $\overline{AE} \cong \overline{UO}$	CPCF Theorem

33. $DE = 12$, $AD = 19$, $BD = 24$ **35.** m$\angle DCB = 130$, m$\angle ADC =$ m$\angle ABC = 50$ **37.** Yes **39.** not necessarily **41.** $\angle A$ **43.** $\overline{EF}$, $\overline{ED}$, $\overline{DF}$ **45.** $\triangle ARP \cong \triangle ABP$ by the HL Congruence Theorem. So, by the CPCF Theorem, $RP = PB$. **47.** $\triangle XBD \cong \triangle XBA \cong \triangle XBC$ by the AAS Congruence Theorem, so $DB = AB = CB$. **49.** Sample: See below. **51.** (Art is enlarged in size) See below. **53.** parallelogram **55.** $AB = CD$, $CD = EF$ by the Properties of a Parallelogram Theorem. Therefore, $AB = EF$ by Transitive Property of Equality.

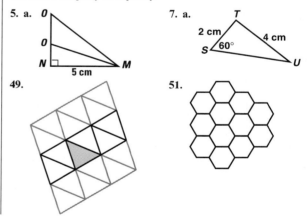

REFRESHER (p. 433)

A. **1.** $12 + 6x$ **2.** $4x - xy$ **3.** $ah + a\ell$ **4.** $\frac{1}{2}bw + \frac{1}{2}bt$ **5.** $ab - 2b^2$ **6.** $2xy^2 + 2x^2y$ **B.** **7.** $x(a + 2)$ **8.** $3c(3c - 4)$ **9.** $4(b - 5h)$ **10.** $\pi(r^2 + h^2)$ **11.** $\frac{1}{2}h(b + 1)$ **12.** $6ar(2r + 1)$ **C.** **13.** $y^2 + 6y + 8$ **14.** $3ab - 21b + 2a - 14$ **15.** $4r^2 - t^2$ **16.** $s^2 + 8s + 16$ **17.** $m^2 + 2mn + n^2$ **18.** $\frac{1}{2}ey + \frac{1}{2}fy + \frac{1}{2}eh + \frac{1}{2}fh$ **D.** **19.** 1.41 **20.** 6.93 **21.** 6.40 **22.** 2.06 **E.** **23.** 10 **24.** $2\sqrt{10}$ **25.** $3\sqrt{2}$ **26.** $6\sqrt{5}$ **F.** **27.** $x = \pm 5$ **28.** $y = \pm 4$ **29.** $z = \pm\sqrt{20} = \pm 2\sqrt{5} \approx \pm 4.47$ **30.** $r = \sqrt{\frac{3}{\pi}} \approx 0.98$

LESSON 8-1 (pp. 436–440)
13. 1.25 ft **15. a.** $8s + 18m + 2\ell$ **b.** 1300 meters **17. a. See below.**
b. $x + y = 8$ ($x > 0$ and $y > 0$) **c.** Sample: $x = 4.5$, $y = 3.5$
19. (d) **21.** Yes **23.** $\approx .79$ inch **25.** 1.5 **27.** No; Sample
counterexample: If $x = -1$, $(x + 1)(2x - 3) = 0$, but $2x^2 - 3 = -1$.
29. a. ≈ 63 mph **b.** ≈ 55 mph **c.** Sample: one route has more
stops or slower speed limits.

17. a.

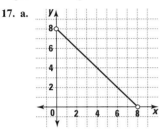

LESSON 8-2 (pp. 441–446)
9. a. 100 units^2 **b.** k^2 units^2
11. $\frac{1}{2}$ yard **13. a.** 288 sq in.
b. 2 sq ft **c.** 144 **15.** 6000
17. 46 cm, 46 cm, 185 cm
19. a. 230 m **b.** 6.25 m
21. $x = \pm 4$ **23. See right.**
$ABCB'$ is a rectangle.

23.

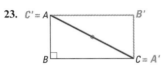

LESSON 8-3 (pp. 447–452)
7. 4096 dots/in^2 **9.** ≈ 66.5 km^2 **11.** about 34% **13.** farm in U.S.
15. a. Its area is multiplied by 100. **b.** Its perimeter is multiplied
by 10. **17.** 56 units

19. Conclusions	Justifications
1. $\triangle PQA \cong \triangle DRA$	AAS Congruence Theorem (Given)
2. $\overline{PA} \cong \overline{DA}$	CPCF Theorem
3. $\triangle PAD$ is isosceles.	definition of isosceles $\triangle$

21. False

LESSON 8-4 (pp. 453–458)
11. 9000 ft^2 **13.** 3 in. **15. a.** 32 m **b.** 48 m^2 **17.** 23 units^2
19. a. 286,650 square miles; answers may vary by
± 4900 square miles. **b.** 266,400 square miles; answers may vary
by ± 1225 square miles. **21.** 124 units^2

23. Conclusions	Justifications
0. $\angle ABD \cong \angle BDC$	Given
1. $\overline{DC} \parallel \overline{AB}$	AIA $\cong \Rightarrow \parallel$ lines
2. $ABCD$ is a trapezoid.	definition of trapezoid

25. (a)

LESSON 8-5 (pp. 459–464)
13. a., b. See below. c. rectangle **d.** 160 units^2 **e.** Yes **15.** $\frac{1}{2}bc$ units^2
17. a. Sample Argument:

Conclusions	Justifications
1. $AF = BC$	Isosceles Trapezoid Theorem
2. $\angle F = \angle C$	definition of isosceles trapezoid
3. $m\angle AEF = 90$ $m\angle BDC = 90$	definition of perpendicular
4. $m\angle AEF = m\angle BDC$	Substitution
5. $\triangle AEF \cong \triangle BDC$	AAS Congruence Theorem (Steps 1, 2, 4)
6. $FE = DC$	CPCF Theorem

b. 52.5 units^2 **19.** (b) **21.** Place a
grid over the shape. Add the number
of squares entirely inside the figure to
half the number of squares partially
covering the figure, and multiply the
result by the area of each square.
23. 108 **25.** $3\sqrt{3}$

13. a., b.

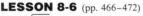
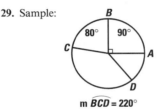

LESSON 8-6 (pp. 466–472)
19. $\sqrt{109} \approx 10.4$ miles **21.** $\sqrt{60} \approx 7.75$ feet **23.** 100 **25.** 270
units^2 **27.** 4.5 units^2 **29. See below.**

29. Sample:

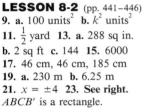

m $\overset{\frown}{BCD} = 220°$

LESSON 8-7 (pp. 474–478)
11. $10\pi \approx 31.4$ m **13.** $\frac{110}{\pi}$ or about 35 seconds **15.** 20 units
17. $A = \frac{1}{2}h(b_1 + b_2)$ **19.** $p = 4s$ **21.** 2000 units^2 **23. a.** Its
perimeter is multiplied by 6. **b.** Its area is multiplied by 36.
25. (d)

LESSON 8-8 (pp. 480–484)
7. a. 11,300 sq meters **b.** 377 m **9. a.** What is the area of Farmer
Bob's field? **b.** $90,000\pi$ m$^2 \approx 282,600$ m^2 **11. a.** $\frac{1}{16}$ **b.** $\frac{15}{16}$
13. a. 72° **b.** $6\pi \approx 18.8$ units **15.** about 466 **17. a.** $18x^2$ units^2
b. $(13 + \sqrt{97})x \approx 22.8x$ units **19.** 7.1 units **21.** 48 units^2
23. 38 cm

CHAPTER 8 PROGRESS SELF-TEST (pp. 488–489)
1. $2(40 + 15) = 110$ units **2.** $\frac{1}{2} \cdot 80 \cdot 210 = 8400$ units^2
3. $A = \frac{1}{2}h(a + c)$ **4.** $\frac{q}{6}$ **5.** $\ell w = 200$ m $= 25w$; $w = 8$ m
6. $48 = \frac{1}{2} \cdot 6 \cdot (9 + b)$; $16 = 9 + b$; $b = 7$ units **7. a.** $C = 12\pi''$;
$A = \pi \cdot 6^2 = 36\pi$ in^2 **b.** 38''; 113 in^2 **8. a.** $40\pi \cdot \frac{45}{360} = 5\pi \approx$
15.7 units **b.** $400\pi \cdot \frac{45}{360} = 50\pi \approx 157$ units^2 **9.** $\sqrt{41^2 - 9^2} =$
$\sqrt{1681 - 81} = \sqrt{1600} = 40$ **10.** $20 + 21 + \sqrt{20^2 + 21^2} =$
$20 + 21 + 29 = 70$ **11. a.** Yes **b.** Because $121 + 3600 = 3721 =$
61^2, meeting the condition of Pythagorean Converse Theorem.
12. $\left(27 + \frac{21}{2}\right) \cdot 25 = 937.5$ sq miles (Answers may slightly vary.)
13. $12^2 - 36\pi \approx 30.90$ units^2 **14. a.** $P = 2(\ell + w)$;
$P' = 2(4\ell + 4w) = 4 \cdot 2(\ell + w) = 4 \cdot P$; its perimeter is
multiplied by 4. **b.** $A = \ell w$; $A' = (4\ell)(4w) = 16 \ell w$; $A' = 16A$;
its area is multiplied by 16. **15.** $16 \cdot 2 + 21 \cdot 2 = 74''$
16. $1.8^2 + b^2 = 5^2$; $b^2 = 21.76$; $b \approx 4.7$ meters **17.** $80^2\pi \approx$
20,100 square miles **18.** $\frac{9}{3} \cdot \frac{15}{3} = 3 \cdot 5 = 15$ sq yd **19.** $\left(\frac{2640}{4}\right)^2 =$
$660^2 = 435,600$ sq ft **20.** $\frac{\pi \cdot 6^2}{\pi \cdot 18^2} = \frac{36\pi}{324\pi} = \frac{1}{9}$ **21.** Area $= \frac{1}{2}bh =$
$\frac{1}{2} \cdot 9 \cdot 1 = 4.5$ units^2 **22.** Area $= 9 \cdot 11 - (3 \cdot 4 + 3 \cdot 3) =$
$99 - 21 = 78$ units^2 **23.** Sample: Greek, Chinese, Babylonian

The chart below keys the **Progress Self-Test** questions to the objectives in the **Chapter Review** on pages 490–493 or to the **Vocabulary** (Voc.) on page 487. This will enable you to locate those **Chapter Review** questions that correspond to questions students missed on the **Progress Self-Test**. The lesson where the material is covered is also indicated on the chart.

Question	1	2	3	4	5	6	7	8	9	10
Objective	A	C	C	A	C	C	F	F	D	A, D
Lesson	8-1	8-4	8-5	8-1	8-2	8-5	8-7, 8-8	8-7, 8-8	8-6	8-1, 8-6
Question	11	12	13	14	15	16	17	18	19	20
Objective	E	B	G	A, C	H	H	J	I	H, I	J
Lesson	8-6	8-3	8-2, 8-8	8-1, 8-2	8-1	8-6	8-8	8-2	8-1, 8-2	8-8
Question	21	22	23							
Objective	K	K	L							
Lesson	8-4	8-2	8-6							

CHAPTER 8 REVIEW (pp. 490–493)

1. 32 units **3.** 235 meters **5.** 10 cm **7.** 12.5 units and 25 units **9.** 13,950 m^2 (Answers may slightly vary.) **11.** 37,500 sq miles (Answers may slightly vary.) **13.** 4.55 cm^2 **15.** 72 units^2 **17.** $\approx$ 153.2 units^2 **19.** 3.5 units **21.** 12 units **23.** $\sqrt{5} \approx 2.24$ units **25.** $\approx 22.2x$ units **27.** 75 cm **29.** No **31.** Yes **33. a.** 20π units, 100π units^2 **b.** 62.83 units, 314.16 units^2 **35.** $\approx 6.4x$ meters

37. a. 41.9 units **b.** $\approx$ 39.27 units^2 **39.** Area of trapezoid: $A = \frac{1}{2}h(b_1 + b_2)$, but in a parallelogram, $b_1 = b_2 = b$, so $A = \frac{1}{2}h(b + b) = \frac{1}{2} \cdot h \cdot 2b = hb$. **41.** $49\pi - 98$ **43.** $3\pi x^2$ units^2 **45.** $\approx$ 4.4 ft **47.** $8k$ units **49.** 14 minutes **51.** 44,100 m^2 **53.** 780,000 sq ft **55.** 28.3 km^2 **57.** $\frac{100}{\pi} \approx 31.8$ ft **59.** 2025 units^2 **61.** 32 units^2 **63.** (a) **65.** True

LESSON 9-1 (pp. 496–499)

9. Yes **11.** Yes **13.** Because a plane has no thickness. **15. a.** True **b.** False **17.** A circle is the set of all points in a plane at a certain distance (its radius) from a certain point (its center). **19.** $x = 67.5$, $y = 112.5$

LESSON 9-2 (pp. 500–504)

11. Sample: the floor, north wall, and east wall **13.** Sample: If two planes are perpendicular to the same plane, then they are parallel. No. **15.** Sample: ℓ and m are skew lines. **See below.** **17. a.** 90° **b.** 30° **19.** No; by Intersecting Planes Assumption, their intersection is a line. **21. a.** 1599 units^2 **b.** about 49.2 units **23. See below.**

15.

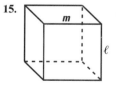

23. Sample:

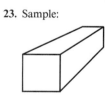

LESSON 9-3 (pp. 505–510)

13. Sample: solid right prism (often hexagonal) or a solid right circular cylinder **15.** Sample: **See right.** **17. a.** 5 units **b.** 13 units **c.** 30 units^2 **19. a.** 60 units **b.** If one edge has length s, then the length of all edges is $12s$. **21.** Sample: **See right.** **23.** infinitely many **25. See right.**

15.

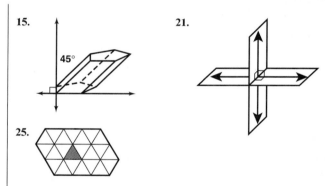

21.

25.

LESSON 9-4 (pp. 511–516)

11. the height of the pyramid, slant height, length of a lateral edge **13. a.** n **b.** $n + 1$ **c.** n **d.** $2n$ **15.** Sample: a platform that elephants stand on **17. a.** $2\pi'' \approx 6.28''$ **b.** $1.0''$ exactly **19.** Sample: **See below.** **21.** In a right cylinder, the direction of the translation is perpendicular to the base; in an oblique cylinder, it is not. **23.** 150 units^2 **25. a.** 240 units^2 **b.** about 18.5 units

19.

LESSON 9-5 (pp. 517–524)

13. a. Sample: **See page 893. b.** Sample: **See page 893. c.** Part **a** is a rectangle; part **b** is a rectangle. **15. a.** Sample: **See page 893. b.** Sample: **See page 893. c.** part **a** is a circle and part **b** is an ellipse **17.** sphere **19. a.** Reflexive Property

of Congruence **b.** definition of sphere **c.** HL Congruence Theorem **d.** CPCF Theorem **21. a.** 8 units **b.** 136 units² **23.** 376 units²
25. The point Q is the reflection image of P over a line ℓ when P is not on ℓ if and only if ℓ is the perpendicular bisector of $\overline{PQ}$.

13. a.   **13. b.**

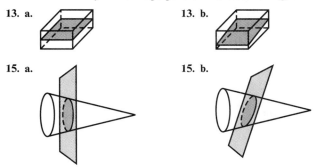

15. a.   **15. b.**

LESSON 9-6 (pp. 525–530)
11. See below. 13. False. A symmetry plane parallel to the bases will have a cross section congruent to the bases. **15. a.** Sample: See below. **b.** Sample: See below. **c.** Part **a** is a circle; part **b** is an ellipse. **17. a.** Sample: See below. **b.** 12 units **19.** box without its top **21.** 240 units² **23.** This has more information than is necessary.

11.   **15. a.**

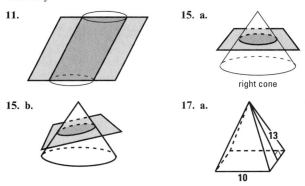

right cone

15. b.   **17. a.**

13

10

LESSON 9-7 (pp. 531–535)
11. ≈ 25 ft **13. See below. 15. a.** 2 **b. See below. 17.** 6 units **19.** one **21.** about 270 mi **23.** a correspondence between two sets of points such that (1) each point in the preimage set has a unique image, and (2) each point in the image set has exactly one preimage

13.

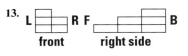
L   R F   B
front   right side

15. b.

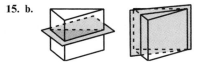

LESSON 9-8 (pp. 537–542)
9. regular octahedron **11.** (d) **13.** Sample: See below. **15.** about 56.5 square inches **17. See below. 19. See below. 21. a.** $V = 9$, $E = 16$, $F = 9$ **b.** $V = 16$, $E = 24$, $F = 10$ **c.** $V = n + 1$, $E = 2n$, $F = n + 1$ **d.** $V = 2n$, $E = 3n$, $F = n + 2$ **e.** (c)
f. pyramid: $(n + 1) + (n + 1) - 2n = 2n + 2 - 2n = 2$ prism: $(n + 2) + (2n) - (3n) = 3n + 2 - 3n = 2$ **23.** about 24.3 inches
25. $t = 95$

13.   **17.**

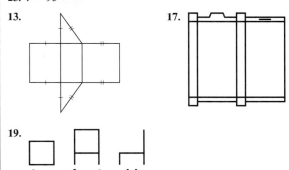

19.

top   front   side

LESSON 9-9 (pp. 544–551)
11. all of them **13.** Sample: See below. **15.** Sample: See below.
17. Sample: See below. **19. See below. 21. See below.**
23. lateral surface of cylinder **25. a.** $\sqrt{136} \approx 11.7$ units
b. $12\sqrt{2} \approx 17.0$ units **c.** $\sqrt{172} \approx 13.1$ units **d.** $\frac{1}{2} \cdot 10 \cdot 6\sqrt{2} = 30\sqrt{2} \approx 42.4$ units²

13.

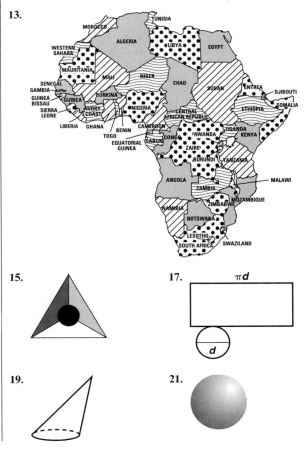

15.   **17.** πd

d

19.   **21.**

CHAPTER 9 PROGRESS SELF-TEST (pp. 556–557)

1. See below. **2.** Sample: See below. **3.** See below. **4.** Sample: The endpoint of one leg does not lie in the plane determined by the endpoints of the other three legs. **5. a.** Sample: See below.
b. Sample: See right. **c.** Part a is a square. Part b is a quadrilateral.
6. 4 **7. a.** See right. **b.** a great circle **8.** See right. **9. a–c.** See right. **10.** infinitely many **11.** circumference $= 2\pi r = 28\pi$;
$\frac{28\pi}{2\pi} = r$; $r = 14$ cm **12. a.** $\sqrt{34^2 - 16^2} = 30$ cm **b.** $10^2\pi = 100\pi \approx 314.16$ cm^2 **13. a.** Area($EGHJ$) $= 9 \cdot 22 = 198$ in^2
b. Area($\triangle EFG$) $= \frac{1}{2} \cdot EF \cdot FG = \frac{1}{2} \cdot 4 \cdot \sqrt{9^2 - 4^2} = 2 \cdot \sqrt{65} \approx 16.12$ in^2 **14.** a rectangular solid **15.** a sphere; it is a surface
16. a. 2 stories **b.** 4 sections **c.** all sections of the left side
17. regular octagonal prism **18.** Sample: See right. **19.** right prism with bases shaped like a parallelogram **20.** (c) and (d) **21.** Sample: See right.

1.

2.

3.

5. a.

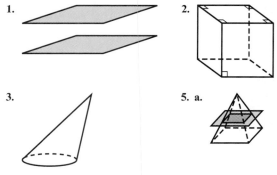

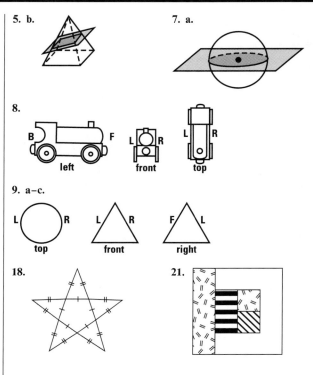

5. b.

7. a.

8.

left   front   top

9. a–c.

L ◯ R   L △ R   F △ L
top   front   right

18.

21.

The chart below keys the **Progress Self-Test** questions to the objectives in the **Chapter Review** on pages 558–561 or to the **Vocabulary** (Voc.) on page 555. This will enable you to locate those **Chapter Review** questions that correspond to questions students missed on the **Progress Self-Test.** The lesson where the material is covered is also indicated on the chart.

Question	1	2	3	4	5	6	7	8	9	10
Objective	A	A	A	F	B	G	B	C	C	G
Lesson	9-2	9-3	9-4	9-1	9-5	9-6	9-5	9-7	9-7	9-6
Question	**11**	**12**	**13**	**14**	**15**	**16**	**17**	**18**	**19**	**20**
Objective	D	D	D	H	H	E	E	J	J	K
Lesson	9-5	9-4	9-3	9-3	9-5	9-7	9-7	9-8	9-8	9-9
Question	**21**									
Objective	I									
Lesson	9-9									

CHAPTER 9 REVIEW (pp. 558–561)

1. See page 895. **3.** See page 895. **5.** See page 895. **7.** See page 895. **9. a.** Sample: See page 895. **b.** Sample: See page 895. **c.** Part a is a circle, and part b is an ellipse. **11. a.** Sample: See page 895.
b. Sample: See page 895. **c.** Part a is a regular pentagon and part b is a pentagon. **13. a.–c.** See page 895. **15. a–c.** See page 895.
17. a. 36 units **b.** 48 units^2 **19. a.** $9\pi \approx 28.3$ cm^2 **b.** $\sqrt{32} \approx 5.7$ cm
21. a. See page 895. **b.** $49\pi \approx 153.9$ mm^2 **23. a.** 34 units

b. 350 units^2 **25. a.** 2 **b.** 2 **c.** back left-hand side **27.** solid right cone **29.** a line **31.** Yes; if a line is $\perp$ to two different lines in space at the same point, then it is $\perp$ to the plane containing those lines. **33.** not necessarily **35. a.** Yes **b.** infinitely many **37.** 9 **39.** solid sphere **41.** rectangular solid **43.** Sample: See page 895. **45.** Sample: See page 895. **47.** Sample: See page 895. **49.** Sample: See page 895. **51. a.** 12 units **b.** $20\pi \approx 62.8$ units **53.** Sample: Connected parts of countries may appear disconnected on different gores. **55.** Distances are not preserved.

1.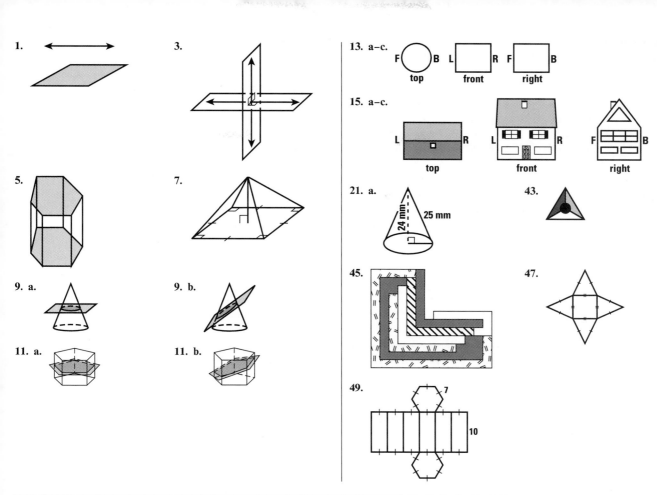

3.

13. a–c.

15. a–c.

5.

7.

21. a.

43.

9. a.

9. b.

45.

47.

11. a.

11. b.

49.

LESSON 10-1 (pp. 564–568)
13. a. right triangular prism **b.** 108 cm^2 **c.** 120 cm^2
15. $2(LW + LH + WH)$ **17.** $100\pi \approx 314.16$ cm^2
19. a. 214 units^2 **b.** 856 units^2 **c.** 4 **d.** 278, 1112, 4 **21.** Sample:
See below. **23.** Find the point of intersection of either the
perpendicular bisectors of any two sides or the bisectors of any two
angles.

21.

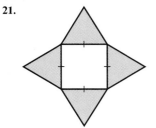

LESSON 10-2 (pp. 570–575)
13. a. Sample: **See right.** **b.** 1000 units^2 **c.** 1100 units^2
15. a. Sample: **See right.** **b.** $175\pi \approx 550$ units^2
c. $224\pi \approx 704$ units^2 **17.** 1332 sq in. = 9.25 sq ft **19.** Sample:
See right. **21.** Uniqueness Property - every polygonal region has a
unique area; Rectangle Formula - the area of a rectangle with
dimensions ℓ and w is ℓw; Congruence Property - congruent figures
have the same area; Additive Property - if A and B are
non-overlapping regions, than Area$(A \cup B)$ = Area(A) + Area(B).
23. a. $x = 61$ **b.** $y = \frac{64}{3}$ **c.** $z = 4$

13. a.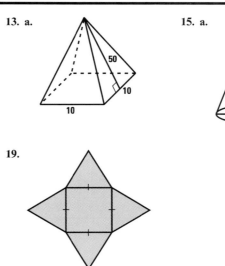

15. a.

19.

LESSON 10-3 (pp. 576–581)
15. 12 inches **17. a.** area of a base **b.** The area of the base is ℓw,
so the formula $V = Bh$ is equivalent to $V = \ell wh$. **19. a.** $\frac{1}{16}$ **b.** $\frac{1}{64}$
21. a. 3 **b.** 9 **c.** 27 **23. a.** $156\pi \approx 490$ in^2 **b.** 4 square feet
25. a. $5\sqrt{11} \approx 16.6$ ft^2 **b.** $84 + 24\sqrt{11} \approx 163.60$ ft^2

LESSON 10-4 (pp. 582–586)

15. $x + 3$ **17.** $33\frac{1}{3}$ cm **19.** Yes, if s is the side length of the original cube and $s = 20$, then $s \cdot (s + 5) \cdot (s - 4) = s^3$. **21.** 1.46 **23.** ≈ 1.8 cm^3 **25.** 2.88 units^2

LESSON 10-5 (pp. 587–592)

13. $\pi r^2 h$ **15.** 9.5 m^2 **17. a.** Yes **b.** For the second glass, $V \approx 2.7\pi h$. For the first glass $V \approx 1.3\pi h$. **19.** The first cylinder has four times the volume of the second cylinder. **21.** $4x^2 + 3xy + 28x - 10y^2 - 35y$ **23.** 20 cm^3 **25.** $A = \frac{1}{2}h(b_1 + b_2)$; $A = bh$; $A = \ell w$; $A = s^2$

LESSON 10-6 (pp. 593–597)

11. a. L.A. $= \frac{1}{2}\ell(2\pi r) = \pi r\ell$ **b.** S.A. $= \pi r\ell + \pi r^2 = \pi r(\ell + r)$ **13. a.** $192 + 192x$ units^2 **b.** S.A. $=$ L.A. $+ 2B$ **15. a.** $144wz$ units^3 **b.** $V = Bh$ **17.** 54 cm **19.** The volume is decreased by $8k^2$. See below. **21.** $\pi(R^2 - r^2)$ **23.** $-\frac{b}{a}$

19.

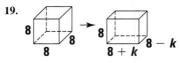

LESSON 10-7 (pp. 599–603)

11. $V = \frac{\pi r^2 h}{3}$ **13.** The volume is multiplied by 31.8. **15. a.** $\frac{72\pi}{3} \approx 75$ cm^3 **b.** 14 **c.** $3\pi\sqrt{73} \approx 81$ cm^2 **17.** at least 817 square inches **19.** volume $= 216$ units^3; S.A. $= 174 + 8\sqrt{117} \approx 261$ units^2 **21.** $\approx 26\%$

LESSON 10-8 (pp. 604–608)

7. 4 m **9. a.** $48\pi \approx 151$ cm^3 **b.** 37.5% **11.** $V = \frac{\pi}{6}d^3$ units^3 **13.** ≈ 449 units^3 **15.** cones, pyramids **17.** $2xyz + 8xy + 6yz + 24y$ **19.** 2.7

21.

Conclusions	Justifications
1. $AB = CB$	definition of equilateral $\triangle$
2. $OB = OB$	Reflexive Property of Equality
3. $OA = OC$	definition of sphere
4. $\triangle OBC \cong \triangle OBA$	SSS Congruence Theorem
5. $m\angle OCB = m\angle OAB$	CPCF Theorem

23. $4\pi r^2$

LESSON 10-9 (pp. 609–612)

7. $\approx 1.83\%$ **9.** $\approx \$464,704$ **11.** $36\pi \approx 113$ m^2 **13. a.** ≈ 26 square inches **b.** ≈ 13 square inches **15.** 24 units **17.** The shorter jar holds twice as much jam. **19.** 65.0 units **21. a.** 2250 cm^2 **b.** 7200 cm^2

CHAPTER 10 PROGRESS SELF-TEST (p. 616)

1. a. L.A. $= ph = (16 + 20 + \sqrt{20^2 - 16^2}) \cdot 30 = (36 + \sqrt{144}) \cdot 30 = 48 \cdot 30 = 1440$ units^2 **b.** Volume $= Bh = \frac{1}{2} \cdot (16 \cdot \sqrt{20^2 - 16^2} \cdot 30 = (\frac{1}{2} \cdot 16 \cdot \sqrt{144}) \cdot 30 = 2880$ units^3

2. Because $ph = 14\pi h = 84\pi$, $h = \frac{84\pi}{14\pi} = 6$ inches

3. Volume $= 400 = \ell wh = \ell \cdot 5 \cdot 10 = 50\ell$; $\ell = \frac{400}{50} = 8$ cm

4. a. L.A. $= \frac{1}{2}\ell p = \frac{1}{2} \cdot 26 \cdot (4 \cdot 20) = 1040$ units^2 **b.** $h = \sqrt{26^2 - 10^2} = 24$; $V = \frac{1}{3} \cdot 400 \cdot h = \frac{1}{3} \cdot 400 \cdot 24 = 3200$ units^3 **5.** Volume $= \frac{1}{3}Bh = \frac{1}{3} \cdot 8^2\pi \cdot 15 = 320\pi \approx 1005.3$ units^3 **6.** Volume $= 96$ cm$^3 = \frac{1}{3}Bh = \frac{1}{3}B \cdot 18 = 6B$; $B = \frac{96}{6} = 16$ cm^2 **7.** S.A. $= 4\pi \cdot 19^2 \approx 4536$ units^2 $V = \frac{4}{3}\pi \cdot 19^3 \approx 28731$ units^3 **8.** S.A. $= 100\pi$ sq units $= 4\pi r^2$; $r^2 = \frac{100\pi}{4\pi} = 25$; $r = \sqrt{25} = 5$ units **9.** $\sqrt[3]{400} \approx 7$

10. L.A. $= \frac{1}{2}\ell p = \frac{1}{2} \cdot \ell \cdot 2\pi r = \pi r\ell$ **11. a.** S.A. $= 6 \cdot (9t)^2 = 6 \cdot 81t^2 = 486t^2$ units^2 **b.** $V = (9t)^3 = 729t^3$ units^3 **12.** $7^3 = 343$; so the volume of the larger box is 343 times that of the smaller one. **13.** Let $e =$ diameter of Earth; then the diameter of Jupiter $= 11e$. S.A. of Earth $= 4\pi r^2 = 4\pi(\frac{e}{2})^2 = \pi e^2$. S.A. of Jupiter $= 4\pi r^2 = 4\pi(\frac{11e}{2})^2 = 121\pi e^2$. Jupiter's area is 121 times that of Earth. **14. a.** Yes **b.** They have congruent bases and heights, and both volumes are $64\pi h$ units^3. **15.** Volume(Prism) $= Bh$; Volume(Pyramid) $= \frac{1}{3}Bh$; the volume of the pyramid is one-third the volume of the prism. **16.** $2(0.9 \cdot 0.7 + 0.9 \cdot 0.2 + 0.7 \cdot 0.2) = 2 \cdot 0.95 = 1.9$ m^2 **17.** L.A. $= \frac{1}{2}p\ell = \frac{1}{2} \cdot 8\pi \cdot 18 = 72\pi \approx 226$ in^2 **18.** $V = \frac{d^3}{6}\pi \approx \frac{(485)^3}{6}\pi \approx 59,734,000$ miles^3 **19.** $V = Bh = \pi r^2 h = 9 \cdot 20\pi \approx 565$ cm^3 **20.** $(3x + 1)(2x + 8) = 3x \cdot 2x + 3x \cdot 8 + 1 \cdot 2x + 1 \cdot 8 = 6x^2 + 24x + 2x + 8 = 6x^2 + 26x + 8$ **21.** $(y + 2)(x + 1)(z + 6)$; $xyz + yz + 2xz + 2z + 6xy + 6y + 12x + 12$

The chart below keys the **Progress Self-Test** questions to the objectives in the **Chapter Review** on pages 617–619 or to the **Vocabulary** (Voc.) on page 615. This will enable you to locate those **Chapter Review** questions that correspond to questions students missed on the **Progress Self-Test**. The lesson where the material is covered is also indicated on the chart.

Question	1	2	3	4	5	6	7	8	9	10
Objective	A	A	A	B	B	B	D	D	C	F
Lesson	10-1, 10-5	10-1	10-3	10-2, 10-7	10-7	10-7	10-8, 10-9	10-9	10-3	10-6

Question	11	12	13	14	15	16	17	18	19	20
Objective	F	E	E	G	F	H	H	I	I	J
Lesson	10-6	10-4	10-9	10-5	10-6	10-1	10-2	10-8	10-5	10-4

Question	21									
Objective	J									
Lesson	10-4									

CHAPTER 10 REVIEW (pp. 617–619)
1. a. $72\pi \approx 226$ units^2 **b.** $104\pi \approx 327$ units^2 **c.** $144\pi \approx 452$ units^3
3. 720 units^3 **5.** 150 units^2 **7. a.** $3e$ units **b.** $27e^3$ units^3
9. S.A. ≈ 155 units^2; $V \approx 128.3$ units^3 **11. a.** 50 units
b. 8000 units^2 **c.** 14,400 units^3 **13.** 200 units^2 **15.** ≈ 6.6 ft **17.** 30
19. 5.31 **21.** S.A. $= 20{,}736\pi \approx 65{,}144$ units^2; $V = 497{,}664\pi \approx$
1,563,458 units^3 **23.** 6 units **25. a.** It is multiplied by 9. **b.** It is
multiplied by 27. **27.** The new volume is 4 times as large.

29. a. $2s\ell$ **b.** $s^2 + 2s\ell$ **31. a.** $\pi r^2 + \pi r\ell$ **b.** $\frac{1}{3}\pi r^2 \sqrt{\ell^2 - r^2}$
33. Plane sections at any level other than the base do not have the
same area. **35. a.** No **b.** Plane sections at any level do not have
the same area. **37.** 320 square inches **39.** $\approx 14{,}142$ cubits^2
41. $\approx 1.13 \cdot 10^8$ km^2 **43.** 15.625 in^3 **45.** ≈ 282.7 m^3
47. $\approx 167{,}000$ cubits^3 **49.** ≈ 73.6 cm^3 **51.** $20xy + 15x + 8y + 6$
53. $(2x + 7)(x + 12) = 2x^2 + 31x + 84$ **55.** $(a + 15)(b + 9)(c + 8)$;
$abc + 8ab + 9ac + 72a + 15bc + 120b + 135c + 1080$

LESSON 11-1 (pp. 622–627)
11. Pete Sampras is world-class. **13.** No conclusion can be made.
15. Thayer is 16 or older. **17.** Nothing can be concluded. **19.** Yes
21. $GH^2 = FG^2 + FH^2$, but $FH^2 = HI^2 + FI^2$, so by substituting,
$GH^2 = FG^2 + HI^2 + FI^2$. But $FI^2 = IE^2 + EF^2$, so by substituting,
$GH^2 = FG^2 + HI^2 + IE^2 + EF^2$. **23.** Sample: **See below.**
25. a. $\frac{2}{5}$ **b.** $\frac{b}{a}$ **c.** -1

23.

LESSON 11-2 (pp. 628–633)
15. a. $x \neq 3$ **b.** Detachment and Contrapositive **17.** Yes; Let $a =$
Joanne apologized and $d =$ Joanne went on her date. Her mother
said not-$a \Rightarrow$ not-d. It is true, so $d \Rightarrow a$ is also true, and d is true,
so you can conclude a: Joanne apologized. **19.** $\triangle ABC$ is not
equilateral. **21.** From a true conditional $p \Rightarrow q$ and a statement or
given information p, you may conclude q. **23.** $\sqrt{130}$ **25.** (b)
27. 12, -22 **29.** $a^2 + b^2$; $a^2 + b^2$; 2; $a^2 + b^2$.

LESSON 11-3 (pp. 634–638)
9. $\overline{BC}$ and $\overline{AD}$ are not parallel. **11.** Catherine Voila: secretary,
Edgar Guinness: manager, Wilbur Farmer: teller, Marjorie Landis:
guard, Shirley Edwards: bookkeeper **13.** Converse: If I eat my hat,
then my cousin is a good cook. Inverse: If my cousin is not a good
cook, then I will not eat my hat. Contrapositive: If I don't eat my
hat, then my cousin is not a good cook. **15.** Diagonals in a square
are perpendicular.
17.

Conclusions	Justifications
1. $AC = BC$, $DC = EC$	definition of isosceles $\triangle$
2. $\angle ACD \cong \angle BCE$	Vertical Angles Theorem
3. $\triangle ACD \cong \triangle BCE$	SAS Congruence Theorem (steps 1 and 2)

19. a. $y = 11$ **b.** $x = h$ **c.** $(h, 11)$ **21.** $4z$

LESSON 11-4 (pp. 639–644)
9. Suppose that a triangle has two right angles. Then the sum of
the interior angles is bigger than 180°, which contradicts the
Triangle-Sum Theorem. So the supposition is false. Therefore, a
triangle cannot have two right angles. **11.** Suppose that x is on
the $\perp$ bisector of $\overline{PQ}$. By the Perpendicular Bisector Theorem,
$PX = QX$, which contradicts the given information. So the
supposition is false. Therefore, x is not on the $\perp$ bisector of $\overline{PQ}$.
13. Suppose that two different lines intersect in more than one
point, for instance, two points P and Q. This contradicts the
Unique Line Assumption of the Point-Line-Plane Postulate. So the
assumption is false. Therefore, two different lines intersect in at
most one point. **15.** Isobel: $5'11''$ and brunette; Mary: $5'10''$ and

black; Ruth: $5'8''$ and auburn; Marcia: $5'7''$ and blond; Grace: $5'6''$
and red **17.** $\triangle ABC$ is not congruent to $\triangle DEF$. **19. a.** If a surface
is not a box, then it is not a prism. **b.** If a surface is a box, then it
is a prism. **c.** If a surface is not a prism, then it is not a box.
d. its converse and inverse (The original statement is not true.)
21. True

LESSON 11-5 (pp. 645–650)
9.

Conclusions	Justifications
1. The slopes of $\overline{EF}$ and $\overline{GH}$ are 1. The slopes of $\overline{EH}$ and $\overline{FG}$ are -1.	definition of slope
2. $\overline{EF} \parallel \overline{GH}$, $\overline{EH} \parallel \overline{FG}$	Parallel Lines and Slopes Theorem
3. $EFGH$ is a parallelogram.	definition of parallelogram

11. The slope of $\overline{EG} = \frac{-7}{3}$, and the slope of $\overline{FH} = \frac{-3}{7}$. $\frac{-7}{3} \cdot \frac{-3}{7} =$
$1 \neq -1$, so $\overline{EG}$ is not perpendicular to $\overline{FH}$. **13. a.** Sample: **See
below. b.** The slope of $\overline{BD}$ is 1, and the slope of $\overline{AC}$ is -1.
Therefore, the product of the slopes is -1. By the Perpendicular
Lines and Slopes Theorem, $\overline{AC} \perp \overline{BD}$. **15.** Either $\sqrt{39{,}600} =$
199 or $\sqrt{39{,}600} \neq 199$. Suppose $\sqrt{39{,}600} = 199$. Then, squaring
both sides: $39{,}600 = 199^2$. However, by the definition of square,
$199^2 = 199 \cdot 199 = 39601$. This contradicts $39{,}600 = 199^2$.
By the Law of Indirect Reasoning, the supposition is false. Thus
$\sqrt{39{,}600} \neq 199$. **17.** Julie walks to school. **19. a.** ≈ 785 sq ft
b. $\approx 25{,}656$ bushels **21. a.** 450 units **b.** 450 units **23.** cannot be
simplified

13. a.

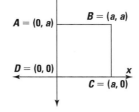

LESSON 11-6 (pp. 651–655)
11. a. See page 898. b. $\sqrt{9.49} \approx 3.08$ miles **13.** Given:
Parallelogram $WXYZ$ in the convenient position as shown below,
with $a > 0$. **See page 898 for drawing.** To Prove: $WX = YZ$ and
$WZ = YX$. According to the Distance Formula, $WX = a$, $YZ =$
$\sqrt{(a + b - b)^2 + (c - c)^2} = \sqrt{a^2} = a$, $WZ =$
$\sqrt{(0 - b)^2 + (0 - c)^2} = \sqrt{b^2 + c^2}$, and $XY =$
$\sqrt{(a + b - a)^2 + (c - 0)^2} = \sqrt{b^2 + c^2}$. Thus, by the Transitive
Property of Equality, $WX = YZ$ and $WZ = YX$. Therefore, each
pair of opposite sides are congruent. **15.** The slope of $\overline{PQ} =$
$\frac{2z - 4z}{7z - 3z} = -\frac{1}{2}$ and the slope of $\overline{QR} = \frac{-8z - 2z}{2z - 7z} = 2$ by the definition
of slope. $-\frac{1}{2} \cdot 2 = -1$. So $\overline{PQ} \perp \overline{QR}$ by the Perpendicular Lines and

Slopes Theorem. Thus, $\triangle PQR$ is a right triangle by the definition of right triangle. **17.** Richard Nixon did not win the majority of electoral votes. **19.** The area of the triangle is one-half the area of the parallelogram. **21.** $\frac{a - b}{2}$

11. a.

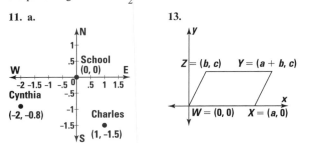

13.

LESSON 11-7 (pp. 656–660)

9. a. $(-1, 0)$ **b.** $\sqrt{2}$ **c.** Sample: $(0, 1)$ **11. a.** $(-6, -2)$ **b.** 1
c. Sample: $(-6, -3)$ **13.** 22π units **15. a. See right. b.** $(2, 0)$
c. $(x - 2)^2 + (y + 1)^2 = 1$ **d.** π square units **17.** $41x$ **19.** Given:
$Q = (9a, 4b)$; $R = (6a, 2b)$, $S = (a, -7b)$, and $T = (-a, -14b)$.
To prove: $QRST$ is a trapezoid.

Conclusions	Justifications
1. slope of $\overline{TQ} =$ $\frac{-14b - 4b}{-a - 9a} = \frac{-18b}{-10a} \frac{9b}{5a}$	definition of slope
2. slope of $\overline{SR} =$ $\frac{-7b - 2b}{a - 6a} = \frac{-9b}{-5a} \frac{9b}{5a}$	definition of slope
3. $\overline{TQ} \parallel \overline{SR}$	Parallel Lines and Slopes Theorem
4. $QRST$ is a trapezoid.	definition of trapezoid

21. By the definition of isosceles triangle, $YX = ZX$. V is the midpoint of $\overline{XY}$, Therefore, $XV = VY = \frac{1}{2}XY$, $XW = WZ = \frac{1}{2}XZ$. Therefore, by the Transitive Property of Equality, $XV = XW$. By the definition of isosceles triangle, $\triangle XVW$ is isosceles.
23. They are collinear.

15. a.

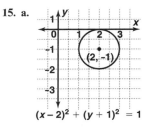

$(x - 2)^2 + (y + 1)^2 = 1$

LESSON 11-8 (pp. 662–667)
13. $BC = 6$ **15.** $\overline{AB} \parallel \overline{NC}$ and $\overline{AN} \parallel \overline{CB}$ by the Midpoint Connector Theorem. Therefore, $ANCB$ is a parallelogram.
17. a. $(1986, 12.25)$ **b.** In 1986 one might estimate that there were 12.25 million African-American, Hispanic, and Asian families in the United States. **19.** Either the diagonals of $MNOP$ bisect each other or they do not. Suppose they do. Thus, $\overline{MO}$ and $\overline{NP}$ have the same midpoint. The midpoint of $\overline{MO} = \left(\frac{8 + -4}{2}, \frac{0 + 6}{2}\right) = \left(\frac{4}{2}, \frac{6}{2}\right) = (2, 3)$. The midpoint of $\overline{NP} = \left(\frac{-8 + 4}{2}, \frac{0 + 6}{2}\right) = \left(\frac{-4}{2}, \frac{6}{2}\right) = (-2, 3)$. These midpoints are not the same. Thus the supposition has led to a false conclusion. Consequently, by the Law of Indirect Reasoning, the diagonals of $MNOP$ do not bisect each other. **21. a.** $(1, -2.5)$ **b.** 8.5 units **c.** $(x - 1)^2 + (y + 2.5)^2 = 72.25$ **23.** $(x + 2)^2 + y^2 = 25$ **25.** Birthday cake does not agree with me. **27.** Sample: **See below.**

27.

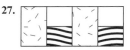

LESSON 11-9 (pp. 668–674)
11. a. $A = (10, 0, 6)$, $B = (0, 0, 6)$, $C = (0, 1, 6)$, $E = (10, 0, 0)$, $G = (0, 1, 0)$, $H = (10, 1, 0)$ **b.** 60 units^3 **c.** 152 units^2
13. $\overline{AB}$: $(3, -.5, 1)$; $\overline{BC}$: $(-3.5, 4, -1.5)$; $\overline{AC}$: $(-4.5, 3.5, 4.5)$
15. $x^2 + y^2 + z^2 = 25$ **17. a.** $(14, 5)$ **b.** $\sqrt{109} \approx 10.44$ units
19. a. $\sqrt{10}$ **b.** $(x - 3)^2 + (y + 2)^2 = 10$ **21.** p
23. a. $x = \frac{55}{2} = 27.5$ **b.** $y = 1.15$ **c.** $z = 3.9$

CHAPTER 11 PROGRESS SELF-TEST (p. 679)

1. $PQ = \sqrt{(30 - 7)^2 + (-7 - 15)^2} = \sqrt{1013} \approx 31.8$ units
2. $M\left(\frac{30 + 7}{2}, \frac{-7 + 15}{2}\right) = M(18.5, 4)$ **3.** $RS =$
$\sqrt{(8 - 3)^2 + (4 - 4)^2} = \sqrt{5^2 + 0^2} = 5$; $RT =$
$\sqrt{(11 - 3)^2 + (8 - 4)^2} = \sqrt{8^2 + 4^2} = \sqrt{80}$; $ST =$
$\sqrt{(11 - 8)^2 + (8 - 4)^2} = \sqrt{3^2 + 4^2} = 5$. So the perimeter of
$\triangle RST = RS + RT + ST = 5 + \sqrt{80} + 5 \approx 18.94$ units.
4. a. See page 899. b. $AO = \sqrt{(5 - 0)^2 + (-6 - 0)^2 + (8 - 0)^2} =$
$\sqrt{125} \approx 11.18$ units **c.** $\left(\frac{5 + 0}{2}, \frac{-6 + 0}{2}, \frac{8 + 0}{2}\right) = (2.5, -3, 4)$
5. a. $\angle A$ is a right angle. **b.** Law of Ruling Out Possibilities
6. a. The volume of figure Q is the area of its base times its height.
b. Law of Detachment **7. a.** If a figure is a polygon, then it is a hexagon. False **b.** If a figure is not a hexagon, then it is not a polygon. False **c.** If a figure is not a polygon, then it is not a hexagon. True **8.** Either $\triangle ABC$ has all angles with measures under 60 or it does not. Suppose that there exists a $\triangle ABC$ where $m\angle A < 60$, $m\angle B < 60$, and $m\angle C < 60$. Then by the Addition Property of Inequality, $m\angle A + m\angle B + m\angle C < 180$. But this

contradicts the Triangle-Sum Theorem. So by the Law of Indirect Reasoning, there is no triangle with three angles all with measures less than 60. **9.** Either $\sqrt{80} = 40$ or $\sqrt{80} \neq 40$. Suppose $\sqrt{80} = 40$. Then squaring both sides, $80 = 1600$. This is a false statement. By the Law of Indirect Reasoning, the supposition $\sqrt{80} = 40$ is false. So $\sqrt{80} \neq 40$. **10.** $HM = \sqrt{6^2 + 8^2} = 10$; $RB = \sqrt{(16 - 10)^2 + (8 - 0)^2} = 10$; $HR = \sqrt{(16 - 6)^2 + (8 - 8)^2} = 10$; $MB = \sqrt{10^2 + 0^2} = 10$ So, $HM = RB = HR = MB$. Therefore, $RHMB$ is a rhombus.
11. By the Midpoint Formula, the midpoint of $\overline{XZ}$ is $\left(\frac{2a + 2b}{2}, \frac{0 + 2c}{2}\right) = (a + b, c)$, and the midpoint of $\overline{YW}$ is $\left(\frac{2a + 2b + 0}{2}, \frac{2c + 0}{2}\right) = (a + b, c)$. So the two midpoints are identical. **12.** $\overline{DF} \parallel \overline{BE}$ by the Midpoint Connector Theorem; $\overline{EF} \parallel \overline{BD}$ by the Midpoint Connector Theorem. By the definition of a parallelogram, $BDFE$ is a parallelogram. **13.** $AE = EB = DF = \frac{1}{2}AB = 5.5$; $EF = BD = DC = \frac{1}{2}BC = 11.15$

14. Let b = someone is a baby, h = someone is happy, and t = someone is teething. The statements translate to (1) $b \Rightarrow h$, (2) $t \Rightarrow b$, (3) Nate is not-h. By the Law of the Contrapositive, (1) implies not-$h \Rightarrow$ not-b and (2) implies not-$b \Rightarrow$ not-t. Applying the Law of Transitivity, not-$h \Rightarrow$ not-t. With this statement and the Law of Detachment, we can conclude Nate is not teething.
15. Let q = Queen, k = King, j = Jack, a = Ace and $>$ mean older than. The statements can be interpreted (1) $q > k$, (2) $j > q$, (3) $a > j$. By the Transitive Property of Inequality, $a > j > q > k$. Thus we conclude that the Ace is from the 17th century, the Jack is from the 18th century, the Queen is from the 19th century and the King is from the 20th century. **16.** If we consider Selkirk to be the origin, then Goodland is $G = (-12, 60)$ and Garden City is $C = (34, -36)$. So the flying distance $GC = \sqrt{46^2 + 96^2} = \sqrt{11332} \approx 106.5$ miles. **17.** $\sqrt{7^2 + 12^2 + 17^2} = \sqrt{482} \approx 21.95''$
18. See right. 19. $x^2 + (y + 19)^2 + (z - 4)^2 = 36$ **20.** Sample: $(-a, 0), (a, 0), (b, c), (-b, c)$

4. a.

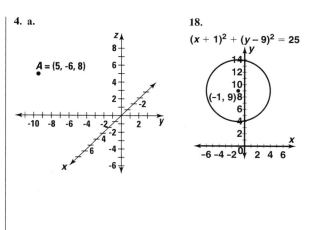

18.
$(x + 1)^2 + (y - 9)^2 = 25$

The chart below keys the **Progress Self-Test** questions to the objectives in the **Chapter Review** on pages 680–683 or to the **Vocabulary** (Voc.) on page 678. This will enable you to locate those **Chapter Review** questions that correspond to questions students missed on the **Progress Self-Test**. The lesson where the material is covered is also indicated on the chart.

Question	1	2	3	4	5	6	7	8	9	10
Objective	A	A	A	C	D	D	E	F	F	G
Lesson	11-6	11-8	11-6	11-9	11-3	11-1	11-2	11-4	11-4	11-6
Question	11	12	13	14	15	16	17	18	19	20
Objective	G	B	B	H	H	I	I	J	J	K
Lesson	11-8	11-8	11-8	11-1, 11-2	11-1	11-6	11-9	11-7	11-9	11-5

CHAPTER 11 REVIEW (pp. 680–683)

1. $\sqrt{136} \approx 11.66$ units **3.** $(-2, -6)$ **5.** 19.49 units **7.** $\sqrt{x^2 + y^2}$ **9.** $WX = 41$, $VX = 82$, $VZ = 80$ **11.** Applying the Midpoint Connector Theorem to $\triangle BCD$, $\overline{EF} // \overline{DB}$. Thus, $BDEF$ is a trapezoid by the definition of trapezoid. **13. See right. 15. a.** $(-1, -1, 3.5)$ **b.** ≈ 9.64 units **17. a.** $LOVE$ is a trapezoid. **b.** Law of Detachment **19. a.** $x = 11$ **b.** Law of Ruling Out Possibilities **21. a.** (d) **b.** (b) **23. a.** If $x^2 = 9$, then $x = 3$. **b.** If $x \neq 3$, then $x^2 \neq 9$. **c.** If $x^2 \neq 9$, then $x \neq 3$. **d.** original, contrapositive **25. a.** All people in the U.S. live in New York. **b.** If a person is not a New Yorker, then that person does not live in the U.S. **c.** If a person does not live in the U.S., then that person is not a New Yorker. **d.** original, contrapositive **27.** Either quadrilateral $ABCD$ has four acute angles or it does not. Suppose that in quadrilateral $ABCD$, $\angle A$, $\angle B$, $\angle C$, and $\angle D$ are acute. Thus, $m\angle A < 90$, $m\angle B < 90$, $m\angle C < 90$, and $m\angle D < 90$. So $m\angle A + m\angle B + m\angle C + m\angle D < 360$, which contradicts the Quadrilateral-Sum Theorem. So by the Law of Indirect Reasoning, the supposition is false. Thus a quadrilateral cannot have four acute angles. **29.** Either $\sqrt{2} = \frac{239}{169}$ or $\sqrt{2} \neq \frac{239}{169}$. Suppose $\sqrt{2} = \frac{239}{169}$. Then $2 = \left(\frac{239}{169}\right)^2 = \frac{57,121}{28,561}$. Then $2 \cdot 28,561 = 57,121$; so $57,122 = 57,121$. This is a false conclusion; so by the Law of Indirect Reasoning, the supposition is false. So $\sqrt{2} \neq \frac{239}{169}$. **31.** Either $ABCD$ is a trapezoid or it is not. Suppose $ABCD$ is a trapezoid. Then by the definition of trapezoid, either $\overline{AB} // \overline{CD}$ or $\overline{AD} // \overline{BC}$. Slope of $\overline{AB} = \frac{10}{8} = \frac{5}{4}$; slope of $\overline{CD} = \frac{12}{14} = \frac{6}{7}$. So, by the Parallel Lines and Slopes Theorem, $\overline{AB}$ is not parallel to $\overline{CD}$.

Then by the Law of Ruling Out Possibilities $\overline{AD} // \overline{BC}$ is the other possibility. Slope of $\overline{AD} = \frac{-5}{-1} = 5$; slope of $\overline{BC} = \frac{3}{-5} = -\frac{3}{5}$. So by the Parallel Lines and Slopes Theorem, $\overline{AD}$ is not parallel to $\overline{BC}$. Therefore, the supposition is false and by the Law of Indirect Reasoning, $ABCD$ is not a trapezoid. **33.** $AB = \sqrt{145}$; $AC = \sqrt{145}$. Since $AB = AC$, $\triangle ABC$ is isosceles by definition of isosceles triangle. **35.** Slope of $\overline{XY} = \frac{-2}{7}$; slope of $\overline{YZ} = \frac{10}{3}$; slope of $\overline{ZW} = \frac{-2}{7}$; slope of $\overline{WX} = \frac{10}{3}$. By the Parallel Lines and Slopes Theorem, $\overline{XY} // \overline{ZW}$ and $\overline{YZ} // \overline{WX}$. Thus, $XYZW$ is a parallelogram by the definition of a parallelogram. **37.** Slope of $\overline{WY} = \frac{s - 0}{s - 0} = \frac{s}{s} = 1$; slope of $\overline{XZ} = \frac{s - 0}{0 - s} = \frac{s}{-s} = -1$. Since $1 \cdot -1 = -1$, by the Perpendicular Lines and Slopes Theorem, $\overline{WY} \perp \overline{XZ}$. **39.** I have not finished my homework. **41.** No bat can live on the moon. **43.** All the names on this list are melodious. **45.** $\sqrt{146} \approx 12.1$ miles **47.** 28 inches **49.** $(x - 8)^2 + (y + 1)^2 = 225$ **51. See below. 53.** $(x - 4)^2 + (y + 3)^2 + z^2 = 100$ **55.** Sample: $(-a, 0), (a, 0), (0, b)$ **57.** Sample: $(-a, 0), (a, 0), (0, b), (0, -c)$

13.

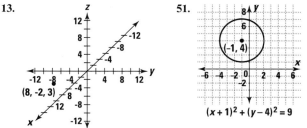

51.
$(x + 1)^2 + (y - 4)^2 = 9$

LESSON 12-1 (pp. 686–690)

9. a. (–6, 21) **b.** $OA = \sqrt{53}$, $AA' = 2\sqrt{53}$, $OA' = 3\sqrt{53}$. Thus, $OA + AA' = OA'$ **c.** By the Properties of S_k Theorem (a), the slopes of OA and OA' are the same. Also, $OA' > OA$. Thus, A is between O and A'. **11. See below. 13.** 144 inches or 12 feet **15.** $P' = (0, 0)$, $R' = (2a, 0)$, $Q' = (2b, 2c)$ **17.** Every isometry preserves Angle measure, Betweenness, Collinearity (lines), and Distance (lengths of segments). **19.** E, $G^{\#}$, B, D **21.** $x = 13.5$ **23.** $AB = 36$

11.

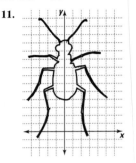

LESSON 12-2 (pp. 692–697)

13. a. expansion **b.** 1.4 **15.** $\frac{3}{5}$ **17.** (0, 0), 2 **19. See below.**
21. a. slope of $\overline{PQ} = \frac{8 - 2}{-4 - 9} = \frac{6}{-13}$, slope of $\overline{QR} = \frac{-10 - 8}{35 - -4} = -\frac{18}{39} = -\frac{6}{13}$, so P, Q, and R are collinear since $\overline{PQ} \,/\!/\, \overline{QR}$ and they share Q. **b.** $PQ = \sqrt{205}$; $QR = \sqrt{1845} = 3\sqrt{205}$; $PR = \sqrt{820} = 2\sqrt{205}$. Since $PQ + PR = QR$, P is between Q and R. **c.** $P' = (36, 8)$, $Q' = (-16, 32)$, $R' = (140, -40)$
d. slope of $\overline{P'Q'} = \frac{32 - 8}{-16 - 36} = \frac{24}{-52} = -\frac{6}{13}$, slope of $\overline{Q'R'} = \frac{-40 - 32}{140 - -16} = \frac{-72}{156} = -\frac{6}{13}$, so P', Q', and R' are collinear since $\overline{P'Q'} \,/\!/\, \overline{Q'R'}$ and they share Q'. **e.** $P'Q' = \sqrt{3280} = 4\sqrt{205}$, $Q'R' = \sqrt{29520} = 12\sqrt{205}$, and $P'R' = \sqrt{13120} = 8\sqrt{205}$. Since $P'Q' + P'R' = Q'R'$, P' is between Q' and R'. **23.** Yes; A to D to C to A to B to C **25. a.** $z = 0.5$ **b.** $M = \pm\sqrt{30} \approx \pm 5.48$

19.

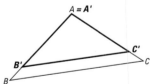

LESSON 12-3 (pp. 698–704)

13. a. $\frac{5}{8} = 0.625$ **b.** M **c.** 7.5 units **d.** 20.8 units **15. See right.**
17. a. 1.4 **b.** 16.8 cm **c.** 60 cm^2 and 117.6 cm^2 **d.** 1.96 **e.** 1.4
19. See right. $k \approx .7$ **21.** $k = \frac{1}{2}$, $Q' = (-4.5, 6)$, $R = (10, 26)$, $T' = (0, -4)$ **23.** composites of 2 reflections are translations and rotations; the other isometry is glide reflections **25.** $k = \frac{2}{b}$

15.

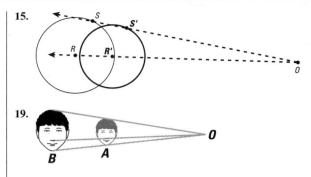

19.

LESSON 12-4 (pp. 705–710)

13. $x = \frac{ac}{b}$ **15.** $\approx \$4.16$ **17.** 112.5 mm **19.** $S(A) = (8, -3)$, $S(B) = (5, 4)$, $AB = \sqrt{6^2 + 14^2} = \sqrt{232} = 2\sqrt{58}$; length of $S(\overline{AB}) = \sqrt{3^2 + 7^2} = \sqrt{58}$; $\sqrt{58} = \frac{1}{2} \cdot 2\sqrt{58}$ **21.** 10 in.
23. a. Since $GHJKL$ is regular, $\overline{GL} \cong \overline{HJ}$. Since $\triangle FGL$ and $\triangle HIJ$ are regular polygons, they are equilateral (definition of regular polygon) and so $\overline{FG} \cong \overline{FL} \cong \overline{GL}$ and $\overline{HI} \cong \overline{JI} \cong \overline{HJ}$. Thus, by the Transitive Property of Congruence, all six of these segments are congruent. So, by the SSS Congruence Theorem, $\triangle FGL \cong \triangle HIJ$. **b.** 168

LESSON 12-5 (pp. 711–716)

15. 6 units **17.** 810″ or 67.5 ft **19. a.** False; for example: under a rotation of magnitude other than 0°, 180°, or –180°, the image of a line intersects its preimage. **b.** True, since if any two figures do not intersect, their images cannot intersect. (Here is a proof of the contrapositive: If their images intersected, then each point of intersection would have a preimage on both figures, so the preimages would have had to intersect.) **21.** 2.5
23. $\frac{244\pi}{3} \approx 256$ cm^3

LESSON 12-6 (pp. 717–723)

11. a. $V = 3025\pi$ units^3; S.A. $= 792\pi$ units^2 **b.** 27, 9
c. $V = 81,675\pi$ units^3; S.A. $= 7128\pi$ units^2 **13. a.** 337.5 in^3
b. ≈ 29.63 in^3 **15. a.** 3 or $\frac{1}{3}$ **b.** 9 or $\frac{1}{9}$ **17.** $\frac{3}{2}$ **19.** $1\frac{1}{2}''$
21. $33\frac{1}{3}\%$

LESSON 12-7 (pp. 724–728)

13. 15.625 kg **15.** ≈ 62 cm **17.** ≈ 123 lb **19.** 11.2 units^2
21. Two figures are similar if and only if there is a composite of size changes and reflections mapping one onto the other.
23.

Conclusions	Justifications
1. $\overline{FK} \cong \overline{KJ}$	definition of midpoint
2. $\angle FKG \cong \angle KJI$, $\angle KFG \cong \angle JKI$	$/\!/$ Lines $\Rightarrow$ corr. $\angle$s $\cong$
3. $\triangle FGK \cong \triangle KIJ$	ASA Congruence Theorem (steps 1 and 2)

CHAPTER 12 PROGRESS SELF-TEST (p. 732)

1. Draw rays from O through A, B, and C. Measure $\overline{OA}$ and calculate $\frac{3}{4} \cdot OA$ to determine OA'. Mark A' on ray $\overline{OA}$. Draw parallel segments through A' for $\overline{A'B'}$, $\overline{A'C'}$, and draw $\overline{B'C'}$. **See page 901. 2. See page 901. 3.** Measure $\overline{FE}$. Calculate $1.5 \cdot FE$ for length of $\overline{FE'}$. Measure $\overline{FD}$. Calculate $1.5 \cdot FD$ for FD'.

Use $\overline{E'D'}$ as radius and draw circle E'. **See page 901. 4. See page 901.** $OP = 5.6$ cm, $OP' = 3.4$; so the magnitude is $\frac{3.4}{5.6}$, about 0.6
5. m$\angle AUQ = 37$; $DA = \frac{3}{4} \cdot 27 = 20.25$; $OU = \frac{4}{3} \cdot 24 = 32$
6. $\frac{WY}{VZ} = \frac{WX}{XV}$; $\frac{4}{12} = \frac{6}{XV}$; $XV = 18$. **7. a.** contraction **b.** X
8. (a) **9.** Two figures F and G are similar if and only if there is a composite of size changes and reflections mapping F onto G.

10. area($\triangle P'Q'R'$) = $5^2 \cdot$ area($\triangle PQR$) = 25 $\cdot$ area($\triangle PQR$)
11. $\frac{V}{729} = \left(\frac{1}{3}\right)^3$; $V = \frac{729}{27} = 27$ cm^3 **12. a.** $\frac{QR}{VT} = \frac{OS}{VU}, \frac{x}{3x} = \frac{OS}{6y}$.
Thus, $QS = 2y$. **b.** $\frac{QR}{VT} = \frac{RS}{TU}, \frac{x}{3x} = \frac{2x}{TU}$. Thus, $TU = 6x$.
13. $\frac{3}{25} = \frac{5}{x}$; $3x = 125$; $x = \frac{125}{3} = 41\frac{2}{3} \approx 41.7$ cm **14.** $\frac{41}{28} = \frac{100}{x}$;
$x \approx 68$ minutes. **15.** $\left(\frac{4}{12}\right)^3 = \left(\frac{1}{3}\right)^3 = \frac{5}{w}$; $w = 5 \cdot 3^3$ or 135 lb
16. $\frac{2 \text{ m}}{0.4 \text{ m}} = \frac{33 \text{ cm}}{y \text{ cm}}$; $2y = 13.2$; $y = 6.6$ cm **17.** 6^3 or 216, 6^2 or 36
18. $S_{\frac{1}{3}}(A) = (2, -2)$, $S_{\frac{1}{3}}(B) = (3, 0)$, $S_{\frac{1}{3}}(C) = \left(-2, \frac{10}{3}\right)$ **See right.**
19. $S_8(P) = (40, -8)$, $S_8(Q) = (8, 16)$; The distance between $S_8(P)$
and $S_8(Q) = \sqrt{(40 - 8)^2 + (-8 - 16)^2} = \sqrt{32^2 + (-24)^2} =$
$\sqrt{1600} = 40$; $PQ = \sqrt{(5 - 1)^2 + (-1 - 2)^2} = \sqrt{4^2 + (-3)^2} =$
$\sqrt{25} = 5$. Therefore, the distance between $S_8(P)$ and $S_8(Q)$ is
$8 \cdot PQ$. **20.** $K' = (0, 4b)$, $I' = (4a, 0)$, $T' = (0, -4c)$,
$E' = (-4a, 0)$

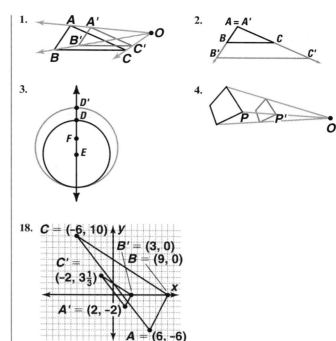

1.

2.

3.

4.

18.

The chart below keys the **Progress Self-Test** questions to the objectives in the **Chapter Review** on pages 733–735 or to the **Vocabulary** (Voc.) on page 731. This will enable you to locate those **Chapter Review** questions that correspond to questions students missed on the **Progress Self-Test**. The lesson where the material is covered is also indicated on the chart.

Question	1	2	3	4	5	6	7	8	9	10
Objective	A	A	A	C	B	B	C	C	Voc.	D
Lesson	12-2	12-2	12-2	12-2	12-5	12-5	12-2	12-3	12-5	12-6
Question	11	12	13	14	15	16	17	18	19	20
Objective	D	B	E	E	F	E	F	G	G	G
Lesson	12-6	12-5	12-4	12-4	12-7	12-5	12-7	12-1	12-1	12-1

CHAPTER 12 REVIEW (pp. 733–735)
1. See below. 3. See below. 5. See right. 7. m$\angle H$ = 100,
$PE = 4.1\overline{6}$, $OU = 7.92$ **9.** $VU = 12$ units **11. a.** $\frac{2}{3}y$ **b.** $3z$
13. a. N **b.** 71 **c.** 44 **15. See right. 17. a.** contraction **b.** True
19. area($\triangle A'B'C'$) = 121 $\cdot$ area($\triangle ABC$) **21.** $\approx$ 35.6 cm^2
23. 6.25" **25.** $\approx$ 20.2" **27.** 3.2" **29.** $\approx$ 12.3 kg **31.** 144
33. 512, 64 **35. See right. 37.** $A' = (6k, -5k)$, $B' = (-10k, 0)$,
$C' = (-3k, 8k)$, $D' = (12k, 20k)$ **39.** $S(E) = \left(\frac{1}{2}a, \frac{1}{2}b\right)$,
$S(H) = \left(0, \frac{1}{2}c\right)$, $S(G) = \left(-\frac{1}{2}a, -\frac{1}{2}b\right)$

1.

3.

5.

15.

35.
$C = (-3, 8)$
$C' = (-1.5, 4)$
$B' = (-5, 0)$
$B = (-10, 0)$
$D = (12, 20)$
$D' = (6, 10)$
$A = (6, -5)$
$A' = (3, -2.5)$

LESSON 13-1 (pp. 738–743)

9. a. $\triangle DEF \sim \triangle UVW$ **b.** $\frac{10}{33}$ or $\frac{33}{10}$ **c.** $UV = 7.\overline{57} \approx 7.6$; $VW = 9.\overline{69} \approx 9.7$ **11. a.** Yes **b.** Using the Pythagorean Theorem, $AC = 39$ and $DF = 32$. Since $\frac{24}{39} = \frac{32}{52} = \frac{40}{65}$, $\triangle ABC \sim \triangle EFD$ by the SSS Similarity Theorem.

13.

Conclusions	Justifications
1. $WX = \frac{1}{2}WY$ $WV = \frac{1}{2}WZ$	definition of midpoint
2. $XV = \frac{1}{2}YZ$	Midpoint Connector Theorem
3. $\frac{WX}{WY} = \frac{1}{2}$, $\frac{WV}{WZ} = \frac{1}{2}$, $\frac{XV}{YZ} = \frac{1}{2}$	Multiplication Property of Equality
4. $\frac{WX}{WY} = \frac{WV}{WZ} = \frac{XV}{YZ}$	Transitive Property of Equality
5. $\triangle WXV \sim \triangle WYX$	SSS Similarity Theorem (step 4)

15. $\frac{1}{5}$ **17.** surface area of a box with sides ℓ, w, h **19.** volume of a cylinder of height h and base of radius r **21. a.** 180 **b.** $\frac{360}{41} \approx 8.78$ **23.** (b) **25.** $\pm 10\sqrt{2}$

LESSON 13-2 (pp. 744–750)

13. a. No **b.** $\frac{12}{24} \neq \frac{5}{7} \neq \frac{13}{25}$ **15. a.** Yes **b.** SSS Similarity Theorem **c.** $\triangle XVW \sim \triangle AZY$ **17.** $\angle ABC \cong \angle EDC$, and $\angle BAC \cong \angle DEC$ from the // Lines $\Rightarrow$ AIA $\cong$ Theorem. Thus, $\triangle CED \sim \triangle CAB$ by the AA Similarity Theorem. **19.** $AB = 1.5$ **21.** $n = 48$

23.

Conclusions	Justifications
0. $\frac{a}{b} = \frac{c}{d}$	Given
1. $ad = bc$	
$\frac{d}{c} = \frac{b}{a}$	Multiplication Prop. of =
2. $\frac{b}{a} = \frac{d}{c}$	Symmetric Prop. of =

24. front **25.** $25\sqrt{2} \approx 35.4$ meters

LESSON 13-3 (pp. 751–757)

11. a. 7.5 **b.** 9 **c.** supplementary **d.** $\frac{BD}{DA} = \frac{FA}{CF}$ **13.** $x = 125$ m; $y = 100$ m **15.** No judgement can be made. **17.** Yes, $\triangle STO \sim \triangle XYZ$ by the SSS Similarity Theorem.

19.

Conclusions	Justifications
1. $\angle ABE \cong \angle CBD$	definition of angle bisector
2. $\triangle ABE \sim \triangle CBD$	AA Similarity Theorem (step 1 and given)

21. (a) **23.** No, the three angle measures are approximately 30, 70, and 30.

LESSON 13-4 (pp. 759–764)

9. $x = \frac{16}{5} = 3.2$, $y = \frac{9}{5} = 1.8$, $h = \frac{12}{5} = 2.4$ **11. a.** part (1) **b.** ≈ 17.8 feet **13.** $a = 3\sqrt{10}$, $b = 3\sqrt{15}$, $CD = 3\sqrt{6}$

15.

Conclusions	Justifications
1. $\overline{WX} \parallel \overline{ZY}$	definition of trapezoid
2. $\angle XWU \cong \angle ZYU$ $\angle WXU \cong \angle YZU$	// lines $\Rightarrow$ AIA $\cong$ Theorem
3. $\triangle WXU \sim \triangle YZU$	AA Similarity Theorem

17. a. No **19. a.** 3 **b.** 3 **c.** 4 **21.** It is circular reasoning to use a theorem to prove itself.

LESSON 13-5 (pp. 765–770)

15. $7\sqrt{3} \approx 12.12$ **17.** 60 **19.** $OP = 4.5$ **21.** $y = 9$; $x = \sqrt{80} \approx 8.94$ **23.** $\triangle CBA$, $\triangle BPA$, $\triangle CPB$ (and also $\triangle ADC$)

LESSON 13-6 (pp. 772–776)

11. $\frac{9}{40} = .225$ **13. a.** $\angle 4$ **b.** $\angle 1$ **15. a.** $\frac{600}{\sqrt{3}} \approx 346.41$ units **b.** $\frac{10000}{\sqrt{3}} \approx 5773.50$ units^2 **17.** 24 **19. a.** No. Suppose a trapezoid is as below, where no two sides of $\overline{AB}$, $\overline{BC}$, $\overline{DC}$, and $\overline{AD}$ are equal. Therefore no two of the triangles are congruent. **See below. b.** Yes. Use the trapezoid in part **a.** Because $\overline{AD} \parallel \overline{BC}$, $\angle DAO \cong \angle BCO$ and $\angle ADO \cong \angle CBO$ (// Lines $\Rightarrow$ AIA $\cong$ Theorem). Therefore, by the AA Similarity Theorem, $\triangle ADO \sim \triangle CBO$. **21. See below.**

19. a.

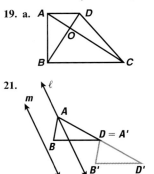

21.

LESSON 13-7 (pp. 777–783)

13. a. Sample: $\sin B \approx \frac{14}{34} \approx .412$; $\sin B' \approx \frac{10}{24} \approx .417$ **b.** Yes; they probably will not be equal, due to measurement error, but very close. **15.** ≈ 20.1 ft **17.** ≈ 6.76 feet **19.** $\sqrt{2} - 1 \approx 0.414$ **21. a.** converse: If m$\angle ABC =$ m$\angle DEF$, then $\triangle ABC \sim \triangle DEF$. inverse: If $\triangle ABC$ is not similar to $\triangle DEF$ with vertices in that order, then m$\angle ABC \neq$ m$\angle DEF$. contrapositive: If m$\angle ABC \neq$ m$\angle DEF$, then $\triangle ABC$ is not similar to $\triangle DEF$ with vertices in that order. **b.** The contrapositive is true. **23. See below.**

23.

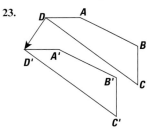

LESSON 13-8 (pp. 784–788)

11. a. ≈ 59.44 cm^2 **b.** ≈ 19 cm^2 **13.** 63 km **15.** $\sin B = \frac{11}{\sqrt{170}}$; $\cos B = \frac{7}{\sqrt{170}}$; $\tan B = \frac{11}{7}$ **17.** $\frac{60}{13}$; $\frac{144}{13}$; $\frac{25}{13}$ **19.** Rudolph is both a y and z, but not an x.

CHAPTER 13 PROGRESS SELF-TEST (pp. 792–793)

1. $\frac{BD}{BC} = \frac{BC}{AB}$ by the Right-Triangle Altitude Theorem; $\frac{2}{6} = \frac{6}{AB}$; $AB = \frac{36}{2} = 18$ **2.** $\frac{75}{ZX} = \frac{ZX}{ZW}$ by the Right-Triangle Altitude Theorem; $75 \cdot ZW = ZX^2 = YZ^2 - XY^2$; $75 \cdot ZW = 75^2 - 60^2$; $ZW = 27$ **3.** $\frac{VW}{VY} = \frac{VX}{VZ}, \frac{11}{24} = \frac{VX}{30}, \frac{330}{24} = VX; VX = 13.75$ **4.** $\frac{VW}{VW + WY} = \frac{WX}{YZ}, \frac{8}{20} = \frac{VW}{15 + VW}; VW = 10$ **5. a.** $BC = AB = 7$ **b.** $AC = 7\sqrt{2}$ **6.** $h = 15\sqrt{3} \approx 25.98$ **7.** Yes; because $\frac{9}{12} = \frac{12}{16} = \frac{15}{20}$, the triangles are similar by the SSS Similarity Theorem. **8.** Area$(\triangle TRI) = \frac{1}{2} \cdot TR \cdot RI \cdot \sin 55° \approx \frac{1}{2} \cdot 36 \cdot 11 \cdot .81915 \approx 162$ units^2 **9.** Since $\overleftrightarrow{AB} \parallel \overleftrightarrow{DE}$, $\angle A \cong \angle E$, $\angle B \cong \angle D$ because of the $\parallel$ Lines $\Rightarrow$ AIA $\cong$ Theorem. Therefore, $\triangle ABC \sim \triangle EDC$ by the AA Similarity Theorem. **10.** ≈ 0.855 **11.** $\frac{\sqrt{3}}{2}$ **12.** $\tan B = \frac{AC}{BC}$ **13.** $\angle 3$ **14.** $\cos A = \frac{AC}{AB} = \sin B = \frac{9}{11}$ **15.** $\sin E = \frac{DF}{DE} = \frac{14}{50} = \frac{7}{25} = .280$ **16.** $\frac{165 \text{ m}}{150 \text{ m}} = \frac{x \text{ m}}{200 \text{ m}}; x = \frac{33000}{150} = 220$ m **17.** $(\sin 80) \cdot 15 \approx 0.98 \cdot 15 \approx 14.8$ ft **18.** $\frac{100 \text{ cm}}{21 \text{ cm}} = \frac{x \text{ m}}{67 \text{ m}}$; $x \approx 319$ m **19. a.** See below. **b.** $\sqrt{6.2^2 + 4.8^2} \approx 7.84$ km **20.** height of tree $= 5 + 40 \cdot \tan 35 \approx 5 + 0.70 \cdot 40 \approx 33$ ft

19. a.

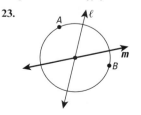

The chart below keys the **Progress Self-Test** questions to the objectives in the **Chapter Review** on pages 794–797 or to the **Vocabulary** (Voc.) on page 791. This will enable you to locate those **Chapter Review** questions that correspond to questions students missed on the **Progress Self-Test**. The lesson where the material is covered is also indicated on the chart.

Question	1	2	3	4	5	6	7	8	9	10
Objective	B	B	A	A	C	C	F	D	F	E
Lesson	13-4	13-4	13-3	13-3	13-5	13-5	13-1	13-8	13-2	13-7

Question	11	12	13	14	15	16	17	18	19	20
Objective	E	G	D	G	D	H	I	H	J	I
Lesson	13-7	13-6	13-6	13-7	13-7	13-3	13-7	13-2	13-8	13-6

CHAPTER 13 REVIEW (pp. 794–797)

1. a. 10.5 **b.** 2.5 **3.** No; since $\frac{2}{4} \neq \frac{6}{9}$. **5.** 8 **7.** 6 **9.** ≈ 14.42 **11.** $AC = 4\sqrt{3} \approx 6.93$; $BC = 8$ **13.** $6\sqrt{3} \approx 10.39$ **15. a.** $7\sqrt{2} \approx 9.90$ **b.** 7 **c.** $7\sqrt{3} \approx 12.12$ **d.** 14 **17.** Sample: m$\angle A \approx 35$, $\tan A \approx .69$, $\sin A \approx .60$, and $\cos A \approx .82$. **19.** $\angle 1$ **21.** $\frac{24}{26} = \frac{12}{13}$ **23.** ≈ 34.4 units^2 **25.** .843 **27.** $\frac{1}{2}$ **29.** 1 **31. a.** Yes **b.** by the SSS Similarity Theorem **33.** Yes, by the AA Similarity Theorem

35.

Conclusions	Justifications
1. $\angle BAE \cong \angle DCE$	$\parallel$ Lines $\Rightarrow$ AIA $\cong$ Theorem
$\angle ABE \cong \angle CDE$	
2. $\triangle ABE \sim \triangle CDE$	AA Similarity Theorem (step 1)

37. $\frac{AC}{AB}$ **39.** $\frac{BC}{AC}$ **41.** sine **43.** $\frac{AC}{BC}$ **45.** 15 m **47. a.** $\frac{50}{3} \approx 16.67$ units **b.** $70 - \frac{70}{3} \approx 46.67$ units **49.** ≈ 365 feet **51.** ≈ 11.1 meters **53.** northern component: ≈ 57 knots; eastern component: ≈ 18.5 knots

LESSON 14-1 (pp. 800–805)

15.

Conclusions	Justifications
1. $OB = OC$, $OA = OD$	definition of circle
2. $\triangle AOB \cong \triangle DOC$	SSS Congruence Theorem (step 1 and given)
3. m$\angle AOB = $ m$\angle DOC$	CPCF Theorem
4. m$\angle AOB = $ m$\overarc{AB}$, m$\angle DOC = $ m$\overarc{CD}$	definition of arc measure
5. m$\overarc{AB} = $ m$\overarc{CD}$	Transitive Property of Equality

17. a., b. See below. **c.** at the center of the circle **d.** Chord-Center Theorem (part 4) **19. a.** 6 units **b.** $6x$ units **21.** ≈ 1.87 m^3 **23.** 84°

17. a., b.

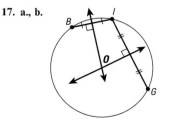

LESSON 14-2 (pp. 806–812)

13. a. n **b.** $\frac{n}{2}$ **15.** 53°
17.

Conclusions	Justifications
1. m$\angle AXM = $ m$\angle BXN$	Vertical Angles Theorem
2. m$\angle AMB = $ m$\angle ANB$	If two inscribed angles intercept the same arc, then they have the same measure
3. $\triangle AXM \sim \triangle BXN$	AA Similarity Theorem (steps 1 and 2)

19. a. 72° **b.** $500 \sin 36° \approx 294$ units **21.** $\sqrt{700} \approx 26.46$ yards **23.** Line m is the perpendicular bisector of the segment with endpoints B and r$_\ell(A)$. **See below. 25.** 200

23.

LESSON 14-3 (pp. 813–817)

7. See below. 9. Sample: **See below. 11. a.** All have measure 30.
b. True **13. a.** $m\angle A = m\angle CDA = 30$, $m\angle ACD = 120$,
$m\angle ADB = 90$, $m\angle BCD = m\angle BDC = m\angle CBD = 60$ **b.** x units
c. $7\sqrt{3} \approx 12.12$ units **15.** 10 units **17.** $625\pi \approx 1963.5$ square feet

7.

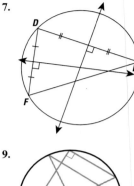

9.

LESSON 14-4 (pp. 818–822)

9. 79 **11.** 143° **13.** ≈ 15.6 ft **15.** $2 \cdot \sqrt{107.25} \approx 20.71$ inches
17. 112° 20′

LESSON 14-5 (pp. 823–828)

11. a., b. The center of the circle is on all of them. **See below.**
13. a. A plane is tangent to a sphere if and only if it is
perpendicular to the radius at the radius's endpoint on the sphere.
b. Yes

15.

Conclusions	Justifications
1. $\overline{PT} \perp \overline{CT}$	Radius-Tangent Theorem
2. $\triangle PTC$ is a right triangle.	definition of right triangle
3. $d^2 + r^2 = (r + h)^2$	Pythagorean Theorem
4. $d^2 + r^2 = r^2 + 2rh + h^2$	Distributive Property
5. $d^2 = 2rh + h^2$	Addition Property of Equality
6. $d = \sqrt{2rh + h^2}$	definition of square root

17. See right. 19. a. 65 **b.** 92.5 **c.** 87.5 **d.** 22.5 **21.** $m\angle G = 81$;
$BC = 15$; $GJ = 16$

11. a., b. The center of the circle is on all of them.

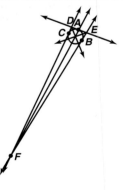

17.

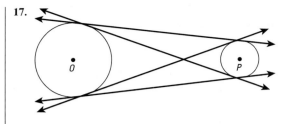

LESSON 14-6 (pp. 829–834)

13. 49° **15.** Draw $\overline{AB}$ and $\overline{AD}$. By the Quadrilateral-Sum Theorem,
$m\angle A + m\angle B + m\angle C + m\angle D = 360$. By the Radius-Tangent
Theorem and substitution, $m\angle A + 90 + m\angle C + 90 = 360$, so
$m\angle A + m\angle C = 180$. But $m\angle A = m\overarc{BD}$, so $m\overarc{BD} + m\angle C = 180$.
17. Let $\overline{BC}$ be a diameter of $\odot O$. By the Radius-Tangent Theorem,
each tangent is perpendicular to $\overline{BC}$. By the Two-Perpendiculars
Theorem, the two tangents are parallel. **19.** 6 **21.** $\frac{x}{4} = \frac{10}{q}$, $\frac{4}{q} = \frac{x}{10}$,
$\frac{10}{x} = \frac{q}{4}$

LESSON 14-7 (pp. 836–841)

11. $8 \pm \sqrt{39} \approx 1.76$ or 14.24 units **13.** Yes; since D, C, and P
are collinear, and B, A, and P are collinear, then D, C, P, B, and A
are coplanar. Points D, C, A, and B lie on the circle which is the
intersection of that plane and the sphere. Apply the Secant Length
Theorem to this circle. **15.** 135° **17.** 42.5 **19.** ≈ 38.4 m

LESSON 14-8 (pp. 842–847)

13. a. 540 ft^2 **b.** $\frac{2025}{\pi} \approx 645$ ft^2 **15.** There is only so much area
that can be covered by an animal skin. The circular base allows for
the largest floor area per animal skin. **17.** ≈ 12.4 units **19.** 90
21. a. $426\frac{2}{3} \approx 426.7$ ft^3 **b.** ≈ 27.3 ft^3 **23. a.** $\frac{1372}{3}\pi$ cm$^3 \approx 1436.8$ cm^3
b. 196π cm$^2 \approx 615.8$ cm^2

LESSON 14-9 (pp. 848–852)

11. a. $36\pi \approx 113$ m^2 **b.** sample: radius of 2 m, height of 9 m;
surface area: $44\pi \approx 138$ m^2 **c.** sample: radius of 6 m, height of 3 m;
surface area: $36\pi + 6\pi\sqrt{45} \approx 240$ m^2 **d.** less **13.** $\frac{\pi\sqrt{x^3}}{6}$ units^3
15. a. circle: $\frac{2304}{\pi} \approx 733$ in^2; square: 576 in^2 **b.** square
17. $m\angle A = 39.5$, $m\angle B = 98$, $m\angle C = 42.5$ **19.** Sample: **See
below. 21.** The distance is from 277 to 339 miles.

19.

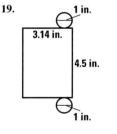

CHAPTER 14 PROGRESS SELF-TEST (pp. 856–857)

1. See right. m$\angle AOB = 120°$ $BC = 210 \sin 60° = 105\sqrt{3}$
Thus, $AB = 2BC = 210\sqrt{3} \approx 363.7$ units **2.** See right.
3. Draw MK, $MK = 60$. Because $\triangle MKL$ is a 45-45-90 triangle,
$KL = \dfrac{60}{\sqrt{2}} = 30\sqrt{2}$. Thus the perimeter of $JKLM = 4 \cdot 30\sqrt{2}$
$= 120\sqrt{2} \approx 169.7$ units. **4.** m$\angle B = \frac{1}{2}$ m$\widehat{DC} = \frac{1}{2} \cdot 80 = 40$.
5. $100 = \frac{1}{2}(80 + $ m$\widehat{AB})$; $200 = 80 + $ m$\widehat{AB}$; m$\widehat{AB} = 120$
6. m$\angle R = \frac{1}{2}($m$\widehat{VT} - m\widehat{US}) = \frac{1}{2}(110 - 30) = \frac{1}{2}(80) = 40$.
7. $QD \cdot QA = CQ \cdot BQ$; $QD \cdot 19 = 40 \cdot 38$; $OD = 80$ units
8. $WV \cdot WZ = WX \cdot WY$; $WV \cdot 10 = 12 \cdot 28$; $WV = 33.6$;
$ZV = 23.6$ **9.** $\ell \perp \overline{OQ}$ since a tangent is perpendicular to a radius
through the point of tangency. $\overline{XY} \perp \overline{OQ}$ because a radius that
bisects a chord is perpendicular to it. Thus ℓ // $\overline{XY}$ because two
lines perpendicular to the same line are parallel. **10. a.** True
b. Sample: $\angle O$ is a 90° angle since m$\widehat{UT} = 90°$ (given). $\angle PTO$ and
$\angle PUO$ are 90° angles since a tangent is perpendicular to a radius
through the point of tangency. So $\angle TPU$ is a 90° angle by the
Quadrilateral-Sum Theorem. So $PUOT$ is a rectangle. Furthermore,
$OU = OT$ since both are radii of $\odot O$. Since opposite sides of a
rectangle are congruent, $PUOT$ has four sides of equal length.
Therefore, by the definition of a square, $PUOT$ is a square.
11. Let $x = $ m$\widehat{TNU}$. m$\angle P = \frac{1}{2}($m$\widehat{TNU} - m\widehat{TU})$
$= \frac{1}{2}(x - (360 - x)) = x - 180$. So $x = $ m$\angle P + 180 = 92 + 180$
$= 272$.

12.

Conclusions	Justifications
0. $\overline{BC}$ // $\overline{AD}$	Given
1. m$\angle CBD = $ m$\angle ADB$	// lines $\Rightarrow$ AIA = Theorem
2. m$\angle CBD = \frac{1}{2}$ m$\widehat{CD}$	
m$\angle ADB = \frac{1}{2}$ m$\widehat{AB}$	Inscribed Angle Theorem
3. m$\widehat{AB} = $ m$\widehat{CD}$	Transitive Property of Equality

13. a. a circle **b.** $2\pi r = 30$; $r = \dfrac{15}{\pi}$; $\pi r^2 = \dfrac{225}{\pi} \approx 71.6$ cm^2
14. The cube, because of all solids with the same volume, the
sphere has the least surface area. **15. a.** Stand on the part of the
circle drawn, or outside the circle and below $\overleftrightarrow{AB}$. See right.
b. $\dfrac{30}{\tan 32°} \approx 48$ ft **16.** 100ft $= \dfrac{100 \text{ ft}}{5280 \text{ ft}}$ mi $\approx .0189$ mi;
(view to horizon)^2 + (Earth's diameter)^2 = (Earth's diameter + tower)^2;
view to horizon $= \sqrt{(3960 + 0.0189)^2 - 3960^2} \approx 12.2$ miles

1.

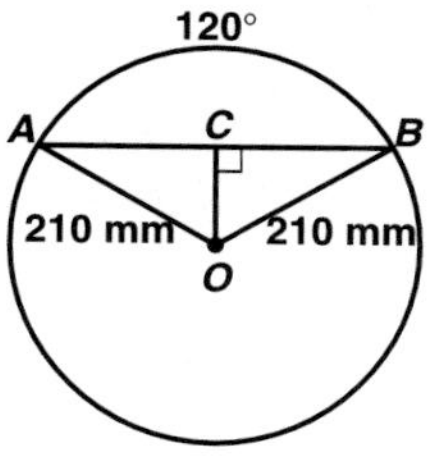

2.

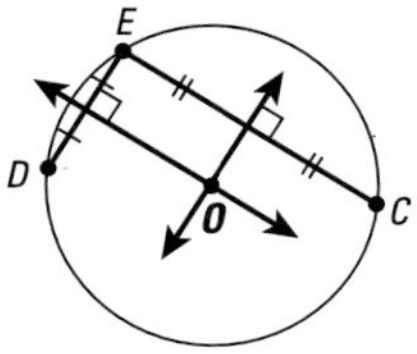

15. a.

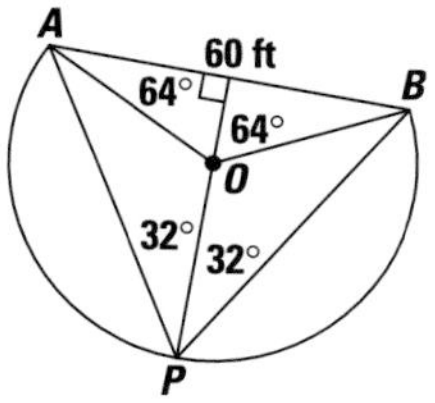

The chart below keys the **Progress Self-Test** questions to the objectives in the **Chapter Review** on pages 858–861 or to the **Vocabulary** (Voc.) on page 855. This will enable you to locate those **Chapter Review** questions that correspond to questions students missed on the **Progress Self-Test.** The lesson where the material is covered is also indicated on the chart.

Question	1	2	3	4	5	6	7	8	9	10
Objective	A	D	A	B	C	C	E	E	F	F
Lesson	14-1	14-3	14-1	14-2	14-4	14-4	14-7	14-7	14-1, 14-5	14-1, 14-5

Question	11	12	13	14	15	16
Objective	C	G	K	H	I	J
Lesson	14-6	14-2	14-8	14-9	14-3	14-5

CHAPTER 14 REVIEW (pp. 858–861)

1. ≈ 57.6 mm **3. a.** ≈ 17.0 units **b.** ≈ 164.4 units² **5. a.** 45°
b. ≈ 91.8 units **7.** 106° **9.** 81 **11.** 120 **13.** m∠EFD =
m∠BFC = 77.5; m∠EBD = m∠ECD = 62.5; m∠BAC = 47.5;
m∠CEB = m∠CDB = 15; m∠BFE = m∠CFD = 102.5;
m∠ABD = m∠ACE = 117.5 **15.** 40° **17. See right.**
19. See right. **21.** 10 units **23.** 21 units **25.** Chord-Center
Theorem, part (1) **27.** m$\widehat{YZ}$ = m$\widehat{XY}$ = m$\widehat{XZ}$, so ZX = ZY = XY
by the Arc-Chord Congruence Theorem. Therefore, by the
definition of equilateral △, △XYZ is equilateral. **29.** m∠BAD = 15,
m∠DAC = 15, m∠BAC = 30, m∠ABD = 90, m∠BDC = 150,
m∠ACD = 90, m∠ADB = 75 and m∠ADC = 75. **31.** ∠A and
∠D are right angles, so they intercept semicircles. **33.** By the
Inscribed Angle Theorem, m∠C = $\frac{1}{2}$ m$\widehat{BD}$. By the Tangent-Chord
Theorem, m∠ABD = $\frac{1}{2}$ m$\widehat{BD}$. By Transitivity of Equality,
m∠ABD = m∠C. **35. a.** a circle **b.** ≈ 100.3 ft **37.** a sphere
39. a. a sphere **b.** ≈ 31.3 cu ft **41. a.** All points shown in the
diagram on ⊙O. **See right. b.** ≈ 99.9 yards **43.** ≈ 61.8 km
45. ≈ 2.74 miles **47. a.** a sphere **b.** Sample: the container could
easily roll off a shelf. **49.** Sample: **See right.**

17.

19.

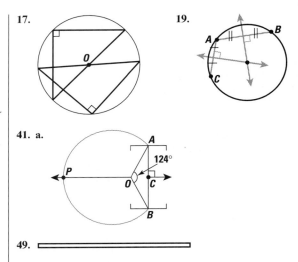

41. a.

49.

906

acre　A unit of area often used to measure plots of land. One acre is equivalent to exactly 43,560 square feet or about 4047 square meters. (451)

acute angle　An angle whose measure is greater than 0 and less than 90. (137)

adjacent angles　Two nonstraight and nonzero angles with a common side interior to the angle formed by the noncommon sides. (139)

adjacent sides　See *consecutive sides*.

adjacent vertices　See *consecutive vertices*.

algorithm　A finite sequence of steps leading to a desired end. (168)

alternate exterior angles　Angles formed by two lines and a transversal whose interiors are not between the two lines and are on different sides of the transversal. (267)

alternate interior angles　Angles formed by two lines and a transversal whose interiors are partially between the two lines and on different sides of the transversal. (265)

altitude　In a triangle or trapezoid, the segment from a vertex perpendicular to the line containing the opposite side; also, the length of that segment. In a prism or cylinder, the distance between the bases. In a pyramid or cone, the length of a segment from the vertex perpendicular to the plane of the base. Also called *height*. (454, 460, 506, 513)

analytic geometry　A geometry in which points are represented by coordinates and where algebraic methods of reasoning are utilized. (621)

angle　The union of two rays (its sides) that have the same endpoint (its **vertex**). (124)

angle bisector　The ray with points in the interior of an angle that forms two angles of equal measure with the sides of the angle. (127)

angle of depression　An angle measured from the horizontal plane downward from an observer's eye to a given point below the plane. (782)

angle of elevation　An angle measured from the horizontal plane upward from an observer's eye to a given point above the plane. (504, 773)

angle of inclination　See *angle of elevation*.

antecedent　The "if" clause of a conditional. Also called *hypothesis*. (69)

apothem　The perpendicular segment from the center to the side of a regular n-gon to one of its sides.

arc　A path from one point (node) of a network to another point (its endpoints or vertices). A part of a circle connecting two points (its endpoints) on the circle. (23, 132)

area　The number of nonoverlapping unit squares or parts of unit squares that can be fit into a region. (441)

arithmetic mean　The result of adding the n numbers in a data set and dividing the sum by n. Also called the *average* or *mean*. (759)

automatic drawer　Computer software or calculator that enables geometric figures to be constructed from input by the user. (101)

auxiliary figure　A figure that is added to a given figure, often to aid in completing proofs. (276)

average　See *arithmetic mean* or *mean*.

axis　A line of reference in a coordinate system. Plural: **axes**. (17, 669)

axis of a cone　The line through the cone's vertex and the center of its base. (512)

base　See *cylindric solid*. See *conic solid*.

base angles of an isosceles triangle　Two angles of an isosceles triangle whose vertices are the endpoints of a base of the triangle. (309)

base angles of a trapezoid　Two angles whose vertices are the endpoints of a base of the trapezoid. (318)

base of an isosceles triangle　The side opposite the vertex angle. (309)

base of a trapezoid　Either of two parallel sides of a trapezoid. (318)

base of a triangle　The side of a triangle to which an altitude is drawn. (454)

betweenness of numbers　A number is between two others if it is greater than one of them and less than the other. (46)

betweenness of points　A point is between two other points on the same line if its coordinate is between their coordinates. (46)

biconditional statement　A statement that includes a conditional and its converse. It may be written in the form $p \Rightarrow q$ and $q \Rightarrow p$, or $p \Leftrightarrow q$, or "p if and only if q." (82)

bilateral symmetry　A space figure has bilateral symmetry if and only if there is a plane over which the reflection image of the figure is the figure itself. (526)

binomial　An algebraic expression with exactly two terms. (583)

bisector　A point, line, ray, or plane which divides a segment, angle, or figure into two parts of equal measure. (85)

bisector of an angle　The ray in the interior of an angle that divides the angle into two angles of equal measures. (127)

bisector of a segment　A line, ray, or segment which intersects a segment at its midpoint but does not contain the segment. (168)

box　A right prism whose faces are rectangles. (505)

capacity　See *volume*.

Cartesian plane　Name given to the plane containing points identified as ordered pairs of real numbers. Also called *coordinate plane*. (17)

center See *circle, rotation, size change, sphere.*

center of a regular polygon The point equidistant from the vertices. (343)

center of symmetry For a rotation-symmetric figure, the center of a rotation that maps the figure onto itself. (335)

central angle of a circle An angle whose vertex is the center of the circle. (132)

centroid of a triangle The point at which all medians of a triangle intersect. (315)

characteristics The designated qualities of the defined term in a conditional. (83)

chord A segment whose endpoints are on a given circle. (347)

circle The set of points in a plane at a certain distance (its **radius**) from a certain point (its **center**). (83)

circularity The "circling back" that sometimes occurs when one tries to define basic terms; returning to the word which one is trying to define. (36)

circumference The perimeter of a circle, which is the limit of the perimeters of inscribed polygons. (473)

circumscribed circle A circle which contains all the vertices of a given polygon. (790)

circumscribed polygon A polygon whose sides are tangent to a given circle. (790, 828)

clockwise orientation The orientation "walking" around a figure keeping its interior to the right. (193)

clockwise rotation The direction in which the hands move on a nondigital clock, designated by a negative magnitude. (134)

coincident lines Lines that contain exactly the same points. (44)

collinear points Points that lie on the same line. (36)

common tangent A line which is tangent to two or more distinct circles. (827)

compass An instrument for drawing circles. (168)

complementary angles Two angles the sum of whose measures is 90. (138)

composite of two transformations The transformation $t_2 \circ t_1$, under which the image of a point or figure F is $t_2(t_1(F))$. (204)

composition The operation, denoted by the symbol $\circ$, of combining two transformations S followed by T by mapping each point or figure F onto $T(S(F))$. (204)

concentric circles Two or more circles that lie in the same plane and have the same center. (133)

conclusion The "then" clause of a conditional. The result of a deduction in a proof. Also called *consequent.* (69, 257)

concurrent Two or more lines that have a point in common. (222)

conditional A statement of the form If . . . then . . . . (69)

cone The surface of a conic solid whose base is a circle. (511)

congruence transformation A transformation that is a reflection or composite of reflections; also called an *isometry.* (229)

congruent figures A figure F is congruent to a figure G if and only if G is the image of F under a reflection or a composite of reflections. (228, 526)

conic section The intersection of a plane with the union of two right conical surfaces that have the same vertex and whose edges are opposite rays; either an ellipse, a parabola, or a hyperbola. (521)

conic solid The set of points between a given point (its **vertex**) and all points of a given region (its **base**), together with the vertex and the base. (511)

conic surface The boundary of a conic solid. (511)

conjecture An educated guess or opinion. (108)

consecutive angles In a polygon, two angles whose vertices are endpoints of the same side. (318)

consecutive sides In a polygon, two sides with an endpoint in common. (96)

consecutive vertices In a polygon, endpoints of a side. (96)

consequent The "then" clause of a conditional, also called the *conclusion.* (69)

construction A drawing which is made using only an unmarked straightedge and a compass following certain prescribed rules. (168)

continuous A figure made up of points with no space between them. (13)

contraction A size change with magnitude between 0 and 1. (693)

contradiction A situation in which there exist contradictory statements. (640)

contradictory statements Two statements that cannot both be true at the same time. (640)

contrapositive A conditional resulting from negating and switching the antecedent and consequent of the original conditional. (629)

convenient location A general location for a figure on a coordinate plane in which its key points are described with the fewest possible variables. (646)

converse The conditional statement formed by switching the antecedent and consequent of a given conditional. (76)

convex polygon A polygon whose polygonal region is a convex set. (97)

convex set A set in which every segment that connects points of the set lie entirely in the set. (66)

coordinate The number or numbers associated with the location of a point on a line, a plane, or in space. (13, 17, 668)

coordinate axes A pair of perpendicular coordinatized lines in a plane that intersect at the point with coordinate 0; three mutually perpendicular coordinatized lines in space that are concurrent at the point with coordinate 0. (17, 668)

coordinate geometry See *analytic geometry.*

coordinate plane See *Cartesian plane.*

coordinatized line A line on which every point is identified with exactly one number and every number is identified with a point on the line. (13)

coplanar Figures that lie in the same plane. (36)

corollary A theorem that is easily proven from another theorem. (312)

corresponding angles A pair of angles in similar locations when two lines are intersected by a transversal. (155)

corresponding parts Angles or sides that are images of each other under a transformation. (244)

cosine of an angle The ratio $\frac{\text{leg adjacent to the angle}}{\text{hypotenuse}}$ in a right triangle. Abbreviated *cos*. (777)

counterclockwise orientation The orientation "walking" around a figure keeping its interior to the left. (193)

counterclockwise rotation The direction opposite that which the hands move on a nondigital clock, designated by a positive magnitude. (134)

counterexample A specific case of a conditional for which the antecedent is true but the consequent is false. An example which shows a conjecture to be false. (71)

cube A box whose dimensions are all equal. (506)

cube root, $\sqrt[3]{\ }$ A real number x is the cube root of a real number y, written $x = \sqrt[3]{y}$, if and only if $x^3 = y$. (579)

cylinder The surface of a cylindric solid whose base is a circle. (507)

cylindric solid The set of points between a region (its **base**) and its translation image in space, including the region and its image. (506)

decagon A polygon with ten sides. (96)

deduction The process of making justified conclusions. (258, 621)

definition A description that clearly and uniquely specifies an object or a class of objects. (81)

degree measure of a major arc *ACB* of ⊙*O* $360 - m\overarc{AB}$. (133)

degree measure of minor arc *AB* of ⊙*O* The measure of the central angle *AOB*. (133)

degree A unit of measure used for the measure of an angle, arc, or rotation. (125)

dense A set with the property that between two elements of the set there is at least one other element of the set. (13)

diagonal of a polygon A segment connecting nonconsecutive vertices of the polygon. (96)

diameter of a circle or a sphere A segment connecting two points on the circle or sphere and containing the center of the circle or sphere; also, the length of that segment. (84, 518)

dihedral angle The angle formed by the union of two half-planes with the same edge. (503)

dimensions of a box The lengths of three edges of the box which meet at the same vertex. (505)

dimensions of a rectangle The lengths of two sides of the rectangle which meet at a single vertex. (437)

direction of translation The direction given by the vector from any preimage point to its image point in a translation. (205, 217)

directly congruent Figures which are congruent and have the same orientation. (229, 526)

direct reasoning (proofs) Reasoning (proofs) using the Law of Detachment and/or the Law of Transitivity. (639)

discrete figure A figure made up of points with space between them. (8)

discrete geometry The study of points as dots separated from each other, and lines made up of these points. (8)

distance between parallel planes The length of a segment perpendicular to the planes with an endpoint in each plane. (502)

distance between two parallel lines The length of a segment that is perpendicular to the lines with an endpoint on each of the lines. (196)

distance between two points The absolute value of the difference of their coordinates on a coordinatized line. (13)

distance from a point to a line The length of the perpendicular segment connecting the point to the line. (502)

distance from a point to a plane The length of the perpendicular segment from the point to the plane. (502)

dodecagon A polygon with twelve sides. (341)

dodecahedron A polyhedron with twelve faces. (537)

edge Any side of a polyhedron's faces. (506)

elevations Planar views of three-dimensional figures given from the top, front, or sides. (531)

ellipse The conic section formed by a plane which intersects only one of the right conical surfaces. (522)

empty set See *null set*.

endpoint See *arc*; *segment*; *ray*.

ends of a kite The common vertices of the equal sides of the kite. (324)

equiangular polygon A polygon with all angles of equal measure. (312)

equidistant At the same distance. (82, 271)

equilateral polygon A polygon with all sides of equal length. (342)

equilateral triangle A triangle with all three sides of equal length. (97)

Euclidean geometry The collection of propositions about figures which includes or from which can be deduced those given by the mathematician Euclid around 250 B.C. (41)

even node A node which is the endpoint of an even number of arcs in a network. (24)

expansion A size change with magnitude greater than one. (693)

extended ratio A sequence of three or more numbers representing the relative sizes of the numbers (or quantities). (282)

exterior angle An angle formed by two lines and a transversal whose interior contains no points between the two lines. An angle which forms a linear pair with an angle of a given polygon. (265, 416)

exterior of a circle The set of points at the distance greater than the radius from the center of the circle. (96)

exterior of a figure When a figure separates the plane into two parts, one bounded and one not, the unbounded part. (96)

exterior of an angle A non-zero angle separates the plane into two sets of points. If the angle is not straight, the non-convex set is the exterior of the angle. (125)

extremes of a proportion The first and fourth terms of the proportion. (706)

face of polyhedron Any of the polygonal regions that form the surface of the polyhedron. (505, 512, 537)

figure A set of points.

flip See *reflection*.

flow chart A diagram that shows a step-by-step progression through a procedure or system. (109)

45-45-90 triangle An isosceles right triangle whose angles measure 45, 45, and 90. (765)

fundamental region A region which is used to tessellate a plane. (398)

generalization A statement that applies to all situations of a particular type. (108)

geometric mean Given numbers a, b, and g, g is the geometric mean of a and b if and only if $\frac{a}{g} = \frac{g}{b}$. (759)

glide reflection The composite of a reflection and a translation parallel to the reflecting line; also called *walk*. (223)

gores A set of tapered sections of a net of a spherical object. (544)

grade The slope of a road, often represented as a percent or ratio. (155)

graph A picture of numbers on a number line or coordinate system. See also *network*. (13, 17, 668)

graph theory The geometry of networks. (23)

great circle of sphere The intersection of a sphere and a plane that contains the center of the sphere. (518)

half-turn A rotation of 180° about a point.

height See *altitude*.

hemisphere The half of a sphere on one side of a great circle. (518)

heptagon A polygon with seven sides. Also called *septagon*. (96)

hexagon A polygon with six sides. (96)

hexahedron A polyhedron with six faces. (537)

hidden lines Lines in a picture or a three-dimensional figure that cannot be seen, but which are marked as dashed or shaded lines so as to show existence or give a feeling of depth. (32)

hierarchy A diagram that shows how various figures or ideas are related, often with a downward direction that moves from more general to more specific. (97)

horizontal component of a vector The first component in the ordered pair description of a vector, indicating its magnitude along the x-axis of the coordinate plane. (218)

horizontal line A line with an equation $y = k$ on the coordinate plane. (18)

hyperbola The conic section formed by a plane which intersects both of the right conical surfaces. (522)

hypotenuse The longest side of a right triangle; the side opposite the right angle. (390)

hypothesis The "if" clause of a conditional, also called the *antecedent*. An assumption used as the basis for an investigation or argument. (69)

icosahedron A polyhedron with twenty faces. (537)

identity transformation A transformation that maps each point onto itself. (693)

if and only if statement A statement consisting of a conditional and its converse. Also called *biconditional*. (82)

if-then statement See *conditional*.

image The result of applying a transformation to an original figure or preimage. (134)

image of a figure The set of all images of points of a given figure. (192)

included angle Of two consecutive sides of a polygon, the angle of the polygon whose vertex is the common point of the sides. (372)

included side Of two consecutive angles of a polygon, the side of the polygon which is on both the angles. (373)

indirect reasoning (proofs) Reasoning (proofs) using the Law of the Contrapositive, the Law of Ruling Out Possibilities, or the Law of Indirect Reasoning. (641)

initial point The beginning point of a vector. (217)

inscribed angle in a circle An angle whose vertex is on the circle and whose sides each intersect the circle at a point other than the vertex. (807)

inscribed circle In a polygon, a circle which is tangent to each side of the polygon. (790)

inscribed polygon In a circle, a polygon whose vertices all lie on the circle. (790)

instance of a conditional A specific case in which the antecedent (*if* part) of the conditional is true and its consequent (*then* part) is also true. (70)

intercepted arc   An arc of the circle in the interior of an angle. (800)

interior angles   Angles formed by two lines and a transversal whose interiors are partially between the lines; the angles of a polygon. (265, 417)

interior of a circle   The set of points at a distance less than the radius from the center of the circle. (85)

interior of a figure   When a figure separates the plane into two parts, one bounded and one not, the bounded part. (96)

interior of an angle   A nonzero angle separates the plane into two sets of points. If the angle is not straight, the convex set is the interior of the angle. (125)

intersecting planes   Two planes that contain the same line. (498)

intersection of two sets   The set of elements which are in both the sets. (87)

inverse   A conditional resulting from negating the antecedent and consequent of the original conditional. (629)

isometry   A transformation that is a reflection or a composite of reflections. Also called *congruence transformation* or *distance-preserving transformation*. (222)

isosceles trapezoid   A trapezoid with a pair of base angles equal in measure. (318)

isosceles triangle   A triangle with at least two sides equal in length. (97)

justification   A definition, postulate, or theorem which enables a conclusion to be drawn. (151)

kite   A quadrilateral with two distinct pairs of consecutive sides of the same length. (317)

lateral area   The sum of the areas of the lateral faces of a solid. (565)

lateral edge of conic solid   Any segment connecting the vertex of the solid to a point on its base. (507, 512)

lateral face of a polyhedron   Any face other than a base. (507, 512)

lateral surface of a conic solid   The surface of the solid excluding its base. (506, 512)

latitude   Of a point P on Earth with Earth's center being C, the measure of the arc $\overset{\frown}{PQ}$, where Q is the intersection of the equator and the plane through P and the North and South Poles. The latitude is North or South according as P is between Q and the North or South Pole. Points with the same latitude lie on the same circle of latitude. (21, 52)

lattice point   A point in the coordinate plane or in space with integer coordinates. (452)

leg   Either side of a right triangle that is on the right angle. (390)

leg adjacent to an angle   The side of the right triangle which is on the acute angle and is not the hypotenuse. (773)

leg opposite an angle   The side of the right triangle which is not on the acute angle. (773)

length   The distance between two points measured along the segment or an arc joining them.   A dimension of a rectangle or rectangular solid. (48, 437, 577)

limit   The value to which the terms of an infinite sequence get closer and closer as one goes further out in the sequence. (449, 474)

line   An undefined geometric term. See *Point-Line-Plane Postulate*. (42, 497)

linear pair   Two adjacent angles whose noncommon sides are opposite rays. (139)

line of reflection   The line over which a preimage is reflected. Also called *reflecting line* or *mirror*. (184)

line of sight   An imaginary line connecting the eye of the viewer to points on an object. (31)

line perpendicular to a plane   A line perpendicular to every line in the plane which passes through the point of intersection. (501)

line segment   See *segment*.

line symmetry   See *reflection-symmetric figure*.

longitude   Of a point P on Earth, the measure of the minor arc $\overset{\frown}{PQ}$ on its circle of latitude, where Q is the intersection of the circle with the prime meridian, the semicircle from the North Pole to the South Pole containing an observatory in Greenwich, England. (283, 524)

magnitude of rotation   In a rotation, the amount that the preimage is turned about the center of rotation, measured in degrees from $-180°$ (clockwise) to $180°$ (counterclockwise), $\pm m\angle POP'$, where P' is the image of P under the rotation and O is its center. (134)

magnitude of size transformation   In a size change, the factor by which length of the preimage is changed, $\frac{A'B'}{AB}$, where A' and B' are the images of A and B under the transformation. Also called a *size change factor* or *scale factor of size transformation*. (687)

magnitude of translation   The distance between any point and its image. (205)

major arc AB of $\odot O$ The points of $\odot O$ that are on or in the exterior of $\angle AOB$. (133)

mapping   See *transformation*.

matrix   A rectangular array of rows and columns. (6)

mean   The sum of a set of numbers divided by the number of numbers in the set. Also called *average* or *arithmetic mean*. See also *geometric mean*. (461, 662, 759)

means of a proportion   The 2nd and 3rd terms of a proportion. (706)

measure   The dimension or amount of something, usually in a system of units. (48, 125)

medial triangle   The union of segments that join the midpoints of the sides of a triangle. (665)

median of a triangle   The segment connecting a vertex of the triangle to the midpoint of the opposite side. (308)

menu A list of options from which a user of a computer can select an operation for the computer to perform. (102)

Mercator projection A two-dimensional map of the Earth's surface named for Gerhardus Mercator, the Flemish cartographer who first created it in 1569. (545)

midpoint of a segment The point on the segment equidistant from the segment's endpoints. (81)

minor arc *AB* of ⊙*O* The points of ⊙*O* that are on or in the interior of ∠*AOB*. (132)

_n_-fold rotation symmetry A figure has *n*-fold rotation symmetry, where *n* is a positive integer, when a rotation of magnitude $\frac{360}{n}$ maps the figure onto itself, and no larger value of *n* has this property. (336)

_n_-gon A polygon with *n* sides. (96)

negation of a statement A statement (called *not-p*) that is true whenever statement *p* is false and is false whenever statement *p* is true. (628)

net A two-dimensional figure that can be folded on its segments or curved on its boundaries into a three-dimensional surface. (538)

network A union of points (its **vertices** or **nodes**) and segments (its **arcs**) connecting them. Also called *graph*. (23)

node An endpoint of an arc in a network. See also *vertex*. (23)

nonagon A polygon with nine sides. (96)

nonconvex set A set in which at least one segment that connects points within the set has points that lie outside of the set. (66)

non-Euclidean geometry A geometry in which the postulates are not the same as those in Euclidean geometry. (278, 282)

nonoverlapping regions Regions that do not share interior points. (384, 443)

null set The set with no elements. Also called *empty set*. (89)

number line A line on which points are identified with real numbers. (12)

oblique cone A cone whose axis is not perpendicular to its base. (512)

oblique figure A 3-dimensional figure in which the plane of the base(s) is not perpendicular to its axis or to the planes of its lateral surfaces. (507, 512)

oblique line A line that is neither horizontal nor vertical. (8)

obtuse angle An angle whose measure is greater than 90 and less than 180. (137)

obtuse triangle A triangle with an obtuse angle. (142)

octagon A polygon with eight sides. (96)

octahedron A polyhedron with eight faces. (537)

odd node A node which is the endpoint of an odd number of arcs in a network. (24)

one-dimensional A space in which all points are collinear. (36)

one-step proof A justified conclusion of a conditional requiring a single definition, theorem, or postulate. (151)

opposite faces A pair of faces of a polyhedron whose planes are parallel to each other. (505)

oppositely congruent Figures which are congruent and have opposite orientation. (229, 526)

opposite rays $\overrightarrow{AB}$ and $\overrightarrow{AC}$ are opposite rays if and only if *A* is between *B* and *C*. (48)

ordered pair The pair of numbers (*a*, *b*) identifying a point in a two-dimensional coordinate system. (17)

ordered pair description of a vector The description of a vector as the ordered pair (*a*,*b*) where *a* is the **horizontal component** and *b* is the **vertical component**. (218)

ordered triple The triple of numbers (*a*, *b*, *c*) identifying a point in a three-dimensional coordinate system. (668)

order _n_ rotation symmetry See *n-fold rotation symmetry*.

orientation The order of the designation of the vertices of a polygon, either *clockwise* or *counterclockwise*. (193)

overlapping figures Figures which have interior points in common. (384)

overlapping triangles Triangles that have interior points in common. (384)

pairing The assignment of two teams or individuals to compete against one another in a tournament. (349)

parabola The conic section formed by a plane parallel to an edge of the conical surface. (522)

paragraph proof A form of written proof in which conclusions and justifications are incorporated into sentences. (264)

parallel lines Two coplanar lines which have no points in common or are identical. (43)

parallelogram A quadrilateral with two pairs of parallel sides. (316)

parallel planes Planes which have no points in common or are identical. (502)

pentagon A polygon with five sides. (96)

perimeter The length of the boundary of a closed region. (435)

perimeter of a polygon The sum of the lengths of its sides. (436)

perpendicular Two segments, rays, or lines such that the lines containing them form a 90° angle. (161)

perpendicular bisector method A method for finding the center of a circle that involves drawing perpendicular bisectors of two chords. (813)

perpendicular bisector of a segment In a plane, the line containing the midpoint of the segment and perpendicular to the segment. In space, the plane that is perpendicular to the segment and contains the midpoint of the segment. (168, 525)

perpendicular planes Planes whose dihedral angle is a right angle. (503)

perspective drawing A drawing of a figure made to look as it would in the real world. (30)

pi, π The ratio of circumference to the diameter of a circle. (475)

picture angle of a camera lens An angle measure indicating how wide a field of vision can be captured in one photo. (806)

pixel A dot on a TV or computer screen or other monitor. (6)

plane An undefined geometric term. See *Point-Line-Plane Postulate*. (17, 496)

plane coordinate geometry The study of points as ordered pairs of numbers. (17, 19)

plane figure A set of points that are all in one plane. (36)

plane geometry The study of figures which lie in the same plane. (35, 38)

plane section The intersection of a three-dimensional figure with a plane. (520)

point An undefined geometric term. See *Point-Line-Plane Postulate*. (41, 497)

point of tangency The point at which a tangent intersects the curve (circle) or curved surface (sphere). (659, 823)

polygon The union of three or more coplanar segments (its **sides**) such that each segment intersects exactly two others, one at each of its endpoints (its **vertices**). (95)

polygonal region The union of a polygon and its interior. (96)

polyhedron A three-dimensional surface which is the union of polygonal regions (its **faces**) and which has no holes. Plural: **polyhedra**. (537)

polynomial An expression that is the sum or difference of two or more terms. (583)

postulate A statement assumed to be true. Also called *axiom*. (41)

power of a point for ⊙O For any secant through *P* intersecting ⊙O at *A* and *B*, the product *PA* · *PB*. (838)

preimage The original figure in a transformation. (134)

preserved property Under a transformation, a property which, if present in a preimage, is present in the image. (191, 699)

prism The surface of a cylindric solid whose base is a polygon. (507)

proof A sequence of justified conclusions, leading from what is given or known to a final conclusion. (150)

proportion A statement that two ratios are equal. (706)

proportional numbers Four numbers that form a true proportion. (706)

protractor A tool commonly used to measure angles. (121)

pyramid The surface of a conic solid whose base is a polygon. (511)

Pythagorean triple A set of three numbers that can be the lengths of the sides of a right triangle. (470)

quadrilateral A polygon with four sides. (96)

radius of circle or sphere A segment connecting the center of a circle or a sphere with a point on that circle or sphere; also, the length of that segment. Plural: **radii**. (83, 517)

ratio A quotient of quantities with the same units. (705)

ratio of similitude In similar figures, the ratio of a distance or length in an image to the corresponding distance or length in a preimage. (714)

ray The ray with **endpoint *A*** containing *B*, denoted $\overrightarrow{AB}$, is the union of $\overline{AB}$ and the set of all points for which *B* is between each of them and *A*. (47)

rectangle A quadrilateral with four right angles. (316)

rectangular solid The union of a box and its interior. (505)

reflecting line The line over which a preimage is reflected. Also called *line of reflection* or *mirror*. (184)

reflecting plane The plane over which a preimage is reflected. Also called *mirror*. (526)

reflection A transformation in which each point is mapped onto its reflection image over a line or plane. (186, 526)

reflection image of a figure The set of all reflection images of the points of the original figure. (192)

reflection image of a point *P* over a line *m* If *p* is not on *m*, the reflection image of *p* is the point *Q* such that *m* is the perpendicular bisector of $\overline{PQ}$. If *P* is on *m*, the reflection image is point *P* itself. (185)

reflection image of a point *A* over a plane *M* If *A* is not on *M*, the reflection image is the point *B* such that *M* is the perpendicular bisector of $\overline{AB}$. If *A* is on *M*, the reflection image is point *A* itself. (525)

reflection-symmetric figure A figure *F* for which there is a reflection r_m such that $r_m(F) = F$. (300, 526)

region The union of a polygon or circle and its interior. More generally, any connected two-dimensional figure that has an area. (96)

regular polygon A convex polygon whose angles are all congruent and whose sides are all congruent. (341)

regular polyhedron A convex polyhedron whose faces are all congruent regular polygons and the same number of edges intersect at each of its vertices. (537)

regular prism A right prism whose base is a regular polygon. (508)

regular pyramid A right pyramid whose base is a regular polygon and whose lateral faces are congruent isosceles triangles. (512)

resolution The capability of distinguishing individual parts of an entire object or picture. (6)

rhombus A quadrilateral with four sides of equal length. (316)

right angle An angle whose measure is 90. (137)

right angle method A method for finding the center of a circle that involves constructing two inscribed right triangles. The intersection of their hypotenuses is the center. (813)

right cone A cone whose axis is perpendicular to the plane of the circular base. (512)

right cylinder A cylinder formed when the direction of translation of the base is perpendicular to the plane of the base. (507)

right prism A prism formed when the direction of translation of the base is perpendicular to the plane of the base. (507)

right pyramid A pyramid whose base is a regular polygon in which the segment connecting its vertex to the center of its base is perpendicular to the plane of the base. (512)

right triangle A triangle which contains a right angle. (142, 453)

rotation The composite of two reflections over intersecting lines; the transformation "turns" the preimage onto the final image about a fixed point (its **center**). Also called *turn*. (134, 211)

rotation-symmetric figure A figure F for which there is a composite of reflections $r_m \circ r_l$ such that $r_m(r_l(F)) = F$. (335)

round-robin tournament A tournament in which each competitor plays each other exactly once. (347)

scale factor of size transformation See *magnitude of size transformation.*

scalene triangle A triangle with no two sides of the same length. (97)

secant to a circle A line that intersects a circle in two distinct points. (819)

sector of a circle The figure bounded by two radii and the included arc of the circle. (480)

segment The set consisting of the distinct points A and B (its **endpoints**) and all points between A and B. Also called *line segment.* (47)

semicircle An arc of a circle whose endpoints are the endpoints of a diameter of the circle. (133)

side of an angle See *angle.*

side of a polygon One of the segments whose union is a polygon; also, the length of that segment. (96)

similar figures Two figures A and B for which there is a similarity transformation mapping one onto the other, written $A \sim B$. (711)

similarity transformation A composite of size changes and reflections. (712)

simple closed curve A figure that is closed and does not intersect itself. (112)

sine of an angle The ratio $\frac{\text{leg opposite the angle}}{\text{hypotenuse}}$ in a right triangle. Abbreviated *sin.* (777)

size change The transformation S such that, for a given point P and a positive real number k (its **magnitude**) and any point O (its **center**), $S(P) = P'$ is the point on $\overrightarrow{OP}$ with $OP' = k \cdot OP$. The transformation in which the image of (x, y) is (kx, ky). Also called *size transformation* or *dilation.* (693)

size change factor See *magnitude of size transformation.*

skew lines Lines that do not lie in the same plane. (504)

slant height of a regular pyramid The altitude from the vertex on any one of the lateral faces of the pyramid. (513)

slant height of a right cone The length of a lateral edge of the cone. (513)

slide See *translation.*

slope In the coordinate plane, the change in y-values divided by the corresponding changes in x-values. (157)

slope-intercept form of an equation for a line A linear equation of the form $y = mx + b$, where m is the slope and b is the y-intercept. (19)

small circle of a sphere The intersection of the sphere and a plane that does not contain the center of the sphere. (518)

solid The union of a surface and the region of space enclosed by the surface. (505)

solid geometry The study of figures in three-dimensional space. (495)

space The set of all points in a geometry. (36)

space figure A figure whose points do not all lie in a single plane. Also called a *three-dimensional figure.* (37)

sphere The set of points in space at a fixed distance (its **radius**) from a point (its **center**). (517)

square A quadrilateral with four sides of equal length and four right angles. (317)

square root, $\sqrt{\ }$ x is a square root of y if and only if $x^2 = y$. If x and y are positive, $x = \sqrt{y}$. (466)

standard form of an equation for a line An equation for a line in the form $Ax + By = C$, where A and B are not both zero. (18)

straight angle An angle whose measure is 180. (137)

straightedge An instrument for drawing the line through two points which has no marks for determining length. (168)

sufficient condition p is a sufficient condition for q if and only if p implies q. (367)

supplementary angles Two angles the sum of whose measures is 180. (138)

surface The boundary of a three-dimensional figure. (505)

surface area The area of the boundary surface of a three-dimensional figure. (564)

symmetry diagonal (of a kite) The diagonal that connects the ends of the kite. (325)

symmetry line For a figure, a line m such that the figure coincides with its reflection image over m. (300)

symmetry plane For a figure, a plane M such that the figure coincides with its reflection image over M. (526)

synthetic geometry A geometry studied without the use of coordinates. (12)

tangent A line, ray, segment, or plane which intersects a curve or curved surface in exactly one point. (659, 823, 825)

tangent of an angle The ratio $\frac{\text{leg opposite to the angle}}{\text{leg adjacent to the angle}}$ in a right triangle. Abbreviated *tan*. (773)

tangent circles Two circles that have exactly one point in common. (853)

terminal point The endpoint of a vector. (217)

tessellation A covering of a plane with congruent nonoverlapping copies of the same region. (398)

tetrahedron A polyhedron with four faces. (537)

theorem A statement deduced from postulates, definitions, or other previously deduced theorems. (43)

30-60-90 triangle A triangle in which the three angles measure 30, 60, and 90. (766)

3-dimensional coordinate system A system of coordinates used to locate points in space by their distances and directions from three mutually perpendicular lines. (668)

three-dimensional figure A figure whose points do not all lie in a single plane. (37)

transformation A correspondence between two sets of points such that each point in the preimage set has a unique image, and each point in the image set has exactly one preimage. Also called *map*. (186)

transformation image of a figure The set of all images of the points of the figure under the transformation. (186)

translation The composite of two reflections over parallel lines. Also called *slide*. (205)

transversal A line that intersects two or more lines. (155)

trapezoid A quadrilateral with at least one pair of parallel sides. (318)

traversable network A network in which all the arcs may be traced exactly once without picking up the tracing instrument. (24)

triangle A polygon with three sides. (96)

triangulate To split a polygon into nonoverlapping triangles. (284)

trigonometric ratio A ratio of the lengths of the sides in a right triangle. (774)

trisect To divide into three congruent parts. (421)

truncated surface A part of a conic surface including its base, the intersection of a plane parallel to the base with the conic solid, and all points of the surface between these. (515)

truth value The condition of a statement in logic; either true or false. (628)

turn See *rotation*.

two-column proof A form of written proof in which the conclusions are written in one column, the justifications beside them in a second column. (264)

two-dimensional Pertaining to figures that lie in a single plane, or to their geometry. (36)

undefined terms A term used without a specific mathematical definition. (36)

union of two sets The set of elements which are in at least one of the sets. (87)

uniquely determined A situation in which there is exactly one element satisfying given conditions. (276)

unit cube A cube in which every edge has length one unit. (576)

unit square A square in which each side has length one unit. (441)

vanishing line A line containing vanishing points. (31)

vanishing point The point at which several lines of a drawing appear to meet at a distance from the viewer's eye. (30)

vector A quantity that has both magnitude and direction. (217)

velocity The rate of change of distance with respect to time. (784)

vertex See *angle*. See *conic solid*. See *network*. See *polygon*. Plural: **vertices**.

vertex angle The angle included by equal sides of an isosceles triangle. (309)

vertex of a polyhedron Any of the vertices of the faces of the polyhedron. (506, 511, 537)

vertical angles Two nonstraight and nonzero angles whose sides form two lines. (140)

vertical component of a vector The second component in the ordered pair description of a vector, indicating its magnitude along the y-axis of the coordinate plane. (218)

vertical line A line with an equation $x = h$ on the coordinate plane. (18)

volume The number of unit cubes or parts of unit cubes that can be fit into a solid. Also called *capacity*. (576)

walk See *glide reflection*.

width (of a rectangle) A dimension of a rectangle or rectangular solid taken at right angles to the length. (437, 577)

window That part of the plane that shows on the screen of an automatic grapher or automatic drawer. (102)

x-axis The line in the coordinate plane or in space, usually horizontal, containing those points whose second coordinates (and third, in space) are 0. (17, 669)

y-axis The line in the coordinate plane, usually vertical, or in space, containing those points whose first coordinates (and third, in space) are 0. (17, 669)

z-axis The line in a three-dimensional coordinate system containing those points whose first and second coordinates are 0. (668)

zero angle An angle whose measure is zero. (137)

INDEX

INDEX

$>$	is greater than
$<$	is less than
$\geq$	is greater than or equal to
$\leq$	is less than or equal to
$\neq$	is not equal to
$\approx$	is approximately equal to
$\pm$	plus or minus
π	pi
$|x|$	absolute value of x
$\sqrt{n}$	positive square root of n
$\sqrt[3]{n}$	cube root of n
$//$	is parallel to
$\perp$	is perpendicular to
$\cong$	is congruent to
$\sim$	is similar to
r_m	reflection over line m
$r_m(P)$	reflection image of point P over line m
$r(P)$	reflection image of point P
$R(P)$	rotation image of point P
$T(P)$	transformation image of point P
$T_1 \circ T_2$	composite of transformation T_2 followed by T_1
S_k	size change of magnitude k
A'	image of point A
A''	image of point A'
$\Rightarrow$	if-then (implication)
$\Leftrightarrow$	if and only if
$\{ \ldots \}$	set
$\{\}, \emptyset$	empty or null set
$N(E)$	the number of elements in set E
$P(E)$	probability of an event E
$\cap$	intersection of sets
$\cup$	union of sets
$\tan A$	tangent of $\angle A$
$\sin A$	sine of $\angle A$
$\cos A$	cosine of $\angle A$
$\overleftrightarrow{AB}$	line through A and B
$\overrightarrow{AB}$	ray with endpoint at A and containing B
$\overline{AB}$	segment with endpoints A and B
AB	distance from A to B
$\angle ABC$	angle ABC
$m\angle ABC$	measure of angle ABC
$\triangle ABC$	triangle with vertices A, B, C

$ABCD \ldots$	polygon with vertices A, B, C, D,$\ldots$
$\odot O$	circle with center O
$\lnot$	right angle symbol
$n°$	n degrees
$\overset{\frown}{AB}$	minor arc with endpoints A and B
$\overset{\frown}{ADB}$	arc with endpoints A and B containing D
$m\overset{\frown}{AB}$	measure of arc AB in degrees
Area(F)	area of figure F
Volume(F)	volume of figure F
(x, y)	ordered pair x, y
(x, y, z)	ordered triple x, y, z
$\overrightarrow{AB}$	vector with initial point A and terminal point B
v or $\vec{v}$	vector v

Calculator Keys

$\pm$ or $+/-$	opposite
y^x or x^y	powering function
INV, 2nd, or F	second function
EE or EXP	scientific notation
1/x	reciprocal
$\sqrt{}$	square root function
x^2	squaring function
π	pi
INV	inverse function
tan	tangent function
sin	sine function
cos	cosine function

Computer Commands

2*2	$2 \cdot 3$
4 / 3	$4 \div 3$
3 ^ 5	3^5
>=	$\geq$
<=	$\leq$
<>	not equal to
SQR(N)	$\sqrt{n}$
IF . . . THEN	THEN statement to be executed only if IF part is true.

Acknowledgments

4(c) Tim Everitt/Tony Stone Worldwide **4-5(t)** Art Institute of Chicago/Bridgeman Art Library, London/Superstock, Inc. **5(b)** Superstock, Inc. **6** Art Institute of Chicago/Bridgeman Art Library, London/Superstock, Inc. **7(b)** ©1988 Sue Klemens/Stock Boston **7(t)** David R. Frazier **9** David R. Frazier **12** Courtesy of the New Explorers, a production of Kurtis Productions, Ltd. and WTTW/CHICAGO/CARA/University of Chicago, Photo: James S. Sweitzer **13** Bob Daemmrich **15(l)** Robert Frerck/Stock Market **16(b)** Demetrio Carrasco/Tony Stone Worldwide **17** Esbin-Anderson/Image Works **21** Ken Hawkins/Sygma **23(t)** Prenzel/Schmida/Animals Animals **27(b)** Lionel Delevingne/Stock Boston **30(t)** Courtesy of the Regis Collection Minneapolis, MN **31** ©1969 by Butcher Lucas. Distrubuted by King Features Syndicate **35** John Maher/Stock Market **41** Scala/Art Resource **43** "The Elements of Euclid of Megara" **46** Lowell Georgia/SS/Photo Researchers **51(t)** William Johnson/Stock Boston **52(t)** Leo de Wys, Inc. **52-53(b)** Donovan Reese/Tony Stone Worldwide **53(t)** Charly Franklin/FPG International Corp. **53(c)** Superstock, Inc. **62(t)** Superstock, Inc. **62(cl)** Superstock, Inc. **62-63(cr)** Superstock, Inc. **63(t)** Superstock, Inc. **66** Jeffrey Markowitz/Sygma **69** Hugh Norton/Superstock, Inc. **73** Gahan Wilson **76** Superstock, Inc. **79(t)** Boltin Picture Library **83** Superstock, Inc. **85** Superstock, Inc. **86** From The World Book Encyclopedia. ©1995 World Book, Inc. By permission of the publisher **87** Alex S. MacLean/Landslides **90(l)** Frank Siteman/Tony Stone Worldwide **90(r)** Reproduced by Permission MBTA, Robert L. Mabardy, Interim General Manager **92(t)** Superstock, Inc. **92(b)** (c)1994 Universal Press Syndicate **95** Carol M. Highsmith **96** Courtesy Watertown Historical Society **99-t** ©1994 by Rand, McNally **99-b** ©1994 by Rand, McNally **103** Peter Ginter/Image Bank **106** Cameramann International, Ltd. **107** Photofest **112** David Young-Wolff/PhotoEdit **113** Superstock, Inc. **114** Courtesy The Milk Foundation/©1995 National Fluid Milk Processor Promotion Board **122(t)** Superstock, Inc. **122(c)** Superstock, Inc. **122-123(b)** Superstock, Inc. **123(t)** Superstock, Inc. **123(c)** Superstock, Inc. **123(cr)** SF **124(t)** J. Sohm/Image Works **129** Superstock, Inc. **132** Cameramann International, Ltd. **134** B. Daemmrich/Image Works **138** Courtesy of the Arthur M. Sackler Gallery, Smithsonian Institution, Washington, D.C. From the exhibition "A rural Basketmaker in Japan," Kenji Nakamura and Nakamura Yuzu Products **142(b)** Superstock, Inc. **144** Joseph Thorn **150** Eric Sander/Gamma Liaison **152** Fanny Broadcast/Gamma Liaison **155** Jan Kanter **156** Myra Miller/Gamma Liaison **165** ©1994 Streetwise Maps, Inc., Amagansett, NY **166(t)** Zavier Testelin/Gamma Liaison **166(b)** The Dell Big Book Of Crosswords And Pencil Puzzles #5 **170** Superstock, Inc. **172(r)** FourByFive/Superstock, Inc. **172(l)** Superstock, Inc. **173(t)** Superstock, Inc. **173(br)** Superstock, Inc. **182(t)** Superstock, Inc. **182-183(c)** Superstock, Inc. **182(bl)** Superstock, Inc. **182-183(br)** Superstock, Inc. **183(tl)** FourByFive Inc/Superstock, Inc. **183(tr)** M.C. Escher/Cordon Art-Baarn-Holland. All rights reserved. **184(t)** Cameramann International, Ltd. **184(b)** Scott Gibson **188** Scott Gibson **191** FPG International Corp. **195(l)** Charles Gupton/Stock Boston **196** Focus on Sports, Inc. **197** Focus on Sports, Inc. **198** Vandystadt Agence de Presse **199** Bob Daemmrich **201** Superstock, Inc. **203** Myrleen Ferguson/PhotoEdit **210** Tony Stone Worldwide **214** Chris Arend/AlaskaStock **216** Rosemary Finn **220** Tony Freeman/PhotoEdit **222** Alan Oddie/PhotoEdit **225** M.C. Escher/Cordon Art-Baarn-Holland. All rights reserved. **226(b)** Tracy Omar/Washington Park Arboretum **233(t)** Superstock, Inc. **240** M.C. Escher/Cordon Art-Baarn-Holland. All rights reserved. **241** Bill Gallery/Stock Market **242-243(tr)** Superstock, Inc. **242-243(c)** FourByFive Inc/Superstock, Inc. **242(B)** FourByFive Inc./Superstock, Inc. **244** Superstock, Inc. **247** Courtesy Lifetouch ®National School Studios **248** Reprinted with permission from Games World Of Puzzles Magazine (19 West 21st Street, New York, NY 10010) Copyright ©1995 B. & P. Publishing Co, Inc. **263** Stacy Pick/Stock Boston **267** Museum of the American Indian **270** John Lamb/Tony Stone Worldwide **278** Hulton/Bettmann **280(b)** Ted Thai/Sygma **280(t)** J.L. Atlan/Sygma **281** Kevin Galvin/Stock Boston **283** Courtesy Lands' End, Dodgeville, WI. Photo: Gary Comer **285** Jon Gray/Tony Stone Worldwide **287** Raymond B. Barnes/Tony Stone Worldwide **288(t)** Superstock, Inc. **288(b)** FourByFive Inc./Superstock, Inc. **297(l)** Jay Syverson/Stock Boston **297(R)** Cameramann International, Ltd. **298** Superstock, Inc. **298(t)** Superstock, Inc. **298-299(b)** Superstock, Inc. **299(t)** Superstock, Inc. **299(c)** Superstock, Inc. **299(tc)** Courtesy ITT Sheraton Corporation **299(tr)** Courtesy Mcdonald's Corporation **300** Stan Osolinski/Tony Stone Worldwide **303** Superstock, Inc. **307** Christopher Morrow/Stock Boston **309** Superstock, Inc. **314** Bob Torrez/Tony Stone Worldwide **316** Larry Lefever/Grant Heilman Photography **323** Ellis Herwig/Stock Boston **325** Sygma **329** Grant Heilman/Grant Heilman Photography **331** Focus on Sports, Inc. **334(l)** Courtesy National Football League/The Pittsburgh Steelers **334(c)** Courtesy Nissan Motor Corporation U.S.A. **334(r)** Courtesy Minolta corporation **335** Bill Gallery/Stock Boston **336** Courtesy Chrysler Corporation **337** Superstock, Inc. **341** ` FourByFive Inc./Superstock, Inc. **346** The Pepsi Ball Design is a registered trademark for soft drinks, used courtesy of PepsiCo, Inc. **347** Michael Newman/PhotoEdit **350** Focus on Sports, Inc. **352** Superstock, Inc. **353(tr)** Museum of the American Indian **353** Superstock, Inc. **353(bl)** From Inversions by Scott Kim **361(l)** Finn, Rosemary **361(tr)** From "The Complete Stitch Encyclopedia" by Jan Eaton. Reprinted 1987 copyright © Quarto Publishing Ltd. **362-363(br)** FourByFive Inc./Superstock, Inc. **362(t)** Superstock, Inc. **362-363(c)** Superstock, Inc. **362(bl)** Superstock, Inc. **363(t)** Superstock, Inc. **365** Superstock, Inc. **366** Superstock, Inc. **369** "Hidden Pictures" puzzle by Christopher Wray from Highlights For Children, July/August 1988 **370** Robert Fried **371(l)** Jay Wolke **374** Courtesy Divine Word International-Original American training center for Divine Word missionaries, now Techny Towers Conference Center, Techny, IL **376** Superstock, Inc. **380** Jan Kanter **381** Jan Kanter **384** Cameramann International, Ltd. **387** Superstock, Inc. **389** Phyllis Picardi/Stock Boston **390** Phyllis Picardi/Stock Boston **395** Dean Abramson/Stock Boston **396** Jim Corwin/Stock Boston **397** Cameramann International, Ltd. **399** Adam Woolfit/Woodfin Camp & Associates **401** Cameramann International, Ltd. **405** Cameramann International, Ltd. **408** Bob Daemmrich **411** Phil Long/AP/Wide World **412** Brian Smith/Stock Boston **415** Andy Sacks/Tony Stone Worldwide **416** FourByFive Inc./Superstock, Inc. **423(t)** FourByFive Inc./Superstock, Inc. **423(bl)** Superstock, Inc. **423(br)** Superstock, Inc. **424(tl)** FourByFive Inc./Superstock, Inc. **434(t)** Superstock, Inc. **434(bl)** Superstock, Inc. **434-435(b)** Superstock, Inc. **435(t)** Superstock, Inc. **437** Superstock, Inc. **439** Courtesy Commonwealth of Australia **440** Superstock, Inc. **441** Frank Lloyd Wright, American, 1867-1959, Window triptych from the Avery Coonley Playhouse, Riverside, Illinois, clear and colored leaded glass in oak frame, 1912, center panel: 88.9 x 109.2 cm; side panels: each 91.4 x 19.7 cm, Restricted gift of Dr. and Mrs. Edwin J. DeCosta and the Walter E. Heller Foundation, 1986.88, "photograph ©1995, Art Institute of Chicago. All Rights Reserved." **442** Grant Heilman/Grant Heilman Photography **446** Tony Freeman/PhotoEdit **447** Courtesy Millinocket Office of Tourism **451** Ulrike Welsch **453** Steve Rosenthal Architectural Photography **457** D.Fox/Superstock, Inc. **459** Esaias Baitel/Gamma-Liaison **464(bl)** Superstock, Inc. **466** Egyptian Expedition of The Metropolitan Museum of Art, Rogers Fund, 1930. **467** Bettmann Archive **471** Thorn, Joseph **474(t)** Alex MacLean/Landslides **474(b)** Courtesy Studio West **475** Michael Newman/PhotoEdit **477** Phil Moughmer/Third Coast Stock Source, Inc. **480** Steve Rosenthal Architectural Photography **483** ©1995 Bill Amend/Universal Press Syndicate. All Rights Reserved. **484** Superstock, Inc. **485(t)** Superstock, Inc. **485(b)** Superstock, Inc. **486** Superstock, Inc. **489** R.Maiman/Sygma **494-495(t)** Superstock, Inc. **494-c** Superstock, Inc. **494(bl)** Superstock, Inc. **494-495(b)** Superstock, Inc. **495(c)** Superstock, Inc. **496** Cameramann International, Ltd. **501** Superstock, Inc. **503** Steve Rosenthal **509** Cameramann International, Ltd. **510** Jim Harrison/Stock Boston **511(l)** David R. Austen/Stock Boston **511(r)** Superstock, Inc. **513** Cameramann International, Ltd. **516** Superstock, Inc. **517** NASA **518** Courtesy L. Meisel Gallery, New York **519** Superstock, Inc. **525** Rosemary Finn **529** Parkinson, Mary Taylor **531(all)** Courtesy Simmons-Boardman Publishing Corporation **539** Finn, Rosemary **544(t)** Tom Van Sant/Geosphere Project, Santa Monica/SPL/Photo Researchers **545(t)** Special Publication No. 1 of the American Cartographic Assn., Copyright 1986 by the American Congress on Surveying and Mapping **545(b)** Copyright ©1992 Quarto Publishing Pic, Marlboro Books Corp, a division of Barnes & Noble, Inc. **545** Special Publication No. 1 of the American Cartographic Assn., Copyright 1986 by the American Congress on Surveying and Mapping **552(t)** Superstock, Inc. **552(b)** Superstock, Inc. **553** Superstock, Inc. **556** Joseph Thorn **562-563(t)** Superstock, Inc. **562(c)** Superstock, Inc. **562(b)** Superstock, Inc. **563(c)** Superstock, Inc. **563(b)** Superstock, Inc. **564** Paul Chesley/Tony Stone Worldwide **568** The World Of Games, Their Origins and History, How to Play Them, and How to Make Them © 1989 first English language edition by Facts on File Inc., New York and Oxford. All rights reserved. **570** Cameramann International, Ltd. **572** Superstock, Inc. **573** Bob Daemmrich/Stock Boston **575** Jan Kanter **576** Brian Yen **577** Ivory Cats ©Lesley Anne Ivory 1990, Licensee ENESCO Corporation **582** John Coletti/Stock Boston **584** Ivory Cats ©Lesley Anne Ivory 1990, Licensee ENESCO Corporation **587** FourByFive Inc./Superstock, Inc. **594** Everett Collection, Inc. **596** Chuck Keeler/Tony Stone Worldwide **599** FourByFive Inc./Superstock, Inc. **602** Roxanna Marino **609** Richard Pasley/Stock Boston **613** Superstock, Inc. **620(cl)** Superstock, Inc. **620-621(t)** Superstock, Inc. **620(cr)** Superstock, Inc. **620-621(b)** Superstock, Inc. **621(c)** FourByFive/Superstock, Inc. **622** Superstock, Inc. **624** Drawings by John Tenniel **625** Giraudon/Art Resource **626** Focus on Sports, Inc. **627** Superstock, Inc. **628** Everett Collection, Inc. **629** David Young-Wolff/PhotoEdit **631** AP/Wide World **632** Richard Hutchings/PhotoEdit **634** Hazel Hankin/Stock Boston **635** Chris Cole/Tony Stone Worldwide **637** Bob Daemmrich/Stock Boston **639** Billy Barnes/Stock Boston **644** David Young-Wolff/PhotoEdit **650** Roberto De Gugliemo/SPL/Photo Researchers **651** Eric Smith/Liaison (Gamma) **655** Focus on Sports, Inc. **656** Courtesy The House on the Rock **665** Cameramann International, Ltd. **666** Richard Hutchings/PhotoEdit **668** Bob Daemmrich/Stock Boston **674** John Elk/Tony Stone Worldwide **675(l)** Superstock, Inc. **675(tr)** FourByFive/Superstock, Inc. **675(b)** The British Library, London/Superstock, Inc. **676** Superstock, Inc. **682** Culver Pictures Inc. **684(tl)** Superstock, Inc. **684-685(tr)** FourByFive/Superstock, Inc. **684-685(c)** Superstock, Inc. **684-b** FourByFive/Superstock, Inc. **685(b)** Superstock, Inc. **685(r)** Superstock, Inc. **686** Finn, Rosemary **698** Karen Stockwell **701** Joseph Thorn **703** Museum of the American Indian, Smithsonian Institution 2742 **705** C.H.Rose/Stock Boston **710** Richard Pasley/Stock Boston **711** Richard Pasley/Stock Boston **717** Seth Resnick/Stock Boston **720** **722** Ernst Haas/Magnum Photos **724** Bettmann Archive **725** Library of Congress **726** Jacksonville Art Museum, Florida/Superstock, Inc. **727** Everett Collection, Inc. **728** Leo Mason/Sports Illustrated **729(t)** Superstock, Inc. **729(b)** FourByFive/Superstock, Inc. **730(tr)** Superstock, Inc. **730(br)** Superstock, Inc. **736(t)** Superstock, Inc. **736-737(c)** FourByFive/Superstock, Inc. **736(b)** Superstock, Inc. **737(t)** Superstock, Inc. **737(b)** Superstock, Inc. **738** Alan Levenson/Tony Stone Worldwide **742** Superstock, Inc. **744** Sovfoto/Eastfoto **746** Cameramann International, Ltd. **753** Simon Milliken **755** Frank Siteman/Stock Boston **759** Bill Gallery/Stock Boston **763** Bob Daemmrich/Stock Boston **764** Cameramann International, Ltd. **765** John Garrett/Tony Stone Worldwide **767** David R. Frazier **777** Superstock, Inc. **779** Bettmann Archive **784** Will & Deni McIntyre/Tony Stone Worldwide **787** Superstock, Inc. **789(tl)** Superstock, Inc. **789(tr)** Superstock, Inc. **789(bl)** Superstock, Inc. **789(c)** Superstock, Inc. **790** Superstock, Inc. **798(tl)** Superstock, Inc. **798-799(tr)** Superstock, Inc. **798-799(c)** Superstock, Inc. **798(b)** FourByFive/Superstock, Inc. **799(b)** Superstock, Inc. **800** Maxwell Mackenzie/Tony Stone Worldwide **801** Superstock, Inc. **804** Steve Rosenthal **806** Underwood & Underwood/Bettmann Archive **809** Steve Rosenthal **812** Superstock, Inc. **814** Margaret Bourke-White/Life Magazine, Time Warner, Inc. **817** Pascal Quittemelle/Stock Boston **818** Jim Pickerell/Stock Boston **829** Everett Collection, Inc. **829** Lina Cornell/Stock Boston **830** Summit Labs/Lerner Fine Art Collections/Superstock, Inc. **842** Focus on Sports, Inc. **844** The World Of Games, Their Origins and History, How to Play Them, and How to Make Them © 1989 first English language edition by Facts on File Inc., New York and Oxford. All rights reserved. **846(l)** Charles Gupton/Stock Boston **846(r)** George Haling Productions/Photo Researchers **848** Jeff Schultz/AlaskaStock **849** Courtesy Chicago Bridge & Iron Company **851** Rosemary Finn **852** Bettmann Archive **853** Superstock, Inc. **854** Superstock, Inc. **857** Richard Pasley/Stock Boston